MULTISCALE METHODS

Multiscale Methods

Bridging the Scales in Science and Engineering

Edited by
JACOB FISH

OXFORD
UNIVERSITY PRESS

Great Clarendon Street, Oxford OX2 6DP

Oxford University Press is a department of the University of Oxford.
It furthers the University's objective of excellence in research, scholarship,
and education by publishing worldwide in

Oxford New York

Auckland Cape Town Dar es Salaam Hong Kong Karachi
Kuala Lumpur Madrid Melbourne Mexico City Nairobi
New Delhi Shanghai Taipei Toronto

With offices in

Argentina Austria Brazil Chile Czech Republic France Greece
Guatemala Hungary Italy Japan Poland Portugal Singapore
South Korea Switzerland Thailand Turkey Ukraine Vietnam

Oxford is a registered trade mark of Oxford University Press
in the UK and in certain other countries

Published in the United States
by Oxford University Press Inc., New York

© Oxford University Press 2010

The moral rights of the author have been asserted
Database right Oxford University Press (maker)

First Published 2010

All rights reserved. No part of this publication may be reproduced,
stored in a retrieval system, or transmitted, in any form or by any means,
without the prior permission in writing of Oxford University Press,
or as expressly permitted by law, or under terms agreed with the appropriate
reprographics rights organization. Enquiries concerning reproduction
outside the scope of the above should be sent to the Rights Department,
Oxford University Press, at the address above

You must not circulate this book in any other binding or cover
and you must impose the same condition on any acquirer

British Library Cataloguing in Publication Data

Data available

Library of Congress Cataloging in Publication Data

Data available

Typeset by Newgen Imaging Systems (P) Ltd., Chennai, India
Printed in Great Britain
on acid-free paper by
CPI Antony Rowe, Chippenham, Wiltshire

ISBN 978–0–19–923385–4

1 3 5 7 9 10 8 6 4 2

PREFACE
Jacob Fish (Rensselaer)

Small scale features and processes occurring at nanometer and femtosecond scales have a profound impact on what happens at larger space and time scales. For instance, in polymer based nanocomposites, interfacial interactions occur at the molecular scale with bond-stretching vibrations at a time scale of femtoseconds, whereas the corresponding relaxation processes span the time scales of hours and spatial scales of meters. In biological processes (Chapter 15), a wide range of spatial scales govern the behavior, from the electronic structure governed by quantum mechanics to atomic, molecular, cellular, organ, system, and genome entities. Temporal scales range from sub-femtosecond for electronic response to billions of years of evolutionary changes. In protein folding, physical phenomena operate across 12 orders of magnitude in time scales, and 10 orders of magnitude in spatial scales in advanced materials (Chapter 14). In nanoelectromechanical systems (NEMS) discussed in Chapter 13, the characteristic length is often a few nanometers with significant quantum effects, material defects, and surface effects, while the entire system can be of the order of micrometers. Nanoporous catalysts (Chapter 16) can be designed at the molecular level, self-organized into channel systems, and further organized into macroscopic structures for use in chemical reactors. In all these applications the disparity between the fine scale at which physical laws originate and the much larger scale of phenomena we wish to understand has been referred to as a *scale gap* in Chapter 7 and as *a tyranny of scales* in the NSF Blue Ribbon Report.

In view of the increasing need for understanding and controlling the behavior of products and processes at multiple scales, multiscale modeling and simulation has emerged as one of the focal research areas in applied science and engineering. Even with rapidly growing computing power, which enables a detailed modeling of multiple physical processes with increasing fidelity, traditional single scale approaches have proven to be inadequate. Consequently, there is a growing need to develop efficient modeling and simulation approaches for multiscale problems. No single solution or approach has the potential of bridging all length and time scales. The interdisciplinary nature of the field, where multiscale approaches are often material- and application-specific, introduces additional challenges. Finally, communication barriers between mathematicians, chemists, physicists, computer scientists, and engineers often prevent satisfactory progress in the field.

The enormous gains that can be achieved by multiscale analysis have been reported in numerous articles. Over the past five years, multiscale science and engineering has been featured in virtually every conference devoted to computational science and engineering. Multiscale computations have been identified[1] as one of the areas critical to future nanotechnology advances. The FY2004 $3.7-billion-dollar National Nanotechnology Bill[1] states that: "approaches that integrate more than one such technique (... molecular simulations, continuum-based models, etc.) will play an important role in this effort." An NSF Blue Ribbon for Simulation-Based Engineering (SBES)[2] recommended an increase in the NSF budget of $300 million per year on SBES-related disciplines with multiscale science and engineering being one of the major benefactors.

The primary objective of this volume is to present the-state-of-the art in multiscale mathematics, modeling, and simulations and to address the following barriers: What is the information that needs to be transferred from one model or scale to another and what physical principles must be satisfied during the transfer of information? What are the optimal ways to achieve such transfer of information? How to quantify variability of physical parameters at multiple scales and how to account for it to ensure design robustness?

Various multiscale approaches presented in this volume are grouped into two main categories: information-passing and concurrent (a more elaborate classification is given in Chapter 4). These multiscale methods are capable of resolving certain quantities of interest with a significantly lower cost than solving the corresponding fine-scale system. Schematically, a multiscale method has to satisfy the so-called Accuracy and Cost Requirements (ACR) test[3]:

$$\text{Error in quantities of interest} < tol$$
$$\frac{\text{Cost of multiscale solver}}{\text{Cost of fine scale solver}} \ll 1$$

In the concurrent approaches described in Chapters 4–6, various scales are simultaneously resolved, whereas in the information-passing methods (sometimes referred to as hierarchical or sequential methods), the fine scale is modeled and its gross response is infused into the coarse scale. Chapter 4 focuses on linking continuum and discrete scales, whereas in Chapter 5 attention is restricted to concurrent linking of discrete scales. Mathematical analysis of some of the concurrent methods is given in Chapter 6. Loosely speaking, for the concurrent multiscale approach to pass the ACR test, the following conditions must be

---

[1] National Nanotechnology Initiative. Supplement to the President's FY 2004 Budget. National Science and Technology Council Committee on Technology.

[2] J.T. Oden (chair) Simulation Based Engineering Science–An NSF Blue Ribbon Report http://www.nsf.gov/pubs/reports/sbes_final_report.pdf

[3] J. Fish. "Bridging the scales in nano engineering and science", Vol. 8, pp. 577–594, *Journal of Nanoparticle Research*, (2006).

typically satisfied:

(i) the interface (or interphase) between the fine and coarse scales should be properly engineered, and

(iia) the fine scale model should be limited to a small part of the computational domain, or alternatively

(iib) the information-passing multiscale approach of choice should serve as an adequate mechanism for capturing the lower frequency response of the fine-scale system.

The state-of-the-art spatial information-passing method is described in Chapters 1–3 with Chapter 2 devoted to systems with self similar microstructure, Chapter 3 focuses on reducing computational complexity of information-passing methods, and Chapter 1 on mathematical analysis of some of information-passing methods. Progress in the space-time information-passing methods is featured in Chapters 8 and 9. The information-passing multiscale approach is likely to pass the so-called ACR (Accuracy and Cost Requirements) test provided that:

(i) the quantities of interest are limited to or defined only on the coarse scale (provided that these quantities are computable from the fine scale), and

(ii) special features of the fine scale problem, such as scale separation and self-similarity, are taken advantage of.

A unified concurrent-information-passing framework which in principle is not limited by such conditions is provided in Chapter7.

The issue of reliability of multiscale modeling and simulation tools is discussed in Chapters 10 and 11, which describe a hierarchy of multiscale models, i.e. a sequence of mathematical models with increasingly more sophisticated effects, and *a posteriori* model error estimation including uncertainty quantification. Component software that can be effectively combined to address a wide range of multiscale simulations is outlined in Chapter 12.

The volume is intended as a reference book for scientists, engineers, and graduate students in traditional engineering and science disciplines as well as in the emerging fields of nanotechnology, biotechnology, microelectronics, and energy.

CONTENTS

Preface   v
    Jacob Fish (Rensselaer)

List of Contributors   xxi

I   INFORMATION-PASSING MULTISCALE METHODS IN SPACE

1   Mixed multiscale finite element methods on adaptive unstructured grids using limited global information   3
    J. E. Aarnes, Y. Efendiev, T.Y. Hou, and L. Jiang

    1.1   Introduction   3
    1.2   Preliminaries and motivation   5
    1.3   Mixed multiscale finite element method for flow equations   7
    1.4   Numerical results   10
    1.5   Analysis of global mixed MsFEM on unstructured grid   16
        1.5.1   Remarks on time dependent problems   24
    1.6   Conclusions   26
    References   27

2   Formulations of mechanics problems for materials with self-similar multiscale microstructure   31
    R.C. Picu and M.A. Soare

    2.1   Introduction   31
    2.2   The geometry of self-similar structures   35
    2.3   Basic concepts in fractional calculus   36
        2.3.1   Non-local fractional operators   36
        2.3.2   Local fractional operators   37
            2.3.2.1   Fractional differential operators   37
        2.3.3   Fractional integral operators   39
    2.4   Mechanics boundary value problems on materials with fractal microstructure   41
        2.4.1   Iterative approaches   43
        2.4.2   Reformulations of governing equations   44
            2.4.2.1   Method used for "homogeneous" fractal medium   44
            2.4.2.2   Methods using local fractional operators   45
                2.4.2.2.1 General formulation   45
                2.4.2.2.2 Examples   49
    2.5   Closure   53
    References   54

3 N-scale model reduction theory — 57
Jacob Fish and Zheng Yuan

    3.1   Introduction — 57
    3.2   The N-scale mathematical homogenization — 58
    3.3   Residual-free governing equations at multiple scales — 61
    3.4   Multiple-scale reduced order model — 63
    3.5   Reduced order unit cell problems — 70
    3.6   Summary of the computational procedure — 76
    3.7   Example: three scale analysis — 78
    3.8   Appendix — 86
    3.9   Acknowledgment — 87
    References — 87

II CONCURRENT MULTISCALE METHODS IN SPACE

4 Concurrent coupling of atomistic and continuum models — 93
Mei Xu

    4.1   Introduction — 93
    4.2   Classification of methods — 96
    4.3   Mechanical equilibrium methods — 98
        4.3.1   Direct and master-slave coupling — 98
        4.3.2   ONIOM method — 101
        4.3.3   Bridging domain method — 104
            4.3.3.1   Ghost forces — 108
            4.3.3.2   Stability of Lagrange multiplier methods — 111
        4.3.4   Quasicontinuum method — 112
    4.4   Molecular dynamics systems — 113
        4.4.1   Conservation properties of bridging domain method — 115
            4.4.1.1   Conservation of linear momentum — 116
            4.4.1.2   Conservation of energy — 116
        4.4.2   Master-slave and handshake methods — 117
        4.4.3   Bridging scale method — 118
    4.5   Numerical results — 119
        4.5.1   Molecular/continuum dynamics with bridging domain method — 119
            4.5.1.1   One-dimensional studies — 119
            4.5.1.2   Two-dimensional studies — 123
        4.5.2   Fracture of defected graphene sheets by QM/MM and QM/CM methods — 125
        4.5.3   Crack propagation in graphene sheet: comparison with Griffith's formula — 129
    4.6   Acknowledgment — 130
    References — 130

Contents

5	**Coarse-grained molecular dynamics: concurrent multiscale simulation at finite temperature** *Robert E. Rudd*		**134**
5.1	Coupling atomistic and continuum length scales		134
5.2	Coarse graining formalism		137
	5.2.1	Shape functions	138
5.3	The CGMD Hamiltonian		139
	5.3.1	Harmonic crystals	141
	5.3.2	How harmonic CGMD differs from FEM	143
	5.3.3	Anharmonic forces	143
5.4	Non-equilibrium effects		144
5.5	Practical details		146
	5.5.1	Shape functions in reciprocal space	147
5.6	Performance tests		148
	5.6.1	Phonon spectra	148
	5.6.2	The finite temperature tantalum CG spectrum	151
	5.6.3	Dynamics and scattering	155
5.7	Conclusion		160
	References		161

6	**Blending methods for coupling atomistic and continuum models** *P. Bochev, R. Lehoucq, M. Parks, S. Badia, and M. Gunzburger*		**165**
6.1	Introduction		165
	6.1.1	An overview of AtC blending methods	166
	6.1.2	Notation	168
6.2	Atomistic and continuum models		168
	6.2.1	Force-based models	168
		6.2.1.1 The atomistic model	169
		6.2.1.2 The continuum model	170
	6.2.2	Energy-based models	170
		6.2.2.1 The atomistic model	171
		6.2.2.2 The continuum model	171
6.3	Force-based blending		172
	6.3.1	An abstract AtC blending method	172
	6.3.2	Blending functions	174
	6.3.3	Assumption	174
	6.3.4	Enforcing the constraints	174
		6.3.4.1 Blending constraint operators	175
	6.3.5	Consistency and patch tests	177
	6.3.6	Definition [Consistency test problem]	177
	6.3.7	Definition [Patch test problem]	177

	6.3.8	Definition [Passing a patch test problem]	177
	6.3.9	Definition [AtC consistency]	177
	6.3.10	Blended atomistic and continuum functionals	177
	6.3.11	Taxonomy of AtC blending methods	179
		6.3.11.1 Methods of type I	179
	6.3.12	Theorem	180
		6.3.12.1 Methods of type II	180
	6.3.13	Theorem	180
		6.3.13.1 Methods of type III	181
	6.3.14	Theorem	181
		6.3.14.1 Methods of type IV	181
	6.3.15	Summary and comparison of force-based AtC blending methods	181
6.4	Energy-based blending		182
	6.4.1	An abstract AtC blending method	182
	6.4.2	Enforcing the constraints	183
	6.4.3	Taxonomy of AtC blending methods	184
6.5	Generalized continua		185
6.6	Conclusions		186
References			186

III SPACE-TIME SCALE BRIDGING METHODS

7 Principles of systematic upscaling — 193
Achi Brandt

7.1	Introduction	193
	7.1.1 Systematic upscaling (SU)	195
	7.1.2 Difference from *adhoc* multiscale modelling	195
	7.1.3 Other numerical upscaling methods	196
	7.1.4 Features of Systematic Upscaling	198
	7.1.5 Plan of this article	198
7.2	Systematic upscaling (SU): an outline	199
	7.2.1 Local equations and interactions	199
	7.2.2 Coarsening	199
		7.2.2.1 Examples of such fine-to-coarse transformations C	199
	7.2.3 Generalized interpolation	200
	7.2.4 The general coarsening criterion	200
	7.2.5 Experimental results	201
7.3	Derivation of coarse equations	202
	7.3.1 Basic hypothesis: localness of coarsening	202
		7.3.1.1 Dependence table	202
		7.3.1.2 Coarse Hamiltonian	203
	7.3.2 Example	204
7.4	Window developments	205

	7.5	Some special situations	206
	7.6	Extensions in brief	207
		7.6.1 Long-range interactions	207
		7.6.2 Dynamical systems	207
		7.6.3 Stochastic coarsening	208
		7.6.4 Joint H^c	209
		7.6.5 Complex fluids	209
		7.6.6 Low temperatures (example)	209
		7.6.7 Multiscale annealing	209
		7.6.8 Coarse-level computability of fine observables	210
		7.6.9 Determinism and stochasticity	210
		7.6.10 Upscaling from quantum mechanics to molecular dynamics	210
	References		210

8 Equation-free computation: an overview of patch dynamics

G. Samaey, A. J. Roberts, and I. G. Kevrekidis

216

	8.1	Introduction	216
	8.2	Equation-free multiscale framework	217
		8.2.1 Definitions	217
		8.2.2 The coarse time-stepper	219
		8.2.2.1 A remark on notation	220
		8.2.3 The gap-tooth scheme and patch dynamics	220
	8.3	Model homogenization problems	221
		8.3.1 Parabolic homogenization problem	222
		8.3.2 Hyperbolic homogenization problem	223
	8.4	The gap-tooth scheme	223
		8.4.1 Formulation	224
		8.4.1.1 Boundary constraints	225
		8.4.1.2 Initial conditions	226
		8.4.1.3 The algorithm	226
		8.4.2 Consistency and stability	226
		8.4.3 Discussion	227
	8.5	Patch dynamics with buffers	228
		8.5.1 Formulation	228
		8.5.1.1 Initial condition	229
		8.5.1.2 Boundary conditions	230
		8.5.1.3 The algorithm	230
		8.5.2 Numerical illustration	231
		8.5.3 Consistency and stability	232
		8.5.4 Application to advection problems	234
		8.5.4.1 Consistency	234
		8.5.4.2 Third order, upwind biased scheme	235
		8.5.4.3 Non-conservation form	236

8.6	General tooth boundary conditions		238
	8.6.1	Diffusion on Dirichlet teeth	239
	8.6.2	Slow manifold support for tooth dynamics	240
8.7	Conclusions		242
8.8	Acknowledgments		242
References			243

9 On multiscale computational mechanics with time-space homogenization 247
P. Ladevèze, David Néron, Jean-Charles Passieux

9.1	Introduction			247
9.2	The reference problem			249
	9.2.1	State laws		250
	9.2.2	Compatibility conditions and equilibrium equations		251
	9.2.3	Formulation of the reference problem		252
9.3	Reformulation of the problem with structure decomposition			252
	9.3.1	Admissibility conditions for Substructure Ω_E		254
	9.3.2	Interface behavior		255
	9.3.3	Reformulation of the reference problem		256
9.4	Multiscale description in the time-space domain $[0,T] \times \Omega$			256
	9.4.1	A two-scale description of the unknowns		256
	9.4.2	Admissibility of the macro quantities		257
9.5	The multiscale computational strategy			258
	9.5.1	The driving force of the strategy		258
	9.5.2	The local stage at Iteration $n+1$		259
	9.5.3	The linear stage at Iteration $n+1$		260
		9.5.3.1	The micro problems defined over each $[0,T] \times \Omega_E$ and $[0,T] \times \Omega_E$	261
		9.5.3.2	The macro problem defined over $[0,T] \times \Omega$	262
		9.5.3.3	Resolution of the linear stage	262
	9.5.4	Choice of the parameters (**H**,**h**) and convergence of the algorithm		262
	9.5.5	First example		263
9.6	The radial time-space approximation			266
	9.6.1	General properties		268
	9.6.2	Illustration		269
	9.6.3	Practical implementation		269
	9.6.4	Reformulation of the linear stage at Iteration $n+1$		271
		9.6.4.1	Rewriting of a micro problem over $[0,T] \times \Omega_E$	271
		9.6.4.2	Choice of admissible radial time-space functions	272
		9.6.4.3	Definition of the best approximation	273
		9.6.4.4	Practical resolution technique	273

9.6.5 Numerical example of the resolution of a micro problem	274
9.7 Conclusions	277
References	278

IV ADAPTIVITY, ERROR ESTIMATION AND UNCERTAINTY QUANTIFICATION

10 Estimation and control of modeling error: a general approach to multiscale modeling   285
J.T. Oden, S. Prudhomme, P.T. Bauman, and L. Chamoin

10.1 Problem setting	285
10.2 The general theory of modeling error estimation	287
10.2.1 The error estimate	288
10.3 A large-scale molecular statics model	289
10.4 The family of six algorithms governing multiscale modeling of polymer densification	290
10.4.1 Polymerization: kinetic monte carlo method	290
10.4.2 Molecular potentials	292
10.4.3 Densification algorithm: inexact Newton-Raphson with trust region	292
10.4.4 Algorithms for computing surrogate models	292
10.4.4.1 Virtual experiments	293
10.4.4.2 The interface	293
10.4.4.3 Residual force calculation	293
10.4.5 The adjoint problem	293
10.4.6 The goals algorithm	294
10.5 Representative results	296
10.5.1 Verification of error estimator and adaptive strategy	296
10.6 Extensions	299
10.7 Concluding comments	303
10.8 Acknowledgments	303
References	303

11 Error estimates for multiscale operator decomposition for multiphysics models   305
D. Estep

11.1 Introduction	305
11.1.1 Challenges and goals of multiscale, multiphysics models	308
11.1.2 Multiscale, multidiscretization operator decomposition	309
11.2 The key is stability. But what is stability...and stability of what?	313

	11.2.1 Pointwise stability of the Lorenz problem	313
	11.2.2 Classic *a priori* stability analysis	316
	11.2.3 Theorem	318
	11.2.4 Stability for stationary problems	318
	11.2.5 The meaning of stability depends on the information to be computed	320
11.3	The tools for quantifying stability properties: functionals, duality, and adjoint operators	323
	11.3.1 Functionals and computing information	323
	11.3.2 Theorem	325
	11.3.3 Theorem	326
	11.3.4 Theorem	326
	11.3.5 The adjoint operator	326
	11.3.6 Theorem	329
	11.3.7 Four good reasons to use adjoints	329
	11.3.8 Theorem	329
	11.3.9 Theorem	329
	11.3.10 Theorem	329
	11.3.11 Adjoint operators for linear differential equations	330
11.4	*A posteriori* error analysis using adjoints	334
	11.4.1 Discretization of elliptic problems	336
	11.4.2 *A posteriori* analysis for elliptic problems	337
	11.4.3 Theorem	337
	11.4.4 Adjoint analysis for nonlinear problems	339
	11.4.5 Discretization of evolution problems	343
	11.4.6 Analysis for discretizations of evolution problems	345
	11.4.7 Theorem	346
	11.4.8 Theorem	347
	11.4.9 Theorem	347
	11.4.10 General comments on *a posteriori* analysis	350
11.5	*A posteriori* error estimates and adaptive mesh refinement	351
	11.5.1 Adaptive mesh refinement in space	352
	11.5.2 Adaptive mesh refinement for evolutionary problems	356
11.6	Multiscale operator decomposition	358
	11.6.1 Multiscale decomposition of triangular systems of elliptic problems	358
	A linear algebra example	362
	Description of the *a posteriori* analysis	364
	11.6.2 Theorem	366
	11.6.3 Multiscale decomposition of reaction-diffusion problems	367
	A linear algebra example	370
	Description of the hybrid *a posteriori–a priori* error analysis	372
	11.6.4 Theorem	374

		Numerical examples	375
	11.6.5	Multiscale decomposition of a fluid–solid conjugate heat transfer problem	376
		Description of an *a posteriori* error analysis	378
	11.6.6	Theorem	380
		Loss of order and flux correction	382
	11.6.7	Theorem	384
11.7	The effect of iteration		385
11.8	Conclusion		386
References			387

V MULTISCALE SOFTWARE

12 Component software for multiscale simulation   393
M.S. Shephard, M.A. Nuggehally, B. FranzDale, C.R. Picu, J. Fish, O. Klaas, and M.W. Beall

12.1	Introduction		393
12.2	Abstraction of adaptive multiscale simulation		394
	12.2.1	Current multiscale simulation implementations	394
	12.2.2	Multiscale simulation abstraction	395
12.3	Multiscale simulation components and structures		399
	12.3.1	Models	399
		12.3.1.1 Abstract model	400
	12.3.2	Domains	401
	12.3.3	Fields	404
12.4	Component tools for crystaline materials		405
	12.4.1	Defining the grain structure	405
	12.4.2	Defining and storing the atomistic information	406
	12.4.3	A fast atomistic relaxation scheme	408
12.5	A concurrent atomistic/continuum adaptive multiscale simulation tool		410
	12.5.1	Concurrent atomistic/continuum multiscale formulation	410
	12.5.2	Analysis components for multiscale simulation tool	414
		12.5.2.1 Atomistic component	414
		12.5.2.2 Continuum component	414
	12.5.3	Adaptive solution procedure	415
	12.5.4	An example result	416
12.6	Acknowledgments		417
References			417

VI SELECTED MULTISCALE APPLICATIONS

13 Finite temperature multiscale methods for silicon NEMS   425
Z. Tang and N. R. Aluru

13.1	Introduction	425

13.2 Finite temperature QC method 427
   13.2.1 Continuum level description 429
   13.2.2 Atomistic level description 430
   13.2.3 QC method at classical zero temperature 431
   13.2.4 Finite temperature formulation 434
      13.2.4.1 Real space quasiharmonic (QHM) model . . 437
      13.2.4.2 Local quasiharmonic (LQHM) model 438
      13.2.4.3 k-space Quasiharmonic (QHMK) Model . . 439
   13.2.5 Results and discussion 441
      13.2.5.1 Lattice constants 441
      13.2.5.2 Strain and temperature effects on PDOS and Grüneisen parameters 441
      13.2.5.3 Bulk elastic constants 446
      13.2.5.4 Mechanical behavior of nanostructures under external loads 449
13.3 Local phonon density of states approach 454
   13.3.1 Theory 454
      13.3.1.1 Lattice dynamics 454
      13.3.1.2 LPDOS and local thermodynamic properties 455
      13.3.1.3 Phonon GF and the recursion method . . . 456
      13.3.1.4 Local mechanical properties 459
   13.3.2 Semilocal model 460
   13.3.3 Silicon surface models 466
   13.3.4 Results and discussion 467
      13.3.4.1 Thermal and mechanical properties of bulk silicon 467
      13.3.4.2 LPDOS of bulk silicon and nanoscale silicon structures 468
      13.3.4.3 Local thermal properties 470
      13.3.4.4 Mechanical properties 473
13.4 Conclusion 475
References 476

14 Multiscale Materials   481
Sidney Yip

14.1 Materials modeling and simulation (computational materials) . 481
   14.1.1 Characteristic length/time scales 482
   14.1.2 Intellectual merits 483
      14.1.2.1 Exceptional bandwidth 485
      14.1.2.2 Removing empiricism 485
      14.1.2.3 Visual insights 485
14.2 Atomistic measures of strength and deformation 486
   14.2.1 Limits to strength: homogeneous deformation 486
   14.2.2 Soft modes 489

		14.2.3 Microstructural effects	491
		14.2.4 Instability in nano-indentation	493
		14.2.5 Instability in shear deformations—dislocation slip versus twinning	495
		14.2.6 Strain localization	498
	14.3	Atomistic measure of defect mobility	500
		14.3.1 Single dislocation glide in a metal	500
		14.3.2 Dislocation mobility in silicon: kink mechanism	503
		14.3.3 Crack front extension in metal and semiconductor	506
	14.4	Outlook for multiscale materials	510
	14.5	Acknowledgment	510
	References		511
15	**From macroscopic to mesoscopic models of chromatin folding**		**514**
	Tamar Schlick		
	15.1	Introduction	514
	15.2	Chromatin structure	515
	15.3	The first-generation macroscopic chromatin models and results	518
	15.4	The second-generation mesoscopic chromatin model and results	520
	15.5	Future perspective	529
		15.5.1 Linker DNA length effects	529
		15.5.2 Histone variant effects	530
		15.5.3 Histone tail modifications	530
		15.5.4 Higher-order chromatin organization	530
	15.6	Conclusions	531
	15.7	Acknowledgments	531
	References		531
16	**Multiscale nature inspired chemical engineering**		**536**
	Marc-Olivier Coppens		
	16.1	Introduction	536
	16.2	Resolving multiscale patterns in heterogeneous systems	538
		16.2.1 Bubbling gas-fluidized beds of solid particles	538
		16.2.1.1 A fluidized bed as a chaotic system	539
		16.2.1.2 Polydispersity and the Student t-distribution as emergent pattern	540
		16.2.2 Diffusion in porous media	540
	16.3	Imposing multiscale structure using a nature-inspired approach	545
		16.3.1 Hierarchical functional architecture in nature	545
		16.3.2 Nature-inspired fluid distributors and injectors	547

	16.3.3 Optimal pore networks in catalysis	549
	16.3.4 Nature-inspired membranes	551
16.4	Concluding remarks	553
16.5	Acknowledgments	554
References		554

Index   561

LIST OF CONTRIBUTORS

J.E. Aarnes
SINTEF Applied Mathematics, N-0314 Oslo, Norway

Achi Brandt
Department of Computer Science and Applied Mathematics,
Weizmann Institute of Science, Rehovot, Israel.

N.R. Aluru
Department of Mechanical Science and Engineering, Beckman Institute for Advanced Science and Technology, University of Illinois at Urbana-Champaign, Urbana, Illinois, USA

S. Badia
CIMNE, Universitat Politècnica de Catalunya, Jordi Girona 1-3, Edifici C1, 08034 Barcelona, Spain. Supported by the European Community through the Marie Curie contract *NanoSim* (MOIF-CT-2006-039522)

P.T. Bauman
Institute for Computational Engineering and Sciences, The University of Texas at Austin, 1 University Station C0200, Austin, TX 78712, USA

M.W. Beall
Simmetrix Inc., 10 Halfmoon Executive Park Drive, Clifton Park, NY 12065, USA

P. Bochev
Sandia National Laboratories, Applied Mathematics and Applications, P.O. Box 5800, MS 1320, Albuquerque NM 87185. Sandia is a multiprogram laboratory operated by Sandia Corporation, a Lockheed Martin Company, for the United States Department of Energy's National Nuclear Security Administration under Contract DE-AC04-94-AL85000

L. Chamoin
Institute for Computational Engineering and Sciences, The University of Texas at Austin, 1 University Station C0200, Austin, TX 78712, USA

Y. Efendiev
Department of Mathematics, Texas A&M University, College Station, TX 77843-3368

B. FranzDale
Scientific Computation Research Center, Rensselaer Polytechnic Institute 110 8th St. Troy NY 12180

M. Gunzburger
School of Computational Science, Florida State University, Tallahassee
FL 32306–4120. Supported in part by the Department of Energy grant
number DE-FG02-05ER25698

T.Y. Hou
Applied Mathematics 217-50, California Institute of Technology, Pasadena,
CA 91125

Jacob Fish
Multiscale Science and Engineering Center, Rensselaer Polytechnic Institute,
Troy, NY 12180, USA

L. Jiang
Department of Mathematics, Texas A&M University, College Station,
TX 77843–3368

I.G. Kevrekidis
Department of Chemical Engineering and PACM, Princeton University,
Princeton, NJ08544

O. Klaas
Simmetrix Inc., 10 Halfmoon Executive Park Drive, Clifton Park, NY 12065,
USA

P. Ladevèze
LMT-Cachan (ENS Cachan/CNRS/UPMC/PRES UniverSud Paris) 61, avenue
du Président Wilson, F-94235 Cachan Cedex, France
EADS Foundation Chair "Advanced Computational Structural Mechanics"

R. Lehoucq
Sandia National Laboratories, Applied Mathematics and Applications,
P.O. Box 5800, MS 1320, Albuquerque NM 87185. Sandia is a multiprogram
laboratory operated by Sandia Corporation, a Lockheed Martin Company,
for the United States Department of Energy's National Nuclear Security
Administration under Contract DE-AC04-94-AL85000

Marc-Olivier Coppens
Howard P. Isermann Department of Chemical and Biological Engineering,
Rensselaer Polytechnic Institute, 110 8th Street, Troy, NY12180, USA

D. Néron
LMT-Cachan (ENS Cachan/CNRS/UPMC/PRES UniverSud Paris) 61, avenue
du Président Wilson, F-94235 Cachan Cedex, France
EADS Foundation Chair "Advanced Computational Structural Mechanics"

M.A. Nuggehally
Scientific Computation Research Center, Rensselaer Polytechnic Institute 110
8th St. Troy NY 12180

J.T. Oden
Institute for Computational Engineering and Sciences, The University of Texas
at Austin, 1 University Station C0200, Austin, TX 78712, USA

M. Parks
Sandia National Laboratories, Applied Mathematics and Applications,
P.O. Box 5800, MS 1320, Albuquerque NM 87185. Sandia is a multiprogram
laboratory operated by Sandia Corporation, a Lockheed Martin Company,
for the United States Department of Energy's National Nuclear Security
Administration under Contract DE-AC04-94-AL85000

J.-Ch. Passieux
LMT-Cachan (ENS Cachan/CNRS/UPMC/PRES UniverSud Paris) 61, avenue
du Président Wilson, F-94235 Cachan Cedex, France
EADS Foundation Chair "Advanced Computational Structural Mechanics"

C.R. Picu
Scientific Computation Research Center, Rensselaer Polytechnic Institute 110
8th St. Troy NY 12180

R.C. Picu
Department of Mechanical, Aerospace and Nuclear Engineering, Rensselaer
Polytechnic Institute, Troy, NY 12180, USA

S. Prudhomme
Institute for Computational Engineering and Sciences, The University of Texas
at Austin, 1 University Station C0200, Austin, TX 78712, USA

Robert E. Rudd
Lawrence Livermore National Laboratory, Physical Sciences, L–045, Livermore,
CA 94551 USA

A.J. Roberts
Computational Engineering and Science Research Centre, Department of
Mathematics and Computing, University of Southern Queensland, Toowoomba,
Queensland 4352, Australia

G. Samaey
Postdoctoral Fellow of the Research Foundation – Flanders
(FWO – Vlaanderen). Scientific Computing, Department of Computer Science,
K. U. Leuven, Celestijnenlaan 200A, 3001 Leuven, Belgium

M.S. Shephard
Scientific Computation Research Center, Rensselaer Polytechnic Institute 110
8th St. Troy NY 12180

M.A. Soare
Division of Engineering, Brown University, Providence, RI. 629121, USA

Tamar Schlick
Department of Chemistry and Courant, Institute of Mathematical Sciences,
New York University, 251 Mercer Street, New York, New York 10012

Z. Tang
Department of Mechanical Science and Engineering, Beckman Institute for
Advanced Science and Technology, University of Illinois at Urbana-Champaign,
Urbana, Illinois, USA

Sidney Yip
Departments of Nuclear Science and Engineering and Materials Science and Engineering, Massachusetts Institute of Technology, 77 Massachusetts Avenue, Cambridge, MA 02139, USA

Zheng Yuan
Multiscale Science and Engineering Center, Rensselaer Polytechnic Institute, Troy, NY 12180, USA

PART I

INFORMATION-PASSING MULTISCALE METHODS IN SPACE

1

MIXED MULTISCALE FINITE ELEMENT METHODS ON ADAPTIVE UNSTRUCTURED GRIDS USING LIMITED GLOBAL INFORMATION

J. E. Aarnes, Y. Efendiev, T.Y. Hou, and L. Jiang

1.1 Introduction

The high degree of variability and multiscale nature of formation properties such as permeability pose significant challenges for subsurface flow modeling. Geological characterizations that capture these effects are typically developed at scales that are too fine for direct flow simulations. Thus new techniques are required to solve the solution of the flow problems effectively. In practice, upscaling procedures have been commonly applied for this purpose and are effective in many cases (see [27] for reviews and discussion). More recently, a number of multiscale finite element (e.g., [30, 16, 7, 1, 5, 22]) and finite volume [32, 33] approaches have been developed and successfully applied to problems of this type.

Multiscale methods presented to date have applied local as well as global calculations to construct their multiscale basis functions. The importance of global information has been illustrated within the context of upscaling procedures in recent investigations [28, 15, 14] as well as within the context of multiscale methods [22, 35]. These studies have shown that the use of global information in the calculation of the upscaled parameters or multiscale basis functions can significantly improve the accuracy of the resulting coarse model. In [15, 14], the global information was computed at the coarse scale (for the sake of computational efficiency), while in [28] fine scale global information was used. Aarnes *et al.* [1, 5] considered the use of both local and global information in their mixed finite element multiscale procedure and observed improved results when global effects were incorporated. In [22, 4], various MsFEM with global information were studied. In particular, these methods were generalized to consider the problems with multiple global information. We refer to [10, 9, 7, 8, 12, 26, 30, 31, 32, 33] for earlier work on multiscale finite element methods and related techniques.

A multi-phase system in heterogeneous porous media consists of the flow equation (elliptic nature) and the transport equation (hyperbolic nature). Multiscale methods described in this paper are designed to solve the flow equation on the coarse grid. Typical multiscale approaches for a multi-phase system [30, 32, 7, 22] solve the flow equations on a coarse grid, while the transport equation is solved on a fine grid using the reconstructed fine-scale velocity field. To provide a truly coarse-scale model, the upscaling of the transport equation is needed. The latter is often difficult because the upscaling of the transport equation requires non-local information. In the upscaling of the transport equation, the unstructured coarse grid is often needed ([20, 2]). In this paper, we consider an approach which is capable of solving the flow and transport equations on the same coarse grid simultaneously. This provides a robust approach for solving the multi-phase flow and transport system since one can solve the coupled system implicitly in time.

We adopt the approach recently introduced in [2] to upscale the transport equation. In this approach, the coarse grid is generated to follow the streamlines of the corresponding single-phase flow (cf. [20]). More precisely, the coarsening algorithm presented in [2] groups high and low values of the single-phase velocity to generate a non-uniform coarse grid to model the transport. The method is applicable to an arbitrary grid and does not impose any smoothness constraint on the coarse grid. The coarsening algorithm is very simple and essentially involves only two parameters which specify the level of coarsening. Consequently, the algorithm allows the user to use the original geomodel as input for flow simulations and specify the simulation grid dynamically to fit the available computer resources. The resulting coarse grid is highly anisotropic (see Figure 11.2). In [2], the flow equation is solved on the fine grid and a non-uniform coarse grid is used for upscaling of the transport equation. To have a truly coarse scale model where both the flow and the transport equations are solved on the coarse grid, one needs to solve the flow equation on a highly anisotropic coarse grid designed for the upscaling of the transport equation, which is the main focus of this paper.

In this paper, we will develop and analyze a mixed multiscale finite element method that can be used to solve the flow equation on a unstructured coarse grid. The mixed MsFEM we consider is similar to that of [4] which uses a quasi-uniform coarse grid. The analysis presented in [4] is not applicable to the mixed MsFEM we consider here since coarse grid blocks are highly anisotropic. To analyze the convergence of the mixed MsFEM on a highly anisotropic grid, we assume that the coarse grid block is described by a hierarchy of scales, e.g. the rectangular coarse grid block with very different sizes in each spatial direction. We analyze the approximation property of the mixed MsFEM that uses limited global information under some restrictive assumptions. We assume that the global fields can smoothly span the solution. The proof involves only algebraic assumptions about the approximation properties. We further extend these results to parabolic and wave equations.

We perform several numerical experiments to demonstrate the improved accuracy and the robustness of the proposed method. Two types of permeability fields are considered in our numerical experiments. In the first example, we use channelized permeability fields from the Tenth SPE Comparative Project [17]. These permeability fields are channelized and difficult to upscale. In particular, due to the channelized nature of these permeability fields, the non-local effects are important. Some type of limited global information is used to improve the accuracy of the multiscale simulations. When a non-uniform coarsening from [2] is employed, the coarse grid is highly anisotropic. We conduct numerical experiments where integrated responses, such as water-cut, as well as saturation fields are compared between the multiscale and the reference solutions. Our numerical results show that the proposed coarse-scale model can accurately represent the integrated responses. Moreover, we show that our method approximates the saturation reasonably well on the coarse grid and preserves the main features of these fields. Further numerical studies reveal that the errors in the saturation field are mainly due to the upscaling of the saturation equation. In fact, the errors due to the mixed MsFEM for the flow equation are much smaller than those from the upscaling of the transport equation. Our numerical results also indicate that the multiscale method on an unstructured grid obtained using limited global information is more accurate. In our second example, we consider a geological model which is defined on an unstructured fine-scale grid. The fine-scale model is composed of corner-point grid [38] which is often used in petroleum applications. This model is generated with SBED^{TM}, and is courtesy of Alf B. Rustad at STATOIL. The permeability ranges from 0.1 mD to 1.7 D. The porosity is assumed to be constant. We employ a non-uniform coarsening of [2] to generate a coarse grid and perform comparison between coarse-scale and fine-scale results. Our numerical results again show that the proposed method provides a good approximation of the detailed multi-phase system in highly heterogeneous porous media.

This paper proceeds as follows. We present the preliminaries and motivation in Section 1.2. In Section 1.3, we describe our mixed MsFEM. Numerical results are presented in Section 1.4. We analyze the convergence of the mixed MsFEM on an unstructured grid in Section 1.5. We close with some concluding remarks in Section 1.6.

1.2 Preliminaries and motivation

In this chapter, we study mixed multiscale finite element methods for two-phase flow equations described by

$$div(\lambda(S)k\nabla p) = f, \tag{1.1}$$

in the absence of gravity and capillary effects. The total mobility $\lambda(S)$ is given by $\lambda(S) = \lambda_w(S) + \lambda_o(S)$ and f is a source term. Here, $\lambda_w(S) = k_{rw}(S)/\mu_w$ and $\lambda_o(S) = k_{ro}(S)/\mu_o$ where μ_o and μ_w are viscosities of oil and water phases, correspondingly, and $k_{rw}(S)$ and $k_{ro}(S)$ are relative permeabilities of oil and water phases, correspondingly. The transport equation is governed by

$$\frac{\partial S}{\partial t} + div(F) = 0, \tag{1.2}$$

where $F = vf_w(S)$, with $f_w(S)$, the fractional flow of water, given by $f_w = \lambda_w/(\lambda_w + \lambda_o)$, and the total velocity v by:

$$v = v_w + v_o = -\lambda(S)k\nabla p. \tag{1.3}$$

To solve the system (1.1) and (1.2) on the coarse grid, coarse-scale models for both the flow and the transport equations are needed. The upscaling of the transport equation is usually difficult because of strong nonlocal effects [29]. The use of a non-uniform coarse grid alleviates this problem to some extent; however, it results in a highly anisotropic coarse grid. An upscaling technique has been proposed in [2] for the transport equation. This upscaling method was analyzed for the transport equations and some numerical results were presented using a fine-scale (resolved) velocity field [2].

In this paper, we would like to develop a mixed MsFEM for the flow equation on the same unstructured coarse grid that is used to upscale the transport equation using the method of [2]. The main difficulty in analyzing this method is due to the highly anisotropic, unstructured coarse grid that is used for the flow equation. We assume that the coarse grid consists of a connected union of fine grid blocks. For example, in Figure 1.1, we plot a typical coarse grid used in the simulations. In this figure, the underlying fine grid is 60×220 (not shown) with 180 coarse grid blocks following the algorithm of [2]. A random color is assigned to each coarse grid block.

We will use limited global information in our mixed MsFEM. Mixed finite element methods are used because they provide mass conservative approximation of the velocity field. The global information is obtained by solving some simplified global problems such that the solutions of these global problems contain the essential non-local information. The use of limited global information is particularly useful for problems without scale separation as those studied in

Figure 1.1: Schematic description of unstructured coarsening for upscaling of transport equation. A random color is assigned to each coarse grid block. (See Plate 1).

this paper. For problems with scale separation, one can use local information to construct basis functions (see [4]). The underlying assumption for these global fields used in the paper is the following. There exists N global fields $p_1, \ldots, p_N$, such that

$$|p - G(p_1, \ldots, p_N)|_{1,\Omega} \leq C\delta, \qquad (1.4)$$

where δ is sufficiently small, G is sufficiently smooth function, and $p_1, \ldots, p_N$ are solutions of $div(k(x)\nabla p) = 0$ with some prescribed boundary conditions. Here $|.|_{1,\Omega}$ denotes the semi-norm of the standard Sobolev space $H^1(\Omega)$. Here δ is a parameter which measures how well the solution can be represented using some functions $p_1, \ldots, p_N$. For example, in the homogenization setting, δ is related to the smallest scale representing the heterogeneities. In some cases, the solution can be represented exactly with the global fields (see below) and in this case we have $\delta = 0$. In the above assumption (1.4), p_i are solutions of the flow equation with different bounary conditions. We denote the corresponding velocity field by v_i, i.e. $v_i = k\nabla p_i$. Then, the above assumption can be written in the following way. There exist sufficiently smooth scalar functions $A_1(x), \ldots, A_N(x)$, such that the corresponding velocities can be written as

$$\|v - A_1(x)v_1 - \cdots - A_N(x)v_N\|_{0,\Omega} \leq C\delta, \qquad (1.5)$$

where $\|.\|_{0,\Omega}$ denotes $L^2(\Omega)$ norm. We note that in the case of the homogenization problem v_i can be obtained from the solution of the periodic cell problem as discussed in [4].

Next, we briefly discuss the assumption (1.4). In [22], it was shown that for channelized permeability fields, p is a smooth function of the single-phase flow pressure (i.e. $N = 1$), where the single-phase pressure equation is described by $div(k(x)\nabla p) = 0$ using the same boundary conditions as those for the corresponding to two-phase flow. These results are shown under the assumption that λ is a smooth function. In a general setting, it was shown by Owhadi and Zhang [35] that for an arbitrary smooth $\lambda(x)$, the solution is a smooth function of d linearly independent solutions of the single-phase flow equation ($N = d$), where d is the space dimension. These results were shown under some assumptions in the case of $d = 2$ and more restrictive assumptions in the case of $d = 3$ provided that λ is a smooth function. When considering random permeability fields, v_i correspond to the solutions of the flow equation for some selected realizations [3].

1.3 Mixed multiscale finite element method for flow equations

In this section, we describe the MsFEM that employs some global information. Let Ω be a bounded domain in $\mathbb{R}^d$ ($d = 2, 3$). We consider an elliptic equation with a homogeneous Dirichlet boundary condition:

$$\begin{aligned} -div(\lambda(x)k(x)\nabla p) &= f(x) & \text{in } \Omega \\ p &= 0 & \text{on } \partial\Omega, \end{aligned} \qquad (1.6)$$

where $k(x)$ is a heterogeneous field and $\lambda(x)$ is a spatial field. In our analysis, we assume that $\lambda(x)$ is a smooth function.

Let $v = -\lambda(x)k(x)\nabla p$ be the velocity field. To construct the velocity basis functions for our global mixed MsFEM, we assume that the velocity field v can be approximated by some *apriori* defined global velocity fields $v_1, \ldots, v_N$ as in (1.5). This assumption (1.5) will be made more precise in Section 1.5. We note that v_i are, in general, solutions of $div(k\nabla p_i) = 0$, $v_i = -k\nabla p_i$, with some boundary conditions. The assumption (1.5) implies that the velocity field in each coarse grid block can be approximated by a linear combination of *a priori* defined velocity fields.

Next, we construct the multiscale velocity basis functions using the information from $v_1, \ldots, v_N$. Specifically, we construct the basis functions for the velocity field as follows:

$$div(k(x)\nabla \phi_{ij}^K) = \frac{1}{|K|} \quad \text{in } K$$

$$k(x)\nabla \phi_{ij}^K \cdot n_{F_l} = \delta_{jl} \frac{v_i \cdot n_{F_l}}{\int_{F_l} v_i \cdot n_{F_l} ds} \quad \text{on } \partial K \tag{1.7}$$

$$\int_K \phi_{ij}^K dx = 0,$$

where $i = 1, \ldots, N$, $j = 1, \cdots, j_K$ (j_K is the number of edge or face of K), F_l are faces (or edges) of K, and

$$\delta_{jj} = 1, \quad \delta_{jl} = 0 \text{ if } j \neq l.$$

We will omit the subscript F_l in n, if the integral is taken along the F_l. In Figure 1.2, we schematically illustrate the basis function construction.

Let $\psi_{ij}^K = k(x)\nabla \phi_{ij}^K$. We define the finite dimensional space spanned by these basis functions by

$$V_h = span\{\psi_{ij}^K\}.$$

Since the divergence of the velocity basis function in each coarse grid is a constant, we set Q_h to be piecewise constant basis functions which are used to approximate the pressure p.

Let $\{v_h, p_h\}$ be the numerical approximation of $\{v, p\}$ with the basis functions defined previously. The numerical mixed formulation of (1.6) is to find $\{v_h, p_h\} \in V_h \times Q_h$ such that

$$((\lambda k)^{-1} v_h, \sigma_h) - (div \sigma_h, p_h) = 0 \quad \forall \sigma_h \in V_h$$
$$(div v_h, q_h) = (f, q_h) \quad \forall q_h \in Q_h, \tag{1.8}$$

where $(\cdot, \cdot)$ denotes the usual L^2 inner product.

Note that for each edge, we have N basis functions and we assume that $v_1, \ldots, v_N$ are linearly independent in order to guarantee that the basis functions are linearly independent. To ensure the boundary condition in (1.7) is well defined, we assume that $\int_{F_l} v_i \cdot n_{F_l} ds$ is not zero. To avoid the possibility that $\int_{F_l} |v_i \cdot n_{F_l}| ds$ is unbounded, we need to make certain assumption which bound $\int_{F_l} |v_i \cdot n_{F_l}| ds$ from below. More precise formulation of the assumption is given in Section 1.5.

In Section 1.5, the convergence of the mixed MsFEM is analyzed under additional assumptions (see Theorem 1.16). These results are formulated in terms of the volume of the element and the area (or length) of the interfaces. We further discuss various options for coarse grid configuration and the assumptions which guarantee that the method converges. For the convergence of the method, we require that A_i's in (1.5) be approximated by a constant over each coarse grid block. This holds, e.g., in the cases considered in [22] and [35]. In general, this assumption states that the solution within each coarse grid block can be approximated using some pre-defined fields. Thus the main results come with no surprise. However, one needs to exercise caution in formulating the main assumptions and obtaining the quantitative error estimates as it is done in Section 1.5.

Remark 1.1 The representative coarse grid K can be non-convex (c.f., Figure 1.2). The analysis presented in Section 1.5 implies that the global mixed multiscale finite element method works for non-convex meshes. Strongly stretched meshes have an impact on the convergence rate of the pressure following the analysis in Section 1.5.

Remark 1.2 The construction of velocity basis functions in (1.7) and the analysis in Section 1.5 imply that K is not necessarily polygon domain and n_F can be a spatial function.

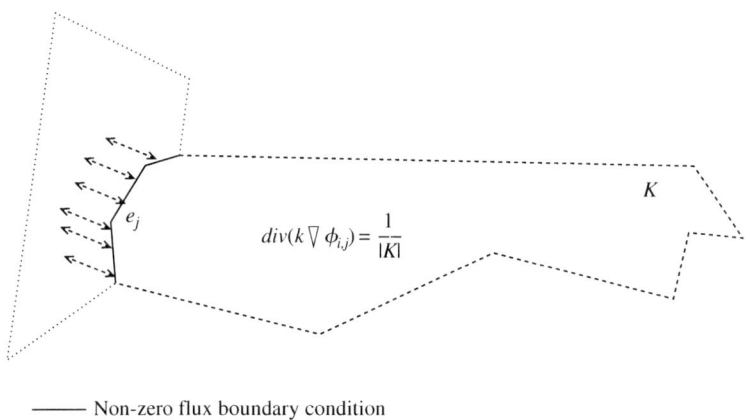

Figure 1.2: Schematic description of a basis function construction.

Remark 1.3 From the velocity basis functions defined in (1.7) and the analysis in Section 1.5, we find that the global mixed multiscale finite element method presented in the paper works for meshes with hanging notes and fine grids which do not necessarily match across coarse grid interfaces.

1.4 Numerical results

In our numerical simulations, we will perform two-phase flow and transport simulations. The equations are given (in the absence of gravity and capillary effects) by (1.1) and (1.2). We take $k_{rw}(S) = S^2$ and $k_{ro}(S) = (1-S)^2$. In the simulations, the multiscale basis functions will be constructed at initial time and will not be changed throughout the simulations. Thus, the pressure equation will be solved on the coarse grid. For the saturation equation, we will use the technique developed in [2]. The main idea of this method is to use an adaptive coarse grid for the saturation equation without changing the form of the equation. In particular, we construct the coarse grid by grouping high and low values of the single-phase velocity field. Extensive numerical results show that the coarse-scale saturation is more accurate when the adaptive coarse grid is used. For the mixed MsFEM, we use a single-phase flow solution as a global field to construct multiscale basis functions, i.e. $N = 1$ in (1.7). The single-phase velocity field is defined by $v_{sp} = -k\nabla p_{sp}$, where p_{sp} satisfies two-phase flow equation with $\lambda = 1$. Note that the single-phase velocity is needed for coarse grid construction following the algorithm in [2]. The steps in the non-uniform coarsening algorithm use the single-phase velocity field (with $\lambda = 1$) and are as follows:

1. Use the logarithm of the velocity magnitude in each cell to segment the cells in the fine grid into ten different bins; that is, each cell c is assigned a number $n(c) = 1, \ldots, 10$ by upper-integer interpolation in the range of $g(c) = \frac{10(\log|v_{sp}(c)| - \min\log|v_{sp}|)}{\max\log|v_{sp}| - \min\log|v_{sp}|}$.
2. Create an initial coarse grid with one block assigned to each connected collection of cells with the same value of $n(c)$.
3. Merge each block with less volume than $V_{\min}$ (a threshold number) with a neighboring block.
4. Refine each block that has more flow than $G_{\max}$ (a threshold number).
5. Repeat Step 3 and terminate.

We employ a backward Euler method for the temporal discretization of the saturation equation where the spatial discretization is a finite-volume method. In the algorithm, we vary V_{min} and G_{max}, while keeping the number of bins fixed $n(c) = 10$.

Figures 1.3(b)–1.3(d) show the logarithm of a velocity field as a piecewise constant function on the fine grid, on a coarse Cartesian grid with 240 blocks, and on a non-uniformly coarsened grid with 239 blocks respectively. We use a permeability field from the Tenth SPE Comparative Project [17] which has a fine resolution 60×220. If we denote the reservoir by Ω, and M is the number

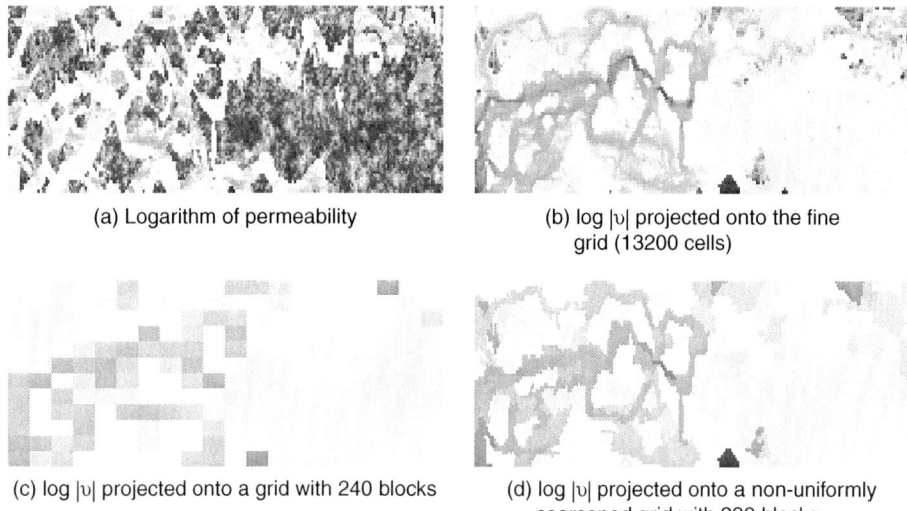

(a) Logarithm of permeability

(b) log |υ| projected onto the fine grid (13200 cells)

(c) log |υ| projected onto a grid with 240 blocks

(d) log |υ| projected onto a non-uniformly coarsened grid with 239 blocks

Figure 1.3: Illustration of the non-uniform coarsening algorithms ability to generate grids that resolve flow patterns and produce accurate production estimates. Results from paper [2]. (See Plate 2).

of cells in the fine grid, then the non-uniform coarse grid is generated under the constraint that each block B satisfies

$$\int_B dx \geq \frac{15}{M} \int_\Omega dx \quad \text{and} \quad \int_B \log|v| dx \leq \frac{75}{M} \int_\Omega \log|v|\, dx.$$

We clearly see that the non-uniformly coarsened grid adapts to the underlying flow pattern. In contrast, the channels with high velocity are almost impossible to detect in Figure 1.3(c). The fact that the coarse grid is capable of resolving the main flow trends leads to improved accuracy in modeled production characteristics. This is illustrated in [2] (see Figure 1.1 in [2]) which shows that the water-cut curves computed by using a non-uniform coarse grid are more accurate than the water-cut curves computed by using the Cartesian coarse grid.

We note that in [2], the fine-scale velocity is used in upscaling of the saturation equation. In this paper, we present the results for a *full* coarse-scale model where both the flow and transport equations are solved on the coarse grid. In this case, the multiscale basis functions are constructed on a highly anisotropic coarse grid which can affect the convergence rate. In our numerical results, we compare the water-cut data as a function of pore volume injected (PVI). The water-cut is defined as the fraction of water in the produced fluid and is given by q_w/q_t, where $q_t = q_o + q_w$, with q_o and q_w being the flow rates of oil and water at the production edge of the model. In particular, $q_w = \int_{\partial\Omega^{out}} f(S) v \cdot n ds$, $q_t = \int_{\partial\Omega^{out}} v \cdot n ds$, where $\partial\Omega^{out}$ is the outer flow boundary. Pore volume injected, defined as $PVI = \frac{1}{V_p} \int_0^t q_t(\tau)d\tau$, with V_p being the total pore volume of the system, provides the dimensionless time for the displacement. We consider a five-spot problem (e.g. [1]), where the water is injected at three corners and the middle of east side of the domain and oil is produced at the middle of the domain.

In our first example, we consider the two dimensional cross sections of permeability fields from the Tenth SPE Comparative Project [17], also known as SPE 10. These permeability fields are highly channelized and difficult to upscale. In particular, due to the channelized nature of these permeability fields, the nonlocal effects are important. In our simulations, we choose the viscosity ratio to be $\mu_o/\mu_w = 5$ and compare the integrated quantities, such as the oil production rate and the total flow rate, as well as the saturation errors at some time instants. Each layer of SPE 10 has 60×220 resolution. We consider layer 65. For our first numerical example, we use 180 coarse grid blocks (we take $V_{min} = 20$ and $G_{max} = 100$), where the fine grid block volume is taken to be 1. In Figure 1.4, the fine-scale permeability and the coarse grid are plotted. In Figure 1.5, we present the water-cut curves. One can observe from this figure that the results are accurate for such high degree of coarsening. Note that the fine-scale system is coarsened by the factor of 75. In Figure 1.6, we present the results for the saturation fields at PVI=1. One can see from this figure that the saturation profile looks realistic when an adaptive coarse grid is used and we preserve the geological realism reasonably well.

Next, we compare the saturation errors. As a measure of comparison, we use the relative L^1 norm when comparing the saturation fields. First, we note that the L^1 relative error for the results presented above with 180 coarse unstructured grid blocks is 19%, while with the structured grid on 10×20 coarse grid is 30%. We note that 19% error is mostly due to the upscaling error in the

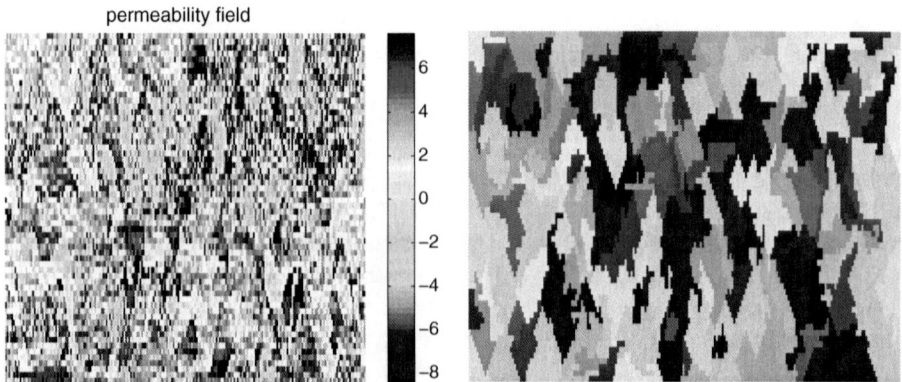

Figure 1.4: 60×220 permeability field and the coarse grid with 180 blocks. A random color is assigned to each coarse grid block. (See Plate 3).

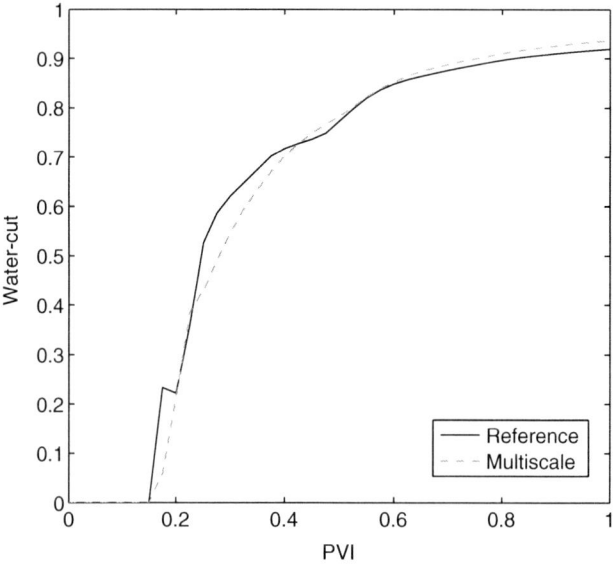

Figure 1.5: Water-cut curves for the reference and multiscale solutions.

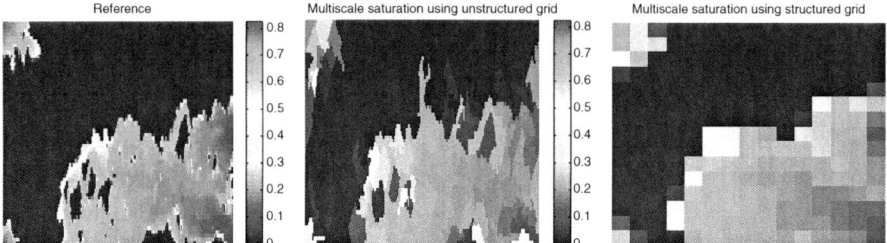

Figure 1.6: Saturation comparisons. (See Plate 4).

saturation. To separate the errors due to the saturation upscaling from those due to the mixed MsFEM on an unstructured grid, we separate two sources of errors for the saturation. In the second and the third columns of Table 1.1, we present the saturation and fractional flow errors by solving the saturation equation on the coarse grid with the velocity obtained from the mixed MsFEM on an unstructured coarse grid. These errors are computed by comparing them with the corresponding saturation and fractional flow which are obtained by solving the saturation equation on the coarse grid with the reference velocity field obtained by solving the fine-scale pressure equation. As we can see, these errors are less than 0.5%, which clearly indicates that the errors in the saturation fields are mostly due to the saturation upscaling. In the fourth and fifth columns of Table 1.1, we compare the saturation errors obtained by using the unstructured

Table 1.1. Relative errors when the transport eq. is solved on the coarse grid (layer = 65)

Unstruct. coarse (number of blocks)	sat err. (due to MsFEM)	water-cut err. (due to MsFEM)	sat. err. (total)	sat. err. (struct. grid) (total)
180	0.0047	0.0030	0.1936	0.3263 (10 × 20)
299	0.0040	0.0023	0.1536	0.2831 (15 × 22)
913	0.0035	0.0012	0.1005	0.2019 (20 × 44)

Table 1.2. Relative errors when the transport eq. is solved on the fine grid (layer = 65)

Unstruct. coarse (number of blocks)	sat. err. (total)	sat. err. (struct. grid) (total)
180	0.0097	0.0130 (10 × 20)
299	0.0080	0.0125 (15 × 22)
913	0.0062	0.009 (20 × 44)

grid with those obtained by using a comparable structured coarse grid. The coarse grid resolution for the structured grid is indicated in the parenthesis. It is clear that the method using an unstructured coarse grid is nearly two times more accurate than that using a structured grid. The issues associated with the transport upscaling are extensively studied in [2]. Though the saturation errors are higher than usual 5%, the coarsening with an unstructured grid better preserves the geological realism in the transport calculations and is needed for the upscaling of the saturation equation. More accurate subgrid capturing mechanisms are needed to improve the saturation upscaling. This issue is beyond the scope of this paper and currently under the investigation.

Next, we compare the saturation errors when the transport equation is solved on the fine grid. One of commonly used methods solves the transport equation on the fine grid, while the flow equation is solved on a coarse grid (e.g. [30, 32, 1]). These methods avoid the issues related to the transport upscaling. In Table 1.2, we present L^1 relative errors for the saturation when different resolutions of the coarse grid are used. In the same table, we show the errors corresponding to the structured grids with comparable number of coarse grid blocks (shown in the parenthesis). We can make two observations from this table. First, the errors are small (less than 1%). Secondly, the mixed MsFEM on an unstructured grid performs better. The latter is due to the fact that the unstructured

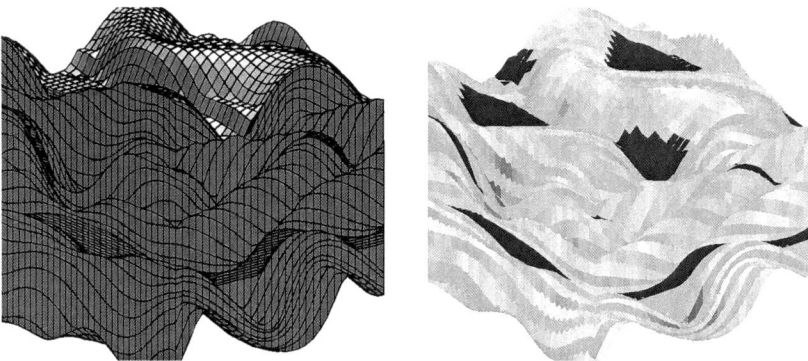

Figure 1.7: A corner-point model with vertical pillars and 100 layers. To the right is a plot of the permeability field on a logarithmic scale. The model is generated with SBED^{TM}, and is courtesy of Alf B. Rustad at STATOIL. (See Plate 5).

grid is constructed using some relevant limited global information which usually increases the accuracy of the method. These issues will be studied in future work.

In our next numerical example, we test the method on a synthetic reservoir with a corner-point grid geometry. The corner-point grid has vertical pillars, as shown in Figure 1.7, 100 layers, and 29629 active cells (cells with positive volume). The permeability ranges from 0.1 mD to 1.7 D and the porosity is assumed to be constant. The corner-point grid (or pillar grid) format [38] is a very flexible grid format that is used in many commercial geomodeling softwares. Essentially a corner-point grid consists of a set of hexahedral cells that are aligned in a logical Cartesian fashion where one horizontal layer in the logical grid is assigned to each sedimentary bed to be modeled. In its simplest form, a corner-point grid is specified in terms of a set of vertical or inclined pillars defined over an areal Cartesian 2-D mesh in the lateral direction. Each cell in the volumetric corner-point grid is restricted by four pillars and is defined by specifying the eight corner points of the cell, two on each pillar.

We consider only 60 vertical layers of the permeability field. The coarse grid is constructed by subdividing the fine-scale model on 30-by-30-by-60 corner-point cells into 202 coarse grid blocks. In Figure 1.8, we plot: coarse grid partitioning (left plot) where a random color is assigned to each coarse grid block; a horizontal slice of coarse partitioning presented on the left plot; and a coarse grid block. In Figure 1.9, we plot the water-cut curves. As we see from this figure, our method provides an accurate approximation of water-cut data. We have observed 17% error in the saturation field. Again, this error is mainly due to the saturation upscaling. The error that is due to the mixed MsFEM is only 2%.

Finally, we would like to note that we have observed similar results for other permeability fields including different layers of SPE 10 as well as 3-D slices of SPE 10 when several layers are used.

Figure 1.8: Left: schematic description of unstructured coarsening (each coarse grid block is assigned a random color). Middle: a horizontal slice of unstructured coarsening presented on the left. Right: a coarse grid block (enlarged). (See Plate 6).

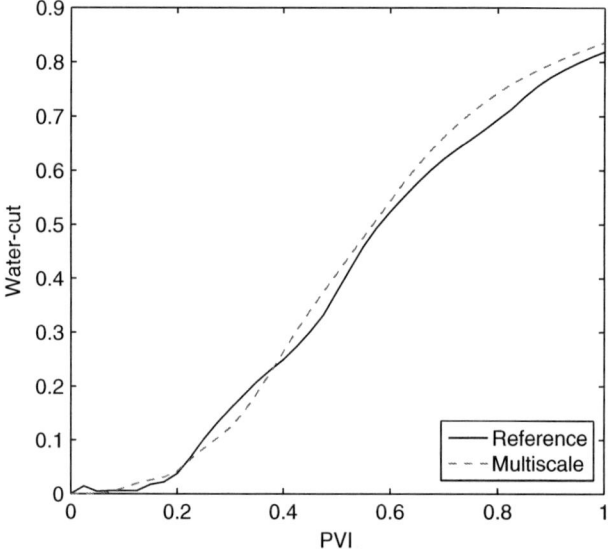

Figure 1.9: Water-cut for reference and multiscale solutions.

1.5 Analysis of global mixed MsFEM on unstructured grid

In this section we study the convergence of the global mixed multiscale finite element method presented in Section 1.3. We assume that $\gamma_{\min}|\xi|^2 \leq \xi^t \lambda(x) k(x) \xi \leq \gamma_{\max}|\xi|^2$ uniformly for any $\xi \in \mathbb{R}^d$ ($d = 2, 3$).

Let $v = -\lambda(x)k(x)\nabla p$ be the velocity. We introduce a finite element partition τ of Ω and let K be a representative element. For the proof of the main convergence results, we will make two main assumptions. First, we will assume that the solution can be represented in each coarse grid block by a family of velocity fields (see (1.9)). Here, we do not assume that the coarse grid is quasi-uniform; however, we assume that the approximating coefficients $A_i(x)$ are smooth (see

(1.9)). We discuss the applicability of this assumption in Remark 1.6. We also need to avoid the possibility that $\int_F |v_i \cdot n_F| ds$ is unbounded. Further assumption is made about the boundedness of the mixed multiscale basis functions.

The proof uses the ideas from [4]. In [4], we have shown the convergence of the method under some assumptions. However, one important assumption used in [4] is that the coarse grid is quasi-uniform. As we mentioned earlier, this assumption is not applicable to the method we consider in this paper since our unstructured grid is highly anisotropic. We need to generalize the previous analysis to allow the use of a high anisotropic coarse grid. Our analysis reveals that if the approximating coefficients A_i vary less within a coarse grid block, then the convergence of the method is improved. Also, if v_i is smoother along the faces F, this further improves the convergence. We introduce additional parameters to quantify the convergence rate.

Assumption A1. *There exist velocity fields $v_1, \cdots, v_N$ and functions $A_1(x), \cdots, A_N(x)$ such that*

$$v(x) = \sum_{i=1}^{N} A_i(x) v_i, \tag{1.9}$$

where $\int_F |v_i \cdot n| ds \leq C |F|^\beta$ ($\beta > 0$) for any face (or edge) F. For any $A_i(x)$ and any $F \subset \partial K$, there exists a small positive number δ_F such that $|A_i(x) - \bar{A}_i| \leq C \delta_F$ in K for some constant C, where $\bar{A}_i = \frac{1}{|F|} \int_F A_i(s) ds$.

Remark 1.4 If $|v_i|$ is bounded along F, then $\beta = 1$. One can use different assumptions by trading-off between the regularities of v_i and A_i, and obtain the convergence rate when $|v_i|$ is not necessarily bounded, though some L^η norm of v_i is bounded along the edge (cf. [4]).

Remark 1.5 For simplicity of the proof, we use the equality (1.9) instead of inequality (1.5). One can also use inequality (1.5) in the proof. This will require an additional assumption on the remainder $v - \sum_{i=1}^{N} A_i(x) v_i$ along faces F's which we would like to avoid.

Remark 1.6 As an example of d global fields in $\mathbb{R}^d$, we use the results in [35]. Let $v_i = -k(x) \nabla p_i$ ($i = 1, \cdots, d$) be defined by the solution of the following elliptic equation:

$$\begin{aligned} div(k(x) \nabla p_i) &= 0 \quad \text{in } \Omega \\ p_i &= x_i \quad \text{on } \partial \Omega, \end{aligned} \tag{1.10}$$

where $x = (x_1, \cdots, x_d)$. The advantage of using the harmonic coordinate $(p_1, \cdots, p_d)$ is that we obtain an enhanced regularity result: $p(p_1, \cdots, p_d) \in W^{2,s}$ ($s \geq 2$) [35]. Consequently, $v = -\lambda(x) k(x) \nabla p = -\sum_i \lambda \frac{\partial p}{\partial p_i} k \nabla p_i := \sum_i A_i(x) v_i$, where $A_i(x) = \lambda \frac{\partial p}{\partial p_i} \in W^{1,s}(\Omega)$.

The mixed formulation of (1.6) is to find $\{v,p\} \in H(div,\Omega) \times L^2(\Omega)$ such that

$$((\lambda k)^{-1}v, \sigma) - (div\sigma, p) = 0 \quad \forall \sigma \in H(div,\Omega)$$
$$(divv, q) = (f, q) \quad \forall q \in L^2(\Omega), \tag{1.11}$$

where $(\cdot, \cdot)$ is the usual L^2-inner product and

$$H(div, \Omega) = \{u(x) \in [L^2(\Omega)]^d : divu \in L^2(\Omega)\}.$$

We define the multiscale velocity basis functions as $\psi_{ij}^K = k(x)\nabla\phi_{ij}^K$ and the corresponding finite element space as

$$V_h = \bigoplus_K \{\psi_{ij}^K\} \bigcap H(div, \Omega).$$

For the boundedness of the velocity basis functions, we make the following assumption:

Assumption A2. $\|\psi_{ij}^K\|_{0,K} \leq C \frac{|K|^{1/2}}{|F_j^K|}$ for all i, j.

Remark 1.7 We note that the lowest order Raviart-Thomas (RT_0) basis functions [13] and the local mixed MsFEM basis functions [16] satisfy the *Assumption A2*. It is easy to check *Assumption A2* when $|v_i|$ is bounded along each F.

Let $Q_h = \oplus_K P_0(K) \subset L^2(\Omega)$ be the finite element space consisiting of piecewise constants basis functions for the pressure. Before we proceed with our convergence analysis, we recall the following lemma.

Lemma 1.8 *[4]* Let v_i be defined in Remark 1.6 and ψ_{ij}^K be defined in (1.7). Then we have

$$v_i|_K \in span\{\psi_{ij}^K\}, \quad i = 1, \ldots, N; \quad j = 1, 2, \ldots j_K.$$

Let

$$X(K) = \left\{ u | u = \sum_{i=1}^{N} a_i(x)v_i \text{ and } \int_F a_i v_i \cdot n_F ds \text{ exists} \right\}$$

be a subspace of $H(div, K)$. We define an interpolation operator $\Pi_h : H(div,\Omega) \cap \prod_{K \in \mathcal{T}_h} X(K) \longrightarrow V_h$ such that in each element K, for any $u = \sum_i a_i(x)v_i$

$$\Pi_h|_K \left(\sum_i a_i(x)v_i \right) = \sum_{i,j} a_{ij}^K \psi_{ij}^K,$$

where $a_{ij}^K = \int_{F_j^K} a_i(x)v_i \cdot n_{F_j^K} ds$. In [4], we have discussed the regularity of $a_i(x)$ and $v_i(x)$ such that $\int_F a_i v_i \cdot n_F ds$ is meaningful. Utilizing the definition of Π_h, we obtain the following lemma.

Lemma 1.9 [4] Let $u = \sum_{i=1}^{N} a_i v_i \in H(div, \Omega) \cap \prod_{K \in T_h} X(K)$ and $q_h \in Q_h$. Then we have $\int_{\Omega} div(u - \Pi_h u) q_h dx = 0$.

From Lemma 1.9, we obtain the following result.

Lemma 1.10 Let v and v_h solve (1.11) and (1.8) respectively. Then

$$\|v - v_h\|_{0,\Omega} \leq (1 + \frac{\gamma_{\min}}{\gamma_{\max}}) \|v - \Pi_h v\|_{0,\Omega}.$$

Proof From equations (1.11) and (1.8), we obtain that

$$\begin{aligned}((\lambda k)^{-1}(v - v_h), \sigma_h) - (div\sigma_h, (p - p_h)) &= 0 \quad \forall \sigma_h \in V_h \\ (div(v - v_h), q_h) &= 0, \quad \forall q_h \in Q_h.\end{aligned} \quad (1.12)$$

Using Lemma 1.9 and the second equation of (1.12), we get

$$(div(\Pi_h v - v_h), q_h) = 0.$$

By taking $q_h = div(\Pi_h v - v_h)$, we have

$$div(\Pi_h v - v_h) = 0.$$

Set $\sigma_h = \Pi_h v - v_h$ in (1.12) and we get

$$((\lambda k)^{-1}(v - v_h), (\Pi_h v - v_h)) = 0. \quad (1.13)$$

Noting that $\gamma_{\max} |\xi|^2 \leq \xi^t (\lambda k)^{-1} \xi \leq \gamma_{\min} |\xi|^2$ for all $\xi \in \mathbb{R}^d$, we have

$$\begin{aligned}\|\Pi_h v - v_h\|_{0,\Omega}^2 &\leq \frac{1}{\gamma_{\max}} ((\lambda k)^{-1}(\Pi_h v - v_h), (\Pi_h v - v_h)) \\ &= \frac{1}{\gamma_{\max}} ((\lambda k)^{-1}(\Pi_h v - v), (\Pi_h v - v_h)) \\ &\quad + \frac{1}{\gamma_{\max}} ((\lambda k)^{-1}(v - v_h), (\Pi_h v - v_h)) \\ &= \frac{1}{\gamma_{\max}} ((\lambda k)^{-1}(\Pi_h v - v), (\Pi_h v - v_h)) \\ &\leq \frac{\gamma_{\min}}{\gamma_{\max}} \|\Pi_h v - v\|_{0,\Omega} \|\Pi_h v - v_h\|_{0,\Omega},\end{aligned} \quad (1.14)$$

where we have used (1.13) in the third step and the Schwarz inequality in the last step. Consequently, it follows

$$\|\Pi_h v - v_h\|_{0,\Omega} \leq \frac{\gamma_{\min}}{\gamma_{\max}} \|\Pi_h v - v\|_{0,\Omega}.$$

We complete the proof by applying the triangle inequality. □

From the above lemma, we obtain the following convergence theorem for the velocity.

Theorem 1.11 *Let v and v_h solve (1.11) and (1.8) respectively. Suppose that Assumption A1 and Assumption A2 hold. Then we have*

$$\|v - v_h\|_{0,\Omega} \leq C\sqrt{\sum_K \sum_j \delta_{F_j^K}^2 |F_j^K|^{2(\beta-1)}|K|}.$$

Proof Let $\beta_{ij}^K = \int_{F_j^K} v_i \cdot n \, ds$ and $A_{ij}^K = \int_{F_j^K} A_i v_i \cdot n \, ds$. A straightforward computation implies that in each K,

$$\begin{aligned}
|A_{ij}^K - \bar{A}_i^j \beta_{ij}^K| &= |\int_{F_j^K} A_i v_i \cdot n \, ds - \bar{A}_i^j \int_{F_j^K} v_i \cdot n \, ds| \\
&= |\int_{F_j^K} (A_i - \bar{A}_i^j) v_i \cdot n \, ds| \\
&\leq C \delta_{F_j^K} |F_j^K|^\beta,
\end{aligned} \qquad (1.15)$$

where we have used *Assumption A1*. By Lemma (1.8), i.e. $v_i|_K = \sum_j \beta_{ij}^K \psi_{ij}^K$, we have

$$\begin{aligned}
\|v - \Pi_h v\|_{0,K} &= \|\sum_{i,j}(A_i(x)\beta_{ij}^K - A_{ij}^K)\psi_{ij}^K\|_{0,K} \\
&\leq \|\sum_{i,j}(A_i(x) - \bar{A}_i^j)\beta_{ij}^K \psi_{ij}^K\|_{0,K} + \|\sum_{i,j}(\bar{A}_i^j \beta_{ij}^K - A_{ij}^K)\psi_{ij}^K\|_{0,K} \\
&\leq \|\sum_{i,j}|A_i(x) - \bar{A}_i^j|\beta_{ij}^K \psi_{ij}^K\|_{0,K} + \|\sum_{i,j}|\bar{A}_i^j \beta_{ij}^K - A_{ij}^K|\psi_{ij}^K\|_{0,K} \\
&\leq C \sum_j \delta_{F_j^K} |F_j^K|^\beta \frac{|K|^{\frac{1}{2}}}{|F_j^K|} \\
&\leq C \sum_j \delta_{F_j^K} |F_j^K|^{\beta-1} |K|^{1/2}
\end{aligned} \qquad (1.16)$$

where we have used (1.15) and *Assumption A2*. Consequently, we get

$$\begin{aligned}
\|v - \Pi_h v\|_{0,\Omega}^2 &\leq \sum_K \|v - \Pi_h v\|_{0,K}^2 \\
&\leq C \sum_K \sum_j \delta_{F_j^K}^2 |F_j^K|^{2(\beta-1)}|K|
\end{aligned} \qquad (1.17)$$

and

$$\|v - \Pi_h v\|_{0,\Omega} \leq C\sqrt{\sum_K \sum_j \delta_{F_j^K}^2 |F_j^K|^{2(\beta-1)}|K|} \qquad (1.18)$$

By using Lemma 1.10, we complete the proof. $\square$

We note that (1.18) is one of our main results. Next, we make some specific assumptions about the coarse mesh. We assume that the coarse mesh is rectangular (or box in 3-D) with the sides having different lengths. The assumption below is used to describe the large anisotropy of the grid and estimate the convergence rate when the sides of the rectangular element differ substantially.

Assumption A3. *Assume each coarse grid block K is rectangular (or box shape in 3-D) with the sides having lengths h_1, h_2 (or h_1, h_2, and h_3 in 3-D). Furthermore, we denote $\delta_{h_i} = \delta_F$ (or $\delta_{h_i h_j} = \delta_F$) for an edge F (or a face F in 3-D).*

For the special case described in *Assumption A3*, Theorem 1.11 implies the following result.

Corollary 1.12 *Under Assumption A1, Assumption A2, and Assumption A3, we have the following convergence estimates:*

$$\|v - v_h\|_{0,\Omega} \leq C \max_{i \in \{1,2\}} \{\delta_{h_i} h_i^{\beta-1}\}$$

in the 2-D case and

$$\|v - v_h\|_{0,\Omega} \leq C \max_{i,j \in \{1,2,3\}} \{\delta_{h_i h_j} (h_i h_j)^{\beta-1}\}$$

in the 3-D case.

Proof We only give the proof for the 3-D case. The proof of (1.16) implies that

$$\|v - \Pi_h v\|_{0,K} \leq C \max_{i,j,l \in \{1,2,3\}} \{\delta_{h_i h_j} (h_i h_j)^{\beta-1/2} h_l^{1/2}\},$$

where *Assumption A1*, *Assumption A2*, and *Assumption A3* have been used. Therefore, we obtain

$$\|v - \Pi_h v\|_{0,\Omega}^2 \leq \sum_K \|v - \Pi_h v\|_{0,K}^2$$
$$\leq C \frac{1}{(h_1 h_2 h_3)} \max_{i,j \in \{1,2,3\}} \{\delta_{h_i h_j}^2 (h_i h_j)^{2\beta-1} h_l\}. \quad (1.19)$$

and

$$\|v - \Pi_h v\|_{0,\Omega} \leq C \max_{i,j \in \{1,2,3\}} \{\delta_{h_i h_j} (h_i h_j)^{\beta-1}\}. \quad (1.20)$$

The proof then follows from Lemma 1.10. This completes the proof. □

Remark 1.13 Define by N_{h_i,h_j,h_l} $(i,j,l \in \{1,2,3\})$ the number of coarse grid blocks with the sides having the lengths $O(h_i)$ and $O(h_j)$ and $O(h_l)$ in 3-D. Then

$$\|v - \Pi_h v\|_{0,\Omega} \leq C \sum_{\{i,j,l\} \in \{1,2,3\}} \max_{i,j,l} \{N_{h_i,h_j,h_l} \delta_{h_i h_j} (h_i h_j)^{\beta-1/2} h_l^{1/2}\}. \quad (1.21)$$

This follows from Corollary 1.12.

Remark 1.14 If Π_h is bounded from $H^1(\Omega)$ to $H(div,\Omega)$, i.e. $\|\Pi_h u\|_{0,\Omega} \leq C\|u\|_{1,\Omega}$ for all u in $H^1(\Omega)$, it follows that [19]

$$\|p - p_h\|_{0,\Omega} \leq C(\|p - P_{Q_h}p\|_{0,\Omega} + \|v - \Pi_h v\|_{0,\Omega}),$$

where P_{Q_h} is L^2 orthogonal projection operator. Therefore the L^2 projection estimate and Theorem 1.11 imply that

$$\|p - p_h\|_{0,\Omega} \leq CH + C\sqrt{\sum_K \sum_j \delta^2_{F_j^K}|F_j^K|^{2(\beta-1)}|K|}$$

provided that $diam(K) \leq CH$ for all K whose chunkiness parameters (see page 99 in [11] for the definition) are bounded below by a constant.

If we would like to estimate the approximation for the pressure p and velocity v simultaneously, we need to assume a discrete inf-sup condition, i.e. for any $q_h \in Q_h$, there exists a constant C such that

$$\sup_{\sigma_h \in V_h} \frac{\int_\Omega div\sigma_h q_h dx}{\|\sigma_h\|_{H(div,\Omega)}} \geq C\|q_h\|_{0,\Omega}. \tag{1.22}$$

Then we have the following stability estimate [13].

$$\|v - v_h\|_{H(div,\Omega)} + \|p - p_h\|_{0,\Omega} \leq C\{\inf_{\sigma_h \in V_h}\|v - \sigma_h\|_{H(div,\Omega)} + \inf_{q_h \in Q_h}\|p - q_h\|_{0,\Omega}\}. \tag{1.23}$$

Remark 1.15 The inf-sup condition (1.22) is investigated in [4] for quasi-uniform meshes.

As a consequence, we obtain the following theorem.

Theorem 1.16 Let $\{v,p\}$ and $\{v_h,p_h\}$ be the solution of (1.11) and (1.8) respectively. Under Assumption A1, Assumption A2 and the discrete inf-sup condition (1.22), we have

$$\|v - v_h\|_{H(div,\Omega)} + \|p - p_h\|_{0,\Omega}$$
$$\leq C(|p|_{1,\Omega} + |f|_{1,\Omega})H + C\sqrt{\sum_K \sum_j \delta^2_{F_j^K}|F_j^K|^{2(\beta-1)}|K|},$$

provided that $diam(K) \leq CH$ for all K whose chunkiness parameters are bounded below by a constant.

Proof For the proof, we need to choose a proper σ_h and a proper q_h such that the right hand side of (1.23) is small.

The second term on the right hand in (1.23) can be easily estimated. In fact, with the choice $q_h|_K = \langle p \rangle_K$, i.e. the average of p in K, we have

$$\inf_{q_h \in Q_h} \|p - p_h\|_{0,\Omega} \leq CH|p|_{1,\Omega}.$$

Next we try to find a $\sigma_h \in V_h$, say $\sigma_h|_K = \Pi_h v$, and estimate the first term on the right hand in (1.23). Invoking Lemma 1.8 and its proof, it follows that in each K

$$v - \sigma_h = v - \Pi_h v$$

$$= \sum_{i,j} (A_i(x)\beta_{ij}^K - \int_{F_j^K} A_i(x)v_i \cdot nds)\psi_{ij}^K. \quad (1.24)$$

Since $\int_K \sum_i div(A_i(x)v_i)dx = f$, the divergence theorem implies

$$\int_{\partial K} \sum_i A_i(x)v_i \cdot nds = f.$$

This gives rise to

$$\|div(v - \Pi_h v)\|_{0,K} = \|f - \sum_{i,j} \int_{F_j^K} A_i(x)v_i \cdot nds \frac{1}{|K|}\|_{0,K}$$
$$= \|f - \langle f \rangle_K\|_{0,K} \leq CH|f|_{1,K}. \quad (1.25)$$

After summing over all K for (1.25), we have

$$\|div(v - \sigma_h)\|_{0,\Omega} \leq CH|f|_{1,\Omega}. \quad (1.26)$$

We complete the proof by using Theorem 1.11. □

Corollary 1.12 and Theorem 1.16 imply the following result.

Corollary 1.17 *If Assumption A1, A2, A3, and (1.22) hold, then we have*

$$\|v - v_h\|_{H(div,\Omega)} + \|p - p_h\|_{0,\Omega} \leq C(|p|_{1,\Omega} + |f|_{1,\Omega})H + C \max_{i \in \{1,2\}} \{\delta_{h_i} h_i^{\beta-1}\}$$

in the 2-D case and

$$\|v - v_h\|_{H(div,\Omega)} + \|p - p_h\|_{0,\Omega}$$
$$\leq C(|p|_{1,\Omega} + |f|_{1,\Omega})H + C \max_{i,j \in \{1,2,3\}} \{\delta_{h_i h_j} (h_i h_j)^{\beta-1}\}$$

in the 3-D case.

Remark 1.18 *If $A_i(x) \in C^1(\prod_{K \in \tau} K) \cap C^0(\Omega)$ in Assumption A1 and v_i are defined such that $\beta = 1$ (e.g. v_i are bounded in $L^\infty(F_l)$ for all F_l), then*

Theorem 1.16 implies that

$$\|v - v_h\|_{H(div,\Omega)} + \|p - p_h\|_{0,\Omega} \le Ch$$

for a quasi-uniform grid with $diam(K) \approx h$.

1.5.1 Remarks on time dependent problems

We can extend the global MsFEM on an unstructured grid to time dependent problems. In this section, we give a brief discussion for linear parabolic equations and wave equations.

For simplicity, we first discuss the following model parabolic equation.

$$\begin{aligned}
\frac{\partial}{\partial t} p - div(k(x)\nabla p) &= f(t,x) \quad \text{in } \Omega_T \\
p &= 0 \quad \text{on } \partial\Omega \times [0,T] \\
p(t=0) &= p_0 \quad \text{in } \Omega.
\end{aligned} \quad (1.27)$$

Let $v = -k(x)\nabla p$ be the velocity, where p solves (1.27). We make the following assumption for v.

Assumption A1T. *There exist velocity fields $v_1, \cdots, v_N$ and functions $A_1(t,x), \cdots, A_N(t,x)$ such that*

$$v(t,x) = \sum_{i=1}^{N} A_i(t,x)v_i, \quad (1.28)$$

where $\int_F |v_i \cdot n| ds \le C|F|^\beta$ for any face (or edge) F. For any $A_i(t,x)$ and any $F \subset \partial K$, $\|A_i(t,x) - \bar{A}_i(t)\|_{L^2(0,T,L^\infty(K))} \le C\delta_F$ for some constant C and all t, where $\bar{A}_i(t) = \frac{1}{|F|} \int_F A_i(t,x) ds$.

Remark 1.19 Let $v_i = -k(x)\nabla p_i$ ($i = 1, \cdots, d$) be defined in (1.10), then $p(t,x) = p(t,p_1,p_2) \in L^2(0,T;W^{2,s}(\Omega))$ ($s > 2$) [36]. Consequently, we have

$$v(t,x) = -k(x)\nabla p = \sum_i \frac{\partial p}{\partial p_i} k \nabla p_i := \sum_i A_i(t,x)v_i,$$

where $A_i(t,x) = \frac{\partial p}{\partial p_i} \in L^2(0,T;W^{1,s}(\Omega))$.

The mixed formulation associated to (1.27) is to find $\{v,p\} : [0,T] \longrightarrow H(div,\Omega) \times L^2(\Omega)$ such that

$$\begin{aligned}
\left(\frac{\partial}{\partial t} p, q\right) + (div v, q) &= (f,q) \quad \forall q \in L^2(\Omega) \\
(k^{-1}v, \sigma) - (div \sigma, p) &= 0 \quad \forall \sigma \in H(div,\Omega) \\
p(0) &= p_0.
\end{aligned} \quad (1.29)$$

The space-discrete mixed formulation is to find $\{v_h, p_h\} : [0, T] \longrightarrow V_h \times Q_h$ such that

$$\left(\frac{\partial}{\partial t} p_h, q_h\right) + (div v_h, q_h) = (f, q_h) \quad \forall q_h \in Q_h$$
$$(k^{-1} v_h, \sigma_h) - (div \sigma_h, p_h) = 0 \quad \forall \sigma_h \in V_h \quad (1.30)$$
$$p_h(0) = p_{0,h}.$$

We define

$$\|v\|_{L^2(0,T;L^2_k(\Omega))}^2 = \int_0^T \int_\Omega v^t \cdot k^{-1}(x) v \, dx \, ds.$$

By using the proof of Theorem 4.2 in [4] and the proof of Theorem 1.16, we conclude the following theorem.

Theorem 1.20 *Let $\{v, p\}$ and $\{v_h, p_h\}$ be the solution of (1.29) and (1.30) respectively. Under Assumption A1T, A2, we have*

$$\|p - p_h\|_{C^0(0,T;L^2(\Omega))} + \|v - v_h\|_{L^2(0,T;L^2_k(\Omega))}$$
$$\leq C |p|_{C^0(0,T;H^1(\Omega))} H + C \sqrt{\sum_K \sum_j \delta_{F_j^K}^2 |F_j^K|^{2(\beta-1)} |K|},$$

provided that $diam(K) \leq CH$ for all K whose chunkiness parameters are bounded below by a constant.

Next we discuss a model wave equation with a homogenous boundary condition

$$\frac{\partial^2}{\partial t^2} p - div k(x) \nabla p = f \quad \text{in } \Omega_T$$
$$p(x, 0) = g_0 \quad \text{in } \Omega \quad (1.31)$$
$$\frac{\partial}{\partial t} p(x, 0) = g_1 \quad \text{in } \Omega$$

Let $v = -k(x) \nabla p$, where p solves (1.31). We make the following observation.

Remark 1.21 Let $v_i = -k(x) \nabla p_i$ ($i = 1, \cdots, d$) be defined in (1.10), then the change of variable implies that $v(t, x) = \sum_i A_i(t, x) v_i$, where $A_i(t, x) = \frac{\partial p}{\partial p_i}$ and $v_i = -k \nabla p_i$. Assume that $f \in L^\infty(L^p(\Omega)) \cap H^1(L^p(\Omega))$, $g_1 \in W^{1,p}(\Omega)$ and $\frac{\partial^2}{\partial t^2} p(0) \in L^p(\Omega)$. Then the proof of Theorem 1.1 in [37] implies that $A_i(t, x) \in L^\infty(W^{1,p}(\Omega))$. Consequently $A_i(t, x) \in L^2(C^{1-\frac{d}{p}}(\Omega))$ if $p > d$ by using the Sobolev embedding theorem.

The mixed formulation of (1.31) is to find $\{p, v\} : [0, T] \longrightarrow L^2(\Omega) \times H(div, \Omega)$ such that

$$\left(\frac{\partial^2}{\partial t^2} p, q\right) + (divv, q) = (f, q) \quad \forall q \in L^2(\Omega)$$
$$(k^{-1}v, \sigma) - (p, div\sigma) = 0 \quad \forall \sigma \in H(div, \Omega)$$
$$(p(0), q) = (g_0, q) \quad \forall q \in L^2(\Omega) \quad (1.32)$$
$$\left((\frac{\partial}{\partial t}p)(0), q\right) = (g_1, q) \quad \forall q \in L^2(\Omega)$$
$$(k^{-1}v(0), \sigma) = -(\nabla g_0, \sigma) \quad \forall \sigma \in H(div, \Omega).$$

The numerical mixed formulation of (1.31) is to find $\{p_h, v_h\} : [0, T] \longrightarrow Q_h \times V_h$ such that

$$\left(\frac{\partial^2}{\partial t^2} p_h, q_h\right) + (divv_h, q_h) = (f, q_h) \quad \forall q_h \in Q_h$$
$$(k^{-1}v_h, \sigma_h) - (p_h, div\sigma_h) = 0 \quad \forall \sigma_h \in V_h$$
$$(p_h(0), q_h) = (g_0, q_h) \quad \forall q_h \in Q_h \quad (1.33)$$
$$\left((\frac{\partial}{\partial t}p_h)(0), q_h\right) = (g_1, q_h) \quad \forall q_h \in Q_h$$
$$(v_h(0), \sigma_h) = (\Pi_h v(0), \sigma_h) \quad \forall \sigma_h \in V_h.$$

By using the proof of Theorem 5.1 in [34] and the proof of Theorem 1.16, we can prove the following theorem.

Theorem 1.22 *Let $\{v, p\}$ and $\{v_h, p_h\}$ be the solution of (1.32) and (1.33) respectively and Assumption A1T, A2 hold. Then we have*

$$\|p - p_h\|_{L^\infty(L^2(\Omega))} + \sup_t \|\int_0^t (v(t) - v_h(t))ds\|_{L^2_k(\Omega)}$$
$$\leq C|p|_{L^\infty(H^1(\Omega))} H + C \sqrt{\sum_K \sum_j \delta_{F_j^K}^2 |F_j^K|^{2(\beta-1)} |K|},$$

provided that $diam(K) \leq CH$ for all K whose chunkiness parameters are bounded below by a constant.

1.6 Conclusions

This paper discusses the mixed multiscale finite element method (MsFEM) on an unstructured coarse grid. Our goal is to develop a coarse-scale model for the two-phase flow and transport. An unstructured coarse grid is often needed to upscale the transport equation. Solving the flow equation on the same coarse grid provides a general robust coarse-scale model for a multi-phase flow and transport. We note that most previous studies use a two-grid approach where the flow equation is solved on the coarse grid while the transport equation is solved on the fine grid.

In this paper, we consider the non-uniform coarsening developed in [2] for solving the transport equation on the coarse grid. The resulting coarse grid is highly anisotropic and is usually not quasi-uniform. We study the convergence of the mixed MsFEM on an unstructured coarse grid following [4] under some restrictive assumptions. We show that the mixed MsFEM using limited global information converges as the coarse mesh approaches to zero.

We present a few representative numerical examples involving highly channelized permeability as well as a 3-D reservoir model using an unstructured fine grid. We compare some integrated responses such as water-cut as well as saturation fields. The use of an unstructured coarse grid provides more accurate results and preserves geological realism in flow and transport simulations. In particular, we observe accurate results for the integrated responses such as water-cut. As for the saturation, we observe that the error is reduced by almost a factor of two when an unstructured coarse grid is used. We show that the error is mostly due to the saturation upscaling. In particular, the error which is introduced by MsFEM is under 1%. Similar results are observed when the transport equation is solved on the fine grid. Finally, we note that it is more appropriate to use an unstructured coarse grid when the underlying fine grid is unstructured. The latter often occurs in petroleum applications.

Although the results presented in this paper are encouraging, there is room for further exploration of some of the underlying approaches. As our intent here was to demonstrate that mixed MsFEM on an unstructured grid could be effectively used in a full coarse-scale model, we did not consider the issue related to the upscaling of the saturation equation. As we noticed that the main source of errors in our coarse-scale simulations is due to the saturation upscaling. The upscaling of the saturation is achieved via the use of a carefully selected coarse grid, however, no subgrid model is used in the upscaling of the transport equations. We plan to investigate the subgrid effects in the saturation equation in the future. This will help to improve further the upscaling result. It may also be worth exploring the impact of the unstructured grid obtained via the single-phase flow on the accuracy of the mixed MsFEM. As we noticed that the mixed MsFEM is more accurate when an unstructured coarse grid is used. Further theoretical investigation of these issues is worth pursuing in future.

References

[1] Aarnes J. (2004). On the use of a mixed multiscale finite element method for greater flexibility and increased speed or improved accuracy in reservoir simulation, *SIAM MMS*, **2**, p. 421–439.

[2] Aarnes J. E., Hauge V. L., and Efendiev Y. (2007). "Coarsening of three-dimensional structured and unstructured grids for subsurface flow", *Advances in Water Resources*, Vol. **30**, Issue 11, November 2007, p. 2177–2193.

[3] Aarnes J. and Efendiev Y. (2008). "Mixed multiscale finite element for stochastic, porous media flows", to appear in *SIAM Sci. Comp*, **30**(5), p. 2319–2339.

[4] Aarnes J., Efendiev Y., and Jiang L. (2008). "Mixed multiscale finite element methods using limited global information", appear in *SIAM MMS*, **7**(2), p. 655–676.

[5] Aarnes J., Kippe V., and Lie K.-A. (2005). "Mixed multiscale finite elements and streamline methods for reservoir simulation of large geomodels", *Advances in Water Resources*, **28**, p. 257–271.

[6] Allaire G. and Brizzi R. (2005). "A multiscale finite element method for numerical homogenization", *SIAM MMS*, **4**(3), p. 790–812.

[7] Arbogast T. (2002). "Implementation of a locally conservative numerical subgrid upscaling scheme for two-phase Darcy flow", *Comput. Geosci.*, **6**, p. 453–481.

[8] Arbogast T., Pencheva G., Wheeler M. F., and Yotov I. (2008). "A multiscale mortar mixed finite element method", *SIAM J. Multiscale Modeling and Simulation*, **6**, p. 319–346.

[9] Babuška I., Caloz G., and Osborn E. (1994). "Special finite element methods for a class of second order elliptic problems with rough coefficients", *SIAM J. Numer. Anal.*, **31**, p. 945–981.

[10] Babuška I. and Osborn E. (1983). "Generalized finite element methods: Their performance and their relation to mixed methods", *SIAM J. Numer. Anal.*, **20**, p. 510–536.

[11] Brenner S. C. and Scott L. R. (2002). *The Mathematical theory of finite element methods*, Springer-Verlag, New York.

[12] Brezzi F. (2000). *Interacting with the subgrid world, in Numerical analysis 1999 (Dundee)*, Chapman & Hall/CRC, Boca Raton, FL, p. 69–82.

[13] Brezzi F. and Fortin M. (1991). *Mixed and hybrid finite element methods*, Springer-Verlag, Berlin – Heidelberg – New York.

[14] Chen Y. and Durlofsky L. J. (2006). "Adaptive local-global upscaling for general flow scenarios in heterogeneous formations", *Transport in Porous Media*, **62**, p. 157–185.

[15] Chen Y., Durlofsky L. J., Gerritsen M., and Wen X. H. (2003). "A coupled local-global upscaling approach for simulating flow in highly heterogeneous formations", *Advances in Water Resources*, **26**, p. 1041–1060.

[16] Chen Z. and Hou T. Y. (2002). "A mixed multiscale finite element method for elliptic problems with oscillating coefficients", *Math. Comp.*, **72**, p. 541–576.

[17] Christie M. and Blunt M. (2001). "Tenth SPE Comparative Solution Project: A comparison of upscaling techniques", *SPE Reser. Eval. Eng.*, **4**, p. 308–317.

[18] Dostert P., Efendiev Y., and Hou T. "Multiscale finite element methods for stochastic porous media flow equations", to appear in *CMMAME*.

[19] Douglas J. and Roberts J. E. (1985). "Global estimates for mixed methods for second order elliptic equations", *Math. Comp.* **44**, p. 39–52.

[20] Durlofsky L. J., Jones, R. C. and Milliken W. J. (1997). "A nonuniform coarsening approach for the scale-up of displacement processes in heterogeneous porous media", *Adv. Water Resources*, **20**, p. 335–347.
[21] E W. and Engquist B. (2003). "The heterogeneous multi-scale methods", *Comm. Math. Sci.*, **1**(1), p. 87–133.
[22] Efendiev Y., Ginting V., Hou T., and Ewing R. (2006). "Accurate multiscale finite element methods for two-phase flow simulations", *J. Comp. Physics.*, **220**(1), p. 155–174.
[23] Efendiev Y., Hou T., and Ginting V. (2004). "Multiscale finite element methods for nonlinear problems and their applications", *Comm. Math. Sci.*, **2**, p. 553–589.
[24] Efendiev Y., Hou T., and Luo W. (2006). "Preconditioning Markov chain Monte Carlo simulations using coarse-scale models", *SIAM J. Sci. Comput.*, **28**, no. 2, p. 776–803
[25] Efendiev Y. R., Hou T. Y., and Wu X. H. (2000). "Convergence of a nonconforming multiscale finite element method", *SIAM J. Num. Anal.*, **37**, p. 888–910.
[26] Efendiev Y. and Pankov A. (2004). "Numerical homogenization of nonlinear random parabolic operators", *SIAM MMS*, **2**(2), p. 237–268.
[27] Gerritsen M. and Durlofsky L. J. (2005). "Modeling of fluid flow in oil reservoirs", *Annual Reviews in Fluid Mechanics*, **37**, p. 211–238.
[28] Holden L. and Nielsen B. F. (2000). "Global Upscaling of Permeability in Heterogeneous Reservoirs: the output least squares (OLS) Method", *Transport in Porous Media*, **40**, p. 115–143.
[29] Hou T. Y., Westhead A., and Yang D. P. (2006). "A Framework for Modeling Subgrid Effects for Two-Phase Flows in Porous Media", *SIAM MMS*, **5**(4), 1087–1127.
[30] Hou T. Y. and Wu X. H. (1997). "A multiscale finite element method for elliptic problems in composite materials and porous media", *Journal of Computational Physics*, **134**, p. 169–189.
[31] Hughes T., Feijoo G., Mazzei L., and Quincy J. (1998). "The variational multiscale method a paradigm for computational mechanics", *Comput. Methods Appl. Mech. Engrg*, **166**, p. 3–24.
[32] Jenny P., Lee, S. H., and Tchelepi H. (2003). "Multi-scale finite volume method for elliptic problems in subsurface flow simulation", *J. Comput. Phys.*, **187**, p. 47–67.
[33] ——— (2005). "Adaptive multi-scale finite volume method for multi-phase flow and transport in porous media", *Multiscale Modeling and Simulation*, **3**, p. 30–64.
[34] Jiang L., Efendiev Y., and Ginting V. "Global multiscale methods for acoustic wave equations with continuum spatial scales", submitted to *SIAM J. Numerical Analysis*.
[35] Owhadi H. and Zhang L. (2007). "Metric based up-scaling", *Com. Pure Appl. Math.*, **60**, p. 675–723.

[36] Owhadi, H. and Zhang, L. "Homogenization of parabolic equations with a continuum of space and time scales", to appear in *SIAM J. Numer. Anal.* 46(2007/08), no. 1, 1–36.
[37] Owhadi H. and Zhang L. (2008). "Numerical homogenization of the acoustic wave equation with a continuum of scales", *Comput. Methods Appl. Mech. Engrg.*, **198**, p. 397–406.
[38] Ponting D. K. (1989). *Corner-point geometry in reservoir simulation.* In: King P. R., editor. Proceedings of the first European Conference on Mathematics of Oil Recovery, Cambridge, (Oxford), Clarendon Press, July 25–27 1989, p. 45–65.
[39] Wu X. H., Parashkevov R., and Stone M. (2007). *Reservoir modeling with global scale-up*, SPE 105237, presented in the 15th Middle East Oil & Gas Show and Conference, March 11–14.

2

FORMULATIONS OF MECHANICS PROBLEMS FOR MATERIALS WITH SELF-SIMILAR MULTISCALE MICROSTRUCTURE

R.C. Picu and M.A. Soare

2.1 Introduction

Materials with hierarchical microstructure are quite often encountered in nature. These are assembled from individual entities which, in turn, are also composed of other identical or dissimilar entities. The hierarchy may be limited to two scales or may be composed of multiple levels extending therefore over a broad range of spatial scale. If the entities observed in the microstructure on various scales are geometrically similar (one can find a scaling mapping that relates them), the hierarchical microstructure is denoted as "self-similar."

Examples of such materials include the trabecular bone, muscles and tendons, the sticky foot of the Gecko lizard, the structural support of various plants and algae, etc. In general, most biological materials, which are made by controlled assembly of molecular components, exhibit hierarchical structures. Self-similarity is usually observed over a limited range of scales and may be deterministic or stochastic in nature. Stochastic self-similarity is generally the rule. The stochastic nature of the structure is related either or both to the dimensions of the building blocks that are replicated from scale to scale, or to their relative position. Figure 2.1(a) shows a schematic representation of a tendon composed mainly from collagen fibers arranged in a hierarchical manner in a non-collagenous matrix [19]. Figure 2.1(b) shows an image of the skeleton of a marine micro-organism, a diatom, which is part of the ocean plankton. This unicellular organism is enclosed in the shell pictured in this image [52]. The shell has pores of various size and exhibits multiple spatial ranges of self-similarity, from 0.6 nm to 6 µm.

Man-made materials belonging to this category include aero-gels, filled polymers in which fillers form fractal aggregates as in tire rubber, and some dendritic structures. In all these materials the amount of geometric detail observed when zooming in increases with decreasing scale of observation. The micro (and nano)-structure may be self-similar from scale to scale (either in a deterministic or a stochastic fashion) or not. Aero-gels are obtained by aggregation of colloidal particles (e.g. base-catalyzed hydrolysis and condensation of silicon tetramethoxide (TMOS) or tetraethoxysilane (TEOS) in alcohol [4, 26] and followed by removal of the solvent through evaporation. These materials usually have low density (as low as $0.09 \, \text{g/cm}^3$) and high porosity, displaying a fractal-mass distribution.

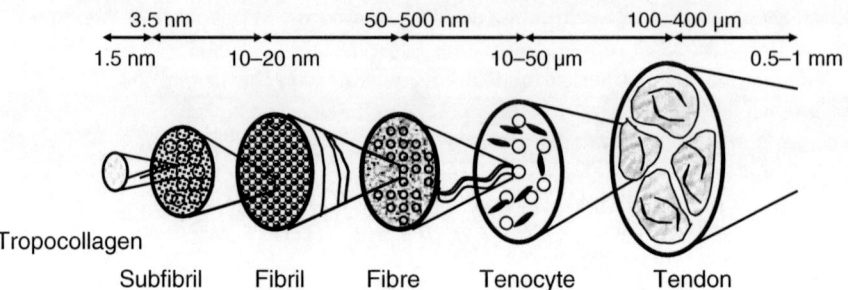

Figure 2.1: a) Schematic representation of the hierarchical organization of a tendon (adapted from [19]).

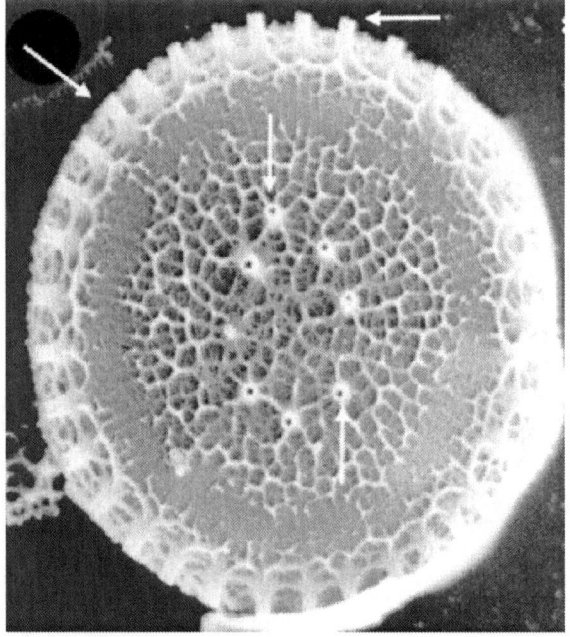

Figure 2.1: b) Image of the skeleton of a marine diaton from the species *Thalassiosira weissflogii*. The diameter of the structure is 0.8 µm [52].

Tire rubber gains desirable properties (stiffness and wear resistance) only after mixing with carbon black nanoparticles. These nanoscale inclusions are known to aggregate in structures that percolate the material and have fractal geometry over a range of scales. This microstructure is the result of a long optimization effort performed mostly by experimental trial and error. Clearly, design assisted by multiscale modeling technologies able to capture the behavior of such network materials would have reduced the time to market.

The self-similar property of the structure of given material may be evidenced by scattering experiments. Figure 2.2 shows a scattering pattern (the scattering intensity versus the scattering vector, k, in units of inverse wavelength of the probing radiation) obtained from a silica aerogel [39]. At long wavelengths, the scattering intensity is independent of the wave-vector indicating that at these length scales the material responds as a continuum. At smaller wave vectors, the discrete nature of the structure becomes evident and the slope of the curve becomes nonzero. A straight line in this diagram indicates self-similarity (the intensity scales as $I(k) \sim k^{-q}$). The exponent q is related to the fractal dimension of the respective object. Most materials exhibit power law scaling over a limited range of scales. In some, multiple scaling ranges are seen (as, for example, in the case of the diatom structure shown in Figure 2.1(b)). In this case, the material has multifractal geometry. At large scattering vectors, k, corresponding to length scales on the order and below 1 nm, the slope of -4 is recovered.

Solving boundary value problems defined for structures made from such materials is not trivial primarily because their constitutive response on the scale on which the problem is defined is unknown. Clearly, the constitutive behavior can be determined by experimentation on the relevant scale. However, this approach has well-known drawbacks as, for example, the lack of relationship

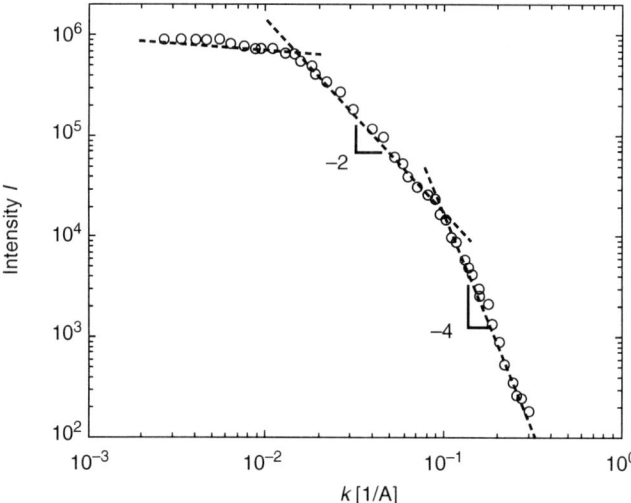

Figure 2.2: Small angle X-ray and light scattering from a silica aero-gel. Three regimes can be distinguished. For large wavelengths (small k), the material behaves as a continuum and the intensity is essentially independent of k; for intermediate k, the self-similar geometry of the structure leads to a slope -2. The usual -4 slope is seen at the largest k

with the microstructure and the fact that performing experiments that provide relevant date may not be always possible. An alternative is to use standard homogenization techniques to determine the effective response starting from the microstructure. As discussed later in this chapter, this meets with significant theoretical difficulties due to the lack of translational symmetry of these geometries and due to the fact that one can seldom assume scale decoupling, a mandatory condition for classical homogenization to apply. If no scale decoupling can be assumed, the microstructure cannot be "smeared out" into an equivalent continuum on the scale of observation. An equivalent statement is that in such materials, inclusions of various sizes are strongly interacting over a broad range of scales (i.e. over the spatial scales covering the range of self-similarity). It should be observed that deterministic self-similar geometries (with a high degree of regularity) have scaling symmetry but no translational symmetry, while stochastic self-similar geometries have scaling symmetry and only an approximate (in the average sense) translational symmetry.

In principle, one may use concurrent multiscale methods to solve this type of problems without scale decoupling. In these methods, various models representing the behavior of the material on various scales are used simultaneously in the problem domain. An example is the use of atomistic models in regions of the model where large field gradients exist, while various forms of continuum are used elsewhere, as in Quasicontinuum [48, 28]. Other types of hybrid discrete-continuum model have been developed to date. Their use in the situation discussed here is problematic due to the large number of scales present in the microstructure.

A completely new approach for such problems was investigated recently. The key idea is to include information about the complex geometry *in the governing equations*. This is opposed to the traditional method of representing the complexity through complicated boundary conditions. To clarify the discussion, let us consider a composite with a very large number of strongly interacting inclusions of various dimensions. Let us consider that a continuum description of the material is adequate both for the matrix and the inclusion materials. To solve mechanics boundary value problems on such domain, one would integrate the governing equations while imposing displacement and traction continuity across all matrix-inclusion interfaces. Hence, explicit representation of the interfaces is required. The solution will be defined over sub-domains, each sub-domain being made from a single material, either matrix or inclusion material. Clearly, when a large number of such interfaces are present, the cost of using this method becomes prohibitive. The new concept discussed here is based on the idea that the geometric complexity of the domain may be incorporated in the governing equations, rather than in the definition of the boundary conditions. This is a revolutionary idea. However, as with any such attempt, reaching a form that is broadly accepted and effectively useful in practice is not straightforward.

In this chapter we review the progress made to date in this direction. The discussion begins with a review of fractal geometry, the geometry that best describes the type of microstructure considered here. Few mathematical results relevant for the various formulations described are then presented, followed by a review of the works performed to date in this direction.

Although stochastic self-similar structures are encountered in nature, the present discussion is focused on deterministic self-similar structures. An approach to address boundary value problems for stochastic structures is presented in [42, 43]. The deterministic structures, although less important in practice, are interesting because they raise theoretical challenges to continuum mechanics and offer broad opportunities for developing new concepts.

2.2 The geometry of self-similar structures

The fractal geometry developed based on the ideas of Mandelbrot (1983) appears adequate to describe certain types of multiscale hierarchical microstructures. A brief overview of few basic notions is presented first.

Let us consider a one-dimensional example: the Cantor set. This set is a fractal embedded in 1D, say in the interval $A = [a, b]$, and is generated by the following geometric iterative procedure: the interval is divided in three segments of equal length and the middle segment is excluded from the set. The procedure is repeated with the end segments. Each such iteration, n, is identified with a "scale." The sets generated from the first three iterations are shown in Figure 2.3. In the following we will denote by F the domains belonging to the fractal inclusions and by A–F the embedding complement.

It is useful to observe that the Cantor set is neither discrete, nor continuum. It has the following properties: [3, 12] it is compact, i.e. it is bounded and closed (the limits of all sets of points from F are included in F), it is perfect, i.e. any point from F is the limit of a set of points from F, and is disconnected, i.e. between any two points of F exists at least a point from A–F. The property of being disconnected may be lost for fractals embedded in Euclidean spaces of dimension larger than one.

Self-similar structures of this type are neither continuum, nor discrete. They are compact and perfect (a characteristic property of a continuum) but any open set from A includes at least an open set containing points from A–F, and the Euclidean dimension of F is zero (which is a characteristic of discrete systems).

Fractals are usually characterized by their fractal dimension. Many such measures have been proposed. One of the most used is the box counting dimension which is determined by covering the set with segments of length ε_n, where n is a natural number representing the iteration step. In the first iteration of the set in Figure 2.3 one needs 2 segments of length $(b-a)/3$. In the n-th iteration,

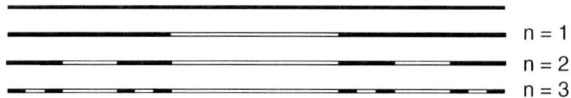

Figure 2.3: Three steps of the iteration leading to a deterministic Cantor fractal set.

$M_n = 2^n$ segments of length $(b-a)/3^n$ are needed. This leads to $\varepsilon_n = (b-a)/3^n$. Denoting by $\varepsilon_n^0 = \varepsilon_n/(b-a)$ a non-dimensional coefficient that decreases to zero as the iteration order increases, the box counting dimension, q, results from the identity $M_n = (\varepsilon_n^0)^{-q}$. Specifically,

$$q = \log(M_n)/\log(1/\varepsilon_n^0) = \log(M_n)/\log[(b-a)/\varepsilon_n] \quad (2.1)$$

In the Cantor set (Figure 2.3) case, $q = \log 2/\log 3$, with $q < 1$, i.e. the fractal dimension is smaller than the dimension of the Euclidean embedding space. Note that the total length corresponding to scale "n" is $L_n = M_n \varepsilon_n = (\varepsilon_n^0)^{-q} \varepsilon_n = (b-a)(\varepsilon_n^0)^{-q}$ [e.g. 13, 41, 42]. Since the parameter ε_n^0 is non-dimensional, L_n has the usual units (e.g. meter) for any n. Generally, if the object is embedded in a space of dimension d, and the topological dimension of the object is D_T, the object is a fractal if its dimension, q, has the property $D_T < q < d$.

It should be noted that the fractal dimension does not fully characterize the geometry and is only an indication of the "irregularity" of the object. Other measures (e.g. the lacunarity) were developed to provide additional information on the multiscale geometry, however, a unique set of quantities of this type probably does not exist, as some problem specificity is always present.

2.3 Basic concepts in fractional calculus

It is useful to review some of the mathematical tools used in this context. Fractional calculus appears to be appropriate for use on geometries with fractal characteristics [51]. This relationship has been identified only recently despite the fact that fractional calculus has its origins more than 100 years ago. It is necessary to define the basic differential and integral operators over such spaces.

2.3.1 Non-local fractional operators

Two equivalent expressions for these operators were initially proposed. One such set is known as the Grunwald–Letnikov operators and is based on the notion of fractional finite differences [17, 24]. Let us consider a function f defined on a one-dimensional domain $[a,b]$, and a real number $q \in (0,1)$. The differential of order q is given by:

$$\frac{d^q}{d(x-a)^q} f = \lim_{h \to 0} \frac{1}{h^q} \sum_{i=0,N} (-1)^i \frac{\Gamma(q+1)}{i!\Gamma(q-i+1)} f(x-ih),$$
$$N = [(x-a)/h], \quad (2.2a)$$

and the integral of order q over $[a,b]$ is given by:

$$I^q([a,x],f) = \lim_{h \to 0} h^q \sum_{i=0,N} \frac{\Gamma(q+i)}{i!\Gamma(q)} f(x-ih). \quad (2.2b)$$

The second set of operators is based on an extension of the multiple integral and are denoted as the Riemann–Liouville operators [16, 18]. The fractional integral of order q is given by:

$$I^q([a,x],f) = \frac{1}{\Gamma(q)} \int_a^x \frac{f(x')}{(x-x')^{1-q}} dx'. \tag{2.3a}$$

and the differential of order q results from (2.3a) as:

$$\frac{d^q}{d(x-a)^q} f = \left(\frac{d}{dx}\right) I^{1-q}([a,x],f) = \frac{1}{\Gamma(1-q)} \left(\frac{d}{dx}\right) \int_a^x \frac{f(x')}{(x-x')^q} dx'. \tag{2.3b}$$

All these operators are nonlocal (derivative at x depends on the choice of the left end of the interval, a) which makes their interpretation difficult. An unusual result is that the derivative of a constant function is not zero.

The non-local fractional operators were extensively used for modeling various physical phenomena as diffusion and transport on porous media [14], relaxation processes of polymers [30, 15], turbulent flows [9], viscous fingering and diffusion limited aggregation [54], for characterization of the rheologic (viscoelastic) behavior of materials [2, 25, 11]; and in fracture mechanics [8, 5]. In general, they were used to represent processes with slow, non-exponential relaxation governed by non-Markovian dynamics (dynamics controlled by spatial and/or temporal correlations). In such cases the system response cannot be represented by a superposition of exponentials (a Prony series), rather by power laws or stretched exponentials, which are natural solutions of fractional PDEs.

2.3.2 Local Fractional Operators

2.3.2.1 Fractional Differential Operators

Local fractional derivatives have been proposed by and Kolwankar and Gangal [22, 23], Kolwankar [21] in the form:

$$D^{qK} f(x_0) = \lim_{x \to x_0} \frac{1}{\Gamma(1-q)} \left(\frac{d}{dx}\right) \int_{x_0}^x \frac{f(x') - f(x_0)}{(x-x')^q} dx' \tag{2.4}$$

An alternate form has been recently used [40] and has the expression:

$$D^q f(x_0) = \lim_{n \to \infty} L^{q-1} \frac{f(x_n) - f(x_0)}{|x_n - x_0|^q \operatorname{sign}(x_n - x_0)} \tag{2.5}$$

The advantage of (2.5) is that its physical meaning is transparent: the derivative of order q is computed in a manner similar to the classical derivative, except that the length of the segment is measured on the fractal set, rather than in the embedding space. $(x_n - x_0)^q / L^{q-1}$ is the distance from x_n to x_0 measured on the fractal set, provided $L = \varepsilon_n$.

Expressions (2.4) and (2.5) differ through a multiplicative constant, specifically, $D^q f(x_0) = (L^{q-1}/\Gamma(1+q)) D^{qK} f(x_0)$. It is also noted that the units of the derivative (2.5) are similar to those of the classical derivative, i.e. 1/length (due to the introduction of parameter L), while those of expression (2.4) are $1/\text{length}^q$.

The elementary function $f : [0,1] \to R$, $f(x) = x^q$ with $0 < q < 1$ is not differentiable classically in $x_0 = 0$ but is fractional differentiable of order q and $D^q x^q|_{x=0} = L^{q-1}$ (with definition (2.5)). On the other hand, at all other points of [0,1], where f is classically differentiable, the local fractional derivative (2.5) vanishes. This is opposed to the non-local form which is non-zero.

It is useful to give an example of a function fractional differentiable at an infinite number of points in an embedding domain. This will also demonstrate, by means of an example, the relationship between the fractal support and the fractional operators. This function is an extension of the "devil staircase" function defined on interval [0,b] with b an arbitrary positive number. It is constructed starting from the deterministic Cantor set. Specifically, consider first the n-th step of the iteration leading to the Cantor set F in Figure 2.3. A continuous function is constructed having linear variation over each segment from $A-F_n$ and power law variation over segments from F_n (Figure 2.5):

$$f_n(x) = \begin{cases} f_n(x_i) + \beta/b(x - x_i) & \text{if } x \in (x_i, x_{i+1}] \subset A - F_n \\ f_n(x_i) + L^{1-q}\gamma/b\,(x - x_i)^q & \text{if } x \in (x_i, x_{i+1}] \subset F_n \end{cases} \quad (2.6)$$

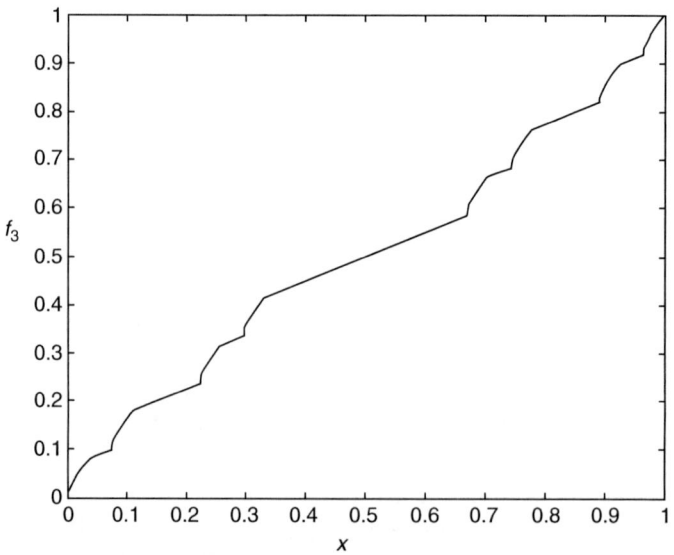

Figure 2.4: A function fractionally differentiable (eq. 2.6) of order $q = \log 2/\log 3$ at points of $n = 3$ order approximation of a Cantor set. Here x is normalized by the length b.

The function of interest here, f, results by taking the limit $n \to \infty$, so $f(x) = \lim_{n \to \infty} f_n(x)$ The set of non-derivability points (in the classical sense) coincides with the points of F. This function has the property:

$$\begin{cases} Df(x) = \beta/b & \text{if } x \in A-F \\ D^q f(x) = \gamma/b & \text{if } x \in F \end{cases}, \quad A = [0,1] \qquad (2.7)$$

i.e. it is a linear function on F and $A-F$ (note that F is defined in the limit $n \to \infty$). By using this procedure, one may define an entire family of power law functions of higher order, starting from (2.7). Several other properties of this operator are discussed in [40].

2.3.3 Fractional Integral Operators

A generalization of the usual integral operator can be obtained starting from Riemann sums [21, 40]. Consider $A = [a, b]$ an arbitrary one-dimensional domain containing a fractal set F and f a real valued function defined on A. A point $x \in [a, b]$ is taken in A and the sub-domain $[a, x]$ is partitioned by a set of points $\{x^m_{i=0\ldots m}\}_{m \geq 1}$. In particular, the partition can be uniform, with the length $\varepsilon_m = x^m_{i+1} - x^m_i$ being independent of i. Obviously, $\varepsilon_m \to 0$ as $m \to \infty$. For any order m of the partition, one may distinguish intervals $[x^m_i, x^m_{i+1}] \subset A-F$ and intervals that contain at least one point of F. The following sum can be evaluated for any m:

$$\Theta_m(f) = \sum_i [x^m_{i+1} - x^m_i] f(x^{*m}_i) + \sum_{\substack{i| \\ [x^m_i, x^m_{i+1}] \cap F \neq \Phi}} L^{1-q}[x^m_{i+1} - x^m_i]^q f(x^{**m}_i)$$

$$- \sum_{\substack{i| \\ [x^m_i, x^m_{i+1}] \cap F \neq \Phi}} L^{1-q}[x^m_{i+1} - x^m_i]^q f(x^{*m}_i) \} \qquad (2.8)$$

The point x^{*m}_i is a point from $A-F$ in the interval $[x^m_i, x^m_{i+1}]$ and x^{**m}_i is a point belonging to F. As with the differential operator of (2.5), L is considered a parameter in this discussion.

If the limit of the above sum for $m \to \infty$ exists and is finite for any partition $\{x^m_{i=0\ldots m}\}_{m \geq 1}$ of the interval $A = [a, x]$ with the norms going to zero, the fractional integral of the function f over $[a, x]$ is defined as:

$$\int_A f(x')d^{Fr}x' = \lim_{m \to \infty} \Theta_m(f) \qquad (2.9)$$

In particular, if the fractal domain is absent ($F = \Phi$) the Riemann integral is recovered (first term in (2.8)). If the integrand is defined strictly on the fractal set F, only the second term appears in the sum (2.8) and the fractional integral operator defined by Kolwankar and Gangal [22, 23] is recovered up to a multiplicative constant. When the integral is considered on the entire domain A containing a fractal F, besides the Riemann sum (first term in (2.8)) the integral on the fractal (the second term in (2.8)) is added. Hence, intervals containing F are counted twice. The third term in (2.8) provides a correction. Note that this third term is well defined since, due to the property of F of being disconnected, any closed interval contains points x_i^{*m} from A–F.

The integral and differential operators defined with (2.5) and (2.9) can be shown to be inverse to each other, in the sense that $\int_A D^{q(x')} f(x') d^{Fr} x' = f(x) - f(a)$, where $q(x')$ is the order of differentiability at x'.

It is both interesting and useful to observe that between the classical and fractional operators so defined exists an approximation relationship. To clarify it, let us partition the domain A in segments of length $\varepsilon_n = |x_{i+1}^n - x_i^n|$ such that each segment exactly overlaps with a "box" used to probe the fractal structure in the n-th step of the fractal generation. Then, the integral over segments belonging to F becomes

$$\int_F f(x') d^{Fr} x' = \lim_{n \to \infty} \left\{ \sum_{F_n} L^{1-q} \varepsilon_n^q f(x_i^{**n}) \right\}, \tag{2.10a}$$

while that over the remaining segments can be written

$$\int_{A-F} f(x') d^{Fr} x' = \lim_{n \to \infty} \left\{ \sum_{A-F_n} \varepsilon_n f(x_i^{*n}) - \sum_{F_n} L^{1-q} \varepsilon_n^q f(x_i^{*n}) \right\}. \tag{2.10b}$$

The fractional integrals are defined only in the limit $n \to \infty$ (limit in which the fractal exists) as in (2.9). However, if one limits attention to a given iteration/scale of the fractal generation procedure (n is given), the integrals in (2.10) become the classical Riemann sums and can be evaluated using standard calculus. These provide a good approximation to the fractional integrals of (2.8) computed with a particular choice of L, i.e. $L = \varepsilon_n$).

$$\int_{F_n} f_n(x) dx \approx \int_F f(x) d^{Fr} x \big|_{L=\varepsilon_n} \tag{2.11a}$$

$$\int_{A-F_n} f_n(x) dx \approx \int_{A-F} f(x) d^{Fr} x \big|_{L=\varepsilon_n} \tag{2.11b}$$

The approximation improves as n increases. The expressions can be used to great advantage in order to predict the solution of a boundary value problem defined on a structure with a finite number of scales, n, from a fictitious solution of the same boundary value problem defined for the infinite fractal, $n \to \infty$. This is discussed further in Section 2.4.2.2.1.

Note that for this L, the expression $L^{1-q}\varepsilon_n^q$ becomes the length of the respective segment measured *on the fractal set*.

As with usual operators, the generalization to multidimensions of the fractional differential and integral operators can be performed by considering (2.5) in each of the principal directions as well as multiple sums in (2.8) and (2.9). Symbolically, one can denote by $\int_A f(\mathbf{x})d^{Fr}\mathbf{x}$ the fractional integral over domain A embedded in arbitrary d-dimensional space. In this case, the parameter ε_n is interpreted as the edge length of the d-dimensional cube used to probe the fractal at step n. The choice of the parameter L in the approximations can be reinterpreted as the condition that the fractal volume computed using the fractional operator should approximate the classical volume of the fractal set at "step n". The condition reads $\int_F 1_F d^{Fr}\mathbf{x}|_{L=\varepsilon_n} = \int_{F_n} 1_{F_n} d\mathbf{x}$.

Under certain conditions (specifically, when (i) the fractal dimension of the projection in direction $\mathbf{e}_j$, at any point, depends only on the x_j coordinate $q_j(\mathbf{x}) = q_j(x_j)$ and (ii) the boundary of the domain is the limit of a set of segments oriented in the reference (i.e. given) frame directions, it is possible to write a flux-divergence theorem as

$$\int_A \sum_{i=1}^d D_i^{q_i(\mathbf{x})} f(\mathbf{x}) d^{Fr}\mathbf{x} = \sum_{i=1}^d \int_{\partial A_{e_i}} f d^{Fr}\Gamma = \int_{\partial A} \sum_{i=1}^d f n_i d^{Fr}\Gamma, \qquad (2.12)$$

where the boundary is defined, as a reunion of segments ∂A_{e_i} oriented normal to $\mathbf{e}_i$, $\partial A = \bigcup_i \partial A_{e_i}$.

2.4 Mechanics boundary value problems on materials with fractal microstructure

Solving boundary value problems on materials with hierarchical self-similar microstructure is notoriously difficult due to the presence of inclusions with a broad range of sizes (size distribution is represented by a power law), which are interacting strongly. This prevents scale decoupling which is a major requirement for all current homogenization procedures. Within the classical theory, such problems can be addressed only numerically. This is also a tedious approach since the discretization must be on the scale of the finest object that needs to be resolved, which becomes impractical very quickly with increasing the number of scales present in the hierarchical microstructure.

In this section we review methods developed to address these difficulties. They are based on the concept that the operators derived from the balance equations have to operate on the field defined over the *entire* problem domain containing heterogeneities, including *at* interfaces. This is opposed to the "patching" method commonly used in which the solution is sought over each homogeneous sub-domain, under the condition of tractions and displacement continuity across the boundaries of the sub-domain. Fractional calculus provides the appropriate tool allowing performing operations on functions not classically differentiable everywhere (i.e. allowing dealing with field discontinuities at interfaces).

Transport [27, 38] over domains with fractal supports have been posed and solved in the past, with or without using concepts from fractional calculus. As it turns out, such problems are significantly simpler than mechanics problems. This is due to the fact that the transport process takes place on the fractal support and hence the solution is sought with the metric of and within the space defined by the structure. With mechanics problems the situation is different since deformation takes place in the embedding space and hence, one has to account for the fractal geometry (with reduced dimensionality) while employing fields defined in the embedding space.

Wave propagation has also been discussed. Strichartz [45, 46], Strichartz and Usher [47] proposed solutions of the wave and heat equations for a particular class of fractals that can be approximated by a sequence of finite graphs ("post-critical finite"). Such an example is the Sierpinsky gasket represented in Figure 2.5. In their approach, the Laplacian operator is redefined based on the Lagrangian (or intrinsic) metric on the fractal space [29]. As inferred earlier by Kigami [20] this operator results as a renormalized limit of graph Laplacians. Details of this construction and of the eigen-functions of the Laplacian on Sierpinsky gasket-like structures can be found in Strichartz [45] and Strichartz and Usher [47]. Vibrations on fractals were studied by Orbach [33] and Alexander and Orbach [1] and by others [31] and a new class of localized vibration modes (fractons) have been identified.

In this chapter we discuss exclusively quasi-static problems. The central problem of interest is that of the deformation of a composite with hierarchical,

Figure 2.5: The Sierpinky gasket.

self-similar (fractal) microstructure. The boundary value problem is defined at the "macroscale," i.e. on the scale of the entire structure to be modeled. The material contains inclusions of many sizes. Deformation takes place in the embedding, Euclidean space. Theoretical and numerical results were obtained by Panagiatopoulos et al. [35], Panagouli [36], Carpinteri and collaborators [6, 7], Tarasov [49, 50], by the present authors [40, 41] and several other groups [37, 34]. These works can be divided broadly in two categories: iterative approaches and approaches based on modifying the governing equations to account for the geometric complexity.

2.4.1 Iterative Approaches

Oshmyan et al. [34] considered the problem of finding the effective moduli for composite structures for which the matrix material is linear elastic, while the inclusions are either rigid or voids and are organized in a Sierpinski carpet. The first three generations of this deterministic geometry are shown in Figure 2.6.

This composite has cubic symmetry. The effective moduli C_{1111}, C_{1122}, C_{1212} are computed for the first generation using the classical finite element method. The procedure is repeated and the effective elastic constants $\{\mathbf{C}^{(n)}\}$ are determined for the next n generations. Due to computational limitations, the analysis was performed only up to the 4th generation ($n = 4$). Renormalization group techniques and a fixed-point theorem [35, 34] were used to extrapolate this information for larger n.

The Poisson ratio $\nu = C_{1122}/C_{1111}$ and the coefficient of anisotropy $\alpha = (C_{1111} - C_{1122})/2C_{1212}$ converge to 0.065 and 4.43 (fixed points of the mapping), respectively, for carpets with voids, and to 0.063 and 3.74, for carpets with rigid inclusions. As expected for both voids and rigid inclusions the scaling of the elastic constants is given by a power law:

$$\mathbf{C}^{(n)} = \mathbf{C}^{(0)} \mathbf{L}^{\beta(n)} \tag{2.13}$$

where L is a normalized characteristic parameter of the structure at step n, $L = 3^n$, and β is an exponent depending on the fractal dimension of the microstructure.

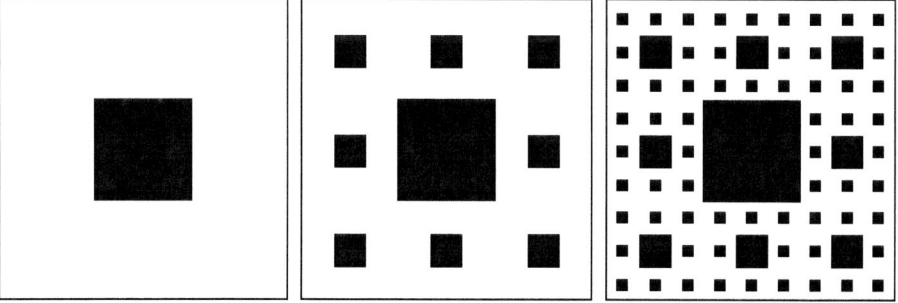

Figure 2.6: The first three generations of the Sierpinki gasket.

The results are in agreement with those previously obtained by Poutet et al. [37] who investigated the scaling properties of porous fractals embedded in 3D, a fractal foam with a fractal box dimension of log(26)/log(3) and the Menger sponge (with a fractal box dimension of log(20)/log(3)). A power law scaling of the equivalent Young moduls of the form $E^{(n)} = \mathrm{E}_0(25/27)^n$ was found for fractal foam and $E^{(n)} = \mathrm{E}_0(2/3)^n$ for the Menger sponge. The results were extrapolated for large n using renormalization arguments starting from the numerical values obtained numerically for the first two iterations.

Dyskin [10] used the differential self-consistent method [44] for media containing self-similar distributions of spherical/ellipsoidal pores or cracks to find the homogenized elastic constants. The author proposes to model such materials by a sequence of continua defined by homogenizing over a sequence of volume elements of various sizes. Specifically, the constitutive behavior of the continuum on scale ε is obtained based on averaged stresses and strains over volume elements of sizeε. Under the restrictive hypothesis that at each scale inhomogeneities (pores/inclusions) of equal size *do not interact*, the elastic constants on scale ε are only functions of the volume fraction of voids/inclusions of smaller size. Under this assumption, the elastic constants follow the scaling of the volume fraction, i.e. a power law $E(\varepsilon) \prec \varepsilon^\beta$.

2.4.2 Reformulations of Governing Equations

The central idea of these attempts is that the complexity of the geometry can be captured in the governing equations, while using simple boundary conditions defined on the largest scale of the problem. The critical issue that needs to be accommodated is the large number density of interfaces (between matrix and inclusions) present; the fields are not classically differentiable at an infinite number of points in the problem domain. The few investigations of this concept to date can be divided based on the type of mathematical tools employed, in methods based on non-local and local fractional operators.

2.4.2.1 Method used for "homogeneous" fractal medium

Tarasov [49, 50] studied porous materials with fractional mass dimensionality. The mass enclosed in a volume of characteristic dimensionε scales as $m(\varepsilon) \prec \varepsilon^{q_m}$. Here q_m is a non-integer number indicating the fractal mass dimension. The author replaces the fractal body with an equivalent continuum having a "fractional measure." The fractional measure (or fractal volume) $\mu_{q_m}(W)$ of a ball W of radius ε is defined such that the mass it contains scales as that of the actual object, i.e.:

$$m(W) = \rho_0 \mu_{q_m}(W) \prec \varepsilon^{q_m} \qquad (2.14)$$

where the density ρ_0 of the homogeneous fractal medium does not depend on the local position (translation and rotation invariant). Then, he defines a fractional integral operator using this measure. The fractional integral of a function f on a fractal volume W of dimension q_m is defined as (Riesz form):

$$I^{q_m} f = \int_W f(\varepsilon) d\mu_{q_m}(W) \qquad (2.15)$$

where the relationship between the differential fractal volume/measure $d\mu_{q_m}$ and the differential Euclidean volume $d^3\varepsilon$ is given by:

$$d\mu_{q_m}(W) = \frac{2^{3-q_m}\Gamma(3/2)}{|\boldsymbol{\varepsilon}|^{3-q_m}\Gamma(q_m/2)} d^3\boldsymbol{\varepsilon} \qquad (2.16)$$

Thus if the ball W of radius $\varepsilon = |\boldsymbol{\varepsilon}|$ contains a fractal set, its fractal volume/measure is $\mu_{q_m}(W) = I^{q_m} 1_W = 4\pi\varepsilon^{q_m}/q_m$. If it does not contain a fractal, the classical volume is recovered $I^{q_m} 1_W = 4\pi\varepsilon^3/3$.

The balance equations for mass, linear and angular momentum conservation are reformulated for the equivalent continuum using this measure.

The fractional operators proposed by Tarasov are a generalization of the classical Riemann–Liouville operator to embedding spaces of arbitrary dimension. The field equations are rewritten for the equivalent "homogenized" continuum and the solutions are obtained in an average sense, without distinguishing between a material point and a pore point.

2.4.2.2 Methods using local fractional operators

2.4.2.2.1 *General formulation* In the framework of solid mechanics, a local version of the fractional operators were initially used by Carpinteri et al. [6, 7] in an attempt to explain size effects of deformation and of fracture processes of heterogeneous materials. Using renormalization group procedures for the fractal like structures, these authors defined new mechanical quantities (such as fractal strain, fractal stress and the corresponding work), which are scale invariant but have unconventional physical dimensions that depend on the fractal dimension of the structure.

The kinematics equations and the principle of virtual work for fractal media embedded in Euclidean spaces were formally presented. They use local fractional operators previously developed by Kolwankar and Gangal [21, 22, 23], which allow writing (formally) the balance equations in a local form. As discussed below, this is not always warranted. These authors went further by postulating the existence of an elastic potential for the fractal microstructure which is then used to define the fractal stress.

This formulation was employed to date only in the rather trivial case of a 1D rigid bar which is allowed to deform only at a set of points that form a Cantor set. In this case, the displacement is represented by the Devil staircase function. This function is piecewise constant (no deformation for the rigid part of the bar) and discontinuous at the Cantor set points.

For structures embedded in multidimensional spaces, the authors suggested a variational formulation of the elasticity problem using their fractional operators and a method to approximate the solution using the Devil's staircase function.

A limitation of this description is that deformation is allowed to take place on the fractal support only. Obviously, this is a very restrictive assumption.

A recent formulation for composite bodies containing fractal inclusions takes a step forward treating the case when both the fractal and its complement deform in the embedding space [40]. The formulation requires the simultaneous use of fractional operators, on the fractal F, and classical operators, on A–F. The structure is referred and its dimension defined relative to a fixed coordinate system denoted as the "reference frame." The method is developed under the following restrictions: (i) the fractal dimension of the projection in each of the reference frame directions $\mathbf{e}_j$, at any point depends only on the X_j coordinate $q_j(\mathbf{X})=q_j(X_j)$; (ii) the fractal box dimension remains unchanged during the deformation (a reasonable assumption under small deformation); (iii) There is no interface sliding between the fractal domain F and its complement.

As in classical continuum, the deformation function $\mathbf{x} = \chi(\mathbf{X}, t)$ mapping the initial configuration ($\mathbf{X}$ is the reference position vector of a generic point from A) into the deformed configuration ($\mathbf{x}$ is the current position vector of the same point) is continuous over the entire domain A. However, while χ is differentiable in the classical sense at all points of A–F, it may be fractional differentiable in one or several directions at points from F. The order of differentiability in a certain direction at a point is given by the fractal dimension in the respective direction [51]. The displacement field is defined as:

$$\mathbf{u}(\mathbf{X}, t) = \chi(\mathbf{X}, t) - \mathrm{id}_A(\mathbf{X}) \tag{2.17}$$

where **id** is the identity function having the derivatives $D_i^{q_i(\mathbf{X})}(\mathrm{id}(\mathbf{X})\mathbf{e}_j) = \delta_{ij}$ $i,j = 1..d$, $q_i(\mathbf{X})$ is the fractal dimension in direction $\mathbf{e}_i$, $q_i(\mathbf{X}) = q_i$ $if\ \mathbf{X} \in F$ and $q_i(\mathbf{X}) = 1$ $if\ \mathbf{X} \in A$–F. Under the hypothesis of small deformations, the strain tensor is defined as:

$$\boldsymbol{\varepsilon} = \frac{1}{2}(\mathbf{H}^T + \mathbf{H}) = \begin{cases} (D_i u_j(\mathbf{X}) + D_j u_i(\mathbf{X}))/2 & if\ \mathbf{X} \in A-F \\ (D_i^{q_i(\mathbf{X})} u_j(\mathbf{X}) + D_j^{q_j(\mathbf{X})} u_i(\mathbf{X}))/2 & if\ \mathbf{X} \in F \end{cases} \tag{2.18}$$

where H is the deformation gradient:

$$H_{ij}(\mathbf{X}, t) = \begin{cases} D_i \chi_j(\mathbf{X}, t) - \delta_{ij} & if\ \mathbf{X} \in A-F \\ D_i^{q_i(\mathbf{X})} \chi_j(\mathbf{X}, t) - \delta_{ij} & if\ \mathbf{X} \in F \end{cases}, \quad i,j = 1\ldots d. \tag{2.19}$$

This expression is similar to that proposed by Carpinteri and collaborators [6, 7], except that the fractional derivatives used here are given by (2.5), i.e. include a parameter L (with physical dimensions of length), and have units similar to those of the classical derivative. (2.18) reduces to the usual strain tensor when **u** is classically differentiable everywhere ($F = \Phi$).

When $F \neq \Phi$, the matrix H does not follow the tensor transformation rule upon rotation of the coordinate system since the fractal dimension q may change as the frame is redefined. For this reason, the quantity (2.18) is named pseudo-strain.

The definition of the complementary stress field is introduced in the context of balance of linear momentum formulated for composite fractal structures. Let P be an arbitrary sub-domain of A at time t in the deformed configuration. Similar to A, P is defined in the d-dimensional Euclidean space and embeds a fractal of smaller dimension. The balance of linear momentum on any such arbitrary sub-domain P at any time t reads:

$$\dot{\mathbf{K}}(P,t) = \mathbf{R}(P,t) + \Im(P,t) \tag{2.20}$$

where $\mathbf{K}(P,t) = \int_P \rho \dot{\mathbf{x}}(\mathbf{x},t) d^{Fr}\mathbf{x}$ is the total linear momentum on P, $\mathbf{R}(P,t) = \int_P \rho \mathbf{b}(\mathbf{x},t) d^{Fr}\mathbf{x}$ is the body force acting on P while $\Im(P,t) = \int_{\partial P} \mathbf{t}(\mathbf{x},\mathbf{n},t) d^{Fr}\Gamma$ is the traction force acting on the boundary ∂P and $\mathbf{t}$ is the boundary traction vector. Assuming that the boundary of P is the limit of a set of segments oriented *exclusively* in the reference frame directions, i.e. $\partial P = \cup_{i=1...d} \partial P_{e_i}$, with $\partial P_{e_i} = \lim_{n\to\infty} \partial P_{e_i}^n$, where $\partial P_{e_i}^n$ is a segment of normal $\mathbf{e}_i$ defined on scale n of the fractal, the traction can be written as:

$$\int_{\partial P} \mathbf{t}(\mathbf{x},\mathbf{n},t) d^{Fr}\Gamma = \sum_{i=1...d} \int_{\partial P_{e_i}} \mathbf{t}(\mathbf{x},\mathbf{e}_i,t) d^{Fr}\Gamma. \tag{2.21}$$

The pseudo-stress field has components:

$$T_{ki}(\mathbf{x},t) = t_k(\mathbf{x},\mathbf{e}_i,t), \quad i,k = 1\ldots d \tag{2.22}$$

Rearranging the terms in (2.20) and using the flux-divergence theorem described in Section 2.3.2, one can write the balance of linear momentum as:

$$\int_P \rho(\ddot{x}_k(\mathbf{x},t) - b_k(\mathbf{x},t)) d^{Fr}\mathbf{x} = \sum_{i=1..d} \int_P D_i^{q_i(\mathbf{x})} T_{ki}(\mathbf{x},t) d^{Fr}\mathbf{x} \quad k=1\ldots d \tag{2.23}$$

Following the same line of thinking, the balance of angular momentum reads:

$$\int_P \in_{ijk} x_j [\rho(\ddot{x}_k(\mathbf{x},t) - b_k(\mathbf{x},t)) - D_l^{q_l(\mathbf{x})} T_{kl}(\mathbf{x},t)] \mathbf{e}_i d^{Fr}\mathbf{x}$$

$$= \int_P \in_{ijk} T_{kj}(\mathbf{x},t) \mathbf{e}_i d^{Fr}\mathbf{x} \tag{2.24}$$

If the integrands of (2.23) and (2.24) are continuous, (2.24) leads further to the symmetry of the pseudo-stress matrix $T_{kj}(\mathbf{x},t) = T_{jk}(\mathbf{x},t)$ $k,j = 1..d$. Note that time differentiation in both (2.23) and (2.24) is classical.

The stress so defined does not rotate following the usual tensorial rule either. The formulation is not frame independent. The fundamental reason for this situation is that the "material point" here is actually an entire structure, which is defined relative to a coordinate system, the reference frame. Hence, the analysis must be performed in this coordinate system, in which the geometry is described. This is not a limitation from a practical point of view.

In order to formulate a boundary value problem for such composite structure, the kinematics (2.18) and equilibrium (2.23, 2.24) must be supplemented with few additional equations. Let us consider a simple quasi-static formulation in the absence of body forces. The balance of linear and angular momentum read:

$$\int_P D_i^{q_i(\mathbf{x})} T_{ki}(\mathbf{x},t) d^{Fr}\mathbf{x} = 0, \quad k=1\ldots d, \quad \int_P \epsilon_{ijk} T_{kj}(\mathbf{x},t) e_i d^{Fr}\mathbf{x} = 0 \quad (2.25)$$

However, one expects some components of the integrand in (2.25) to be discontinuous at interfaces. Under such circumstances, the above non-local forms must be supplemented by the traction continuity condition:

$$(\mathbf{t}^F(\mathbf{x},t) - \mathbf{t}^{A-F}(\mathbf{x},t))|_\Gamma = 0. \quad (2.26)$$

where Γ is the interface between the fractal material and its complement. The displacement continuity is insured by construction.

Also, a constitutive relation between the pseudo-strain (2.18) and pseudo-stress field (2.22) needs to be postulated (just the same with postulating such a relationship linking standard strain and stress). For example, one can assign linear elastic behavior to both matrix and inclusions:

$$\varepsilon_{ij}(\mathbf{x}) = L_{ijkl}(\mathbf{x}) T_{kl}(\mathbf{x}), \quad (2.27)$$

where the compliance L_{ijkl} is different for matrix and inclusions. ε and T stand for the classical strain and stress over A–F and for the pseudo-strain and pseudo-stress over F.

(2.18), (2.25), (2.26) and (2.27), together with a set of displacement, tractions or mixed boundary conditions:

$$u_i = u_i^0 \quad \text{on } \Gamma_u, \qquad T_{ij} n_j = t_i^0 \quad \text{on } \Gamma_t \quad (2.28)$$

where $\Gamma_u \cup \Gamma_t = \partial A$ and $\Gamma_u \cap \Gamma_t = \Phi$, fully define a quasi-static problem for the composite. The formulation corresponds to a structure with an infinite number of scales, an ideal mathematical geometry, while real materials exhibit self-similarity over a finite range of scales only. It seems that (2.18), (2.25)–(2.28), the so-called "trial problem," is practically irrelevant. However, the fictitious solution of this system of equations can be used to approximate solutions on any (finite) scale n. This approximation is based on relations (2.11a,b) between the fractional and classical operators. Specifically, let us denote by $\{\mathbf{u}^n, \mathbf{T}^n\}$

the solution of a boundary value problem formulated on a geometry A_n, which constitutes the n-th iteration in the composite fractal generation. The solution of the "trial problem", $\{\mathbf{u}, \mathbf{T}\}$, is expressed in terms of an arbitrary parameter L that appears in the formulation of the fractional operators. The solution $\{\mathbf{u}^n, \mathbf{T}^n\}$ can be approximated simply by a particularization of L, in the solution $\{\mathbf{u}, \mathbf{T}\}$. The specific value of the parameter L for each iteration step n, is determined from the condition that the volume of the fractal domain F evaluated using the fractional integral, approximates the volume of F_n evaluated by classical integration:

$$\int_F 1_F d^{Fr}\mathbf{x}|L = \varepsilon_n = \int_{F_n} 1_{F_n} d\mathbf{x}. \qquad (2.29)$$

The procedure is exemplified in the next section for two simple problems.

2.4.2.2.2 *Examples* Let us consider two examples: 1D and 2D composite domains for which the first 3 iterations are shown in Figures 2.3 and 2.7, respectively. In the second case, that of the plate, the geometry is defined by partitioning each of the principal axes ("reference frame") in Cantor sets (denoted by F_{x_1} and F_{x_2}). The total Euclidean length of the bar, as well as the edge length of the plate are denoted by b.

The fractal box-dimensions for the bar and plate are $\ln(2)/\ln(3)$ and $1 + \ln(2)/\ln(3)$, respectively. The "inclusions" are represented in black and are characterized by a set of isotropic elastic constants E, ν, while the complement is represented in white and it is characterized by E_0, ν. Both cases of stiffer ($E_0 > E$) and more compliant matrix material ($E_0 < E$) are studied. The aim is to characterize the solution of boundary value problems formulated for composite geometries obtained at any iteration step n (1D traction for the first case, plane stress conditions for the second).

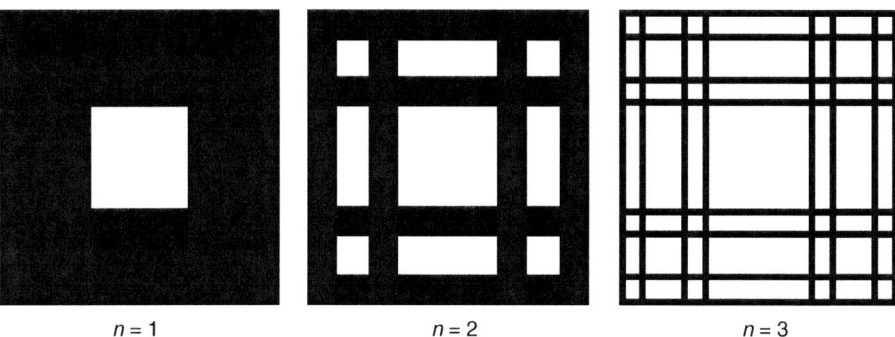

Figure 2.7: The first three iteration of a plate with deterministic, self-similar geometry.

The "trial problem" (2.18, 2.25–2.28) is solved first in terms of L. Depending on the boundary conditions, one can use a primal, dual or mixed variational principle. In the mixed variational formulation [32, 53], the solution $(\mathbf{u},\mathbf{T})$ of the elasticity problem associated with the composite with an infinite number of scales is the unique critical (saddle) point of the functional:

$$U(\mathbf{v},P) = \frac{1}{2}\int_A P_{ij}L_{ijkl}P_{kl}d^{Fr}\mathbf{x}$$

$$-\int_A P_{ij}\varepsilon_{ij}(\mathbf{v})d^{Fr}\mathbf{x} + \int_A \varepsilon_{ij}(\mathbf{v})L^{-1}_{ijkl}\varepsilon_{kl}(\mathbf{v})d^{Fr}\mathbf{x} - \int_{\Gamma_t} v_i t^0 d^{Fr}\Gamma \quad (2.30)$$

over the space of all vector fields satisfying the displacement boundary conditions: $v_i|_{\Gamma_u} = u_i^0$, with the properties $\int_A \mathbf{v}^2 d^{Fr}x < \infty$, $\int_A [Grad(\mathbf{v})]^2 d^{Fr}x < \infty$, and all pseudo-traction fields $\mathbf{P}$ with the property $\int_A \mathbf{P}^2 d^{Fr}x < \infty$. The solution of the variational problem may be obtained by using a finite element procedure. The domain A is discretized in a finite number of elements $A = \bigcup_{p=1,N} A_p$, with a total number of M_u nodes. The solution is approximated using a set of generic shape functions $N_m(\xi)$, $m = 1..M_u$ for the displacement field $\mathbf{u} = \sum_{m=1,M_u} \mathbf{u}^m N_m$ and a corresponding set of shape functions $M_m(\xi)$, $m = 1..M_T$ for the stress field $\mathbf{T} = \sum_{m=1,M_T} \mathbf{T}^m M_m$. Both sets are simultaneously used in (2.30). As usual, this transforms the minimization problem stated above into a system of equations for the unknown coefficients $\{\mathbf{u}^m, \mathbf{T}^m\}$ and any additional ("internal") parameters that may be built into the shape functions. Using the simple matrix notation for the displacement field $\mathbf{u}(\mathbf{x}) = \mathbf{N}(\mathbf{x})\mathbf{X}$ (which leads to a strain field of the form $\boldsymbol{\varepsilon}(\mathbf{x}) = (\nabla \mathbf{N}(\mathbf{x}))\mathbf{X}$), and the stress field $\mathbf{T}(\mathbf{x}) = \mathbf{M}(\mathbf{x})\mathbf{Y}$, the energy functional of the mixed formulation (2.30) can be expressed as:

$$U = \frac{1}{2}\mathbf{Y}^T\mathbf{A}\mathbf{Y} + \mathbf{X}^T\mathbf{B}\mathbf{X} - \mathbf{Y}^T\mathbf{C}\mathbf{X} - \mathbf{X}^T\mathbf{F}^0 \quad (2.31)$$

where:

$$\begin{cases} \mathbf{A} = \int_A \mathbf{M}^T(\mathbf{x})\mathbf{L}(\mathbf{x})\mathbf{M}(\mathbf{x})d^{Fr}\mathbf{x} \\ \mathbf{B} = \int_A (\nabla \mathbf{N}(\mathbf{x}))^T \mathbf{L}^{-1}(\mathbf{x})(\nabla \mathbf{N}(\mathbf{x}))d^{Fr}\mathbf{x} \\ \mathbf{C} = \int_A \mathbf{M}^T(\mathbf{x})(\nabla \mathbf{N}(\mathbf{x}))\,d^{Fr}\mathbf{x} \\ \mathbf{F}^0 = \int_{\Gamma_t} \mathbf{N}^T(\mathbf{x})\mathbf{t}^0(\mathbf{x})d^{Fr}\Gamma \end{cases}$$

One of the key aspects in this formulation is the selection of the shape functions, which incorporate information about the self-similarity of the structure. In the context of modeling localization in 1D bars using fractal concepts, Carpinteri and Chiaia [5] used as shape function the Lebesque Cantor staircase. This function has nonzero fractional derivative at the Cantor set points and zero classical derivatives elsewhere. The staircase function results as a limit of a sequence of functions defined at each step n of the iteration leading to the Cantor set. In the current example, the staircase function is modified, as in example (2.6 and 2.7) of section 2.3.2 $S_{\beta,\gamma}(x) = f(x)$ such that is has nonzero fractional derivatives (γ/b) at all points of the Cantor set (F) and *nonzero* classical derivatives (β/b) outside the fractal set $(A-F)$. The condition $f(b) = 1$ imposes though, a relationship between the two constants: $\gamma = (L/b)^{q-1}(1-\beta)+\beta$, such that the staircase function depends on β, S_β. The formulation for the geometry embedded in the 1D space is straightforward (only the displacement formulation is used in this case and one element requires only one parameter β):

$$u = u^{(1)} N_1 + u^{(2)} N_2 \tag{2.32a}$$

where $N_1, N_2 : [0, b] \to R$, $N_1(x) = 1 - S_\beta(x)$, $N_2(x) = S_\beta(x)$.

For the structure embedded in 2D, the plate is modeled with quadrilateral elements and β may be different for each of the principal directions. One the other hand, the mixed variational form is used and discretization of both displacement and stress fields are needed.

$$\mathbf{u} = \mathbf{NX} \text{ and respectively } \mathbf{T} = \mathbf{MY}$$

$$u = \begin{bmatrix} u_1 \\ u_2 \end{bmatrix}, \quad N = \begin{bmatrix} N_1 & 0 & N_2 & 0 & N_3 & 0 & N_4 & 0 \\ 0 & N_1 & 0 & N_2 & 0 & N_3 & 0 & N_4 \end{bmatrix}, \quad X = \begin{bmatrix} X_1 \\ \vdots \\ X_8 \end{bmatrix}$$

$$\begin{aligned} N_1(x_1, x_2) &= (1 - S_{\beta_1^u}(x_1))(1 - S_{\beta_2^u}(x_2)) \\ N_2(x_1, x_2) &= S_{\beta_1^u}(x_1)(1 - S_{\beta_2^u}(x_2)) \\ N_3(x_1, x_2) &= S_{\beta_1^u}(x_1) S_{\beta_2^u}(x_2) \\ N_4(x_1, x_2) &= (1 - S_{\beta_1^u}(x_1)) S_{\beta_2^u}(x_2) \end{aligned} \tag{2.32b}$$

$$T = \begin{bmatrix} T_{11} \\ T_{22} \\ T_{12} \end{bmatrix}, \quad M = \begin{bmatrix} 1 & S_{\beta_1^T}(x_1) & 1_{F_{x_2}} & 0 & 0 & 0 & 0 \\ 0 & 0 & 0 & 1 & 1_{F_{x_1}} & S_{\beta_2^T}(x_2) & 0 \\ 0 & -S_{\beta_1^T}(x_2) & 0 & 0 & 0 & -S_{\beta_2^T}(x_1) & 1 \end{bmatrix}, \quad Y = \begin{bmatrix} Y_1 \\ \vdots \\ Y_7 \end{bmatrix}$$

where $1_{F_{x_i}}$ is the characteristic function of the fractal set F_{x_i}.

In a first approximation, the "trial problem" for both examples can be solved using a single element.

(i) Let us consider first the case of a 1D bar under zero displacement at one end and simple traction $t(b)$ at the other end. The total potential energy reads $U = (1/2)\int_{[0,b]} T_{11}\varepsilon_{11} d^{Fr}x - tu(b)$ which becomes $U = u^{(2)}2b[(L/b)^{1-q}E(\gamma/b)^2 + E_0(\beta/b)^2(1-(L/b)^{1-q})]/2 - u^{(2)}t$. A straightforward minimization of U with respect to β and $u^{(2)}$ leads to a solution: $u = bt[(L/b)^{1-q}/E + (1-(L/b)^{1-q})/E_0]S(x)$. In order to estimate the solution at step n, one has to take the parameter L equal to the characteristic length $\varepsilon_n = b/3^n$ (2.29). Thus, for geometries corresponding to each step n, the displacement at the end of the bar results equal to the analytical solution obtained directly for step n using the classical formulation: $u_n^{(2)} = bt[(2/3)^n/E + (1-(2/3)^n)/E_0]$. The strain field for the "trial" problem is simply obtained using the fractional differential on F and the classical differential on A–F of the displacement field:

$$\varepsilon(x) = D_x^q u = t/E \text{ for } x \in F, \quad \varepsilon(x) = D_x u = t/E_0 \text{ for } x \in A{-}F$$

This is the expected analytic piecewise constant strain field for F_n and respective A_n–F_n.

(ii) Let us consider now, the case of the composite plate under simple traction conditions: $\mathbf{T}(x_1, b)\mathbf{n_2} = t\mathbf{e_2}$. The solution results as the stationary point $(\mathbf{X}, \mathbf{Y}, \boldsymbol{\beta})$ of the total energy (2.31): $U = \mathbf{Y}^T\mathbf{A}(\beta_i^T)\mathbf{Y}/2 + \mathbf{X}^T\mathbf{B}(\beta_i^u)\mathbf{X} - \mathbf{Y}^T\mathbf{C}(\beta_i^u, \beta_i^T)\mathbf{X} - \mathbf{X}^T\mathbf{F}^0$ under restrictions imposed by the natural boundary conditions. Figure 2.8(a) shows the displacement at the end of the plate in the direction of the applied force for geometries obtained at various iterations n. Figures 2.8(b,c) represent the stress field components in the same direction for the fractal material and its complement. The continuous lines are obtained

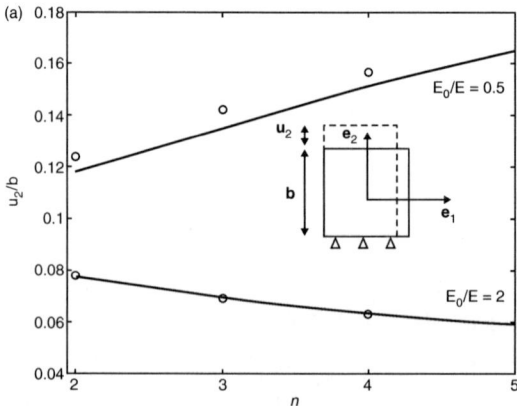

Figure 2.8: (a) Displacement in the vertical direction, normalized by the undeformed plate width, as a function of the fractal approximation step order n. The lines represent the solution obtained by the present method, while the symbols represent the solution obtained using the standard finite element method.

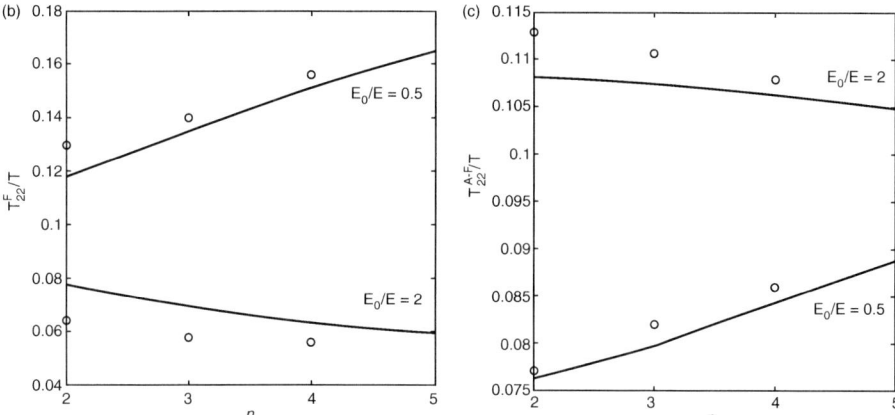

Figure 2.8: (b,c) Pseudo-stress components, in vertical direction (normalized with t/b^2) acting on the fractal material (F) and on its complement (A–F). The lines represent the solutions of the present method (constant, for each step order n), while the symbols represent the similar components (in the average values), obtained using the standard finite element procedure.

from the "trial solution", by selecting the $L = b/3^n$, as imposed by condition (2.29). The symbols represent the same quantities obtained using the standard finite method and a fine mesh at each n. Both cases of stiffer and softer fractal material ($E_0/E = 0.5$ and 2) are considered. As expected, the error decreases as n increases, reaching a value of less than 7% for the stress field at $n = 3$. The "brute force" solution was computed only for $n \leq 4$ as the computational effort increases dramatically with n (the lower bound of the number of elements required is 3^{2n}). The solution obtained using the above procedure presented here requires the same computational cost for any scale n. Thus, it provides a computationally efficient method for characterizing the solution of a boundary value problem for a self-similar structure with a lower scale cut-off.

2.5 Closure

Materials and structures with self-similar multiscale geometry are widespread in nature and exhibit interesting magnetic, transport and mechanical properties. The geometric complexity observed in such materials is large and increases with decreasing scale of observation. Solving boundary value problems on such domains is not trivial and, using existing methods, requires a large computational effort. The approaches presented in this chapter are attempts to address the complexity of the problem in a new way. The concepts discussed here are not yet fully accepted and the formulations required much further work, however, they represent the first steps in a direction that may prove fruitful.

References

[1] Alexander S. and Orbach R. (1982). "Density of states on fractals: 'fractons'", *J. Phys.- Lett.*, **43**, L625–31.

[2] Bagley R. L. and Torvik P. J. (1986). "On the fractional calculus model of viscoelastic behavior", *J. Rheol.*, **30**, p. 133–155.

[3] Barnsley M. F. (1993). *Fractals everywhere*, Academic Press, Cambridge MA.

[4] Beaucage G. (1996). "Small-Angle Scattering from Polymeric Mass Fractals of Arbitrary Mass-Fractal Dimension", *J. Appl. Cryst.*, **29**, p. 134–146.

[5] Carpinteri A. and Chiaia B. (1996). "Power scaling laws and dimensional transitions in solid mechanics Chaos", *Solitons and Fractals*, **9**, p. 1343–1364.

[6] Carpinteri A., Chiaia B., and Cornetti P. (2001). "Static-kinematic duality and the principle of virtual work in the mechanics of fractal media", *Comput. Methods Appl. Mech. Engr.*, **191**, p. 3–11.

[7] Carpinteri A., Chiaia B., and Cornetti P. (2004). "A fractal theory for the mechanics of elastic materials", *Mat Sci and Eng. A*, **365**, p. 235–240.

[8] Cherepanov G. P., Balankin A. S., and Ivanova, V. S. (1995). "Fractal fracture mechanics—a review", *Eng. Frac. Mech.*, **51**(6), p. 997–1033.

[9] Del-Castillo-Negrete D., Carreras B. A., and Lynch V. E. (2005). "Nondiffusive Transport in Plasma Turbulence: A Fractional Diffusion Approach", *Phys. Rev. Lett.*, **94**, p. 065003.

[10] Dyskin A. V. (2005). "Effective characteristics and stress concentrations in materials with self-similar microstructure", *Int. J. Sol. Struc.*, **42**, p. 477–502.

[11] Djordjevic V. D., Jaric J., Fabry B., Fredberg J. J., and Stamenovic, D. (2003). "Fractional derivatives embody essential features of cell rheological behaviour", *Annals of Biomed. Eng.*, **3**, p. 692–699.

[12] Falconer K. J. (1985). *The geometry of fractal sets*, Cambridge University Press, Cambridge.

[13] Feder J. (1988). *Fractals*, Plenum Press, New York.

[14] Giona M. and Roman H. E. (1992). "Fractional diffusion equation on fractals: one-dimensional case and asymptotic behavior", *J. Phys. A. Math. Gen.*, **25**, p. 2093–2105.

[15] Glöcke W. G. and Nonnenmacher T. F. (1993). "Fox function representation of non-debye relaxation processes", *J. Stat. Phys.*, **71**, p. 741–757.

[16] Gorenflo R. and Mainardi F. (1997). *Fractional calculus: integral and differential equations of fractional order*. In: Carpinteri A. and Mainardi F. (Eds.), CISM Lecture Notes Fractals and Fractional Calculus in Continuum Mechanics, Springer Verlag, Wien, NY, 378, p. 223–376.

[17] Grundwald A. K. (1867). "Ueber begrenzte Derivationen und deren Anwendung", *Zeitschrift fur Mathematik und Physik XII* (**6**), p. 441–480.

[18] Hilfer R. (2000). *Applications of fractional calculus in physics*, World Scientific, Singapore.

[19] Kastelic J., Galeski A., and Baer, E. (1978). "The multicomposite structure of tendon", *Connective Tissue Research*, **6**, p. 11–23.
[20] Kigami J. (1989). "A harmonic calculus on the Sierpinski spaces", *Japan J. Appl. Math*, **8**, p. 259–290.
[21] Kolwankar, K. M. (1998). Studies of fractal structures and processes using methods of fractional calculus. PhD thesis, University of Pune, India.
[22] Kolwankar K. M. and Gangal, A. D. (1996). "Fractional differentiability of nowhere differentiable functions and dimensions", *Chaos* **6**, p. 505–524.
[23] Kolwankar K. M. and Gangal, A. D. (1997). Local fractional derivatives and fractal functions of several variables, in the proceedings of the Conference 'Fractals in Engineering', Archanon.
[24] Letnikov A. V. (1868). "Theory of differentiation of arbitrary order", *Mat Sb.*, **3**, p. 1–68.
[25] Mainardi F. (1997). in *CISM Lecture Notes Fractals and Fractional Calculus in Continuum Mechanics*, Eds. Carpinteri A. and Mainardi F. (Springer Verlag, Wien, NY 378), p. 291.
[26] Marliere C., Despetis F., Etienne P., Woignier T., Dieudonne P., and Phalippou J. (2001). "Very large scale structures in sintered silica aerogels as evidenced by atomic force microscopy and ultra-small angle X-ray scattering experiments", *J Non-Crys. Sol.*, **285**, p. 148–153.
[27] Meerschaert M., Benson D., Scheffler H. P., and Baeumer B. (2002). "Stochastic solution of space-time fractional diffusion equations", *Phys. Rev. E,* **65**, p. 1103–1106.
[28] Miller R., Tadmor E. B., Phillips R., and Ortiz M. (1998). "Quasicontinuum simulation of fracture at the atomic scale", *Modelling Simul. Mater. Sci. Eng.*, **6**(5), p. 607–638.
[29] Mosco U. (1997). "Invariant Field Metrics and Dynamical Scalings on Fractals", *Phys Rev. Lett.*, **79**(21), p. 4067–4070.
[30] Nonnemacher T. F. (1990). "Fractional integral and differential equations for a class of Levy-type probability densities", *J. Phys.*, **23**(14), L697S–L700S.
[31] Nakayama T., Yakubo K., and Orbach R. (1994). "Dynamical properties of fractal networks: Scaling, numerical simulations, and physical realizations", *Rev. Mod. Phys.*, **66**, p. 381–443.
[32] Nečas J. and Hlaváček I. (1981). *Mathematical theory of elastic and elastico-plastic bodies: an introduction*, Elsevier, Amsterdam.
[33] Orbach R. (1986). "Dynamics of Fractal Networks", *Science*, **231**(4740), p. 814–819.
[34] Oshmyan V. G., Patlashan S. A., and Timan S.A. (2001). "Elastic properties of Sierpinski-like carpets: Finite-element-based simulation", *Physical Review E*, **64**, 056108: 1–10.
[35] Panagiatopoulos P. D., Mistakidis E. S., and Panagouli O. K. (1992). "Fractal interfaces with unilateral contact and friction conditions", *Comp. Meth. Appl. Mech. Eng.*, **99**, p. 395–412.

[36] Panagouli O. K. (1997). "On the fractal nature of problems in mechanics", *Chaos, Solitons and Fractals*, **8**(2), p. 287–301.

[37] Poutet J., Manzoni D., Chehade F. H., Jacquin C. J., Bouteca M. J., Thovert J. F., and Adler P. M. (1996). "The effective mechanical properties of random porous media", *J.M.P.S.*, **44**(10), p. 1587–1620.

[38] Saichev A.I. and Zaslavsky G.M. (1997). Fractional kinetic equations: solutions and applications", *Chaos*, **7**, p. 753–764.

[39] Schaefer D. W. and Keefer K. D. (1986). "Structure of Random Porous Materials: Silica Aerogel", *Phys. Rev. Letters*, **56**(20), p. 2199–2202.

[40] Soare M. A. (2006). Mechanics of materials with hierarchical fractal structure. Ph.D. Thesis, Rensselaer Polytechnic Institute, Troy, NY.

[41] Soare M. A. and Picu R. C. (2007). "An approach to solving mechanics problems for materials with multiscale self-similar microstructure", *Int. J. Sol. Struct.*, **44**, p. 7877–7890.

[42] Soare M. A. and Picu R.C. (2008a). "Boundary value problems defined on stochastic self-similar multiscale geometries", *Int. J. Numer. Meth. Eng.*, **74**, p. 668–696.

[43] Soare M. A. and Picu R. C. (2008b). "Spectral decomposition of random fields defined over the generalized Cantor set", *Chaos, Solitons and Fractals*, **37**, p. 566–573.

[44] Salganik R. L. (1973). "Mechanics of bodies with many cracks", *Mech. of Solids.*, **8**, p. 135–143.

[45] Strichartz R. S. (1999). "Some properties of Laplacians on fractals", *J. of Functional Analysis*, **164**, p. 181–208.

[46] Strichartz R. S. (2003). "Function spaces on fractals", *J. of Functional Analysis*, **198**, p. 43–83.

[47] Strichartz R. S. and Usher M. (2000). "Splines on fractals", *Math. Proc. Cambridge Philos. Soc.*, **129**, p. 331–360.

[48] Tadmor E. B., Ortiz M., and Phillips R. (1996). "Quasicontinuum analysis of defects in solids", *Phil. Mag. A* **6**(73), p. 1529–1573.

[49] Tarasov V. E. (2005a). "Fractional hydrodynamics equations for fractal media", *Annals of Physics*, **318**(2), p. 286–307.

[50] Tarasov V. E. (2005b). "Continuous medium model for fractal media", *Physics Letters A,* **336**, p. 167–174.

[51] Tatom F. B. (1995). "The Relationship between fractional calculus and fractals", *Fractals*, **3**(1), p. 217–229.

[52] Vrieling E. G., Beelen T. P. M., Sun Q., Hazelaar S., van Santen R. A., and Gieskes W. W. "Ultrasmall, small, and wide angle X-ray scattering analysis of diatom biosilica: interspecific differences in fractal properties", *J. Matl. Chem.*, **14**, p. 1970–1975.

[53] Washizu K. (1982). *Variational Methods in Elasticity and Plasticity*, Pergamon Press, New York.

[54] Witten T. A. and Sander L. M. (1983). "Diffusion-limited aggregation", *Phys. Rev.*, **27**(9), p. 5686–5697.

3
N-SCALE MODEL REDUCTION THEORY
Jacob Fish and Zheng Yuan

3.1 Introduction

Nature and man-made products are replete with multiple scales. Consider for instance the Airbus A380—one of the largest man made creatures. It is 53 m long with wing span of 80 m and height of 24 m. The A380 consists of hundreds of thousands of structural components and many more structural details. Just in the fuselage alone there are more than 750,000 holes and cutouts. In addition to various structural scales, there are numerous material scales. The composite portion of the fuselage, consists of a laminate and a woven or textile composite scale at the two coarsest material scales, followed by a tow microstructure and ending in one or more discrete scales. The metal portion of the airplane, consists of the macroscopic (phemenological plasticity) and the grain (crystal plasticity) scales, followed by a sub-grain (continuously distributed dislocations) scale, and ending with several discrete scales including collections of discrete dislocations and atomistic scale.

Derivation of coarse-scale equations (or so-called upscaling) from finer-scale equations is often facilitated by the method of homogenization [1, 2, 3, 4] for continuous scales and its generalization to discrete scales [5, 6, 7, 8, 9] provided that the scale separation is valid. The enormous gains that can be accrued from deducting coarse-scale behavior directly from the fine scale has been articulated by the Nobel Prize Laureate Richard Feynman [10] 50 years ago, who stated: "What would the properties of materials be if we could really arrange the atoms the way we want them?" To take advantage of the enormous potential offered by this bottom-up approach, computational homogenization approaches [11, 12, 13, 14, 15, 16, 17, 18, 19, 20, 21, 22, 23, 24, 25, 26, 27, 28, 29] and closely related methods [30, 31, 32, 33, 34, 35] emerged in recent years. These methods are either based on the Hill-Mandel relation [36] or on the multiple-scale asymptotic expansions in combination with various numerical methods.

And yet, while computational homogenization approaches can drastically reduce the computational cost of the direct numerical simulation (DNS) they remain computationally prohibitive for large-scale nonlinear problems. This is because a nonlinear unit cell problem for a two-scale problem has to be solved number of times equal to the product of number of quadrature points at a coarse scale times the number of load increments and iterations at the coarse scale. With each additional scale the number of times the finest scale unit cell problem has to be solved is increased by a factor equal to the product of the number of quadrature points, load increments and iterations at the previous scale.

It is therefore desirable to reduce the computational complexity of the fine-scale problem without significantly compromising on the solution accuracy in the quantities of interest, which are often at the system level. This would provide a more realistic bottom-up perspective very much along the lines advocated by Einstein who said that the "everything (model) should be the simplest possible, but no simpler". The method presented in this book chapter is a generalization of the two-scale model reduction approach developed in [37] to arbitrary number of scales.

The book chapter is organized as follows. In Section 3.2, the N-scale (or multiple scale) mathematical homogenization formulation is developed. The governing equations are reformulated in terms of residual-free stresses based on the analytical influence (Green's) functions in Section 3.3. Section 3.4 presents a reduced order model based on the partition-by-partition representation of numerical influence functions. Solution of the system of nonlinear equations corresponding to the unit cell problems at multiple scales is described in Section 3.5. The computational procedure is summarized in Section 3.6. Derivation of the three-scale formulation as a special case of the N-scale formulation as well as validation and verification studies on tube crush tests conclude the book chapter.

3.2 The N-scale mathematical homogenization

The premise of the mathematical homogenization (or any other homogenization type) is that at a certain scale the governing equations including constitutive equations of phases are well understood (or at least better understood than at the coarse scale). Under this assumption, homogenization provides a mathematical framework by which coarse-scale equations can be derived from the well-defined fine-scale equations provided that spatial scales are separable at all relevant scales.

This section focuses on the derivation (upscaling) of the governing equations at multiple spatial scales from the equations on the finest scale of interest. We start with a strong form of the boundary value problem stated at the finest scale of interest

$$\sigma^\zeta_{ij,j}(\mathbf{x}) + b^\zeta_i(\mathbf{x}) = 0 \quad \mathbf{x} \in \Omega \tag{3.1}$$

$$\sigma^\zeta_{ij}(\mathbf{x}) = L^\zeta_{ijkl}(\mathbf{x})[\varepsilon^\zeta_{kl}\mathbf{x} - \mu^\zeta_{kl}(\mathbf{x})] \quad \mathbf{x} \in \Omega \tag{3.2}$$

$$\varepsilon^\zeta_{ij}(\mathbf{x}) = u^\zeta_{(i,j)}(\mathbf{x}) \equiv \frac{1}{2}(u^\zeta_{i,j} + u^\zeta_{j,i}) \quad \mathbf{x} \in \Omega \tag{3.3}$$

$$u^\zeta_i(\mathbf{x}) = \bar{u}_i(\mathbf{x}) \quad \mathbf{x} \in \Gamma_u \tag{3.4}$$

$$\sigma^\zeta_{ij}(\mathbf{x})n_j(\mathbf{x}) = \bar{t}_i(\mathbf{x}) \quad \mathbf{x} \in \Gamma_t \tag{3.5}$$

$$\delta^\zeta_i(\mathbf{x}) \equiv [\![u^\zeta_i(\mathbf{x})]\!] = u^\zeta_i\big|_{S^\zeta_-} - u^\zeta_i\big|_{S^\zeta_+} \quad \mathbf{x} \in S^\zeta \tag{3.6}$$

$$\sigma_{ij}^\zeta n_j\big|_{s_+^\zeta} + \sigma_{ij}^\zeta n_j\big|_{s_-^\zeta} = t_i^\zeta\big|_{s_+^\zeta} + t_i^\zeta\big|_{s_-^\zeta} = 0 \qquad (3.7)$$

where the superscript ζ denotes dependence of response fields on the finest scale heterogeneities. Constitutive relations described in (3.2) are assumed to admit an additive decomposition of the total strains ε_{ij}^ζ into elastic and inelastic components (or more generally stated as eigenstrains μ_{kl}^ζ arising from either inelastic deformation, thermal changes, moisture effects, and phase transformation). S^ζ in (3.6) and (3.7) denotes the interface between the microconstituents with S_+^ζ and S_-^ζ denoting the two sides of the interface. δ_i^ζ in (3.6) is the displacement jump along the interface, and $[\![\cdot]\!]$ is the jump operator. The traction continuity along the interface is given by (3.7).

We now introduce the formalism of the multiple-scale mathematical homogenization. Let $^I\mathbf{x}$ be the position vector at scale I related to the position vector at scale I-1 as

$$^I\mathbf{x} \equiv {}^{I-1}\mathbf{x}/\zeta \quad for \; I = 1, \ldots, N-1 \qquad (3.8)$$

In the case of three scales, the leading order position vectors are often denoted as $^0\mathbf{x} = \mathbf{x}, {}^1\mathbf{x} = \mathbf{y}$, and $^2\mathbf{x} = \mathbf{z}$. The volume of the unit cells at a scale I is denoted by Θ_I and the corresponding interface between the phases in the unit cell is denoted by S_I. Here we focus on a periodic theory (see [38] for nonperiodic case) by which any locally periodic function depends on position vectors at multiple scales

$$\phi^\zeta(\mathbf{x}) \equiv \phi\left({}^0\mathbf{x}, \ldots, {}^{N-1}\mathbf{x}\right) \qquad (3.9)$$

The indirect spatial derivative is given by

$$\phi_{,i}^\zeta(\mathbf{x}) = \sum_{I=0}^{N-1} \zeta^{-I} \phi_{,^I x_i}\left({}^0\mathbf{x}, \ldots, {}^{N-1}\mathbf{x}\right) \qquad (3.10)$$

The N-scale asymptotic expansion of the displacement field can be expressed as

$$u_i\left({}^0\mathbf{x}, \ldots, {}^{N-1}\mathbf{x}\right) = \sum_{I=0}^{N-1} \zeta^I u_i^I\left({}^0\mathbf{x}, \ldots, {}^I\mathbf{x}\right) + \cdots \qquad (3.11)$$

From the indirect spatial derivative (3.10), the N-scale asymptotic expansion of the strain field is

$$\varepsilon_{ij}\left({}^0\mathbf{x}, \ldots, {}^{N-1}\mathbf{x}\right) = \sum_{I=0}^{N-1} \zeta^I \varepsilon_{ij}^I\left({}^0\mathbf{x}, \ldots, {}^{N-1}\mathbf{x}\right) + \cdots \qquad (3.12)$$

where the leading order strain is

$$\varepsilon_{ij}^{0}\left({}^{0}\mathbf{x},\ldots,{}^{N-1}\mathbf{x}\right) = \sum_{I=0}^{N-1} u_{(i,{}^{I}x_{j})}^{I}\left({}^{0}\mathbf{x},\ldots,{}^{I}\mathbf{x}\right) \tag{3.13}$$

The average strain at scale I-1 is obtained by averaging the strain at scale I over the repetitive volume element (or unit cell) Θ_I at scale I

$$^{I-1}\varepsilon_{ij}\left({}^{0}\mathbf{x},\ldots,{}^{I-1}\mathbf{x}\right) \equiv \frac{1}{|\Theta_I|} \int_{\Theta_I} {}^{I}\varepsilon_{ij}\left({}^{0}\mathbf{x},\ldots,{}^{I}\mathbf{x}\right) d\Theta_I \quad for\ I = 1,\ldots,N-1 \tag{3.14}$$

with the strain at the finest scale $(N-1)$ being

$$^{N-1}\varepsilon_{ij}\left({}^{0}\mathbf{x},\ldots,{}^{N-1}\mathbf{x}\right) \equiv \varepsilon_{ij}^{0}\left({}^{0}\mathbf{x},\ldots,{}^{N-1}\mathbf{x}\right) \tag{3.15}$$

The relation between strains at two subsequent scales is

$$^{I}\varepsilon_{ij}\left({}^{0}\mathbf{x},\ldots,{}^{I}\mathbf{x}\right) = {}^{I-1}\varepsilon_{ij}\left({}^{0}\mathbf{x},\ldots,{}^{I-1}\mathbf{x}\right) + u_{(i,{}^{I}x_{j})}^{I}\left({}^{0}\mathbf{x},\ldots,{}^{I}\mathbf{x}\right)$$
$$for\ I = 1,\ldots,N-1 \tag{3.16}$$

The N-scale asymptotic expansion of the stress field is

$$\sigma_{ij}\left({}^{0}\mathbf{x},\ldots,{}^{N-1}\mathbf{x}\right) = \sum_{I=0}^{N-1} \zeta^{I}\sigma_{ij}^{I}\left({}^{0}\mathbf{x},\ldots,{}^{N-1}\mathbf{x}\right) + \cdots \tag{3.17}$$

where the leading order stress is obtained from the constitutive equation (3.2)

$$\sigma_{ij}^{0}\left({}^{0}\mathbf{x},\ldots,{}^{N-1}\mathbf{x}\right) = L_{ijkl}\left({}^{N-1}\mathbf{x}\right)\left[\varepsilon_{kl}^{0}\left({}^{0}\mathbf{x},\ldots,{}^{N-1}\mathbf{x}\right) - \mu_{kl}\left({}^{0}\mathbf{x},\ldots,{}^{N-1}\mathbf{x}\right)\right] \tag{3.18}$$

Similarly to (3.14) the average stress at each scale is

$$^{I-1}\sigma_{ij}\left({}^{0}\mathbf{x},\ldots,{}^{I-1}\mathbf{x}\right) \equiv \frac{1}{|\Theta_I|} \int_{\Theta_I} {}^{I}\sigma_{ij}\left({}^{0}\mathbf{x},\ldots,{}^{I}\mathbf{x}\right) d\Theta_I \quad for\ I = 1,\ldots,N-1 \tag{3.19}$$

with the stress at the finest scale $(N-1)$ of interest being

$$^{N-1}\sigma_{ij}\left({}^{0}\mathbf{x},\ldots,{}^{N-1}\mathbf{x}\right) \equiv \sigma_{ij}^{0}\left({}^{0}\mathbf{x},\ldots,{}^{N-1}\mathbf{x}\right) \tag{3.20}$$

The equilibrium equations at each scale follow from the equilibrium equation (3.1) and indirect spatial differentiation rule (3.10)

$$\boxed{\begin{aligned} O\left(\zeta^{-I}\right): &\quad {}^{I}\sigma_{ij,{}^{I}x_{j}}\left({}^{0}\mathbf{x},\ldots,{}^{I}\mathbf{x}\right) = 0 \quad for\ I = 1,\ldots,N-1 \\ O\left(\zeta^{0}\right): &\quad {}^{0}\sigma_{ij,{}^{0}x_{j}}\left({}^{0}\mathbf{x}\right) + {}^{0}b_{i} = 0 \end{aligned}} \tag{3.21}$$

where ${}^{0}b_i$ is the average body force at the coarsest scale.

3.3 Residual-free governing equations at multiple scales

In this section, we formulate equilibrium equations (3.21) in terms of variables (eigendeformation modes) that *a priori* satisfy equilibrium equations at multiple scales (except for the macro scale). Computation of the residual-free stresses eliminates the need for costly solution of the discretized equilibrium equations at each step.

Using the definitions of strain (3.13) and stress (3.21) in combination with the constitutive relation (3.2), the equilibrium equation at any scale (except for the macro scale) is given by

$$\{^I L_{ijkl}\left(^0\mathbf{x},\ldots,^I\mathbf{x}\right)\left[^{I-1}\varepsilon_{kl}\left(^0\mathbf{x},\ldots,^{I-1}\mathbf{x}\right)+{}^I u_{(k,^I x_l)}\left(^0\mathbf{x},\ldots,^I\mathbf{x}\right)\right.$$
$$\left.-{}^I\mu_{kl}\left(^0\mathbf{x},\ldots,^I\mathbf{x}\right)\right]\}_{,^I x_j}=0 \quad (3.22)$$

Extending the two-scale formulation [37] to multiple scales, the displacement field can be decomposed using the elastic influence function $^I H_{ikl}$, the eigenstrain influence function $^I h^\mu_{ikl}$ and the eigenseparation influence function $^I h^\delta_{i\hat{n}}$ as

$$^I u_i\left(^0\mathbf{x},\ldots,^I\mathbf{x}\right) = {}^I u_i^e\left(^0\mathbf{x},\ldots,^I\mathbf{x}\right) + {}^I u_i^\mu\left(^0\mathbf{x},\ldots,^I\mathbf{x}\right) + {}^I u_i^\delta\left(^0\mathbf{x},\ldots,^I\mathbf{x}\right)$$
$$= {}^I H_{ikl}\left(^0\mathbf{x},\ldots,^I\mathbf{x}\right){}^{I-1}\varepsilon_{kl}\left(^0\mathbf{x},\ldots,^{I-1}\mathbf{x}\right)$$
$$+ \int_{\Theta_I} {}^I h^\mu_{ikl}\left(^0\mathbf{x},\ldots,^I\mathbf{x},^I\hat{\mathbf{x}}\right){}^I\mu_{kl}\left(^0\mathbf{x},\ldots,^I\hat{\mathbf{x}}\right) d\Theta$$
$$+ \int_{S_I} {}^I h^\delta_{i\hat{n}}\left(^0\mathbf{x},\ldots,^I\mathbf{x},^I\hat{\mathbf{x}}\right){}^I\delta_{\hat{n}}\left(^0\mathbf{x},\ldots,^I\hat{\mathbf{x}}\right) dS$$
$$(3.23)$$

where the influence (or Green's) functions $^I H_{ikl}, {}^I h^\mu_{ikl}, {}^I h^\delta_{i\hat{n}}$ are computed by solving a sequence of elastic boundary value problems independent of and prior to solving a nonlinear macro problem. The subscript $\hat{n}$ in the eigenseparation influence function denotes the component in the local Cartesian coordinate system of the interface. The influence functions are chosen to satisfy the equilibrium equation (3.22) for arbitrary $^I\varepsilon_{kl}, {}^I\mu_{kl}, {}^I\delta_{\hat{n}}$, and domains Θ_I, S_I they are defined. The governing equations for the elastic influence functions are

$$\{^I L_{ijkl}\left(^0\mathbf{x},\ldots,^I\mathbf{x}\right)\left[I_{klmn}\left(^0\mathbf{x},\ldots,^I\mathbf{x}\right)+{}^I G_{klmn}\left(^0\mathbf{x},\ldots,^I\mathbf{x}\right)\right]\}_{,^I x_j}$$
$$= 0 \quad {}^I\mathbf{x}\in\Theta_I \quad (3.24)$$

where

$$^I G_{ijkl}\left(^0\mathbf{x},\ldots,^I\mathbf{x}\right) = {}^I H_{(i,^I x_j)kl}\left(^0\mathbf{x},\ldots,^I\mathbf{x}\right) \quad (3.25)$$

The eigenstrain influence functions are computed by solving

$$\{{}^I L_{ijkl}({}^0\mathbf{x},\ldots,{}^I\mathbf{x})[{}^I g^\mu_{klmn}({}^0\mathbf{x},\ldots,{}^I\mathbf{x},{}^I\hat{\mathbf{x}}) - I_{klmn}({}^0\mathbf{x},\ldots,{}^I\hat{\mathbf{x}})\, d\,({}^I\mathbf{x}-{}^I\hat{\mathbf{x}})]\}_{,{}^I x_j}$$
$$= 0 \quad {}^I\mathbf{x}, {}^I\hat{\mathbf{x}} \in \Theta_I \tag{3.26}$$

where

$$ {}^I g^\mu_{ijkl}({}^0\mathbf{x},\ldots,{}^I\mathbf{x},{}^I\hat{\mathbf{x}}) = {}^I h^\mu_{(i,{}^I x_j)kl}({}^0\mathbf{x},\ldots,{}^I\mathbf{x},{}^I\hat{\mathbf{x}}) \tag{3.27}$$

and d is the Dirac delta function.

The governing equations for eigenseparation influence functions are

$$\{{}^I L_{ijkl}({}^0\mathbf{x},\ldots,{}^I\mathbf{x})\,{}^I g^\delta_{kl\hat{n}}({}^0\mathbf{x},\ldots,{}^I\mathbf{x},{}^I\hat{\mathbf{x}})\}_{,{}^I x_j} = 0 \quad {}^I\mathbf{x} \in \Theta_I,\ {}^I\hat{\mathbf{x}} \in S_I$$
$$\text{E.B.C.} \quad \delta_{\hat{n}}({}^I\tilde{\mathbf{x}}) = d\,({}^I\tilde{\mathbf{x}} - {}^I\hat{\mathbf{x}}) \quad {}^I\tilde{\mathbf{x}} \in S_I \tag{3.28}$$

where

$$ {}^I g^\delta_{ij\hat{n}}({}^0\mathbf{x},\ldots,{}^I\mathbf{x},{}^I\hat{\mathbf{x}}) = {}^I h^\delta_{(i,{}^I x_j)\hat{n}}({}^0\mathbf{x},\ldots,{}^I\mathbf{x},{}^I\hat{\mathbf{x}}) \tag{3.29}$$

The elastic stiffness tensor at each scale is given by

$$ {}^{I-1}L_{ijkl}({}^0\mathbf{x},\ldots,{}^{I-1}\mathbf{x}) = \frac{1}{|\Theta_I|}\int_{\Theta_I} {}^I L_{ijmn}({}^0\mathbf{x},\ldots,{}^I\mathbf{x})\,{}^I A_{mnkl}({}^0\mathbf{x},\ldots,{}^I\mathbf{x})\,d\Theta $$

$$ {}^{N-1}L_{ijkl}({}^0\mathbf{x},\ldots,{}^{N-1}\mathbf{x}) \equiv L_{ijkl}({}^0\mathbf{x},\ldots,{}^{N-1}\mathbf{x}) $$

$$ {}^I A_{ijkl} = I_{ijkl} + {}^I G_{ijkl} \tag{3.30}$$

The residual-free stress at scale I is given by

$$\boxed{\begin{aligned}
{}^I\sigma_{ij} = {}^I L_{ijmn}&({}^0\mathbf{x},\ldots,{}^I\mathbf{x}) \\
\times \Big[&{}^I A_{mnkl}({}^0\mathbf{x},\ldots,{}^I\mathbf{x})\,{}^{I-1}\varepsilon_{kl}({}^0\mathbf{x},\ldots,{}^{I-1}\mathbf{x}) \\
&+ \int_{S_I} {}^I g^\delta_{mn\hat{n}}({}^0\mathbf{x},\ldots,{}^I\mathbf{x},{}^I\hat{\mathbf{x}})\,{}^I\delta_{\hat{n}}({}^0\mathbf{x},\ldots,{}^I\hat{\mathbf{x}})\,dS \\
&+ \int_{\Theta_I} {}^I g^\mu_{mnkl}({}^0\mathbf{x},\ldots,{}^I\mathbf{x},{}^I\hat{\mathbf{x}})\,{}^I\mu_{kl}({}^0\mathbf{x},\ldots,{}^I\hat{\mathbf{x}})\,d\Theta - {}^I\mu_{mn}({}^0\mathbf{x},\ldots,{}^I\mathbf{x}) \Big]
\end{aligned}}$$
$$\tag{3.31}$$

The eigenstrain at scale $I-1$ ($I \geq 2$) can be expressed in terms of the eigendeformation (eigenstrain and eigenseparation) at a finer scale I:

$$
\begin{aligned}
&\text{For } I = 2, \ldots, N-1 \\
&{}^{I-1}\mu_{ij}\left({}^{0}\mathbf{x}, \ldots, {}^{I-1}\mathbf{x}\right) \\
&= -\left[{}^{I-1}L_{ijst}\left({}^{0}\mathbf{x}, \ldots, {}^{I-1}\mathbf{x}\right)\right]^{-1} \\
&\quad \cdot \frac{1}{|\Theta_I|}\int_{\Theta_I}\left\{{}^{I}L_{stmn}\left({}^{0}\mathbf{x}, \ldots, {}^{I}\mathbf{x}\right)\left[\begin{array}{l}\int_{S_I}{}^{I}g_{mn\hat{n}}^{\delta}\left({}^{0}\mathbf{x}, \ldots, {}^{I}\mathbf{x}, {}^{I}\hat{\mathbf{x}}\right){}^{I}\delta_{\hat{n}}\left({}^{0}\mathbf{x}, \ldots, {}^{I}\hat{\mathbf{x}}\right)dS \\ + \int_{\Theta_I}{}^{I}g_{mnkl}^{\mu}\left({}^{0}\mathbf{x}, \ldots, {}^{I}\mathbf{x}, {}^{I}\hat{\mathbf{x}}\right){}^{I}\mu_{kl}\left({}^{0}\mathbf{x}, \ldots, {}^{I}\hat{\mathbf{x}}\right)d\Theta \\ -{}^{I}\mu_{mn}\left({}^{0}\mathbf{x}, \ldots, {}^{I}\mathbf{x}\right)\end{array}\right]\right\}d\Theta
\end{aligned}
$$
(3.32)

with eigenstrains at the finest scale of interest denoted as

$$
\begin{aligned}
{}^{N-1}\mu_{ij}\left({}^{0}\mathbf{x}, \ldots, {}^{N-1}\mathbf{x}\right) &\equiv \mu_{ij}\left({}^{0}\mathbf{x}, \ldots, {}^{N-1}\mathbf{x}\right) \\
{}^{N-1}\delta_{\hat{n}}\left({}^{0}\mathbf{x}, \ldots, {}^{N-1}\mathbf{x}\right) &\equiv \delta_{\hat{n}}\left({}^{0}\mathbf{x}, \ldots, {}^{N-1}\mathbf{x}\right)
\end{aligned}
$$
(3.33)

The hierarchical structure of eigendeformations at different scales is shown in Table 3.1 where the terms in the gray background denote independent variables. At the finest scale of interest the eigendeformation (μ, δ) are independent, whereas for the coarser scales the eigenstrain at scale $I-1$ ($I < N$) is derived from the variables at scale I. The eigenseparation, on the other hand, is assumed to be an independent variable at all scales.

3.4 Multiple-scale reduced order model

In this section, we focus on reducing computational complexity of the multiple-scale unit cell problems derived in Section 3.3. This is accomplished by

Table 3.1. The hierarchical structure of eigendeformations

Scale	Eigendeformations	
$N-1$	μ	δ
$N-2$	μ	δ
$\vdots$	$\vdots$	$\vdots$
$N-k$	μ	δ
$\vdots$	$\vdots$	$\vdots$
1	μ	δ

discretizing the eigendeformation and formulating the resulting residual-free reduced order governing equations.

The eigenstrains at each scale are discretized in terms of piecewise constant function $N^{(\alpha_I)}\left(^I\mathbf{x}\right)$ as:

$$^I\mu_{ij}\left(^0\mathbf{x},\ldots,{}^I\mathbf{x}\right) = \sum_{\alpha_I=1}^{n_I} N^{(\alpha_I)}\left(^I\mathbf{x}\right){}^I\mu_{ij}^{(\alpha_I)}\left(^0\mathbf{x},\ldots,{}^{I-1}\mathbf{x}\right) \quad (3.34)$$

where

$$N^{(\alpha_I)}\left(^I\mathbf{x}\right) = \begin{cases} 1 & {}^I\mathbf{x} \in \Theta_I^{(\alpha_I)} \\ 0 & {}^I\mathbf{x} \notin \Theta_I^{(\alpha_I)} \end{cases}$$

$$^I\mu_{ij}^{(\alpha_I)}\left(^0\mathbf{x},\ldots,{}^{I-1}\mathbf{x}\right) = \frac{1}{\left|\Theta_I^{(\alpha_I)}\right|}\int_{\Theta_I^{(\alpha_I)}} {}^I\mu_{ij}\left(^0\mathbf{x},\ldots,{}^I\mathbf{x}\right)d\Theta \quad (3.35)$$

in which the total volume of the unit cell at scale I is partitioned into n_I nonoverlapping subdomains denoted by $\Theta_I^{(\alpha_I)}$. Partitions at various scales are denoted by a superscript enclosed in parenthesis: (α_I), (β_I) denote phase (volume) partitions at scale I; (ξ_I), (η_I) denote the interface partitions at scale I.

The eigenseparation $^I\delta_{\hat{n}}\left(^0\mathbf{x},\ldots,{}^I\tilde{\mathbf{x}}\right)$ at scale I is discretized in terms of C^0 continuous interface partition shape function $N^{(\xi_I)}\left(^I\tilde{\mathbf{x}}\right)$ as

$$^I\delta_{\hat{n}}\left(^0\mathbf{x},\ldots,{}^I\tilde{\mathbf{x}}\right) = \sum_{\xi_I=1}^{m_I} N^{(\xi_I)}\left(^I\tilde{\mathbf{x}}\right){}^I\delta_{\hat{n}}^{(\xi_I)}\left(^0\mathbf{x},\ldots,{}^{I-1}\mathbf{x}\right) \quad (3.36)$$

where $N^{(\xi_I)}\left(^I\tilde{\mathbf{x}}\right)$ is defined by a linear combination of the piecewise linear finite element shape functions over partition ξ_I

$$N^{(\xi_I)}\left(^I\tilde{\mathbf{x}}\right) = \begin{cases} \sum_{a \in S_I^{(\xi_I)}} N_a\left(^I\tilde{\mathbf{x}}\right) & {}^I\tilde{\mathbf{x}} \in S_I^{(\xi_I)} \\ 0 & {}^I\tilde{\mathbf{x}} \notin S_I^{(\xi_I)} \end{cases}$$

$$^I\delta_{\hat{n}}^{(\xi_I)}\left(^0\mathbf{x},\ldots,{}^{I-1}\mathbf{x}\right) = \frac{1}{\left|S_I^{(\xi_I)}\right|}\int_{S_I^{(\xi_I)}} {}^I\delta_{\hat{n}}\left(^0\mathbf{x},\ldots,{}^I\tilde{\mathbf{x}}\right)dS \quad (3.37)$$

in which the total interface at the scale I is divided into m_I partitions denoted by $S_I^{(\xi_I)}$, and $N_a\left({}^I\tilde{\mathbf{x}}\right)$ is a linear shape function associated with a finite element mesh node a along the interface of scale I. Note that in contrast to the volume partitions, surface partitions are overlapping as illustrated in Figure 3.1 for the case of fibrous unit cell. Figure 3.1(a) depicts the unit cell finite element mesh. Figure 3.1(b) shows the two interface partitions. The overlapping areas are shown as shaded areas. Figures 3.1(c) and 3.1(d) illustrate the two interface partition shape functions.

Substituting (3.34) and (3.36) into (3.31) yields the reduced order residual-free stresses for scales ranging from $I = 1$ to $N - 1$

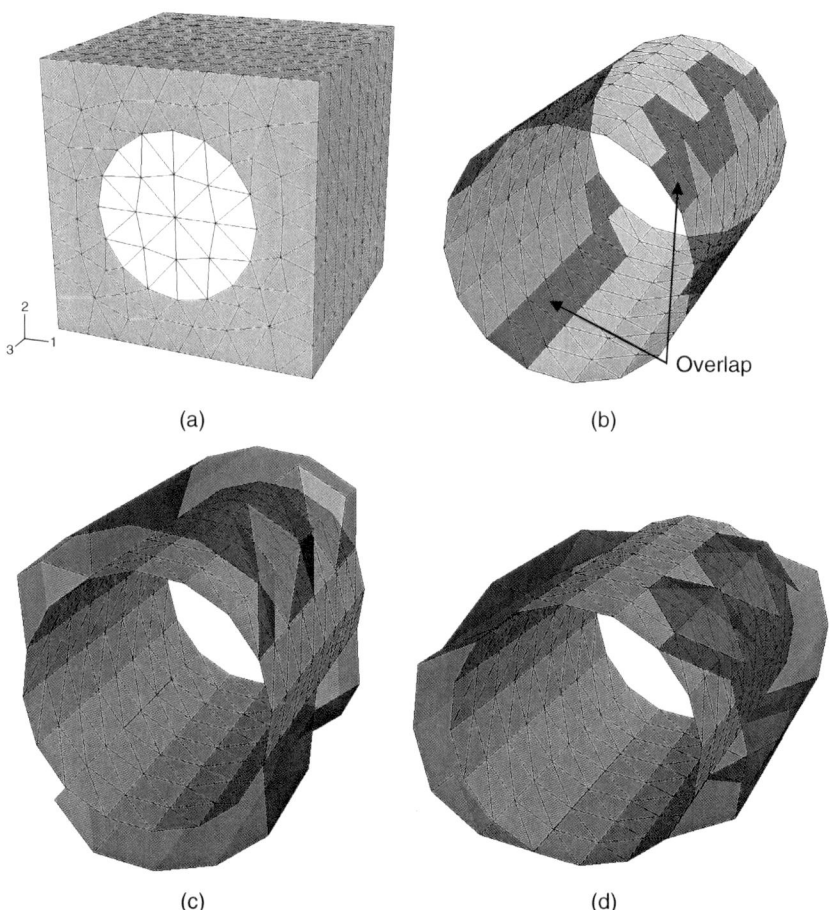

Figure 3.1: Interface partitions: (a) unit cell mesh; (b) interface partitions; (c) shape function for partition 1; (d) shape function for partition 2.

$$^I\sigma_{ij} = {}^I L_{ijmn}\left({}^0\mathbf{x},\ldots,{}^I\mathbf{x}\right) \begin{bmatrix} {}^I A_{mnkl}\left({}^0\mathbf{x},\ldots,{}^I\mathbf{x}\right){}^{I-1}\varepsilon_{kl}\left({}^0\mathbf{x},\ldots,{}^{I-1}\mathbf{x}\right) \\ + \sum_{\xi_I=1}^{m_I} {}^I Q_{mn\hat{n}}^{(\xi_I)}\left({}^0\mathbf{x},\ldots,{}^I\mathbf{x}\right){}^I\delta_{\hat{n}}^{(\xi_I)}\left({}^0\mathbf{x},\ldots,{}^{I-1}\mathbf{x}\right) \\ + \sum_{\alpha_I=1}^{n_I} {}^I S_{mnkl}^{(\alpha_I)}\left({}^0\mathbf{x},\ldots,{}^I\mathbf{x}\right){}^I\mu_{kl}^{(\alpha_I)}\left({}^0\mathbf{x},\ldots,{}^{I-1}\mathbf{x}\right) \end{bmatrix}$$

(3.38)

where

$$^I Q_{ij\hat{n}}^{(\xi_I)}\left({}^0\mathbf{x},\ldots,{}^I\mathbf{x}\right) = \int_{S_I} {}^I g_{ij\hat{n}}^\delta\left({}^0\mathbf{x},\ldots,{}^I\mathbf{x},{}^I\hat{\mathbf{x}}\right) N^{(\xi_I)}\left({}^I\hat{\mathbf{x}}\right) dS$$

$$^I S_{ijkl}^{(\alpha_I)}\left({}^0\mathbf{x},\ldots,{}^I\mathbf{x}\right) = {}^I P_{ijkl}^{(\alpha_I)}\left({}^0\mathbf{x},\ldots,{}^I\mathbf{x}\right) - I_{ijkl}^{(\alpha_I)}\left({}^0\mathbf{x},\ldots,{}^I\mathbf{x}\right)$$

(3.39)

and

$$P_{ijkl}^{(\alpha_I)}\left({}^0\mathbf{x},\ldots,{}^I\mathbf{x}\right) = \int_{\Theta_I} {}^I g_{ijkl}^\mu\left({}^0\mathbf{x},\ldots,{}^I\mathbf{x},{}^I\hat{\mathbf{x}}\right) N^{(\alpha_I)}\left({}^I\hat{\mathbf{x}}\right) d\Theta$$

$$= \int_{\Theta_I^{(\alpha_I)}} {}^I g_{ijkl}^\mu\left({}^0\mathbf{x},\ldots,{}^I\mathbf{x},{}^I\hat{\mathbf{x}}\right) d\Theta \qquad (3.40)$$

$$I_{ijkl}\left(\alpha_I\right)\left({}^0\mathbf{x},\ldots,{}^I\mathbf{x}\right) = \begin{cases} I_{ijkl} & {}^I\mathbf{x} \in \Theta_I(\alpha_I) \\ 0 & {}^I\mathbf{x} \notin \Theta_I(\alpha_I) \end{cases} \qquad (3.41)$$

The reduced order influence functions, $P_{ijkl}(\alpha_I)\left({}^0\mathbf{x},\ldots,{}^I\mathbf{x}\right)$ and ${}^I Q_{ij\hat{n}}(\xi_I)$ $\left({}^0\mathbf{x},\ldots,{}^I\mathbf{x}\right)$, can be computed in two ways: (i) directly from (3.40) and (3.39) in combination with (3.26) and (3.28), or by requiring reduced order stress (3.38) to be residual free.

By Integrating (3.26) over partition ${}^I\hat{\mathbf{x}} \in \Theta_I^{(\alpha_I)}$ and exploiting (3.40) yields the governing equation

$$\left\{ {}^I L_{ijkl}\left({}^0\mathbf{x},\ldots,{}^I\mathbf{x}\right) \left[{}^I P_{klmn}^{(\alpha_I)}\left({}^0\mathbf{x},\ldots,{}^I\mathbf{x}\right) - I_{klmn}^{(\alpha_I)}\left({}^0\mathbf{x},\ldots,{}^I\mathbf{x}\right) \right] \right\}_{,{}^I x_j}$$

$$= 0 \quad {}^I\mathbf{x} \in \Theta_I \qquad (3.42)$$

from which reduced order eigenstrain influence function can be solved for.

Precisely the same expression can be obtained from (3.38) assuming vanishing strain ${}^{I-1}\varepsilon_{ij}$ and partitioned eigenseparation ${}^I\delta_{\hat{n}}^{(\xi_I)}$.

Similarly, the governing equation for the reduced order eigenseparation influence function is given by

$$\left\{ {}^{I}L_{ijkl}\left({}^{0}\mathbf{x},\ldots,{}^{I}\mathbf{x}\right){}^{I}Q_{kl\hat{n}}^{(\xi_I)}\left({}^{0}\mathbf{x},\ldots,{}^{I}\mathbf{x}\right)\right\}_{,{}^{I}x_j} = 0 \quad {}^{I}\mathbf{x} \in \Theta_I$$
(3.43)

E.B.C. $\quad \delta_{\hat{n}}\left({}^{I}\tilde{\mathbf{x}}\right) = N^{(\xi_I)}\left({}^{I}\tilde{\mathbf{x}}\right) \quad {}^{I}\tilde{\mathbf{x}} \in S_I$

In the following, we focus on expressing dependent partitioned eigenstrains in terms of the independent partitioned eigendeformations (eigenstrain at the finest scale and eigenseparation at each scale).

We start by expressing eigenstrain at scale $I-1$ in terms of the partitioned eigendeformation at scale I by substituting (3.34), (3.36), and (3.39) into (3.32), which yields

$$
{}^{I-1}\mu_{ij}\left({}^{0}\mathbf{x},\ldots,{}^{I-1}\mathbf{x}\right) = -\left[{}^{I-1}L_{ijmn}\left({}^{0}\mathbf{x},\ldots,{}^{I-1}\mathbf{x}\right)\right]^{-1}
$$
$$
\times \left\{ \begin{array}{l} \displaystyle\sum_{\xi_I=1}^{m_I} {}^{I-1}F_{mn\hat{n}}^{(\xi_I)}\left({}^{0}\mathbf{x},\ldots,{}^{I-1}\mathbf{x}\right){}^{I}\delta_{\hat{n}}^{(\xi_I)}\left({}^{0}\mathbf{x},\ldots,{}^{I-1}\mathbf{x}\right) \\ + \displaystyle\sum_{\alpha_I=1}^{n_I} {}^{I-1}E_{mnkl}^{(\alpha_I)}\left({}^{0}\mathbf{x},\ldots,{}^{I-1}\mathbf{x}\right){}^{I}\mu_{kl}^{(\alpha_I)}\left({}^{0}\mathbf{x},\ldots,{}^{I-1}\mathbf{x}\right) \end{array} \right\}
$$
(3.44)

where

$${}^{I-1}F_{ij\hat{n}}^{(\xi_I)}\left({}^{0}\mathbf{x},\ldots,{}^{I-1}\mathbf{x}\right) = \frac{1}{|\Theta_I|}\int_{\Theta_I} {}^{I}L_{ijmn}\left({}^{0}\mathbf{x},\ldots,{}^{I}\mathbf{x}\right){}^{I}Q_{mn\hat{n}}^{(\alpha_I)}\left({}^{0}\mathbf{x},\ldots,{}^{I}\mathbf{x}\right) d\Theta$$

$${}^{I-1}E_{ijkl}^{(\alpha_I)}\left({}^{0}\mathbf{x},\ldots,{}^{I-1}\mathbf{x}\right) = \frac{1}{|\Theta_I|}\int_{\Theta_I} {}^{I}L_{ijmn}\left({}^{0}\mathbf{x},\ldots,{}^{I}\mathbf{x}\right){}^{I}S_{mnkl}^{(\alpha_I)}\left({}^{0}\mathbf{x},\ldots,{}^{I}\mathbf{x}\right) d\Theta$$
(3.45)

Then, the partitioned eigenstrain is given by

$${}^{I-1}\mu_{ij}^{(\alpha_{I-1})}\left({}^{0}\mathbf{x},\ldots,{}^{I-2}\mathbf{x}\right)$$
$$= -\left[{}^{I-1}L_{ijmn}\left({}^{0}\mathbf{x},\ldots,{}^{I-2}\mathbf{x}\right)\right]^{-1}$$
$$\times \left\{ \begin{array}{l} \displaystyle\sum_{\xi_I=1}^{m_I} {}^{I-1}F_{mn\hat{n}}^{(\xi_I)}\left({}^{0}\mathbf{x},\ldots,{}^{I-2}\mathbf{x}\right){}^{I}\delta_{\hat{n}}^{(\alpha_{I-1}:\xi_I)}\left({}^{0}\mathbf{x},\ldots,{}^{I-2}\mathbf{x}\right) \\ + \displaystyle\sum_{\alpha_I=1}^{n_I} {}^{I-1}E_{mnkl}^{(\alpha_I)}\left({}^{0}\mathbf{x},\ldots,{}^{I-2}\mathbf{x}\right){}^{I}\mu_{kl}^{(\alpha_{I-1}:\alpha_I)}\left({}^{0}\mathbf{x},\ldots,{}^{I-2}\mathbf{x}\right) \end{array} \right\}$$
(3.46)

where

$$^I\delta_{\hat{n}}^{(\alpha_{I-1}:\xi_I)}\left(^0\mathbf{x},\ldots,^{I-2}\mathbf{x}\right) = \frac{1}{\left|\Theta_{I-1}^{(\alpha_{I-1})}\right|}\int_{\Theta_{I-1}^{(\alpha_{I-1})}} {}^I\delta_{\hat{n}}^{(\xi_I)}\left(^0\mathbf{x},\ldots,^{I-1}\mathbf{x}\right)d\Theta$$

$$^I\mu_{ij}^{(\alpha_{I-1}:\alpha_I)}\left(^0\mathbf{x},\ldots,^{I-2}\mathbf{x}\right) = \frac{1}{\left|\Theta_{I-1}^{(\alpha_{I-1})}\right|}\int_{\Theta_{I-1}^{(\alpha_{I-1})}} {}^I\mu_{ij}^{(\alpha_I)}\left(^0\mathbf{x},\ldots,^{I-1}\mathbf{x}\right)d\Theta$$

(3.47)

Finally, using the above relations recursively, we can express all the dependent partitioned eigenstrains in terms of independent eigendeformations as

$$^{I-1}\mu_{ij}^{(\alpha_{I-1})}(^0\mathbf{x},\ldots,^{I-2}\mathbf{x}) = \sum_{\alpha_I=1}^{n_I}\cdots\sum_{\alpha_{N-1}=1}^{n_{N-1}}\prod_{K=I}^{N-1}{}^{K-1}S_{ijkl}^{(\alpha_K)}\cdot {}^{N-1}\mu_{kl}^{(\alpha_{I-1}:\cdots:\alpha_{N-1})}$$

$$\times (^0\mathbf{x},\ldots,^{I-2}\mathbf{x}) + \sum_{J=I}^{N-1}\left[\sum_{\alpha_I=1}^{n_I}\cdots\sum_{\alpha_{J-1}=1}^{n_{J-1}}\sum_{\xi_J=1}^{m_J}\prod_{K=I}^{J-1}{}^{K-1}S_{ijkl}^{(\alpha_K)}{}^{J-1}Q_{kl\hat{n}}^{(\xi_J)J}\right.$$

$$\left.\times \delta_{\hat{n}}^{(\alpha_{I-1}:\cdots:\alpha_{J-1}:\xi_J)}\left(^0\mathbf{x},\ldots,^{I-2}\mathbf{x}\right)\right]$$

(3.48)

where

$$^{K-1}S_{ijkl}^{(\alpha_K)} \equiv -\left[^{K-1}L_{ijmn}\right]^{-1}{}^{K-1}E_{mnkl}^{(\alpha_K)}; \quad {}^{J-1}Q_{ij\hat{n}}^{(\xi_J)}$$

$$\equiv -\left[^{J-1}L_{ijmn}\right]^{-1}{}^{J-1}F_{mn\hat{n}}^{(\xi_J)}$$

$$^{N-1}\mu_{ij}^{(\alpha_{I-1}:\cdots:\alpha_{N-1})} \equiv \frac{1}{\left|\Theta_I^{(\alpha_{I-1})}\right|}\int_{\Theta_I^{(\alpha_{I-1})}}\cdots\frac{1}{\left|\Theta_{N-1}^{(\alpha_{N-1})}\right|}$$

$$\int_{\Theta_{N-1}^{(\alpha_{N-1})}} {}^{N-1}\mu_{ij}\left(^0\mathbf{x},\ldots,^{N-1}\mathbf{x}\right)d\Theta\cdots d\Theta$$

(3.49)

$$^J\delta_{\hat{n}}^{(\alpha_{I-1}:\cdots:\alpha_{J-1}:\xi_J)} \equiv \frac{1}{\left|\Theta_{I-1}^{(\alpha_{I-1})}\right|}\int_{\Theta_I^{(\alpha_I)}}\cdots\frac{1}{\left|\Theta_{J-1}^{(\alpha_{J-1})}\right|}\int_{\Theta_{J-1}(\alpha_{J-1})}\frac{1}{\left|S_J^{(\xi_J)}\right|}$$

$$\int_{S_J^{(\xi_J)}} {}^J\delta_{\hat{n}}\left(^0\mathbf{x},\ldots,^{J-1}\mathbf{x},{}^J\tilde{\mathbf{x}}\right)dSd\Theta\cdots d\Theta$$

To update the stresses at the macro (coarsest) scale, we express the reduced order macro stress in terms of the independent partitioned eigendeformations as

$$
\begin{aligned}
{}^0\sigma_{ij} = {}&{}^0L_{ijkl} \cdot {}^0\varepsilon_{kl}({}^0\mathbf{x}) + \sum_{\xi_y=1}^{m_y} {}^0F_{ij\hat{n}}^{(\xi_1)1}\delta_{\hat{n}}^{(\xi_1)}({}^0\mathbf{x}) \\
&+ \sum_{\alpha_1=1}^{n_1} \cdots \sum_{\alpha_{N-1}=1}^{n_{N-1}} {}^0E_{ijkl}^{(\alpha_1:\cdots:\alpha_{N-1})} \cdot {}^{N-1}\mu_{kl}^{(\alpha_1:\cdots:\alpha_{N-1})}({}^0\mathbf{x}) \\
&+ \sum_{J=2}^{N-1} \left[\sum_{\alpha_1=1}^{n_1} \cdots \sum_{\alpha_{J-1}=1}^{n_{J-1}} \sum_{\xi_J=1}^{m_J} {}^0F_{ij\hat{n}}^{(\alpha_1:\cdots:\alpha_{J-1}:\xi_J)J}\delta_{\hat{n}}^{(\alpha_1:\cdots:\alpha_{J-1}:\xi_J)}({}^0\mathbf{x}) \right]
\end{aligned}
$$
(3.50)

where

$$
{}^0L_{ijkl} = \frac{1}{|\Theta_1|}\int_{\Theta_1} {}^1L_{ijmn}({}^1\mathbf{x})\,{}^1A_{mnkl}({}^1\mathbf{x})d\Theta
$$

$$
{}^0F_{ij\hat{n}}^{(\xi_1)} = \frac{1}{|\Theta_1|}\int_{\Theta_1} {}^1L_{ijkl}({}^1\mathbf{x})\,{}^1Q_{kl\hat{n}}(\xi_1)({}^1\mathbf{x})d\Theta
$$
(3.51)

The recursive definition of ${}^0E_{ijkl}^{(\alpha_1:\cdots:\alpha_{N-1})}$ is given by

for $I = 1,\ldots,N-2$

$$
{}^{I-1}E_{ijkl}^{(\alpha_I:\cdots:\alpha_{N-1})} = -\frac{1}{|\Theta_I|}\int_{\Theta_I} {}^IL_{ijmn}({}^I\mathbf{x})\,{}^IS_{mnpq}^{(\alpha_I)}({}^I\mathbf{x})\left[{}^IL_{pqst}({}^I\mathbf{x})\right]^{-1}
$$
$$
\times\,{}^IE_{stkl}^{(\alpha_{I+1}:\cdots:\alpha_{N-1})}d\Theta
$$
(3.52)

where the term at the finest scale of interest is

$$
{}^{N-2}E_{ijkl}^{(\alpha_{N-1})} = \frac{1}{|\Theta_{N-1}|}\int_{\Theta_{N-1}} {}^{N-1}L_{ijmn}({}^{N-1}\mathbf{x})\,{}^{N-1}S_{mnkl}^{(\alpha_{N-1})}({}^{N-1}\mathbf{x})d\Theta \quad (3.53)
$$

The recursive definition of ${}^0F_{ij\hat{n}}^{(\alpha_1:\cdots:\alpha_{J-1}:\xi_J)}$ is given as

for $I = 1,\ldots,J-1$

$$
{}^{I-1}F_{ij\hat{n}}^{(\alpha_I:\cdots:\alpha_{J-1}:\xi_J)} = -\frac{1}{|\Theta_I|}\int_{\Theta_I} {}^IL_{ijmn}({}^I\mathbf{x})\,{}^IS_{mnpq}^{(\alpha_I)}({}^I\mathbf{x})\left[{}^IL_{pqst}({}^I\mathbf{x})\right]^{-1}
$$
$$
\times\,{}^IF_{st\hat{n}}^{(\alpha_{I+1}:\cdots:\alpha_{J-1}:\xi_J)}d\Theta
$$
(3.54)

with the finest scale term for the Jth eigendisplacement being

$$^{J-1}F_{ij\hat{n}}^{(\xi_J)} = \frac{1}{|\Theta_J|}\int_{\Theta_J} {}^JL_{ijmn}\left({}^J\mathbf{x}\right) {}^JQ_{mn\hat{n}}^{(\xi_J)}\left({}^J\mathbf{x}\right)d\Theta \qquad (3.55)$$

The salient feature of the method is that all the influence functions can be obtained by approximately solving the corresponding elastic boundary value problem using finite element method independent of and prior to the nonlinear macro analysis. Once the influence functions are determined, all of the coefficient tensors of the reduced-order model at multiple scales ($^0L_{ijkl}$, $^0F_{ij\hat{n}}^{(\xi_1)}$, $^{I-1}E_{ijkl}^{(\alpha_I:\cdots:\alpha_{N-1})}$, and $^{I-1}F_{ij\hat{n}}^{(\alpha_I:\cdots:\alpha_{J-1}:\xi_J)}$) can be calculated. Since these computations are independent of the nonlinear macro analysis, they are precomputed in the preprocessing stage.

3.5 Reduced order unit cell problems

Since all the coefficient tensors of the reduced order model can be precomputed prior to the macro analysis, only independent partitioned eigendeformations $^1\delta_{\hat{n}}^{(\xi_1)}\left({}^0\mathbf{x}\right)$, $^{N-1}\mu_{kl}^{(\alpha_1:\cdots:\alpha_{N-1})}\left({}^0\mathbf{x}\right)$, and $^J\delta_{\hat{n}}^{(\alpha_1:\cdots:\alpha_{J-1}:\xi_J)}\left({}^0\mathbf{x}\right)$ need to be updated at each increment of macro analysis. In this section, we construct the system of nonlinear equations for updating these quantities.

Referring to the reduced order residual-free stress (3.38) for scales $I = 1,\ldots,N-1$, the corresponding strains are given as:

$$^I\varepsilon_{ij}\left({}^0\mathbf{x},\ldots,{}^I\mathbf{x}\right) = {}^IA_{ijkl}\left({}^0\mathbf{x},\ldots,{}^I\mathbf{x}\right) {}^{I-1}\varepsilon_{kl}\left({}^0\mathbf{x},\ldots,{}^{I-1}\mathbf{x}\right)$$

$$+ \sum_{\xi_I=1}^{m_I} {}^IQ_{ij\hat{n}}^{(\xi_I)}\left({}^0\mathbf{x},\ldots,{}^I\mathbf{x}\right) {}^I\delta_{\hat{n}}^{(\xi_I)}\left({}^0\mathbf{x},\ldots,{}^{I-1}\mathbf{x}\right) \qquad (3.56)$$

$$+ \sum_{\alpha_I=1}^{n_I} {}^IP_{ijkl}^{(\alpha_I)}\left({}^0\mathbf{x},\ldots,{}^I\mathbf{x}\right) {}^I\mu_{kl}^{(\alpha_I)}\left({}^0\mathbf{x},\ldots,{}^{I-1}\mathbf{x}\right)$$

Applying $\frac{1}{|\Theta_I^{(\beta_I)}|}\int_{\Theta_I^{(\beta_I)}} \cdot d\Theta$ averaging operator on (3.56) yields

$$^I\varepsilon_{ij}^{(\beta_I)}\left({}^0\mathbf{x},\ldots,{}^{I-1}\mathbf{x}\right) = {}^IA_{ijkl}^{(\beta_I)}\left({}^0\mathbf{x},\ldots,{}^{I-1}\mathbf{x}\right)\cdot {}^{I-1}\varepsilon_{kl}\left({}^0\mathbf{x},\ldots,{}^{I-1}\mathbf{x}\right)$$

$$+ \sum_{\xi_I=1}^{m_I} {}^IQ_{ij\hat{n}}^{(\beta_I,\xi_I)}\left({}^0\mathbf{x},\ldots,{}^{I-1}\mathbf{x}\right)\cdot {}^I\delta_{\hat{n}}^{(\xi_I)}\left({}^0\mathbf{x},\ldots,{}^{I-1}\mathbf{x}\right)$$

$$+ \sum_{\alpha_I=1}^{n_I} {}^IP_{ijkl}^{(\beta_I,\alpha_I)}\left({}^0\mathbf{x},\ldots,{}^{I-1}\mathbf{x}\right)\cdot {}^I\mu_{kl}^{(\alpha_I)}\left({}^0\mathbf{x},\ldots,{}^{I-1}\mathbf{x}\right)$$

$$(3.57)$$

Substituting the corresponding expression for $^{I-1}\varepsilon_{ij}$ into the right hand side of (3.57) yields

$$^{I}\varepsilon_{ij}^{(\beta_I)}\left(^{0}\mathbf{x},\ldots,^{I-1}\mathbf{x}\right)$$

$$= {}^{I}A_{ijmn}^{(\beta_I)}\left(^{0}\mathbf{x},\ldots,^{I-1}\mathbf{x}\right)$$

$$\cdot \left\{ \begin{array}{l} {}^{I-1}A_{mnkl}\left(^{0}\mathbf{x},\ldots,^{I-1}\mathbf{x}\right){}^{I-2}\varepsilon_{kl}\left(^{0}\mathbf{x},\ldots,^{I-2}\mathbf{x}\right) \\ + \displaystyle\sum_{\xi_{I-1}=1}^{m_{I-1}} {}^{I-1}Q_{mn\hat{n}}^{(\xi_{I-1})}\left(^{0}\mathbf{x},\ldots,^{I-1}\mathbf{x}\right){}^{I-1}\delta_{\hat{n}}^{(\xi_{I-1})}\left(^{0}\mathbf{x},\ldots,^{I-2}\mathbf{x}\right) \\ + \displaystyle\sum_{\alpha_{I-1}=1}^{n_{I-1}} {}^{I-1}P_{mnkl}^{(\alpha_{I-1})}\left(^{0}\mathbf{x},\ldots,^{I-1}\mathbf{x}\right){}^{I-1}\mu_{kl}^{(\alpha_{I-1})}\left(^{0}\mathbf{x},\ldots,^{I-2}\mathbf{x}\right) \end{array} \right\}$$

$$+ \sum_{\xi_I=1}^{m_I} {}^{I}Q_{ij\hat{n}}^{(\beta_I,\xi_I)}\left(^{0}\mathbf{x},\ldots,^{I-1}\mathbf{x}\right)\cdot {}^{I}\delta_{\hat{n}}^{(\xi_I)}\left(^{0}\mathbf{x},\ldots,^{I-1}\mathbf{x}\right)$$

$$+ \sum_{\alpha_I=1}^{n_I} {}^{I}P_{ijkl}^{(\beta_I,\alpha_I)}\left(^{0}\mathbf{x},\ldots,^{I-1}\mathbf{x}\right)\cdot {}^{I}\mu_{kl}^{(\alpha_I)}\left(^{0}\mathbf{x},\ldots,^{I-1}\mathbf{x}\right) \qquad (3.58)$$

Further applying $\dfrac{1}{\left|\Theta_{I-1}^{(\beta_{I-1})}\right|}\int_{\Theta_{I-1}^{(\beta_{I-1})}}\cdot d\Theta$ averaging operator on (3.58) yields

$$^{I}\varepsilon_{ij}^{(\beta_{I-1},\beta_I)}\left(^{0}\mathbf{x},\ldots,^{I-2}\mathbf{x}\right)$$

$$= {}^{I}A_{ijmn}^{(\beta_I)}\left(^{0}\mathbf{x},\ldots,^{I-2}\mathbf{x}\right)$$

$$\cdot \left\{ \begin{array}{l} {}^{I-1}A_{ijmn}^{(\beta_{I-1})}\left(^{0}\mathbf{x},\ldots,^{I-2}\mathbf{x}\right){}^{I-2}\varepsilon_{kl}\left(^{0}\mathbf{x},\ldots,^{I-2}\mathbf{x}\right) \\ + \displaystyle\sum_{\xi_{I-1}=1}^{m_{I-1}} {}^{I-1}Q_{mn\hat{n}}^{(\beta_{I-1},\xi_{I-1})}\left(^{0}\mathbf{x},\ldots,^{I-2}\mathbf{x}\right){}^{I-1}\delta_{\hat{n}}^{(\xi_{I-1})}\left(^{0}\mathbf{x},\ldots,^{I-2}\mathbf{x}\right) \\ + \displaystyle\sum_{\alpha_{I-1}=1}^{n_{I-1}} {}^{I-1}P_{mnkl}^{(\beta_{I-1},\alpha_{I-1})}\left(^{0}\mathbf{x},\ldots,^{I-2}\mathbf{x}\right){}^{I-1}\mu_{kl}^{(\alpha_{I-1})}\left(^{0}\mathbf{x},\ldots,^{I-2}\mathbf{x}\right) \end{array} \right\}$$

$$+ \sum_{\xi_I=1}^{m_I} {}^{I}Q_{ij\hat{n}}^{(\beta_{I-1},\beta_I)(\xi_I)}\left(^{0}\mathbf{x},\ldots,^{I-2}\mathbf{x}\right)\cdot {}^{I}\delta_{\hat{n}}^{(\alpha_{I-1},\xi_I)}\left(^{0}\mathbf{x},\ldots,^{I-2}\mathbf{x}\right)$$

$$+ \sum_{\alpha_I=1}^{n_I} {}^{I}P_{ijkl}^{(\beta_{I-1},\beta_I,)(\alpha_I)}\left(^{0}\mathbf{x},\ldots,^{I-2}\mathbf{x}\right)\cdot {}^{I}\mu_{kl}^{(\alpha_{I-1},\alpha_I)}\left(^{0}\mathbf{x},\ldots,^{I-2}\mathbf{x}\right)$$

$$(3.59)$$

Repeating the similar derivation for $^{I-2}\varepsilon_{ij}$ and so on, yields the partitioned strain at the finest scale of interest expressed in terms of macro strain and all the independent partitioned eigendeformations

$$
\begin{aligned}
{}^{N-1}\varepsilon_{ij}^{(\beta_1:\cdots:\beta_{N-1})}({}^0\mathbf{x}) \\
= A_{ijkl}^{(\beta_1:\cdots:\beta_{N-1})} \cdot {}^0\varepsilon_{kl}({}^0\mathbf{x}) + \sum_{\xi_1=1}^{m_1} Q_{ij\hat{n}}^{(\beta_1:\cdots:\beta_{N-1})(\xi_1)} \cdot {}^1\delta_{\hat{n}}^{(\xi_1)}({}^0\mathbf{x}) \\
+ \sum_{\alpha_1=1}^{n_1}\cdots\sum_{\alpha_{N-1}=1}^{n_{N-1}} P_{ijkl}^{(\beta_1:\cdots:\beta_{N-1})(\alpha_1:\cdots:\alpha_{N-1})} \cdot {}^{N-1}\mu_{kl}^{(\alpha_1:\cdots:\alpha_{N-1})}({}^0\mathbf{x}) \\
+ \sum_{J=2}^{N-1}\left[\sum_{\alpha_1=1}^{n_1}\cdots\sum_{\alpha_{J-1}=1}^{n_{J-1}}\sum_{\xi_J=1}^{m_J} Q_{ij\hat{n}}^{(\beta_1:\cdots:\beta_{N-1})(\alpha_1:\cdots:\alpha_{J-1}:\xi_J)} {}^J\delta_{\hat{n}}^{(\alpha_1:\cdots:\alpha_{J-1}:\xi_J)}({}^0\mathbf{x})\right]
\end{aligned}
$$

(3.60)

where

$$A_{ijkl}^{(\beta_1:\cdots:\beta_{N-1})} = \prod_{K=1}^{N-1} {}^K A_{ijkl}^{(\beta_K)} \tag{3.61}$$

with the coefficient tensor related to the eigenstrain at the finest scale being

$$
\begin{aligned}
P_{ijkl}^{(\beta_1:\cdots:\beta_{N-1})(\alpha_1:\cdots:\alpha_{N-1})} &= {}^{N-1}P_{ijkl}^{(\beta_1:\cdots:\beta_{N-1})(\alpha_1:\cdots:\alpha_{N-1})} \\
&+ {}^{N-1}A_{ijmn}^{(\beta_{N-1})} \cdot {}^{N-2}P_{mnpq}^{(\beta_1:\cdots:\beta_{N-2},\alpha_1:\cdots:\alpha_{N-2})} \\
&\cdot {}^{N-2}S_{pqkl}^{(\alpha_{N-1})} + \cdots \\
&+ \prod_{K=2}^{N-1} {}^K A_{ijmn}^{(\beta_K)} \cdot {}^1 P_{mnpq}^{(\beta_1,\alpha_1)} \cdot \prod_{K=2}^{N-1} {}^{K-1}S_{pqkl}^{(\alpha_K)}
\end{aligned}
\tag{3.62}
$$

and the coefficient tensor related to all independent eigenseparation given as

$$Q_{ij\hat{n}}^{(\beta_1:\cdots:\beta_{N-1})(\xi_1)} = \left[\prod_{K=2}^{N-1} {}^K A_{ijkl}^{(\beta_K)}\right] \cdot Q_{kl\hat{n}}^{(\beta_1)(\xi_1)} \tag{3.63}$$

$$\underbrace{Q_{ij\hat{n}}^{(\beta_1:\cdots:\beta_{N-1})(\alpha_1:\cdots:\alpha_{J-2}:\xi_{J-1})}}_{\text{for } J=3,\ldots,N-1}$$

$$= {}^J A_{ijmn}^{(\beta_J:\cdots:\beta_{N-1})} \cdot {}^{J-1}Q_{mn\hat{n}}^{(\beta_1:\cdots:\beta_{J-1})(\xi_{J-1})}$$

$$+ {}^{J}A_{ijst}^{(\beta_J:\cdots:\beta_{N-1})} \cdot {}^{J-1}A_{stmn}^{(\beta_{J-1})} \cdot {}^{J-2}P_{mnpq}^{(\beta_1:\cdots:\beta_{J-2},\alpha_1:\cdots:\alpha_{J-2})} \cdot {}^{J-2}Q_{pq\hat{n}}^{(\xi_{J-1})}$$

$$+ {}^{J}A_{ijst}^{(\beta_J:\cdots:\beta_{N-1})} \cdot {}^{J-1}A_{stvw}^{(\beta_{J-1})} \cdot {}^{J-2}A_{vwmn}^{(\beta_{J-2})} \cdot {}^{J-3}P_{mnpq}^{(\beta_1:\cdots:\beta_{J-3},\alpha_1:\cdots:\alpha_{J-3})}$$

$$\cdot {}^{J-3}S_{pqkl}^{(\alpha_{J-2})} \cdot {}^{J-2}Q_{kl\hat{n}}^{(\xi_{J-1})} + \cdots$$

$$+ {}^{J}A_{ijst}^{(\beta_J:\cdots:\beta_{N-1})} \cdot \prod_{K=2}^{J-1} {}^{K}A_{stmn}^{(\beta_K)} \cdot {}^{1}P_{mnpq}^{(\beta_1,\alpha_1)} \cdot \prod_{K=2}^{J-2} {}^{K-1}S_{pqkl}^{(\alpha_K)} \cdot {}^{J-2}Q_{kl\hat{n}}^{(\xi_{J-1})}$$

(3.64)

$$Q_{ij\hat{n}}^{(\beta_1:\cdots:\beta_{N-1})(\alpha_1:\cdots:\alpha_{N-2}:\xi_{N-1})}$$

$$= {}^{N-1}Q_{ij\hat{n}}^{(\beta_1:\cdots:\beta_{N-1})(\xi_{N-1})}$$

$$+ {}^{N-1}A_{ijmn}^{(\beta_{N-1})} \cdot {}^{N-2}P_{mnpq}^{(\beta_1:\cdots:\beta_{N-2},\alpha_1:\cdots:\alpha_{N-2})} \cdot {}^{N-2}Q_{pq\hat{n}}^{(\xi_{N-1})}$$

$$+ {}^{N-1}A_{ijst}^{(\beta_{N-1})} \cdot {}^{N-2}A_{stmn}^{(\beta_{N-2})} \cdot {}^{N-3}P_{mnpq}^{(\beta_1:\cdots:\beta_{N-3},\alpha_1:\cdots:\alpha_{N-3})} \quad (3.65)$$

$$\cdot {}^{N-3}S_{pqkl}^{(\alpha_{N-2})} \cdot {}^{N-2}Q_{kl\hat{n}}^{(\xi_{N-1})} + \cdots$$

$$+ \prod_{K=2}^{N-1} {}^{K}A_{ijmn}^{(\beta_K)} \cdot {}^{1}P_{mnpq}^{(\beta_1,\alpha_1)} \cdot \prod_{K=2}^{N-2} {}^{K-1}S_{pqkl}^{(\alpha_K)} \cdot {}^{N-2}Q_{kl\hat{n}}^{(\xi_{N-1})}$$

Next, we focus on the traction fields at multiple scales, which are given by

$$\boxed{\begin{aligned}
{}^{I-1}t_{\tilde{m}}^{(\beta_1:\cdots:\beta_{I-2}:\eta_{I-1})}({}^{0}\mathbf{x}) \\
= B_{\tilde{m}kl}^{(\beta_1:\cdots:\beta_{I-2}:\eta_{I-1})} \cdot {}^{0}\varepsilon_{kl}({}^{0}\mathbf{x}) + \sum_{\xi_1=1}^{m_1} W_{\tilde{m}\tilde{n}}^{(\beta_1:\cdots:\beta_{I-2}:\eta_{I-1})(\xi_1)} \cdot {}^{1}\delta_{\hat{n}}^{(\xi_1)}({}^{0}\mathbf{x}) \\
+ \sum_{\alpha_1=1}^{n_1} \cdots \sum_{\alpha_{N-1}=1}^{n_{N-1}} V_{\tilde{m}kl}^{(\beta_1:\cdots:\beta_{I-2}:\eta_{I-1})(\alpha_1:\cdots:\alpha_{N-1})} \cdot {}^{N-1}\mu_{kl}^{(\alpha_1:\cdots:\alpha_{N-1})}({}^{0}\mathbf{x}) \\
+ \sum_{J=2}^{N-1} \left[\sum_{\alpha_1=1}^{n_1} \cdots \sum_{\alpha_{J-1}=1}^{n_{J-1}} \sum_{\xi_J=1}^{m_J} W_{\tilde{m}\tilde{n}}^{(\beta_1:\cdots:\beta_{I-2}:\eta_{I-1})(\alpha_1:\cdots:\alpha_{J-1}:\xi_J)} {}^{J}\delta_{\hat{n}}^{(\alpha_1:\cdots:\alpha_{J-1}:\xi_J)}({}^{0}\mathbf{x}) \right] \\
\text{for } I = 3, \ldots, N
\end{aligned}}$$

(3.66)

$$\begin{aligned}
{}^1 t_{\hat{m}}^{(\eta_1)}\left({}^0\mathbf{x}\right) &= B_{\hat{m}kl}^{(\eta_1)} \cdot {}^0\varepsilon_{kl}\left({}^0\mathbf{x}\right) + \sum_{\xi_1=1}^{m_1} W_{\hat{m}\hat{n}}^{(\eta_1)(\xi_1)} \cdot {}^1\delta_{\hat{n}}^{(\xi_1)}\left({}^0\mathbf{x}\right) \\
&+ \sum_{\alpha_1=1}^{n_1} \cdots \sum_{\alpha_{N-1}=1}^{n_{N-1}} V_{\hat{m}kl}^{(\eta_1)(\alpha_1:\cdots:\alpha_{N-1})} \cdot {}^{N-1}\mu_{kl}^{(\alpha_1:\cdots:\alpha_{N-1})}\left({}^0\mathbf{x}\right) \\
&+ \sum_{J=2}^{N-1}\left[\sum_{\alpha_1=1}^{n_1}\cdots\sum_{\alpha_{J-1}=1}^{n_{J-1}}\sum_{\xi_J=1}^{m_J} W_{\hat{m}\hat{n}}^{(\eta_1)(\alpha_1:\cdots:\alpha_{J-1}:\xi_J)} {}^J\delta_{\hat{n}}^{(\alpha_1:\cdots:\alpha_{J-1}:\xi_J)}\left({}^0\mathbf{x}\right)\right]
\end{aligned} \tag{3.67}$$

where

$$B_{\hat{m}kl}^{(\beta_1:\cdots:\beta_{I-2}:\eta_{I-1})} = \frac{1}{\left|S_{I-1}^{(\eta_{I-1})}\right|}\int_{S_{I-1}^{(\eta_{I-1})}} {}^{I-1}a_{\hat{m}i}\,{}^{I-1}L_{ijmn} A_{mnkl}^{(\beta_1:\cdots:\beta_{I-2})}\,{}^{I-1}\hat{n}_j^d S \tag{3.68}$$

The coefficient tensor related to eigenstrain at the finest scale is given as

$$V_{\hat{m}kl}^{(\beta_1:\cdots:\beta_{I-2}:\eta_{I-1})(\alpha_1:\cdots:\alpha_{N-1})}$$
$$= \frac{1}{\left|S_{I-1}^{(\eta_{I-1})}\right|}\int_{S_{I-1}^{(\eta_{I-1})}} {}^{I-1}a_{\hat{m}i}\,{}^{I-1}L_{ijmn} S_{mnkl}^{(\beta_1:\cdots:\beta_{I-2})(\alpha_1:\cdots:\alpha_{N-1})}\,{}^{I-1}\hat{n}_j^d S \tag{3.69}$$

where

$$\begin{aligned}
S_{ijkl}^{(\beta_1:\cdots:\beta_{I-2})(\alpha_1:\cdots:\alpha_{N-1})} &= {}^{N-1}S_{ijkl}^{(\beta_1:\cdots:\beta_{I-2})(\alpha_1:\cdots:\alpha_{N-1})} \\
&+ {}^{N-1}A_{ijmn}^{(\beta_{I-2})}\cdot {}^{N-2}S_{mnpq}^{(\beta_1:\cdots:\beta_{I-3},\alpha_1:\cdots:\alpha_{N-2})}\cdot {}^{N-2}S_{pqkl}^{(\alpha_{N-1})} \\
&+ \cdots \\
&+ \prod_{K=2}^{I-2}{}^K A_{ijmn}^{(\beta_K)}\cdot {}^1 S_{mnpq}^{(\beta_1,\alpha_1)}\cdot\prod_{K=2}^{N-1}{}^{K-1}S_{pqkl}^{(\alpha_K)}
\end{aligned} \tag{3.70}$$

The coefficient tensor is related to all independent eigenseparations as

$$W_{\hat{m}\hat{n}}^{(\beta_1:\cdots:\beta_{I-2}:\eta_{I-1})(\xi_1)}$$
$$= \frac{1}{\left|S_{I-1}^{(\eta_{I-1})}\right|}\int_{S_{I-1}^{(\eta_{I-1})}} {}^{I-1}a_{\hat{m}i}\,{}^{I-1}L_{ijmn} Q_{mn\hat{n}}^{(\beta_1:\cdots:\beta_{I-2})(\xi_1)}\,{}^{I-1}\hat{n}_j dS \tag{3.71}$$

$$\underbrace{W_{\hat{m}\hat{n}}^{(\beta_1:\cdots:\beta_{I-2}:\eta_{I-1})(\alpha_1:\cdots:\alpha_{J-2}:\xi_{J-1})}}_{for\ J=3,\ldots,N-1}$$

$$= \frac{1}{\left|S_{I-1}^{(\eta_{I-1})}\right|} \int_{S_{I-1}^{(\eta_{I-1})}} {}^{I-1}a_{\hat{m}i}\, {}^{I-1}L_{ijmn} Q_{mn\hat{n}}^{(\beta_1:\cdots:\beta_{I-2})(\alpha_1:\cdots:\alpha_{J-2}:\xi_{J-1})}\, {}^{I-1}\hat{n}_j dS \tag{3.72}$$

$$W_{\hat{m}\hat{n}}^{(\beta_1:\cdots:\beta_{I-2}:\eta_{I-1})(\alpha_1:\cdots:\alpha_{N-2}:\xi_{N-1})}$$

$$= \frac{1}{\left|S_{I-1}^{(\eta_{I-1})}\right|} \int_{S_{I-1}^{(\eta_{I-1})}} {}^{I-1}a_{\hat{m}i}\, {}^{I-1}L_{ijmn} Q_{mn\hat{n}}^{(\beta_1:\cdots:\beta_{I-2})(\alpha_1:\cdots:\alpha_{N-2}:\xi_{N-1})}\, {}^{I-1}\hat{n}_j dS \tag{3.73}$$

where

$$Q_{ij\hat{n}}^{(\beta_1:\cdots:\beta_{I-2})(\xi_1)} = \left[\prod_{K=2}^{I-2} {}^{K}A_{ijkl}^{(\beta_K)}\right] \cdot Q_{kl\hat{n}}^{(\beta_1)(\xi_1)} \tag{3.74}$$

$$\underbrace{Q_{ij\hat{n}}^{(\beta_1:\cdots:\beta_{I-2})(\alpha_1:\cdots:\alpha_{J-2}:\xi_{J-1})}}_{for\ J=3,\ldots,N-1}$$

$$= {}^{J}A_{ijmn}^{(\beta_J:\cdots:\beta_{I-2})} \cdot {}^{J-1}Q_{mn\hat{n}}^{(\beta_1:\cdots:\beta_{J-1})(\xi_{J-1})}$$

$$+ {}^{J}A_{ijst}^{(\beta_J:\cdots:\beta_{I-2})} \cdot {}^{J-1}A_{stmn}^{(\beta_{J-1})} \cdot {}^{J-2}S_{mnpq}^{(\beta_1:\cdots:\beta_{J-2},\alpha_1:\cdots:\alpha_{J-2})} \cdot {}^{J-2}Q_{pq\hat{n}}^{(\xi_{J-1})}$$

$$+ {}^{J}A_{ijst}^{(\beta_J:\cdots:\beta_{I-2})} \cdot {}^{J-1}A_{stvw}^{(\beta_{J-1})} \cdot {}^{J-2}A_{vwmn}^{(\beta_{J-2})} \cdot {}^{J-3}S_{mnpq}^{(\beta_1:\cdots:\beta_{J-3},\alpha_1:\cdots:\alpha_{J-3})}$$

$$\cdot {}^{J-3}S_{pqkl}^{(\alpha_{J-2})} \cdot {}^{J-2}Q_{kl\hat{n}}^{(\xi_{J-1})} + \cdots$$

$$+ {}^{J}A_{ijst}^{(\beta_J:\cdots:\beta_{I-2})} \cdot \prod_{K=2}^{J-1} {}^{K}A_{stmn}^{(\beta_K)} \cdot {}^{1}S_{mnpq}^{(\beta_1,\alpha_1)} \cdot \prod_{K=2}^{J-2} {}^{K-1}S_{pqkl}^{(\alpha_K)} \cdot {}^{J-2}Q_{kl\hat{n}}^{(\xi_{J-1})} \tag{3.75}$$

$$Q_{ij\hat{n}}^{(\beta_1:\cdots:\beta_{I-2})(\alpha_1:\cdots:\alpha_{N-2}:\xi_{N-1})}$$

$$= {}^{N-1}Q_{ij\hat{n}}^{(\beta_1:\cdots:\beta_{I-2})(\xi_{N-1})}$$

$$+ {}^{N-1}A_{ijmn}^{(\beta_{I-2})} \cdot {}^{N-2}S_{mnpq}^{(\beta_1:\cdots:\beta_{I-2},\alpha_1:\cdots:\alpha_{N-2})} \cdot {}^{N-2}Q_{pq\hat{n}}^{(\xi_{N-1})}$$

$$+ {}^{N-1}A_{ijst}^{(\beta_{I-2})} \cdot {}^{N-2}A_{stmn}^{(\beta_{I-3})} \cdot {}^{N-3}S_{mnpq}^{(\beta_1:\cdots:\beta_{I-4},\alpha_1:\cdots:\alpha_{N-3})}$$

$$\cdot {}^{N-3}S_{pqkl}^{(\alpha_{N-2})} \cdot {}^{N-2}Q_{kl\hat{n}}^{(\xi_{N-1})} + \cdots$$

$$+ \prod_{K=2}^{I-2} {}^{K}A_{ijmn}^{(\beta_K)} \cdot {}^{1}S_{mnpq}^{(\beta_1,\alpha_1)} \cdot \prod_{K=2}^{N-2} {}^{K-1}S_{pqkl}^{(\alpha_K)} \cdot {}^{N-2}Q_{kl\hat{n}}^{(\xi_{N-1})} \qquad (3.76)$$

In the above $^{I-1}a_{\hat{m}i}$ is the transformation matrix from the global coordinates system to the local interface coordinate system. All the coefficient tensors in (3.67) have similar expressions without the superscript $(\beta_1 : \cdots : \beta_{I-2})$.

The constitutive relations for all independent eigendeformations are assumed to be known based on the material models of phases and interfaces. The partitioned constitutive relations can be expressed as

$$\boxed{{}^{N-1}\mu_{ij}^{(\alpha_1:\cdots:\alpha_{N-1})}({}^{0}\mathbf{x}) = {}^{N-1}f\left({}^{N-1}\varepsilon_{ij}^{(\alpha_1:\cdots:\alpha_{N-1})}({}^{0}\mathbf{x}), {}^{N-1}\sigma_{ij}^{(\alpha_1:\cdots:\alpha_{N-1})}({}^{0}\mathbf{x})\right)}$$

(3.77)

$$\boxed{{}^{I}t_{\hat{n}}^{(\alpha_1:\cdots:\alpha_{I-1}:\xi_I)}({}^{0}\mathbf{x}) = {}^{I}g\left({}^{I}\delta_{\hat{n}}^{(\alpha_1:\cdots:\alpha_{I-1}:\xi_I)}({}^{0}\mathbf{x})\right) \quad for\ I = 2, ..., N-1}$$

(3.78)

$$\boxed{{}^{1}t_{\hat{n}}^{(\xi_1)}({}^{0}\mathbf{x}) = {}^{1}g\left({}^{1}\delta_{\hat{n}}^{(\xi_1)}({}^{0}\mathbf{x})\right)}$$

(3.79)

Eqs. (3.60), (3.66), and (3.67) along with the partitioned constitutive relations (3.77)–(3.79) form a system of nonlinear equations for the unknowns $^{1}\delta_{\hat{n}}^{(\xi_1)}({}^{0}\mathbf{x})$, $^{N-1}\varepsilon_{kl}^{(\alpha_1:\cdots:\alpha_{N-1})}({}^{0}\mathbf{x})$, and $^{J}\delta_{\hat{n}}^{(\alpha_1:\cdots:\alpha_{J-1}:\xi_J)}({}^{0}\mathbf{x})$. Independent partitioned eigendeformations are obtained by solving the system of equations within each macro increment. Once the independent partitioned eigendeformations have been determined, (3.50) is used to update the macro stress.

3.6 Summary of the computational procedure

The computational procedure consists of first precomputing the coefficient tensors prior to the nonlinear analysis and subsequently solving the nonlinear N-scale reduced-order unit cell problem at each load increment and iteration of the macro problem. In order to solve the N-scale reduced-order unit cell problem and then update the macro-stress, all the coefficient tensors

($^0L_{ijkl}$, $^0F_{ij\hat{n}}^{(\xi_1)}$, $^{I-1}E_{ijkl}^{(\alpha_1:\cdots:\alpha_{N-1})}$, and $^{I-1}F_{ij\hat{n}}^{(\alpha_1:\cdots:\alpha_{J-1}:\xi_J)}$ in (3.50), $A_{ijkl}^{(\beta_1:\cdots:\beta_{N-1})}$, $Q_{ij\hat{n}}^{(\beta_1:\cdots:\beta_{N-1})(\xi_1)}$, $P_{ijkl}^{(\beta_1:\cdots:\beta_{N-1})(\alpha_1:\cdots:\alpha_{N-1})}$, and $Q_{ij\hat{n}}^{(\beta_1:\cdots:\beta_{N-1})(\alpha_1:\cdots:\alpha_{J-1}:\xi_J)}$ in (3.60), $B_{\hat{m}kl}^{(\beta_1:\cdots:\beta_{I-2}:\eta_{I-1})}$, $W_{\hat{m}\hat{n}}^{(\beta_1:\cdots:\beta_{I-2}:\eta_{I-1})(\xi_1)}$, $V_{\hat{m}kl}^{(\beta_1:\cdots:\beta_{I-2}:\eta_{I-1})(\alpha_1:\cdots:\alpha_{N-1})}$, and $W_{\hat{m}\hat{n}}^{(\beta_1:\cdots:\beta_{I-2}:\eta_{I-1})(\alpha_1:\cdots:\alpha_{J-1}:\xi_J)}$ in (3.66)) have to be obtained in advance. From the definitions of each coefficient tensor, it can be seen that all the coefficient tensors are functions of the influence functions. As mentioned in Section 3.4, all the influence functions are obtained by solving the corresponding elastic boundary value problem using finite element method independent of and prior to the nonlinear macro analysis. The sequence of operations in the preprocessing stage is summarized below:

1. Solve the elastic influence function problem (3.24) for each scale (from finest to coarsest) to obtain $^IG_{ijkl}$ and calculate $^IA_{ijkl}$ and $^{I-1}L_{ijkl}$ using (3.30).

2. Solve the reduced order eigenstrain influence function problem (3.42) for each scale to obtain $^IP_{ijkl}^{(\alpha_I)}$ and calculate $^IS_{ijkl}^{(\alpha_I)}$ using (3.39).

3. Solve the reduced order eigenseparation influence function problem (3.43) for each scale to obtain $^IQ_{ij\hat{n}}^{(\xi_I)}$.

4. Calculate the coefficient tensors $^0L_{ijkl}$, $^0F_{ij\hat{n}}^{(\xi_1)}$, $^{I-1}E_{ijkl}^{(\alpha_1:\cdots:\alpha_{N-1})}$, and $^{I-1}F_{ij\hat{n}}^{(\alpha_1:\cdots:\alpha_{J-1}:\xi_J)}$ using (3.51)–(3.55).

5. Calculate the coefficient tensors $A_{ijkl}^{(\beta_1:\cdots:\beta_{N-1})}$, $Q_{ij\hat{n}}^{(\beta_1:\cdots:\beta_{N-1})(\xi_1)}$, $P_{ijkl}^{(\beta_1:\cdots:\beta_{N-1})(\alpha_1:\cdots:\alpha_{N-1})}$, and $Q_{ij\hat{n}}^{(\beta_1:\cdots:\beta_{N-1})(\alpha_1:\cdots:\alpha_{J-1}:\xi_J)}$ using (3.61)–(3.65).

6. Calculate the coefficient tensors $B_{\hat{m}kl}^{(\beta_1:\cdots:\beta_{I-2}:\eta_{I-1})}$, $W_{\hat{m}\hat{n}}^{(\beta_1:\cdots:\beta_{I-2}:\eta_{I-1})(\xi_1)}$, $V_{\hat{m}kl}^{(\beta_1:\cdots:\beta_{I-2}:\eta_{I-1})(\alpha_1:\cdots:\alpha_{N-1})}$, and $W_{\hat{m}\hat{n}}^{(\beta_1:\cdots:\beta_{I-2}:\eta_{I-1})(\alpha_1:\cdots:\alpha_{J-1}:\xi_J)}$ using (3.68)–(3.76).

To this end we focus on the system of nonlinear equations to be solved at each Gauss point in the macro (coarsest) mesh. The number of equations defined by (3.60) is given by

$$n^{(1)} = 6 \left(\sum_{\alpha_1=1}^{n_1} \cdots \sum_{\alpha_k=1}^{n_k|_{\alpha_{k-1}}} \cdots \sum_{\alpha_{N-2}=1}^{n_{N-2}|_{\alpha_{N-3}}} n_{N-1}|_{\alpha_{N-2}} \right) \qquad (3.80)$$

where $n_k|_{\alpha_{k-1}}$ denotes the number of partitions at scale k corresponding to the α_{k-1}th partition at $k-1$ scale. Note that for different α_{k-1} partitions, $n_k|_{\alpha_{k-1}}$ might be different.

The number of equations defined by (3.66) is given as

$$for\ I = 3, \ldots, N$$

$$n_I^{(2)} = 3 \left(\sum_{\alpha_1=1}^{n_1} \cdots \sum_{\alpha_k=1}^{n_k|_{\alpha_{k-1}}} \cdots \sum_{\alpha_{I-2}=1}^{n_{I-2}|_{\alpha_{I-3}}} m_{I-1}|_{\alpha_{I-2}} \right) \tag{3.81}$$

The number of equations defined by (3.67) is given as

$$n^{(3)} = 3m_1 \tag{3.82}$$

The total number of equations (unknowns) for the N-scale reduced-order unit cell problem is given as

$$n = n^{(1)} + \sum_{I=3}^{N} n_I^{(2)} + n^{(3)} \tag{3.83}$$

For example, for the three-scale unit cell problem described in Section 3.7, $n^{(1)}$ is computed as follows:

$$n^{(1)} = 6 \left(\sum_{\alpha_y=1}^{n_y} n_z|_{\alpha_y} \right) = 6\left(1 + 2 + 2 + 2\right) = 42$$

where

α_y	Matrix	Axial tow	Bias tow 1	Bias tow 2
$n_z|_{\alpha_y}$	1	2 (fiber; matrix)	2 (fiber; matrix)	2 (fiber; matrix)

3.7 Example: three scale analysis

In this section, the three-scale formulation is considered as a special case of the N-scale formulation. In the following, the three-scale formulation is derived from the N-scale formulation and then validated against the tube crash tests.

In the three-scale case ($N = 3$), we have three position vectors denoted as $^0\mathbf{x} = \mathbf{x}$, $^1\mathbf{x} = \mathbf{y}$, and $^2\mathbf{x} = \mathbf{z}$. The equilibrium equations for each scale can be directly deduced from (3.21)

$$\boxed{\begin{array}{l} O\left(\zeta^{-2}\right): \ ^2\sigma_{ij,z_j}(\mathbf{x}, \mathbf{y}, \mathbf{z}) = 0 \\ O\left(\zeta^{-1}\right): \ ^1\sigma_{ij,y_j}(\mathbf{x}, \mathbf{y}) = 0 \\ O\left(\zeta^{0}\right): \ ^0\sigma_{ij,x_j}(\mathbf{x}) + {^0b_i} = 0 \end{array}} \tag{3.84}$$

The residual-free stresses at the two finest scales are directly derived from (3.31)

$$
\begin{aligned}
{}^2\sigma_{ij} = {}^2L_{ijmn}\left(\mathbf{x},\mathbf{y},\mathbf{z}\right) \\
\times \begin{bmatrix} {}^2A_{mnkl}\left(\mathbf{x},\mathbf{y},\mathbf{z}\right){}^1\varepsilon_{kl}\left(\mathbf{x},\mathbf{y}\right) \\ + \displaystyle\int_{S_z} {}^2g^\delta_{mn\hat{n}}\left(\mathbf{x},\mathbf{y},\mathbf{z},\hat{\mathbf{z}}\right){}^2\delta_{\hat{n}}\left(\mathbf{x},\mathbf{y},\hat{\mathbf{z}}\right)dS \\ + \displaystyle\int_{\Theta_z} {}^2g^\mu_{mnkl}\left(\mathbf{x},\mathbf{y},\mathbf{z},\hat{\mathbf{z}}\right){}^2\mu_{kl}\left(\mathbf{x},\mathbf{y},\hat{\mathbf{z}}\right)d\Theta - {}^2\mu_{mn}\left(\mathbf{x},\mathbf{y},\mathbf{z}\right) \end{bmatrix}
\end{aligned}
$$

(3.85)

$$
{}^1\sigma_{ij} = {}^1L_{ijmn}\left(\mathbf{x},\mathbf{y}\right) \begin{bmatrix} {}^1A_{mnkl}\left(\mathbf{x},\mathbf{y}\right){}^0\varepsilon_{kl}\left(\mathbf{x}\right) \\ + \displaystyle\int_{S_y} {}^1g^\delta_{mn\hat{n}}\left(\mathbf{x},\mathbf{y},\hat{\mathbf{y}}\right){}^1\delta_{\hat{n}}\left(\mathbf{x},\hat{\mathbf{y}}\right)dS \\ + \displaystyle\int_{\Theta_y} {}^1g^\mu_{mnkl}\left(\mathbf{x},\mathbf{y},\hat{\mathbf{y}}\right){}^1\mu_{kl}\left(\mathbf{x},\hat{\mathbf{y}}\right)d\Theta - {}^1\mu_{mn}\left(\mathbf{x},\mathbf{y}\right) \end{bmatrix}
$$

(3.86)

A single dependent eigenstrain value ${}^1\mu_{ij}$ expressed in terms of independent eigendeformation (${}^2\mu_{ij}$ and ${}^2\delta_{\hat{n}}$) at the finest scale follows from (3.32)

$$
\begin{aligned}
&{}^1\mu_{ij}\left(\mathbf{x},\mathbf{y}\right) \\
&= -\left[{}^1L_{ijst}\left(\mathbf{x},\mathbf{y}\right)\right]^{-1}\frac{1}{|\Theta_z|}\int_{\Theta_z} \\
&\cdot \left\{ {}^2L_{stmn}\left(\mathbf{x},\mathbf{y},\mathbf{z}\right) \begin{bmatrix} \displaystyle\int_{S_z} {}^2g^\delta_{mn\hat{n}}\left(\mathbf{x},\mathbf{y},\mathbf{z},\hat{\mathbf{z}}\right){}^2\delta_{\hat{n}}\left(\mathbf{x},\mathbf{y},\hat{\mathbf{z}}\right)dS \\ + \displaystyle\int_{\Theta_z} {}^2g^\mu_{mnkl}\left(\mathbf{x},\mathbf{y},\mathbf{z},\hat{\mathbf{z}}\right){}^2\mu_{kl}\left(\mathbf{x},\mathbf{y},\hat{\mathbf{z}}\right)d\Theta \\ - {}^2\mu_{mn}\left(\mathbf{x},\mathbf{y},\mathbf{z}\right) \end{bmatrix} \right\} d\Theta
\end{aligned}
$$

(3.87)

The reduced-order residual-free stresses for the two finest scales are directly deduced from (3.38)

$$
{}^2\sigma_{ij} = {}^2L_{ijmn}(\mathbf{x},\mathbf{y},\mathbf{z}) \begin{bmatrix} {}^2A_{mnkl}(\mathbf{x},\mathbf{y},\mathbf{z})\,{}^1\varepsilon_{kl}(\mathbf{x},\mathbf{y}) \\ + \sum_{\xi_z=1}^{m_z} {}^2Q_{mn\hat{n}}^{(\xi_z)}(\mathbf{x},\mathbf{y},\mathbf{z})\,{}^2\delta_{\hat{n}}^{(\xi_z)}(\mathbf{x},\mathbf{y}) \\ + \sum_{\alpha_z=1}^{n_z} {}^2S_{mnkl}^{(\alpha_z)}(\mathbf{x},\mathbf{y},\mathbf{z})\,{}^2\mu_{kl}^{(\alpha_z)}(\mathbf{x},\mathbf{y}) \end{bmatrix} \quad (3.88)
$$

$$
{}^1\sigma_{ij} = {}^1L_{ijmn}(\mathbf{x},\mathbf{y}) \begin{bmatrix} {}^1A_{mnkl}(\mathbf{x},\mathbf{y})\,{}^0\varepsilon_{kl}(\mathbf{x}) \\ + \sum_{\xi_y=1}^{m_y} {}^1Q_{mn\hat{n}}^{(\xi_y)}(\mathbf{x},\mathbf{y})\,{}^1\delta_{\hat{n}}^{(\xi_y)}(\mathbf{x}) \\ + \sum_{\alpha_y=1}^{n_y} {}^1S_{mnkl}^{(\alpha_y)}(\mathbf{x},\mathbf{y})\,{}^1\mu_{kl}^{(\alpha_y)}(\mathbf{x}) \end{bmatrix} \quad (3.89)
$$

The relation between the partitioned eigendeformations follows from (3.48)

$$
{}^1\mu_{ij}^{(\alpha_y)}(\mathbf{x}) = \sum_{\alpha_z=1}^{n_z} {}^1S_{ijkl}^{(\alpha_z)} \cdot {}^2\mu_{kl}^{(\alpha_y:\alpha_z)}(\mathbf{x}) + \sum_{\xi_z=1}^{m_z} {}^1Q_{kl\hat{n}}^{(\xi_z)}\,{}^2\delta_{\hat{n}}^{(\alpha_y:\xi_z)}(\mathbf{x}) \quad (3.90)
$$

The reduced-order macro stress is derived from (3.50)

$$
{}^0\sigma_{ij} = {}^0L_{ijkl} \cdot {}^0\varepsilon_{kl}(\mathbf{x}) + \sum_{\xi_y=1}^{m_y} {}^0F_{ij\hat{n}}^{(\xi_y)}\,{}^1\delta_{\hat{n}}^{(\xi_y)}(\mathbf{x})
$$
$$
+ \sum_{\alpha_y=1}^{n_y}\sum_{\alpha_z=1}^{n_z} {}^0E_{ijkl}^{(\alpha_y:\alpha_z)} \cdot {}^2\mu_{kl}^{(\alpha_y:\alpha_z)}(\mathbf{x}) + \sum_{\alpha_y=1}^{n_y}\sum_{\xi_z=1}^{m_z} {}^0F_{ij\hat{n}}^{(\alpha_y:\xi_2)}\,{}^2\delta_{\hat{n}}^{(\alpha_y:\xi_z)}(\mathbf{x})
$$

(3.91)

The reduced-order unit cell problem can be obtained from (3.60), (3.66), and (3.67)

$$
\begin{aligned}
{}^2\varepsilon_{ij}^{(\beta_y:\beta_z)}(\mathbf{x}) = {} & A_{ijkl}^{(\beta_y:\beta_z)} \cdot {}^0\varepsilon_{kl}(\mathbf{x}) + \sum_{\xi_y=1}^{m_y} Q_{ij\hat{n}}^{(\beta_y:\beta_z)(\xi_y)} \cdot {}^1\delta_{\hat{n}}^{(\xi_y)}(\mathbf{x}) \\
& + \sum_{\alpha_y=1}^{n_y} \sum_{\alpha_z=1}^{n_z} P_{ijkl}^{(\beta_y:\beta_z)(\alpha_y:\alpha_z)} \cdot {}^2\mu_{kl}^{(\alpha_y:\alpha_z)}(\mathbf{x}) \\
& + \sum_{\alpha_y=1}^{n_y} \sum_{\xi_z=1}^{m_z} Q_{ij\hat{n}}^{(\beta_y:\beta_z)(\alpha_y:\xi_z)} {}^2\delta_{\hat{n}}^{(\alpha_y:\xi_z)}(\mathbf{x})
\end{aligned}
\qquad (3.92)
$$

$$
\begin{aligned}
{}^2t_{\hat{m}}^{(\beta_y:\eta_z)}(\mathbf{x}) = {} & B_{\hat{m}kl}^{(\beta_y:\eta_z)} \cdot {}^0\varepsilon_{kl}(\mathbf{x}) + \sum_{\xi_y=1}^{m_y} W_{\hat{m}\hat{n}}^{(\beta_y:\eta_z)(\xi_y)} \cdot {}^1\delta_{\hat{n}}^{(\xi_y)}(\mathbf{x}) \\
& + \sum_{\alpha_y=1}^{n_1} \sum_{\alpha_z=1}^{n_z} V_{\hat{m}kl}^{(\beta_y:\eta_z)(\alpha_y:\alpha_z)} \cdot {}^2\mu_{kl}^{(\alpha_y:\alpha_z)}(\mathbf{x}) \\
& + \sum_{\alpha_y=1}^{n_y} \sum_{\xi_z=1}^{m_z} W_{\hat{m}\hat{n}}^{(\beta_y:\eta_z)(\alpha_y:\xi_z)} {}^2\delta_{\hat{n}}^{(\alpha_y:\xi_z)}(\mathbf{x})
\end{aligned}
\qquad (3.93)
$$

$$
\begin{aligned}
{}^1t_{\hat{m}}^{(\eta_y)}(\mathbf{x}) = {} & B_{\hat{m}kl}^{(\eta_y)} \cdot {}^0\varepsilon_{kl}(\mathbf{x}) + \sum_{\xi_y=1}^{m_y} W_{\hat{m}\hat{n}}^{(\eta_y)(\xi_y)} \cdot {}^1\delta_{\hat{n}}^{(\xi_y)}(\mathbf{x}) \\
& + \sum_{\alpha_y=1}^{n_y} \cdots \sum_{\alpha_z=1}^{n_z} V_{\hat{m}kl}^{(\eta_z)(\alpha_y:\alpha_z)} \cdot {}^2\mu_{kl}^{(\alpha_y:\alpha_z)}(\mathbf{x}) \\
& + \sum_{\alpha_y=1}^{n_y} \sum_{\xi_z=1}^{m_z} W_{\hat{m}\hat{n}}^{(\eta_y)(\alpha_y:\xi_z)} {}^2\delta_{\hat{n}}^{(\alpha_y:\xi_z)}(\mathbf{x})
\end{aligned}
\qquad (3.94)
$$

with the partitioned constitutive relation following from (3.77)–(3.79)

$$
{}^2\mu_{ij}^{(\alpha_y:\alpha_z)}(\mathbf{x}) = {}^2f\left({}^2\varepsilon_{ij}^{(\alpha_y:\alpha_z)}(\mathbf{x}),{}^2\sigma_{ij}^{(\alpha_y:\alpha_z)}(\mathbf{x})\right) \qquad (3.95)
$$

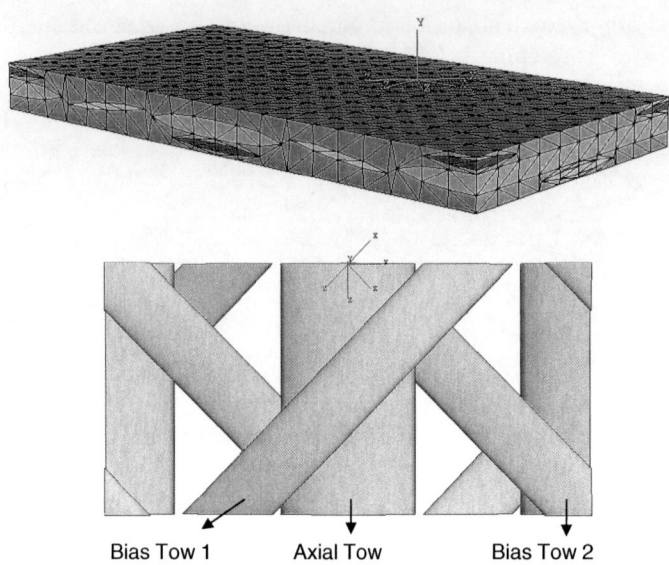

Figure 3.2: Braid architecture considered for validation.

$$^2 t_{\hat{n}}^{(\alpha_y : \xi_z)} (\mathbf{x}) = {}^2 g \left({}^2 \delta_{\hat{n}}^{(\alpha_y : \xi_z)} (\mathbf{x}) \right) \tag{3.96}$$

$$^1 t_{\hat{n}}^{(\xi_y)} (\mathbf{x}) = {}^1 g \left({}^1 \delta_{\hat{n}}^{(\xi_y)} (\mathbf{x}) \right) \tag{3.97}$$

We now focus on the validation of the three-scale reduced-order formulation on the tube crush tests of the braided composite. We consider a triaxially-braided composite unit cell with 45° angle of bias tows as shown in Figure 3.2.

In the unit cell model we identify bias tows with plus/minus (clockwise/counterclockwise) 45° angles with respect to the axial tow as two different inclusion phases. Hence there are four phases (matrix, axial tow, bias tow 1, and bias tow 2) and three interfaces associated with each inclusion phase.

Since each tow in the unit cell consists of thousands of filaments of carbon fiber (axial tow: 80k[1] Fortafil 511; bias tow: 12k Grafil 34-700) and epoxy matrix (Ashland Hetron 922), we have a three-scale representation of the composite: fibrous unit cell model at the micro scale, braided unit cell model (constructed in the previous step) at the meso scale, and tube model at the macro scale. Figure 3.3 depicts the unit cell model at the micro scale. We assume that the three tow phases in the meso scale have the same fibrous unit cell model.

---

[1] An 80 k tow nominally contains 80,000 individual fibers.

The next step is to identify elastic properties of phases in the micro and meso scales. There are three phases: fiber phase, matrix phase in the tow, and matrix phase outside the tow. The identified phase properties (see [39] for parameter identification process) are given in Table 3.2. The identified macro properties of the braided composite are given in Table 3.3.

We employ one partition per phase and assume no interface failure. This results in seven phase partitions as shown in Table 3.4. There are 42 (7 partitions

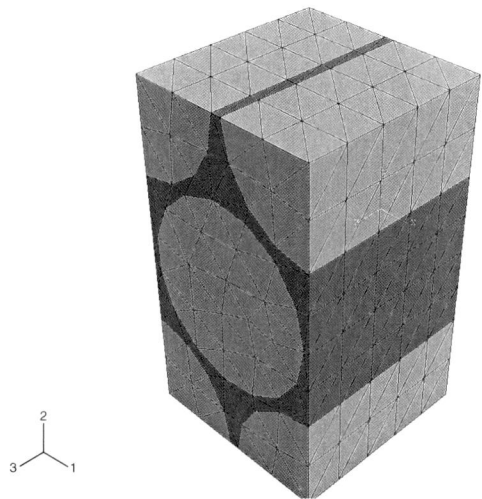

Figure 3.3: Unit cell model at the micro scale.

Table 3.2. Identified micro and meso properties

	E (GPa)	v
Fiber	448	0.1
Matrix in the tow	2.94	0.35
Matrix outside the tow	2.94	0.35

Table 3.3. Identified macro elastic properties

	Experiment	Simulation
$\bar{E}_A$ (GPa)	60.3±2.95	60.33
$\bar{E}_T$ (GPa)	8.7±0.24	8.78

time 6 modes) eigenstrain components that need to be updated within each macro increment. The computational complexity is comparable to the phemenological single scale model with 42 state variables.

The constitutive behavior of phases at the meso scale is modeled using continuum damage mechanics (see Appendix for details). Model parameters including the ultimate strength of matrix, S_{matrix}, the ultimate strength of fiber, S_{fiber}, the compression factor of matrix, C_{matrix}, and the buckling strength of fiber S_{tow}^b are calibrated using inverse methods [39]. The values of the identified parameters and their comparison to the experimental values are summarized in Table 3.5.

Once the model parameters have been identified, we proceed with a crush simulation of a circular tube and compare the simulation result to the experimental data [40]. The macro model of the circular tube is shown in Figure 3.4. The inner diameter of the tube is 2.35 inch and there two layers of braided composite. An

Table 3.4. Phase partitions

Meso partition	Micro partition	Partition ID
Matrix	-	1
Bias Tow 1	Matrix	2
	Fiber	3
Bias Tow 2	Matrix	4
	Fiber	5
Axial Tow	Matrix	6
	Fiber	7

Table 3.5. Calibration of inelastic parameters

Step #	Active Parameter	Type of the Test	Overall Ultimate Strength of Composite (MPa)	
			Experiment	Simulation
1	S_{matrix}	Transverse Tension	65.2	64.7
2	S_{tow}	Longitudinal Tension	654.3	653.0
3	C_{matrix} & S_{tow}^b	Longitudinal Compression	375.8	380.1

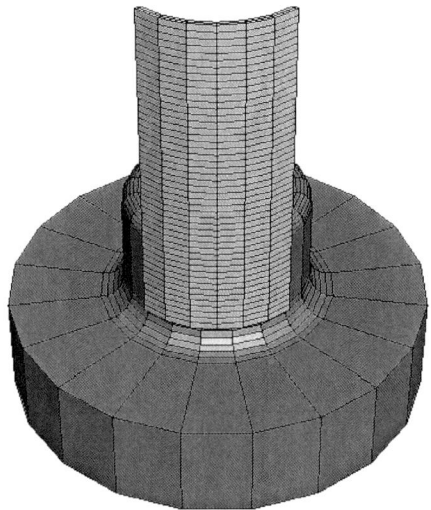

Figure 3.4: Circular tube model.

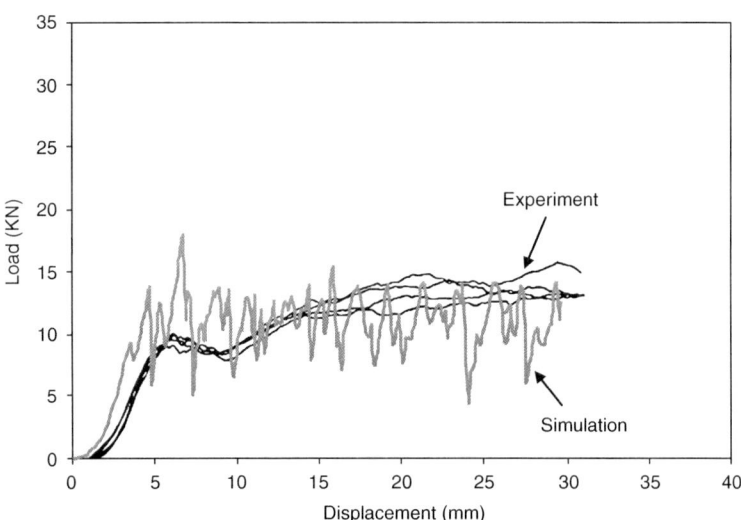

Figure 3.5: Comparison of the simulation and experimental results for circular tube subjected to quasi static loading.

initiator plug with 5/16 inch fillet radius is considered and quasi-static loading (load rate = 0.5 in/min) is assumed.

Figure 3.5 compares the three-scale reduced-order simulations with the experimental results. Good agreement with the experiment can be observed. The high oscillated profile may be attributed to the fact that we haven't accounted for interface failure.

3.8 Appendix

Continuum damage mechanics with isotropic damage law is employed to model constitutive behavior of phases. By this approach, the constitutive relation is given by

$$\sigma_{ij} = (1 - w_{\text{ph}}) L_{ijkl} \varepsilon_{kl} \qquad (3.98)$$

where $w \in [0, 1]$ is a damage state variable. Thus the partitioned eigenstrain is defined as

$$\mu_{ij}^{(\alpha_y)} = w_{\text{ph}}^{(\alpha_y)} \varepsilon_{ij}^{(\alpha_y)} \qquad (3.99)$$

The damage state variable $w_{\text{ph}}^{(\alpha_y)}$ is taken to be a piecewise-continuous function of damage equivalent strain $\kappa_{\text{ph}}^{(\alpha_y)}$. The evolution of phase damage may be expressed as

$$w_{\text{ph}}^{(\alpha_y)} = \begin{cases} 0 & \kappa_{\text{ph}}^{(\alpha_y)} \leq {}^1\kappa_{\text{ph}}^{(\alpha_y)} \\ \Phi\left(\kappa_{\text{ph}}^{(\alpha_y)}\right) & {}^1\kappa_{\text{ph}}^{(\alpha_y)} < \kappa_{\text{ph}}^{(\alpha_y)} \leq {}^2\kappa_{\text{ph}}^{(\alpha_y)} \\ 1 & \kappa_{\text{ph}}^{(\alpha_y)} > {}^2\kappa_{\text{ph}}^{(\alpha_y)} \end{cases} \qquad (3.100)$$

where ${}^1\kappa_{\text{ph}}^{(\alpha_y)}$ and ${}^2\kappa_{\text{ph}}^{(\alpha_y)}$ are model parameters corresponding to the initial and fully damaged state, respectively.

The damage equivalent strain $\kappa_{\text{ph}}^{(\alpha_y)}$ is assumed to be a function of the principal strain

$$\kappa_{\text{ph}}^{(\alpha_y)}(t) = \max \left\{ \sqrt{\sum_{I=1}^{3} \left\langle \varepsilon_I^{(\alpha_y)}(\tau) \right\rangle^2}, \tau > t \right\} \qquad (3.101)$$

where

$$\langle x \rangle = \begin{cases} x & x \geq 0 \\ Cx & x < 0 \end{cases} \qquad (3.102)$$

and C is the compression factor (one of inelastic material parameters).

${}^1\kappa_{\text{ph}}^{(\alpha_y)}$ can be expressed in terms of material strength S (another inelastic material parameter) and stiffness E

$${}^1\left(\kappa_{\text{ph}}^{(\alpha_y)}\right) = \frac{S}{E} \qquad (3.103)$$

Based on the selection of the damage evolution function, $^2\kappa_{ph}^{(\alpha_y)}$ can be expressed in terms of physical-based material parameters (E, S and G- the total strain energy). For various evolution functions $\Phi\left(\kappa_{ph}^{(\alpha_y)}\right)$ see [39].

Similarly, eigenseparation are modeled using damage mechanics based cohesive law

$$t_{\hat{n}}^{(\xi_y)} = \left(1 - w_{int}^{(\xi_y)}\right) K^{(\xi_y)} \delta_{\hat{n}}^{(\xi_y)} \quad (3.104)$$

where $K^{(\xi_y)}$ is the interface stiffness. In the normal direction, $w_{int}^{(\xi_y)}$ is considered only when $\delta_{\hat{n}}^{(\xi_y)} > 0$.

The interface damage state variable $w_{int}^{(\xi_y)}$ is assumed to be a function of damage equivalent displacement $\kappa_{int}^{(\xi_y)}$ defined as

$$K_{int}^{(\xi_y)}(t) = \max\left\{\sqrt{\left\langle\delta_{\hat{n}}^{\xi_y}(\tau)\right\rangle_+^2 + \left(\delta_{\hat{t}2}^{(\xi_y)}(\tau)\right)^2}, \tau < t\right\} \quad (3.105)$$

where

$$\langle x \rangle_+ = \begin{cases} x & x \geq 0 \\ 0 & x < 0 \end{cases} \quad (3.106)$$

The evolution of interface damage can be expressed as

$$w_{int}^{(\xi_y)} = \begin{cases} 0 & \kappa_{int}^{(\xi_y)} \leq {}^1\kappa_{int}^{(\xi_y)} \\ \Phi\left(\kappa_{int}^{(\xi_y)}\right) & {}^1\kappa_{int}^{(\xi_y)} < \kappa_{int}^{(\xi_y)} \leq {}^2\kappa_{int}^{(\xi_y)} \\ 1 & \kappa_{int}^{(\xi_y)} > {}^2\kappa_{int}^{(\xi_y)} \end{cases} \quad (3.107)$$

3.9 Acknowledgment

The financial supports of National Science Foundation under grants CMS-0310596, 0303902, 0408359, Rolls-Royce contract 0518502, Automotive Composite Consortium contract 606-03-063L, and AFRL/MNAC MNK-BAA-04-0001 contract are gratefully acknowledged.

References

[1] Hill R. (1948). "A theory of the yielding and plastic flow of anisotropic metals", *Proc. Roy. Soc. Lond.*, **A193**, p. 281–297.

[2] Babuska I. (1975). *Homogenization and Application. Mathematical and Computational Problems*. In Numerical Solution of Partial Differential Equations – III, SYNSPADE, ed. B Hubbard, Academic Press.

[3] Benssousan A., Lions J. L., and Papanicoulau G. (1978). *Asymptotic Analysis for Periodic Structures*, North-Holland.
[4] Sanchez-Palencia E. (1980). *Non-homogeneous media and vibration theory*, Lecture notes in physics, Vol. 127, Springer-Verlag. Berlin.
[5] Chen W. and Fish J. (2006). "A Mathematical Homogenization Perspective of Virial Stress", *International Journal for Numerical Methods in Engineering*, **67**, p. 189–207.
[6] Chen W. and Fish J. (2006). "A generalized space-time mathematical homogenization theory for bridging atomistic and continuum scales", *International Journal for Numerical Methods in Engineering*, **67**, p. 253–271.
[7] Fish J., Chen W., and Tang, Y. (2005). "Generalized Mathematical Homogenization of Atomistic Media at Finite Temperatures," *International Journal for Multiscale Computational Engineering*, Vol. **3**, Issue 4.
[8] Fish J., Chen W., and Li, R. (2007). "Generalized mathematical homogenization of atomistic media at finite temperatures in three dimensions", *Comp. Meth. Appl. Mech. Engng.*, Vol. **196**, p. 908–922.
[9] Li A., Li R., and Fish J. (2008). "Generalized Mathematical Homogenization: From Theory to Practice", to appear in *Comp. Meth. Appl. Mech. Engng.*
[10] Feynman R. P. (1959). There's Plenty of Room at the Bottom in 29th Annual Meeting of the American Physical Society. California Institute of Technology.
[11] Terada K. and Kikuchi N. (1995). "Nonlinear homogenization method for practical applications", *Computational Methods in Micromechanics*, Ghosh S. and Ostoja-Starzewski M. (eds.), ASME, New York, Vol. AMD-212/MD-62, p. 1–16.
[12] Terada K. and Kikuchi N. (2001). "A class of general algorithms for multiscale analysis of heterogeneous media", *Comput. Methods Appl. Mech. Eng.*, **190**, p. 5427–5464.
[13] Matsui K., Terada K., and Yuge K. (2004). "Two-scale finite element analysis of heterogeneous solids with periodic microstructures", *Computers and Structures*, **82** (7–8), p. 593–606.
[14] Smit R. J. M., Brekelmans W. A. M., and Meijer H. E. H. (1998). "Prediction of the mechanical behavior of nonlinear heterogeneous systems by multilevel finite element modeling", *Comput. Methods Appl. Mech. Eng.*, **155**, p. 181–192.
[15] Miehe C. and Koch A. (2002). "Computational micro-to-macro transition of discretized microstructures undergoing small strain", *Arch. Appl. Mech.*, **72**, p. 300–317.
[16] Kouznetsova V., Brekelmans W.-A., and Baaijens F.P.-T. (2001). "An approach to micro-macro modeling of heterogeneous materials", *Comput. Mech.*, **27**, p. 37–48.
[17] Feyel F. and Chaboche J.-L. (2000). "FE2 multiscale approach for modeling the elastoviscoplastic behavior of long fiber sic/ti composite materials", *Comput. Methods Appl. Mech. Eng.*, **183**, p. 309–330.

[18] Ghosh S., Lee K., and Moorthy S. (1995). "Multiple scale analysis of heterogeneous elastic structures using homogenization theory and Voronoi cell finite element method", *Int. J. Solids Struct.*, **32**, p. 27–62.

[19] Ghosh S., Lee K., and Moorthy S. (1996). "Two scale analysis of heterogeneous elasticplastic materials with asymptotic homogenization and Voronoi cell finite element model". *Comput. Methods Appl. Mech. Engng.*, **132**, p. 63–116.

[20] Michel J.-C., Moulinec H., and Suquet P. (1999). "Effective properties of composite materials with periodic microstructure: a computational approach", *Comput. Methods Appl. Mech. Eng.*, **172**, p. 109–143.

[21] Geers M. G. D., Kouznetsova V., Brekelmans W. A. M. (2001). "Gradient-enhanced computational homogenization for the micro-macro scale transition", *J. Phys. IV*, Vol. **11**, p. 145–152.

[22] McVeigh C., Vernerey F., Liu W. K., and Brinson L. C. (2006). "Multiresolution analysis for material design", *Comput. Methods Appl. Mech. Engrg.*, Vol. **195**, p. 5053–5076.

[23] Miehe, C. (2002). "Strain-driven homogenization of inelastic microstructures and composites based on an incremental variational formulation", *International Journal for Numerical Methods in Engineering*, **55**, p. 1285–1322.

[24] Fish J., Shek K., Pandheeradi M., and Shephard M. S. (1997). "Computational Plasticity for Composite Structures Based on Mathematical Homogenization: Theory and Practice", *Comp. Meth. Appl. Mech. Engng.*, Vol. **148**, p. 53–73.

[25] Fish J. and Shek K. L. (1999). "Finite Deformation Plasticity of Composite Structures: Computational Models and Adaptive Strategies", *Comp. Meth. Appl. Mech. Engng.*, Vol. **172**, p. 145–174.

[26] Fish J. and Yu Q. (2001). "Multiscale Damage Modeling for Composite Materials: Theory and Computational Framework", *International Journal for Numerical Methods in Engineering*, Vol. **52**, No 1-2, p. 161–192.

[27] Clayton J. D. and Chung P. W. (2006), "An atomistic-to-continuum framework for nonlinear crystal mechanics based on asymptotic homogenization", *J. Mech. Physics of Solids*, **54**, p. 1604–1639.

[28] Fish J. and Shek K. L. (1999). "Finite Deformation Plasticity of Composite Structures: Computational Models and Adaptive Strategies", *Comp. Meth. Appl. Mech. Engng.*, Vol. **172**, p. 145–174.

[29] Yuan Z. and Fish J. (2007). "Towards Realization of Computational Homogenization in Practice", to appear in *International Journal for Numerical Methods in Engineering*.

[30] Zohdi T. I., Oden J. T., and Rodin G. J. (1996). "Hierarchical modeling of heterogeneous bodies". *Computer Methods in Applied Mechanics and Engineering*, Vol. **138**, p. 273–298.

[31] Feyel F. and Chaboche J. L. (2000). "Fe2 multiscale approach for modelling the elastoviscoplastic behaviour of long fibre sic/ti composite materials", *Computer Methods in Applied Mechanics and Engineering*, Vol. **183**, p. 309–330.

[32] E W. and Enquist B., et al., (2003). "Heterogeneous multiscale method: A general methodology for multiscale modeling", *Physical Review B*, **67**(9), p. 1–4.

[33] Hou T. Y. and Wu X. H. (1997). "A multiscale finite element method for elliptic problems in composite materials and porous media", *Journal of Computational Physics*, Vol. **134**(1), p. 169–189.

[34] Loehnert S. and Belytschko T. (2007). "A multiscale projection method for macro/microcrack simulations", *International Journal for Numerical Methods in Engineering*, Vol. **71**, Issue 12, p. 1466–1482.

[35] Belytschko T., Loehnert S., and Song J. (2008). "Multiscale aggregating discontinuities: A method for circumventing loss of material stability", *International Journal for Numerical Methods in Engineering*, Vol. **73**, Issue 6, p. 869–894.

[36] Hill R. (1963). "Elastic properties of reinforced solids: some theoretical principles", *Journal of the Mechanics and Physics of Solids*, **11**, p. 357–372.

[37] Oskay C. and Fish J. (2007). "Eigendeformation-Based Reduced Order Homogenization", *Comp. Meth. Appl. Mech. Engng.*, **196**, p. 1216–1243.

[38] Fish J. and Fan R. (2008). "Mathematical Homogenization of Nonperiodic Heterogeneous Media Subjected to Large Deformation Transient Loading", accepted in *International Journal for Numerical Methods in Engineering*.

[39] Yuan Z. (2008). *Multiscale Design System*, PhD thesis, Rensselaer Polytechnic Institute, Troy, NY, January 2008.

[40] Beard S. J. (2001). *Energy Absorption of Braided Composite Tubes*. Thesis. Stanford University.

PART II

CONCURRENT MULTISCALE METHODS IN SPACE

4
CONCURRENT COUPLING OF ATOMISTIC AND CONTINUUM MODELS

Mei Xu

4.1 Introduction

This chapter describes the concurrent coupling of atomistic methods with continuum mechanics. Such models are useful in the study of phenomena such as fracture and dislocation dynamics, where molecular mechanics and/or quantum mechanics models are required for phenomena such as bond breaking, but the relevant configuration is far too large to permit a completely atomic description. To make such problems computationally tractable, the quantum/atomistic model must be limited to small clusters of atoms in the vicinity of a domain of interest where such high resolution models are necessary and a continuum method should be used for the rest of the domain, see e.g. Khare *et al.* [1, 2]. In the remainder of the domain, where no bond breaking occurs, the continuum description permits much less expensive computations with sufficient accuracy to describe the behavior of the system.

This chapter describes two types of concurrent models

1. Mechanical equilibrium models
2. Molecular dynamics models

Equilibrium models are used to determine the energy landscape of a system, i.e., the surface that gives the energy at 0 K. They provide valuable information about the system such as energy differences due to configurational changes, activation barriers, etc. The adjective *mechanical* should be carefully noted. Equilibrium is often used in molecular mechanics to describe thermodynamic equilibrium in atomistic systems, but thermodynamic equilibrium also includes thermal and chemical in addition to mechanical equilibrium. Mechanical equilibrium can be thought of as molecular statics, in contrast to molecular dynamics.

We will describe five concurrent coupling methods

1. Direct and master-slave coupling
2. The ONIOM method
3. The bridging-domain method
4. The quasicontinuum method
5. The bridging scale method

Although this list is not comprehensive, it provides a birdseye view of concurrent methods. We will first give a brief review of the literature. This is followed by a taxonomy of multiscale methods, so that the place of concurrent methods can be seen. Several concurrent methods are then described. Similarities and differences of the methods will be described, along with their advantages and disadvantages. The description is by no means comprehensive and emphasizes the bridging domain method.

Among the first atomistic/continuum models are those in Baskes et al. [3] and Mullins and Dokainish [4]. The latter studied a plane crack in alpha iron. In their model, the finite element mesh was coarser than the lattice constant (the lattice constant is the distance between nuclei). The displacements of the atoms not coincident with finite element nodes were prescribed by using the finite element interpolation (the term nuclei is more precise since atoms include the electrons, but we will use both interchangeably). It can be said that the atoms were slaved to the finite element mesh. This procedure is often called a master-slave approach in the finite element literature, see Belytschko et al. [5], and we will use this nomenclature here and describe the method.

Kohlhoff et al. [6] similarly coupled finite elements with an atomistic model in a method called FEAt, but they used a nonlocal elasticity theory for the continuum in the overlapping domain to link with the nonlocal lattice theories. The model was applied to the cleavage fracture of two b.c.c. crystalline materials, tungsten and iron. The FEAt method introduces the concept of a "pad" which overlaps the finite element model. However, the energy of the underlying finite element model is not included in the total energy. Its purpose is to provide the displacements of the pad atoms. At the atomistic/finite element interface, the nodes and atoms are coincident to ensure the compatibility between the two models.

The quasicontinuum method (QC) (Tadmor et al. [8]) is aimed at making the transition to atomistic models more seamless. The entire model is viewed as an atomistic model; finite element interpolation and the Cauchy-Born rule are used to eliminate atoms where atomic resolution is not needed. Accurate computation of the energy in the *atomistic* subdomain requires that the atoms be coincident with the finite element nodes. Representative atoms are selected in subdomains to eliminate the unnecessary atomistic degrees of freedom, and the mechanical variables, e.g. displacements and forces, of atoms in the subdomains are calculated using finite element interpolation. This almost bypasses the distinction between atomistic and continuum domains. Ghost forces due to the mismatch in the transition zone were later taken care of by the developers of the method [9]. The convergence, accuracy and the effect of tolerances were evaluated in [10]. Several special purpose variants of the method have been developed. For example, Knap and Ortiz [10] formulated a cluster summation rule for the QC method, which made it possible to apply the QC model to more complex problems. Rudd and Broughton [11] developed a similar method, called the coarse-grained molecular dynamics (CGMD) method. The coincidence of the atoms and nodes was required and the equations of motion were directly obtained from a statistical coarse graining procedure.

Shilkrot *et al.* [7] developed the coupled atomistic/discrete-dislocation (CADD) method, which combines the QC method with continuum defect models. It added the capability to pass dislocations between the atomistic and the continuum models. Closed form solutions and superposition methods were used for this purpose, and the *pad* atoms were used in the CAAD method too, but they were not slaved to the continuum nodes. The role of the *pad* was to eliminate the free surface when dealing with the atomistic model.

Considerable interest in multiscale methods has been aroused by the pioneering work of Abraham *et al.* [12] and Broughton *et al.* [13], in which a quantum model was added in the bond-breaking subdomain. Their method is called macroscopic-atomistic-ab initio dynamics (MAAD). In their atomistic/continuum coupling, the two models are overlapped and the contribution of each model to the total Hamiltonian is taken to be the average of the two Hamiltonians (i.e. the energies). The coupling procedure is only described briefly. They called the overlapping domain the *handshake* domain, and this name has become widely used. In their models, the finite element size at the atomistic/continuum interface was the same as the lattice constant.

Recently, several methods have been developed for equilibrium solutions and dynamics that do not require coincident nodes and atoms. Belytschko and Xiao [22] have developed an overlapping domain decomposition approach, called the bridging domain method, in which the energies of the atomistic and continuum models are scaled in the overlapping domain. Compatibility between the two models is enforced by Lagrange multipliers. A similar method was developed for molecular dynamics in Xiao and Belytschko [21]. A comparison of two types of coupling with bridging domain methods, L^2 and H^1 coupling, is given by Guidault and Belytschko [23]. Fish *et al.* [24] have developed a similar method in which the forces instead of the energies are scaled, which permits the continuum to be an irreversible material.

Wagner and Liu [14] developed a bridging scale method (BSM) for coupling molecular dynamics with continuum mechanics. In their method, the continuum and atomistic models are initially applied on the entire domain. Then the fine-scale degrees of freedom are eliminated from the coarse-scale domain by a time history kernel, i.e. a Green's function. In the fine-scale region a projection of the fine-scale solution onto the coarse-scale model is used to link the models. A Laplace transform was applied to derive the time history kernel function which corresponds to the missing degrees of freedom in the coarse-scale region. Park *et al.* [37] extended the method to two dimensions by computing the kernel function numerically. The performance of BSM is studied by Farrell *et al.* [16].

In all concurrent coupling methods for dynamics, a general issue is the spurious reflection of waves at the fine-scale domain boundary due to the difference in the cutoff frequencies of the models. Several silent boundary conditions have been developed to solve this problem. Cai *et al.* [17] introduced a time-dependent memory kernel to the equations of motion in the fine-scale. It gave the exact boundary conditions for the microscale simulation but was generally computationally expensive. E and Huang [18] proposed a matching boundary condition by optimizing the reflection term which was added to the displacement field. The time history kernel of the bridging scale method [14] serves a similar purpose. In the bridging domain method [21], a diagonalized constraint matrix was used to eliminate such reflections. To and Li [19] applied a perfectly matched layer [20] to the overlapping region to absorb high frequency waves leaving the atomistic domain.

4.2 Classification of methods

Before describing concurrent methods in more detail, we will first give an overview of different multiscale methods. Multiscale methods are often classified as either hierarchical or concurrent. But this classification is too simple to reflect the subtleties of different variants of multiscale analysis methods. Table 4.1 gives a classification of multiscale methods, and a schematic of the various methods is shown in Figure 4.1.

In Figure 4.1 the fine-scale models are on the left and the coarse-scale models on the right. The essence of the procedures is the same for any scales that are linked, whether it is quantum to continuum, microstructure of a composite to an engineering structure, or an atomistic (molecular) model to a continuum. Furthermore, coarse-graining is not restricted to link only two scales: it is feasible to couple three, four, or even more scales.

Hierarchical methods are the most widely used and computationally the most efficient, see Figure 4.1(a). In these methods, the response of a representative volume element at the fine-scale is first computed over a range of expected inputs $\mathcal{E}$, and from these a stress-strain law is extracted. The stress-strain law can be a constitutive equation, with parameters determined by the fine-scale solutions, a numerical data base, or a neutral network-based law. For linear response, this process can be simplified since there is a robust theory of homogenization whereby linear constants can be effectively extracted, sometimes even without recourse to numerical micromodels.

Table 4.1. Classification of multiscale methods illustrated for Atomistic/ Continuum coupling

Multiscale techniques	Information transfer
Hierarchical	$\Omega^{atomic} \to \Omega^{continuum}$
Hybrid hierarchical-semiconcurrent	$\Omega^{atomic} \to \Omega^{continuum}$ if $\mathbf{E} \in \mathcal{E}_I$
	$\Omega^{atomic} \leftrightarrow \Omega^{continuum}$ if $\mathbf{E} \notin \mathcal{E}_I$
Semiconcurrent	$\Omega^{atomic} \leftrightarrow \Omega^{continuum}$
Concurrent	$\Omega^{atomic} \Leftrightarrow \Omega^{continuum}$

($\mathbf{E}$ =input, $\mathcal{E}_I$ =domain of input, $\to$, $\leftrightarrow$: weak coupling, $\Leftrightarrow$: strong coupling)

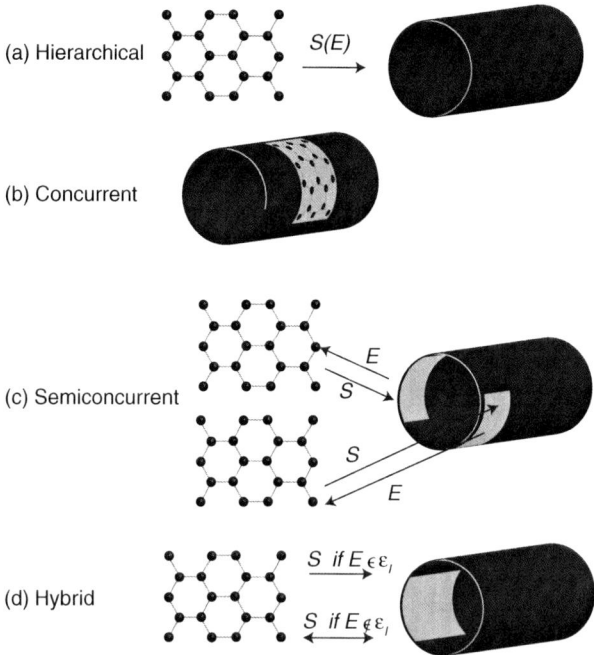

Figure 4.1: Schematic of hierarchical, concurrent, semiconcurrent and hybrid hierarchical-semiconcurrent coupling methods; $E =$ strain, $S =$ stress.

For severely nonlinear problems, hierarchical models become more problematical, particularly if the fine-scale response is path dependent. It should be noted that when failure occurs, in many circumstances standard hierarchical models are invalid and cannot be used, see Song *et al.* [15].

Concurrent methods are those in which the fine-scale model is embedded in the coarse-scale model and is directly and intimately coupled to it, as shown in Figure 4.1(b). Both compatibility and momentum balance are enforced across the interface. For example, in Khare *et al.* [1, 2] electronic structure models solved by quantum mechanics (in the bond-breaking subdomain) are linked to atomistic and continuum models. Similarly the MAAD method [12], the BSM [14], and the bridging domain method [22] are concurrent. Such models are quite effective when the subdomain, where a higher order theory is required, is small compared to the domain of the problem. In Khare *et al.* [1, 2], the QM model, which is the highest order theory, was restricted to the area around a slit where bond breaking was likely to occur; some of these models are described in Section 4.5. In the study of fracture, fine-scale models can be inserted in *hot spots* where strains or stresses become large. These *hot spots* can be identified on the fly or by a previous run.

Semiconcurrent methods are a class of methods in which a fine-scale model is calculated concurrently with a coarser-grained model, as shown in Figure 4.1(c).

They are distinguished from concurrent methods by the fact that the fine-scale model is not coupled directly with the coarse-grained model; compatibility and momentum balance are only satisfied approximately. Examples of such methods are the FE^2 methods of Feyel and Chaboche [28], and their extensions to gradient models in Kouznetsova *et al.* [29]. The reduction in computational effort from a fully concurrent approach is not great. The appeal of these methods lies more in the fact that the fine-scale and coarse-scale models can be run by separate software and the avoidance of numerical problems associated with rapid changes in element size (the lattice constant can be viewed as an element size). These difficulties include

- Ill-conditioning of the equilibrium equations
- Spurious reflections off the interface between fine-grain and coarse-grain models

The fourth type of methods are hybrid hierarchical/semiconcurrent methods, Figure 4.1(d). In these methods, a hierarchical approach is used as long as the requested information is available. When the requested information is not available, i.e. when the strain is outside the strain domain $\mathcal{E}$ that has been exercised by the fine-scale model, a direct computation is then invoked. If the fine-scale response is history dependent, it would be necessary to reconstruct at least an approximate history.

4.3 Mechanical equilibrium methods

In this section we describe several concurrent methods for mechanical equilibrium problems, often called 0 K methods; we will often simply use the term equilibrium methods. In this description and subsequent descriptions of the method, we denote the complete domain in the initial configuration by Ω_0 and its boundaries by Γ_0; Γ_0 consists of traction boundaries Γ_0^t and the displacement boundaries Γ_0^u. The domain is subdivided into the subdomains treated by continuum mechanics (CM), Ω_0^C, by molecular mechanics (MM), i.e. by atomistic methods, Ω_0^M, and by quantum mechanics (QM), Ω_0^Q; $\Omega_0^M \cup \Omega_0^Q$ is the domain encompassed by the atoms of the model. The intersection of Ω_0^C and Ω_0^M is denoted by Ω_0^{MC} and is often called the handshake domain; Γ_0^α denotes the edges of the continuum domain. The position of atom I in the reference configuration, or more precisely the nuclei of the atom I, is given by $\mathbf{X}_I$ and its displacement by $\mathbf{d}_I$. Sometimes we use a superscript A to indicate molecular mechanics related variables instead of M.

4.3.1 *Direct and master-slave coupling*

A common method of coupling molecular mechanics with continuum mechanics is what we call master-slave coupling. It can also be used for QM/MM and QM/CM coupling. This is a well-known method in finite elements (see [5], p 168) based on energy. In one dimension, a model with master-slave coupling is illustrated in Figure 4.2. As indicated by the dashed lines, the atoms and coincident nodes

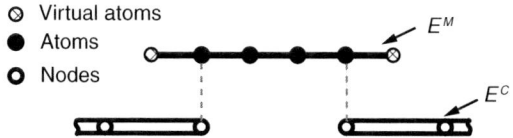

Figure 4.2: Schematic of direct coupling method.

are constrained to move together. In master-slave coupling, the nodes need not be coincident. We will describe this coupling in more detail.

The total energy of the coupled system is

$$E\left(\mathbf{u}, \mathbf{d}^M\right) = E^C\left(\mathbf{u}\right) + E^M\left(\mathbf{d}^M\right) - E^{\text{ext}} \qquad (4.1)$$

where E^M and E^C are the energies of the molecular and continuum subdomains, respectively, and E^{ext} is the work of external forces. The vector $\mathbf{u}$ is the displacement field in the continuum and $\mathbf{d}^M$ is a vector of discrete molecular displacements. The continuum energy is given by

$$E^C\left(\mathbf{u}\right) = \int_{\Omega_0^C} w_C\left(\mathbf{F}\right) d\Omega_0^C \qquad (4.2)$$

where w_C is the strain energy density and $\mathbf{F}$ is the deformation gradient: $\mathbf{F} = \mathbf{I} + \frac{\partial \mathbf{u}}{\partial \mathbf{X}}$.

If a finite element method is used for the continuum, then the displacement field is given by

$$\mathbf{u}\left(\mathbf{X}\right) = \sum_{I \in \mathcal{S}} N_I\left(\mathbf{X}\right) \mathbf{u}_I \qquad (4.3)$$

where $N_I\left(\mathbf{X}\right)$ are the shape functions expressed as functions of the material coordinates $\mathbf{X}$, $\mathbf{u}_I$ are the nodal displacements, and $\mathcal{S}$ is the set of all FEM nodes.

The potential energy of the bond $I - J$ is denoted by $w_{IJ}^M = w_M(\mathbf{x}_I, \mathbf{x}_J)$ for a two-body potential, where $\mathbf{x}_I = \mathbf{X}_I + \mathbf{d}_I$ is the current position of atom I. The total potential energy of the molecular mechanics subdomain Ω^M is then

$$E^M = \frac{1}{2} \sum_{I,J \in \mathcal{M}} w_{IJ}^M \qquad (4.4)$$

where $\mathcal{M}$ is a set of atoms in the atomistic model.

Master-slave methods can be classified as follows:

- Methods with atoms coincident with continuum nodes on the interface, Figure 4.3.
- Methods where atoms are not coincident with nodes on the interface, Figure 4.4.

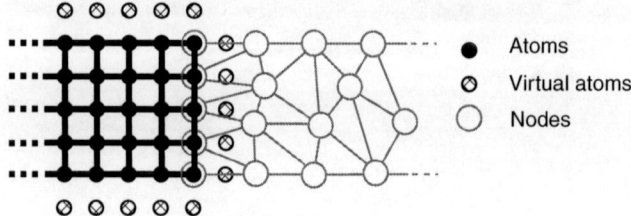

Figure 4.3: Master-slave model with nuclei coincident with nodes at the coupling interface.

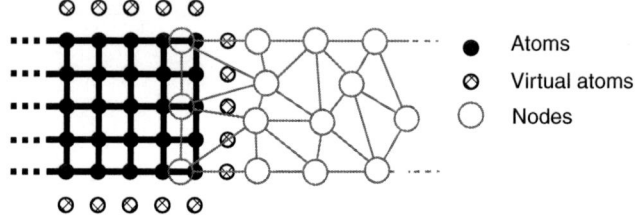

Figure 4.4: Master-slave model with nuclei not coincident with nodes at the coupling interface.

An illustration of a model with coincident atoms/nodes is shown in Figure 4.3. As can be seen, to reduce the nodal spacing to the lattice constant, it is necessary to construct rather complicated meshes. This is awkward in two dimensions and can be a severe obstacle in three dimensions.

In models with coincident coupling, the atomistic and nodal displacements at the interface are taken to be equal. Therefore, we define a common vector $\mathbf{d} = \{\mathbf{d}_I\}_{I=1}^n$, where n is the number of independent nodes and atoms. Compatibility is enforced by letting

$$\mathbf{d}_I = \begin{cases} \mathbf{d}_I^M & \mathbf{X}_I \in \Omega_0^M \\ \mathbf{d}_I^M = \mathbf{u}_I & \mathbf{X}_I \in \Omega_0^M \cap \Omega_0^C \\ \mathbf{u}_I & \mathbf{X}_I \in \Omega_0^C \end{cases} \quad (4.5)$$

The discretized energy of the model can then be written as

$$E(\mathbf{d}) = E^C(\mathbf{d}) + E^M(\mathbf{d}) - E^{\text{ext}} \quad (4.6)$$

Equilibrium configurations correspond to the stationary points of (4.6), i.e., the configurations where the derivatives with respect to $\mathbf{d}_I$ vanish. This gives

$$0 = \frac{\partial E^M}{\partial \mathbf{d}_I} + \frac{\partial E^C}{\partial \mathbf{d}_I} - \frac{\partial E^{\text{ext}}}{\partial \mathbf{d}_I} \quad (4.7)$$

These derivatives correspond to nodal and atomistic forces, so the above can be written as

$$0 = \mathbf{f}_I^M + \mathbf{f}_I^C - \mathbf{f}_I^{\text{ext}} \tag{4.8}$$

where from (4.2) and (4.3) it follows that

$$\mathbf{f}_I^C = \int_{\Omega_0^C} \frac{\partial w_C(\mathbf{F})}{\partial \mathbf{F}} \frac{\partial \mathbf{F}}{\partial \mathbf{d}_I} d\Omega_0^C = \int_{\Omega_0^C} \mathbf{P} \frac{\partial N_I}{\partial \mathbf{X}} d\Omega_0^C \tag{4.9}$$

where $\mathbf{P}$ is the first Piola-Kirchhoff stress and $\mathbf{f}_I^{\text{ext}}$ are external forces.

The positions of atoms which lie inside Ω_0^C, denoted as virtual atoms in Figure 4.3, are obtained by the finite element interpolation (4.3),

$$\mathbf{d}_J^M = \sum_{I \in \mathcal{S}} N_I(\mathbf{X}_J) \mathbf{u}_I \tag{4.10}$$

Models with noncoincident nodes and atoms at the interface, as shown in Figure 4.4, are treated by master-slave methods by letting the atoms be slaves. The displacements of the atoms on the interface are obtained by (4.10). The governing equations are identical to those described for coincident coupling. However, the forces on the slave atoms are added to the continuum nodes by the usual master-slave procedure. This can be seen by rederiving (4.6) for nodes on the interface:

$$\frac{\partial E^M}{\partial \mathbf{u}_I} = \sum_J \frac{\partial E^M}{\partial \mathbf{d}_J^M} \frac{\partial \mathbf{d}_J^M}{\partial \mathbf{u}_I} = \mathbf{f}_I^M N_J(\mathbf{X}_I) \tag{4.11}$$

where the last step follows from the definition of nodal forces and (4.10).

4.3.2 ONIOM method

In the ONIOM method, first developed by Svensson et al. [32], the higher order method is superimposed on the lower order method on the subdomain in which higher accuracy is needed. For QM/MM coupling, the ONIOM method is basically a direct coupling method. Its advantage lies in the ease with which it can be used to couple models that reside in two different software systems. Since the QM and MM methods that are needed for a particular model are often in different software packages, this often proves useful. The ONIOM scheme is illustrated in one dimension in Figure 4.5 for a quantum/molecular model. As can be seen, the model consists of three layers in the subdomain where the higher order theory is used:

- The layer Q modeled by the higher order theory, in this case an electronic structure method (quantum mechanics)
- The bottom layer M which extends throughout the entire computational domain, in this case an atomistic model

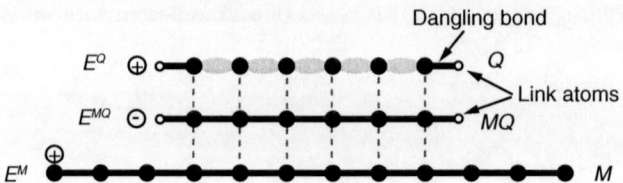

Figure 4.5: Schematic of the ONIOM method for quantum/molecular coupling. The signs indicate the sign of the corresponding terms in energy (force) expression; the vertical lines indicate that the atoms in the three models are constrained to move identically.

- The overlaid (middle) layer MQ which cancels the energy of the lower order method (molecular mechanics) on Ω^Q

The middle layer is sometimes called fictitious, but both M and MQ are fictitious on Ω^Q. They are only retained to simplify the coupling.

In the ONIOM method, the displacements of corresponding nuclei in difference layers are equal (except the link atoms) and will here be denoted by $\mathbf{d}_I$.

The energy of the system as treated by ONIOM is given by

$$E(\boldsymbol{\alpha}, \mathbf{d}) = E^Q(\boldsymbol{\alpha}, \mathbf{d}) + E^M(\mathbf{d}) - E^{MQ}(\mathbf{d}) - E^{\text{ext}}(\mathbf{d}) \qquad (4.12)$$

as is readily apparent from Figure 4.5, where $\mathbf{d}$ is the matrix of atomic displacements and $\boldsymbol{\alpha}$ the electronic solution parameters. The equilibrium equations for a system with external loads $\mathbf{f}_I^{\text{ext}}$ on atom I is then

$$0 = \frac{\partial E}{\partial \mathbf{d}_I} = \frac{\partial E^Q}{\partial \mathbf{d}_I} + \frac{\partial E^M}{\partial \mathbf{d}_I} - \frac{\partial E^{MQ}}{\partial \mathbf{d}_I} - \mathbf{f}_I^{\text{ext}} \qquad (4.13)$$

This equation can be written as

$$0 = \mathbf{f}_I^Q + \mathbf{f}_I^M - \mathbf{f}_I^{MQ} - \mathbf{f}_I^{\text{ext}} = \mathbf{f}_I^{\text{int}} - \mathbf{f}_I^{\text{ext}} \qquad (4.14)$$

Generally, in the ONIOM coupling of QM/MM models, the atomic displacements $\mathbf{d}_I$ are obtained by minimizing the total energy with respect to the parameters of the electronic structure $\boldsymbol{\alpha}$ and $\mathbf{d}_I$. This involves two nested iterative procedures. First the energy must be minimized with respect to the parameters of the electronic structure. In the outer loop, the positions of the nuclei are updated to bring them towards equilibrium.

An interesting aspect of ONIOM is that the quality of the coarse-scale model underlying the fragment has no effect on the results. Therefore, models that are quite inaccurate for the phenomena under study can be used. For example, in the study of fracture with MM/QM ONIOM schemes, the MM model can be quite inaccurate near the fracture strain. However, it is crucial that in the range of behavior that occurs at the QM/MM interface, the models are matched carefully. For example, in Khare et al. [1], it was reported that a standard Brenner MM model for carbon coupled with either a DFT or PM3 quantum model often fractured at the interface. Therefore, they proposed a scaling of the MM potential to provide better results. A major source of error in these models is the "dangling" bonds at the boundary of the model (in Figure 4.5 the two ends of the model). Electronic structure calculations can be very inaccurate at non-periodic boundaries when the QM model is terminated, especially for covalent bonds. One common remedy is to cap the dangling bonds with hydrogen atoms; these are called link atoms.

The ONIOM scheme can also be used for coupling an atomistic (molecular) or quantum model with a finite element model of a continuum. This procedure is schematically shown in Figure 4.6 for a MM/CM model; the model consists of three layers

1. The low order model (in this case the continuum model)
2. A fictitious continuum model that overlays the fragment to be modeled by atomistics
3. The atomistic (molecular) model at the top in Figure 4.6

In this type of application, the nodal spacing of the continuum model need not coincide with the atomic spacing (lattice constant). Instead, the atomistic displacements can be computed by using the finite element interpolation. This is of advantage since the finite element continuum model need not be refined to the atomistic lattice at the coupling interface. Of course, if the disparity between the lattice constant and element size is too large, substantial error can be engendered.

The formulation framework is identical to that for QM/MM coupling by ONIOM. The energy of the continuum/atomistic system is

$$E(\mathbf{d}) = E^C(\mathbf{d}) + E^M(\mathbf{d}) - E^{CM}(\mathbf{d}) - E^{\text{ext}}(\mathbf{d}) \qquad (4.15)$$

where the energies are associated with the models as shown in Figure 4.6 and $\mathbf{d}$ is a vector of nodal displacements.

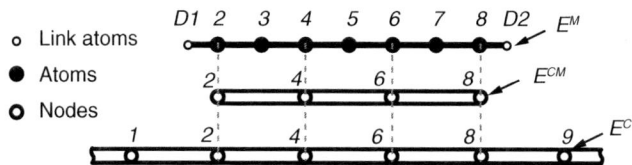

Figure 4.6: ONIOM scheme for continuum (C)- molecular (M) coupling.

The equations of equilibrium are then

$$0 = \frac{\partial E}{\partial \mathbf{d}_I} = \frac{\partial E^C}{\partial \mathbf{d}_I} + \frac{\partial E^M}{\partial \mathbf{d}_I} - \frac{\partial E^{CM}}{\partial \mathbf{d}_I} - \mathbf{f}_I^{\text{ext}}$$

We note that as before, the derivatives with respect to the nodal displacements are nodal forces, so

$$0 = \mathbf{f}_I^C + \mathbf{f}_I^M - \mathbf{f}_I^{CM} - \mathbf{f}_I^{\text{ext}} \qquad (4.16)$$

As in QM/MM coupling, the major difficulty occurs on the edge of the atomistic domain, since the behavior of the outermost bonds (the far left and far right bonds in Figure 4.5) can be polluted by incorrect placement of the link atoms D_1 and D_2.

4.3.3 Bridging domain method

In the following, we describe the bridging domain method for coupling continuum models with molecular models in mechanical equilibrium problems. The bridging domain method is in essence an overlapping domain decomposition scheme where compatibility in the overlapping (handshaking) domain is enforced by Lagrange Multipliers. The advantage of this method is that the atomic nuclei need not be coincident with the nodes of the continuum mesh. Furthermore, the method is generally used with a linear scaling of the energies in the handshaking domain, with the atomistic (continuum) energy dominant near the purely atomistic (continuum) domain. This enables the method to alleviate the errors that arise from dropping atomistic energies due to far field atoms and provides a gradual transition from the molecular model to the continuum model. An example of a concurrent bridging domain model of graphene is shown in Figure 4.7.

The total potential energy of the molecular mechanics subdomain Ω^M is given by (4.4) and the total potential energy of the continuum is given by (4.2).

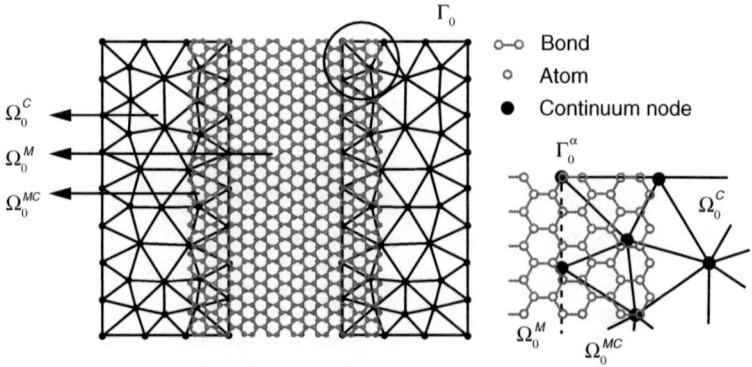

Figure 4.7: Schematic of bridging domain method; from [21] and [22].

In expressing the total internal potential energy of the system, we employ an energy scaling function $\alpha(\mathbf{X})$ in the overlapping subdomain. The energy scaling function is defined as

$$\alpha(\mathbf{X}) = \begin{cases} 0 & \text{in } \Omega_0^M \setminus \Omega^{MC} \\ l(\mathbf{X})/l_0(\mathbf{X}) & \text{in } \Omega_0^{MC} \\ 1 & \text{in } \Omega_0^C \setminus \Omega_0^{MC} \end{cases} \quad (4.17)$$

where $l(\mathbf{X})$ is the least square projection of $\mathbf{X}$ onto Γ_0^α, shown in Figure 4.8, $l_0(\mathbf{X})$ is defined by the distance between the overlapping domain boundaries Γ_0^α on the orthogonal projection of $\mathbf{X}$. The scaling parameter vanishes at the edge of the continuum and is unity at the other edge of Ω_0^{MC}; Ω_0^{MC} should include the last line of atoms.

The total potential energy is given by

$$E_\alpha^C(\mathbf{u}) + E_\alpha^M(\mathbf{d}) = \int_{\Omega_0^C} \alpha(\mathbf{X}) w_C(\mathbf{F}) d\Omega_0^C + \frac{1}{2} \sum_{I,J \in \mathcal{M}} (1 - \alpha_{IJ}) w_{IJ}^M \quad (4.18)$$

where $\alpha_{IJ} = \frac{1}{2}(\alpha_I + \alpha_J)$, $\alpha_I = \alpha(\mathbf{X}_I)$, and $\mathbf{d}$ is the vector of atomic displacements. The external potential is scaled similarly, so

$$E_\alpha^{\text{ext}} = \int_{\Omega_0^C} \alpha \rho_0 \mathbf{b} \cdot \mathbf{u} d\Omega_0^C + \int_{\Gamma_0^t} \alpha \rho_0 \bar{\mathbf{t}} \cdot \mathbf{u} d\Gamma_0^t + \sum_{I \in \mathcal{M}} (1 - \alpha_I) \bar{\mathbf{f}}_I^{\text{ext}} \cdot \mathbf{d}_I \quad (4.19)$$

where k is the body force. $\bar{\mathbf{t}}$ is the traction boundary condition, and $\bar{\mathbf{f}}_I^{\text{ext}}$ is any external force applied to atom I.

Let $\mathcal{M}^{MC} = \{I | \mathbf{X}_I \in \Omega_0^{MC}\}$ be the set of atoms initially in the handshaking region. In the initial version of the bridging domain method [22], displacement compatibility in the handshaking region Ω_0^{MC} was enforced by the constraint

$$g_I = ||\mathbf{u}(\mathbf{X}_I) - \mathbf{d}_I||^2 = \sum_{i=1}^{ND} [u_i(\mathbf{X}_I) - d_{iI}]^2 = 0, \forall I \in \mathcal{M}^{MC} \quad (4.20)$$

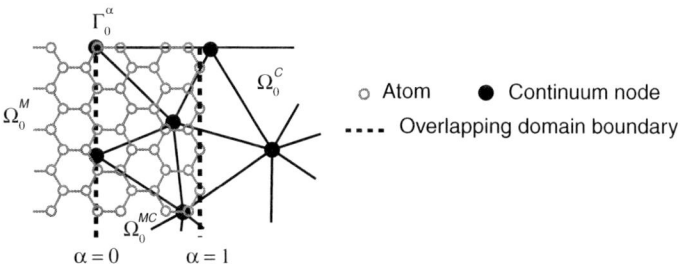

Figure 4.8: Overlapping subdomain of the atomistic and continuum models.

where ND denotes number of dimensions. Later versions of the bridging domain method [21, 30] have used the following stronger constraint

$$g_{iI} = u_i(\mathbf{X}_I) - d_{iI} = 0, \forall I \in \mathcal{M}^{MC} \,\&\, i = 1..ND \tag{4.21}$$

We note that by (4.20) a single constraint is applied per atom in $\mathcal{M}^{MC}$ whereas by (4.21) one constraint is applied for each component of the displacement of each atom. The use of constraint (4.20) rather than (4.21) results in a system of equations with fewer Lagrange multipliers but it is quite peculiar and not recommended. For this reason (4.21) will be used in the following derivation.

We first show how the constraint is applied by the Lagrange multiplier method; then we will add the modifications needed for the augmented Lagrangian method.

In the Lagrange multiplier method, the problem is posed as follows: find the stationary point of

$$W_L = E_\alpha^C + E_\alpha^M - E_\alpha^{\text{ext}} + \boldsymbol{\lambda}^T \mathbf{g} \tag{4.22}$$

where $\boldsymbol{\lambda} = \{\lambda_{iI}\}$ is the vector of Lagrange multipliers and $\mathbf{g} = \{g_{iI}\}$. The Lagrange multiplier λ_{iI} is the generalized force that enforces the constrain g_{iI} given by (4.21).

To develop the corresponding discrete equations, we use a finite element method in Ω_0^C and molecular mechanics in Ω_0^M. The displacements in Ω_0^C are given by (4.3). The Lagrange multiplier field is expressed in terms of shape functions denoted by $N_I^\lambda(\mathbf{X})$:

$$\boldsymbol{\lambda}(\mathbf{X}, t) = \sum_{I \in \mathcal{S}^\lambda} N_I^\lambda(\mathbf{X}) \bar{\boldsymbol{\lambda}}_I(t) \tag{4.23}$$

where $\mathcal{S}^\lambda$ is the set of all nodes of the Lagrange multiplier mesh. Generally, the shape functions for the Lagrange multiplier field will differ from those for the displacements, $N_I(\mathbf{X})$, and they must satisfy the LBB conditions. The Lagrange multiplier field is usually constructed by an auxiliary finite element triangulation of the intersection (overlapping) domain, as shown in Figure 4.9. To distinguish the Lagrange multiplier $\boldsymbol{\lambda}_I$ in (4.22) from the nodal values of the field, we use $\bar{\boldsymbol{\lambda}}_I$ to denote the Lagrange multipliers at the Lagrange multiplier nodes.

The discrete equilibrium equations are then obtained by substituting (4.2–4) and (4.23) into W_L (4.22) and setting the derivatives of W_L with respect to $\mathbf{u}_I$,

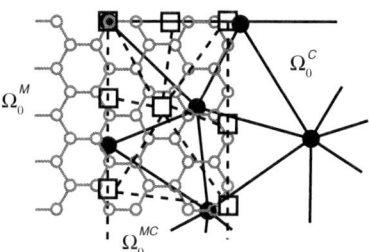

Figure 4.9: Lagrange multiplier interpolation in the overlapping region.

$\mathbf{d}_I$, and $\bar{\boldsymbol{\lambda}}_I$ to zero, which yields the following equations

$$\frac{\partial W_L}{\partial u_{iI}} = \left(F_{iI}^{\text{int}} - F_{iI}^{\text{ext}}\right) + \sum_{J \in \mathcal{M}^{MC}} \lambda_{iJ} N_{IJ} = 0, \forall I \in \mathcal{S} \quad (4.24)$$

$$\frac{\partial W_L}{\partial d_{iI}} = \left(f_{iI}^{\text{int}} - f_{iI}^{\text{ext}}\right) - \lambda_{iI} = 0, \forall I \in \mathcal{M} \quad (4.25)$$

$$\frac{\partial W_L}{\partial \bar{\lambda}_{iI}} = \sum_{J \in \mathcal{M}^{MC}} N_{IJ}^{\lambda} g_{iJ} = 0, \forall I \in \mathcal{S}^{\lambda} \quad (4.26)$$

where $N_{IJ} = N_I(\mathbf{X}_J)$, $N_{IJ}^{\lambda} = N_I^{\lambda}(\mathbf{X}_J)$, $\lambda_{iI} = \lambda_i(\mathbf{X}_I)$, and

$$F_{iI}^{\text{int}} = \int_{\Omega_0^C} \alpha \frac{\partial w_C(\mathbf{F})}{\partial u_{iI}} d\Omega_0^C \quad (4.27)$$

$$f_{iI}^{\text{int}} = \sum_{I,J \in \mathcal{M}} (1 - \alpha_{IJ}) \frac{\partial w_{IJ}^M}{\partial d_{iI}} \quad (4.28)$$

$$F_{iI}^{\text{ext}} = \frac{\partial E^{\text{ext}}}{\partial u_{iI}} \quad (4.29)$$

$$f_{iI}^{\text{ext}} = \frac{\partial E^{\text{ext}}}{\partial d_{iI}} \quad (4.30)$$

The augmented Lagrangian method can be developed by adding a penalty to (4.22) as follows:

$$W_{AL} = E_{\alpha}^C + E_{\alpha}^M - E_{\alpha}^{\text{ext}} + \boldsymbol{\lambda}^T \mathbf{g} + \frac{1}{2} p \mathbf{g}^T \mathbf{g} \quad (4.31)$$

where p is the penalty parameter; if $p = 0$, (4.31) is identical to (4.22).

The discrete equations for the augment Lagrangian method form of the bridging domain method are obtained by inserting (4.3) and (4.23) into (4.31) and setting the derivatives of W_{AL} with respect to $\mathbf{u}_I$, $\mathbf{d}_I$, and $\bar{\boldsymbol{\lambda}}_I$ to zero. This gives

$$\frac{\partial W_{AL}}{\partial u_{iI}} = \frac{\partial W_L}{\partial u_{iI}} + p \sum_{J \in \mathcal{M}^{MC}} g_{iJ} N_{IJ} = 0, \forall I \in \mathcal{S} \tag{4.32}$$

$$\frac{\partial W_{AL}}{\partial d_{iI}} = \frac{\partial W_L}{\partial d_{iI}} - p\, g_{iI} = 0, \forall I \in \mathcal{S} \tag{4.33}$$

$$\frac{\partial W_{AL}}{\partial \bar{\lambda}_{iI}} = \frac{\partial W_L}{\partial \bar{\lambda}_{iI}} = 0, \forall I \in \mathcal{S}^\lambda \tag{4.34}$$

We have found that often, the bridging domain method works best when the Lagrange multiplier fields are Dirac delta functions at the locations of the nuclei. This choice of Lagrange multipliers enforces compatibility exactly with the finite element approximation for the continuum. We can write this approximation for the Lagrange multipliers as

$$\boldsymbol{\lambda}(\mathbf{X}) = \sum_{I \in \mathcal{M}^{MC}} \hat{\boldsymbol{\lambda}}_I \delta(\mathbf{X} - \mathbf{X}_I) \tag{4.35}$$

where $\hat{\boldsymbol{\lambda}}_I$ are the unknown values of the Lagrange multipliers and $\delta(\bullet)$ is the Dirac delta function. For the strict Lagrange multiplier method, the discrete equations are

$$\frac{\partial W_{AL}}{\partial u_{iI}} = \left(F_{iI}^{\text{int}} - F_{iI}^{\text{ext}}\right) + \sum_{J \in \mathcal{M}^{MC}} \hat{\lambda}_{iJ} N_{IJ}$$

$$+ p \sum_{J \in \mathcal{M}^{MC}} g_{iJ} N_{IJ} = 0, \forall I \in \mathcal{S} \tag{4.36}$$

$$\frac{\partial W_{AL}}{\partial d_{iI}} = \left(f_{iI}^{\text{int}} - f_{iI}^{\text{ext}}\right) - \hat{\lambda}_{iI} - p\, g_{iI} = 0, \forall I \in \mathcal{M} \tag{4.37}$$

$$\frac{\partial W_{AL}}{\partial \hat{\lambda}_{iI}} = g_{iI} = 0, \forall I \in \mathcal{S}^\lambda \tag{4.38}$$

4.3.3.1 *Ghost forces*
In an ideal coupling scheme the forces acting on the atoms in the handshaking domain should be exactly those that would occur if the whole domain were modeled by atomistics. The difference between the ideal forces acting on the atoms and the actual multiscale model forces are termed *ghost forces*. In many multiscale methods the ghost forces occur, even for linear homogeneous deformations

and harmonic atomistic potentials—a study of the solutions to this problem for the various quasicontinuum methods can be found in Curtin and Miller [31].

In this subsection will we address the issue of ghost forces in the bridging domain method. We will show that for a linear elastic continuum and a molecular model with nearest neighbor interaction and a harmonic potential, ghost forces are eliminated when the pointwise Lagrange multiplier approximation (4.35), which we will also call strict coupling, is used; however, when the Lagrange multipliers are interpolated by finite element shape functions as in (4.23) ghost forces exist.

Consider the one-dimensional lattice depicted in Figure 4.10(a) with a harmonic potential with constant k. This is equivalent to a linear elastic constitutive model for the continuum with the elastic modulus $Y = ak$, where a is the lattice constant. For this simple molecular system a consistent multiscale method (i.e. one without ghost forces) should satisfy the patch test, i.e. be able to reproduce a constant strain field exactly.

Let $\beta_i = \beta(X_i) = 1 - \alpha(X_i)$ and $\beta_{ij} = (\beta_i + \beta_j)/2$. For the coupled model shown in Figure 4.10a, the weight function β has a value of unity at atom s and decreases linearly to zero at atom j. The coupling domain spans only one

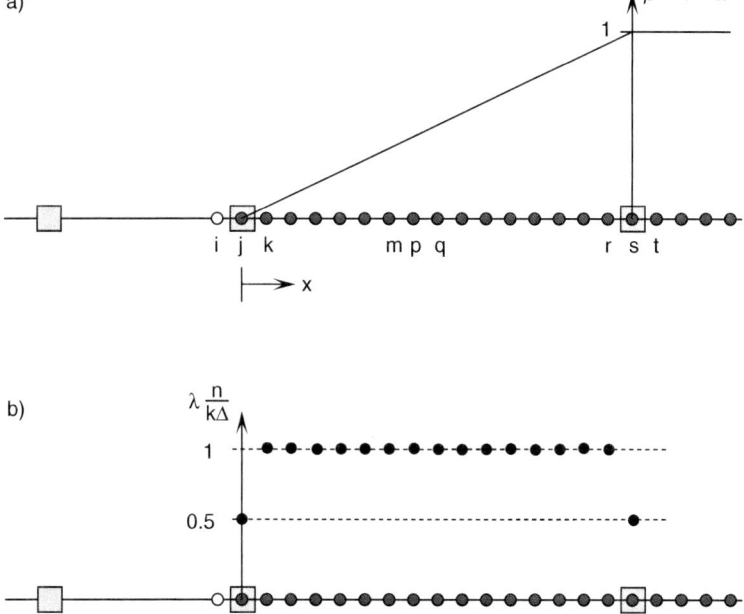

Figure 4.10: Illustration of the bridging domain in one-dimension. Dark circles represent atoms, white atoms represent the virtual atoms and grey squares represent FEM nodes.

element and so the energy scaling function is given by

$$\beta(\bar{x}) = \frac{\bar{x}}{h_e} = \frac{\bar{x}}{a(n-1)} \tag{4.39}$$

where h_e is the element length, a is the lattice spacing, n is the number of atoms in the handshaking domain. Note that both atoms j and s are assumed to be in the coupling domain.

Consider the uniform straining of the lattice in Figure 4.10(a), such that each atomistic bond is elongated by Δ. For simplicity we will only study the Lagrange multiplier method with $p = 0$. The equilibrium equations, given by (4.26) or (4.38), for atoms j, p, and s are, respectively, given by

$$r_j = \mathsf{k}\Delta\left(\beta_{ij} - \beta_{jk}\right) + \lambda\left(\mathbf{X}_j\right) = -\mathsf{k}\Delta\beta_{jk} + \lambda\left(\mathbf{X}_j\right) \tag{4.40}$$

$$r_p = \mathsf{k}\Delta\left(\beta_{mp} - \beta_{pq}\right) + \lambda\left(\mathbf{X}_p\right) \tag{4.41}$$

$$r_s = \mathsf{k}\Delta\left(\beta_{rs} - \beta_{st}\right) + \lambda\left(\mathbf{X}_s\right) = \mathsf{k}\Delta\left(\beta_{rs} - 1\right) + \lambda\left(\mathbf{X}_s\right) \tag{4.42}$$

where r_i is the residual force on atom i and is equal to the ghost force of atom i. For the bridging domain method to satisfy the patch test, the residual at all atoms must be zero, which gives

$$\lambda\left(\mathbf{X}_j\right) = \frac{1}{2}\frac{\mathsf{k}\Delta}{(n-1)} \qquad \lambda\left(\mathbf{X}_p\right) = \frac{\mathsf{k}\Delta}{(n-1)} \tag{4.43}$$

$$\lambda\left(\mathbf{X}_s\right) = \frac{1}{2}\frac{\mathsf{k}\Delta}{(n-1)} \tag{4.44}$$

The discrete Lagrange multipliers required to satisfy the patch test are shown in Figure 4.10(b). We can see from this figure and equations (4.43–4.44) that in general a linear Lagrange multiplier field cannot provide the Lagrange multipliers needed to satisfy the patch test. Therefore, ghost forces will always exist in the bridging domain method when the Lagrange multipliers are interpolated by shape functions as in (4.23); however, when strict Lagrange multipliers (4.35) are used no ghost forces occur.

The conclusions of the previous paragraph are verified numerically in the following example. Consider a one-dimensional domain of length 128, with the origin of the coordinate system located at the left end. The atomistic subdomain is $0 \le x \le 80$ and the continuum subdomain is $64 \le x \le 128$, so the handshaking domain is $64 \le x \le 80$. The lattice spacing is taken to be $a = 1$, and a uniform finite element mesh is used with the element size selected so that the handshaking domain is spanned exactly by a single element, so $h_e = 16a$. The atom at $x = 0$ is fixed and a displacement of magnitude 1 is applied to the node at $x = 128$.

In Figure 4.11 the value of the Lagrange multipliers at each atom in the handshaking domain are shown for the bridging-domain method with weak coupling (4.23) and strict coupling (4.35). The values are normalized by $\frac{\mathsf{k}\Delta}{(n-1)}$ with

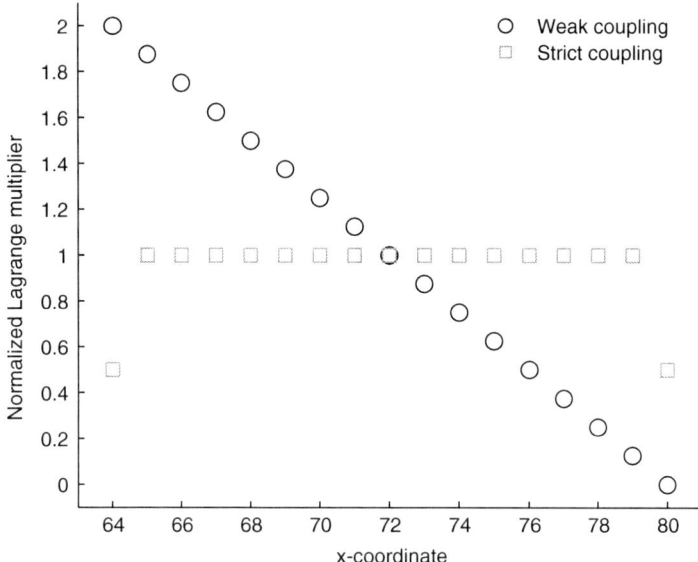

Figure 4.11: Lagrange multipliers solution from the bridging–domain method for a patch test on a one-dimensional lattice.

$n = 16$. The normalized L^2 error norm in the displacements given by

$$\text{Normalized } L^2 \text{ error norm} = \frac{1}{n^A} \sqrt{\sum_{I=1}^{n^A} \frac{(d_I - u^{\text{exact}}(X_I))^2}{(u^{\text{exact}}(X_I))^2}} \qquad (4.45)$$

Where n^A is the number of atoms. For strict coupling the error is zero to within machine precision and therefore the strict coupling method passes the patch test. We can observe that the Lagrange multipliers required to satisfy the patch test, as shown in Figure 4.11, are those given by equations (4.43–4.44). The normalized L^2 error norm is 1.9×10^{-4} for the bridging domain method with weak coupling (4.23) and so it does not pass the patch test. This is because a linear field cannot exactly interpolate the required Lagrange multipliers given by (4.43–4.44). We note that although the bridging domain method with weak coupling does not exactly satisfy the patch test, the error in the solution is small.

4.3.3.2 Stability of Lagrange multiplier methods

The development of molecular/continuum methods parallels recent work in the finite element community on the *gluing* of disjoint meshes in an overlapping subdomain. Ben Dhia and Rateau [26, 27] were the first to study these methods and developed a new method which is discussed below. They called these methods Arlequin methods. They showed that for a constant scaling in the overlapping domain, such as in the original handshake methods of Abraham et al. [12], instabilities appear in the Lagrange multiplier problem. In [27], a comparison of two coupling methods was made. In an atomistic/continuum coupling, the energy of the coupled model is given by

$$E = E^A(\alpha) + E^C(\alpha) + (\boldsymbol{\lambda}, \delta \mathbf{u}) \qquad (4.46)$$

where E is the total energy, $E^A(\alpha)$ is the scaled atomistic energy, $E^C(\alpha)$ is the scaled continuum energy, and $(\lambda, \delta u)$ is a scalar product of the Lagrange multiplier field λ and the difference in displacements, as in the bridging domain method [22, 21]. The two types of coupling studied by Ben Dhia were called L^2 and H^1 couplings in [27] and are given, respectively, by

$$(\boldsymbol{\lambda}, \delta \mathbf{u}) = (\boldsymbol{\lambda}, \delta \mathbf{u})_{\mathcal{L}^2(\Omega_B)} = \int_{\Omega_B} \boldsymbol{\lambda} \cdot \delta \mathbf{u} d\Omega \equiv \int_{\Omega_B} \lambda_i \delta u_i d\Omega \qquad (4.47)$$

$$(\boldsymbol{\lambda}, \delta \mathbf{u}) = (\boldsymbol{\lambda} \delta \mathbf{u})_{\mathcal{H}^1(\Omega_B)} = \int_{\Omega_B} \left(\boldsymbol{\lambda} \cdot \delta \mathbf{u} + l^2 \epsilon(\boldsymbol{\lambda}) : \epsilon(\delta \mathbf{u}) \right) d\Omega$$

$$\equiv \int_{\Omega_B} \left(\lambda_i \delta u_i + l^2 \lambda_{i,j} \delta u_{i,j} \right) d\Omega \qquad (4.48)$$

It was shown in [27] that for the linear scaling of the energy, the L^2 coupling used in the bridging domain method is also stable. These conclusions also appear to apply to atomistic/continuum coupling.

4.3.4 *Quasicontinuum method*

The quasicontinuum method, as presented in Tadmor *et al.* [8], is a landmark work in atomistic/continuum coupling. It was one of the first works to introduce the concept of using the Cauchy-Born rule in the continuum part of the model, so that the two models are compatible. Many of the difficulties in atomistic/continuum coupling, such as internal modes, are also dealt with in the subsequent papers.

From a coupling point of view, the quasicontinuum method [8] is basically a direct atomistic-continuum coupling scheme where selected atoms are coincident with finite element nodes. Therefore, as in any direct coupling method, the mesh size must be decreased to the atomistic lattice spacing wherever atomistic potentials are to be used, which complicates mesh generation. For example, for a problem with several interacting dislocations in three dimensions, meshing could be quite difficult.

The pioneering contribution of the method is the seamless way it transitions from an atomistic description to a continuum description based on the Cauchy-Born rule. This concept has been widely adopted in other concurrent atomistic/continuum methods such as the bridging domain method.

It is also of interest that the quasicontinuum method avoids ill-posedness of the continuum model, i.e. in areas of bond breaking, for in these areas the method transitions to atomistic potentials, which are called nonlocal models. This process introduces a length scale wherever there is bond breaking. Thus, although an unstable process is modeled, the system remains well posed.

4.4 Molecular dynamics systems

Most of the methods for coupling atomistic to continua are also applicable to dynamics. The major difficulty that arises in extending these methods to dynamics is that when the cutoff frequency of the continuum model is below that of the molecular dynamics model, spurious reflections occur at the interface. This relationship of cutoff frequencies is almost always the case, since there is no benefit in using continuum model unless the internodal spacing is greater than the lattice constant. The cutoff frequency of a continuum model with low order elements is given by

$$\omega_{max}^C = \frac{c}{h_{min}} \tag{4.49}$$

where c is the wavespeed and h_{min} is the smallest distance between nodes. It can be shown with the Cauchy-Born rule that the atomistic cutoff frequency is of a crystal lattice

$$\omega_{max}^A = \frac{c}{l} \tag{4.50}$$

where l is the lattice constant. Phonons in the atomistic domain with frequencies above ω_{max}^C will not be able to penetrate into the continuum model. This results in spurious reflections of high frequency phonons, which need to be mitigated. The phonons above the atomic cutoff frequency represent thermodynamic energy that is not represented in the continuum model. It can be treated by linking the energy equation to a thermostat. One such approach is discussed in [21].

We next consider the concurrent coupling of molecular dynamics with continuum mechanics by the bridging domain method. For simplicity, we consider only pair potentials. The Hamiltonians of the atomistic and continuum subdomains are

$$H_M = \frac{1}{2} \sum_{I,J \in \mathcal{M}} w_{IJ}^M + \frac{1}{2} \sum_{I \in \mathcal{M}} M_I^M \dot{d}_{iI} \dot{d}_{iI} \tag{4.51}$$

$$H_C = \int_{\Omega_0^C} w_C(\mathbf{F}) d\Omega_0^C + \frac{1}{2} \sum_{I \in \mathcal{S}} M_I^C \dot{u}_{iI} \dot{u}_{iI} \tag{4.52}$$

where M^M denotes the mass of atoms and M^C the mass of the nodes in the lumped mass matrix. Indicial notation convention applies to lower-case indices, so repeated indices indicate summation.

To derive the equations of motion, it is convenient to transform Hamiltonian to the corresponding Lagrangian L. The same energy scaling function as for equilibrium systems (4.17) is used in the bridging domain. Lagrange multipliers are used to enforce the constraints between the atoms and the continuum mesh. In this case, the pointwise Lagrange multiplier field is used, so

$$L = \frac{1}{2} \sum_{I \in \mathcal{M}} (1 - \alpha_I) M_I^M \dot{d}_{iI} \dot{d}_{iI} - \sum_{I \in \mathcal{M}} \sum_{J \in \mathcal{M}, J > I} w_{IJ}^M (1 - \alpha_{IJ})$$

$$+ \frac{1}{2} \sum_{I \in \mathcal{S}} \alpha_I M_I^C \dot{u}_{iI} \dot{u}_{iI} - \int_{\Omega_0^C} \alpha w_C d\Omega_0^C + G \quad (4.53)$$

where

$$G = \sum_{J \in \mathcal{M}} \lambda_{iJ} \left(\sum_{I \in \mathcal{S}} N_{IJ} u_{iI} - d_{iJ} \right) \quad (4.54)$$

After this transformation, the equations of motion can be obtained by using Lagrange's equation

$$\frac{d}{dt} \left(\frac{\partial L}{\partial \dot{q}_i} \right) - \frac{\partial L}{\partial q_i} = 0 \quad (4.55)$$

where $\mathbf{q} = [\{d_{iI}\}, \{u_{iI}\}, \{\lambda_{iI}\}]$. This yields

$$\bar{M}_I^M \ddot{d}_{iI} + f_{iI}^{\text{int}} + \frac{\partial G}{\partial d_{iI}} = 0 \quad i = 1 \ldots n_{SD} \quad (4.56)$$

$$\bar{M}_I^C \ddot{u}_{iI} + F_{iI}^{\text{int}} + \frac{\partial G}{\partial u_{iI}} = 0 \quad i = 1 \ldots n_{SD} \quad (4.57)$$

$$\sum_{J \in \mathcal{S}} N_{JI} u_{iJ} - d_{iI} = 0 \quad i = 1 \ldots n_{SD} \quad (4.58)$$

where $\bar{M}_I^C$ is a diagonalized mass matrix.

$$\bar{M}_I^M = (1 - \alpha_I) M_I^M \quad (4.59)$$

$$\bar{M}_I^C = \sum_{J \in \mathcal{S}} \int_{\Omega_0^C} \alpha \rho_0 N_I N_J d\Omega_0^C \quad (4.60)$$

The equations are integrated by a predictor-corrector central difference method (which is equivalent to the Verlet method) for time integration. The acceleration at time step n can be expressed as

$$\ddot{d}^n = \frac{d^{n+1} - 2d^n + d^{n-1}}{\Delta t^2} \quad (4.61)$$

$$\ddot{u}^n = \frac{u^{n+1} - 2u^n + u^{n-1}}{\Delta t^2} \quad (4.62)$$

Substituting (4.61) and (4.62) into (4.56) and (4.57), respectively, we get the displacements at step $n + 1$

$$d_{iI}^{n+1} = -\frac{f_{iI}^{int}}{\bar{M}_I^M} + 2d_{iI}^n - d_{iI}^{n-1} + \frac{\Delta t^2 \sum_{J \in \mathcal{M}} \lambda_{iJ}^n \delta_{IJ}}{\bar{M}_I^M} \qquad (4.63)$$

$$u_{iI}^{n+1} = -\frac{F_{iI}^{int}}{\bar{M}_I^M} + 2u_{iI}^n - u_{iI}^{n-1} - \frac{\Delta t^2 \sum_{J \in \mathcal{M}} \lambda_{iJ}^n N_{IJ}}{\bar{M}_I^C} \qquad (4.64)$$

The equations for the Lagrange multipliers can be obtained by substituting the above displacements into the constraint (4.58)

$$\sum_{L \in \mathcal{M}} \lambda_{iL}^n A_{IL} = -\frac{f_{iI}^{int}}{\bar{M}_I^M} + \frac{2d_{iI}^n - d_{iI}^{n-1}}{\Delta t^2} + \sum_{J \in \mathcal{S}} N_{JI} \left(\frac{F_{iJ}^{int}}{\bar{M}_J^C} - \frac{2u_{iJ}^n - u_{iJ}^{n-1}}{\Delta t^2} \right) \qquad (4.65)$$

where

$$A_{IL} = -\sum_{J \in \mathcal{S}} \frac{N_{JI} N_{JL}}{\bar{M}_J^C} - \frac{\delta_{IL}}{\bar{M}_I^M} \qquad (4.66)$$

is called the consistent constraint matrix. These Lagrange multipliers are then used in (4.56) and (4.57) for a corrector calculation. A diagonalized constraint matrix **B** is used in the implementation

$$B_{IJ} = \begin{cases} 0 & I \neq J \\ \sum_{L \in \mathcal{M}} A_{IL} & I = J \end{cases} \qquad (4.67)$$

because that is essential for suppressing spurious reflections. This is discussed further in Section 4.5.

4.4.1 *Conservation properties of the bridging domain method*

In the following we show that the conservation of both linear momentum and the energy of the system in the bridging domain method. The conservation of angular momentum is shown in [33]. The work of Hairer et al. [34] shows that for the central difference method (symplectic Verlet), the energy is conserved in time if the ordinary differential equations in time are conservative. Therefore we only examine the time derivatives of the conservation variables.

The analysis shows that the bridging domain method exactly conserves linear momentum and energy. This presents the dilemma: how can the method suppress spurious reflections? As will be shown with numerical results, the answer lies in the diagonalization, which renders the method to be dissipative and remarkably effective in suppressing spurious reflections.

4.4.1.1 *Conservation of linear momentum*
The system linear momentum p_i is

$$p_i = p_i^M + p_i^C = \sum_{I \in \mathcal{M}} \bar{M}_I^M \dot{d}_{iI} + \sum_{J \in \mathcal{S}} \bar{M}_J^C \dot{u}_{iJ} \qquad (4.68)$$

where $\bar{M}_I^M$ and $\bar{M}_I^C$ are defined in (4.59) and (4.60), respectively. The rate of change of the total linear momentum of the discrete system is given by

$$\frac{dp_i}{dt} = \frac{d}{dt}\left(\sum_{I \in \mathcal{M}} \bar{M}_I^M \dot{d}_{iI} + \sum_{J \in \mathcal{S}} \bar{M}_J^C \dot{u}_{iJ}\right) \qquad (4.69)$$

Substituting (4.56–4.60) into (4.69) gives

$$\frac{dp_i}{dt} = \sum_{I \in \mathcal{M}} \lambda_{iI} - \sum_{I,J \in \mathcal{M}} \frac{\partial w_{IJ}^M}{\partial d_{iI}} \alpha_{IJ} - \sum_{I \in \mathcal{M}} \lambda_{iI} \sum_{J \in \mathcal{S}} N_{JI}$$

$$- \sum_{I \in \mathcal{S}} \int_{\Omega_0^C} (1-\alpha) \frac{\partial N_I}{\partial X_j} P_{ji} d\Omega_0^C \qquad (4.70)$$

By the partition of unity property of finite element interpolants, $\sum_{J \in \mathcal{S}} N_J(\mathbf{X}) = 1$ for $\forall \mathbf{X}$; thus, the first and the third terms cancel. The fourth term vanishes, which can easily be seen if the summation is pulled into the integral and the partition-of-unity property of the shape functions is invoked. This partition-of-unity property has also been used to show the conservation properties of meshfree methods in Belytschko et al. [35]. The second term vanishes because of its antisymmetry on I and J; for example, w_{IJ}^M is a function of $||\mathbf{x}_I - \mathbf{x}_J||$ for a two body potential, so

$$\sum_{I,J \in \mathcal{M}} \frac{\partial w_{IJ}^M}{\partial d_{iI}} = \frac{1}{2} \sum_{I,J \in \mathcal{M}} \left(\frac{\partial w_{IJ}^M}{\partial d_{iI}} + \frac{\partial w_{IJ}^M}{\partial d_{iJ}}\right) = 0 \qquad (4.71)$$

It can be shown that this term also vanishes for other potentials, such as the angle bending potential. So $dp_i/dt = 0$, which shows that linear momentum is conserved. The conservation property is independent of the form of the scaling function in the coupling domain.

4.4.1.2 *Conservation of energy*
The time derivative of the total energy reads

$$\dot{E}^{tot} = \dot{T}^M + \dot{W}^M + \dot{T}^C + \dot{W}^C \qquad (4.72)$$

where

$$\dot{T}^M = \sum_{I \in \mathcal{M}} \bar{M}_I^M \ddot{d}_{iI} \dot{d}_{iI} \tag{4.73}$$

$$\dot{W}^M = \sum_{I \in \mathcal{M}} \sum_{J \in \mathcal{M} > I} \left(\frac{\partial w_{IJ}^M}{\partial d_{iI}} \dot{d}_{iI} + \frac{\partial w_{IJ}^M}{\partial d_{iJ}} \dot{d}_{iJ} \right) (1 - \alpha_{IJ}) \tag{4.74}$$

$$\dot{T}^C = \sum_{I \in \mathcal{S}} \bar{M}_I^C \ddot{u}_{iI} \dot{u}_{iI} \tag{4.75}$$

$$\dot{W}^C = \int_{\Omega_0^C} \alpha \frac{\partial w_C}{\partial \mathbf{F}} : \dot{\mathbf{F}} d\Omega_0^C = \int_{\Omega_0^C} \alpha \mathbf{P} : \dot{\mathbf{F}} d\Omega_0^C \tag{4.76}$$

Substituting (4.56–4.57) into (4.73–4.76) gives

$$\dot{E}^{tot} = -\sum_{I \in \mathcal{M}} (f_{iI}^{int} + \frac{\partial G}{\partial d_{iI}}) \dot{d}_{iI} - \sum_{I \in \mathcal{S}} (F_{iI}^{int} + \frac{\partial G}{\partial u_{iI}}) \dot{u}_{iI}$$

$$- \sum_{I \in \mathcal{M}} \sum_{J \in \mathcal{M} > I} \left(\frac{\partial w_{IJ}^M}{\partial d_{iI}} \dot{d}_{iI} + \frac{\partial w_{IJ}^M}{\partial d_{iJ}} \dot{d}_{iJ} \right) (1 - \alpha_{IJ}) + \int_{\Omega_0^C} \alpha \mathbf{P} : \frac{\partial N_I}{\partial \mathbf{X}} d\Omega_0^C \dot{u}_{iI}$$

$$= \sum_{I \in \mathcal{M}} \frac{\partial G}{\partial d_{iI}} \dot{d}_{iI} - \sum_{I \in \mathcal{S}} \frac{\partial G}{\partial u_{iI}} \dot{u}_{iI} = -\dot{G} \tag{4.77}$$

Since G is the constraint term which vanishes by the Karush-Kuhn-Tucker condition, the energy is conserved.

4.4.2 Master-slave and handshake methods

The master-slave method can be applied to MM/CM coupling by either edge-to-edge to overlaid domain decomposition. In either case, the position of any atom which is common to the continuum and atomistic domains is given by

$$\mathbf{d}_J = \sum N_I(\mathbf{X}_J) \mathbf{u}_I \quad \text{or} \quad \mathbf{d} = \mathbf{N}\mathbf{u}^C \tag{4.78}$$

The nodal forces and mass of the atoms are distributed to the nodes of the continuum elements or edges that are overlaid by the atoms. Using standard master-slave coupling formulae based on energy (see [5]), these additional forces and masses are

$$\mathbf{f}_I = \sum_J N_I(\mathbf{X}_J) \mathbf{f}_J^M \quad \text{or} \quad \mathbf{f} = \mathbf{N}^T \mathbf{f}^M \tag{4.79}$$

$$\mathbf{M}_{IJ} = \sum_{I,J} N_I(\mathbf{X}_K) M_{KK}^M N_K(\mathbf{X}_I) \quad \text{or} \quad \mathbf{M} = \mathbf{N}^T \mathbf{M}^M \mathbf{N} \tag{4.80}$$

The resulting equations can then be solved by explicit or implicit methods.

In the handshake method, the nodes of the two coupled systems are coincident. Therefore the same Lagrangian as in (4.53) can be used with the constraint term G omitted. Generally, a constant energy scaling function α is used in the overlapping domain, i.e. handshake domain. The overlapping nodes are often coupled with dampers which suppress some of the spurious reflections at the interface.

4.4.3 Bridging scale method

In the bridging-scale method as originated in Wagner and Liu [14], the atomic displacements are decomposed into coarse-scale and fine-scale displacements. The decomposition of the matrix of atomistic displacements is

$$u_I^A = N_{IJ} d_J^C + \tilde{u}_I^A \quad \text{or} \quad \mathbf{u}^A = \underbrace{\mathbf{N} \mathbf{d}^C}_{\bar{\mathbf{u}}} + \tilde{\mathbf{u}}^A \tag{4.81}$$

where $\bar{\mathbf{u}}$ is the coarse-scale field, which, as indicated, is usually a finite element continuum field; $\mathbf{d}^c$ are the nodal displacements of a coarse-scale mesh. The notation is the opposite of that in previous sections to conform to [14], so we have added superscripts A and C for atoms and continua, respectively, for clarity.

The key concept in their work is that the coarse-scale displacements are obtained by a least square projection weighted by the atomic masses. This projection is

$$\mathbf{d}^C = \mathbf{M}^{-1} \mathbf{N}^T \mathbf{M}^A \mathbf{u}^A \tag{4.82}$$

where $\mathbf{M}$ is the matrix of the coarse-scale masses, given by

$$\mathbf{M} = \mathbf{N}^T \mathbf{M}^A \mathbf{N} \tag{4.83}$$

which is the mass that would be obtained by a master-slave relationship, see (4.80). It can be seen from (4.82) that the nodal displacements of the finite element mesh can then be obtained at any time from the atomistic displacements.

Combining (4.81) and (4.82) gives

$$\mathbf{u}^A = \mathbf{N} \mathbf{M}^{-1} \mathbf{N}^T \mathbf{M}^A \mathbf{u} + \tilde{\mathbf{u}}^A = \mathbf{P} \mathbf{u}^A + \tilde{\mathbf{u}}^A \tag{4.84}$$

$$\mathbf{P} = \mathbf{N} \mathbf{M}^{-1} \mathbf{N}^T \mathbf{M}^A \tag{4.85}$$

It follows from the above that

$$\tilde{\mathbf{u}}^A = (\mathbf{I} - \mathbf{P}) \mathbf{u}^A \tag{4.86}$$

In the implementation of Wagner and Liu [14] and the subsequent improvements [36, 37], this projection is apparently only used at the interface. By using this concept, they are able to obtain an equation for the fine-scale (i.e. relative acceleration) at the interface without any Lagrange multipliers. They use the approach of Adelman and Doll [38].

The approach has been improved and then streamlined by updating the interface atoms by Green's functions for lattices. Green's functions have been developed for various lattices [36, 37]. The method also entails the computation of Fourier transforms at the interface.

To and Li [19] and Li et al. [39] have enhanced the method by joining the coarse-scale to the fine-scale on an overlapping domain by the perfectly matched layer (PML) method. The perfect matched layer method is very efficient in suppressing spurious reflections. Comparisons of the Li and To PML method with the bridging domain method are given in Section 4.5.

4.5 Numerical results

This section describes some models and solutions obtained by concurrent multiscale methods. The first set of examples is for coupling of molecular dynamics with continuum dynamics by the bridging domain method. The second set are coupled QM/MM, QM/CM, and QM/MM/CM equilibrium solutions of graphene sheets with a defect.

4.5.1 *Molecular/continuum dynamics with bridging domain method*

To examine the performance of the bridging domain method, we investigate the reflection and the transmission properties at the molecular-continuum interface. We will also check how well the conservation properties we have proven hold when the equations are integrated numerically. For these purposes, we consider several one- and two-dimensional problems. In all problems, we evaluate the conservation properties by measuring the change of the conserved variable normalized by its initial absolute value, except in the cases where the initial value is zero. For example, for energy, the error e_{en} is given by

$$e_{en} = \frac{E(t) - E_0}{E_0} \quad (4.87)$$

where E_0 is the initial energy of the system. In these examples, all of the energy and momentum is imparted to the system by initial conditions in the velocities and displacements. There are no external forces applied. The central difference method (symplectic Verlet) is used for time integration in all examples.

4.5.1.1 *One-dimensional studies*

In the first set of examples, we consider a one-dimensional model as shown in Figure 4.12. The domain consists of 400 equally spaced atoms with a lattice constant $r_0 = 1.23\,\text{Å}$, and 400 elements of length $h = n_1 r_0$, so n_1 is the number of bonds in an element. We denote the number of elements in the coupling subdomain by n_2; n_1 and n_2 are varied in the examples to determine the effect of the element size and overlaid domain size on the performance of the bridging domain method. No dummy atoms are introduced at the interface between the atomistic and continuum model, because the scaled masses of these atoms vanish according to (4.59). The motions of these atoms are determined by the compatibility constraints.

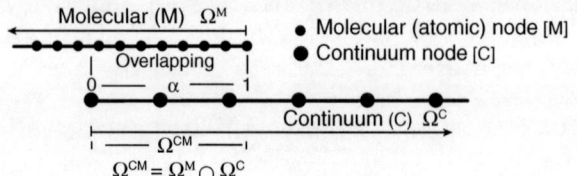

Figure 4.12: Schematic of one dimensional coupling model.

The Lennard-Jones potential is chosen to represent the atomic interaction with the nearest neighbour assumption:

$$w_{IJ}^M = 4\beta_1 \left[\left(\frac{\beta_2}{r_{IJ}}\right)^{12} - \left(\frac{\beta_2}{r_{IJ}}\right)^6 \right] \quad (4.88)$$

where $r_{IJ} = \| \mathbf{x}_J - \mathbf{x}_I \|$ is the bond length between the nearest neighboring atoms, i.e. $|I - J| = 1$. β_1 and β_2 are parameters. In these examples, $\beta_1 = 0.2$eV and $\beta_2 = 1.1$Å. This corresponds to a Young's modulus $E_Y = 11.66$eV/Å^3. The atomic mass $M^M = 1.036 \times 10^{-4}$eV·ps^2/Å^2. The initial displacements and velocities for the atoms with $X \leq 100$ Å are obtained from linear combinations of the following solutions of the wave equation by letting $t = 0$.

$$u(X,t) = (a_1 + a_1 \cos(2\pi a_2 (X - ct) + \pi))(1 + a_3 \cos(2\pi X)) \quad (4.89)$$

$$\dot{u}(X,t) = ca_1 \sin(2\pi a_2 (X - ct) + \pi)(1 + a_3 \cos(2\pi X)) \quad (4.90)$$

$$u(X,0) = \dot{u}(X,0) = 0 \quad \text{for} \quad X > 100\text{Å} \quad (4.91)$$

where $c = (E/\rho)^{1/2} = 372.82$Å/ps is the wavespeed, $a_1 = 6.17 \times 10^{-4}$, $a_2 = 0.01$, $a_3 = 0.1$. The initial conditions include waves at four different frequencies: $\omega_1 = 2\pi c$, $\omega_2 = 2\pi a_2 c$, $\omega_3 = \omega_1 + \omega_2$, $\omega_4 = \omega_1 - \omega_2$.

Figure 4.13 shows the linear momentum as a function of time; we do not use the error measure (4.87) because the initial momentum is zero. As can be seen, the momentum is exactly conserved until the wave reaches the overlapping subdomain. At that point, there is a moderate fluctuation in the total momentum, but it is restored perfectly after the wave passes. We have no explanation for this temporary violation of momentum conservation.

The effectiveness of the bridging domain method can be seen from Figure 4.14 which shows the reflection and transmissivity factors for the interface for various combinations of overlapping elements and atoms per element. In these figures, the cutoff frequencies of the finite element and atomistic models are indicated by vertical lines. The cutoff frequency of the finite element model is given by (4.49). In the atomistic model, the cutoff frequency is given by $\omega_{cut}^M = (k/m)^{1/2}$, where k is the tangent of the force-elongation curve of the bond; for a Lennard-Jones potential, $k = 36 \times 2^{2/3} \beta_1/\beta_2^2$. An ideal coupling scheme would suppress all reflections

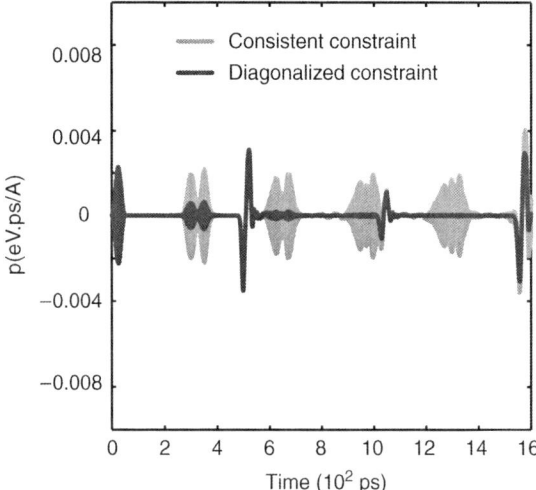

Figure 4.13: History of linear momentum of one dimensional coupled model with $n_1 = n_2 = 3$ for consistent and diagonalized constraints methods.

of phonons between the cutoff frequencies of the atomistic model and the continuum model (frequencies above the atomistic frequency are due to aliasing and can be ignored). It is noted that the coupling should transmit all phonons with frequencies below the cutoff frequency of the continuum, so ideally the transmission factor should be unity below the continuum cutoff frequency. Both the diagonalized and consistent constraints show nearly perfect transmission of these phonons. However, the consistent form exhibits substantial reflection of phonons with frequencies between the cutoff frequencies of the atomistic and continuum models. These spurious reflections are far less severe in the diagonalized form.

Figure 4.14f shows the reflectivity and transmissivity of the interface for the To et al. [19] method based on PML. It can be seen that its performance is comparable to that of the bridging domain method with diagonalized constraints.

Table 4.2 shows the energy dissipation of the high frequency ($\omega > \omega_{\text{cut}}^C$) and the low frequency ($\omega < \omega_{\text{cut}}^C$) incident waves for the diagonalized constraint matrix in one dimension. It is observed that more than 90% of the high frequency energy emanating from the atomistic domain is eliminated, but the low frequency dissipation is less than 1%. The diagonalized constraint matrix damps out the high frequency of the incident wave and effectively eliminates the spurious reflection at the boundary of different scales.

When the diagonalized constraint matrix is used instead of the consistent constraint matrix, compatibility inside the coupled subdomain is not satisfied exactly. Therefore $\dot{G}$ no longer vanishes, and energy is not conserved. The computations show that the time derivative of G is always positive, which implies energy is dissipated when the wave goes through the coupling domain (Figure 4.15)

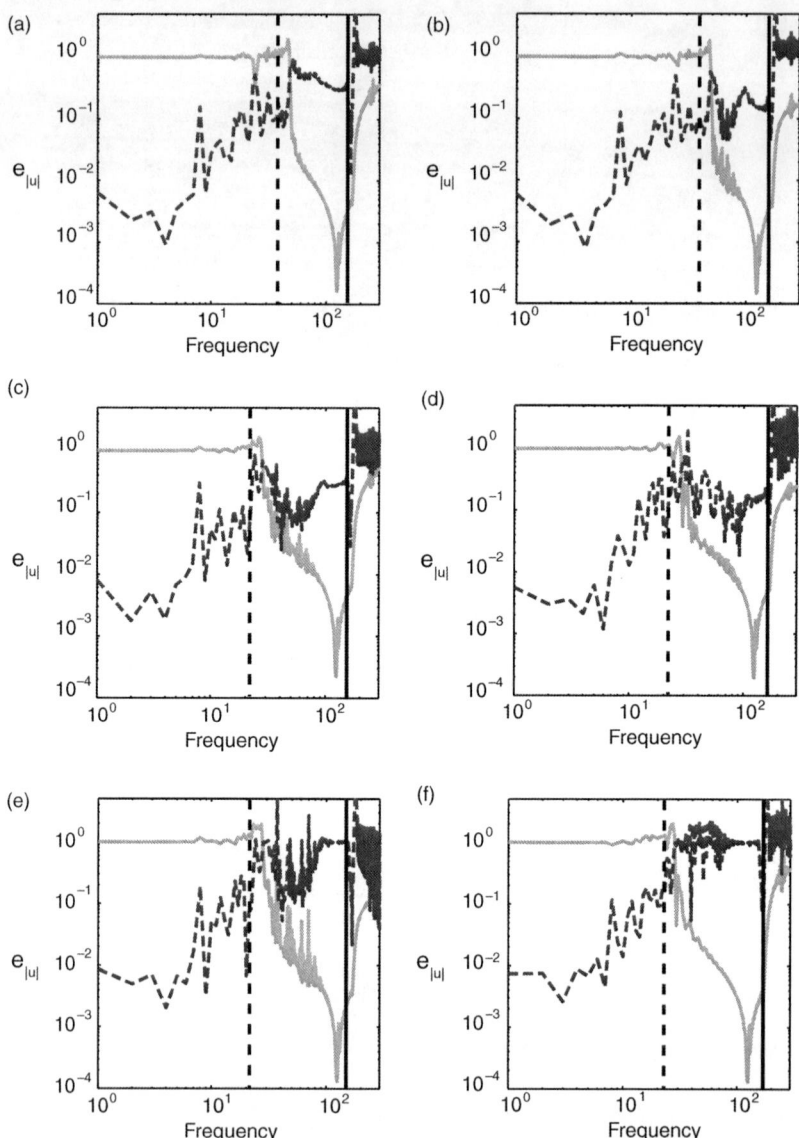

Figure 4.14: Reflectivity (dashed grey line) and transmissivity (grey line) for n_2 overlaid elements with n_1 atoms per element. (a) $n_1 = 4$, $n_2 = 3$, diagonalized constraint; (b) $n_1 = 4$, $n_2 = 6$, diagonalized constraint; (c) $n_1 = 7$, $n_2 = 3$, diagonalized constraint; (d) $n_1 = 7$, $n_2 = 6$, diagonalized constraint; (e) $n_1 = 7$, $n_2 = 3$, consistent constraint; (f) To *et al.* [20]. (The dashed and solid vertical lines are the cutoff frequencies of the continuum and the atomic models, respectively.)

Table 4.2. Energy dissipation of low and high frequency waves for n_1 overlaid elements with $n_2 = 3$

	$n_1 = 7$	$n_1 = 4$
Low Frequency	0.9%	0.2%
High Frequency	92%	91%

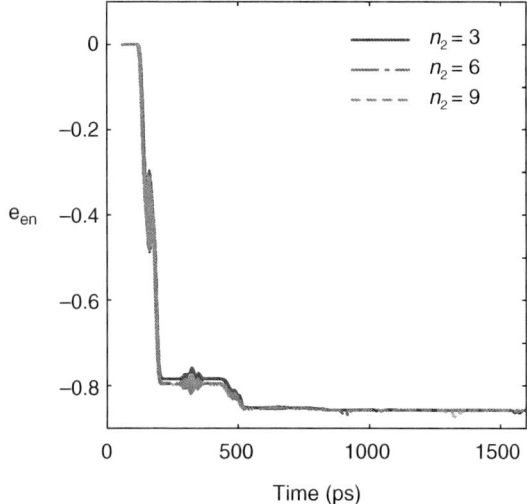

Figure 4.15: Error in total energy of coupled model for diagonalized constraints methods.

according to (4.77). The energy loss increases as the element size increases relative to the lattice constant.

4.5.1.2 Two-dimensional studies

In the two-dimensional example, we consider the square lattice (shown in Figure 4.16) of 38,000 nodes and 40,401 atoms. No dummy atoms are introduced at the interface between the atomistic and continuum model for the same reason as in one-dimensional problems. The lattice constants of the atomic model are $r_0 = 1.23 \, \text{Å}$ and $\theta_0 = 90°$. Harmonic potentials are used for both stretch and angle bending

$$W^M = \sum_{I \in \mathcal{M}} \sum_{J \in \mathcal{M} > I} \frac{1}{2} k_r (r_{IJ} - r_0)^2 + \sum_L \frac{1}{2} k_\theta (\theta_L - \theta_0)^2 \qquad (4.92)$$

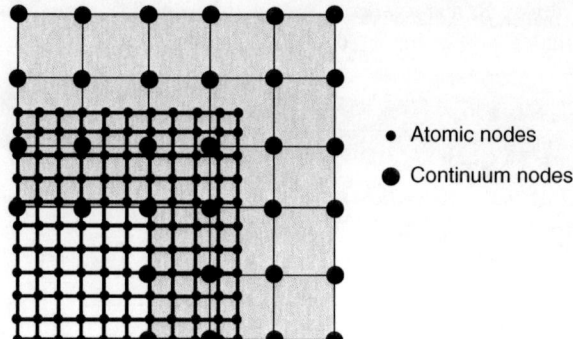

Figure 4.16: Schematic of two dimensional coupled model (in calculations, more elements and nodes were used).

In the above θ_L is the bond angle, $k_r = 14.418\text{eV}/\text{Å}^2$ and $k_\theta = 5.618\text{eV}/\text{rad}^2$ are the stretch and angle bending constants, respectively, and L is summed over the number of unique three-body interactions. The internal force of atom I, $\mathbf{f}_I^{int}$, is given by the derivative of the potential function with respect to the atomic coordinates.

$$f_{iI}^{int} = -\frac{\partial W^M}{\partial x_{iI}} = -\sum_{J \in \mathcal{M}} k_r(r_{IJ} - r_0)\frac{\partial r_{IJ}}{\partial x_{iI}} - \sum_L k_\theta(\theta_L - \theta_0)\frac{\partial \theta_L}{\partial x_{iI}} \quad (4.93)$$

We consider the model shown in Figure 4.16; four-fold symmetry is used. We initiate a cylindrical wave at the bottom left in the atomistic model with the initial conditions

$$u_1(\mathbf{X}_I) = \big[a_1 + a_1 \cos(2\pi a_2 r_I + \pi)\big]\big[1 + a_3 \cos(2\pi r_I)\big] \cos\gamma$$

$$u_2(\mathbf{X}_I) = \big[a_1 + a_1 \cos(2\pi a_2 r_I + \pi)\big]\big[1 + a_3 \cos(2\pi r_I)\big] \sin\gamma$$

$$\dot{u}_1(\mathbf{X}_I) = 2\pi c a_1 \sin(2\pi a_2 r_I + \pi)\big[1 + a_3 \cos(2\pi r_I)\big] \cos\gamma$$

$$\dot{u}_2(\mathbf{X}_I) = 2\pi c a_1 \sin(2\pi a_2 r_I + \pi)\big[1 + a_3 \cos(2\pi r_I)\big] \sin\gamma \quad (4.94)$$

for the atoms with $r_I \leq 100\,\text{Å}$, where $r_I = ||\mathbf{x}_I||$ is the distance of atom I to the origin, and γ is the angle between the positive x direction and $\mathbf{x}_I$. The initial displacements and velocities vanish for the rest of the atoms. The values of a_1, a_2, and a_3 are the same as in the one-dimensional examples. The incident wave contains four different frequencies ω_1 to ω_4.

The linear momentum of the system is perfectly conserved for both the consistent and diagonalized constraints. There is no temporary loss of linear momentum as observed in the one-dimensional examples when the wave reaches the overlapping domain. The initial total angular momentum is zero, and remains so when using the consistent constraint. For the diagonalized constraint, the angular momentum in the coupling domain oscillates about zero with an amplitude of 10^{-3}.

The error in energy is shown in Figure 4.17. The energy is conserved with the consistent constraint. For the diagonalized constraint, about 45% of the energy is dissipated after the wave passes the coupling domain; most of this energy is in the high frequency part of the spectrum.

4.5.2 Fracture of defected graphene sheets by QM/MM and QM/CM methods

Here we describe coupled QM/MM and QM/CM calculations of defected graphene sheets. The coupled method uses the modified Tersoff-Brenner (MTB-G2) potential for the molecular mechanics model and a linear elastic finite element model for continuum mechanics subdomain. The quantum mechanics calculations were made with the semiempirical method PM3 [41]. MTB-G2 [42, 43, 44, 45] is a modified version of the standard reactive empirical bond order (REBO) [44] potential, in which the cutoff function is removed and interatomic interactions are included only for those atom pairs that are less than 2.0 Å apart in the initial unstrained configuration. This was found [42, 45] to be necessary to prevent spuriously high values of the fracture strength due to the cutoff function.

Three coupling methods were used to calculate the stress-strain curves and fracture stresses of a graphene sheet containing defects. In these calculations, the end carbon atoms of the graphene sheet were displaced with increments of 0.5% of the sheet length until the sheet fractured. At each strain increment,

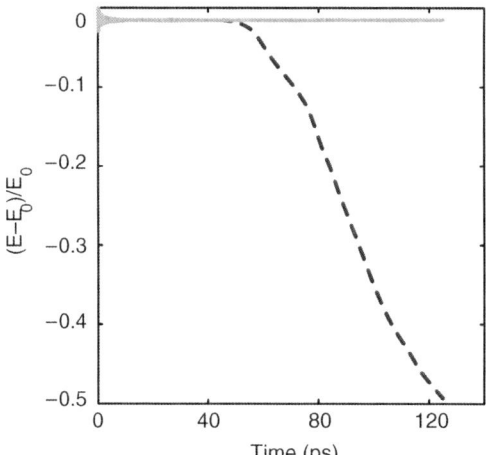

Figure 4.17: Rate change in total energy for a two dimensional example with consistent (solid line) and diagonalized (dashed line) constraints ($n_1 = 4$, $n_2 = 1$).

the atomistic configuration was optimized with the carbon atoms at each edge constrained to lie within a plane. Once the geometry was optimized at a given applied strain, the tensile stress in the sheet was calculated by taking derivatives of the energy using finite differences with respect to the strain. An effective thickness of 3.4 Å was used for the stress calculations; this corresponds to the interlayer spacing in graphite, and is widely used for stresses on graphene sheets.

We first consider a graphene sheet containing an asymmetric two-atom vacancy [46, 47]. It was stretched in the direction perpendicular to the zigzag edge. The coupled models of this system for the two coupling methods considered are shown in Figures 4.18–4.19. The sheet was 27 Å by 21 Å, and contained 248 carbon atoms with 64 carbon atoms in the QM domain and 38 carbon atoms in the overlapping domain for the modified ONIOM (called QtMMOD) scheme, (see [30]). The results were compared to those calculated by pure QM calculations. In addition to the two QM/MM models, a QM/CM model (shown in Figure 4.18) using triangular elements was also used.

The stress-strain curves for the three methods are shown in Figure 4.20. The stresses calculated by the QtMMOD and ONIOM coupling methods are almost indistinguishable. Both results match quite well with the benchmark pure QM results although there are small differences near the fracture strain. The QtMMOD and ONIOM methods calculated fracture stress values of 87.4 GPa and 87.2 GPa, respectively, in comparison to 88.1 GPa obtained from the strictly QM calculations. The stress-strain curve for the QM/CM coupled calculations matches the stresses from the other methods quite well and yields a fracture stress value of 87.3 GPa. However, small differences are observed, which may

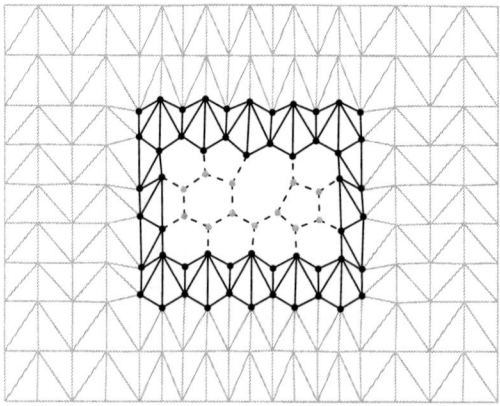

Figure 4.18: A QM/CM model of a graphene sheet containing an asymmetric two-atom vacancy defect. Grey (ball and stick) atoms in the center constitute the strictly QM region, dark (ball and stick) atoms constitute the overlapping region, and grey (stick) region represents the strictly CM region.

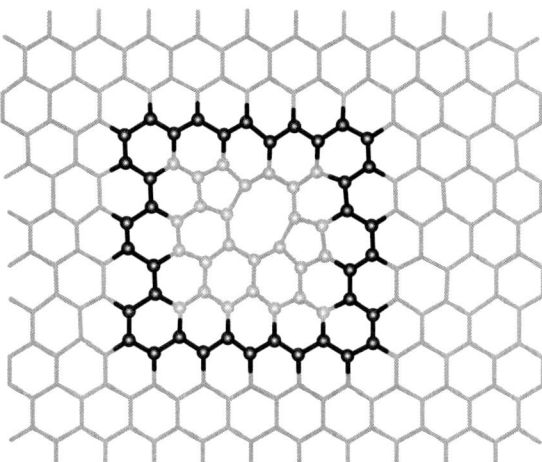

Figure 4.19: A QtMMOD model of a graphene sheet containing an asymmetric two-atom vacancy defect. Grey (ball and stick) atoms constitute the strictly QM region, dark (ball and stick) atoms constitute the overlapping region, and grey (stick) atoms constitute the strictly MM region.

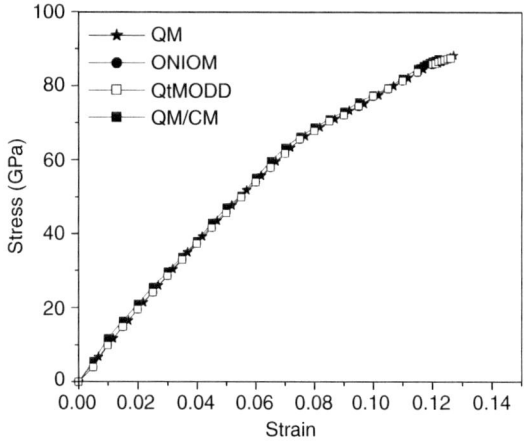

Figure 4.20: Stress-strain curves obtained by the ONIOM, QtMMOD, QM/CM coupling, and pure QM methods for the graphene sheet models shown in Figures 4.3–4.5.

be due to several reasons, such as the fact that a finite element model does not reflect the physics near the defect as accurately as the MM method. Nevertheless, the QM/CM model is still sufficiently accurate to be useful for studies of large systems, such as calculations of the fracture stress of a graphene sheet containing

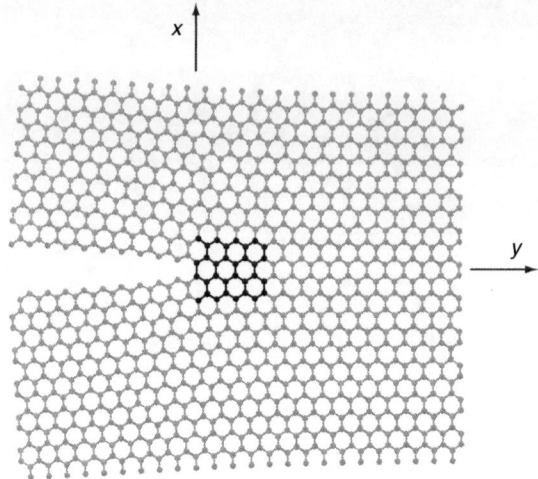

Figure 4.21: QM/MM model of a graphene sheet with a crack. Dark atoms represent the QM subdomain and the grey atoms represent additional atoms treated at the MM level.

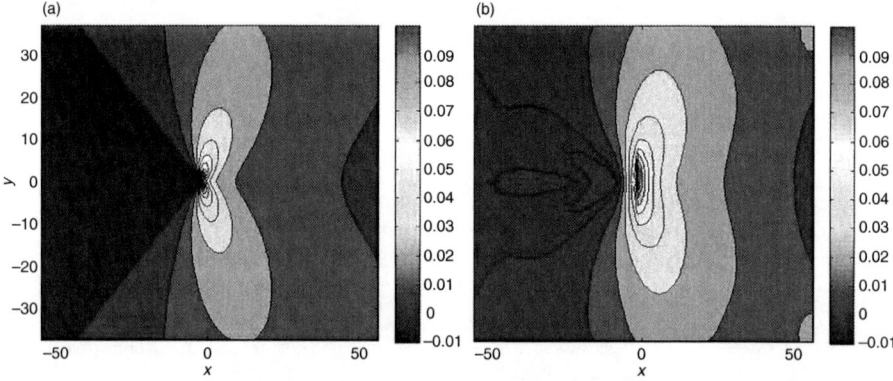

Figure 4.22: ϵ_y obtained from (a) elasticity solution and (b) MLS method for the sheet shown in Figure 4.21.

a crack [1]. In such models, the region near the crack tip can be modeled quantum mechanically and the CM model can be applied elsewhere. Figure 4.22 shows the comparison of elasticity solution of the strain ϵ_y at stress intensity factor K = 3 MPa$\sqrt{\text{m}}$ with the results calculated by the QM/MM methods. The strains were computed from the configuration by the MLS scheme [48, 49]. The agreement is quite good, although the atomistic strains are not as localized near the crack tip as in the elastic solution.

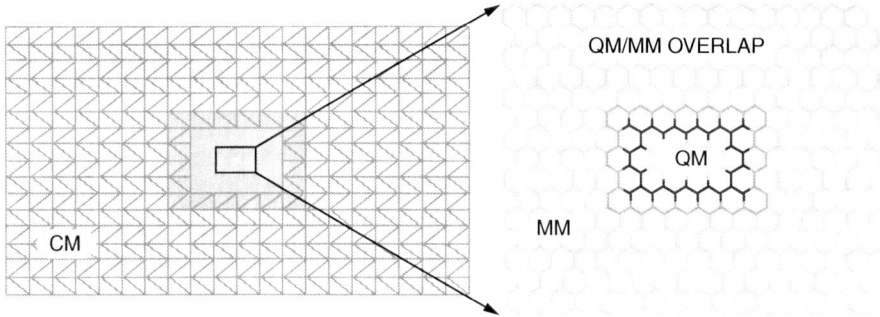

Figure 4.23: QM/MM/CM model of a graphene sheet containing a slit-like defect.

4.5.3 Crack propagation in a graphene sheet: comparison with Griffith's formula

In this example we consider the fracture strength of a graphene sheet with slit defects of various lengths and compare the calculated strengths to those predicted by Griffith's formula for a crack in an infinite sheet. Slit defects were created by removing four rows of atoms and then saturating the dangling bonds by hydrogen atoms [1, 43]. In order to be able to compare with the results for an infinite sheet, very large sheets were required, so we used a QM/MM/CM concurrent calculation [2], where the QM/MM coupling was obtained by using the modified ONIOM scheme, and the bridging domain method was used to link the MM model to a finite element model. The PM3 method [41] was used for the quantum calculation.

A representation of the QM/MM/CM model for one slit size is shown in Figure 4.23. The dimensions of the full sheet are 393.5 Å by 411.8 Å, the QM model consists of 76 atoms around the slit. The atomistic region is 115.5 Å by 82.3 Å and located in the center of the sheet.

In Figure 4.24 the numerical results are compared with the results of the Griffith formula, which is given by

$$\sigma = \sqrt{\frac{2Y\gamma}{\pi a}}, \tag{4.95}$$

where σ is the Griffith formula stress, Y is the Young's modulus, a is the half length of the slit, and γ is the surface energy density. We calculated γ as the difference between the PM3 energy of a geometry optimized pristine square sheet and the energy of the two fragments after they are separated to a large distance, and subsequently optimized to yield their equilibrium geometries, divided by twice the surface area. The factor of two is included because two surfaces are produced by fracture.

The QM/MM/CM results agree well with the prediction of the Griffith formula and decrease approximately as the inverse of the square root of the crack

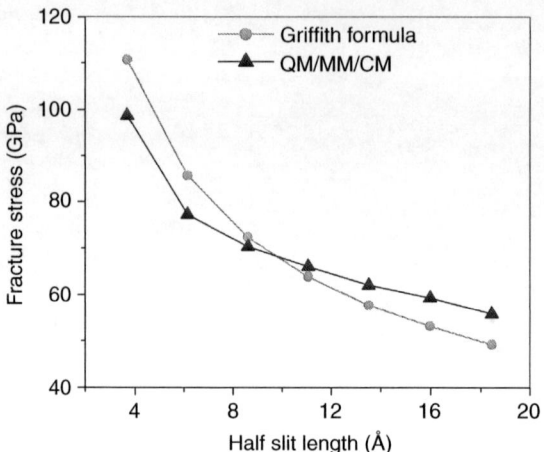

Figure 4.24: Results of the Griffith formula versus QM/MM/CM fracture strengths for a graphene sheet containing a slit-like defect.

length. The Griffith stress is a lower bound on the fracture stress, but the approximate stress estimates of the Griffith formula need not be, and as seen in Figure 4.24, the Griffith formula results fall below the numerical ones for slits longer than 20 Å. Note that on the whole, remarkably good agreement is obtained between the Griffith formula and the QM/CM/MM calculation, which indicates that the Griffith formula is also applicable to nanoscale fracture.

4.6 Acknowledgment

The support of the Army Research Office and Office of Naval Research is gratefully acknowledged.

References

[1] Khare R., Mielke S. L., Paci J. T., Zhang S., Ballarini R., Schatz G. C., and Belytschko T. (2007). "Coupled quantum mechanical/molecular mechanical modeling of the fracture of defective carbon nanotubes and graphene sheets", *Phys. Rev. B*, **75**, p. 075412.

[2] Khare R., Mielke S. L., Schatz G. C., and Belytschko T. (2008). "Multiscale coupling schemes spanning the quantum mechanical, atomistic forcefield, and continuum regimes", *Comput. Methods Appl. Mech. Engrg*, **197**, p. 3190–3202.

[3] Baskes M. I., Melius C. F., and Wilson W. D. (1980). "Solubility and diffusivity of hydrogen and helium at dislocations and in the stress field near a crack tip", *Journal of Metals*, **32**, p. 14.

[4] Mullins M. and Dokainish M. A. (1982). "Simulation of the (001) plane crack in α-iron employing a new boundary scheme", *Phil. Mag. A*, **46**, p. 771–787.

[5] Belytschko T., Liu W. K., and Moran B. (2001). *Nonlinear Finite Elements for Continua and Structures*, John Wiley & Sons Inc.
[6] Kohlhoff S., Gumbsch P., and Fischmeister H. F. (1991). "Crack propagation in bcc crystals studied with a combined finite-element and atomistic model". *Phil. Mag. A*, **64**, p. 851–878.
[7] Shilkrot L. E., Curtin W. A., and Miller R. E. (2002). "A coupled atomistic/continuum model of defects in solids", *J. Mech. Phys. Solids*, **50**, p. 2085–2106.
[8] Tadmor E. B., Ortiz M., and Phillips R. (1996). "Quasicontinuum analysis of defects in solids", *Phil. Mag. A*, **73**, p. 1529–1563.
[9] Shenoy V. B., Miller R., Tadmor E. B., Rodney D., Phillips R., and Ortiz M. (1998). "An adaptive methodology for atomic scale mechanics: the quasicontinuum method", *J. Mech. Phys. Sol.*, **47**, p. 611–642.
[10] Knap J. and Ortiz M. (2001). "An analysis of the quasicontinuum method", *J. Mech. Phys. Sol.*, **49**, p. 1899–1923.
[11] Rudd R. E. and Broughton J. Q. (1998). "Coarse-grained molecular dynamics and the atomic limit of finite elements", *Phys. Rev. B*, **58**, R5893–R5896.
[12] Abraham F. F., Broughton J. Q., Bernstein N., and Kaxiras E. (1998). "Spanning the length scales in dynamic simulation", *Computers in Physics*, **12**, p. 538–546.
[13] Broughton J. Q., Abraham F. F., Bernstein N., and Kaxiras E. (1999). "Concurrent coupling of length scales: methodology and application", *Phys. Rev. B*, **60**, p. 2391–2403.
[14] Wagner G. J. and Liu W. K. (2003). "Coupling of atomistic and continuum simulations using a bridging scale decomposition", *J. Comput. Physi.*, **190**, p. 249–274.
[15] Song J.-H., Loehnert S., and Belytschko T. (2008). "Multi-scale aggregating discontinuities: a method for circumventing loss of material stability". *Int. J. Numer. Mech. Engng*, **73**, p. 869–894.
[16] Farrell D. E., Park H. S., and Liu W. K. (2007). "Implementation aspects of the bridging scale method and application to intersonic crack propagation", *Int. J. Numer. Meth. Engng*, **71**, p. 583–605.
[17] Cai W., Koning M., Bulatov V. V., and Yip S. (2000). "Minimizing boundary reflections in coupled-domain simulation", *Phys. Rev. Lett.*, **85**, p. 3213–3216.
[18] E W. and Huang Z. (2001). "Matching conditions in atomistic-continuum modeling of materials", *Phys. Rev. Lett.*, **87**, p. 135501.
[19] To A. and Li S. F. (2005). "Perfectly matched multiscale simulations", *Phys. Rev. B*, **72**, p. 035414.
[20] Berenger J. P. (1994). "A perfectly matched layer for the absorption of electromagnetic waves", *J. Comput. Phys.*, **114**, p. 185–200.
[21] Xiao S. P. and Belytschko T. (2004). "A bridging domain method for coupling continua with molecular dynamics", *Comput. Meth. Appl. Mech. Eng.*, **193**, p. 1645–1669.

[22] Belytschko T. and Xiao S. P. (2003). "Coupling methods for continuum model with molecular model", *Int. J. Numer. Meth. Engng*, **1**, p. 115–126.
[23] Guidault P. A. and Belytschko T. (2007). "On the L^2 and the H^1 couplings for an overlapping domain decomposition method using Lagrange multipliers", *Int. J. Numer. Meth. Engng*, **70**, p. 322–350.
[24] Fish J., Nuggehally M. A., Shephard M. S., Picu C. R., Badia S., Parks M. L., and Gunzburger M. (2007). "Concurrent AtC coupling based on a blend of the continuum stress and the atomistic force", *Comput. Methods Appl. Mech. Engng*, **196**, p. 4548–4560.
[25] Li X. and E W. (2005). "Multiscale modeling of the dynamics of solids at finite temperature". *J. Mech. Phys. Solids*, **53**, p. 1650–1685.
[26] Dhia H. B. and Rateau G. (2001). "Mathematical analysis of the mixed Arlequin method". *Comptes Rendus de l'Academie des Sciences Series I Mathematics*, **332**, p. 649–654.
[27] Dhia H. B. and Rateau G. (2005). "The Arlequin method as a flexible engineering desing tool", *Int. J. Numer. Mech. Engng*, **62**, p. 1442–1462.
[28] Feyel F. and Chaboche J. L. (2000). "FE^2 multiscale approach for modelling the elastoviscoplasitic behavior of long fibre SiC/Ti composite materials", *Comput. Methods Appl. Mech. Engng*, **183**, p. 309–330.
[29] Kouznetsova V., Brekelmans W. A. M., and Baaijens F. P. T. (2001). "An approach to micro-macro modeling of heterogeneous materials", *Comput Mech*, **27**, p. 37–48.
[30] Zhang S., Khare R., Lu Q., and Belytschko T. (2007). "A bridging domain and strain computation method for coupled atomistic/continuum modelling of solids", *Int. J. Numer. Mech. Engng*, **70**, p. 913–933.
[31] Curtin W.A. and Miller R. E. (2003). "Atomistic/continuum coupling in computational materials science", *Modelling and Simulation in Materials Science and Engineering*, **11**, R33–R68.
[32] Svensson M., Humbel S., Froese R. D. J., Matsubara T. Sieber S., and Morokuma K. (1996). "ONIOM: a multilayered integrated MO+MM method for geometry optimizations and single point energy predictions. A test for Diels-Alder reactions and $Pt(P(t-Bu)_3)_2 + H_2$ oxidative addition", *J. Phys. Chem.*, **100**, p. 19357–19363.
[33] Xu M. and Belytschko T. (2008). "Conservation properties of the bridging domain method for coupled molecular/continuum dynamics", *Int. J. Numer. Meth. Engng*, **76**, p. 278–294.
[34] Hairer E., Norsett S. P., and Wanner G. (2002). *Solving Ordinary Differential Equations* Springer.
[35] Belytschko T., Krongauz Y., Dolbow J., and Gerlach C. (1998). "On the completeness of meshfree particle methods", *Int. J. Numer. Mech. Engng*, **43**, p. 785–819.
[36] Karpov E. G., Wagner G. J., and Liu W. K. (2005). "A Green's function approach to deriving non-reflecting boundary conditions in molecular dynamics simulation", *Int. J. Numer. Mech. Engng*, **62**, p. 1250–1262.

[37] Park H. S., Karpov E. G., Klein P. A., and Liu W. K. (2005). "The bridging scale for two-dimensional atomistic/continuum coupling", *Philosophical Magazine*, **85**, p. 79–113.

[38] Adelman S. A. and Doll J. D. (1974). "Generalized langevin equation approach for atom/solid-surface scattering-collinear atom/harmonic chain model", *J. Chem. Phys.*, **61**, p. 4242–4245.

[39] Li S., Liu X., Agrawal A., and To A. (2006). "Perfectly matched multiscale simulations for discrete lattice systems: Extension to multiple dimensions", *Phys. Rev. B*, **74**, p. 045418.

[40] Khare R., Mielke S. L., Paci J.T., Zhang S., Ballarini R., Schatz G. C., and Belytschko T. (2007). "Coupled quantum mechanical/molecular mechanical modeling of the fracture of defective carbon nanotubes and graphene sheets", *Phys. Rev. B*, **75**, p. 75412.

[41] Stewart J. J. P. (1989). "Optimization of parameters for semiempirical methods", *J. Comput. Chem.*, **10**, p. 209–220.

[42] Belytschko T., Xiao, S. P., Schatz G. C., and Ruoff R. S. (2002). "Atomistic simulations of nanotube fracture", *Phys. Rev. B*, **65**, p. 235430.

[43] Zhang S., Mielke S. L., Khare R., Troya D., Ruoff R. S., Schatz G. C., and Belytschko T. (2005). "Mechanics of defects in carbon nanotubes: Atomistic and multiscale simulations", *Phys. Rev. B*, **71**, p. 115403.

[44] Brenner D. W., Shenderova O. A., Harrison J. A., Stuart S. J., Ni, B., and Sinnott B. (2002). "A second-generation reactive empirical bond order (REBO) potential energy expression for hydrocarbons", *J. Phys.: Condens. Matter*, **14**, p. 783–802.

[45] Shenderova O. A., Brenner D. W., Omeltchenko A., Su X., and Yang L. H. (2000). "Atomistic modeling of the fracture of polycrystalline diamond", *Phys. Rev. B*, **61**, p. 3877–3888.

[46] Mielke S. L., Troya D., Zhang S., Li J. L., Xiao S. P., Car R., Ruoff R. S., Schatz G. C., and Belytschko T. (2004). "The role of vacancy defects and holes in the fracture of carbon nanotubes", *Chemi. Physi. Lett.*, **390**, p. 413–420.

[47] Krasheninnikov A. V., Nordlund K., Sirvio M., Salonen E., and Keinonen J. (2001). "Formation of ion-irradiation-induced atomic-scale defects o walls of carbon nanotubes", *Phys. Rev. B*, **63**, p. 245405.

[48] Belytschko T., Lu Y. Y., and Gu L. (1994). "Element-free Galerkin methods", *Int. J. Numer. Mech. Engng.*, **37**, p. 229–256.

[49] Belytschko T., Krongauz Y., Organ D., Fleming M., and Krysl P. (1996). "Meshless methods: An overview and recent developments", *Comput. Methods Appl. Mech. Engng.*, **139**, p. 3–47.

5

COARSE-GRAINED MOLECULAR DYNAMICS: CONCURRENT MULTISCALE SIMULATION AT FINITE TEMPERATURE[1]

Robert E. Rudd

5.1 Coupling atomistic and continuum length scales

Coarse-Grained Molecular Dynamics (CGMD) is a computational technique designed to provide an efficient simulation of processes at the atomic scale in heterogeneous systems [1, 2, 3, 4, 5, 6]. The heterogeneity of materials is interesting. It is interesting from both scientific and practical perspectives. Real materials have structure in their electronic bonds, in the position and arrangement of their atoms such as in crystal lattices containing one or more elements in ordered, partially ordered, or random arrays, in the configuration of the defects both in the atomic configuration of the defects and how the defects are arrayed, in the nature of the grains and the structure of the boundaries that separate them, in higher order patterns in the microstructure, and so on. The nature of the structures at different length scales is one of the principal characteristics available to the materials scientist to improve mechanical properties of solid systems, just as it unavoidably limits materials performance in other cases. For example, aluminum alloys are strengthened to the level needed for structural applications through an aging process in which nanoscale structures such as Guinier-Preston zones precipitate and form impediments to dislocation flow [7]. Carbide precipitates play a similar role at longer length scales in steels [8]. Indeed, much of metallurgy deals with optimizing microstructures for improved properties. On the other hand, the motion of point defects, surface material, and dislocations limits the quality factor of nanoscale mechanical resonators [9, 10]. Metal failure processes typically start at material heterogeneities, whether through interface decohesion or fracture of brittle components [11]. Furthermore, much of nanoscience is influenced by the heterogeneous properties of the surface since the surface area to volume ratio is so large for nanoscale systems [12, 13, 14], and the influence can be advantageous or not. For these and countless other cases inhomogeneous materials are of interest, and computer modeling tools are needed that can model parts of the system fully resolved to the level of atoms as well as relatively large volumes of the surrounding material—preferably at a lower computational cost.

---

[1]The work in this chapter was performed under the auspices of the United States Department of Energy by Lawrence Livermore National Laboratory under contract DE-AC52-07NA27344.

In heterogeneous systems, not all parts are equal. Some regions require higher resolution modeling than others.

Indeed, the practitioner of molecular dynamics (MD) [15] often wants to model an interesting part of a system fully resolved at the atomic level, but must contend with modeling the surrounding medium in order to have the correct boundary conditions for the region of interest. Dislocation core structures are a well documented example where the atomic structure is affected by how boundary conditions are set many lattice spacings away [16]. Dislocation motion is a good example of a process that not only has interesting features at the atomic scale coupled to larger volumes of the system but is also thermally activated or thermally damped. Nanoscale mechanical resonators [10, 17] and void growth [18] are other examples of multiscale systems where temperature is important. Including the surrounding medium in a conventional MD simulation may incur a substantially greater computational cost, even though the sub-region of interest couples to only a small fraction of the degrees of freedom in the surroundings. There is a strong motivation to generalize molecular dynamics to be able to run with only the relevant degrees of freedom at finite (nonzero) temperature.

Concurrent multiscale models of materials (see for example, [1, 2, 6, 19, 20, 21, 22, 23, 24, 25, 26]) couple an atomistic description of some parts of the system to a more coarse-grained description of others. The two resolutions co-exist in a single simulation, controlled by an irregularly sized mesh as is typical of finite element modeling (FEM) [27, 28]. Unlike conventional finite element modeling, when the mesh size is taken down to the atomic level, a concurrent multiscale model becomes atomistic, and the mesh nodes behave as atoms interacting via interatomic bonds rather than a conventional finite element stiffness. The notion that an atomistic model could be coupled to FEM with gains in computational efficiency was introduced by Kohlhoff *et al.* [19] in a study of fracture. The first technique to combine atomistics with something like finite element modeling while systematically deriving the properties of the coarse-meshed region from the atomistic model was the Quasicontinuum Technique of Tadmor *et al.* [20]. It relied on molecular statics. The Coupling of Length Scales approach [21] successfully coupled finite temperature molecular dynamics to a conventional finite element model.

The goal of Coarse-Grained Molecular Dynamics [1, 2, 3, 4, 5, 6] is to provide a formalism within which a concurrent multiscale model is developed derived entirely from the underlying finite temperature molecular dynamics model using a mesh to control the amount of coarse graining in different regions of space, as shown in Figure 5.1. Thus, it is an "effective theory" in terminology borrowed from the renormalization group community where, as in the Quasicontinuum Technique, the concurrent multiscale model is derived entirely from the atomistics without any continuum-level input. By contrast, in a hybrid model, such as the Coupling of Length Scales, existing atomistic and continuum models are linked to run simultaneously. The advantage of formulating a multiscale model

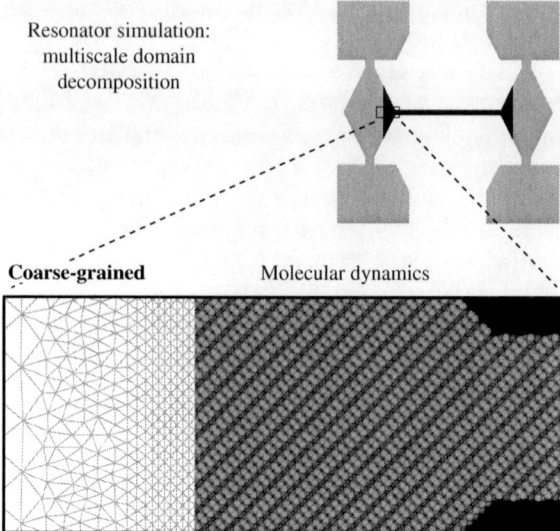

Figure 5.1: Schematic diagram of a coarse-grained simulation of a NEMS silicon microresonator [10, 29]. The coarse-grained (CG) region comprises most of the volume, but the molecular dynamics (MD) region contains most of the simulated degrees of freedom. Each sphere shown in the MD region represents an atom. Note that the CG mesh is refined to the atomic scale where it joins with the MD lattice [6].

based on an underlying MD model in an effective theory is that the crossover from a continuum description appropriate for a very coarse mesh to the correct physics in the atomic limit of the fully resolved mesh arises naturally from the formulation of the model. It is not *ad hoc*. In a hybrid scheme based on FEM, for example, reducing the mesh size down to the atomic scale does not lead to physical interatomic forces. The conventional theory of continuum elasticity does not have a length scale, and its implementation in FEM certainly does not have any special behavior at the length scale of the atomic lattice. An effective theory does not need to take the continuum limit, so it does not loose the memory of the lattice length scale. In CGMD when the mesh is taken from a coarse size down to the point where it is commensurate with the lattice, the atomic forces are recovered naturally.

Section 5.2 describes the formalism of finite temperature coarse graining of a solid MD system. This discussion is at a general level. Section 5.3 develops the theory further making specific assumptions about the form of the MD Hamiltonian. Section 5.4 describes some extensions of CGMD to coarse-graining nonequilibrium systems. Then Section 5.5 deals with the specific implementation needed to obtain reasonable performance on a computer, single processor desktop, or supercomputer. Section 5.6 describes some applications and performance figures. Finally, we conclude in Section 5.7 with an outlook for future developments of CGMD.

5.2 Coarse graining formalism

We begin the discussion of CGMD at the most general level: the development of a formalism for coarse graining of a finite temperature MD system on an irregular mesh. Coarse graining involves the systematic removal of degrees of freedom in the theoretical description of a physical system. A prescription for how those degrees of freedom are removed while effectively retaining their contribution to the equations of motion (or equations of equilibrium for static systems) is the fundamental construct of a coarse-graining formulation. The effect of what are consequently internal degrees of freedom appears through sub-grid contributions to the system energy and forces. In CGMD, the fundamental principle is that the internal degrees of freedom are in thermal equilibrium consistent with the CG fields. This principle is formulated mathematically in this section. It is different than the fundamental principle of the Quasicontinuum Technique [20] in which the atoms are placed exactly where the CG fields specify they should go according to the Cauchy-Born rule (at least at the representative atoms). In contrast, for nonzero temperatures thermal fluctuations are important, and the *ansatz* that the internal degrees of freedom are in local thermal equilibrium provides a means to formulate the multiscale model.

In CGMD we focus on a solid system, crystalline, or amorphous, whose energy is given by the MD Hamiltonian H_{MD}. It is a sum of potential and kinetic energies depending on the relative atomic positions and atomic momenta, respectively. For convenience, we label the atoms with Greek letters, so $\mathbf{x}_\mu$ gives the position of atom μ, and $\mathbf{x}_{\mu 0}$ gives its equilibrium position in a reference state. We also have a mesh with nodes located at the zero-displacement positions $\mathbf{x}_j$. Continuum fields are to be represented on the mesh by interpolation between values given at the nodes. The most important of these fields for our purposes is the continuum displacement field $\mathbf{u}(\mathbf{x})$ that gives the continuum limit of the atomic displacements $\mathbf{u}_\mu = \mathbf{x}_\mu - \mathbf{x}_{\mu 0}$. Introducing a shape function $N_j(\mathbf{x})$ for each node j that gives the support for the field around node j, the interpolated field is given by

$$\mathbf{u}(\mathbf{x}) = \sum_j \mathbf{u}_j N_j(\mathbf{x}) \qquad (5.1)$$

where $\mathbf{u}_j$ is the nodal displacement.

The starting point for our coarse-graining procedure is the link between the atomistic and coarse-grained variables

$$\mathbf{u}_j = \sum_\mu f_{j\mu} \mathbf{u}_\mu, \qquad (5.2)$$

where $f_{j\mu}$ is a weighting function, related to the shape function $N_j(\mathbf{x})$ in a way that is described below. To begin, it is enough to know that each nodal displacement is a linear combination of the atomic displacements, a certain local weighted average. An analogous expression relates the nodal velocities to the atomic momenta $\mathbf{p}_\mu$:

$$\dot{\mathbf{u}}_j = \sum_\mu f_{j\mu}\, \mathbf{p}_\mu/m_\mu, \tag{5.3}$$

where m_μ is the mass of atom μ.

Now we formulate the correspondence principle discussed above. Specifically, we define the coarse-grained (CG) energy as the average energy of the canonical ensemble on this constrained phase space:

$$E(\mathbf{u}_k, \dot{\mathbf{u}}_k) = \langle\, H_{MD}\, \rangle_{\mathbf{u}_k, \dot{\mathbf{u}}_k} \tag{5.4}$$

$$= \int d\mathbf{x}_\mu d\mathbf{p}_\mu\; H_{MD}\, e^{-\beta H_{MD}}\, \Delta/Z, \tag{5.5}$$

$$Z(\mathbf{u}_k, \dot{\mathbf{u}}_k) = \int d\mathbf{x}_\mu d\mathbf{p}_\mu\; e^{-\beta H_{MD}}\, \Delta, \tag{5.6}$$

$$\Delta = \prod_j \delta\!\left(\mathbf{u}_j - \sum_\mu \mathbf{u}_\mu f_{j\mu}\right) \delta\!\left(\dot{\mathbf{u}}_j - \sum_\mu \frac{\mathbf{p}_\mu f_{j\mu}}{m_\mu}\right), \tag{5.7}$$

where $\beta = 1/(k_B T)$ is the inverse temperature, k_B is the Boltzmann constant, Z is the partition function, and $\delta(\mathbf{u})$ is a three-dimensional delta function. The delta functions enforce the mean field constraint (5.2). Again we use Latin indices, $j, k, \ldots$, to denote mesh nodes and Greek indices, $\mu, \nu, \ldots$, to denote atoms. The energy (5.5) is computed below in (5.28).

This formulation assumes that the internal degrees of freedom evolve adiabatically as the CG fields evolve. It is possible to formulate CGMD as a constant temperature simulation, in which the internal energy (5.4) is replaced by a free energy [6], but it will not be discussed in detail here.

5.2.1 Shape functions

The basis of shape functions $\{N_j(\mathbf{x})\}_{j=1}^{N_{node}}$ is used to interpolate continuum fields on the mesh, as mentioned above, and they determine the weights $f_{i\mu}$ that appear in (5.2) and (5.3). The shape functions need to have the following properties:

$$N_j(\mathbf{x}_k) = \delta_{jk}, \tag{5.8}$$

$$\sum_{j=1}^{N_{node}} N_j(\mathbf{x}) = 1, \tag{5.9}$$

$$N_j(\mathbf{x}) \in C^0 \cap H^1(\Omega). \tag{5.10}$$

where Ω is the spatial extent of the system. The first property is that the functions are normalized and local on the mesh nodes, $\mathbf{x}_k$. Thus, $N_j(\mathbf{x})$ is equal to one at $\mathbf{x} = \mathbf{x}_{j0}$, and it vanishes as it reaches any of the other nodes. Locality makes the interpretation of the fields more natural and physical, and it keeps the bandwidth of the system of equations of motion to a minimum. The second property is that the shape functions form a partition of unity. This property ensures that uniform displacement is represented, so that the internal modes do not include this zero-energy mode. The third property states that the functions are continuous C^0, and their derivatives are square-integrable even though they may not be continuous: technically, they belong to the Sobolev space H^1. The elastic energy is proportional to an integral of the square of the strain ($\varepsilon_{ab} = \frac{1}{2}(\partial_a u_b + \partial_b u_a)$), so the third property ensures that the elastic energy is well-behaved in the continuum limit. Additional considerations may come into play, such as the need to refine the mesh onto a particular crystal lattice at the MD/CG interface as discussed in Rudd [30].

Now we derive the weights $f_{j\mu}$. They are determined by the relationship of the atomic positions to the CG displacement field. Given a set of atomic displacements we can find the displacement field represented on the CG mesh that best fits the atomic data in the least-squares sense:

$$\chi^2 = \sum_{\mu} \left| \mathbf{u}_\mu - \sum_{j} \mathbf{u}_j N_{j\mu} \right|^2, \tag{5.11}$$

where $N_{j\mu} = N_j(\mathbf{x}_{\mu 0})$. The χ^2 error is minimized by nodal displacements $\mathbf{u}_j$ of the form given in (5.2) with the weighting function $f_{j\mu}$ given in terms of the shape function:

$$f_{j\mu} = \sum_{k} \left(\sum_{\nu} N_{j\nu} N_{k\nu} \right)^{-1} N_{k\mu}. \tag{5.12}$$

The relationship between the nodal displacements and the atomic displacements (5.2) shows that many atomic configurations project to the same set of nodal displacements, if the number of atoms exceeds the number of nodes as is typically the case. Because of the appearance of a matrix inverse in (5.12), the support of $f_{j\mu}$ extends farther from node j than that of $N_{j\mu}$. On the other hand, in regions where the atoms and nodes are in one-to-one correspondence, $N_{j\mu}$ and $f_{j\mu}$ are delta functions, $\delta_{j\mu}$, and conventional MD is recovered. We note that recently this relationship introduced in CGMD has arisen in the bridging scale, and other L^2 projection techniques [25].

5.3 The CGMD Hamiltonian

The CG energy (5.5) is a powerful construct for the development of concurrent multiscale models. In practice this Hamiltonian must be further refined to provide the computationally tractable model needed for applications. Fortunately, it may be computed in closed form using analytic techniques in the case of a harmonic lattice. Two different but equivalent expressions have been derived [1, 6]. We just give an indication of the derivation here, and refer the reader to Rudd and Broughton [6] for the details.

We begin with the MD model underlying the CGMD system. The Hamiltonian that expresses the energy of the system of atoms is quadratic in the atomic momenta, and typically higher order in the atomic positions:

$$H_{MD} = \sum_\mu \frac{\mathbf{p}_\mu^2}{2m_\mu} + \sum_\mu U_\mu \tag{5.13}$$

$$= \sum_\mu \frac{\mathbf{p}_\mu^2}{2m_\mu} + \sum_\mu E_\mu^{\text{coh}} + \sum_{\mu,\nu} \frac{1}{2} \mathbf{u}_\mu \cdot D_{\mu\nu} \mathbf{u}_\nu + \ldots \tag{5.14}$$

where U_μ is the potential energy of atom μ (a function of the atomic positions of atom μ and its neighbors), E_μ^{coh} is the cohesive energy, and $D_{\mu\nu}$ is the matrix of potential energy coefficients (the dynamical matrix in real space). It acts as a tensor on the components of the displacement vector at each site. To simplify the calculations we re-express the CG energy (5.5) using a parametric derivative of the log of the constrained partition function (5.6),

$$E(\mathbf{u}_k, \dot{\mathbf{u}}_k) = -\partial_\beta \log Z(\mathbf{u}_k, \dot{\mathbf{u}}_k; \beta), \tag{5.15}$$

This expression enables us to focus on the calculation of the CGMD partition function Z.

Below, we consider the harmonic (quadratic) and anharmonic (higher order) parts of the potential energy; first, we consider the kinetic energy which is exactly quadratic, at least in typical, unconstrained systems, and calculate the kinetic part of the CGMD partition function Z_{kin}:

$$Z_{kin}(\dot{\mathbf{u}}_k) = \int d\mathbf{p}_\mu \, e^{-\beta \sum p_\mu^2/(2m)} \Delta_K, \tag{5.16}$$

$$\Delta_K = \prod_j \delta\left(\dot{\mathbf{u}}_j - \sum_\mu \frac{\mathbf{P}_\mu f_{j\mu}}{m}\right), \tag{5.17}$$

where we have assumed that all of the atomic masses are equal to m, for simplicity. The case of multiple masses is described in the references [1, 6].

The first approach to calculate this integral [1] is to express the delta functions in a Fourier representation

$$\delta(z) = \int_{-\infty}^{\infty} \frac{d\lambda}{2\pi} e^{i\lambda z}. \tag{5.18}$$

Substituting this expression for the delta functions in Z_{kin} (5.16) we have

$$Z_{kin}(\dot{\mathbf{u}}_k) = \int dp \, d\lambda \, e^{-\beta \frac{\mathbf{P}_\mu \cdot \mathbf{P}_\mu}{2m} + i\lambda_j \cdot \left(\dot{\mathbf{u}}_j - \frac{\mathbf{P}_\mu f_{j\mu}}{m}\right)} \tag{5.19}$$

where $dp = (dp)^{3N_{\text{atom}}}$, $d\lambda = \left(\frac{d\lambda}{2\pi}\right)^{3N_{\text{node}}}$, and repeated indices are summed. This Gaussian integral can be calculated analytically, as can be seen by completing the square in the argument of the exponential, or equivalently by re-expressing the integral in terms of the variables $\tilde{\mathbf{p}}_\mu$ and $\tilde{\lambda}_j$:

$$\tilde{\mathbf{p}}_\mu = \mathbf{p}_\mu + i\lambda_j f_{j\mu}/\beta, \tag{5.20}$$

$$\tilde{\lambda}_j = \lambda_j - i\beta \dot{\mathbf{u}}_k M_{kj} \tag{5.21}$$

where we have been led to introduce the CG mass matrix M_{jk}:

$$M_{jk} = \left(f_{j\mu} m_\mu^{-1} f_{k\mu}\right)^{-1} \tag{5.22}$$

$$= m\, N_{j\mu} N_{k\mu}. \tag{5.23}$$

The first expression involves the inverse of an $N_{\text{node}} \times N_{\text{node}}$ matrix. The second expression is simpler and more intuitive. The mass is evenly spread between neighboring nodes similar to the consistent mass matrix of FEM [28, 31], but it only applies in the case when all of the atomic masses are equal. Making the substitutions (5.20) and (5.21) into (5.19) and computing the Gaussian integrals, we have

$$Z_{kin}(\dot{\mathbf{u}}_k) = C_{kin} \beta^{-3(N_{\text{atom}}-N_{\text{node}})/2}\, e^{-\frac{1}{2}\beta \dot{\mathbf{u}}_j M_{jk} \dot{\mathbf{u}}_k} \tag{5.24}$$

where C_{kin} is a constant independent of β, and hence irrelevant to the CGMD energy (5.15).

5.3.1 Harmonic crystals

A similar treatment of the contribution coming from the potential energy for a harmonic crystal (or the leading quadratic part of the potential energy more generally) to the partition function, Z_{pot}, leads to

$$Z_{\text{pot}}(u_k;\beta) = C_{pot}\, \beta^{-3(N_{\text{atom}}-N_{\text{node}})/2}\, e^{-\frac{1}{2}\beta u_j K_{jk} u_k} \tag{5.25}$$

In this expression, the CG stiffness matrix appears:

$$K_{jk} = \left(f_{j\mu} D_{\mu\nu}^{-1} f_{k\nu}\right)^{-1} \tag{5.26}$$

which involves two matrix inverses.

The full CG energy for a harmonic solid of N_{atom} atoms coarse-grained to N_{node} nodes is then calculated from (5.15) to be

$$E(\mathbf{u}_k, \dot{\mathbf{u}}_k) = -\partial_\beta \log\left(Z_{kin} Z_{pot}\right) \tag{5.27}$$

$$= U_{\text{int}} + \frac{1}{2} \sum_{j,k} \left(M_{jk}\, \dot{\mathbf{u}}_j \cdot \dot{\mathbf{u}}_k + \mathbf{u}_j \cdot K_{jk} \mathbf{u}_k\right), \tag{5.28}$$

where the contribution of the internal degrees of freedom is

$$U_{\text{int}} = N_{\text{atom}} E^{\text{coh}} + 3(N_{\text{atom}} - N_{\text{node}}) kT. \tag{5.29}$$

The equations of motion are the Euler-Lagrange equations for this Hamiltonian:

$$M_{ij}\ddot{u}_j^a = -K_{ik}^{ab}u_k^b \tag{5.30}$$

subject to a set of boundary and initial conditions where we have written the spatial indices a and b explicitly.

Remarkably, there is another way to write the CGMD mass and stiffness matrices that does not require two inverses [6]. The derivation of this form does not involve the Fourier representation of the constraint. Instead, the atomic displacements are explicitly separated into the CG mean field and fluctuations about it. The separation is accomplished through the use of projection matrices

$$P_{\mu\nu}^{CG} \equiv f_{j\mu}(f_{j\lambda}f_{k\lambda})^{-1}f_{k\nu} \tag{5.31}$$

$$= N_{j\mu}(N_{j\lambda}N_{k\lambda})^{-1}N_{k\nu} \tag{5.32}$$

$$= N_{j\mu}f_{j\nu} \tag{5.33}$$

$$P_{\mu\nu}^{\perp} \equiv \delta_{\mu\nu} - P_{\mu\nu}^{CG} \tag{5.34}$$

where the matrix $P_{\mu\nu}^{CG}$ has been constructed to project onto the CG field: $P_{\mu\nu}^{CG}\mathbf{u}_\nu = \mathbf{u}_j N_{j\mu}$. Inserting the identity matrix in the form $\delta_{\mu\nu} = P_{\mu\nu}^{CG} + P_{\mu\nu}^{\perp}$ into the expressions for the partition function and some algebra and Gaussian integrals [6] leads to the same form of the CG energy (5.28), but with different expressions for the mass and stiffness matrices:

$$K_{jk} = N_{j\mu}D_{\mu\nu}N_{k\nu} - D_{j\mu}^{\times}\tilde{D}_{\mu\nu}^{-1}D_{k\nu}^{\times} \tag{5.35}$$

$$M_{jk} = N_{j\mu}m_\mu N_{k\mu} - M_{j\mu}^{\times}\tilde{M}_{\mu\nu}^{-1}M_{k\nu}^{\times} \tag{5.36}$$

where the other block components of the matrices are defined as

$$D_{j\mu}^{\times} = P_{\mu\rho}^{\perp}D_{\rho\nu}N_{j\nu} \tag{5.37}$$

$$\tilde{D}_{\mu\nu} = P_{\mu\rho}^{\perp}D_{\rho\sigma}P_{\sigma\nu}^{\perp} + \varepsilon P_{\mu\nu}^{CG} \tag{5.38}$$

We do not currently know of a direct way to show the equivalence of (5.26) and (5.35), but we have checked that they are numerically equal. Each form has its preferred uses. In the calculation of the CGMD spectra and any other application in which the stiffness matrix in reciprocal (Fourier) space is needed, (5.26) is advantageous. It is formally simpler, but it suffers from requiring two inverses and from formal singularities due to the zero modes of the potential energy coefficient matrix. Both of these drawbacks disappear in reciprocal space, where the matrix with its nodal indices Fourier transformed, known as the dynamical matrix, is diagonal. On the other hand, (5.35) is well defined, and it only requires one inverse. Typically, the inverse in the atomic indices is taken in reciprocal space making use of the perfect crystal space group symmetry, whereas the second inverse (5.26) is in the nodal indices for which reciprocal space offers an advantage only in special cases where the mesh is uniform. For irregular meshes, (5.35) is preferred.

5.3.2 How harmonic CGMD differs from FEM

In the CG region, CGMD for a harmonic crystal is a generalization of an FEM implementation of linear elasticity insofar as the equations of motion (5.30) are linear and similar in form to the conventional FEM equations of motion. In FEM, the mass and stiffness matrices are given by:

$$M_{ij}^{FEM} = \int d^3x\, \rho(\mathbf{x})\, N_i(\mathbf{x})\, N_j(\mathbf{x}) \qquad (5.39)$$

$$K_{ij}^{ab\,FEM} = C_{acbd} \int d^3x\, \partial_c N_i(\mathbf{x})\, \partial_d N_j(\mathbf{x}) \qquad (5.40)$$

[28] where $N_i(\mathbf{x})$ is the continuum shape function for node i and we have written the dependence on the spatial indices $a, \ldots, d$ explicitly. It is the continuum elastic constant tensor C_{acbd} that enter the expression for the stiffness matrix. These elastic constants are related to the infinite wavelength limit of the dynamical matrix. The CGMD stiffness matrix goes beyond this convention continuum elasticity and includes dispersion in the stiffness that arises from the atomic interactions. In the theory of linear continuum elasticity, the acoustic wave frequency and wavenumber obey $\omega = ck$, where c is the wavespeed; for MD and CGMD $\omega = \omega(k)$, a nonlinear function that agrees with the linear elasticity result for small wavenumber but then curves over to have zero slope at the boundary of the Brillouin zone. This nonlinear dispersion encodes the length scale of the lattice and is part of what allows CGMD to transition smoothly to MD based on interatomic forces as the mesh size is reduced to the atomic scale. This point will be clearer when we consider the CGMD spectrum below.

5.3.3 Anharmonic forces

Both of the expressions for the CGMD Hamiltonian discussed in Section 5.3.1 assume that the underlying MD model is harmonic, i.e. it is quadratic in the momenta and displacements. The MD kinetic energy is always quadratic in the momenta, unless inertia is neglected and then it is even lower order in the momenta. In either case, the derivation given in the previous section needs at most a minor modification to be completely general. The same may not be said of the dependence of the MD Hamiltonian on the displacements. Most interatomic force laws are nonlinear and the corresponding interatomic potentials are not quadratic. They have infinitely strong repulsive forces as the interatomic separation goes to zero to represent the hard-core repulsion of the inner electron shells, and the forces go smoothly to zero at long distances (either at a specified cutoff or at infinity). Quadratic potentials do not have these properties.

However, for small displacements from equilibrium, the potential may be approximated as quadratic. Such an approximation is possible even at finite temperature or under pressure where the lattice constant and even the equilibrium structure may be different than they are at zero temperature and zero pressure [1, 6]. The result is a quasi-harmonic model, such as is familiar from crystal thermodynamics [32]. In such a model the matrix of potential energy coefficients $D_{\mu\nu}$

(or its Fourier transform, the dynamical matrix) becomes a function of the temperature and pressure. This approach automatically includes thermal expansion and the thermal softening of elastic constants through the zero-displacement reference state. Apart from the temperature-dependent stiffness matrix, the derivation of the CGMD Hamiltonian remains the same.

It is also possible to expand the interatomic potentials in a Taylor series, and use that series to derive higher-order expressions for the CGMD Hamiltonian. The fourth-order Hamiltonian is derived in Rudd and Broughton [6]. While this expression provides insight into how CGMD works for more general potentials, it is very complicated for practical uses. The quasi-harmonic approach is better in practice.

5.4 Non-equilibrium effects

In developing CGMD the coarse-grained prescription is based on the ansatz that the internal degrees of freedom are in thermal equilibrium with the modes represented on the mesh. This ansatz is an improvement over eliminating thermal fluctuations entirely, but it is clearly an approximation. Molecular dynamics is often used to study nonequilibrium phenomena, and in processes such as shock waves even the shortest wavelength modes in the system are driven out of equilibrium because of the abrupt rise in the wave. This kind of nonequilibrium phenomenon does not cause a problem in CGMD provided it is contained in the atomically resolved part of the system; however, if it propagates into the CG region problems may ensue.

Firstly, the behavior of the nonequilibrium process may be incorrect in the CG region. In the case of a shock wave, an obvious problem occurs if the mesh is too coarse to represent the abrupt rise of the shock. Even if it is not too coarse, the entropy generated at the short wavelengths affects the form of the wave front, and so the wave profile may not be correct due to the assumption of equilibrium in the internal modes.

Secondly, the transition of a nonequilibrium process from an atomistic or fine-grained region into a coarse-grained region (or vice versa) may not be handled correctly. The example that has been discussed widely for many concurrent multiscale models is acoustic wave propagation and the unphysical reflection of acoustic waves from CG regions or boundaries [1, 4, 21, 25, 33, 34, 35]. In models like CGMD that conserve energy, when a short wavelength mode impinges on a region of the mesh that is too coarse to represent it, the energy must go somewhere. The result is that the wave reflects from the CG region [1]. The same phenomenon occurs in conventional MD simulations in which waves incident on the boundaries of the simulation box must go somewhere. With periodic boundary conditions they wrap around so that outgoing waves become incoming waves; with rigid boundaries, the waves reflect. Either case results in unphysical incoming waves. This wave reflection effect may have significant consequences to a particular simulation [36, 37] or it may be benign, or completely irrelevant to the quantities of interest. Because graded meshes induce wave dispersion that breaks up shock waves, they already mitigate some of the worst effects [2], and may be sufficient for particular applications including crack propagation [21].

In principle, the coarse-grained system should not be a Hamiltonian system [4]. Energy is not conserved exactly because there can be energy exchange between the CG and internal modes. This phenomenon is best known in the context of Brownian motion, where a large particle immersed in a bath of much smaller molecules is observed to be subject to random and dissipative forces. The resulting motion is described by Langevin dynamics rather than Hamiltonian dynamics.

The equations of motion for CGMD with random and dissipative forces may be derived using projection operator techniques [38]. The result is a generalized Langevin equation

$$M_{ij}\ddot{\vec{u}}_j = -K_{ik}\vec{u}_k + \int_{-\infty}^{t} dt'\, \eta_{ik}(t-t')\dot{\vec{u}}_k(t') + \vec{F}_i(t). \tag{5.41}$$

where M_{ij} is a mass matrix, K_{ik} is a stiffness matrix, $\eta_{ik}(\Delta t)$ is a memory function, and $\vec{F}_i$ is a random force that is linked in magnitude and correlation time to $\eta_{ik}(\Delta t)$ through a fluctuation-dissipation theorem [4]. The details of how M_{ij} and K_{ik} are calculated in CGMD are presented in Section 5.3. Here we briefly discuss some interesting features of $\eta(t)$ and $F(t)$ for coarse-grained systems.

If the forces in the CG region are approximately harmonic, the expression for η_{ik} may be derived in terms of the inverse Laplace transform of a certain spectral function. In practice, we have used a lumped mass approximation for M_{ij} to allow explicit integration of the equations of motion [2, 15].

While Langevin dynamics and memory functions provide a solution to the wave reflection problem, the most general Langevin formulation of CGMD has proved very expensive computationally. In particular, if the goal is to construct a massively parallel computer code that is capable of handling millions of atoms in 3D in the MD region, then the Langevin formulation suffers from the additional memory requirements to store the CG variable histories over a correlation time (the maximum Δt such that $\eta_{ik}(\Delta t)$ in (5.41) is not negligible) and the additional inter-processor communication requirements arising from the spatial range of η_{ik}. Powerful simplifications may be possible, but at least in the forms tried to date the Langevin formulation has been difficult to implement in a scalable code, and we do not discuss it in detail here.

One particularly intriguing result was presented in Rudd [4]. In CGMD, because long wavelength elastic waves are able to propagate out onto the coarse mesh, the memory kernel does not need to absorb them. In the case of absorbing boundary conditions at a fixed boundary in MD, the absorption of these long wavelength, low frequency modes requires $\eta_{ij}(\Delta t)$ in (5.41) to be long ranged in both time and space. In CGMD its range is shorter in both time and space, considerably reducing the overhead associated with the memory needed to save the history, and the communication needed to pass the requisite spatial data in a parallel simulation. Another intriguing result is that the memory kernel can act as a propagator, absorbing waves as they impinge on a coarse-grained region and then (if desired) recreating them on the far side as they pass back out into another fine-grained region [4].

5.5 Practical details

Many of the details of how CGMD is implemented are explained in Rudd and Broughton [6]. CGMD is run much the way conventional MD would be. The forces in the CG region are determined by the CGMD stiffness matrix and the nodal displacements; the forces in the MD region are computed from the full interatomic potential in order to capture anharmonic effects. Using the corresponding accelerations, the velocity Verlet time integrator is used to evolve the system in time [15]. The same time step is used throughout the simulation, even though the time step could in principle be lengthened in the CG region since the natural frequency is lower. The CG region entails relatively little computational expense, and a uniform time step works well.

We do not alter the CG mesh during the run. There is no adaptive mesh refinement. The mesh refinement technique would need to be able to position atoms quickly without introducing heat to the system. Since we focus on solids at finite temperature without a thermostat, it is very important to have minimal heat generation from any numerical (unphysical) process, and we have not found a suitable algorithm.

The stiffness matrix K_{ij} is computed once at the start of a simulation. Since the mesh does not change, K_{ij} remains unaltered during the subsequent dynamics. It does not matter whether atoms vibrate across cell boundaries, as long as the crystal lattice topology does not change and diffusion is negligible. The elements of K_{ij} decrease exponentially with distance from the diagonal, and in practice it is necessary to truncate the stiffness matrix in order to control the memory and CPU requirements for simulating large systems with irregular meshes. Doing so allows the simulation of billion atom systems (greatly coarse-grained) on desktop workstations without approximation beyond those presented here.

For $T \neq 0$ the finite-temperature matrix of potential energy coefficients should be used for $D_{\mu\nu}$. This quasiharmonic approximation ensures self-consistent thermodynamics. For example, in ergodic systems the time average of the kinetic energy term in the CG energy (5.28) is related to the temperature through an equipartition expression. While the harmonic approximation may be good in peripheral regions, it is typically not appropriate for the regions that are fully resolved at the atomistic level (the MD region). We have shown that the CGMD and MD equations of motion agree in regions where the mesh coincides with the atomic sites. In these regions, the full MD potential may, and should, be employed, so that effects such as diffusion and dislocation dynamics are allowed.

Each of the expressions for the stiffness matrix, (5.26) and (5.35), involves averaging an underlying matrix $D_{\mu\nu}$ that is the size of the full system of atoms.

In order to simulate large CG regions, an approach that is more efficient that averaging in real space is needed. One approach is to make use of the long-range order in a single crystal in computing the stiffness matrix. The eigenstates of the matrix of potential energy coefficients $D_{\mu\nu}$ are plane waves. For monatomic lattices they correspond to the normal modes of the system, the longitudinal and transverse phonons in the acoustic and optical branches that are familiar from lattice dynamics [32, 39]. In reciprocal space, where the basis elements are exactly these plane waves, the dynamical matrix is diagonal. The inverse of the dynamical matrix is then trivial to compute. It also resolves the technical difficulty that the matrix of potential energy coefficients is singular (its inverse does not exist) due to the three zero energy modes corresponding to uniform translation of the entire system. In reciprocal space the dynamical matrix naturally factorizes into a direct product of the three zero modes with $\mathbf{k} = 0$ and all of the other modes with nonzero eigenvalues. The explicit formula for the stiffness matrix is given in Rudd and Broughton [6]. It also describes at length how the zero-energy center-of-mass modes should be treated in order to achieve an efficient and numerically stable algorithm.

5.5.1 Shape functions in reciprocal space

An extensive discussion of various shape functions is given in [6]. We do not repeat it here. However, some of the analysis we describe below relies on having explicit expressions for the CGMD Hamiltonian in the coarse-grained reciprocal space.

In order to make use of the reciprocal space representation of $D_{\mu\nu}$, it is necessary to have the Fourier transform of the shape functions. The Fourier transform can be computed numerically, of course, using fast Fourier transform (FFT) techniques. In some cases it is also possible to compute the Fourier transform analytically. In this section, we derive the atomic-index Fourier transform of the linear interpolation function in one dimension. The result is immediately applicable to the 4-node square element and the 8-node brick element in two and three dimensions, respectively.

Consider the symmetric linear interpolation function on a regular mesh in one dimension:

$$N_j(x) = \begin{cases} 1 - \left|\frac{x-x_j}{x_{j+1}-x_j}\right| & \text{for } |x - x_j| < x_{j+1} - x_j \\ 0 & \text{otherwise.} \end{cases} \quad (5.42)$$

Let a be the lattice constant, and $N_{per} = (x_{j+1} - x_j)/a$, the number of atoms per mesh cell. It is shown in Rudd and Broughton [6] that well known formulae for summing geometric series can be used to write an analytic expression for the shape function in reciprocal space:

$$N_j(k) = \frac{1}{N_{per}} \frac{\sin^2(kaN_{per}/2)}{\sin^2(ka/2)} e^{ikx_j} \quad (5.43)$$

where N_{per} is the number of lattice sites per CG cell. In higher dimensions, N_{per} would be replaced by N_{per}^x, N_{per}^y, and N_{per}^z. This result applies to the regular CG lattice. Using this formula, we have derived analytic formulae for the spectrum of elastic waves within CGMD, as discussed below. The corresponding formula for a general one-dimensional CG lattice is much more complicated, and using a numerical FFT to calculate it is generally recommended.

5.6 Performance tests

Coarse-grained molecular dynamics has been tested in various ways. In addition to straightforward application of CGMD to conduct simulations, we have also developed a detailed analysis to gain some insight into how CGMD works. In every case we take as the ideal a fully-resolved MD simulation or the pure MD result. The goal is to reproduce the MD result for the observables of interest at a significantly reduced computational expense. We give a brief overview of the detailed analysis here. Specifically we consider the behavior of long wavelength elastic waves: Are their frequency-wavelength relationships correct? Is the scattering of waves off a region of increasing mesh size minimal?

5.6.1 Phonon spectra

The CGMD phonon spectrum offers a good first test of the model. Consider a regular, but not necessarily commensurate CG mesh. The CGMD equations of motion (5.30) for the harmonic system are

$$M_{ij}\,\ddot{u}_j^a = -K_{il}^{ab}\,u_l^b,$$

where $\ddot{u}_j^a$ is the nodal acceleration of the jth node in the ath direction. Substitution of a plane-wave normal mode $u_j^a(t) = u_0^a e^{i\mathbf{k}\cdot\mathbf{x}_j - i\omega t}$ produces the secular equation

$$M(\mathbf{k})\,\omega^2\,\delta_{ab} = K^{ab}(\mathbf{k}) \tag{5.44}$$

where $M(\mathbf{k})$ and $K^{ab}(\mathbf{k})$ are the Fourier transform of the mass and stiffness matrices, respectively. If the crystal lattice has $N_{unitAtom}$ atoms per unit cell, then $M(\mathbf{k})$ is an $N_{unitAtom} \times N_{unitAtom}$ matrix, and $K(\mathbf{k})$ is a $3N_{unitAtom} \times 3N_{unitAtom}$ matrix. For simplicity, we have written the form for the case when there is a single atom per unit cell. For such a monatomic lattice the form of the mass matrix may be simplified further:

$$m^2\omega^2\,\delta_{ab} = \left[M\,(ND^{-1}N^T)_{ab}^{-1}\right](\mathbf{k}) \tag{5.45}$$

$$\omega = \frac{1}{m}\left\{\text{Eigenvalue}\left[M\,(ND^{-1}N^T)^{-1}\right](\mathbf{k})\right\}^{1/2} \tag{5.46}$$

where we have used (5.23) and (5.26). The three eigenvalues of the 3×3 matrix correspond to the three acoustic phonon branches. Were there multiple atoms per primitive unit cell, the MD spectrum would include three acoustic branches and

$3(N_{unitAtom} - 1)$ optical branches. Whether or not CGMD is then constructed to retain the long wavelength optical phonons is a choice that depends on the application.

When the mesh nodes form a sublattice of the equilibrium atomic lattice (i.e. the case of a uniform commensurate mesh), we may go to reciprocal space and the formulae simplify considerably. The CGMD spectrum (5.46) may be computed in closed form for a monatomic solid with a commensurate CG mesh in one dimension, making use of the Fourier transform of the linear shape function (5.43). The normal modes are plane waves both on the underlying ring of atoms and on the CG mesh. The wave vector **k** provides a suitable index for both. The nonzero terms of the dynamical matrix are of the form: $D_{\mu\mu} = 2K, D_{\mu,\mu\pm 1} = -K$. Figure 5.2 shows the resulting phonon spectra in four cases: exact, CGMD, consistent, mass FEM- and lumped-mass FEM.[2]

We presented the analytic expression for the spectrum with nearest neighbor interactions and linear interpolation in [1],

$$\omega(k) = 2\sqrt{\frac{K}{m}} \left(\frac{\sum_p \sin^{-4}\left(\frac{1}{2}ka + \frac{\pi p}{N_{per}}\right)}{\sum_p \sin^{-6}\left(\frac{1}{2}ka + \frac{\pi p}{N_{per}}\right)} \right)^{1/2} \tag{5.47}$$

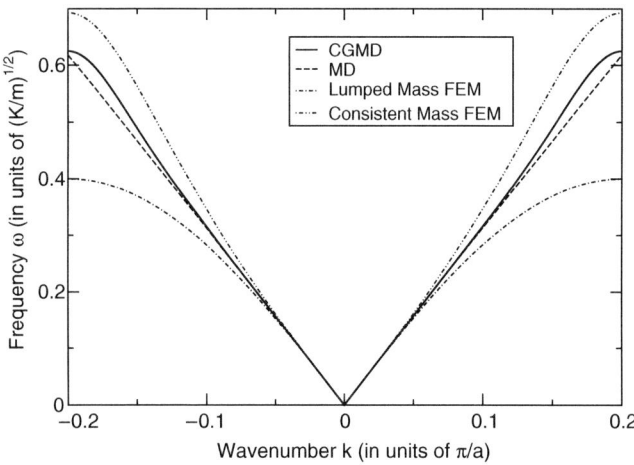

Figure 5.2: The spectrum of elastic waves in a 1D ring, as given by MD, CGMD, lumped-mass FEM, and consistent-mass FEM [1]. The case plotted is with 5 atoms per coarse-grained cell, and the entire CG Brillouin zone is shown, $[-\pi/5a, \pi/5a]$. In these long wave modes, CGMD provides a better approximation to the fully-resolved MD than either implementation of finite elements.

---

[2]The FEM lumped mass matrix is a diagonal approximation of the true consistent mass matrix (cf. [28]). The consistent mass matrix is sometimes called the distributed mass matrix, as we have done in previous publications.

where the sums over p run from 0 to $N_{per} - 1$, $N_{per} = N_{atom}/N_{node}$ is the number of atoms per cell, and K is the nearest-neighbor spring constant. This formula shows the contribution of many modes of the underlying crystal to each CGMD mode, resulting from the choice of interpolation functions which have many normal mode components. Near the center of the CG Brillouin zone, a single mode ($p = 0$) dominates the sums (5.47), reflecting the fact that long wavelength modes are well represented on the CG mesh. Near the boundary of the CG zone $[k \approx N_{node}\pi/(Na)]$, many modes contribute. Many modes are needed because periodicity in the CG reciprocal space forces the slope of the spectrum to zero, and the modes act in concert to keep the CGMD spectrum close to the true spectrum which is not smooth at the boundary. As given in [1], the formula for the FEM spectrum with the consistent mass matrix is

$$\omega^{\text{dist}}(k) = 2\sqrt{\frac{K}{m}} \frac{1}{N_{per}} \frac{\sin\left|\frac{1}{2}kN_{per}a\right|}{\sqrt{1 - \frac{2}{3}\sin^2\left(\frac{1}{2}kN_{per}a\right)}} \qquad (5.48)$$

and the formula for the exact MD spectrum is

$$\omega^{\text{MD}}(k) = 2\sqrt{\frac{K}{m}} \sin\left|\frac{1}{2}ka\right|. \qquad (5.49)$$

The coarse-grained mass and stiffness matrices conspire to produce a Padé approximant of the true spectrum, and thereby achieve the $\mathcal{O}(k^4)$ improved relative error, compared to the $\mathcal{O}(k^2)$ relative error of the two FEM spectra. A similar analytic formula for the spectrum of 3D systems is derived in Rudd and Broughton [6].

Figure 5.2 shows that for the most basic test case CGMD gives a better approximation to the true phonon spectrum than the two kinds of FEM do. All three do a good job at the longest wavelengths, as expected, but CGMD offers a higher order of accuracy. The relative error for CGMD is $\mathcal{O}(k^4)$ while that of the two versions of FEM is only $\mathcal{O}(k^2)$. At shorter wavelengths, there are significant deviations from the exact spectrum. The worst relative error of CGMD is about 6%, better by more than a factor of three than that for FEM. This improvement is made possible by the longer-ranged interactions of CGMD as compared to FEM. The continuity condition satisfied by linear interpolation is enough to ensure that the hydrodynamic modes ($k \sim 0$) are well modeled, but the lack of continuity of the derivatives shows up as error in the spectrum of the modes away from the zone center. This error vanishes for the smooth, nonlocal basis consisting of the longest wavelength normal modes. It turns out that the CGMD error at the CG zone boundary is relatively small (less than 1%) for technical reasons. Also note that even though the number of atoms varies from cell to cell in the incommensurate mesh, the CGMD spectrum is free of anomalies. Other computations have shown that CGMD with linear interpolation is well behaved on irregular meshes, as well.

5.6.2 The finite temperature tantalum CG spectrum

It was shown in [1] that the CGMD phonon spectrum is closer to the true spectrum than that of FEM for a one-dimensional chain of atoms with nearest-neighbor interactions. We have also investigated the fidelity of CGMD spectra in various 3D systems including solid argon and tantalum [2, 6]. CGMD has proven to provide a high quality representation of the part of the MD spectrum that is supported on the mesh i.e., the modes whose wavelengths are twice the mesh spacing or greater. The case of solid argon is presented in detail in [6], including explicit formulae for the CG elastic wave spectra for MD, CGMD, and FEM. The three spectra were compared on fully resolved MD systems, moderately coarse-grained systems with a few atoms per mesh cell, and greatly coarse-grained systems with many atoms per cell.

We have made comparisons on more common materials, too. We consider the case of the phonon spectra of tantalum at room temperature here. Tantalum was chosen to demonstrate CGMD in a more open crystal structure (bcc) and in a system using many-body interatomic potentials. We have used the Finnis-Sinclair many-body potential for tantalum [40] with the improved Ackland-Thetford core repulsion [41]. The FEM stiffness matrix was constructed using the elastic constants for this potential: $C_{11} = 266.0$ GPa, $C_{12} = 161.2$ GPa, and $C_{44} = 82.4$ GPa. We have calculated the finite temperature dynamical matrix in a conventional MD simulation consisting of 2000 atoms in a lattice of $10\times10\times10$ bcc unit cells with periodic boundary conditions. The system is equilibrated to $T = 300 \pm 0.1$K and $P = 0 \pm 10^{-3}$ GPa through scaling of the box size and velocities every 100 time steps until the target temperature and pressure were attained, and then an additional 5000 steps without rescaling to ensure equilibration. The equilibrium lattice constant at this temperature was found to be 3.3129 Å, expanded by 0.2% from the $T = 0$K value of 3.3058 Å.

Subsequently, the dynamical matrix was calculated every 1000 time steps averaged over every atom in the simulation. In principle we are computing an ensemble average, which we have implemented by averaging over the equivalent lattice sites of the system and over multiple time steps (relying on ergodicity for the equivalence of ensemble and temporal averages in equilibrium). In all, we have averaged over a total of ten snapshots of the system and imposed the cubic (O_h) point group symmetry by averaging over the point group operations. The range of the dynamical matrix includes out to the sixth nearest-neighbor shell in tantalum at T=300K. In contrast, the range of the pairwise functions entering the potential includes the first and second nearest-neighbor shells.

The spectra have been computed for three levels of coarse-graining: the atomic limit (one atom per cell or no coarse-graining), a slight coarse-graining (eight atoms per cell), and a case approaching the continuum limit (32,678 atoms per cell). The elastic wave spectra are plotted in Figures 5.3, 5.4, and 5.5. The first figure shows the spectrum of elastic waves on a fully-refined mesh. The CGMD and MD spectra agree exactly and are overlapping. The second figure shows the spectrum on a mesh consisting of eight atoms per rhombohedral cell,

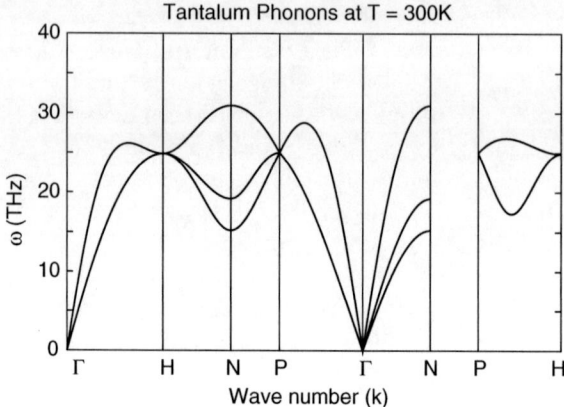

Figure 5.3: The room temperature phonon spectra for solid tantalum on a mesh with no coarse-graining are shown as plots of wave frequencies vs. wave number along various high symmetry lines through the Brillouin zone [6]. The CGMD phonon spectrum agrees exactly with the MD spectrum when the mesh is refined to the atomic limit. It is common practice to leave the gap between the second N and P points since that part of the spectrum is already represented to the left.

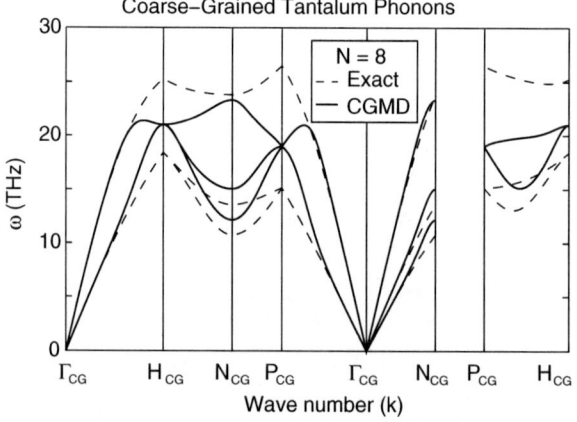

Figure 5.4: The room temperature phonon spectra for solid tantalum on a mesh with 8 atoms per mesh cell are shown as plots of wave frequencies vs. wave number along various high symmetry lines through the Brillouin zone [6]. The CGMD phonon spectrum agrees very well with the MD spectrum in the limit of long wavelengths (near the Γ point), and it agrees reasonably well throughout the coarse-grained Brillouin zone.

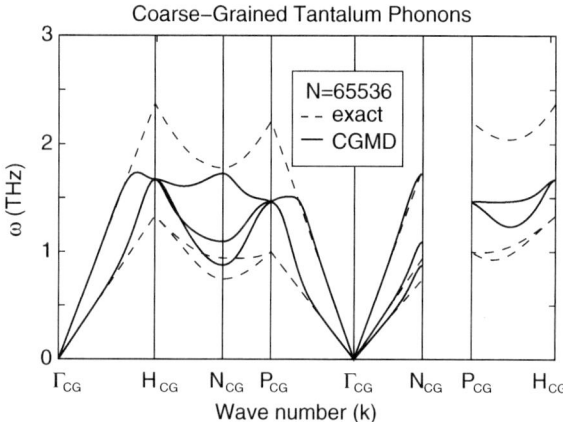

Figure 5.5: The room temperature phonon spectra for solid tantalum on a mesh with $32 \times 32 \times 32$ bcc unit cells (65,536 atoms) per mesh cell are shown as plots of wave frequencies vs. wave number along various high symmetry lines through the Brillouin zone. The CGMD phonon spectrum agrees very well with the MD spectrum in the limit of long wavelengths (near the Γ point), with somewhat greater errors near the coarse-grained Brillouin zone boundary than in the cases of less coarse-graining. The performance is still superior to FEM.

and the third figure shows the case of 32,678 atoms also in a rhombohedral cell, as described in [6]. In these cases with degrees of freedom removed, the MD and CGMD spectra do not agree exactly, but the results are quite good. CGMD performs better than either of the two FEM models (data not shown) in reproducing the MD spectra. The performance is similar in quality to the case of solid argon, shown in [6] with the FEM results. The many-body potential and the more open crystal structure do not have a significant impact on the quality of the results.

Elastic wave spectra are of interest because they provide a means of quantifying the small-amplitude dynamics of the system. They represent the energetics of every normal mode of vibration of a system of atoms. In coarse-grained dynamics, the wave spectra provide an excellent way to quantify the fidelity of the coarse-grained model. Since the normal modes of a crystal are plane waves, they are uniquely identified by a wave number $\mathbf{k}$ and a branch index, for example, indicating a transverse optical mode or a longitudinal acoustic mode. The normal modes correspond to a lattice of points in reciprocal space ($\mathbf{k}$-space) inside a bounded region known as the Brillouin zone. It is not possible to plot $\omega(\mathbf{k})$ for $\mathbf{k}$ throughout the three dimensional Brillouin zone, so typically $\omega(\mathbf{k})$ is plotted along lines, in particular high symmetry lines, through the 3D Brillouin zone (see for example Fig. 22.13 in [42] or Section 11.6 in [43]).

Consider the spectra in the atomic limit shown in Figure 5.3. The CGMD spectrum agrees precisely with the true spectrum. Of the two FEM spectra (data not shown), the lumped-mass spectrum is closer to the true spectrum. This is sensible because the mass is localized to the nodes in the atomic limit, since each node represents one atom. Overall, the lumped-mass frequencies are lower than the true frequencies, whereas the consistent-mass frequencies are higher. This ordering remains true regardless of the level of coarse-graining.

In the continuum limit shown in Figure 5.5, the CGMD spectrum no longer agrees exactly with the true spectrum, but it is still a better approximation than either FEM spectrum (data not shown). In the continuum limit, the consistent mass produces the better spectrum of the two finite element cases. This again makes sense, because the mass is becoming more evenly spread throughout the CG cell. Still, the CGMD spectrum is significantly better than the consistent mass FEM spectrum.

The intermediate case is shown in Figure 5.4. Already the consistent-mass FEM is in better agreement with the MD spectra than the lumped-mass FEM is (data not shown). It is remarkable that cells containing as few as eight atoms are beginning to exhibit continuum behavior. This one-two-many qualitative dependence is typical of large-N expansions, where the large-N limit quickly becomes a good approximation to the real system behavior, and even for N as low as two or three it is a good approximation. The CGMD spectra are again in better agreement with the MD spectra than either of the FEM spectra are.

The shape of each FEM spectrum is the same regardless of the level of coarse-graining. Continuum elasticity is scale invariant, and the changes in FEM spectra are a simple rescaling of frequency and wave number. The scale, N_{per}, enters the FEM consistent mass through a prefactor N_{per}^3, scaling the frequency and the factor of N_{per} in the argument of the cosine scaling the wave number. The same scaling of the wave number is present in the stiffness matrix, but its prefactor goes like N_{per}^3 rather than N_{per}^3. As a result, the frequency ($\sim \sqrt{K/M}$) and the wave number scale like $1/N_{per}$. In the linear portion of the spectra near $k = 0$, the two effects cancel, but the spectra are modified significantly near the zone boundary. Thus, when we discuss the comparison of the MD spectra with the FEM spectra and find better agreement with the lumped-mass FEM for small cells and better agreement with consistent-mass FEM for large cells, it is not that the FEM spectra are changing. The shape of the MD spectra are changing from discrete atomic behavior to continuum behavior, since the CG Brillouin zone retains an ever smaller piece of the full Brillouin zone as the level of coarse-graining increases, and the dispersion curve becomes ever straighter. It is the natural scale dependence of true lattice dynamics. CGMD has this scale dependence arise naturally, as well, and so it is able to a large extent to track the changes in the MD spectra with cell size.

It is not obvious, but even in the long wavelength limit (near the Γ point), the CGMD spectrum is better than either of the two FEM spectra. The relative error for the CGMD spectrum is of order $\mathcal{O}(k^4)$, whereas it is $\mathcal{O}(k^2)$ for the two FEM cases. This improved error was demonstrated above with the formulae for the 1D frequencies, and it continues to hold in 3D. An order $\mathcal{O}(k^2)$ relative error is the naive expectation, since the phonon dispersion relation is linear, with the leading corrections of order $\mathcal{O}(k^3)$ due to symmetry. The higher-order error for CGMD is due to a subtle cancellation between the mass and stiffness matrices. This cancellation can be seen from the formula for the general 1D CGMD spectra (5.47) and its 3D generalization

$$\omega_{ab}^2(\mathbf{k}) = \frac{1}{m} \prod_{b=1}^{3} \frac{\sum_{p_b=1}^{N_{per}} \sin^{-4}(\frac{1}{2}\mathbf{a}_b \cdot \mathbf{k}_{p_b})}{\sum_{p_b=1}^{N_{per}} \sin^{-4}(\frac{1}{2}\mathbf{a}_b \cdot \mathbf{k}_{p_b}) D_{ab}^{-1}(\mathbf{k}_{p_b})} \quad (5.50)$$

where $\mathbf{k}_{p_b} = \mathbf{k} + \frac{2\pi p_b}{N_{per}^b a}\mathbf{b}_b$ and $\mathbf{b}_b$ is the reciprocal lattice basis element. Coarse-grained molecular dynamics naturally produces a subtle cancellation to attain a higher-order accuracy.

5.6.3 Dynamics and scattering

Variation in the mesh size is what makes CGMD a multiscale technique. For example, in studies of crack propagation it is advantageous to introduce a coarse-grained model of the far-field regions away from the crack [21]. The error associated with a varying mesh spacing must be assessed. It is important in a multiscale crack simulation that elastic waves generated at the crack tip do not reflect from the coarse-grained region and perturb the crack propagation [36, 37]. CGMD offers the chance to allow the longest wavelength modes to propagate much farther into the periphery within incurring a commensurately greater computational expense. Other coarse-grained models of the periphery, such as hybrid FEM/MD schemes, may offer the same advantage. In energy-conserving CGMD, the short wavelength modes are reflected from the CG region, but this process is sufficiently dispersive that shock waves are smoothed out and the potential wave-reflection pathologies are averted. The unphysical wave reflection may also produce a nonzero Kapitza resistance at the interface, which can lead to an unphysical temperature gradient across the interface. Of course, a stationary system started in thermal equilibrium remains at a constant, uniform temperature given a reasonable measure of temperature in the CG region, but a system driven out of equilibrium may exhibit unphysical gradients on time scales shorter than the relaxation time.

Given the potential problems associated with wave reflection, we have developed a methodology to quantify the accuracy. The natural measure of the ability of waves to propagate from an atomistic region into a CG region is the S-matrix of scattering theory, or in one dimension, the transmission and reflection coefficients, $\mathcal{T}$ and $\mathcal{R}$, respectively. The basic approach to scattering problems is to look for solutions of the equations of motion of the form of an incoming plane wave and an outgoing spherical wave,

$$\mathbf{u}(\mathbf{r},t) \sim \frac{1}{(2\pi)^{3/2}}\left[e^{i\mathbf{k}\cdot\mathbf{r}-iw\omega t} + f_{\mathbf{k}}(\hat{\mathbf{k}})\frac{e^{ikr-iw\omega t}}{r}\right] \quad (5.51)$$

where this asymptotic form of the displacement field holds well outside the scattering region. The S-matrix and the scattering cross-section may be determined from $f_{\mathbf{k}}(\hat{\mathbf{k}})$. For CGMD, the scattering region is the region where the stiffness matrix differs from the MD dynamical matrix. A tremendous amount of theoretical analysis has been developed for scattering problems [44]. In lattice dynamics, scattering is complicated by the anisotropy of the lattice. The asymptotic form given above (5.51) is only applicable to isotropic wave propagation, but the formalism is readily generalized [6].

We focus on the one-dimensional case, for which the analysis is more straight-forward, and consider the scattering properties of CGMD and FEM, for comparison. The reflection coefficients are computed in the same way for both. The asymptotic region is described by harmonic MD on a regular lattice, and the normal modes are the well-known plane wave solutions. As in (5.51), the asymptotic form of the displacements is known for each frequency:

$$u_j(t) = \begin{cases} A\left(e^{ikx_j - i\omega t} + r\, e^{-ikx_j - i\omega t}\right) & \text{for } j \leq 1 \\ A\, t\, e^{ikx_j - i\omega t} & \text{for } j \geq N \end{cases} \quad (5.52)$$

where the reflection and transmission coefficients are given by $R = |r|^2$ and $T = |t|^2$, respectively. We have assumed that the scattering region is contained between x_1 and x_N, in the sense that these points bound the CG region of the mesh and are separated by more than the range of the MD potential from any coarse-grained cell. This requirement guarantees that the form of $u_j(t)$ in the relation (5.52) is a strict equality and not just an asymptotic relation like (5.51). The wave number k is determined by the frequency, $m\omega^2 = D(k)$, where $D(k)$ is the MD dynamical matrix. The leading coefficient A just determines the amplitude and is irrelevant for our purposes, so we set $A = 1$. In principle, the displacement field could have components with many different frequencies, but since the problem is linear, we may restrict to a single frequency without loss of generality.

The equations of motion (5.30) are given by

$$M_{ij}\ddot{u}_j(t) = -K_{il}u_l(t)$$

and we look for solutions with angular frequency ω,

$$u_j(t) = e^{ikx_j - i\omega t} + v_j e^{-i\omega t} \quad (5.53)$$

such that the asymptotic boundary conditions (5.52) are satisfied:

$$v_j = \begin{cases} r\, e^{-ikx_j} & \text{for } j \leq 1 \\ (t-1)\, e^{ikx_j} & \text{for } j \geq N \end{cases} \quad (5.54)$$

where again these boundary conditions are a strict equality. The equations of motion for v_j are

$$\left(K_{ij} - \omega^2 M_{ij}\right) v_j = -\left(K_{ij} - \omega^2 M_{ij}\right) e^{ikx_j} \quad (5.55)$$

In principle, there are many ways to solve the equations of motion (5.55) with the boundary conditions (5.54); in practice, we have found this problem to be rather subtle. The approach we have taken is based on the observation that the solution in the outer regions can be related to the solution at the boundary points

$$v_{1-n} = e^{inka} v_1 \quad (5.56)$$

$$v_{N+n} = e^{inka} v_N \quad (5.57)$$

where $n \geq 0$. Here a is the lattice constant. Using this trick, the problem is reduced to the calculation of $v_1, \ldots, v_N$ using a set of N equations as described in [6]. Then the scattering coefficients are determined by $\mathcal{R} = |r|^2$ and $\mathcal{T} = |t|^2$ with

$$r = e^{ikx_1} v_1 \quad (5.58)$$

$$t = e^{-ikx_N} v_N + 1 \quad (5.59)$$

which follow from (5.54).

In Figure 5.6 we plot the reflection coefficient $\mathcal{R}(k)$ for scattering from a CG region of 72 nodes representing 652 atoms in the middle of an infinite harmonic

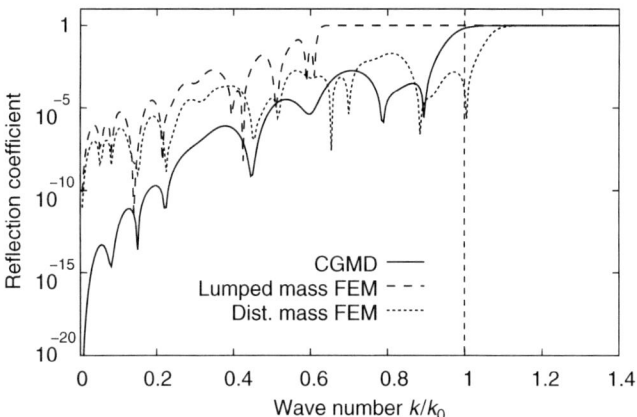

Figure 5.6: A comparison of the reflection of elastic waves from a CG region in three cases: CGMD and two varieties of FEM [1]. Note that the reflection coefficient is plotted on a log scale. The dashed line marks the natural cutoff $[k_0 = \pi/(N_{max}a)]$, where N_{max} is the number of atoms in the largest cells. The bumps in the curves are scattering resonances. Note that at long wavelengths CGMD offers significantly suppressed scattering.

chain of atoms. The reflection coefficients for CGMD, lumped-mass FEM, and consistent-mass FEM are plotted. The lattice used for the calculation is plotted in Figure 13 of Rudd and Broughton [6]. The cell size increases smoothly in the CG region, as it should, to a maximum of $N_{max} = 20$ atoms per cell. In all three cases shown $\mathcal{R}$ becomes exponentially small in the long wavelength limit, and it goes to unity as the wavelength becomes smaller than the mesh spacing—a coarse mesh cannot support short wavelength modes. The threshold occurs approximately at $k = \pi/(N_{max}a)$, the natural cutoff corresponding to a wavelength of $\lambda = 2N_{max}a$. The threshold for CGMD takes place almost exactly at this point because the CGMD phonon frequencies are more accurate than those of the two versions of FEM. According to three-dimensional scattering theory in the limit of wavelengths much larger than the size of the scattering region, the scattering cross-section should vary like $\sigma \sim k^4$. This favorable transmission of long wavelengths is the well known Rayleigh scattering that gives us a blue sky [45]. In these one-dimensional scattering calculations, the trend is for $\mathcal{R} \sim k^\beta$ where the exponent β is roughly $\beta \approx 4 \pm 1$ for FEM and $\beta \approx 8 \pm 2$ for CGMD. We have hypothesized in [6] that the difference is due to the improved accuracy of CGMD at long wavelengths. Using the Born approximation, the scattering strength should go roughly like $r \sim k^\gamma$, where γ is the order of the error, or $\gamma = 2$ in FEM and $\gamma = 4$ in CGMD. Then the reflection coefficient would go like $\mathcal{R} = |r|^2 \sim k^{2\gamma}$, giving $\beta \approx 4$ for FEM and $\beta \approx 8$ for CGMD in rough agreement with the numerical solution; however, this hypothesis has not been proved mathematically and the resonance structure of the scattering curve leads to large uncertainties in the β fit. If we fit to the tops of the peaks at long wavelengths in scattering from an abruptly changing mesh, the uncertainty is reduced to $\beta \approx 4 \pm 0.2$ and 8 ± 0.2 for FEM and CGMD, respectively.

The plot in Figure 5.6 shows a number of interesting features. A series of bumps is visible in each of the curves in the transmissible region. The features are the result of weak scattering resonances, wavelengths at which the scattering cross-section is increased because the frequency of the incoming wave is in resonance with an internal mode of the scattering region. Most of these bumps are not visible when the reflection coefficient is plotted on a linear scale, as in Fig. 2 of Rudd and Broughton [1]. The log scale used in Figure 5.6 brings out these features in regions with extremely low scattering, where the features are more of academic than practical interest. Of course, they are much more peaked on a linear scale, where their width is an indication of the lifetime of the state. The curvature of the peaks in the log-linear plot is low, indicating short-lived resonances. The height of the peaks is an indication of the scattering strength. If a peak were high and narrow, it would indicate a strongly scattering localized mode, which would be pathological behavior in a concurrent multiscale simulation. In general, weak scattering with broad resonances (if any) is desirable. The wave reflections in CGMD are weaker and the resonances much less strong than in FEM, although the shoulder in the reflection coefficient for consistent mass FEM is 10% higher in wavenumber than CGMD because its frequencies are higher.

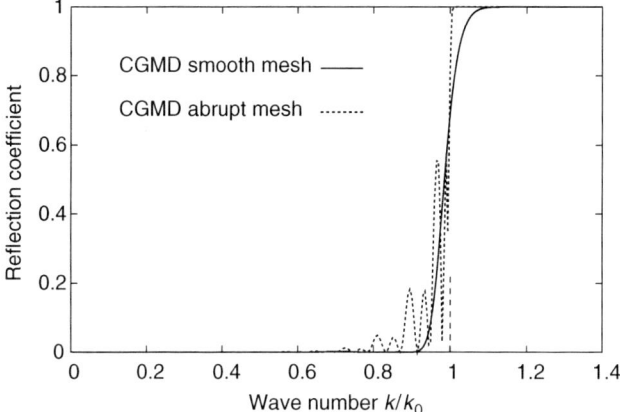

Figure 5.7: A comparison of the reflection of elastic waves from a CG region for a mesh with smoothly varying cell size and one with an abrupt change in cell size, both computed in CGMD [6]. The dashed line marks the natural cutoff $[k_0 = \pi/(N_{max}a)]$. Note that on the linear scale the resonances are not visible for the smooth mesh, but are quite pronounced for the abruptly changing mesh.

It is also interesting to consider what kind of mesh transition between the MD and CG regions is optimal. We have calculated the scattering on many different coarse-grained regions. The general features of the reflection coefficient curves remain the same as the mesh is varied, but the details change. One of the most pronounced changes happens if the mesh varies too abruptly. For a rapidly varying mesh, strong scattering resonances may occur near the threshold, even for CGMD, as shown in Figure 5.7. Unlike the previous figure, this plot has a linear scale so the features are quite strong. The mesh used for the previous comparison of FEM and CGMD was generated using a tanh function for the cell size to ensure smoothness. In contrast, we have plotted in Figure 5.7 the reflection coefficient for a mesh with an abrupt transition to CG region of the same size but consisting almost exclusively of cells of size $N_{max} = 20$, and the abrupt change in mesh size leads to much stronger resonances. The increased reflection of waves with k at resonance could lead to an unphysical size scale, and smooth meshes should be used to prevent this undesirable behavior. Apart from ensuring smoothness, we have not optimized the mesh, and it may be possible to reduce the scattering further still through a more optimized mesh.

In applications like the crack propagation problem, it may be important to consider nonlinear effects as well. In the anharmonic MD crystal, waves of sufficiently large amplitude will steepen into shock waves. The wave velocity increases with the pressure, so that a wavefront with a slow rise to a high pressure will steepen into a step-like shock wave in which the abruptness of the rise is ultimately limited by dissipative processes at the front and lattice effects. So a crack will see acoustic radiation from its own periodic image (or mirror image for hard boundaries), and the compressive waves generated at the crack tip may evolve into shock waves that have a strong impact on the crack propagation. The reflection coefficient does not give the complete picture of this process. It is a property of the system in the small amplitude, harmonic limit, and as such does not give any information about the behavior of shock waves. Shock waves have an abrupt rise and hence have power at short wavelengths localized at the wavefront. When a shock wave is incident on the CG mesh, the short wavelengths are reflected. Since the mesh spacing increases gradually, this reflection disperses the power at the front. The shortest waves are reflected first, then the next shortest and so on. The shock wave is dispersed and much of the power flows out to the CG mesh, so the reflected wave is a low amplitude wave that does not steepen into a shock wave. Thus, while some short wavelength components are reflected, they are not shock waves and do not appear to have an appreciable impact on the processes in the MD region. The majority of the power is carried out into the CG region, effectively delaying its return to the MD region by the transit time across the CG region. The MD/CG interface has an effective Kapitza resistance so the MD region will gradually heat up if no thermostat is used, but typically this heating is minor.

This dispersion and delay in wave reflection due to the CG region is the way CGMD and FEM/MD hybrid methods solve the reflection problem. Several other solutions have been proposed that make use of absorbing boundaries [4, 25, 33, 34, 35], as discussed briefly in Section 5.4. Those techniques have much lower scattering of short wavelengths and hence a lower effective Kapitza resistance at zero temperature. They also involve considerable computer memory usage and considerable coding overhead. At this point it is clear that several approaches exist that solve the wave reflection problem in principle, but it is not yet clear which will offer the ease of use and scalability that will be demanded for widespread use in large-scale simulation.

5.7 Conclusion

Coarse-grained molecular dynamics is a computational technique to couple an MD region seamlessly to a mesh-based coarse-grained region at finite temperature. Both regions of the system are simulated concurrently in a single run. The result is that a large system may be simulated with far fewer degrees of freedom than in a conventional MD simulation, with a corresponding reduction in the required computational resources. This saving has the potential to be quite dramatic. In a simulation we have conducted in the past year for a molten metal system for which a version CGMD has not yet been developed, we used conventional MD with many-body potentials to simulate a cubic micron of material with 90 billion atoms for over one nanosecond of simulated time [46]. The computational cost of this simulation was extraordinary. It required about 200,000 processors and ran for over a month. As we are driven to perform simulations of larger systems, the advantages of multiscale techniques become very compelling.

In practice the formulation of CGMD relies heavily on the properties of a crystal lattice, and it is therefore suited to solids. The formulation discussed here is based on a Hamiltonian, a conserved energy for the system, and is free from ghost forces. We have applied CGMD to three-dimensional systems with interatomic potentials that are many-body in nature and extend well beyond nearest neighbors. We have performed a variety of tests on CGMD, including the elastic wave spectra and the wave scattering properties described in this chapter. In the process, it has been possible to develop analytic formulae for some of the properties that provide a detailed insight into how CGMD, and more generally concurrent multiscale models, work. We have begun to explore extensions of CGMD that work for nonequilibrium systems and systems in which the exchange of energy between the coarse-grained and internal modes is important.

Much remains to be done with CGMD, and we are actively developing the model. CGMD includes temperature through equilibrium thermodynamics, but we have not addressed the question of heat flow in the coarse-grained degrees of freedom. In the nanoscale systems typically treated by MD, much of the elastic wave energy is in ballistic waves; however, in CGMD the CG regions may be larger than the phonon mean free path, so that the elastic wave energy should be diffusive. In effective theories the transition from ballistic to diffusive transport should occur naturally, a feature that would be very nice. At this point the theory remains to be developed, and it would be a natural extension of CGMD.

Many aspects of the numerical efficiency of CGMD could still be improved. A controlled means of bandwidth reduction for the CGMD stiffness matrix is needed. Also an efficient and consistent treatment of wave absorption is an open challenge. We have not discussed the implementation of CGMD on parallel platforms; the decomposition of the MD region into parallel domains is straightforward, but the decomposition of the CG region is less obvious and is linked to the question of the stiffness bandwidth. More fundamentally, CGMD has been based on coherent interfaces between the MD and CG regions. While atoms at the interface can move, the topology of the crystal lattice at the interface is assumed to be unchanging. This assumption precludes interesting effects such as defect motion into or out of the CG region. Other groups are pursuing multiscale techniques that include such effects. It would be very interesting to have a finite temperature methodology that is based on an effective theory for a coarse-grained solid including fluxes across the interface. Of course, there are many ways to take the concurrent multiscale techniques forward, and we can expect many exciting developments in the coming years.

References

[1] Rudd, R. E. and Broughton, J. Q. (1998). "Coarse-grained molecular dynamics and the atomic limit of finite elements", *Phys. Rev. B*, **58**, R5893–R5896.

[2] Rudd R. E. and Broughton J. Q. (2000). "Concurrent coupling of length scales in solid state systems", *Phys. Stat. Sol. (b)*, **217**, p. 251–291.

[3] Rudd R. E. (2001). "Concurrent multiscale modeling of embedded nanomechanics", In *Mat. Res. Soc. Symp. Proc.*, Volume **677**, Warrendale, PA, p. A1.6.1–12. Materials Research Society.
[4] Rudd R. E. (2002). "Coarse-grained molecular dynamics: Dissipation due to internal modes", In *Mat. Res. Soc. Symp. Proc.*, Volume **695**, Warrendale, PA, p. T10.2.1–6. Materials Research Society.
[5] Rudd R. E. (2004). "Coarse-grained molecular dynamics for computer modeling of nanomechanical systems", *Intl. J. on Multiscale Comput. Engin.*, **2**, p. 203–220.
[6] Rudd R. E. and Broughton J. Q. (2005). "Coarse-grained molecular dynamics: Nonlinear finite elements and finite temperature", *Phys. Rev. B*, **72**, 144104 [32 pages].
[7] Rudd R. E., Mason D. R., and Sutton A. P. (2007b). "Lanczos and recursion techniques for multiscale kinetic monte carlo simulations", *Prog. Mater. Sci.*, **52**, p. 319–332.
[8] Argon A. S. (2008). *Strengthening Mechanisms in Crystal Plasticity*. Oxford University Press, Oxford.
[9] Cleland A. N. and Roukes M. L. (1996). "Fabrication of high frequency nanometer scale mechanical resonators from bulk Si crystals", *Appl. Phys. Lett.*, **69**, p. 2653–2655.
[10] Rudd R. E. and Broughton J. Q. (1999a). "Atomistic simulation of MEMS resonators through the coupling of length scales", *J. Modeling and Sim. of Microsys.*, **1**, p. 29–38.
[11] McClintock F. A. and Argon A. S. (1966). *Mechanical Behavior of Materials*, Addison-Wesley, Reading, MA.
[12] Shchukin V. A. and Bimberg D. (1999). "Spontaneous ordering of nanostructures on crystal surfaces", *Rev. Mod. Phys.*, **71**, p. 1125–1171.
[13] Rudd R. E., Briggs G. A. D., Sutton A. P., Medeiros-Ribeiro, G., and Williams, R. S. (2003). "Equilibrium model of bimodal distributions of epitaxial island growth", *Phys. Rev. Lett.*, **90**, 146101 [4 pages].
[14] Rudd R. E., Briggs G. A. D., Sutton A. P., Medeiros-Ribeiro G., and Williams R. S. (2007a). "Equilibrium distributions and the nanostructure diagram for epitaxial quantum dots", *J. Comput. Theor. Nanosci.*, **4**, p. 335–350.
[15] Allen M. P. and Tildesley D. J. (1987). *Computer Simulation of Liquids*, Clarendon Press, Oxford.
[16] Woodward C. and Rao S. I. (2001). "Ab-initio simulation of isolated screw dislocations in bcc Mo and Ta", *Philos. Mag. A*, **81**, p. 1305–1316.
[17] Chu M., Rudd R. E., and Blencowe M. P. (2007). The role of reconstructed surfaces in the intrinsic dissipative dynamics of silicon nanoresonators. *arXiv:0705.0015*.
[18] Seppälä E. T., Belak J., and Rudd R. E. (2004). "Effect of stress-triaxiality on void growth in dynamic fracture of metals: a molecular dynamics study", *Phys. Rev. B*, **69**, 134101 [19 pages].
[19] Kohlhoff S., Gumbsch P., and Fischmeister H. F. (1991). "Crack propagation in b.c.c. crystals studied with a combined finite-element and atomistic model", *Philos. Mag.*, **64**, p. 851–878.

[20] Tadmor E. B., Ortiz M., and Phillips R. (1996). "Quasicontinuum analysis of defects in solids", *Philos. Mag.*, **73**, p. 1529–1563.
[21] Broughton J. Q., Abraham F. F., Bernstein N., and Kaxiras, E. (1999) "Concurrent coupling of length scales: Methodology and application", *Phys. Rev. B*, **60**, p. 2391–2403.
[22] Curtarolo S. and Ceder G. (2002). 'Dynamics of an inhomogeneously coarse grained multiscale system", *Phys. Rev. Lett.*, **88**, 255504 [4 pages].
[23] Miller R. E. and Tadmor E. B. (2002). "The quasicontinuum method: overview, applications and current directions", *J. Comput.-Aided Mater. Design*, **9**, p. 203–239.
[24] Curtin W. A. and Miller R. E. (2003). "Atomistic/continuum coupling in computational materials science", *Modelling Simul. Mater. Sci. Eng.*, **11**, R33–R68.
[25] Park H. S. and Liu W. K. (2004). "An introduction and tutorial on multiple-scale analysis in solids", *Comput. Meth. Appl. Mech. Engng*, **193**, p. 1733–1772.
[26] Dupuy L. M., Tadmor E. B., Miller R. E., and Phillips R. (2005). "Finite-temperature quasicontinuum: Molecular dynamics without all the atoms", *Phys. Rev. Lett.*, **95**, 060202 [4 pages].
[27] Zienkiewicz O. C. and Taylor R. L. (1991). *The Finite Element Method* (4th edn), McGraw-Hill, New York.
[28] Hughes T. J. R. (1982). *The Finite Element Method: Linear Static and Dynamic Finite Element Analysis*, Dover, Mineola.
[29] Rudd R. E. and Broughton J. Q. (1999b). *Coupling of length scales and atomistic simulation of MEMS resonators*. In Proc. DTM '99, Paris, France, March–April 1999, SPIE Vol. **3680**, p. 104–113. SPIE.
[30] Rudd R. E. (2000) *The atomic limit of finite elements in the simulation of micro-resonators*. In Proc. of the 3rd Intl. Conf. on Modeling and Simulation of Microsystems (MSM2000), San Diego, CA, March 27–29, 2000, Boston, MA, p. 465–468. Computational Publications.
[31] Archer J. S. (1963). "Consistent mass matrix for distributed systems". *Proc. ASCE*, **89**(ST4), p. 161–178.
[32] Wallace D. C. (1972). *Thermodynamics of Crystals*. Dover, Mineola.
[33] Adelman S. A. and Doll J. D. (1974). "Generalized Langevin equation approach for atom/solid-surface scattering: Collinear atom/harmonic chain model", *J. Chem. Phys.*, **61**, p. 4242–4245.
[34] Cai W., de Koning M., Bulatov V. V., and Yip S. (2000). "Minimizing boundary reflections in coupled-domain simulations", *Phys. Rev. Lett.*, **85**, p. 3213–3216.
[35] Weinan E. and Huang Z. (2001). "Matching conditions in atomistic-continuum modeling of materials", *Phys. Rev. Lett.*, **87**, 135501 [4 pages].
[36] Abraham F. F., Brodbeck D., Rafey R. A., and Rudge W. E. (1994). "Instability dynamics of fracture: A computer simulation investigation", *Phys. Rev. Lett.*, **73**, p. 272–275.
[37] Holian B. L. and Ravelo R. (1995). "Fracture simulations using large-scale molecular dynamics", *Phys. Rev. B*, **51**, p. 11275–11288.

[38] Zwanzig R. (2001). *Nonequilibrium Statistical Mechanics*, Oxford University Press, Oxford.
[39] Born M. and Huang K. (1954). *Dynamical Theory of Crystal Lattices*, Clarendon Press, Oxford.
[40] Finnis M. W. and Sinclair J. E. (1984). "A simple empirical N-body potential for transition metals", *Philos. Mag.*, **50**, p. 45–55.
[41] Ackland G. J. and Thetford R. (1987). "An improved N-body semi-empirical model for body-centred cubic transition metals", *Philos. Mag.*, **56**, p. 15–30.
[42] Ashcroft N. W. and Mermin N. D. (1976). *Solid State Physics*, Saunders College Press, Philadelphia.
[43] Inui T., Tanabe Y., and Onodera Y. (1996). *Group Theory and its Applications in Physics*, Springer-Verlag, Berlin.
[44] Taylor J. R. (1983). *Scattering Theory*, Krieger, Malabar, FL.
[45] Strutt J. W. (Lord Rayleigh) (1871). "On the light from the sky, its polarisation and colour", I. *Philos. Mag.*, **41**, p. 107–120.
[46] Glosli J. N., Caspersen K. J., Richards D. F., Rudd R. E., Streitz F. H., and Gunnels J. A. (2007). *Micron-scale simulations of kelvin-helmholtz instability with atomistic resolution.* In Proc. Supercomputing 2007 (SC07), Reno, NV, Nov. 2007.

6

BLENDING METHODS FOR COUPLING ATOMISTIC AND CONTINUUM MODELS

P. Bochev, R. Lehoucq, M. Parks, S. Badia, and M. Gunzburger

6.1 Introduction

In this paper we review recent developments in blending methods for atomistic-to-continuum (AtC) coupling in material statics problems. Such methods tie together atomistic and continuum models by using a bridge domain that connects the two models. There are several reasons why AtC coupling methods (of which blending methods are a subset) are important and have been subject to an increased interest in the recent years. Despite tremendous increases in computational power, fully atomistic simulations on an entire model domain remain computationally infeasible for many applications of interest. As a result, attention has focused on hybrid schemes where in all regions with well-behaved solutions, the atomistic (microscopic) model is replaced by a (macroscopic) continuum model enabling a more efficient computational scheme (see [14, 29, 9] for general information). The main challenge is the synthesis of the two distinct models in a manner that minimizes, or altogether eliminates, undesirable artifacts such as ghost forces, unphysical solutions let alone supporting mathematical analysis. Notable AtC coupling methods include the quasicontinuum method [39], the bridging scale decomposition [41], and [24] where atomistic and continuum models are overlapped (see the latter two references, and those mentioned above for numerous citations to the literature). The numerical analysis of AtC methods has lagged in comparison to the number of methods proposed; see the recent papers [2, 3, 17, 31, 26, 27] for analyses of the quasicontinuum method.

Blending methods couple atomistic/continuum modes via a dedicated *blending*, or bridge, region inserted between the atomistic and continuum subdomains. The atomistic and continuum models are tied together by using a suitable "continuity" condition (or balance law) for the atomistic and continuum positions or displacements in this region. A complicating factor is that the atomistic and (classical) continuum elastic models rely on nonlocal and local models of force interaction, respectively. In the (classical) elastic context, the "local force" is that exerted on a body by contact forces occurring on the surface (of the body). In contrast, in the atomistic model, forces are summed from atoms separated by a finite distance. The incompatibility arising from coupling local and nonlocal force models is intrinsic. The goal of our paper is threefold. First, we review how blending approaches attempt to ameliorate the various negative effects by relying on an interface between the two models. Second, we discuss the beginnings of a numerical analysis by discussing consistency of the resulting numerical schemes. Third, we suggest how the intrinsic limitation associated with coupling nonlocal and local mechanical theories may be avoided by replacing the classical elastic theory with a nonlocal elastic theory.

The use of a bridge domain in blending methods bears a resemblance to conventional overlapping domain decomposition schemes (see [33, 40] for discussions and references to the literature). However, blending methods are more complicated because they combine two mathematically distinct descriptions of the material response, and their main goal is to *reconcile* these descriptions over a transitional region. To fix the main ideas below we briefly summarize the basic concepts of AtC blending methods.

6.1.1 An overview of AtC blending methods

Suppose that we want to find the deformed configuration at zero temperature of a material that occupies a bounded region Ω in a two- or three-dimensional Euclidean space. Assume that we have two different mathematical models of this material: an *atomistic* and *continuum* described by atomistic and continuum operators $\mathcal{L}^a$ $\mathcal{L}^c$, respectively. Finally, assume that the problem configuration is such that

- the atomistic model is valid throughout Ω
- solving the atomistic problem on all of Ω is prohibitively expensive
- the continuum model is valid in a subregion $\Omega' \subset \Omega$
- the continuum operator $\mathcal{L}^c$ approximates well (in a suitable sense) the atomistic operator in Ω', but is not valid[1] in $\Omega_a = \Omega \setminus \Omega'$.

The last two assumptions define the scope of hybrid methods: they are unnecessary if the continuum model remains valid throughout Ω, and they are not appropriate unless atomistic and continuum models are "close" on some part of the domain.

In particular, an AtC blending method partitions Ω' into nonoverlapping continuum region Ω_c and a blending region Ω_b so that $\Omega = \Omega_a \cup \Omega_b \cup \Omega_c$; see Figure 6.1. The basic idea is to avoid the computational expense associated with solving the atomistic equations on all of Ω by using instead

- the continuum model in Ω_c (where it is assumed valid)
- the atomistic model in Ω_a (where the continuum model is assumed invalid)
- a blending of the two models over the bridge region Ω_b.

The cost of an ensuing computational method is minimized when Ω_a and Ω_b are small relative to Ω_c.

A typical AtC blending method has four main ingredients:

- blending functions θ_a and θ_c that form a partition of unity on Ω
- an operator $\mathcal{L}_\theta^a$ acting on $\Omega_a \cup \Omega_b$, and such that $\mathcal{L}_\theta^a|_{\Omega_a} = \mathcal{L}^a$

---

[1] By this we mean that the physical phenomena in Ω' cannot be modeled well by $\mathcal{L}^c$.

Introduction

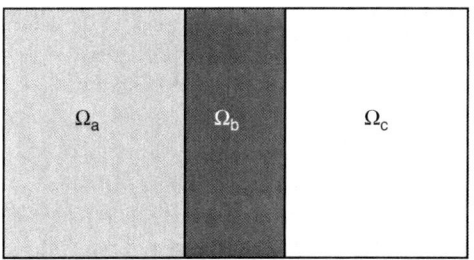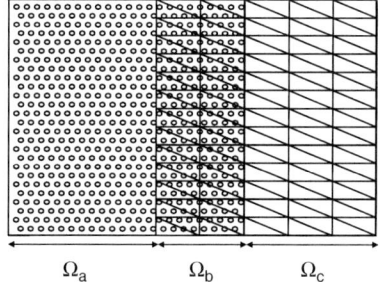

Figure 6.1: Typical domain configurations for AtC blending methods. The left plot depicts the original problem domain Ω and its partition into an atomistic domain Ω_a, continuum domain Ω_c, and a bridge domain Ω_b. The right plots depict the atomic lattice in $\Omega_a \cup \Omega_b$ and a finite element mesh in $\Omega_b \cup \Omega_c$ for the continuum model.

- an operator $\mathcal{L}_\theta^c$ acting on $\Omega_b \cup \Omega_c$, and such that $\mathcal{L}_\theta^c|_{\Omega_c} = \mathcal{L}^c$
- an operator $\mathcal{C}$ acting on Ω_b.

Operators $\mathcal{L}_\theta^a$ and $\mathcal{L}_\theta^c$ are *blended* versions of the original atomistic and continuum operators $\mathcal{L}^a$ and $\mathcal{L}^c$, defined using the blending functions. Their purpose is to avoid "duplication" of the material response in the bridge region that results from simply superimposing the two models in Ω_b. $\mathcal{C}$ is a constraint operator that enforces the "continuity" between atomistic and continuum solutions in the bridge region.

AtC blending methods are a relatively recent development that is driven by simulation needs in nanotechnology and material sciences. These simulation needs have resulted in the creation of many *ad hoc* AtC blending methods that are often loosely defined and focused on specific problems. This makes the numerical analysis of blending methods difficult. Depending on the atomistic and continuum models used in the AtC method $\mathcal{L}^a$ and $\mathcal{L}^c$ may correspond to a force equilibrium or an energy minimization principle. In the first case these operators are related to Newton's second law and we talk about *force-based* AtC blending. Examples of force-based blending methods derived from mechanical arguments are given in [4, 5, 20]. In the second case AtC blending is *energy-based*. A representative example of an energy-based method, defined by blending of atomistic and continuum energy functionals, can be found in [8]. One-dimensional results and analysis for blending harmonic potentials and linear elasticity via the Arlequin method [16] is the subject of [6, 34]. The overlapping AtC method presented in [24] can also be considered a blending method, and demonstrates how ghost forces can be eliminated by considering a patch test. We also remark that blending methods have been introduced within the context of meshfree methods [22] where the atomistic region is replaced by a region discretized with a meshfree method.

Our paper reviews an abstract framework for force- or energy-based AtC blending methods that includes precise notions of patch and consistency tests. Our discussion on energy-based blending methods is new work. The framework allows us to identify four general classes of AtC methods, explains the origin of so-called *ghost forces*, and the satisfaction of Newton's third law.

We have organized the paper as follows. The model atomistic and continuum problems are introduced in Section 6.2. Section 6.3 discusses force-based AtC blending methods and their taxonomy, and states formal definitions of consistency and patch test. Energy-based AtC blending methods and their taxonomy are briefly discussed in Section 6.4. Our conclusions are summarized in Section 6.6.

6.1.2 *Notation*

The following notation is used throughout the paper. Double fonts ($\mathbb{A}$) denote sets of atoms, except for $\mathbb{R}^d$ and $\mathbb{R}$ that stand for d-dimensional and 1-dimensional Euclidean spaces, respectively. Standard upper-case fonts (A) are used for atomistic and continuum spaces, calligraphic fonts ($\mathcal{L}$) for operators and functionals, lower-case bold letters ($\boldsymbol{a}$) for vector-valued continuum functions, lower-case bold Greek letters ($\boldsymbol{\alpha}$) for vector-valued atomistic (discrete) functions and Lagrange multipliers.

A superscript in the space designation indicates the support of the functions in this space and a subscript shows the type of boundary constraint imposed on its elements. For example, the elements of X^{ab} are supported on $\Omega_a \cup \Omega_b$, the elements of X_0 vanish on the boundary, and X_D is an affine space whose elements are subject to an inhomogeneous boundary condition. A lack of superscript, e.g., X, indicates that the elements of the space are supported on all of Ω. Likewise, absence of a subscript means that the space is not constrained by boundary conditions.

Dual spaces are denoted by $(\cdot)'$. In general, discrete and continuous L^2 inner products, L^2 norms, and duality pairings are denoted by $(\cdot,\cdot)$, $\|\cdot\|$, and $\langle\cdot,\cdot\rangle$, respectively. Additional notation will be introduced as needed.

6.2 Atomistic and continuum models

This section provides a brief summary of the basic atomistic and continuum material statics models that will be used in the paper. Because our main focus is on mathematical aspects of AtC blending, rater than material modeling, we intentionally choose the simplest possible material models.

6.2.1 *Force-based models*

Force-based atomistic and continuum models are derived by equilibrating internal and applied forces at each atom or material point. Material properties in this case are encoded in terms of *internal force* operators. For continuum materials internal forces are usually described by partial differential operators, although other (non-local, integral) choices are also possible, as discussed in Section 6.5.

6.2.1.1 The atomistic model

Let $\mathbb{P}$ denote an undeformed (or reference) lattice of $|\mathbb{P}|$ identical particles located in a bounded, Euclidean region $\Omega \subset \mathbb{R}^d$. The spatial position vectors of the particle $\alpha \in \mathbb{P}$ in the undeformed and deformed configurations are $\boldsymbol{x}_\alpha$ and $\boldsymbol{q}_\alpha$, and $\boldsymbol{\psi}_\alpha = \boldsymbol{q}_\alpha - \boldsymbol{x}_\alpha$ is the displacement vector of α. The set of all possible atomistic displacements is a finite dimensional space X whose elements $\boldsymbol{\phi} \in X$ are sets of properly ordered $|\mathbb{P}| \times d$ scalar values ϕ_α^i for $\alpha \in \mathbb{P}$ and $i = 1, \ldots, d$.

Let $\mathbb{D} \subset \mathbb{P}$ denote the subset of particles whose positions in the deformed configuration are fixed. We define the affine space

$$X_D := \{\boldsymbol{\phi} \in X \mid \boldsymbol{\phi}_\alpha = \boldsymbol{\psi}_\alpha^\mathbb{D} \ \forall \alpha \in \mathbb{D}\},$$

where $\boldsymbol{\psi}_\alpha^\mathbb{D}$ is the given, fixed displacement vector, and the subspace

$$X_0 := \{\boldsymbol{\phi} \in X \mid \boldsymbol{\phi}_\alpha = \boldsymbol{0} \ \forall \alpha \in \mathbb{D}\}$$

of X in which all particles from $\mathbb{D}$ have zero displacements. These spaces are atomistic counterparts of Sobolev spaces constrained by inhomogeneous and homogeneous Dirichlet boundary conditions for a continuum problem.

The strong form of the lattice statics problem consists of finding an equilibrium (deformed) configuration $\{\boldsymbol{\psi}_\alpha\}_{\alpha \in \mathbb{P} \setminus \mathbb{D}}$ which satisfies the force-balance problem

$$\left(\mathcal{L}^a(\boldsymbol{\psi})\right)_\alpha + \boldsymbol{\chi}_\alpha = \boldsymbol{0} \qquad \forall \alpha \in \mathbb{P} \setminus \mathbb{D} \tag{6.1a}$$

$$\boldsymbol{\psi}_\alpha = \boldsymbol{\psi}_\alpha^\mathbb{D} \qquad \forall \alpha \in \mathbb{D} \tag{6.1b}$$

In (6.1a) $(\mathcal{L}^a(\cdot))_\alpha$ and $\boldsymbol{\chi}_\alpha$ are the internal and external forces acting on the particle α, respectively. Therefore, the atomistic problem (6.1) is simply Newton's second law for a system of particles interacting via the force operator $\mathcal{L}^a$ and the applied forces $\boldsymbol{\chi}$, and constrained to satisfy the "boundary condition" (6.1b). Note that $\mathcal{L}^a : X \to X$ is generally a non-linear operator.

A *weak formulation* of (6.1) is as follows: find $\boldsymbol{\psi} \in X_D$ such that

$$\mathcal{B}^a(\boldsymbol{\psi}, \boldsymbol{\phi}) = \mathcal{G}^a(\boldsymbol{\phi}) \qquad \forall \boldsymbol{\phi} \in X_0, \tag{6.2}$$

where

$$\mathcal{B}^a(\boldsymbol{\psi}, \boldsymbol{\phi}) = (\mathcal{L}^a(\boldsymbol{\psi}), \boldsymbol{\phi}) \quad \text{and} \quad \mathcal{G}^a(\boldsymbol{\phi}) = -(\boldsymbol{\chi}, \boldsymbol{\phi}) \quad \text{for } \boldsymbol{\psi} \in X, \boldsymbol{\phi} \in X_0. \tag{6.3}$$

Equation (6.2) is the principle of virtual work. While it is similar in form to variational equations arising from continuum models, it is important to keep in mind that force balance equations (6.1a) are inherently *non-local* in nature, i.e., in general, the force acting on a particle $\alpha \in \mathbb{P} \setminus \mathbb{D}$ depends on the displacements of many other particles separated by a finite distance.

6.2.1.2 The continuum model

Let $\boldsymbol{x}$, $\boldsymbol{q}$, and $\boldsymbol{u} = \boldsymbol{q} - \boldsymbol{x}$ be the spatial position vectors in the undeformed and deformed continuum configurations, and the continuum displacement vector, respectively. The strong form of the continuum model is

$$\mathcal{L}^c(\boldsymbol{u}) = \boldsymbol{f} \quad \text{in } \Omega, \tag{6.4a}$$

$$\boldsymbol{u} = \boldsymbol{u}^{\partial \Omega} \quad \text{on } \partial\Omega, \tag{6.4b}$$

where $\mathcal{L}^c$ denotes a (possibly nonlinear) differential operator, $\boldsymbol{f}$ the external force, $\partial\Omega$ the boundary of Ω, and $\boldsymbol{u}^{\partial\Omega}$ the prescribed boundary data.[2] In what follows we assume that (6.4) is a *local* model in the sense that the stress at a point $\boldsymbol{x} \in \Omega$ depends only on the values of $\boldsymbol{u}$ and $\nabla \boldsymbol{u}$ at that point. We also assume that there exist differential operators $\mathcal{L}_S^c(\cdot)$ and $\mathcal{L}_E^c(\cdot)$ such that

$$\langle \mathcal{L}^c(\boldsymbol{u}), \boldsymbol{v} \rangle = \int_\Omega \mathcal{L}_S^c(\boldsymbol{u}) : \mathcal{L}_E^c(\boldsymbol{v}) d\boldsymbol{x}$$

for all smooth functions $\boldsymbol{u}$ and $\boldsymbol{v}$ with $\boldsymbol{v} = \boldsymbol{0}$ on $\partial\Omega$; $(\cdot) : (\cdot)$ denotes the scalar tensor product operation. To state the weak form of (6.4) let Y denote a Hilbert space, defined with respect to Ω, and such that for any $\boldsymbol{v} \in Y$, $\mathcal{L}_S^c(\boldsymbol{v})$ and $\mathcal{L}_E^c(\boldsymbol{v})$ are meaningful in L^2 sense. We define the affine subspace and subspace

$$Y_D := \{\boldsymbol{v} \in Y \mid \boldsymbol{v} = \boldsymbol{u}^{\partial\Omega} \text{ on } \partial\Omega\} \quad \text{and} \quad Y_0 := \{\boldsymbol{v} \in Y \mid \boldsymbol{v} = \boldsymbol{0} \text{ on } \partial\Omega\},$$

respectively, and the functionals

$$\mathcal{B}^c(\boldsymbol{u}, \boldsymbol{v}) := \int_\Omega \mathcal{L}_S^c(\boldsymbol{u}) : \mathcal{L}_E^c(\boldsymbol{v}) d\boldsymbol{x} \quad \text{and} \quad \mathcal{G}^c(\boldsymbol{v}) = \langle \boldsymbol{f}, \boldsymbol{v} \rangle \quad \text{for } \boldsymbol{u} \in Y, \, \boldsymbol{v} \in Y_0. \tag{6.5}$$

Then, a *weak formulation* of (6.4) is: given $\boldsymbol{f} \in (Y_0)'$, find $\boldsymbol{u} \in Y_D$ such that

$$\mathcal{B}^c(\boldsymbol{u}, \boldsymbol{v}) = \mathcal{G}^c(\boldsymbol{v}) \quad \forall \, \boldsymbol{v} \in Y_0. \tag{6.6}$$

Note that, as a rule, the operator $\mathcal{L}_E^c(\cdot)$ is linear but, in general, $\mathcal{L}_S^c(\cdot)$ is nonlinear.

6.2.2 Energy-based models

Energy-based models in material statics postulate that the equilibrium (deformed) configuration of an atomistic or a continuum system minimizes the potential energy of the system. For these models material properties are encoded in terms of *potential energy* functionals.

---

[2] For simplicity, we consider only Dirichlet boundary conditions.

6.2.2.1 The atomistic model

We retain the notation from the force-based atomistic model. In particular, $\mathbb{P}$ is an undeformed lattice of identical particles in Ω and ψ_α is the displacement of the particle α. The (atomistic) potential energy of the particle system is given by the expression

$$\mathcal{E}^a(\psi) = \sum_{\alpha \in \mathbb{P}} \left(W_\alpha^a(\psi) + \chi_\alpha \cdot \psi_\alpha \right), \tag{6.7}$$

where $W_\alpha^a(\cdot)$ and χ_α denote the potential energy associated with the particle α, and the external force applied to that particle. Note that W_α^a may depend on the displacements of all the particles ψ_β, $\beta \in \mathbb{P}$, although in many cases it will depend only on the displacement of the particles within some ball $\mathcal{B}_\alpha = \{\mathbf{x} \in \Omega : |\mathbf{x} - \mathbf{x}_\alpha| \leq r\}$ for some given $r > 0$. As an example, we could set W_α^a to the Leonard-Jones potential.

The equilibrium deformed atomistic configuration ψ is characterized by

$$\psi = \arg \min_{\phi \in X_D} \mathcal{E}^a(\phi) \tag{6.8}$$

As a result, $\mathbf{u}$ is subject to the first-order necessary optimality condition

$$\delta_\psi \mathcal{E}^a(\psi) \equiv \lim_{\epsilon \to 0} \frac{d}{d\epsilon} \mathcal{E}^a(\psi + \epsilon \phi) = 0 \quad \forall \phi \in X_0. \tag{6.9}$$

This problem can be further replaced by a weak variational formulation that has the same form as (6.2).

6.2.2.2 The continuum model

Suppose that Ω is occupied by a continuum material with a prescribed boundary displacement $\mathbf{u}^{\partial\Omega}$, and whose potential energy at a point x in the deformed configuration is given by $W^c(\mathbf{u})$. Then, the total potential energy associated with this material is

$$\mathcal{E}^c(\mathbf{u}) = \int_\Omega \left(W^c(\mathbf{u}) + \mathbf{f} \cdot \mathbf{u} \right) d\Omega, \tag{6.10}$$

where $\mathbf{f}$ is the external volumetric force applied at each point in Ω. The equilibrium deformed configuration is characterized by an energy principle similar to (6.8),

$$\mathbf{u} = \arg \min_{\mathbf{v} \in X_D} \mathcal{E}^c(\mathbf{v}), \tag{6.11}$$

and is subject to an analogous first-order optimality condition:

$$\delta_\mathbf{u} \mathcal{E}^c(\mathbf{u}) \equiv \lim_{\epsilon \to 0} \frac{d}{d\epsilon} \mathcal{E}^c(\mathbf{u} + \epsilon \mathbf{v}) = 0 \quad \forall \mathbf{v} \in Y_0. \tag{6.12}$$

This problem also can be replaced by a weak variational equation that assumes the same form as (6.6). As an example, we could set $W^c(\boldsymbol{u}) = \sigma(\boldsymbol{u}) : \varepsilon(\boldsymbol{u})$ where σ and ε denote the work conjugate stress and strain tensors, respectively. Then, minimizers of (6.11) solve the weak equation: seek $\boldsymbol{u} \in Y_D$ such that

$$\int_\Omega \sigma(\boldsymbol{u}) : \varepsilon(\boldsymbol{v}) d\Omega = \int_\Omega \boldsymbol{f} \cdot \boldsymbol{v} d\Omega \quad \forall \boldsymbol{v} \in Y_0. \tag{6.13}$$

With $\mathcal{L}_S^c(\cdot) = \sigma(\cdot)$ and $\mathcal{L}_E^c(\cdot) = \varepsilon(\cdot)$, the abstract weak problem (6.6) reduces to (6.13).

6.3 Force-based blending

This section reviews an AtC blending method to merge force-based atomistic and continuum material models. The blending process is effected by using the weak forms (6.2) and (6.6) of the models. General definitions of consistency and patch tests for AtC methods are also reviewed.

6.3.1 An abstract AtC blending method

Let $\mathbb{A}$, $\mathbb{B}$, and $\mathbb{C}$ denote the particles associated with Ω_a, Ω_b, and Ω_c, respectively; particles on the interfaces between Ω_b and the other two subdomains are assigned to $\mathbb{B}$. We introduce the atomistic spaces

$$X_0^{ab} := \{(\phi_\alpha)|_{\alpha \in \mathbb{A} \cup \mathbb{B}} \mid \phi \in X_0\} \quad \text{and} \quad X_D^b := \{(\phi_\alpha)|_{\alpha \in \mathbb{B}} \mid \phi_\alpha \in X_D\},$$

the continuum subspace

$$Y_0^{bc} := \{\boldsymbol{v}|_{\Omega_b \cup \Omega_c} \mid \boldsymbol{v} \in Y_0\}$$

and the continuum affine subspaces

$$Y_D^{bc} := \{\boldsymbol{v}|_{\Omega_b \cup \Omega_c} \mid \boldsymbol{v} \in Y_D\} \quad \text{and} \quad Y_D^b := \{\boldsymbol{v}|_{\Omega_b} \mid \boldsymbol{v} \in Y_D\}.$$

A force-based AtC blending method has the following key ingredients:

1. *atomistic and continuum blending functions* θ_a and θ_c, respectively, such that $\theta_a \geq 0$, $\theta_c \geq 0$, $\theta_a = 1$ in Ω_a, $\theta_c = 1$ in Ω_c and $\theta_a + \theta_c = 1$ in Ω;
2. a *constraint operator* $\mathcal{C}(\cdot, \cdot) : X_D^b \times Y_D^b \to Q$, where Q is a function space whose definition depends on the particular nature of the constraints.
3. *blended atomistic functionals* $\mathcal{B}_\theta^a(\cdot, \cdot; \theta_a) : X_D \times X_0^{ab} \to \mathbb{R}$ and $\mathcal{G}_\theta^a(\cdot; \theta_a) : X_0^{ab} \to \mathbb{R}$ such that

    $$\mathcal{B}_\theta^a(\psi, \phi; 1) = \mathcal{B}^a(\psi, \phi) \quad \text{and} \quad \mathcal{G}_\theta^a(\phi; 1) = \mathcal{G}^a(\phi) \quad \text{for all } \{\psi, \phi\} \in X_D \times X_0^{ab};$$

4. *blended continuous functionals* $\mathcal{B}_\theta^c(\cdot, \cdot; \theta_c) : Y_D^{bc} \times Y_0^{bc} \to \mathbb{R}$ and $\mathcal{G}_\theta^c(\cdot; \theta_c) : Y_0^{bc} \to \mathbb{R}$ such that

    $$\mathcal{B}_\theta^c(\boldsymbol{u}, \boldsymbol{v}; 1) = \mathcal{B}^c(\boldsymbol{u}, \boldsymbol{v}) \quad \text{and} \quad \mathcal{G}_\theta^c(\boldsymbol{v}; 1) = \mathcal{G}^c(\boldsymbol{v}) \quad \text{for all } \{\boldsymbol{u}, \boldsymbol{v}\} \in Y_D^{bc} \times Y_0^{bc};$$

Using these definitions, an abstract, force-based AtC blending method can be expressed in the following form: find $\{\boldsymbol{\psi}, \boldsymbol{u}\} \in X_D \times Y_D^{bc}$ such that

$$\begin{cases} \mathcal{B}_\theta^a(\boldsymbol{\psi}, \boldsymbol{\phi}; \theta_a) + \mathcal{B}_\theta^c(\boldsymbol{u}, \boldsymbol{v}; \theta_c) = \mathcal{G}_\theta^a(\boldsymbol{\phi}; \theta_a) + \mathcal{G}_\theta^c(\boldsymbol{v}; \theta_c) \ \forall \{\boldsymbol{\phi}, \boldsymbol{v}\} \in X_0^{ab} \times Y_0^{bc} \\ \qquad\qquad\qquad\qquad \text{subject to} \\ \boldsymbol{\psi}_\alpha = \boldsymbol{u}(\boldsymbol{x}_\alpha) \ \forall \alpha \in \mathbb{C} \setminus (\mathbb{C} \cap \mathbb{D}) \quad \text{and} \quad \mathcal{C}(\boldsymbol{\psi}, \boldsymbol{u}) = 0 \ \forall \{\boldsymbol{\psi}, \boldsymbol{u}\} \in X_D^b \times Y_D^b \end{cases}$$
(6.14)

The first equation in (6.14) is a weak *blended force-balance* equation. The second equation is a constraint which states that atomistic displacements in Ω_c are slaved to the continuum displacements, and the last equation is a constraint that ties the two models together over the blend region. A particular AtC blending method results from choosing the four key ingredients above, i.e., the blending functions and definitions of the functionals $\mathcal{B}_\theta^a$, $\mathcal{B}_\theta^c$, $\mathcal{G}_\theta^c$, $\mathcal{G}_\theta^a$, and $\mathcal{C}$.

Note that in (6.14) the atomistic test functions from X_0^{ab} are supported only on $\mathbb{A} \cup \mathbb{B}$ and the continuum test functions from Y_0^{bc} are supported only on $\Omega_b \cup \Omega_c$. As a result, away from the bridge region the equations in the AtC blending method default to their atomistic and continuum definitions, respectively.

Due to the non-local nature of the atomistic model, the particles included in the force balance equations in $\Omega_a \cup \Omega_b$ will interact with at least some of the particles in Ω_c. This fact is reflected in the choice of X_D as a trial space for the atomistic part of the solution in the AtC blending method. Of course, in (6.14) we never solve for the displacements of the particles in Ω_c, instead, whenever their displacements are needed we approximate them by the continuum displacement at the location of the particle according to the first constraint in (6.14). This requires us to assume that certain types of materials and behaviors, such as multilattice materials or phase transitions, are not present in Ω_c.

Remark 6.1 Restricting the atomistic trial space in (6.14) to particles in $\Omega_a \cup \Omega_b$ only, will result in neglecting the forces acting on the particles in $\Omega_a \cup \Omega_b$ due to the particles in Ω_c. This gives rise to what is known as the *ghost force* effect [25, 28]. The AtC blending method (6.14) mitigates the ghost force effect in two ways. First, owing to the choice of the atomistic trial space, interactions between the particles in $\Omega_a \cup \Omega_b$ and Ω_c are included in the problem. Second, because the atomistic blending function θ_a has small values near Ω_c, the errors caused by the use of approximate slaved atomistic displacements will be greatly reduced.

An equally important, but much less discussed issue, is the inconsistency that occurs in the continuum model at the interface between Ω_a and Ω_b. Simply restricting the weak continuum equation (6.6) to $\Omega_b \cup \Omega_c$ forces the *unphysical* natural boundary condition $\mathcal{L}_S^c(\boldsymbol{u}) \cdot \mathbf{n} = \mathbf{0}$ along that interface. In the blended method (6.14) the adverse effects from this artificial boundary condition are greatly reduced or altogether removed thanks to the fact that $\theta_c = 0$ on the interface between Ω_a and Ω_b.

6.3.2 Blending functions

With this section we begin to examine the four key ingredients of (6.14). The atomistic and continuum blending functions θ_a and θ_c are the first ingredient of a blending method. Their main purpose is to ensure that atomistic and continuum forces are *blended* rather than *superimposed* in the bridge region. In addition, judicious choice of θ_a can help reduce the ghost forces and errors due to slaving of the atomistic displacements in (6.14); see Remark 6.1. Likewise, with a suitable choice of θ_c one can avoid the imposition of an artificial natural boundary condition on the interface between Ω_a and Ω_b.

It is clear that in practice, to define θ_a and θ_c, it suffices to pick a single blending function θ such that $0 \leq \theta \leq 1$ in Ω, $\theta = 1$ in Ω_a, and $\theta = 0$ on Ω_c. Then, one can set $\theta_a = \theta$ and $\theta_c = 1 - \theta$. A key requirement is that θ is small near the interface between Ω_b and Ω_c (so that θ_a is small there) and that it be close to one near the interface between Ω_a and Ω_b (so that θ_c is small there.) Methods for constructing the blending function θ are discussed in [4].

To achieve these desirable properties the blending functions have to be at least of class C^0. The continuum blending function θ_c will also be required to satisfy the following property.

6.3.3 Assumption

For every $v \in Y$, we have that $\theta_c v \in Y$.

6.3.4 Enforcing the constraints

Recall that the blended weak problem (6.14) is subject to two constraints. The first one is simply a slaving condition which postulates that *atomic* displacements in the *continuum* region are slaved to the continuum displacements. It can be trivially imposed by simply substituting the appropriate atomistic displacements in the weak equations by the associated continuum values. Thus, in what follows we shall assume that this constraint had already been enforced in the blended problem.

The second constraint is non-trivial in the sense that in the bridge region the atomistic and continuum models coexist and their displacements must be reconciled in a physically meaningful sense. As a result, the operator $\mathcal{C}$ in (6.14) can have a rather general form which may make direct imposition of this constraint more difficult. We shall discuss some specific examples of this operator after examining two possible strategies for its enforcement.

The first strategy is to use Lagrange multipliers, which leads to the mixed problem [11]: find $\{\psi, u\} \in X_D \times Y_D^{bc}$, and $\boldsymbol{\lambda} \in Q'$ such that

$$\begin{cases} \mathcal{B}_\theta^a(\psi, \phi; \theta_a) + \mathcal{B}_\theta^c(u, v; \theta_c) + \langle (\delta_\psi \mathcal{C}(\psi, u))\phi, \boldsymbol{\lambda} \rangle + \langle (\delta_u \mathcal{C}(\psi, u))v, \boldsymbol{\lambda} \rangle \\ \quad = \mathcal{G}_\theta^a(\phi; \theta_a) + \mathcal{G}_\theta^c(v; \theta_c) \qquad \forall \{\phi, v\} \in X_0^{ab} \times Y_0^{bc} \\ \langle \mathcal{C}(\psi, u), \boldsymbol{\mu} \rangle = 0 \qquad \forall \boldsymbol{\mu} \in Q' \end{cases} \quad (6.15)$$

where $\delta_\psi \mathcal{C}(\cdot,\cdot)$ and $\delta_u \mathcal{C}(\cdot,\cdot)$ are the Gâteaux derivatives of $\mathcal{C}(\cdot,\cdot)$ with respect to ψ and u, respectively and Q' is a suitable Lagrange multiplier space. Since $\phi \in X_0^{ab}$ and $v \in Y_0^{bc}$ are independent of each other, (6.15) assumes the form

$$\begin{cases} \mathcal{B}_\theta^a(\psi,\phi;\theta_a) + \langle (\delta_\psi \mathcal{C}(\psi,u))\phi, \lambda \rangle = \mathcal{G}_\theta^a(\phi;\theta_a) \ \forall \phi \in X_0^{ab} \\ \mathcal{B}_\theta^c(u,v;\theta_c) + \langle (\delta_u \mathcal{C}(\psi,u))v, \lambda \rangle = \mathcal{G}_\theta^c(v;\theta_c) \ \forall v \in Y_0^{bc} \\ \langle \mathcal{C}(\psi,u), \mu \rangle = 0 \qquad\qquad \forall \mu \in Q'. \end{cases} \quad (6.16)$$

The coupling of the atomistic and continuum models in (6.16) is effected solely through the Lagrange multiplier terms in the first two equations and the constraint equation.

The second approach is to impose the constraint operator directly on the approximating spaces. Let

$$\begin{aligned} Z_D^b &= \{\psi \in X_D^b, u \in Y_D^b \mid \mathcal{C}(\psi,u) = 0\} \\ Z_0^b &= \{\psi \in X_0^b, u \in Y_0^b \mid \mathcal{C}(\psi,u) = 0\} \end{aligned} \quad (6.17)$$

where X_0^b and Y_0^b have the obvious definitions. Then, (6.14) assumes the following form: find $\{\psi,u\} \in (X_D^a \times Y_D^c) \oplus (X_D^b \times Y_D^b) \cap Z_D^b$ such that

$$\begin{cases} \mathcal{B}_\theta^a(\psi,\phi;\theta_a) + \mathcal{B}_\theta^c(u,v;\theta_c) \\ \quad = \mathcal{G}_\theta^a(\phi;\theta_a) + \mathcal{G}_\theta^c(v;\theta_c) \quad \forall \{\phi,v\} \in (X_0^a \times Y_0^c) \oplus (X_0^b \times Y_0^b) \cap Z_0^b. \end{cases} \quad (6.18)$$

The coupling between the atomistic and continuum models is now accomplished implicitly, owing to the fact that the test functions ϕ and v are forced to satisfy the constraint. Among other things this implies that the split of (6.15) into (6.16) no longer possible.

Remark 6.2 The two approaches to enforce the constraints are mathematically equivalent in the sense that the solutions obtained using either one are identical. When a basis for Z_0 can be easily constructed, the advantage of (6.18) is that discretization of this problem has much fewer degrees of freedom than discretization of (6.16). This method is also more convenient from an analysis point of view because it does not require an inf-sup stability condition. □

6.3.4.1 Blending constraint operators

Because the purpose of $\mathcal{C}$ is to tie together the atomistic and continuum parts of the solution in the bridge region, definition of this operator is very important for the quality of the AtC blending method. In particular, simply slaving the displacements to each other, which was appropriate in Ω_c, may not be enough to obtain good coupling of the atomistic and continuum models. As a result, $\mathcal{C}$ can have a rather general form. In this section we focus primarily on constraint operators of the form

$$\mathcal{C}_a(\boldsymbol{\psi}) - \mathcal{C}_c(\boldsymbol{u}) = 0 \quad \text{for } \{\boldsymbol{\psi}, \boldsymbol{u}\} \in X_D^b \times Y_D^b,$$

where $\mathcal{C}_a(\cdot) : X_D^b \to Q$ and $\mathcal{C}_c(\cdot) : Y_D^b \to Q$ are linear operators. Two useful classes of constraint operators result from setting $\mathcal{C}_a$ or $\mathcal{C}_c$ to identity. In the first case, the constraint assumes the form

$$\boldsymbol{\psi} = \Pi_{\mathbb{B}}(\boldsymbol{u}), \qquad (6.19)$$

where $\Pi_{\mathbb{B}} : Y_D^b \mapsto X_D^b$ is an *expansion* operator and $Q = X_D^b$. This operator slaves *particle displacements* to the continuous displacement field. As a result, atomistic degrees of freedom can be eliminated from Ω_b. The simplest example[3]

$$\boldsymbol{\psi}_\alpha = \boldsymbol{u}(\boldsymbol{x}_\alpha) \qquad \forall \alpha \in \mathbb{B} \setminus (\mathbb{B} \cap \mathbb{D}) \qquad (6.20)$$

embodies the physical assumption that continuous and atomistic deformation fields agree. This is precisely the case for a Cauchy-Born deformation [18, 10].

Setting $\mathcal{C}_c$ to identity gives a complementary class of constraint operators

$$\boldsymbol{u} = \pi_b(\boldsymbol{\psi}), \qquad (6.21)$$

where $\pi_b : X_D^b \mapsto Y_D^b$ is a *compression* operator and $Q = Y_D^b$. In this case the *continuous displacement field* is slaved to particle displacements. This enables elimination of continuum degrees of freedom from Ω_b in a discretized AtC model.

Remark 6.3 It is important to note that using (6.20) to eliminate atomistic degrees of freedom from Ω_b does not delete any of the atomistic force balance equations in Ω_b from the AtC blending method (6.14), as is done in the quasi-continuum method [39]. Instead, using (6.20) means that $\boldsymbol{\phi}$ is constrained to satisfy $\boldsymbol{\phi}_\alpha = \boldsymbol{v}(\boldsymbol{x}_\alpha)$ for all $\alpha \in \mathbb{B} \setminus (\mathbb{B} \cap \mathbb{D})$, where $\boldsymbol{v}$ is the continuum test function. □

More complex types of constraints, where $\mathcal{C}_a$ and/or $\mathcal{C}_c$ are not necessarily the identity operators can also be defined. We mention two examples which use subdivision $\{\Omega_{b,j}\}_{j=1}^J$ of Ω_b into J nonoverlaping, covering subdomains, i.e., $\Omega_{b,j} \cap \Omega_{b,k} = \emptyset$ whenever $j \neq k$ and $\cup_{j=1}^J \Omega_{b,j} = \Omega_b$. This subdivision induces a partition of $\{\mathbb{B}_j\}_{j=1}^J$ of $\mathbb{B}$, where $\alpha \in \mathbb{B}_j$ whenever $\boldsymbol{x}_\alpha \in \Omega_{b,j}$. Let $|\Omega_{b,j}|$ and $|\mathbb{B}_j|$ denote the volume of $\Omega_{b,j}$ and the number of particles located in $\Omega_{b,j}$, respectively. Then,

$$\frac{1}{|\Omega_{b,j}|} \int_{\Omega_{b,j}} \boldsymbol{u} \, d\boldsymbol{x} = \frac{1}{|\mathbb{B}_j|} \sum_{\alpha \in \mathbb{B}_j} \boldsymbol{\psi}_\alpha \qquad \text{for } j = 1, \ldots, J.$$

---

[3] A non-linear version of (6.20); see [8], is $|\boldsymbol{\psi}_\alpha - \boldsymbol{u}(\boldsymbol{x}_\alpha)| = 0 \quad \forall \alpha \in \mathbb{B} \setminus (\mathbb{B} \cap \mathbb{D})$, where $|\cdot|$ is the Euclidean norm in $\mathbb{R}^d$. The advantage of this operator is that it requires fewer Lagrange multipliers (one instead of d per particle) than (6.20). On the other hand, (6.20) is much easier to implement.

defines a set of constraints that is less stringent than (6.20). The integral can be approximated by a simple average to obtain another version of this operator:

$$\sum_{\alpha \in \mathbb{B}_j} u(x_\alpha) = \sum_{\alpha \in \mathbb{B}_j} \psi_\alpha \qquad \text{for } j = 1, \ldots, J. \tag{6.22}$$

In either case, we have that $Q = \mathbb{R}^{Jd}$.

Constraints such as (6.22) involve *linear combinations* of displacements and are difficult to enforce on the trial spaces. In this case, the Lagrange multiplier approach is more useful. For the constraint equations (6.22), one defines the Lagrange multipliers to be piecewise constant functions with respect to the subdivision $\{\Omega_{b,j}\}_{j=1}^{J}$ of Ω_b.

6.3.5 Consistency and patch tests

We conclude with formal definitions of consistency and patch tests introduced in [5].

6.3.6 Definition [Consistency test problem]

The set $\{\chi, \psi^\mathbb{D}; f, u^{\partial \Omega}\}$ is called a consistency test problem *if the solutions $\widetilde{\psi}$ and $\widetilde{u}$ of the global problems (6.2) and (6.6), respectively,[4] are such that $\mathcal{C}(\widetilde{\psi}, \widetilde{u}) = 0$ holds on Ω.*

6.3.7 Definition [Patch test problem]

A consistency test problem is called patch test problem *if the continuous component $\widetilde{u}$ of $(\widetilde{\psi}, \widetilde{u})$ is such that $\mathcal{L}_S^c(\widetilde{u})$ is constant, i.e., $\widetilde{u}$ is a constant stress solution.*

The following definitions formalize the notion of *passing a patch test* and a definition of *consistency* for an AtC coupling method.

6.3.8 Definition [Passing a patch test problem]

Assume that $\{\chi, \psi^\mathbb{D}; f, u^{\partial \Omega}\}$ is a patch test problem with solution $(\widetilde{\psi}, \widetilde{u})$. An AtC coupling method passes a patch test if $(\widetilde{\psi}, \widetilde{u})$ satisfies the AtC coupled problem (6.16) or (6.18).

6.3.9 Definition [AtC consistency]

An AtC coupling method is consistent *if, for any consistency test problem, the pair $(\widetilde{\psi}, \widetilde{u})$ satisfies the coupled AtC system.*

Atomistic problems with Cauchy-Born solutions (see [18, 10]) are a physical example of consistency test problems. From the previous definitions, one can easily infer that consistency implies passage of the patch test problem. However, the converse statement is not true.

6.3.10 Blended atomistic and continuum functionals

Assuming that blending functions, constraint operators and a method for their enforcement have already been chosen, all that remains to be done to obtain an

---

[4] Recall that $\{\chi, \psi^\mathbb{D}; f, u^{\partial \Omega}\}$ provides the data for these two problems.

AtC method is to define the blended functionals $\mathcal{B}_\theta^c$, $\mathcal{B}_\theta^a$, $\mathcal{G}_\theta^c$, and $\mathcal{G}_\theta^a$ appearing in (6.14).

There are two possible ways to obtain these functionals from the original atomistic and continuum functionals (6.3) and (6.5). The first one is to use $\mathcal{B}^a(\psi,\phi)$ and/or $\mathcal{B}^c(\boldsymbol{u},\boldsymbol{v})$ directly without changing their definitions. We call this approach *external* blending because it preserves the internal definitions of force balance from (6.3) and (6.5). The second possibility is to define $\mathcal{B}_\theta^c$ and $\mathcal{B}_\theta^a$ by modifying $\mathcal{B}^a(\psi,\phi)$ and $\mathcal{B}^c(\boldsymbol{u},\boldsymbol{v})$ over Ω_b. We call this approach *internal* blending because it modifies the internal definition of the force balance in (6.3) and (6.5).

For the atomistic blended functional these choices are

$$\mathcal{B}_\theta^a(\psi,\phi;\theta_a) = \begin{cases} \mathcal{B}^a(\psi,\boldsymbol{\Theta}_a\phi) = (\mathcal{L}^a(\psi),\boldsymbol{\Theta}_a\phi) & \Leftarrow \text{external} \\ \text{or} & \\ (\mathcal{L}_\theta^a(\psi;\theta_a),\phi) & \Leftarrow \text{internal} \end{cases} \quad (6.23)$$

for all $\psi \in X_D$ and $\phi \in X_0^{ab}$, where $\boldsymbol{\Theta}_a$ is a diagonal weighting matrix whose diagonal values are equal to θ_a evaluated at the corresponding particle positions:

$$(\boldsymbol{\Theta}_a)_{\alpha\beta}^{ij} = \delta_{ij}\delta_{\alpha\beta}\theta_a(\boldsymbol{x}_\alpha), \quad \text{for } i,j=1,\ldots,d,\ \alpha,\beta \in \mathbb{P}.$$

For examples of how $\mathcal{L}_\theta^a(\psi;\theta_a)$ may be defined we refer to [4].

For the continuum blended functional the choices are

$$\mathcal{B}_\theta^c(\boldsymbol{u},\boldsymbol{v};\theta_c) = \begin{cases} \mathcal{B}^c(\boldsymbol{u},\theta_c\boldsymbol{v}) = \int_{\Omega_b\cup\Omega_c} \mathcal{L}_S^c(\boldsymbol{u}) : \mathcal{L}_E^c(\theta_c\boldsymbol{v})d\boldsymbol{x} & \Leftarrow \text{external} \\ \text{or} & \\ \int_{\Omega_b\cup\Omega_c} \theta_c \mathcal{L}_S^c(\boldsymbol{u}) : \mathcal{L}_E^c(\boldsymbol{v})d\boldsymbol{x} & \Leftarrow \text{internal} \end{cases} \quad (6.24)$$

for all $\boldsymbol{u} \in Y_D^{bc}$ and $\boldsymbol{v} \in Y_0^{bc}$.

Remark 6.4 The difference between internal and external blending can be further appreciated by examining the strong forms of the differential operators in (6.24). Suppose that $\mathcal{L}^c = -\Delta$ is the Laplace operator. The blended versions of $\mathcal{L}^c$ are

$$\mathcal{B}_\theta^c(u,v;\theta_c) = \begin{cases} \int_{\Omega_b\cup\Omega_c} \nabla(u)\cdot\nabla(\theta_c v)d\boldsymbol{x} & \Leftarrow \text{external} \\ \text{or} & \\ \int_{\Omega_b\cup\Omega_c} \theta_c\nabla(u)\cdot\nabla(v)d\boldsymbol{x} & \Leftarrow \text{internal} \end{cases} \quad (6.25)$$

The strong forms of the externally and internally blended operators are

$$\mathcal{L}_E^c = -\theta_c\Delta \quad \text{and} \quad \mathcal{L}_I^c = -\nabla\cdot\theta_c\nabla,$$

respectively. We see that external blending only scales $\mathcal{L}^c$ without changing its definition. In contrast, internal blending modifies $\mathcal{L}^c$ in the blend region which effectively changes the response of the continuum material there.

For either one of the choices in (6.23) and (6.24), the blended linear data functionals appearing in (6.14) are defined[5] as

$$\mathcal{G}_\theta^a(\phi;\theta_a) = \mathcal{G}^a(\Theta_a\phi) = -(\chi,\Theta_a\phi) = -(\Theta_a\chi,\phi) \quad \forall \phi \in X_0^{ab} \quad (6.26)$$

and

$$\mathcal{G}_\theta^c(v;\theta_c) = \mathcal{G}^c(\theta_c v) = \langle f, \theta_c v\rangle = \langle \theta_c f, v\rangle = \int_{\Omega_b \cup \Omega_c} \theta_c f \cdot v\, dx \quad \forall v \in Y_0^{bc}, \quad (6.27)$$

respectively.

6.3.11 Taxonomy of AtC blending methods

Because external and internal blendings can be applied independently to the continuum and atomistic problems there are four possible ways to define the blended AtC functional $\mathcal{B}^a(\psi,\phi;\theta_a)+\mathcal{B}_\theta^c(u,v;\theta_c)$. These choices are summarized in Table 6.1.

Below we review each one of the four types ot AtC blending methods and comment on their properties.

6.3.11.1 Methods of type I

For external atomistic and internal continuum blending the abstract AtC method (6.14) assumes the form: find $\{\psi, u\} \in X_D \times Y_D^{bc}$ such that

$$\begin{cases} (\mathcal{L}^a(\psi),\Theta_a\phi) + \int_{\Omega_b \cup \Omega_c} \theta_c \mathcal{L}_S^c(u) : \mathcal{L}_E^c(v)\, dx \\ \qquad = -(\chi,\Theta_a\phi) + \int_{\Omega_b \cup \Omega_c} \theta_c f \cdot v\, dx \quad \forall \{\phi,v\} \in X_0^{ab} \times Y_0^{bc} \quad (6.28) \\ \mathcal{C}(\psi,u) = 0 \quad \forall \{\psi,u\} \in X_D^b \times Y_D^b \end{cases}$$

For this method, we have the following result [5].

Table 6.1. Force-based AtC blending methods classified by blending types

Type of the method	Type of the blending	
	Atomistic model	Continuum model
I	external	internal
II	internal	internal
III	external	external
IV	internal	external

---

[5] We can restrict the integrals in (6.24) and (6.27) to $\Omega_b \cup \Omega_c$ because, by the definition of the test space Y_0^{bc}, the continuum test function v is supported only within that subregion.

6.3.12 Theorem

Methods of Type I are inconsistent and do not pass the patch test.

This conclusion also extends to discretizations of Type I methods where the continuum part is approximated by, e.g., finite elements.

6.3.12.1 Methods of type II

For internal atomistic and continuum blending the abstract AtC method (6.14) assumes the form: find $\{\psi, u\} \in X_D \times Y_D^{bc}$ such that

$$\begin{cases} (\mathcal{L}_\theta^a(\psi;\theta_a), \phi) + \int_{\Omega_b \cup \Omega_c} \theta_c \mathcal{L}_S^c(u) : \mathcal{L}_E^c(v) dx \\ \qquad = -(\chi, \Theta_a \phi) + \int_{\Omega_b \cup \Omega_c} \theta_c f \cdot v dx \quad \forall \{\phi, v\} \in X_0^{ab} \times Y_0^{bc} \quad (6.29) \\ \mathcal{C}(\psi, u) = 0 \qquad \forall \{\psi, u\} \in X_D^b \times Y_D^b \end{cases}$$

Consistency of Type II methods depends on the definition of $\mathcal{L}_\theta^a(\cdot; \theta_a)$. The following theorem [5] gives an abstract consistency condition for this operator.

6.3.13 Theorem

Under Assumption 6.3.3, a sufficient condition for Type II methods to be consistent is that for any consistency test problem solution $\{\widetilde{\psi}, \widetilde{u}\}$ the identity

$$\begin{aligned}(\mathcal{L}_\theta^a(\widetilde{\psi};\theta_a), \phi) &- (\mathcal{L}^a(\widetilde{\psi}), \Theta_a \phi) \\ &= -\int_{\Omega_b \cup \Omega_c} \Big(\theta_c \mathcal{L}_S^c(\widetilde{u}) : \mathcal{L}_E^c(v) - \mathcal{L}_S^c(\widetilde{u}) : \mathcal{L}_E^c(\theta_c v)\Big) dx \end{aligned} \quad (6.30)$$

holds for all $\{\phi, v\} \in X_0^{ab} \times Y_0^{bc}$. Type II methods pass the patch test if (6.30) is satisfied for patch test solutions.

Remark 6.5 It is not hard to see that the reason Type I methods fail to be consistent is that for any solution of a consistency test problem the atomistic terms in (6.28) vanish but the continuum terms do not. By using internal atomistic and continuum blending, Type II methods make it possible to "cancel" consistency errors by using a suitably defined $\mathcal{L}_\theta^a$. Indeed, the right hand side in (6.30) and the term $(\mathcal{L}^a(\widetilde{\psi}), \Theta_a \phi)$ on the left hand side are completely defined once θ_c and θ_a have been selected. Then, one can choose $\mathcal{L}_\theta^a$ so that (6.30) holds.

In [4], a specific choice for $\mathcal{L}_\theta^a(\cdot; \theta_a)$ is defined that satisfies (6.30) for a particular set of one-dimensional patch test problems. The choice can be mechanically justified as a *blended force balance*.

It is worth pointing out that among the four methods discusssed here, Type II methods are the only one that can preserve the symmetry of the underlying atomistic and continuum problems. This has a positive effect on their stability and leads to symmetric linear systems that are easier to solve than the nonsymmetric linear systems generated by the other three methods.

6.3.13.1 Methods of type III

For external atomistic and continuum blending the abstract AtC method (6.14) assumes the form: find $\{\psi, u\} \in X_D \times Y_D^{bc}$ such that

$$\begin{cases} (\mathcal{L}^a(\psi), \Theta_a \phi) + \int_{\Omega_b \cup \Omega_c} \mathcal{L}_S^c(u) : \mathcal{L}_E^c(\theta_c v) dx \\ \quad = -(\chi, \Theta_a \phi) + \int_{\Omega_b \cup \Omega_c} \theta_c f \cdot v dx \quad \forall \{\phi, v\} \in X_0^{ab} \times Y_0^{bc} \quad (6.31) \\ \mathcal{C}(\psi, u) = 0 \quad \forall \{\psi, u\} \in X_D^b \times Y_D^b \end{cases}$$

Type III methods can be interpreted as *residual blending methods* because they merge the continuum residual $\mathcal{L}^c u - f$ with the atomistic residual $\mathcal{L}^a(\psi) - \chi$. This fact endows Type III methods with an attractive feature that is not shared by the other three AtC blending methods: for a consistency problem solution, the atomistic and continuum terms in (6.29) separately cancel out and so, this method is intrinsically consistent. This observation is formalized in the following theorem [5].

6.3.14 Theorem

Under Assumption 6.3.3, Type III methods are consistent and pass the patch test.

6.3.14.1 Methods of type IV

For internal atomistic and external continuum blending the abstract AtC method (6.14) assumes the form: find $\{\psi, u\} \in X_D \times Y_D^{bc}$ such that

$$\begin{cases} (\mathcal{L}_\theta^a(\psi; \theta_a), \phi) + \int_{\Omega_b \cup \Omega_c} \theta_c \mathcal{L}_S^c(u) : \mathcal{L}_E^c(v) dx \\ \quad = -(\chi, \Theta_a \phi) + \int_{\Omega_b \cup \Omega_c} \theta_c f \cdot v dx \quad \forall \{\phi, v\} \in X_0^{ab} \times Y_0^{bc} \quad (6.32) \\ \mathcal{C}(\psi, u) = 0 \quad \forall \{\psi, u\} \in X_D^b \times Y_D^b \end{cases}$$

The types of the atomistic and the continuum blending in Type IV methods are reversed with respect to Type I methods. As a result, these methods can be thought of as "dual" to Type I. The duality of the two classes is further underscored by the fact that in Type IV methods the continuous terms cancel for any consistency problem solution but the atomistic terms do not; exactly the opposite was true for Type I methods. Clearly, Type IV methods are also inconsistent.

6.3.15 Summary and comparison of force-based AtC blending methods

Consistency properties of the four types of AtC blending methods are compared and contrasted in Table 6.2.

AtC blending methods of Types I and II have appeared previously. For example, according to our classification scheme, the AtC blending method described in [8] is of Type I; see equations (10) and (11) of that paper. An example of a

Table 6.2. Atomistic and continuous contributions in AtC methods at a consistency problem solution $\{\widetilde{\psi},\widetilde{u}\}$

Type of the method	$\mathcal{B}_\theta^a(\widetilde{\psi},\phi) - \mathcal{G}_\theta^a(\phi)$	$\mathcal{B}_\theta^c(\widetilde{u},v) - \mathcal{G}_\theta^c(v)$	Consistent?
I	$= 0$	$\neq 0$	No
II	$\neq 0$	$\neq 0$	Depends on $\mathcal{L}_{\theta_a}^a$
III	$= 0$	$= 0$	Yes
IV	$\neq 0$	$= 0$	No

Type II method can be found in [4, 20]. We also remark that Methods of types I, III, and IV do not satisfy Newton's third law over the blend region, and so these methods do not lead to a symmetric formulation.

We also contrast AtC blending with the quasicontinuum method [39]. In a local quasicontinuum method, the Cauchy-Born hypothesis [10] is used to eliminate degrees of freedom in a particle model, lessening the computational complexity. The local quasicontinuum approximation has no direct relation to blending. In certain circumstances, the local/nonlocal interface arising in the quasicontinuum method can be viewed as the blending approach of Type II methods with a $d-1$ dimensional interface; see [14]. Furthermore, the forces in the quasicontinuum method are derived from a global energy functional and obey Newton's third law (or equivalently, the conservation of linear momentum).

6.4 Energy-based blending

Obviously we can always tie together energy based atomistic and continuum models by blending weak forms of their associated first-order optimality systems. This *optimize and then blend* approach reduces formulation of AtC methods for energy-based models to the force-based setting from the last section. In this section we shall consider a bona fide energy-based blending that can be described as *blend and then optimize* approach.

6.4.1 An abstract AtC blending method

In the *blend and then optimize* approach the first two key blending ingredients, i.e., the blending functions and the constraint operator $\mathcal{C}$, are shared with the force-based approach but the last two are different and are derived from the atomistic (6.7) and continuum (6.10) potential energy functionals. Specifically, instead of $\mathcal{B}_\theta^a(\cdot,\cdot;\theta_a)$ and $\mathcal{B}_\theta^c(\cdot,\cdot;\theta_c)$ we now consider:

- a *blended atomistic potential energy functional*

$$\mathcal{E}_\theta^a(\psi;\theta_a) = \sum_{\alpha\in\mathbb{A}\cup\mathbb{B}} \left(W_{\alpha;\theta_a}^a(\psi) + \chi_{\alpha;\theta_a}\cdot\psi_\alpha \right) \quad \text{such that} \quad \mathcal{E}_\theta^a(\psi;1) = \mathcal{E}^a(\psi);$$

- a *blended continuum potential energy functional*

$$\mathcal{E}_\theta^c(\boldsymbol{u};\theta_c) = \int_{\Omega_b \cup \Omega_c} \Big(W_{\theta_c}(\boldsymbol{u}) + \mathbf{f}_{\theta_c} \cdot \boldsymbol{u}\Big) d\Omega \quad \text{such that} \quad \mathcal{E}_\theta^c(\boldsymbol{u};1) = \mathcal{E}^c(\boldsymbol{u}).$$

Using these definitions, an abstract, energy-based AtC blending method can be stated as the following constrained optimization problem (compare with (6.14)):

$$\begin{cases} \text{minimize } \mathcal{E}_\theta(\boldsymbol{\psi}, \boldsymbol{u}; \theta_a, \theta_c) = \mathcal{E}_\theta^a(\boldsymbol{\psi}; \theta_a) + \mathcal{E}_\theta^c(\boldsymbol{u}; \theta_c) \\ \text{subject to} \\ \boldsymbol{\psi}_\alpha = \boldsymbol{u}(\boldsymbol{x}_\alpha) \ \forall \alpha \in \mathbb{C} \backslash (\mathbb{C} \cap \mathbb{D}) \quad \text{and} \quad \mathcal{C}(\boldsymbol{\psi}, \boldsymbol{u}) = 0 \ \forall \{\boldsymbol{\psi}, \boldsymbol{u}\} \in X_D^b \times Y_D^b \end{cases} \quad (6.33)$$

The blended atomistic-continuum potential energy $\mathcal{E}_\theta(\boldsymbol{\psi}, \boldsymbol{u}; \theta_a, \theta_c)$ describes a hybrid model that is neither continuous nor atomistic. However, owing to the definition of blended atomistic and continuum energies, $\mathcal{E}_\theta$ has the following additive property:

$$\mathcal{E}_\theta(\boldsymbol{u}, \boldsymbol{\psi}; \theta_a, \theta_c)$$

$$= \sum_{\alpha \in \mathbb{A}} \Big(W_\alpha^a(\boldsymbol{\psi}) + \boldsymbol{\chi}_\alpha \cdot \boldsymbol{\psi}_\alpha\Big) \qquad \rightarrow \text{atomistic energy in atomistic region}$$

$$+ \int_{\Omega_c} \Big(W^c(\boldsymbol{u}) + \mathbf{f} \cdot \boldsymbol{u}\Big) d\Omega \qquad \rightarrow \text{continuum energy in continuum region}$$

$$+ \sum_{\alpha \in \mathbb{B}} \Big(W_{\alpha;\theta_a}^c(\boldsymbol{\psi}) + \boldsymbol{\chi}_{\alpha;\theta_a} \cdot \boldsymbol{\psi}_\alpha\Big) \rightarrow \text{atomistic energy in bridge region}$$

$$+ \int_{\Omega_b} \Big(W_{\theta_c}^c(\boldsymbol{u}) + \mathbf{f}_{\theta_c} \cdot \boldsymbol{u}\Big) d\Omega \qquad \rightarrow \text{continuum energy in bridge region}$$

The first and the second terms are simply the atomistic and continuum potential energies for the atomistic and continuum subdomains, respectively. The third and fourth terms together represent the blended energy in the bridge region. Different definitions of $W_{\theta_c}^c$, $W_{\alpha;\theta_a}^a$, etc., will result in different blending methods.

6.4.2 Enforcing the constraints

Similarly to the weak force-based weak problem (6.14) the blended optimization problem (6.33) is subject to two constraints. The first one is the usual slaving condition which ties atomistic displacements in Ω_c to the continuum displacements. As in the force-based case, this constraint can be trivially imposed by simply substituting the appropriate atomistic displacements in the energy functional by the associated continuum values. Thus, in what follows we shall assume that this constraint had already been enforced in (6.33).

The second, blending, constraint in (6.33) serves the same purpose as in the force-based blending: its role is to tie together the atomistic and continuum solutions over the bridge region. Examples of such operators were given in Section 6.3.4.1; the same blending constraints can also be applied in energy-based methods. To enforce these constraints we again have a choice of two different strategies.

The first one uses Lagrange multipliers to replace the constrained optimization problem (6.33) by the unconstrained problem of finding the saddle-point $\{\{\boldsymbol{\psi}, \boldsymbol{u}\}, \boldsymbol{\lambda}\} \in \{X_D \times Y_D^{bc}\} \times Q'$ of the Lagrangian functional

$$\mathcal{L}(\boldsymbol{u}, \boldsymbol{\psi}, \vec{\lambda}; \theta_a, \theta_c) = \mathcal{E}_\theta(\boldsymbol{u}, \boldsymbol{\psi}; \theta_a, \theta_c) + \langle C(\boldsymbol{u}, \boldsymbol{\psi}), \boldsymbol{\lambda} \rangle \,.$$

According to the Lagrange multiplier rule, we can now take *independent variations* in each of $\boldsymbol{\psi}$, $\boldsymbol{u}$, and $\boldsymbol{\lambda}$ to obtain the optimality system

$$\begin{cases} \delta_\psi \mathcal{E}_\theta(\boldsymbol{u}, \boldsymbol{\psi}; \theta_a, \theta_c) + \langle \delta_\psi C(\boldsymbol{\psi}, \boldsymbol{u}), \boldsymbol{\lambda} \rangle = 0 \\ \delta_u \mathcal{E}_\theta(\boldsymbol{u}, \boldsymbol{\psi}; \theta_a, \theta_c) + \langle \delta_u C(\boldsymbol{\psi}, \boldsymbol{u}), \boldsymbol{\lambda} \rangle = 0 \\ C(\boldsymbol{u}, \boldsymbol{\psi}) = 0 \end{cases} \quad (6.34)$$

Note that since $\mathcal{E}_\theta(\boldsymbol{u}, \boldsymbol{\psi}; \theta_a, \theta_c)$ is a sum of functionals of $\boldsymbol{u}$ only and of $\boldsymbol{\psi}$ only, that $\delta_u \mathcal{E}_\theta$ depends only on $\boldsymbol{u}$ and $\delta_\psi \mathcal{E}_\theta$ depends only on $\boldsymbol{\psi}$. Thus, coupling is effected solely by the Lagrange multiplier terms which mirrors the force-based setting. In particular, the structure of (6.34) closely resembles that of (6.16).

The second strategy to enforce the constraints, used in force-based blending methods, is also applicable to energy-based methods. We remind that in this approach the constraints are imposed on the solution spaces, which transforms (6.33) into the following unconstrained minimization problem: find[6] $\{\boldsymbol{\psi}, \boldsymbol{u}\} \in (X_D^a \times Y_D^c) \oplus (X_D^b \times Y_D^b) \cap Z_D^b$ such that

$$\mathcal{E}_\theta(\boldsymbol{u}, \boldsymbol{\psi}; \theta_a, \theta_c) \le \mathcal{E}_\theta(\boldsymbol{v}, \boldsymbol{\phi}; \theta_a, \theta_c) \quad \forall \{\boldsymbol{\phi}, \boldsymbol{v}\} \in (X_0^a \times Y_0^c) \oplus (X_0^b \times Y_0^b) \cap Z_0^b \,. \quad (6.35)$$

Problem (6.35) is the counterpart of the blended weak problem (6.18).

6.4.3 Taxonomy of AtC blending methods

In many ways formulation of energy-based AtC blending methods closely follows that of force-based methods, and gives rise to problems with like structures. In Section 6.3.11 we identified four different types of force-based AtC blending methods. Roughly speaking, these four types corresponded to the number of possible ways to allocate the blending functions between test and trial functions in (6.14). We called the blending "external" when the blending function was assigned to the test function and "internal" otherwise.

---

[6] The spaces Z_D^b and Z_0^b were defined in (6.17).

In energy-based blending there are no test and trial functions but analogues of internal and external blending still exists. Since external blending is not supposed to change the mechanical definition of the potential energy, in the present case it corresponds to a *weighting* of the energy functional. For the atomistic potential energy we thus have the choice of

$$\mathcal{E}_\theta^a(\psi;\theta_a) = \begin{cases} \sum_{\alpha\in\mathbb{A}\cup\mathbb{B}} \theta_a(\boldsymbol{x}_\alpha)\Big(W_\alpha^a(\psi) + \chi_\alpha \cdot \psi_\alpha\Big) & \Longleftarrow \text{external} \\ \text{or} & \\ \sum_{\alpha\in\mathbb{A}\cup\mathbb{B}} \Big(W_{\alpha;\theta_a}^c(\psi) + \chi_{\alpha;\theta_a} \cdot \psi_\alpha\Big) & \Longleftarrow \text{internal} \end{cases} \quad (6.36)$$

Similarly, for the continuum blended potential energy the choices are

$$\mathcal{E}_\theta^c(\boldsymbol{u};\theta_c) = \begin{cases} \int_{\Omega_b\cup\Omega_c} \theta_c(\boldsymbol{x})\Big(W^c(\boldsymbol{u}) + \mathbf{f}\cdot\boldsymbol{u}\Big) d\Omega & \Longleftarrow \text{external} \\ \text{or} & \\ \int_{\Omega_b\cup\Omega_c} \Big(W_{\theta_c}^c(\boldsymbol{u}) + \mathbf{f}_{\theta_c}\cdot\boldsymbol{u}\Big) d\Omega & \Longleftarrow \text{internal} \end{cases} \quad (6.37)$$

It follows that formally, energy-based blending can also result in four different types of AtC blending methods. Perhaps the two most important cases are the analogues of Type II and Type III methods, in which the type of the blending is the same for the atomistic and continuum energies. The overlapping domain decomposition coupling method in [8] is an example of a Type II blending method.

It is not clear whether or not analogues of Type I and Type IV methods would be useful at all in the energy-based coupling context. One concern is that by mixing different blending strategies it may be very difficult to ensure that the energy in the blend regions is not over or under-counted.

6.5 Generalized continua

The intrinsic incompatibility of coupling mechanical models with local/nonlocal force interaction suggests that we consider generalizations of classical continuum mechanics as described in [7, 13, 19, 35, 36, 37], or continuum realizations of molecular dynamics [1, 42, 12, 43, 30]. We refer to these classes of methods as generalized continua. They are motivated, in large part, by introducing a length-scale (absent in classical elasticity) by augmenting the displacement field with supplementary fields (e.g., rotations) that provide information about fine-scale kinematics, by using higher-order gradients of the displacement field, by averaging local strains and/or stresses, or by introducing a notion of a field into molecular dynamics. We also mention the papers [15, 23] where variational principles for the generalized continua are described useful for finite element based discretizations, and [21] where the classical theory is augmented with internal bonds.

The blending methods reviewed in this paper may also be applied for use with generalized continua, and require modification to (6.4), and the equations that follow. In particular, $\mathcal{L}^c(\boldsymbol{u}), \mathcal{L}^c_S(\cdot)$ and $\mathcal{L}^c_E(\cdot)$ may not correspond to a differential operator, and a continuum notion of the force operator $\mathcal{L}^a$ introduced in (6.1) may need modification.

A generalized continua can fulfill two roles within a blending method. The first is to replace the classical elastic theory, the second is as an intermediatory between molecular dynamics and classical elasticity. The second approach becomes viable if there is theoretical justification linking the generalized continua with classical elasticity, especially if the force interaction model is nonlocal.

For instance, the recent paper [38] explains that if the underlying deformation is sufficiently smooth, then peridynamics [36, 37] is asymptotically equivalent [38] to classical elasticity as the length-scale decreases, and has a symbiotic relationship with molecular dynamics. As a result, a discretized peridynamic model can be implemented using an off-the-shelf molecular dynamic code as described in [32]. Future work blends molecular dynamics and peridynamics to enable multiscale materials modeling.

6.6 Conclusions

In this paper we reviewed a novel mathematical framework for encoding and classifying force or energy-based AtC blending methods. In particular, we identified four key ingredients of such methods and two possible blending types that can be applied in the force-based or energy-based settings. We also stated formal definitions of consistency and patch tests for AtC blending methods, and explained how the ghost force effects and unphysical interface boundary conditions are mitigated in such methods.

Based on the type of blending applied to the atomistic and continuum models, AtC blending methods can be divided into four different categories. Only one of these categories, Type II methods, leads to AtC formulations that are simultaneously consistent, symmetric and do not violate Newton's third law over the bridge region.

Equally important for the AtC blending methods is the choice of the blending constraint operator whose purpose is to reconcile atomistic and continuum displacements in the bridge region so as to impose a suitable notion of "continuity" in the coupled solution. The choice of how to enforce these constraints does not modify the final result, but has important implications for the implementation of AtC methods. We have considered two different choices: classical Lagrange multipliers, and restricted AtC spaces whose elements explicitly satisfy the constraints.

References

[1] Arndt M. and Griebel M. (2005). "Derivation of higher order gradient continuum models from atomistic models for crystalline solids", *Multiscale Mod. Simul.*, **4**(2), p. 531–562.

[2] Arndt M. and Luskin M. (2007). "Goal-oriented atomistic-continuum adaptivity for the quasicontinuum approximation", *International Journal for Multiscale Computational Engineering*, **5**, p. 407–415.

[3] Arndt M. and Luskin M. (2008). "Error estimation and atomistic-continuum adaptivity for the quasicontinuum approximation of a Frenkel-Kontorova model", *SIAM J. Multiscale Modeling & Simulation*, **7**, p. 147–170.

[4] Badia S., Bochev P., Fish J., Gunzburger M., Lehoucq R., Nuggehally M., and Parks M. (2007). "A force-based blending model for Atomistic-to-Continuum coupling", *International Journal for Multiscale Computational Engineering*, **5**(5), p. 387–406.

[5] Badia S., Parks M., Bochev P., Gunzburger M., and Lehoucq R. (2008). "On atomistic-to-continuum coupling by blending", *Multiscale Modeling & Simulation*, **7**(1), p. 381–406.

[6] Bauman P. T., Dhia H. B., Elkhodja, N., Oden J. T., and Prudhomme S. (2008). "On the application of the Arlequin method to the coupling of particle and continuum models", *Computational Mechanics*. Article in Press.

[7] Bažant, Zdeněk P. and Jirásek M. (2002). "Nonlocal integral formulations of plasticity and damage: Survey of progress", *J. of Eng. Mech.*, **128**, p. 1119–1149.

[8] Belytschko T. and Xiao S. P. (2003). "Coupling methods for continuum model with molecular model", *International Journal for Multiscale Computational Engineering*, **1**(1), p. 115–126.

[9] Blanc, Xavier, Bris, Claude Le, and Lions, P.-L. (2007). "Atomistic to continuum limits for computational materials science", *Math. Model. Num. Anal.*, **41**(2), p. 391–426.

[10] Born M. and Huang K. (1954). *Dynamical Theory of Crystal Lattices*, Clarendon Press.

[11] Brezzi F. and Fortin M. (1991). *Mixed and hybrid finite element methods*, Springer, Verlag.

[12] Chen Y. and Lee J. (2005). "Atomistic formulation of a multiscale field theory for nano/micro solids", *Philosophical Magazine*, **85**(33–35), p. 4095–4126.

[13] Chen Y., Lee J. D., and Eskandarian A. (2004). "Atomistic viewpoint of the applicability of microcontinuum theories", *Int. J. of Solids and Structures*, **41**, p. 2085–2097. 10.1016/j.ijsolstr.2003.11.030.doi

[14] Curtin W. A. and Miller R. E. (2003). "Atomistic/continuum coupling methods in multi-scale materials modeling", *Modeling and Simulation in Materials Science and Engineering*, **11**(3), R33–R68.

[15] de Sciarra, Francesco Marotti (2008). "Variational formulations and a consistent finite-element procedure for a class of nonlocal elastic continua", *International Journal of Solids and Structures*, **45**, p. 4184–4202.

[16] Dhia H. B. and Rateau G. (2005). "The Arlequin method as a flexible engineering design tool", *International Journal for Numerical Methods in Engineering*, **62**, p. 1442–1462.

[17] Dobson M. and Luskin M. (2008). "Analysis of a force-based quasicontinuum approximation", *Mathematical Modelling and Numerical Analysis*, **42**, p. 113–139.

[18] E W. and Ming P. (2007). "Cauchy–Born rule and the stability of crystalline solids: Static problems", *Archives for Rational Mechanics and Analysis*, **183**(2), p. 241–297.

[19] Eringen, A. Cemal (2002). *Nonlocal continuum field theories*, Springer-Verlag, NewYork, Inc.

[20] Fish J., Nuggehally M. A., Shephard M. S., Picu C. R., Badia S., Parks M. L., and Gunzburger M. (2007). "Concurrent AtC coupling based on a blend of the continuum stress and the atomistic force", *Computer Methods in Applied Mechanics and Engineering*, **196**, p. 4548–4560.

[21] Gao, Huajian and Klien, Patrick (1998). "Numerical simulation of crack growth in an isotropic solid with randomized internal cohesive bonds", *J. Mech. Phys. Solids*, **46**, p. 187–218.

[22] Huerta A., Belytschko T., Fernández-Médez S., and Rabczuk, Timon (2004). Meshfree methods. In *Encyclopedia of Computational Mechanics* (ed. Stein E., de Borst R., and Hughes T. J. R.), Volume 1 of *Fundamentals*, Chapter 10, Wiley.

[23] Kirchner N. and Steinmann P. (2007). "Mechanics of extended continua: modeling and simulation of elastic microstretch materials", *Comput. Mech.*, **40**.

[24] Klein P. A. and Zimmerman J. A. (2006). "Coupled atomistic–continuum simulations using arbitrary overlapping domains", *Journal of Computational Physics*, **213**, p. 86–116.

[25] Knap J. and Ortiz M. (2001). "An analysis of the quasicontinuum method", *J. Mech. Phys. Solids*, **49**, p. 1899–1923.

[26] Lin P. (2003). "Theoretical and numerical analysis for the quasi-continuum approximation of a material particle model", *Mathematics of Computation*, **72**, p. 657–675.

[27] Lin P. (2007). "Convergence analysis of a quasi-continuum approximation for a two-dimensional material without defects", *SIAM Journal on Numerical Analysis*, **45**(1), p. 313–332.

[28] Miller R. and Tadmor E. (2002). "The quasicontinuum method: Overview, applications and current directions", *J. Comput. Aided Mater. Des.*, **9**, p. 203–239.

[29] Miller R. E. and Tadmor E. B. (2007, November). Hybrid continuum mechanics and atomistic methods for simulating materials deformation and failure. In *Multiscale Modeling in Advanced Materials Research* (ed. J. J. de Pablo and W. A. Curtin), Volume 32 of *MRS Bulletin*, p. 920–926.

[30] Murdoch A. I. and Bedeaux D. (1994). "Continuum equations of balance via weighted averages of microscopic quantities", *Proc. Roy. Soc. London A*, **445**, p. 157–179.

[31] Ortner, Christoph and Süli, Endre (2008). "Analysis of a quasicontinuum method in one dimension", *M2AN Math. Model. Numer. Anal.*, **42**(1), p. 57–91.

[32] Parks M. L., Bochev P. B., and Lehoucq R. B. (2008). "Connecting atomistic-to-continuum coupling and domain decomposition", *Multiscale Modeling & Simulation*, **7**(1), p. 362–380.

[33] Parks M. L., Lehoucq R. B., Plimpton S. J., and Silling S. A. (2008). Implementing peridynamics within a molecular dynamics code. Technical Report SAND2007–7957J, Sandia National Laboratories, Albuquerque, New Mexico 87185 and Livermore, California 94550, **179**(11), p. 777–783.

[34] Prudhomme S., Dhia H. B., Bauman P. T., Elkhodja N., and Oden, J. Tinsley (2008). "Computational analysis of modeling error for the coupling of particle and continuum models by the Arlequin method", *Computer Methods in Applied Mechanics and Engineering*, **197**(41–42), p. 3399–3409.

[35] Rogula D. (1982). Introduction to nonlocal theory of material media. In *Nonlocal theory of material media* (ed. D. Rogula), Number 268 in CISM Courses and Lectures, p. 125–222. Springer Verlag, Wien.

[36] Silling S. A. (2000). "Reformulation of elasticity theory for discontinuities and long-range forces", *J. Mech. Phys. Solids*, **48**, p. 175–209.

[37] Silling S. A., Epton M., Weckner O., Xu J., and Askari E.(2007). "Peridynamic states and constitutive modeling", *J. Elasticity*, **88**, p. 151–184.

[38] Silling, S. A. and Lehoucq, R. B. (2008). "Convergence of peridynamics to classical elasticity theory", *J. Elasticity*, **93**(1), p. 13–37.

[39] Tadmor E. B., Ortiz M., and Phillips R. (1996). "Quasicontinuum analysis of defects in solids". *Philosophical Magazine A*, **73**, p. 1529–1563.

[40] Toselli, Andrea and Widlund, Olof B. (2005). *Domain Decomposition Methods–Algorithms and Theory*, Volume 34 of *Springer Series in Computational Mathematics*. Springer.

[41] Wagner G. J. and Liu W. K. (2003). "Coupling of atomic and continuum simulations using a bridging scale decomposition", *Journal of Computational Physics*, **190**, p.249–274.

[42] Zhou Min (2005). "Thermomechanical continuum representation of atomistic deformation at arbitrary size scales", *Proceedings of The Royal Society A*, **461**, p. 3437–3472. 10.1098/rspa.2005.1468.

[43] Zhou, Min and McDowell, David L. (2002). "Equivalent continuum for dynamically deforming atomistic particle systems", *Philosophical Magazine A*, **82**, p. 2547–2574.

PART III

SPACE-TIME SCALE BRIDGING METHODS

7
PRINCIPLES OF SYSTEMATIC UPSCALING
Achi Brandt

7.1 Introduction

Despite their dizzying speed, modern supercomputers are still incapable of handling many most vital scientific problems. This is primarily due to *the scale gap*, which exists between the microscopic scale at which physical laws are given and the much larger scale of phenomena we wish to understand.

This gap implies, first of all, a huge number of *variables* (e.g., atoms, or gridpoints, or pixels), and possibly even a much larger number of *interactions* (e.g., one force between every pair of atoms). Moreover, computers simulate physical systems by moving a *few variables at a time*; each such move must be extremely small, since a larger move would have to take into account all the motions that should in parallel be performed by all other variables. Such a computer simulation is particularly incapable of moving the system across large-scale *energy barriers*, which can each be crossed only by a large, *coherent* motion of very many variables.

This type of obstacle makes it impossible, for example, to calculate the properties of nature's building blocks (elementary particles, atomic nuclei, etc.), or to computerize chemistry and materials science, so as to enable the design of materials, drugs, and processes, with enormous potential benefits for medicine, biotechnology, nanotechnology, agriculture, materials science, industrial processing, etc. With current common methods the amount of computer processing often increases so steeply with problem size, that even much faster computers will not do.

Past studies have demonstrated that scale-born slownesses can often be overcome by multiscale algorithms. Such algorithms have first been developed in the form of fast *multigrid solvers* for discretized PDEs [1, 2, 3, 4, 13, 14, 15]. These solvers are based on two processes: (1) classical *relaxation* schemes, which are generally slow to converge but fast to *smooth* the error function; (2) approximating the smooth error on a *coarser grid* (typically having twice the mesh size), by solving on that grid equations which are derived from the PDE and from the fine-grid residuals; the solution of these coarse-grid equations is obtained by using recursively the same two processes. As a result, large-scale changes are effectively calculated on correspondingly coarse grids, based on information gathered from finer grids. Such multigrid solvers yield *linear complexity*, i.e., the solution work is proportional to the number of variables in the system.

In many years of research, the range of applicability of these methods has steadily grown to cover most major types of linear and nonlinear large systems of equations appearing in science and engineering. This has been accomplished by extending the concept of 'smoothness' in various ways, finally to stand generally for any solution component which cannot be determined locally, and by correspondingly diversifying the types of coarse representations, to include, for instance, grid-free solvers (algebraic multigrid [7, 8, 9, 16]), non-deterministic problems ([10, 11, 12, 20, 21]) and multiple coarse-level representations for wave equations [5].

It has been shown (see survey [29]) that the inter-scale interactions can indeed eliminate all kinds of scale-associated difficulties, such as: slow convergence (in minimization processes, PDE solvers, etc.); critical slowing down (in statistical physics); ill-posedness (e.g., of inverse problems); conflicts between small-scale and large-scale representations (e.g., in wave problems, bridging the gap between wave equations and geometrical optics); numerousness of long-range interactions (in many body problems or integral equations); the need to produce many fine-level solutions (e.g., in optimal control, design, and data assimilation problems), or a multitude of eigenfunctions (e.g., in calculating electronic structures), or very many fine-level independent samples (in statistical physics); etc. Since the local processing (relaxation, etc.) in each scale can be done in parallel at all parts of the domain, the multiscale algorithms, based on such processing, proved ideal for implementation on massively parallel computers.

To obtain even further generality, there emerge however two basic reasons to go much beyond these multigrid methods. First, they cannot perform well for *highly nonlinear cases*, where configurations cannot be decomposed into weakly-interacting local and non-local parts. Second, for many systems, even attaining linear complexity is not good enough, since the number of variables is huge. Such systems on the other hand are typically *highly repetitive*, in the sense that the same small set of governing equations (or Hamiltonian terms) keeps repeating itself throughout the physical domain. This opens the way to the possibility of having, at the coarse level too, a small set of governing equations that are valid everywhere, and that can be derived from fine-level processing conducted only in some small representative 'windows' (see below).

These two basic reasons point in fact in the same direction. Instead of relaxing the given system of equations so as to obtain a smooth error that can be approximated on a coarse level, one should use coarse-level variables that are little sensitive to relaxation (e.g., representing chosen *averages*, rather than a subset of individual fine-level values) and that represent the *full* solution rather than the correction to any given current approximation. Such coarse variables can be chosen (as described below) so that the coarse-level equations can be derived just by local processing. We use the term *'upscaling'* for this type of direct (full-solution) transition from a fine level to a coarser one. Such a transition is valid even in those highly nonlinear cases where all scales interact with each other so strongly that correction-based multi-leveling is inapplicable.

In fact, upscaling, under the name 'renormalization', was first introduced into exactly those systems where all scales interact most strongly: systems of statistical mechanics at the critical temperature of phase transition. The *renormalization group* (RG) method (see, e.g., [17, 18, 19, 65, 66]) was developed contemporaneously with, but independently of the multigrid method, its chief purpose having been to investigate the behavior of such critical systems at the limit of very large scales. The RG method has thus mainly focused on analyzing, theoretically and computationally, the fixed point of the group of successive renormalization steps, and various universal asymptotic power laws associated with it. Little has been done to upscale systems without a fixed point, which is the prevalent situation in many practical problems. This is related to the fact that the RG computational efficiency remained very limited, due to the lack of a systematic coarse-to-fine transition, which is needed either for accelerating simulations at all levels (as in multigrid solvers) and/or for confining them to small representative windows (as described below).

7.1.1 *Systematic Upscaling (SU)*

Building on the complementary advantages of multigrid and RG described above, *Systematic Upscaling (SU)* is a methodical derivation of equations (or statistical rules) that govern a given physical system at increasingly larger scales, starting at a microscopic scale where first-principle laws are known, and advancing, scale after scale, to obtain suitable variables and operational rules for processing the system at much larger scales. Unlike classical RG, the SU algorithms include repeated coarse-to-fine transitions, which are essential for (1) testing the adequacy of the set of coarse-level variables (thus providing a general tool for constructing that set); (2) accelerating the finer-level simulations; and, most importantly (3) confining those simulations to small representative subdomains (called *windows*) within the coarser-level simulations. (SU was described briefly in [68] and at length in [69].)

7.1.2 *Difference from adhoc multiscale modelling*

Upscaling should not be confused with various methods of *multiscale modelling* (also called 'multiscale simulation') being developed in several fields (e.g., materials science). Those methods study a physical system by employing several different *adhoc* models, each describing a very different scale of the system. Their basic approach is the *fine-to-coarse parameter passing*, in which data obtained from simulating a finer-scale model, often coupled with experimental observations, are used to determine certain parameters of a larger-scale model, regarding the latter as a coarse-graining of the former. The most basic feature missing in this approach (as in RG) is the accurate transition from coarse levels back to finer ones, and the use of this transition as a systematic vehicle for choosing an adequate set of coarse variables, and for iterating between the levels to accelerate fine-level calculations or restrict them to small windows. Successful as multiscale modelling methods are in many cases, they lack generality, are often inapplicable (especially when the scales are not widely separated) or inaccurate (based on questionable large-scale models), and involve much slowdown due to large-scale gaps. SU, by contrast, inherits from multigrid and RG general and methodical procedures to construct and iteratively employ all intermediate scales and thus, attain slowness-free efficiency and fully-controlled coarse-level accuracy.

7.1.3 Other numerical upscaling methods

The research on numerical upscaling, i.e., precise numerical derivation of coarse-level (e.g., macroscopic) equations from fine-level laws, has had of course a long history. In particular it has been active in computational mechanics for at least 30 years. Many systems of 'homogenization', i.e., rigorous derivation of macroscopic (continuum or discretized-continuum) equations from microscopic (either continuum, or discrete continuum, or atomistic) laws have been advanced, first in the engineering literature and then in more rigorous mathematical analyses.

Most widely developed are methods of *asymptotic expansion*, in particular multiple-scale asymptotic expansion for *periodic* heterogeneous structures; see, e.g., [57, 58, 59]. For elastodynamics of composite materials, high-order methods were developed, introducing effects such as polarization, dispersion, and attenuation of a *single-frequency* stress wave [56], [60]. Such expansions for *initial-boundary-value* problems in periodic media were developed by Jacob Fish and his group [62, 63, 64], who then extended the theory to discrete-state (atomistic) fine-scale models [42], and then also to finite temperatures [43].

Asymptotic expansions, not assuming fine-scale periodicity but relying on vast-scale separations, were developed for singularly perturbed systems of differential equations based on rigorously derived averaging principles, both for deterministic and stochastic problems ([46, 47, 48, 49] for example; see comprehensive survey in [67]). The usefulness of these theories was limited since 'it is often impossible, or impractical, to obtain the reduced equations in closed form. This has motivated the development of algorithms such as projective and coarse projective integration [50, 51, 52] within the so-called equation-free framework [53, 54]. In this framework, short bursts of appropriately initialized fine-scale simulations are used to estimate on demand the numerical quantities required to perform scientific computing tasks with coarse-grained models (time derivatives, the action of (slow) Jacobians, and, for the case of *stochastic* coarse-grained models, the local effective noise drift and diffusivity [55]).' (Quoted from [45].)

The developed method is in fact not entirely free of any equation. It actually tends to be rather similar to the *Heterogeneous Multiscale Method* (HMM) [44], which numerically solves continuum equations *of a known form* but with certain unknown local data (e.g., the stress—in a fluid dynamics macroscopic model), finding the latter by performing atomistic simulations on tiny subdomains, with constraints (e.g., the local average gradient) supplied by the macroscopic model.

The 'equation-free' and HMM methods require rederivation of the macroscopic quantities by running the microscopic model at each spatial gridpoint within each time step (with grid and time step resolution fine enough for accurate macroscopic simulation), so they are not exactly 'upscaling' methods in the strict sense defined above. More important, such schemes would usually require *double* scale separation as well as extensive work at each fine-level simulation. Indeed, if the fine (temporal or spatial) step is h_f, and the scale of desired coarse (macroscopic) resolution is h_c, there usually exists also at least one intermediate scale $h_i \gg h_f$ at which the equations homogenize (and equilibrate, in the stochastic case). Now h_i must of course be much smaller than h_c for this multiscale processing to pay off. Moreover, one would typically need at least $O(h_i^2/h_f^2)$ time steps (or iterations, in static or equilibrium problems) at each fine-level simulation to reliably calculate the coarse coefficients in the HMM methods, and sometimes even up to $O(h_c^2/h_f^2)$ such steps to capture the coarse-level derivatives needed in the equation-free framework (unless one uses multigrid-like accelerations, which means introducing more computational levels between h_f and h_i, or sometimes even between h_f and h_c).

A general approach for deriving coarse-level from fine-level equations, which does not require scale separation, is based on interpolation. This was originally introduced in the multigrid framework, including in particular its nonlinear (FAS) version [2]. The approach has been applied to the case of atomistic (not PDE) fine level, for example in the framework of the *quasi-continuum (QC)* method, used in particular to derive coarse equations at subdomains where the continuum equations break down (e.g., at the tip of a crack).

It is worth noting that early versions of the QC method had previously appeared in the multigrid literature: the method of local refinements of the PDE discretization in which each finer level locally corrects the equations of the next coarser level is part of the FAS approach (see, e.g., Sections 7–9 in [2]), and the particular situation in which the finest level is atomistic, is described for example in Section 1.1 of [11].

As shown, for example by Fish and Chen [61], the efficiency of interpolation-based methods depends critically on the interpolation scheme. This issue has in fact a long history in the multigrid literature, starting with [7, 8, 9, 16, 37, 41], and all later papers on Algebraic Multigrid (AMG). Still more general and accurate methods for deriving interpolations, based on relaxed vectors and Bootstrap AMG (see Section 17.2 in [29]), or on adaptive smooth aggregation [40], were developed in recent years.

The interpolation-based multiscale methods were extended to simple finite-temperature models in statistical mechanics [10, 12, 38, 39], (reviewed in Section 13.1 of [29]), showing the feasibility of calculating *thermodynamic limits* (observable averages at system sizes tending to infinity) at '*statistically optimal' efficiency* (achieving accuracy ε in $O(\varepsilon^{-2})$ random number generations!). However, attaining similar performance for the highly nonlinear models of statistical physics (involving inseparable interaction scales) by interpolation-based methods proved impossible (as explained in Section 13.1 of [29]), which gave rise to the Renormalization Multigrid (RMG) method ([6] or Section 13.2 of [29]), the forerunner of the SU methods described here.

7.1.4 Features of Systematic Upscaling

In the SU method, no scale separation is assumed; in fact, a small ratio between successive scales can often be essential: it ensures slowdown-free computations that at each scale can be confined to certain representative windows, each containing a moderate, bounded number of variables.

Notice that SU does not depend on having a coarse level of continuum type (see examples below). Moreover, unlike other approaches, SU is being developed also for *non-local* interactions (see Section 7.6.1 below). Note as well that the SU coarsening can also be used for vast *acceleration* of fine-level calculations, sometimes even by just one modest coarsening step (e.g., with just a 1:3 coarsening ratio: see Section 7.2.5 below).

A basic feature of SU is that coarsening is derived *once for all*. That is, the life of each window is limited: once it has accumulated enough statistics (translated into coarser-level equations) its task is over. The total size-times-duration of all the windows serving a given system at a given level depends only on the number of *different* local situations that can arise, not on the overall size and duration of the system. This implies enormous potential savings not only in computer resources, but also in human effort because coarsened equations developed by one can be used by others, possibly without ever returning to the fine level. One may, for example, just develop the first level of coarsening a macromolecule (or even just part of it), thereby already providing it with much faster simulations; another investigator can then build on this the next coarser level (or add the coarsening of another part), and so on. *The upscaling of an important physical system can thus become a gradual and collaborative systematic endeavor.*

The development of SU methods should provide necessary tools for surmounting extreme computational bottlenecks in many areas of science and engineering, such as: statistical mechanics, especially at phase transition; elementary-particle physics; electronic structure of molecular systems (for deriving the interatomic force fields); molecular dynamics of fluids, condensed matter, nano structures, and macromolecules, including proteins and nucleic acids; materials science; turbulent fluid dynamics; global optimization of systems with multiple-scale energy barriers.

7.1.5 Plan of this article

Section 7.2 outlines the basic procedures of the SU methodology, in terms of examples from partial differential equations, molecular and macromolecular dynamics, and statistical dynamics. It also reports some preliminary results. Section 7.3 gives more precise details of the derivation of coarse equations, using for illustration one particular example (polymer in equilibrium). Section 7.4 describes the creation of windows and the algorithmic flow between windows at various levels. Section 7.5 mentions some special situations, such as boundaries. Section 7.6 is a brief survey of possible extensions of the SU methodology to deal with long-range interactions, time-dependent systems, complex fluids, low temperatures, global optimization in systems with multiscale attraction basins, and transition from quantum mechanics to molecular dynamics.

7.2 Systematic upscaling (SU): an outline

7.2.1 Local equations and interactions

Computationally we will always deal with a discrete system, whose n variables (or unknowns) $u_1, u_2, \ldots, u_n$ will typically be either the discrete values of functions (grid values, or finite elements, etc.), or the locations of particles. An equation in the d-dimensional physical space (usually $1 \leq d \leq 4$) is called *local* if it involves only $O(1)$ neighboring unknowns. A discretized partial differential equation, for example, is a system of local equations. Similarly, an 'interaction', i.e., an additive term in an energy functional or Hamiltonian H, is called *local interaction* if it involves only $O(1)$ neighboring variables. In equilibrium calculations we will usually assume H to already include the $(k_B T)^{-1}$ factor, so that the probability density $P(u)$ of each configuration $u = (u_1, u_2, \ldots, u_n)$ is proportional to $\exp(-H(u))$.

For simplicity of discussion we describe SU first for systems which are *stationary or at equilibrium*, and such that their equations or interactions are *local*. We will however point out in Section 7.6 extensions to long-range equations or interactions, and to dynamic and non-equilibrium systems.

7.2.2 Coarsening

Similar to multigrid, SU is based on two processes: The usual *local processing* (relaxation in deterministic problems, Monte Carlo (MC) in stochastic ones) and repeated *coarsening*, creating increasingly coarser descriptions of the same physical system, with interscale interactions to be described below. At each coarsening stage, one constructs from a current level of description (the fine level) a coarser level, employing the following general principles.

To each fine-level configuration $u = (u_1, \ldots, u_n)$ one defines (using the criterion below) a unique coarse-level configuration $Cu = u^c = (u_1^c, \ldots, u_m^c)$, which is a vector with a reduced number of variables; typically the scale ratio is $1.5 < n/m < 10$ (except of course for problems where the equations homogenize, or a renormalization fixed-point emerges, already at some intermediate scale. Introducing additional computational scales is then needed only between the finest scale and that intermediate scale.) How to choose this fine-to-coarse transition C is of course a central question, which we discuss in detail below, after giving several examples and defining first the coarse-to-fine transition.

7.2.2.1 Examples of such fine-to-coarse transformations C

(i) For discretized continuous (e.g., PDE) problems—each coarse variable is usually an average of several neighboring fine variables. (But see Section 7.5 for an example where averaging works only up to a certain scale, above which more elaborate variables should be added.)

(ii) For a simple polymer, which consists of a chain of n atoms at the three-dimensional locations $(u_1, u_2, \ldots, u_n)$—each coarse-level 'atom' location u_j^c is at the average location of q, say, consecutive real atoms:

$$u_j^c = \frac{1}{q}(u_{qj-q+1} + u_{qj-q+2} + \cdots + u_{qj}), \ (j = 1, \ldots, m; \ m = \frac{n}{q}). \quad (7.1)$$

(iii) For a simple atomistic fluid, described by the positions u in space of its n molecules—the coarse level variables are defined at the points of a lattice placed over the flow domain, with each variable u_j^c summarizing a property of the set of molecules around that lattice point (e.g., their total mass, or density, total dipole moment, etc.). Or each u_j^c may be a *vector* which summarizes *several* such properties. At lower temperatures, as the fluid starts to solidify, additional types of coarse variables must enter, accounting for larger-scale order parameters (see Section 7.6.6).

(iv) For a lattice of Ising spins—each coarse variable is again an Ising spin, standing for the *sign* of the sum of a block of fine-level spins.

7.2.3 Generalized interpolation

To any given coarse configuration $U = (U_1, \ldots, U_m)$, there are generally many fine-level configurations u which are *compatible* with U (i.e., such that $Cu = U$). The interpolation (transition from U to one specific fine configuration u) is created by *compatible Monte Carlo* (CMC) (or compatible relaxation, in the deterministic case), i.e., by the local processing, restricted to configurations compatible with U. The interpolation is completed once the CMC has practically reached its equilibrium (or the compatible relaxation has converged).

For instance, in the case of polymer (Example (ii) above), if the coarse variables are defined by (7.1), each step in a CMC would offer a simultaneous change of two consecutive atomic positions, u_k and u_{k+1}, such that $u_k + u_{k+1}$ is kept unchanged ($qj - q + 1 \leq k \leq qj - 1;\ 1 \leq j \leq m$).

The CMC interpolation was first introduced for the case of lattice Ising spins, establishing its fast equilibration [6].

7.2.4 The general coarsening criterion

The fine-to-coarse transformation C is said to be adequate *if (and to the extent that) a compatible Monte Carlo that equilibrates fast (or a compatible relaxation that converges fast) is available.*

In our polymer example, for instance, if the polymer force field includes the usual bond length, bond angle, and dihedral (torsion) interactions, it turns out that the coarsening (7.1) is adequate for $q = 2$ or 3, but a larger q yields a much slower CMC equilibration. (This result makes sense: at coarsening ratio $q \leq 3$, the fixed coarse values u_j^c implicitly nearly fix the dihedral values, which determine the large-scale polymer configuration.)

In the framework of Algebraic Multigrid (AMG), the criterion of fast convergence of compatible relaxation as a tool for choosing the coarse variables (introduced in [36]) is already used by several groups. (Particularly useful is the fact that the rate of convergence of (suitably arranged) compatible relaxation is an accurate predictor of the obtainable multigrid convergence rate; so accurate in fact that it can be used to design and debug the AMG solver.)

A major problem in coarsening any system is to find a suitable set of coarse variables. The above criterion gives a general and very effective tool for developing such a set. Importantly, a coarsening that satisfies that criterion practically implies local dependence of every fine variable on neighboring coarse variables, and hence the theoretical possibility to construct, just by local processing, a set of *'equations'* (in the form of numerical tables) that will govern simulations at the coarse level. How to go about this construction in practice is discussed in Section 7.3 below.

Moreover, in highly repetitive systems (defined above), this local construction of the coarse equations need not be done everywhere: the coarse-level equations can iteratively be derived by comparing coarse-level with fine-level simulations, where the latter are performed only in some relatively small *windows* (subdomains, on the boundaries of which the fine level is kept compatible with the coarse level. See Section 7.4 below for details.)

Thus, the fine-level simulations supply (or correct) the equations (or Hamiltonian) of the next coarser level. On the other hand, the coarse level selects the windows where these fine-level simulations should take place. A window is opened over a region where the coarse level has detected local relations in a range for which fine level computations in previous windows could not supply accurate equations (see more in Section 7.4). Iterating back and forth between windows at all the levels quickly settles into a self-consistent multilevel equilibrium and compatibility; as in the multigrid, if the coarsening ratio n/m is not large, no slowdown should occur. More important, at each level the computations need extend only over a collection of small windows, whose number depends on the diversity of local situations, not on the size of the entire system.

7.2.5 *Experimental results*

The four simple examples mentioned above have already revealed the very high potential of the SU approach. For instance, in Example (ii), in which conventional simulations run into extreme slowdowns, even the single coarsening level (1), with $q = 3$, already accelerates the simulation by at least two orders of magnitude, while accurately reproducing all the relevant statistics (using the coarse Hamiltonian described in Section 7.3 below. This result was obtained in computations devised by Dr Dov Bai). The reason for this acceleration is that this first coarsening already effectively averages over the attraction basins caused by the local minima of the fine-level dihedral interactions.

Also, as mentioned, wide experience has already been accumulated confirming the effectiveness of the above general coarsening criterion in the special case of Algebraic Multigrid. The criterion has also been essential in developing coarse variables for high-Reynolds flows (see Section 7.5 below).

7.3 Derivation of coarse equations

7.3.1 Basic hypothesis: localness of coarsening

The *solution* to a system each of whose equations is local cannot be determined locally: it depends on *all* equations, near and far. However, *what can essentially be determined just from local information are coarser-level equations (or interactions)*. More precisely: provided the coarse set of variables is adequate (satisfying the above general coarsening criterion), a coarse system of local equations (interactions) equivalent to the fine-level system (in the sense that a coarse solution/equilibrium would yield the fine solution/equilibrium by a brief compatible local processing) is obtainable locally (i.e., by processing only a fine-level neighborhood comparable in size to the typical distance between coarse variables) with an error that decreases exponentially as a function of the total work, assuming this work is invested to suitably enlarge the number of variables involved in each coarse equation and increase the size of the fine-level neighborhood and the number of iteration involved in the local processing. This hypothesis has emerged from the long and diverse experience with both RG and multigrid solvers.

The actual derivation of the coarse equations, incorporating RG-type techniques, is based on fine-level simulations in representative regions. The simulations give a sequence of fine-level configurations u, which is readily translated into a sequence of coarse configurations $u^c = Cu$. There exist several approaches as to how and in what form to derive governing coarse-level rules from such a sequence. We briefly describe two basic approaches with which some experience has already been gained (recommending mainly the second of them).

7.3.1.1 Dependence table

In this (older) approach, the sequence of coarse configurations calculated by fine-level simulations is used to accumulate statistics of the dependence of each coarse variable (the *'pivot'*) on a certain set of neighboring coarse variables (the *'neighborhood'*). For this purpose the set of possible neighborhood configurations is partitioned into bins (described below). In fully deterministic problems, the average value of the pivot in each bin is accumulated and then tabulated. From such a table, the pivot value for any given neighborhood configuration can be interpolated, which is all one needs in order to operate (e.g., perform Gauss-Seidel relaxation) at the coarse level. In stochastic problems, additional statistics (e.g., variance) of the pivot over each neighborhood bin are tabulated, enough to enable accurate Monte Carlo simulations at the coarse level. Successful experience with simple versions of this approach, including the cases of Examples (iii) and (iv) above, are reported in [6, 24, 25] and [26].

The partition into bins should be done in terms of neighborhood functionals $S_1, S_2, \ldots$, roughly ordered in decreasing order of their influence on the pivot. For example, S_1 can be the sum of the values at the nearest neighbors to the pivot; S_2—the sum of the next-nearest neighbors; S_3—the variance in the set of nearest-neighbor values; etc. The space $S = (S_1, S_2, \ldots)$ is divided into (very roughly equiprobability) bins. The number of bins need not be very large; it only needs to allow sufficiently accurate interpolation to any probable value $s \in S$. The interpolation is usually done by a polynomial whose integral over each bin near s yields the statistics accumulated at that bin; the interpolation accuracy can of course deteriorate with decreasing probability of s.

The construction of the space of functional S and its binning should be guided by physical insight. It need not be very particular: many different choices would be adequate, due to the interdependence between neighborhood functionals. Still, the construction will usually be cumbersome. Therefore, for most models, the next approach seems preferable, so we describe it in greater detail.

7.3.1.2 Coarse Hamiltonian

In this approach the sequence of coarse configurations is used to calculate averages of many coarse-level observables $O_1, O_2, \ldots, O_\mu$. The average of O_i is denoted by $\langle O_i \rangle_f$, subscript f indicating that this average of the coarse observable has been calculated in the *fine*-level simulations. The coarse level itself is intended to be governed by a (yet to be calculated) Hamiltonian-like functional $H^c(u^c)$, that is, the probability of a coarse configuration u^c will be proportional to $\exp(-H^c(u^c))$. For any given approximate H^c, one can run simulations at the coarse level during which $\langle O_i \rangle_c$, the average of O_i according to H^c, can be calculated. Our first aim is to construct H^c such that $\langle O_i \rangle_c = \langle O_i \rangle_f$, $(i = 1, \ldots, \mu)$.

For this purpose H^c is written in the general form

$$H^c(u^c) = \sum_{k=1}^{K} a_k H_k(u^c) , \qquad (7.2)$$

where each H_k is a known functional of u^c (see examples below) and $\{a_k\}$ is a set of coefficients that need to be found. A crude approximation to H^c, possibly with a reduced number of terms (reduced K), can inexpensively be obtained from small-scale fine-level calculations assuming independence of various quantities (as in the example below). The approximation is then improved in a few Newton-like iterations, during which additional terms may be added to H^c (increasing K). Specifically, in each iteration H^c is changed by adding to it $\delta H^c = \sum_k \delta a_k H_k$. Using the first-order relation (used already by K.G. Wilson and R.H. Swendsen; cf., e.g., [27])

$$\delta \langle O \rangle = \langle O \rangle \langle \delta H^c \rangle - \langle O \cdot \delta H^c \rangle , \qquad (7.3)$$

one gets a system of μ equations

$$\sum_{k}^{K} (\langle O_i \rangle \langle H_k \rangle - \langle O_i H_k \rangle) \delta a_k = \langle O_i \rangle_f - \langle O_i \rangle_c , \; (i = 1, \ldots, \mu) \qquad (7.4)$$

where $\langle O_i \rangle$, $\langle H_k \rangle$, and $\langle O_i H_k \rangle$ are averages calculated during the fine-level simulations. (Corresponding averages are also calculated in the coarse-level simulation, for a purpose explained below.) The corrections $\{\delta a_k\}$ at each iteration are calculated to satisfy (7.4) best, in a least-square sense. We choose for this purpose enough observables: $\mu > K$. Usually, the set of observables $\{O_i\}_i$ will include $H_1, H_2, \ldots, H_K$ and some others. Then H^c is replaced by $H^c + \delta H^c$ and if this change is not small enough, another iteration is performed with the new H^c.

These Newton-like iterations converge fast. Having converged, the relation $\langle O_i \rangle_c = \langle O_i \rangle_f$ is satisfied for $i = 1, \ldots, \mu$, but possibly not for observables which were not included in the process. To check the accuracy of the calculated H^c, we therefore calculate the discrepancy $\langle O \rangle_f - \langle O \rangle_c$ for many additional observables, in particular for all the 'second moment', or product, observables $O_i H_k$, $(i = 1, \ldots, \mu; \ k = 1, \ldots, K)$, whose averages, needed in (7.4), have anyway been calculated. An observable O for which the discrepancy $\langle O \rangle_f - \langle O \rangle_c$ remains particularly large can be added to the list of Hamiltonian terms (increasing K) to facilitate decrease of this discrepancy in the next iterations. Due to the interdependence of observables, obtaining in this way accurate reproduction of the observables that have had the largest discrepancies will usually cause the discrepancies to sharply decrease in other observables as well, including still-higher-moment observables. The iterations continue as long as all such observables exhibit satisfactorily small discrepancies.

7.3.2 Example

A test of the coarse Hamiltonian approach has been carried out with the polymer case (Example (ii) in Section 7.2.2.1), employing a united-atom model of polymethylene taken from [28], using the coarsening (7.1) with $q = 3$. (A detailed description, but with $q = 4$, has appeared in [22].) The given fine-level Hamiltonian in this case is the sum of bond lengths, bond angles, bond dihedrals (torsions), and Lennard-Jones interactions. Similarly, the first approximation to H^c is chosen in the form

$$H^c(u^c) = \sum_i F_1(|r_i - r_{r+1}|) + \sum_i F_2(\theta_i) + \sum_i F_3(\tau_i) + \sum_{|i-j|>2} F_4(|r_i - r_j|), \quad (7.5)$$

where $r_i = u_i^c$ is the location of the i-th coarse 'atom', $|r_i - r_{i+1}|$ is the distance between two successive coarse 'atoms', θ_i is the angle (r_{i-1}, r_i, r_{i+1}), τ_i is the torsion $(r_{i-1}, r_i, r_{i+1}, r_{i+2})$, i.e., the angle between the plane spanned by (r_{i-1}, r_i, r_{i+1}) and the one spanned by (r_i, r_{i+1}, r_{i+2}), and F_4 is a Lennard-Jones-like interaction. Each of the initially-unknown functions F_ℓ can be expanded in the form

$$F_\ell(\xi) = \sum_j a_{\ell,j} w_j(\xi), \quad (\ell = 1, 2, 3, 4) \quad (7.6)$$

with unknown coefficients $a_{\ell,j}$ and known basis function $w_j(\xi)$; e.g., local basis functions (one-dimensional finite elements). Upon collecting (over the relevant

$\sum_i$ in (7.5)) all the Hamiltonian terms that include the same unknown $a_{\ell,j}$, the coarse Hamiltonian (7.5) obtains the general form (7.2).

A reasonable first approximation to F_4 is the given Lennard-Jones interactions, multiplied by q^2. A first approximation to F_1 (similarly: F_2, F_3) can be calculated by fine-level simulation of a rather short polymer chain (e.g., $n = 24$ and $m = 8$) from the bare distribution of the distances $\{|r_i - r_{i+1}|\}_{i=2}^{m-2}$, omitting the exceptional distances at the ends ($i = 1$ and $i = m - 1$), for which a different function F_1 may separately be calculated. These first approximations ignore all correlations between the internal coordinates ($\{|r_i - r_{i+1}|\}, \{\theta_i\}$ and $\{\tau_i\}$). The iterations described above will then automatically correct for those correlations, introducing on the way some new explicit correlation terms into H^c (increasing K by adding 'second-moment' terms as described above). With the improved H^c one can then make simulations at the coarse level with much longer chains (e.g., $n = 120$) and further improve H^c by making additional iterations, now making the fine-level simulations in *windows within that longer chain* (see Section 7.4). Then H^c can be similarly used to derive Hamiltonians at still coarser levels, with simulations on still longer chains. The longer chains may produce new situations (e.g., contact points due to folding) that require some new, window-within-window calculations at all finer levels to further correct H^c. The formulation is very flexible, allowing the introduction of new Hamiltonian terms to account for new situations (see next).

7.4 Window developments

Usually the computation starts in a rather small finest-level domain, with some artificial boundary conditions. For example, in the case of atomistic fluids (Example (iii) above), one can start with several hundreds (in two-dimensional problems) or several thousands (in 3D) particles, with periodic boundary conditions, where the period is chosen so that the fluid has its known average density. Similarly, in the case of discretized PDE problems (e.g., a fluid satisfying discretized Navier-Stokes equations), one can start with a grid of several hundreds (in 2D) to several thousands (in 3D) gridpoints, with periodic boundary conditions. In the polymer case, one can start with a short chain, as mentioned above.

From the computations in those small domains, one derives the first approximation to the first-coarse-level equations (or Hamiltonian), as in the example above. Being coarser, the new level then allows inexpensive calculations in a much larger domain. Note, however, that these computations in a larger domain are likely to encounter situations not accounted for by the initial small-domain fine-level simulations. For instance, the density fluctuations in the large domain can be much larger than in the small domain, creating regions with densities outside the ranges that have been simulated (or that have accumulated enough statistics) at the fine level. Similarly, in the polymer example, the short initial chain may well have lacked folding situations typical to longer chains. To obtain more accurate coarse equations in such new situations, more fine-level small-domain calculations should be done.

Some of the needed additional calculations could in fact have been done in advance. Namely, *several* small-domain calculations could initially be launched, independent of each other, to simulate for example several different average densities, or several different mixtures of atomic species or chain elements. However, it is difficult to anticipate in advance all the different local situations that will arise. So, generally, much of the needed additional fine-level simulations will be carried out 'upon demand', in regions where the coarse level encounters new situations, and they will be carried out as fine-level *windows* within the coarse-level simulation. This has also the advantage of giving these simulations more realistic boundary conditions (for fluids), or more realistic folding situations (for polymers), etc.

A window is created by replacing a certain coarse subdomain (where new situations have arisen) by a fine-level patch (including again several hundred to several thousand degrees of freedom). On the boundary of that patch the fine-level simulations are kept compatible with the coarse simulations using the generalized interpolation described above. As in Section 7.3 above, the fine simulation will give rise to (or correct) the coarse equations needed at the refined subdomain.

The coarse-level equations derived in the window can of course be used outside that window's subdomain, wherever similar conditions exist. Also, the window may be shut off (returning to pure coarse-level simulations at that subdomain) as soon as it has accumulated enough statistics to make the derived course equations as accurate as desired.

The process can of course be recursive: the coarse-level simulations can be used to construct equations for a still coarser level, which will be simulated in a still larger domain, possibly creating, upon encountering additional local conditions, new windows of the first-coarse level, with new fine-level windows inside them, and so on.

7.5 Some special situations

Various special situations require special or modified coarse-level equations (or additional terms in H^c). In the case of PDEs, special coarse equations would usually be needed near boundaries, with different equations near different types of fine-level boundary conditions.

In some cases, new types of variables need be introduced to satisfy the above general coarsening criterion. For example, in Navier-Stokes simulations of incompressible two-dimensional fluids, at a fine enough scale suitable coarsening can be defined in terms of averages of the vorticity function ω. But this coarsening is no longer adequate for large time steps and/or large spatial meshsizes that do not sufficiently resolve the rotation of the flow in strong vortexes. It has been shown that suitable coarsening (satisfying the above coarsening criterion) at such scales can be constructed by decomposing the vorticity function into the sum of idealized vortexes (radial local functions that satisfy the steady-state

Euler equations) and the averages of a remaining low-vorticity part (work in progress, in collaboration with Jim McWilliams and Boris Diskin).

In the polymer case described above, each of the coarse-level functions F_l should have a separate special expansion (7.6) for the internal coordinates near the ends of the chain. Hence these coordinates will have different sets of unknowns a_{lj}, which thus cannot be collected together with the terms arising away from the ends. Also, at large scales, special situations arise when the polymer folds upon itself, bringing into proximity atoms that are not neighbors along the chain. Special terms should then be added to H^c that depend on the distances between such atoms and on local angles created by them.

7.6 Extensions in brief

Various important extensions of the upscaling techniques, to diverse physical situations, can be developed. The following principal directions have already been preliminarily considered.

7.6.1 Long-range interactions

Long-range interactions (e.g., between electrostatic charges) can each be decomposed into the sum of a smooth interaction and a local one ('smooth' and 'local' being meant on the scale of the next coarse level; all familiar physical interactions can be decomposed this way; see [23] and examples in [30, 32] and [33]). To any desired accuracy, the smooth part can directly be represented at the coarse-level, e.g., by aggregated charges and dipoles moving with the coarse level 'atoms' (in Example (ii) above) or by (high order) adjoint interpolation of charges to the coarse-level lattice (in Example (iii)). The local part is essentially transferred, together with all other local interactions, using the fine/coarse iterations described above. Effectively, the amount of work invested per charge involves only calculating its local interactions, and, even more importantly, only charges within selected windows need be treated. It can be shown that this is possible due to the smoothness of the non-local interactions, which makes them minimally sensitive to the local MC moves, especially when the latter are explicitly designed to conserve certain moments of the charge distribution.

7.6.2 Dynamical systems

Generally, for time-dependent systems, the equilibrium coarsening criterion of Section 7.2.4 is replaced by the requirement that the fine-level configuration (if its evolution is stable) or its ensemble statistics (in the case of instability) at any given time can fully be recovered from the coarse configurations at a small number of previous time steps. Dependence tables (e.g., in kinetic Monte Carlo computations) have been derived in the form of flux dependence on both current-time and previous-time neighboring coarse variables [34]. A general computational criterion has been formulated for the size of the time steps to increase with the spatial coarsening level, so as to maintain full efficiency.

A Hamiltonian-like functional that governs every time step can also be developed analogously to the one described in Section 7.3.1.2.

For Hamiltonian systems (e.g., corresponding to Examples (ii) and (iii) in Section 7.2.2.1), the multiscale structure allows a natural combination of temperature-accurate statistical simulation at small scales with time-accurate dynamics at large scales. Assuming that after any given time interval the fine-scale degrees of freedom settle into a local equilibrium slave to the coarse-level averages (where that level increases with the size of the time interval), the general equilibrium criterion for choosing the coarse variables (see Section 7.2.4) still directly applies. Large time steps, based on implicit discretization of Newton law, can then be made, using a multigrid-like solver where the relaxation at fine levels is replaced by CMC (cf. Section 14.8 in [29]). This approach yields two benefits in performing very large time steps. Firstly, it allows much easier handling of local minima. Secondly, it avoids the killing of highly-oscillatory modes (those vibrations that are not resolved by the time step), which would occur if the implicit equations of a large time step were imposed at all scales. Instead, these modes assume stochastic amplitudes according to their equilibrium probability distribution. The desired temperature is introduced very directly in this way, avoiding the need for fabricating Langevin stochastic forces.

Another possible approach, for fluids, is to first develop at equilibrium a coarse-level Hamiltonian $H^c(u^c)$ such that the relation $Probability(u^c) \sim \exp(-H^c(u^c)/k_B T)$ will simultaneously hold for a full range of temperatures T. (This can be achieved by adding several moments $(H^c)^m$ to the list of observables (O_i) used in (7.4), and constructing a *joint Hamiltonian* (see below) for different temperatures in the range of interest.) Then H^c can be used in Newtonian dynamics at the coarse level, where effective coarse-level masses (and their possible dependence on the coarse coordinates) are determined by comparing (in windows of fine-level dynamic simulations) coarse-level accelerations with the gradient of H^c.

Still another approach for fluids is a Boltzmann-type upscaling in the 6D space of positions and velocities. Starting simulations at the individual-particle level, increasingly coarser spatial levels will describe velocity distributions at progressively higher resolutions.

7.6.3 Stochastic coarsening

Our studies (e.g., [34]) have shown that averaging upon coarsening should often better be stochastic. The added stochasticity is important to create smoother coarse dynamics, hence simpler dependence table or easier H^c, especially for a fine level with discrete-state (e.g., integer-valued) variables or highly oscillating Hamiltonian. One general way is to modify a deterministic averaging (or anterpolation—the adjoint of interpolation) by adding to each coarse variable a small increment, where the field of increments is in equilibrium governed by a Hamiltonian-like functional H_p. A corresponding CMC process has been developed, and the general coarsening criterion then effectively checks that H_p has been properly designed, i.e., so that it prohibits increment fields that correspond to long-range moves.

7.6.4 Joint H^c

The same coarse functional H^c should sometimes simultaneously satisfy $\{<O_i>_c = <O_i>_f\}_i$ for several different MC situations, such as: (a) under different external fields; (b) at different temperatures (cf. Section 7.6.2); (c) in different energy basins (cf. Section 7.6.6). Generally, this can be achieved by adding in (7.2) terms H_k that are particularly sensitive to the differences between the different simulated situations.

7.6.5 Complex fluids

More elaborate coarse Hamiltonians are needed for fluids with more complex molecules of one or several species, such as water with methanol, or glycerol, etc. A gradual construction can then be planned, starting for example with H^c constructed for atomistic equilibrium in a periodic domain containing only two molecules. Adding then to the simulation one molecule at a time, the coefficients of H^c are updated by (7.4), with additional terms H_k being introduced that correspond at each iteration to correlation observables that are still ill-approximated.

7.6.6 Low temperatures (example)

At high temperatures, the coarse variables for a simple fluid in equilibrium are gridpoint values, each standing for some local averaging of $m(\mathbf{x})$, the masses m of particles at positions $\mathbf{x} = (x_1, x_2, x_3)$ near the gridpoint. At lower temperatures, as the fluid starts to crystallize, roughly with periods $u^{(\ell)} = (u_1^{(\ell)}, u_2^{(\ell)}, u_3^{(\ell)})$, ($\ell = 1, 2, 3$), say, three new coarse-level fields should enter, standing for local averaging of $\exp(2\pi i w^{(\ell)} \mathbf{x}) * m(\mathbf{x})$, ($\ell = 1, 2, 3$), where $\mathbf{w}^{(\ell)} \cdot \mathbf{u}^{(\ell)} \simeq \delta_{k\ell}$. If the crystal is perfect and $\mathbf{w}^{(\ell)}$ are exactly known, these three coarse fields will be constants. When $w^{(\ell)}$ are only approximate, these fields will oscillate smoothly. Similar averaging at the next coarser levels will then describe these oscillations, effectively correcting the erroneous $\mathbf{w}^{(\ell)}$. If the crystal is not perfect, meaningful averaging of this type will persist only up to a certain scale; usually, the lower the temperature the larger that scale (see Section 14.7.3 in [29]).

Also upon lowering the temperature, energy barriers emerge at increasingly larger scales. By insisting on constructing, level after level, *joint* H^c (see Section 7.6.4), statistically correct transitions between different energy basins can be simulated efficiently.

7.6.7 Multiscale annealing

Thus, as a system is gradually cooled, increasingly larger-scale degrees of freedom are identified. This identification of increasingly larger collective moves makes such a computation extremely more effective than simple simulated annealing [35] for *minimizing* the energy (the limit $T \to 0$), especially in the physically common situation of multiscale nested attraction basins. The multiscale annealing may provide efficient solvers to very difficult global optimization problems. (See more in Section 18.2 of [29], and in [31].)

7.6.8 Coarse-level computability of fine observables

Often, an observable of interest is not directly expressed in terms of the coarse-level variables. We have developed a general procedure (similar to Section 7.3.1.2) for computing a functional dependence of a quantity of interest upon the coarse variables, based on suitable statistics accumulated during the fine-level simulations.

7.6.9 Determinism and stochasticity

The discussion above is written mainly in terms of stochastic systems, but can be extended to deterministic ones. Moreover, a stochastic system at the fine level often yields a deterministic system at large enough scales. The opposite exists too: a deterministic fluid flow at the viscous scale acquires stochastic features at the large scales reigned by turbulence. Similarly, a particle system governed by Newtonian dynamics can give rise to various stochastic developments at different scales. The coarsening approaches described above can accommodate such transitions.

7.6.10 Upscaling from quantum mechanics to molecular dynamics

The electronic distribution of a molecular system with N valence electrons can be approximately computed by solving the Kohn-Sham equations. (This involves the calculation of N eigenfunctions of a Schroedinger operator, whose potential depends on the eigenfunctions. A very efficient multigrid is developed for this task; see Section 9 in [29].) It is estimated that such calculations can solve sufficiently large systems to enable upscaling to molecular dynamics (MD) or molecular equilibrium (ME), ie., derivation of the MD or ME force fields. The coarse-level variables in this upscaling are quite obviously the nuclear positions. The force field tables for ME calculations, for example, can iteratively be derived by comparing the ME simulations with Kohn-Sham solutions, where the latter need be computed only in relatively small windows. This is similar to the derivation of the *coarse-level* macromolecular force field described in Section 7.3.1.2.

An even more intriguing possibility is the attempt to use the SU approach to derive the molecular force fields directly from the fundamental, high-dimensional equations of quantum mechanics, using Feynman's path integrals, as sketched in [70].

References

[1] Brandt A. (1973). Multi-level adaptive technique (MLAT) for fast numerical solutions to boundary value problems. In *Proc. 3rd Int. Conf. on Numerical Methods in Fluid Mechanics* (Cabannes H. and Temam R., eds), p. 82–89. Lecture Notes in Physics, 18, Springer-Verlag.

[2] Brandt A. (1977). "Multi-level adaptive solutions to boundary value problems", *Math. Comp.*, **31**, p. 333–390.
[3] Brandt A. (1982). Guide to multigrid development. In *Multigrid Methods* (Hackbusch W. and Trottenberg U., eds.), Springer-Verlag, p. 220–312.
[4] Brandt A. (2007). *Multigrid Techniques*: *1984 Guide, with Applications to Fluid Dynamics*, 1984, 191 pages, ISBN-3-88457-081-1; GMD-Studien Nr. 85. To appear in SIAM Classics in Applied Mathematics series.
[5] Brandt A. and Livshits I. (1997). "Wave-ray multigrid method for standing wave equations", *Electronic Trans. Num. An.*, **6**, p. 162–181.
[6] Brandt A. and Ron D. (2001). "Renormalization multigrid (RMG): Statistically optimal renormalization group flow and coarse-to-fine Monte Carlo acceleration", Gauss Center Report WI/GC–11, June 1999. *J. Stat. Phys.*, **102**, p. 231–257.
[7] Brandt A., McCormick S., and Ruge J. (1982). Algebraic multigrid (AMG) for automatic multigrid solution with application to geodetic computations, Institute for Computational Studies, POB 1852, Fort Collins, Colorado.
[8] Brandt A., McCormick S., and Ruge J. (1984). Algebraic multigrid (AMG) for sparse matrix equations. In *Sparsity and its Applications* (Evans, D.J., ed.), Cambridge University Press, Cambridge, p. 257–284.
[9] Brandt A. (1986). Algebraic multigrid theory: The symmetric case, in *Preliminary Proc. Int. Multigrid Conf.*, Copper Mountain, Colorado, April 6–8, 1983; *Appl. Math. Comp.*, **19**, p. 23–56.
[10] Brandt A., Ron D., and Amit D. J. (1986). Multi-level approaches to discrete-state and stochastic problems. In *Multigrid Methods, II* (Hackbusch, W. and Trottenberg, U., eds.), Springer-Verlag, p. 66–99.
[11] Brandt A. (1992). "Multigrid methods in lattice field computations", *Nucl. Phys. B* (Proc. Suppl.), **26**, p. 137–180.
[12] Brandt A., Galun M., and Ron D. (1994). "Optimal multigrid algorithms for calculating thermodynamic limits", *J. Stat. Phys.*, **74**, p. 313–348.
[13] Hackbusch W. (1985). *Multigrid Methods and Applications*, Springer, Berlin.
[14] Trottenberg U, Oosterlee C. W., and Schüller A. (2000). *Multigrid*, Academic Press, London.
[15] Briggs W. L., Henson V. E., and McCormick S. F. (2000). *A Multigrid Tutorial*, 2nd Ed., SIAM.
[16] Ruge J. and Stüben K. (1987). Algebraic multigrid. In *Multigrid Methods* (McCormick S. F., ed.), SIAM, Philadelphia, p. 73–130.
[17] Wilson K. G. (1983). "The Renormalization Group and Critical Phenomena", *Rev. Mod. Phys.*, **55**, p. 583–600. (1982, Nobel Prize Lecture)
[18] Swendsen R. H. (1979). "Renormalization group Monte Carlo methods", *Phys. Rev. Lett.*, **42**, p. 859–862.
[19] Swendsen R.H. (1979). "Monte Carlo renormalization-group studies of the $d = 2$ Ising model", *Phys. Rev. B*, **20**, p. 2080–2087.
[20] Mack G. and Pordt A. (1985). Convergent perturbation expansions for Euclidean quantum field theory, *Comm. Math. Phys.*, **97**, p. 267. Also:

Mack G. in: Nonperturbative Quantum Field Theory, G. t'Hooft *et al.*, eds. Plenum Press, New York, 1988, p. 309.

[21] Goodman J. and Sokal A. D. (1986). "Multigrid Monte Carlo methods for lattice field theories", *Phys. Rev. Lett.*, **56**, p. 1015–1018.

[22] Bai D. and Brandt A. (2001). Multiscale computation of polymer models. In: Brandt A., Bernholc J., and Binder K. (Eds.), *Multiscale Computational Methods in Chemistry and Physics*. NATO Science Series: Computer and System Sciences, Vol. **177**, IOS Press, Amsterdam, p. 250–266.

[23] Brandt A. (1991). "Multilevel computations of integral transforms and particle interactions with oscillatory kernels", *Comp. Phys. Comm.*, **65**, p. 24–38.

[24] Brandt A. and Iliyn V. (2001). Multilevel approach in statistical physics of liquids. In: Brandt A., Bernholc J., and Binder K. (Eds.), *Multiscale Computational Methods in Chemistry and Physics*. NATO Science Series: Computer and System Sciences, Vol. 177, IOS Press, Amsterdam, p. 187–197.

[25] Brandt A. and Ilyin V. (2003). "Multilevel Monte Carlo methods for studying large-scale phenomena in fluids", *J. of Molecular Liquids*, **105**, p. 253–256.

[26] Shmulyian S. (1999). *Toward Optimal Multigrid Monte Carlo Computation in Two-Dimensional $O(N)$ Non-Linear σ-Models*, Ph.D. Thesis, Weizmann Institute of Science.

[27] Gupta R. (1987). "Open problems in Monte Carlo Renormalization Group: Application to critical phenomena", *J. Appl. Phys.*, **61**, p. 3605–3611.

[28] Paul W., Yoon D. Y., and Smith G. D. (1995). "An optimized united atom model for simulations of polymethylene melts", *J. Chem. Phys.*, **103**, p. 1702–1709.

[29] Brandt A. (2001). Multiscale scientific computation: review 2001. In Barth T. J., Chan T. F., and Haimes,, R. (eds.): *Multiscale and Multiresolution Methods: Theory and Applications*, Springer Verlag, Heidelberg, p. 1–96. Available in www.wisdom.weizmann.ac.il/∼achi/review00.ps.

[30] Brandt A. and Venner C. H. (1998). Multilevel evaluation of integral transforms with asymptotically smooth kernels, Gauss Center Report WI/GC–2, The Weizmann Institute of Science, Rehovot, April 1995; *SIAM J. Sci. Comp.*, **19**, p. 468–492.

[31] Brandt A. and Ron D. (2003). Multigrid solvers and Multilevel Optimization Strategies. In: *Multilevel Optimization and VLSICAD*, (Cong J. and Shinnerl J. R., eds.), Kluwer Academic Publishers, Boston, p. 1–69.

[32] Ramirez I. H. (2005). *Multilevel Multi-Integration Algorithm for Acoustics*, Ph.D. Thesis, University of Twente, Enschede, The Netherlands.

[33] Suwan I. (2005). *Multiscale Methods in Molecular Dynamics*, Ph.D. Thesis, Feinberg Graduate School, Weizmann Institute of Science, Rehovot, Israel.

[34] Saad N. (2005). *Multiscale Algorithm for Time-Dependent One-Dimensional System*, Feinberg Graduate School, Weizmann Institute of Science, Rehovot, Israel.

[35] Kirkpatrick S., Gelatt C. D., Jr., and Vecci M. P. (1983). "Optimization by simulated annealing", *Science*, **220**, p. 671.
[36] Brandt A. (2000). General highly accurate algebraic coarsening schemes. Gauss Center Report WI/GC-13, May 1999. *Electronic Trans. Num. Anal.*, **10**, p. 1–20.
[37] Alcouffe R. E., Brandt A., Dendy J. E., Jr., and Painter J. W. (1981). "The multigrid methods for the diffusion equation with strongly discontinuous coefficients", *SIAM J. Sci. Stat. Comp.*, **2**, p. 430–454.
[38] Brandt A. and Galun M. (1996). "Optimal multigrid algorithm for the massive Gaussian model and path integrals", *J. Stat. Phys.*, **82**, p. 1503–1518.
[39] Brandt A. and Galun M. (1997). "Optimal multigrid algorithms for variable-coupling isotropic Gaussian models", *J. Stat. Phys.*, **88**, p. 637–664.
[40] Brezina M., Falgout R., MacLachlan S., Manteuffel T., McCormick S., and Ruge J. (2005). "Adaptive smoothed aggregation (αSA) multigrid", *SIAM Review*, **47**, p. 317–346.
[41] Dendy J. E., Jr. (1982). "Black box multigrid". *J. Comp. Phys.*, **48**, p. 366–386.
[42] Chen W. and Fish J. (2006). "A generalized space-time mathematical homogenization theory for bridging atomistic and continuum scales", *International J. Num. Meth. in Eng.*, **67**, p. 253–271.
[43] Fish J., Chen W., and Li R. (2006). "Generalized mathematical homogenization of atomistic media at finite temperatures in three dimensions", *Comput. Methods Appl. Mech. Engr.*, doi:10.1016/j.ema.2006.08.001.
[44] E W. and Engquist B. (2002). "The heterogeneous multi-scale methods", *Commun. Math. Sci.*, **1**, p. 87–132.
[45] Givon D., Kevrekidis I. G., and Kupferman R. (2006). "Strong convergence of projective integration schemes for singularly perturbed stochastic differential systems", *Comm. Math. Sci.*, **4**, p. 707–729.
[46] van Kampen N. G. (1985). "Elimination of fast variables", *Phys. Rep.*, **124**, p. 69–160.
[47] Skorokhod A. V. (1987). *Asymptotics Method in the Theory of Stochastic Differential Equations*, AMS, Providence, RI.
[48] Veretennikov A. Yu. (1991). "On the averaging principle for systems of stochastic differential equations", *Math. USSR Sborn.*, **69**, p. 271–284.
[49] Freidlin M. I. and Wentzell A. D. (1984). *Random Perturbations of Dynamical Systems*, Springer, New York.
[50] Gear C. W., Kevrekidis I. G., and Theodoropoulos C. (2002). "'Coarse' integration/bifurcation analysis via microscopic simulators: micro-Galerkin methods", *Comp. Chem. Engr.*, **26**, p. 941–963.
[51] Gear C. W. and Kevrekidis I. G. (2003). "Projective methods for stiff differential equations: problems with gaps in their eigenvalue spectrum", *SIAM J. Sci. Comp.*, **24**, p. 1091–1106.

[52] Rico-Martinez R., Gear C. W., and Kevrekidis I. G. (2004). "Coarse projective KMC integration: forward/reverse initial and boundary value problems", *J. Comp. Phys.*, **196**, p. 474–489.

[53] Kevrekidis I. G., Gear C. W., Hyman J. M., Kevrekidis P. G., Runborg O., and Theodoropoulos, K. (2003). "Equation-free coarse-grained multi-scale computation: enabling microscopic simulators to perform system-level tasks", *Comm. Math. Sci.*, **1**, p. 715–762.

[54] Kevrekidis I. G., Gear C. W., and Hummer G. (2004). "Equation-free: the computer assisted analysis of complex multiscale systems", *AIChE J.*, **50**, p. 1346–1354.

[55] Hummer G. and Kevrekidis I. G. "Coarse molecular dynamics of a peptide fragment: free energy, kinetics and long time dynamics computations", *J. Chem. Phys.*, **118**, p. 10762–10773.

[56] Ting T. C. T. (1980). "Dynamic response of composites", *Appl. Mech. Rev.*, **33**, p. 1629–1635.

[57] Benssousan A., Lions J. L., and Papanicoulau G. (1978). *Asymptotic Analysis for Periodic Structures*, North Holland, Amsterdam.

[58] Bakhvalov N. S. and Panasenko G. P. (1989). *Homogenization: Averaging Processes in Periodic Media*, Kluwer, Dordrecht.

[59] Sanchez-Palencia E. (1980). *Non-homogeneous Media and Vibration Theory*, Springer, Berlin.

[60] Boutin C. and Auriault J. L. (1993). "Rayleigh scattering in elastic composite materials", *Int. J. Engng. Sci.*, **12**, p. 1669–1689.

[61] Fish J. and Chen W. (2004). "Discrete-to-continuum bridging based on multigrid principles", *Computer Methods in Applied Mechanics and Engineering*, **193**, p. 1693–1711.

[62] Fish J., Chen W., and Nagai G. (2000). "Nonlocal dispersive model for wave propagation in heterogeneous media: one-dimensional case", *International Journal for Numerical Methods in Engineering*, **54**, p. 331–346.

[63] Fish J., Chen W., and Nagai G. (2004). "Nonlocal dispersive model for wave propagation in heterogeneous media: multidimensional case", *International Journal for Numerical Methods in Engineering*, **54**, p. 347–363.

[64] Fish J. and Chen W. (2004). "Space-time multiscale model for wave propagation in heterogeneous media", *Comp. Meth. Appl. Mech. Engng.*, **193**, p. 4837–4856.

[65] Kadanoff L. (2002). *Statistical Physics, Statics, Dynamics and Renormalization*, World Scientific, Singapore.

[66] Fisher M. E. (1998). "Renormalization group theory: Its basis and formulation in statistical physics", *Rev. Mod. Phys.*, **70**(2), p. 653–681.

[67] Givon D., Kupferman R., and Stuart A. (2004). "Extracting macroscopic dynamics: model problems and algorithms", *Nonlinearity*, **17**, R55-R127.

[68] Brandt A. (2005). "Multiscale solvers and systematic upscaling in computational physics", *Computer Physics Communication*, **169**, p. 438–441.

[69] Brandt A. (2006). *Methods of Systematic Upscaling*, Report MCS06–05, Department of Computer Science and Applied Mathematics, Weizmann Institute of Science, Rehovot, Israel.//
[70] Zlochin M. and Brandt A. (2005). *Systematic Upscaling for Feynman Path Integrals*, Progress Report MCS05-10, Department of Computer Science and Applied Mathematics, Weizmann Institute of Science, Rehovot, Israel.

8

EQUATION-FREE COMPUTATION: AN OVERVIEW OF PATCH DYNAMICS

G. Samaey, A. J. Roberts, and I. G. Kevrekidis

8.1 Introduction

In many contemporary problems in science and engineering, one is interested in system behavior on macroscopic scales, while a constitutive equation at this scale cannot be obtained in closed form without simplifying assumptions that are hard to justify. Instead, a microscopic model (possibly of a different type) is available, which describes the system in full detail; however, the associated computational cost can be prohibitive for macro-scale simulations. For time-dependent problems of this type, an "equation-free" framework has recently been proposed, e.g. [1, 2].

Equation-free computation aims at extracting macroscopic information (such as steady states or time derivatives), using only simulations with a given microscopic model (or simulation code) in small subsets of the space-time domain (the so-called patches). This chapter reviews the patch dynamics scheme. When the unavailable macroscopic model is a partial differential equation (PDE), this scheme, as originally proposed, estimates the time derivative of a "method of lines" (finite difference) spatial discretization by performing a local simulation with the (known) microscopic model in small patches around the macroscopic grid points. In the closely related heterogeneous multiscale method, a solver for the unavailable macroscopic moded is formulated and supplemented with local microscopic simulations [34]. As such, both methods can be considered as *information passing* methods in the spirit of the classification of this book: macroscopic information (in casu: a finite difference approximation of spatial derivatives) is used to initialize simulations with the microscopic model; and, in turn, macroscopic information (in casu: the macroscopic time derivative) is estimated from these simulations.

When implementing patch dynamics schemes, one needs to provide initial and boundary conditions for the microscopic simulations, and choose the size of the patches and the simulated microscopic time. These choices need to be made based on: (1) the capabilities of the available microscopic simulation code; (2) the accuracy of the macroscopic information obtained from the microscopic simulations; and (3) knowledge of the properties of the (unknown) macroscopic equation. We discuss these issues throughout the chapter.

Section 8.2 introduces equation-free methods, the gap-tooth scheme and patch dynamics. In this chapter we use the classical homogenization problem to illustrate the convergence and efficiency of the gap-tooth and patch dynamics schemes. Section 8.3 gives the basic properties of the model problems. Subsequently, Section 8.4 describes the gap-tooth scheme and discusses its convergence for a class of model homogenization problems. The gap-tooth scheme as formulated in Section 8.4 is based on symmetric interpolation; this type of interpolation automatically results in central finite difference schemes for the macroscopic equation. Moreover, it requires one to impose *constant macroscopic gradient* boundary conditions on the patches; this may not always be possible in practice. Section 8.5 describes the patch dynamics scheme. Here, we investigate to what extent (and at what computational cost) one can provide the patches with *arbitrary* boundary conditions. To this end, the patches are surrounded with *buffer regions*, which shield the region of interest from boundary artifacts; the size of the buffer region determines the additional computational cost. We also generalize the initialization of the patches, such that *any* finite difference scheme can be approximated, beyond the central schemes, as in Section 8.4. Section 8.6 follows a different approach: we retain the advantage of using the microscopic simulation code with standard boundary conditions, while avoiding the introduction of buffer regions. This is achieved by updating the boundary conditions on the patches after *every microscopic time step*; the values for these boundary conditions are obtained via center manifold theory, implemented through computer algebra. Finally, Section 8.7 contains conclusions and an outlook for future research.

8.2 Equation-free multiscale framework

In the equation-free framework one assumes that a macroscopic, effective model exists, but that it cannot be obtained *explicitly* from the available microscopic model. When this is the case, one can construct a so-called *coarse time-stepper* to approximate a time-stepper for the unknown macroscopic equation. This time-stepper can subsequently be used to estimate macroscopic time derivatives.

A motivation for the development of this framework originates from the observation that, in research on numerical methods for the study of dynamical systems and bifurcations, methods have been developed that can compute steady states and periodic solutions, as well as their stability and bifurcations using only repeated calls to an already available simulation code. Well known examples of such methods are the recursive projection method [3], and its generalization as the Newton–Picard method [4, 5]. When replacing the macroscopic simulation code with a coarse time-stepper, all these methods become available to study *coarse* steady states, periodic solutions, and other coarse features of microscopic simulation codes.

This section briefly reviews the basic setup and introduces our notation.

8.2.1 Definitions

Consider an abstract microscopic evolution law (and corresponding time-stepper)

$$\partial_t u(\mathbf{x}, t) = f(u(\mathbf{x}, t)), \quad u(\mathbf{x}, t + \mathrm{d}t) = s(u(\mathbf{x}, t), \mathrm{d}t), \tag{8.1}$$

in which $u(\mathbf{x}, t)$ represents the microscopic state variables, $\mathbf{x}$ and t are the microscopic independent variables, and dt is the size of the microscopic time-step. We assume that an equivalent macroscopic model exists, but cannot be obtained in closed form. We denote this model (and time-stepper) by

$$\partial_t U(\mathbf{X},t) = F(U(\mathbf{X},t)), \quad U(\mathbf{X}, t+\delta t) = S(U(\mathbf{X},t), \delta t), \qquad (8.2)$$

in which $U(\mathbf{X},t)$ denotes the set of macroscopic state variables, $\mathbf{X}$ and t are the macroscopic independent variables, and δt is the size of the coarse time-step.

As an illustrative example [6], consider the microscopic model to be the stochastic differential equation (SDE) for a particle in a potential but forced by a fast time Ornstein–Uhlenbeck process:

$$\begin{aligned} dx(t) &= -\partial_x V(x)dt + \frac{1}{\epsilon}\sigma y(t)dt\,, \\ dy(t) &= -\frac{y}{\epsilon^2}dt + \frac{1}{\epsilon}dW_y(t)\,, \end{aligned} \qquad (8.3)$$

in which $x(t), y(t) \in \mathbb{R}$, $V(x)$ is a potential function, $\epsilon \ll 1$, and $W_y(t)$ is a standard Brownian motion on $\mathbb{R}$. One can think of $x(t)$ as being the position of the particle, and $y(t)$ as a fast internal variable. For our purposes, the macroscopic model of interest is the *Fokker–Planck equation* that the describes the evolution of the probability density of particles $\bar{\rho}(x,t)$:

$$\partial_t \bar{\rho}(x,t) - \partial_x(\partial_x V(x)\bar{\rho}(x,t)) = \frac{\sigma^2}{2}\partial_x^2 \bar{\rho}(x,t), \qquad (8.4)$$

corresponding to the following effective SDE for the behavior of a particle,

$$dx(t) = -\partial_x V(x)dt + \sigma dW_x(t),$$

where $W_x(t)$ is again a standard Brownian motion.

Then $u(\mathbf{x},t)$ consists of the position and internal state of all particles in the system, that is,

$$u(\mathbf{x},t) \equiv u(t) = \{x_j(t), y_j(t)\}_{j=1}^M,$$

with M the number of particles in the system. The corresponding macroscopic variable $U(\mathbf{X},t)$ is the density $\bar{\rho}(x,t)$; thus $\mathbf{X} = x$.

In the equation-free framework, one constructs a coarse time-stepper for the variables $U(\mathbf{X},t)$, which we denote as

$$\bar{U}(\mathbf{X},t+\delta t) = \bar{S}(\bar{U}(\mathbf{X},t), \delta t), \qquad (8.5)$$

where δt denotes the size of the coarse time-step, and the bars have been introduced to emphasize that the time-stepper for the macroscopic variables is only an *approximation* of a time-stepper for (8.2), since this equation is not explicitly known.

To define a coarse time-stepper (8.5), we introduce two operators that make the transition between microscopic and macroscopic variables. We define a *lifting operator*,

$$\mu : U(\mathbf{X}, t) \mapsto u(\mathbf{x}, t) = \mu(U(\mathbf{X}, t)), \qquad (8.6)$$

which maps macroscopic to microscopic variables. The corresponding *restriction operator*

$$\mathcal{M} : u(\mathbf{x}, t) \mapsto U(\mathbf{X}, t) = \mathcal{M}(u(\mathbf{x}, t)) \qquad (8.7)$$

computes the value of the macroscopic variable from the microscopic state. The restriction operator can often be determined as soon as the macroscopic variables are known. For instance, when the microscopic model consists of an evolving ensemble of many particles, the restriction typically consists of the computation of the low order moments of the particle distribution (density, momentum, energy).

The construction of the lifting operator is usually more involved. Again taking the example of a particle model, we need to define a mapping from a few low order moments to an initial condition for each of the particles. We know that the higher order moments of the resulting particle distribution should be functionals of the low order ones; but unfortunately these functionals are unknown (since the macroscopic evolution law is also unknown). Several approaches have been suggested to address this problem. One could for instance initialize the higher order moments randomly. This introduces a *lifting error*, and one then relies on the separation of time scales to ensure that the higher order moments relax quickly to a functional of the low order moments (*healing*) [7, 8, 9]. In some cases this approach produces inaccurate results [10]. To initialize the higher order moments correctly, one should perform a simulation of the microscopic system with the additional constraint that the low order moments should be kept fixed. How this can be done using only a time-stepper for the original microscopic system has been explained and analyzed for lattice Boltzmann problems and singularly perturbed systems [11, 12, 13, 14]. When the particles are interacting, as in molecular dynamics simulations, the problem becomes more involved. In such cases, one has the additional complication that particles cannot be positioned too closely to each other; see e.g. [15, 16] for particle insertion in coupled atomistic-continuum simulations. We refer to [17] for the study of a lifting procedure for molecular dynamics simulations of dense fluids.

8.2.2 *The coarse time-stepper*

Given an initial condition for the macroscopic variables $\bar{U}(\mathbf{X}, t^*)$ at some time t^*, we construct the coarse time-stepper (8.5) in the following way.

1. **Lifting** Using the lifting operator (8.6), create appropriate initial conditions $\bar{u}(\mathbf{x}, t^*)$ for the microscopic time-stepper (8.1), consistent with the macroscopic variables $\bar{U}(\mathbf{X}, t^*)$.
2. **Simulation** Use the time-stepper (8.1) to compute the microscopic state $\bar{u}(\mathbf{x}, t)$ for $t \in [t^*, t^* + \delta t]$.
3. **Restriction** Obtain the macroscopic state $\bar{U}(\mathbf{X}, t^* + \delta t)$ from the microscopic state $\bar{u}(\mathbf{x}, t^* + \delta t)$ using the restriction operator (8.7).

Assuming $\delta t = k \, \mathrm{d}t$, this can be written as

$$\bar{U}(\mathbf{X}, t + \delta t) = \bar{S}(\bar{U}(\mathbf{X}, t), \delta t) = \mathcal{M}(s^k(\mu(\bar{U}(\mathbf{X}, t)), \mathrm{d}t)), \tag{8.8}$$

where we represent the k microscopic time-steps by a superscript on s. If the microscopic model is stochastic, one may need to perform multiple replica simulations to get an accurate averaged result.

So far, we have only added a computational wrapper around the microscopic code. As a consequence, if there is a big separation in time scales, any feasible coarse time-step δt will still be small compared to the macroscopic time scales of interest. However, as already mentioned, the coarse time-stepper is used as a *building block* to create computationally efficient numerical schemes. It can, for instance, be used as input for time-stepper-based bifurcation algorithms to perform a bifurcation analysis for the unavailable macroscopic equation [2, 8, 18, 19]. This approach has been used in a number of applications [9, 20], and also empowers one to perform more general system level tasks, such as coarse control and optimization [21].

One accelerates time integration using so-called *coarse projective integration* methods [22, 23]. These algorithms use a few coarse time-steps of size δt to obtain an estimate of the time derivative of the macroscopic variables (the unknown right-hand side of (8.2)). This time derivative estimate is subsequently used to take a large step of size $\Delta t \gg \delta t$, which should be determined by accuracy and stability considerations for the unknown macroscopic equation only. Coarse projective integration has been used in a number of applications [e.g. 24, 25, 26].

8.2.2.1 A remark on notation
Throughout this text, we use $\mathrm{d}t$ to denote a microscopic time-step, and δt for a coarse time-step. As outlined above, δt corresponds to a *time integration interval* for the microscopic simulations. We use $\Delta t \gg \delta t$ to denote a macroscopic time-step; for example, in the context of projective integration. Correspondingly, we use Δx and $\mathrm{d}x$ to denote a macroscopic, respectively microscopic, spatial mesh width.

8.2.3 The gap-tooth scheme and patch dynamics
This chapter considers situations where the macroscopic model (8.2) is assumed to be representable in principle by a PDE in one space dimension, so $\mathbf{X} = x$, say

$$\partial_t U(x, t) = F(U(x, t), \partial_x U(x, t), \ldots, \partial_x^d U(x, t)), \tag{8.9}$$

in which $U(x,t) \in \mathbb{R}^n$ denotes the macroscopic state as a function of space and time and d is the order of the PDE (the highest derivative that appears in the equation). For this type of problems, the *gap-tooth scheme* was proposed [1]; it can be generalized in several space dimensions. A number of small intervals (the *teeth*) are introduced, separated by large gaps; they qualitatively correspond to mesh points for a traditional discretization of the unavailable equation. In higher space dimensions, these intervals would become *boxes* around the coarse mesh points, a term that we will also use throughout this text. Given an initial condition for the macroscopic unknowns at some time t^*, a coarse time-δt map is constructed as follows.

1. **Lifting** Using the lifting operator (8.6), create appropriate initial conditions $\bar{u}_i(\mathbf{x}, t^*)$ for the microscopic time-stepper (8.1) *in each small box* around the mesh point x_i, consistent with the *spatial profile* of the macroscopic solution.

2. **Simulation** Use the time-stepper (8.1) to compute the microscopic state $\bar{u}_i(\mathbf{x}, t)$ *in each box* for $t \in [t^*, t^* + \delta t]$ *using appropriate boundary conditions*.

3. **Restriction** Obtain the macroscopic state *in each box* from the microscopic state $\bar{u}_i(\mathbf{x}, t + \delta t)$ using the restriction operator (8.7).

The result is a coarse time-stepper as above; however, the simulations are only performed in a small portion of the spatial domain. Note that the lifting step becomes more involved because, in addition to the average value of the macroscopic unknowns in the small interval, we also need to initialize some approximation of the spatial derivatives of the macroscopic solution to capture correctly the evolution equation (8.9). An additional (and crucial) difficulty is the imposition of boundary conditions on each small box, since each box is supposed to mimic local evolution of the microscopic problem *as if it were embedded in a larger domain*.

One combines the gap-tooth scheme with projective integration to form *patch dynamics* [1]. The gap-tooth scheme has been used with lattice Boltzmann simulations of the Fitzhugh–Nagumo dynamics [1, 27] and with particle-based simulations of the viscous Burgers equation [28].

8.3 Model homogenization problems

This section briefly reviews basic elements of standard homogenization theory [29]. This theory starts from a partial differential equation in which one of the coefficients depends on a small-scale parameter ϵ, and analytically derives a (macroscopic) equation, in which the dependence on ϵ has been eliminated. In our scenario, this macroscopic equation is assumed to be unknown; it is used for analysis purposes only.

8.3.1 Parabolic homogenization problem

As a *microscopic problem*, we consider the parabolic partial differential equation,

$$\partial_t u_\epsilon(x,t) = \partial_x \left(a\left(x/\epsilon\right) \partial_x u_\epsilon(x,t)\right) + r(u_\epsilon(x,t)),$$
$$u_\epsilon(x,0) = u^0(x) \in L^2([0,1]), \quad (8.10)$$
$$u_\epsilon(0,t) = u_\epsilon(1,t) = 0,$$

where $a(y) = a(x/\epsilon)$ is uniformly elliptic and periodic in y, and ϵ is a small parameter. We choose homogeneous Dirichlet boundary conditions for simplicity.

On the macroscopic scale we are interested in the *effective, homogenized* partial differential equation, in which the small-scale parameter ϵ has been eliminated. According to classical homogenization theory [29], the solution of (8.10) can be written as an asymptotic expansion in ϵ,

$$u_\epsilon(x,t) = U(x,t) + \sum_{i=1}^{\infty} \epsilon^i \left(u_i(x, x/\epsilon, t)\right), \quad (8.11)$$

where the functions $u_i(x,y,t) \equiv u_i(x, x/\epsilon, t)$, $i = 1, 2, \ldots$ are periodic in y. Here, $U(x,t)$ is the solution of the *homogenized equation*

$$\partial_t U(x,t) = \partial_x \left(a^* \partial_x U(x,t)\right) + r(U(x,t)),$$
$$U(x,0) = u^0(x) \in L^2([0,1]), \quad (8.12)$$
$$U(0,t) = U(1,t) = 0.$$

Here, the constant effective coefficient

$$a^* = \int_0^1 a(y) \left(1 - \frac{d}{dy}\chi(y)\right) dy, \quad (8.13)$$

and $\chi(y)$ is the periodic solution of the so-called *cell problem*

$$\frac{d}{dy}\left(a(y) \frac{d}{dy}\chi(y)\right) = \frac{d}{dy}a(y). \quad (8.14)$$

The solution of (8.14) is only defined up to an additive constant, so we impose the extra condition

$$\int_0^1 \chi(y) dy = 0.$$

In one space dimension, an explicit formula is known for a^* [29]:

$$a^* = \left[\int_0^1 \frac{1}{a(y)} dy\right]^{-1}. \quad (8.15)$$

These asymptotic expansions have been rigorously justified by Bensoussan et al. [29], see also [30]. Under the smoothness assumptions made on $a(x/\epsilon)$, one obtains *strong* convergence of $u_\epsilon(x,t)$ to $U(x,t)$ as $\epsilon \to 0$ in $L^2([0,1]) \times C([0,T))$. Indeed,

$$\|u_\epsilon(x,t) - U(x,t)\|_{L_2([0,1])} \leq C_0 \epsilon, \tag{8.16}$$

uniformly in t.

8.3.2 Hyperbolic homogenization problem

We consider the following hyperbolic partial differential equation in one space dimension,

$$\partial_t u_\epsilon(x,t) + \partial_x \left[c\left(x/\epsilon\right) u_\epsilon(x,t) \right] = 0,$$
$$u_\epsilon(x,0) = u^0(x) \in L^2([0,1]), \quad \partial_x u_\epsilon(0,t) = 0, \tag{8.17}$$

where $c(y) = c(x/\epsilon) > 0$ is periodic in y and ϵ is a small parameter. We choose a homogeneous Neumann boundary condition for simplicity.

As in the previous section, we are interested in an effective, homogenized partial differential equation on a macroscopic scale, where the dependence on the small scale parameter has been eliminated. According to classical homogenization theory [29, 30], the solution of (8.17) converges *weakly* in the limit of $\epsilon \to 0$ to the solution of

$$\partial_t U(x,t) + \partial_x \left[c^* U(x,t) \right] = 0,$$
$$U(x,0) = u^0(x) \in L^2([0,1]), \quad \partial_x U(0,t) = 0, \tag{8.18}$$

which describes the evolution of the averaged, effective behaviour. As in the parabolic case, the effective coefficient is the harmonic average,

$$c^* = \left[\int_0^1 \frac{1}{c(y)} dy \right]^{-1}. \tag{8.19}$$

8.4 The gap-tooth scheme

The gap-tooth scheme was originally proposed by Kevrekidis et al. [1]; the key idea was that, when the macroscopic behavior is sufficiently smooth in space, one reduces the computational complexity by only performing microscopic simulations in a number of small subdomains, which are coupled via interpolation. This scheme has been analyzed by Kevrekidis et al. [1] for parabolic problems using a symmetric interpolating polynomial in the special case where $a(x/\epsilon) = a^* = 1$ and without reaction. In this case, the microscopic model and the macroscopic model are the same pure diffusion equation, and there is no dependence on a small-scale parameter ϵ. Then the gap-tooth scheme is equivalent to a standard second order finite difference/forward Euler scheme for the diffusion equation when constant gradient boundary conditions are used for the boxes.

We have since then extended the analysis for the gap-tooth scheme in several ways [31, 32]. We derived a formulation which approximates finite difference schemes with higher order accuracy in space, and we analyzed the convergence of this generalized scheme for a one-dimensional parabolic homogenization problem with non-linear reaction. This analysis shows that the gap-tooth scheme approximates a finite difference scheme for the correct effective equation in the presence of microscopic scales.

Because of the constant gradient boundary conditions, this original formulation is restricted to the situation where the macroscopic equation is of second order (a diffusion equation); due to the symmetric interpolation, it also automatically leads to central finite difference schemes for the unknown macroscopic equation. Although this is sufficient for the illustrative parabolic model problem, some generalizations are necessary to apply the gap-tooth scheme to a broader class of problems. Clearly, if required by the properties of the macroscopic equations, one can also use non-symmetric interpolation. In Section 8.5, we will present a more general formulation that can be used to mimic *any* finite difference scheme, i.e. in which the finite difference approximation of each derivative can be chosen independently. This leads to schemes that can also be applied for advection problems and higher order equations.

8.4.1 Formulation

Suppose we want to obtain the solution of the *unknown* equation (8.12) on the interval $[0, 1]$, using an equidistant macroscopic mesh $\Pi(\Delta x) := \{0 = x_0 < x_1 = x_0 + \Delta x < \cdots < x_N = 1\}$. We assume that the equation is of second order, that is $d = 2$. (A strategy to obtain this information is given in [33].) Given equation (8.12), we could define a space-time discretization, for example, a forward Euler/spatial finite difference scheme, as

$$U^{n+1} = S(U^n, \delta t) = U^n + \delta t \cdot \left(a^* D^2(U^n) + r(U^n)\right), \tag{8.20}$$

where the numerical solution is denoted by $U^n = (U_0^n(t_n), \ldots, U_N^n(t_n))^T$, with $U_i^n \approx U(x_i, t_n)$, and $D^2(U^n)$ is a suitable finite difference approximation of the second order spatial derivative, which will be specified below.

Since equation (8.12) is not known explicitly, we use equation (8.20) for analysis purposes only. We refer to it as the *comparison scheme*. Instead, we propose the following (gap-tooth) scheme. We consider a small interval (box, *tooth*) of size $h \ll \Delta x$ around each mesh point, and define the discrete solution as being the average of the microscopic solution in these small boxes,

$$\bar{U}_i(t) = \mathcal{S}_h(u^\epsilon)(x_i, t) = (1/h) \int_{x_i - h/2}^{x_i + h/2} u^\epsilon(\xi, t) d\xi, \quad i = 1, \ldots, N-1. \tag{8.21}$$

For the definition of $\bar{U}_0$ and $\bar{U}_N$, the integration bounds can be adjusted to ensure that $u^\epsilon(x, t)$ is defined everywhere. The solution $\bar{U}(t) \in \mathbb{R}^{N+1}$ is defined as $\bar{U}(t) = (\bar{U}_0(t), \ldots, \bar{U}_N(t))^T$, and we denote an approximation of $\bar{U}(t)$ at $t = t_n$ as $\bar{U}^n$.

We construct a time δt-map for $\bar{U}(t)$ by solving the microscopic problem (8.10) in each box, subject to the following boundary constraints and initial condition.

8.4.1.1 Boundary constraints

Each box should provide information on the evolution of the global problem at that location in space. It is therefore crucial that the (artificially imposed) boundary conditions are chosen to emulate the correct behaviour in a larger domain. Roberts and Kevrekidis [32] argued that high order accuracy could be obtained for the gap-tooth scheme for whatever boundary condition the microscopic simulator required, see section 8.6. Here, we impose a fixed macroscopic gradient at the boundaries of each small box during a time interval of length δt. We determine the value of this gradient via a polynomial approximation of the macroscopic concentration profile $U(x,t)$, based on the (given) box averages $\bar{U}_i^n$, $i = 0, \ldots, N$. Thus, we approximate

$$U(x,t_n) \approx p_i^k(x,t_n), \quad x \in [x_i - h/2, x_i + h/2],$$

where $p_i^k(x,t_n)$ denotes a polynomial of (even) degree k. We require that the approximating polynomial has the same box averages as the initial condition in box i and in $k/2$ boxes to the left and to the right:

$$\frac{1}{h} \int_{x_{i+j}-h/2}^{x_{i+j}+h/2} p_i^k(\xi,t_n)d\xi = \bar{U}_{i+j}^n, \quad j = -k/2, \ldots, k/2. \quad (8.22)$$

One can easily check that

$$S_h(p_i^k)(x,t_n) = \sum_{j=-k/2}^{k/2} \bar{U}_{i+j}^n L_{i,j}^k(x), \quad L_{i,j}^k(x) = \prod_{\substack{l=-k/2 \\ l \neq j}}^{k/2} \frac{(x - x_{i+l})}{(x_{i+j} - x_{i+l})}, \quad (8.23)$$

where $L_{i,j}^k(x)$ denotes a Lagrange polynomial of degree k. The derivative of this approximating polynomial determines the value of the gradient at the boundary of the box:

$$\partial_x p_i^k \big|_{x_i \pm h/2} = s_i^\pm. \quad (8.24)$$

If we did have an equation for the macroscopic behaviour, we would use these slopes as Neumann boundary conditions. Here, we use these derivatives to constrain the *average* gradient of the microscopic solution $\bar{u}_i^\epsilon(x,t)$ in box i over one small-scale period around the end points:

$$\begin{aligned} \frac{1}{\epsilon} \int_{x_i-h/2-\epsilon/2}^{x_i-h/2+\epsilon/2} \partial_\xi \bar{u}_i^\epsilon(\xi,t)d\xi &= s_i^-, \\ \frac{1}{\epsilon} \int_{x_i+h/2-\epsilon/2}^{x_i+h/2+\epsilon/2} \partial_\xi \bar{u}_i^\epsilon(\xi,t)d\xi &= s_i^+. \end{aligned} \quad (8.25)$$

Note that we approximate a box average in a box of size $h \gg \epsilon$, while we average for boundary condition purposes over a length scale ϵ that is characteristic for the microscopic model. Hence, we replace each boundary condition and its effect on the simulation by an algebraic constraint.

8.4.1.2 Initial conditions

For time integration, we must impose an initial condition $\bar{u}_i^\epsilon(x, t_n)$ in each box $[x_i - h/2, x_i + h/2]$, at time t_n. We require $\bar{u}_i^\epsilon(x, t_n)$ to satisfy the boundary conditions and the given box average. We choose a quadratic polynomial $\bar{u}_i^\epsilon(x, t_n)$, centered around the coarse mesh point x_i,

$$\bar{u}_i^\epsilon(x, t_n) \equiv A(x - x_i)^2 + B(x - x_i) + C. \tag{8.26}$$

Using the constraints (8.25) in the limit for $\epsilon \to 0$ and requiring

$$\frac{1}{h} \int_{x_i - h/2}^{x_i + h/2} \bar{u}_i^\epsilon(\xi, t_n) d\xi = \bar{U}_i^n,$$

we obtain

$$A = \frac{s_i^+ - s_i^-}{2h}, \quad B = \frac{s_i^+ + s_i^-}{2}, \quad C = \bar{U}_i^n - \frac{h}{24}(s_i^+ - s_i^-). \tag{8.27}$$

8.4.1.3 The algorithm

A complete *gap-tooth* algorithm to proceed from $\bar{U}^n$ to $\bar{U}^{n+1}$ is given below.

1. **Lifting** At time t_n, construct the initial condition $\bar{u}_i^\epsilon(x, t_n)$, $i = 0, \ldots, N$, using the box averages $\bar{U}_i^n$ ($i = 0, \ldots, N$) as defined in (8.27).
2. **Evolution** Compute $\bar{u}_i^\epsilon(x, t)$ by solving the equation (8.10) until time $t_{n+1} = t_n + \delta t$ with the boundary constraints (8.25).
3. **Restriction** Compute the box averages $\bar{U}_i^n$, $i = 1, \ldots, N$, at time t_{n+1},

$$\bar{U}_i^{n+1} = \frac{1}{h} \int_{x_i - h/2}^{x_i + h/2} \bar{u}_i^\epsilon(\xi, t_{n+1}) d\xi.$$

This algorithm amounts to a 'coarse-to-coarse' time-δt map:

$$\bar{U}^{n+1} = \bar{S}(\bar{U}^n, \delta t; h), \tag{8.28}$$

where we explicitly identify the dependence on the box width h.

8.4.2 Consistency and stability

Samaey et al. [31] proved the following consistency estimate:

Theorem

Consider the finite difference scheme (8.20) and the gap-tooth timestepper (8.28). When $U_i^n = \bar{U}_i^n$, the local error is bounded by

$$\|U_i^{n+1} - \bar{U}_i^{n+1}\| \leq C(\bar{U}^n) \left(\frac{\epsilon}{\sqrt{h}} + \delta t^2 + \delta t h^2 \right),$$

in which the constant $C(\bar{U}^n)$ depends on $\bar{U}^n$.

The accuracy is affected by an error in the restriction (which is of order $\epsilon/\sqrt{h}$), and a difference in the approximation of the nonlinear reaction term. While in the finite difference scheme a pointwise value is taken and kept constant over δt, the gap-tooth scheme takes the average reaction term over a small box of size h and time interval of size δt.

To obtain convergence, one also needs stability. However, in the nonlinear setting considered here, one has to be careful about the notion of stability. In particular, since the error constants depend on $\bar{U}^n$ and its divided differences, we can only establish stability for numerical solutions for which these constants remain bounded, which is exactly the property that we would like to prove. In recent work, E and Engquist [34] state that the heterogeneous multiscale method is stable if the corresponding macroscopic scheme is stable. In order to obtain this result, they need to assume that the numerical solution remains in a class K, which is defined as the class of discrete functions on the numerical grid (x_i, t_n), $i = 0, \ldots N$, $t_n = n\delta t$, $n = 0, \ldots, T/\delta t$, with bounded finite differences [35],

$$K = \{\{U^n\} : \|D^k(U^n)\| \leq C_k \text{ for } k \leq d, \ n\delta t \leq T\},$$

where d is the highest spatial derivative that occurs in the finite difference scheme, $D^k(U^n)$ is the finite difference operator of order k, and C_k is independent of δt and Δx.

We have the following convergence theorem [34, Theorem 5.5].

Theorem

Assume $\{\bar{U}^n\}, \{U^n\} \in K$. *Then, the gap-tooth scheme (8.28) is convergent if the finite difference time-stepper (8.20) is stable. Moreover, if* $U^0 = \bar{U}^0 = \mathcal{S}_h(u^0)(x)$, *then the error with respect to the homogenized solution* $U(x,t)$ *of (8.12) is bounded by*

$$\|\bar{U}_i^n - U(x_i, t_n)\| \leq C(\bar{U}^n)\left(h^2 + \frac{\epsilon}{\sqrt{h\delta t}} + \delta t + \Delta x^k\right). \tag{8.29}$$

8.4.3 Discussion

This section analyzed a simple formulation of the gap-tooth scheme for a class of parabolic homogenization problems. We showed that the method approximates a finite difference scheme of arbitrary (even) order for the homogenized equation when we appropriately constrain the microscopic problem in the boxes. This formulation has a number of drawbacks, which will be addressed in the next section which discusses the patch dynamics scheme with buffers.

Because the box initial conditions are constructed using a symmetric interpolating polynomial, we automatically approximate a *central* finite difference scheme. Our analysis revealed that the presence of microscopic scales introduces an error term that grows with decreasing δt, which means that the optimal accuracy will always be limited by the size of the small scale parameter ϵ.

As already mentioned, we also intend to use the gap-tooth scheme when the microscopic model is not a PDE, but some other microscopic model; for example, kinetic Monte Carlo. In this case, several complications arise. First, imposing macroscopically inspired boundary conditions (of the type discussed above) is highly nontrivial. Indeed, for instance a particle microscopic model typically only has knowledge about particle positions and velocities, and some artificial dynamics will need to be introduced to keep a macroscopic gradient fixed, see e.g. [36] for a control-based strategy. Moreover, it is possible that one already has a microscopic simulation code available, containing a number of standard, *built-in* boundary conditions, that has already been thoroughly tested. These codes do not usually allow us to perform microscopic simulations subject to externally specified macroscopic boundary conditions, unless serious modifications to the code are made – something we would like to avoid if possible.

8.5 Patch dynamics with buffers

This section discusses the patch dynamics scheme with buffers. In patch dynamics, the gap-tooth scheme as discussed previously is used in conjunction with projective integration; via a gap-tooth step of size δt, we estimate the macroscopic time derivative, which is subsequently used to take a macroscopic step of size $\Delta t \ll \delta t$.

The gap-tooth scheme itself is modified in two ways: (1) the initial conditions are based on a Taylor-like polynomial, in which the spatial derivatives are replaced by suitably chosen finite difference approximations on the macroscopic grid; and (2) we introduce buffer regions around the boxes, and impose *arbitrary* boundary conditions on the buffers. (This comes at a, possibly large, computational cost.) We analyze the effect of the size of the buffer regions on the accuracy of the method, and show that these modifications allow us to also approximate higher order or advection dominated macroscopic equations.

8.5.1 Formulation

We devise a scheme for the evolution of the effective behaviour $U(x,t)$ of the general homogenization problem

$$\partial_t u_\epsilon = f(u_\epsilon, \partial_x u_\epsilon, \ldots, \partial_x^d u_\epsilon, t; \epsilon), \tag{8.30}$$

where ∂_t denotes again the time derivative, and ∂_x^k denotes the kth spatial derivative ($k = 1, \ldots, d$, where $k = 1$ is usually omitted). We assume that a time integration code for this equation has already been written and is available with a number of *standard* boundary conditions, such as no-flux or Dirichlet. Further, we assume that the (unavailable) macroscopic equation is of the form

$$\partial_t U = F(U, \partial_x U, \ldots, \partial_x^d U, t). \tag{8.31}$$

Suppose we want to obtain the solution of (8.31) on the interval $[0,1]$, using an equidistant, macroscopic mesh $\Pi(\Delta x) := \{0 = x_0 < x_1 = x_0 + \Delta x < \cdots < x_N = 1\}$. Given equation (8.31), we define a method-of-lines space discretization,

$$\partial_t U_i(t) = F(U_i(t), D^1(U_i(t)), \ldots, D^d(U_i(t)), t), \quad i = 0, \ldots, N. \tag{8.32}$$

where $U_i(t) \approx U(x_i, t)$ and $D^k(U_i(t))$ denotes a suitable finite difference approximation for the kth spatial derivative. We subsequently discretize equation (8.32) in time using a time integration method of choice, for example, forward Euler. We denote the resulting time-stepper as

$$U^{n+1} = S(U^n, t_n; \Delta t) = U^n + \Delta t\, F(U^n, t_n), \quad (8.33)$$

where $U^n = (U_0(t_n), \ldots, U_N(t_n))^T$ and Δt denotes the macroscopic time-step. We suppress the dependence of $F(U^n, t_n)$ on the spatial derivatives for notational convenience. Note that, although we have used the forward Euler scheme here for concreteness, in principle any time discretization method can be used to solve equation (8.32).

Since equation (8.31) is assumed not to be known explicitly, we use (8.33) for analysis purposes only. We construct a (patch dynamics) scheme to approximate (8.33). To this end, consider a small interval (box, *tooth*) of size $h \ll \Delta x$ around each mesh point, and define the discrete solution $\bar{U}(t) = (\bar{U}_0(t), \ldots, \bar{U}_N(t))^T \in \mathbb{R}^{N+1}$ as being the average of the microscopic solution in the small boxes,

$$\bar{U}_i(t) = \mathcal{S}_h(u_\epsilon)(x_i, t) = \frac{1}{h} \int_{x_i - h/2}^{x_i + h/2} u_\epsilon(\xi, t)\, d\xi, \quad i = 0, \ldots, N. \quad (8.34)$$

We denote an approximation of $\bar{U}(t)$ at $t = t_n$ as $\bar{U}^n$.

A patch dynamics scheme is now constructed as follows. Introduce a larger *buffer* box of size $H > h$ around each mesh point. In each box of size H, perform a time integration over a time interval of size δt using the microscopic model (8.30), and restrict to macroscopic variables. Use the results to estimate the macroscopic time derivative. We provide each microscopic simulation with the following initial and boundary conditions.

8.5.1.1 Initial condition
We define the initial condition by constructing a local Taylor expansion, based on the (given) box averages $\bar{U}_i^n$, $i = 0, \ldots, N$, at mesh point x_i and time t_n:

$$\bar{u}_\epsilon^i(x, t_n) = \sum_{k=0}^{d} D_i^k(\bar{U}^n) \frac{(x - x_i)^k}{k!}, \quad x \in [x_i - \frac{H}{2}, x_i + \frac{H}{2}], \quad (8.35)$$

where d is the order of the macroscopic equation (8.31). The coefficients $D_i^k(\bar{U}^n)$, $k > 0$, are the same finite difference approximations for the kth spatial derivative that would be used in the comparison scheme (8.32), whereas $D_i^0(\bar{U}^n)$ is chosen such that

$$\frac{1}{h} \int_{x_i - h/2}^{x_i + h/2} \bar{u}_\epsilon^i(\xi, t_n)\, d\xi = \bar{U}_i^n. \quad (8.36)$$

8.5.1.2 *Boundary conditions*

The time integration of the microscopic model in each box should provide information on the evolution of the *global* problem at that location in space. Crucially, the boundary conditions are chosen such that the solution inside each box evolves *as if it were embedded in the larger domain*. To this end, we introduce a larger box of size $H > h$ around each macroscopic mesh point. The simulation is subsequently performed using any of the *built-in* boundary conditions of the microscopic code. Lifting and (short-term) evolution (using *arbitrary* available boundary conditions) are performed in the larger box; yet the restriction is done by processing the solution (here taking its average) over the inner, small box only. The goal of the additional computational domains, the *buffers*, is to buffer the solution inside the small box from the artificial disturbance caused by the (repeatedly updated) boundary conditions. This is accomplished over *short enough* time intervals, provided the buffers are *large enough*; analyzing the method is tantamount to making these statements quantitative.

The idea of a buffer region was also introduced in the multiscale finite element method of Hou (oversampling) [37] to eliminate boundary layer effects; also Hadjiconstantinou makes use of overlap regions to couple a particle method with a continuum code [38].

8.5.1.3 *The algorithm*

The complete algorithm to obtain an estimate of the macroscopic time derivative at time t_n is given below.

1. **Lifting** At time t_n, construct the initial condition $\bar{u}_\epsilon^i(x, t_n)$, $i = 0, \ldots, N$, using the box averages $\bar{U}_i^n$ defined in (8.35).
2. **Simulation** Compute the box solution $\bar{u}_\epsilon^i(x, t)$, $t > t_n$, by solving equation (8.30) in the interval $[x_i - H/2, x_i + H/2]$ with *some* boundary conditions up to time $t_{n+\delta} = t_n + \delta t$. The boundary conditions can be anything that the microscopic code allows.
3. **Restriction** Compute the average $\bar{U}_i^{n+\delta} = (1/h) \int_{x_i-h/2}^{x_i+h/2} \bar{u}_\epsilon^i(\xi, t_{n+\delta}) d\xi$ over the *inner, small box only*.
4. **Estimation** We estimate the time derivative at time t_n as

$$\bar{F}^d(\bar{U}^n, t_n; h, \delta t, H) = \frac{\bar{U}^{n+\delta} - \bar{U}^n}{\delta t}, \qquad (8.37)$$

where the superscript d denotes the highest spatial derivative initialized in the lifting step. We also make explicit the dependence of the estimate on H and δt.

Since the first three steps constitute a gap-tooth time-step, we call the estimator (8.37) a *gap-tooth time derivative estimator*. It can be used in any ODE time integration code. For example, a forward Euler patch dynamics scheme would be

$$\bar{U}^{n+1} = \bar{U}^n + \Delta t \, \bar{F}^d(\bar{U}^n, t_n; h, \delta t, H). \qquad (8.38)$$

We emphasize (again) that an initialization according to equation (8.35) has the important advantage that one chooses a suitable finite difference approximation for each derivative independently. This property is crucial, and empowers us to approximate advection-dominated dynamics effectively.

8.5.2 Numerical illustration

We illustrate the theory with a diffusion homogenization problem. Consider the model problem (8.10) with

$$a(x/\epsilon) = 1.1 + \sin(2\pi x/\epsilon), \quad \epsilon = 10^{-5}, \qquad (8.39)$$

as a microscopic problem on the domain $[0,1]$ with homogeneous Dirichlet boundary conditions and initial condition $u(x,0) = 1 - 4(x-1/2)^2$. The corresponding macroscopic equation is given by equation (8.12), with $a^* = 0.45825686$. This problem has also been used as a model example in [31, 39]. To solve this microscopic problem, we use a second order, finite difference discretization with mesh width $\delta x = 10^{-7}$ and lsode [40] as time-stepper. The concrete gap-tooth scheme for this example is again defined by taking second order central finite differences.

We compare a gap-tooth step with $h = 2 \cdot 10^{-3}$ and $\Delta x = 10^{-1}$ with the reference estimator $a^* D^2(\hat{U}^n)$, in which the effective diffusion coefficient is known to be $a^* = 0.45825686$. Figure 8.1 shows the error with respect to the finite difference time derivative as a function of (a) H and (b) δt. The convergence is in agreement with Theorem 8.5.3. We see that smaller values of δt result in larger values for the optimal error, but the convergence towards this optimal error is faster.

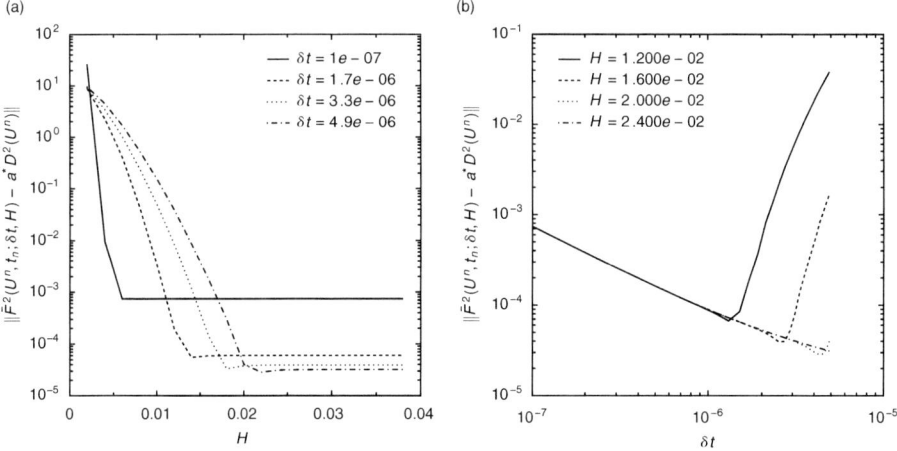

Figure 8.1: Error of the gap-tooth estimator $\bar{F}^2(U^n, t_n; \delta t, H)$ with respect to the finite difference time derivative $a^* D^2(U^n)$ on the same mesh. (a) Error with respect to H for fixed δt; (b) Error with respect to δt with fixed H.

8.5.3 Consistency and stability

For the effective equation (8.12), one writes a finite-difference/forward Euler time-stepper as

$$\begin{aligned} U^{n+\delta} &= S(U^n, t_n; \delta t) \\ &= U^n + \delta t\, F(U^n, D^1(U^n), D^2(U^n), t_n) \\ &= U^n + \delta t\, [a^* D^2(U^n)]. \end{aligned} \qquad (8.40)$$

We compare the gap-tooth time-derivative estimator with the effective time derivative. For concreteness, we impose Dirichlet boundary conditions at the boundaries of the boxes, which will clearly introduce artifacts on the estimated time derivative. The subsequent theorem shows that these artifacts can be made arbitrarily small by increasing the buffer size H [41].

Theorem

Let $\bar{F}^2(\bar{U}^n, t_n; h, \delta t, H)$ be a gap-tooth time-stepper for the homogenization problem (8.10). Then, assuming $U^n = \bar{U}^n$,

$$\left\| \bar{F}^2(\bar{U}^n, t_n; \delta t, H) - a^* D^2(U^n) \right\| \le$$

$$C_4 \underbrace{\frac{\epsilon}{\sqrt{h\delta t}}}_{\text{micro-scales}} + C_5 \underbrace{\left(1 + \frac{h^2}{\delta t}\right)}_{\text{averaging}} \underbrace{\left[1 - \exp\left(-a^* \pi^2 \frac{\delta t}{H^2}\right)\right]}_{\text{boundary conditions}}. \qquad (8.41)$$

The bound (8.41) shows the main consistency properties of the gap-tooth estimator. The error decays exponentially as a function of buffer size, but the optimal accuracy of the estimator is limited by the presence of the microscopic scales. Therefore, we need to make a tradeoff to determine an optimal choice for H and δt. The smaller δt, the smaller H can be used to reach optimal accuracy (and thus the smaller the computational cost), but smaller δt implies a larger optimal error.

Samaey et al. [41] showed numerically that the convergence result does not depend crucially on the type of boundary conditions. For example, for no-flux boundary conditions, we obtain qualitatively the same result. However, if we know how the macroscopic solution behaves at the boundaries of the boxes, we can use this knowledge to eliminate the buffers, as we did in Section 8.4 for diffusion problems.

As in Section 8.4, E and Engquist provide a stability result ([34] Theorem 5.5). However, it is also illustrative to investigate the stability numerically by computing the eigenvalues of the time-derivative estimator as a function of H. We define the concrete patch dynamics scheme to be a forward Euler scheme,

$$\bar{U}^{n+1} = U^n + \Delta t\, \bar{F}^2(\bar{U}^n, t_n; \delta t, H), \qquad (8.42)$$

with the box initialization defined by (8.35) with second order, central finite differences. In this case, the comparison finite difference scheme for the macroscopic equation is

$$U^{n+1} = U^n + \Delta t\, F(U^n, t_n) = U^n + a^*\Delta t\, \frac{U_{i+1}^n - 2U_i^n + U_{i-1}^n}{\Delta x^2}. \quad (8.43)$$

The time derivative operator $F(U^n, t_n)$ in the comparison scheme (8.43) has eigenvalues

$$\lambda_k = -\frac{4a^*}{\Delta x^2} \sin^2(\pi k \Delta x), \quad (8.44)$$

which, using the forward Euler scheme as time-stepper, results in the stability condition

$$\max_k |1 + \lambda_k \Delta t| \leq 1 \quad \text{or} \quad \frac{\Delta t}{\Delta x^2} \leq \frac{1}{2} a^*.$$

As the operator $\bar{F}^2(U^n, t_n; \delta t, H)$ is linear, we can interpret its evaluation as a matrix-vector product. We can therefore use any matrix-free linear algebra technique to compute the eigenvalues of $\bar{F}^2(U^n, t_n; \delta t, H)$, for example, Arnoldi iteration [42]. We choose to compute $\bar{F}^2(U^n, t_n; \delta t, H)$ and $F(U^n, t_n)$ on the domain $[0, 1]$ with Dirichlet boundary conditions, on a mesh of width $\Delta x = 0.05$ and with an inner box width of $h = 2\cdot 10^{-3}$. We choose $\delta t = 5\cdot 10^{-6}$ and compute the eigenvalues of $\bar{F}^2(U^n, t_n; \delta t, H)$ as a function of H as shown in Figure 8.2. When the buffer size is too small, the eigenvalues of the gap-tooth estimator are closer to 0 than the corresponding eigenvalues of the finite difference scheme. This is because the microscopic simulation approaches a steady state quickly (due to the Dirichlet boundary conditions), instead of following the true system evolution in a larger domain. With increasing buffer size H, the eigenvalues of

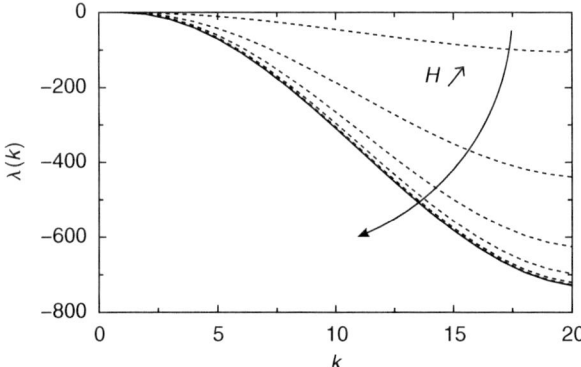

Figure 8.2: Spectrum of the estimator $\bar{F}^2(U^n, t_n; \delta t, H)$ (dashed) for the model equation (8.10) for $H = 2\cdot 10^{-3}, 4\cdot 10^{-3}, \ldots, 20\cdot 10^{-3}$ and $\delta t = 5\cdot 10^{-6}$, and the eigenvalues (8.44) of $F(U^n, t_n)$ (solid).

$\bar{F}^2(U^n, t_n; \delta t, H)$ approximate those of $F(U^n, t_n)$, which is an indication of consistency for larger H. Since all eigenvalues are negative and the most negative eigenvalue for $\bar{F}^2(U^n, t_n; \delta t, H)$ is always smaller in absolute value than the corresponding eigenvalue of $F(U^n, t_n)$, the patch dynamics scheme is always stable if the comparison scheme is stable.

8.5.4 Application to advection problems

We now illustrate the numerical properties of the method. Samaey et al. [41] showed in detail the numerical properties of the patch dynamics scheme for diffusion problems, and also showed how the scheme can be applied when the macroscopic behavior is of fourth order. Here, we show that the scheme can also be used when the macroscopic behavior is advective; in this case, one only needs to choose appropriate finite difference formulae in the initialization.

Consider equation (8.17) with

$$c(x/\epsilon) = \frac{1}{3 + \sin(2\pi x/\epsilon)}, \quad \epsilon = 10^{-5}. \quad (8.45)$$

The effective equation is then given by (8.18) with $c^* = 1/3$. The available microscopic simulation code is an upwind/forward Euler time-stepper on a grid with size $\delta x = 5 \cdot 10^{-10}$ and a time-step $dt = 5 \cdot 10^{-11}$. We take the size of the small boxes to be $h = 5 \cdot 10^{-4}$.

We first investigate how the accuracy of the scheme is influenced by the buffer size H and the gap-tooth time-step δt. Once a good set of method parameters is found, we perform a long-term simulation. We construct patch dynamics schemes to mimic the upwind, third order, upwind biased spatial discretization, and central, fourth order, spatial discretization.

8.5.4.1 Consistency

To determine the buffer size H and the gap-tooth time-step δt, we perform a numerical simulation for this model on the domain $[-H/2, +H/2]$, with $H = h + 5i \cdot 10^{-9}$ for $i = 1, \ldots, 20$ on the time interval $[0, \delta t]$ with $\delta t = j \cdot 10^{-9}$, $j = 1, \ldots, 100$, and the linear initial condition

$$u_\epsilon(x, 0) = D^1 x + D^0 = 3.633x + 0.9511.$$

The results are shown in Figure 8.3(a). Notice two differences with respect to the parabolic dynamics. First, we do not need very large buffer regions. Indeed, the advective nature of equation (8.17) ensures that information travels with finite speed. The consequence is that, as soon as the time-step is too short for the boundary information to reach the interior of the domain, the buffer size H will not have any influence on the accuracy of the result.

The second difference is that the error decreases monotonically with decreasing δt, whereas the theoretical result for diffusion indicates an error term of the form $O(\epsilon/\delta t)$. This discrepancy is due to additional numerical inaccuracies during the restriction step, which are caused by the weak convergence towards the

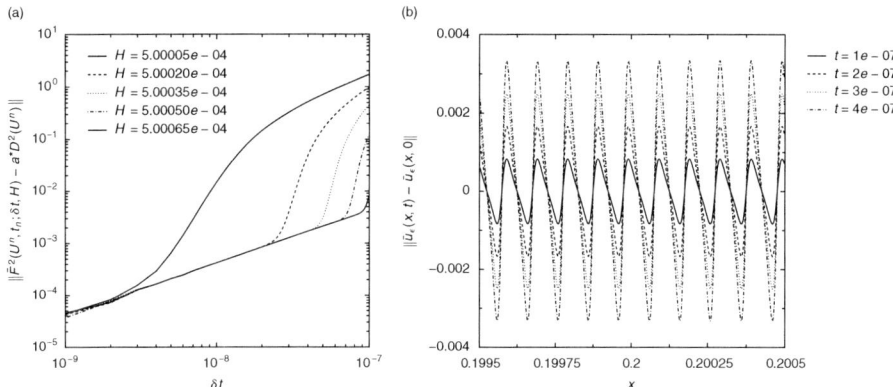

Figure 8.3: (a) Error of the gap-tooth estimator with respect to the macroscopic time derivative c^*D^1. (b) Difference between the solution inside the box at time t and the initial condition at time $t = 0$.

homogenized equation in the hyperbolic case. Figure 8.3 shows how $u_\epsilon(x,t)$ varies as a function of time: the microscopic solution develops oscillations which grow in amplitude with time. Recall that the macroscopic quantity at time $t = \delta t$ is computed as the spatial average of the solution $u_\epsilon(x, \delta t)$ over a box of size h. We need to approximate this spatial average using a quadrature formula, in which we can only use the solution on the numerical grid points as quadrature points. Thus, we may expect a decrease of accuracy in the computation of the box average for increasing values of δt. We numerically verified this intuitive reasoning by increasing ϵ. The box solution then becomes less oscillatory, and we observed that the accuracy of the restriction was increased.

Based on these results, we choose $H = h + 10^{-7}$ and $\delta t = 5 \cdot 10^{-9}$. Since our macroscopic schemes will use $\Delta x = O(10^{-2})$ and $\Delta t = O(10^{-2})$, the method results in gains of the order of 100 in space and 10^6 in time. However, for realistic microscopic problems, part of this spectacular gain will be lost because we need to initialize the microscopic system consistently (the lifting step) using only a few low-order moments, which may require additional microscopic simulations [11, 12, 13].

8.5.4.2 Third order, upwind biased scheme

As an illustration, we design a patch dynamics algorithm to mimic the third order, upwind biased scheme as a spatial discretization, which we combine with the classical fourth order Runge–Kutta time integration method. (Samaey et al. [52] give other examples.) In this case, the macroscopic time derivative is

$$F(U_i^n, t_n) = \frac{c^*}{\Delta x}\left(-\frac{1}{6}U_{i-2}^n + U_{i-1}^n - \frac{1}{2}U_i^n - \frac{1}{3}U_{i+1}^n\right). \tag{8.46}$$

The Runge–Kutta method requires some auxiliary evaluations of the time derivative operator:

$$k_1 = F(U_i^n, t_n),$$
$$k_2 = F(U_i^n + \frac{\Delta t}{2} k_1, t_n + \frac{\Delta t}{2}),$$
$$k_3 = F(U_i^n + \frac{\Delta t}{2} k_2, t_n + \frac{\Delta t}{2}), \quad (8.47)$$
$$k_4 = F(U_i^n + \Delta t\, k_3, t_n + \Delta t);$$

and the time-stepper $U^{n+1} = S(U^n, t_n; \Delta_t)$ is then defined as

$$U^{n+1} = U^n + \Delta t \left(\frac{1}{6} k_1 + \frac{1}{3} k_2 + \frac{1}{3} k_3 + \frac{1}{6} k_4 \right). \quad (8.48)$$

The corresponding patch dynamics scheme is defined by the algorithm in Section 8.5.1, where the initial condition (8.35) is defined by taking $d = 1$ and

$$D_i^1(\bar{U}^n) = \frac{1}{\Delta x} \left(\frac{1}{6} \bar{U}_{i-2}^n - \bar{U}_{i-1}^n + \frac{1}{2} \bar{U}_i^n + \frac{1}{3} \bar{U}_{i+1}^n \right),$$
$$D_i^0(\bar{U}^n) = \bar{U}_i^n. \quad (8.49)$$

The resulting time derivative estimator is subsequently used inside the fourth order Runge–Kutta method.

We perform a numerical simulation on a macroscopic mesh with size $\Delta x = 2 \cdot 10^{-2}$ and $\Delta t = 2 \cdot 10^{-2}$. The results are shown in Figure 8.4. The patch dynamics scheme clearly has the same properties as the comparison finite difference scheme. It is less diffusive than the upwind scheme, but some artificial oscillations are introduced. Figure 8.4(a) shows the L_2-error of patch dynamics with respect to the finite difference scheme, and with respect to an 'exact' solution of the effective equation, which was obtained using the upwind scheme on a very fine mesh with $\Delta x = 10^{-4}$ and $\Delta t = 10^{-4}$. Again, the error in approximating the exact solution is completely dominated by the error of the macroscopic scheme, whereas the errors due to estimation are negligible.

8.5.4.3 Non-conservation form

As a second example, we consider equation (8.17) with

$$c(x, x/\epsilon) = \frac{1}{3 + \sin(2\pi x/\epsilon) + \sin(2\pi x)}, \quad \epsilon = 10^{-5}. \quad (8.50)$$

The effective equation is then (8.18) with $c^*(x) = 1/[3+\sin(2\pi x)]$. The available microscopic simulation code is an upwind/forward Euler time-stepper on a grid with size $dx = 5 \cdot 10^{-10}$ and a time-step $dt = 5 \cdot 10^{-11}$. We take the size of the small boxes to be $h = 5 \cdot 10^{-4}$.

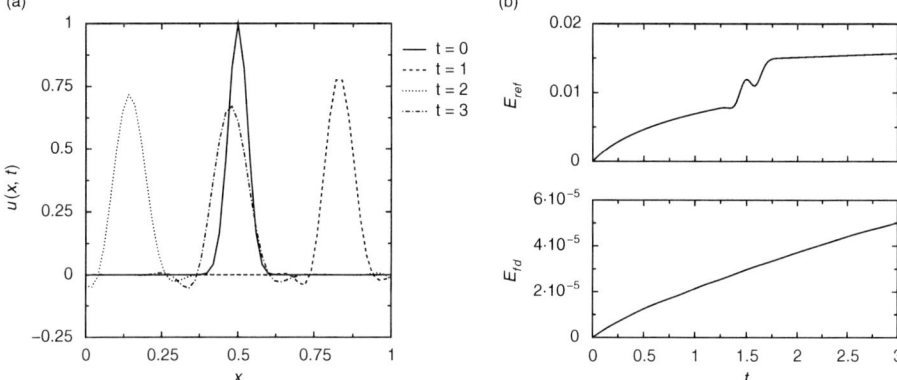

Figure 8.4: (a) Snapshots of the solution of the homogenization advection equation (8.17) with coefficient (8.45) using the upwind biased patch dynamics scheme at certain moments in time. (b) L_2-error with respect to the 'exact' solution of the effective equation (8.18) (top) and the finite difference comparison scheme (8.48)–(8.46) (bottom). The total error is dominated by the error of the finite difference scheme.

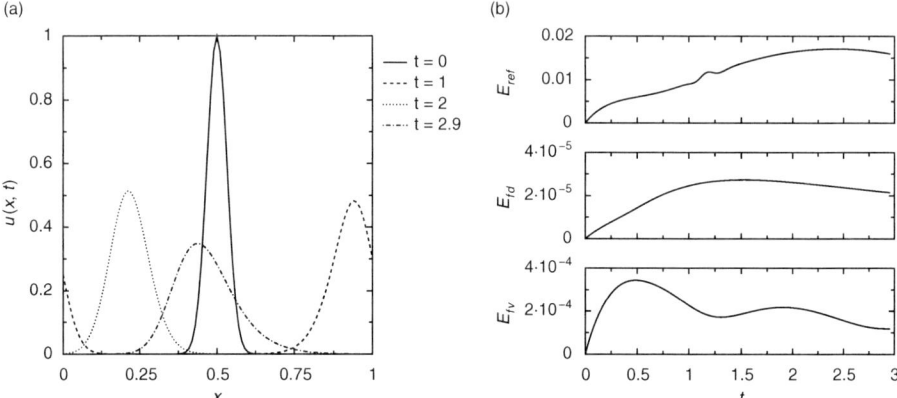

Figure 8.5: (a) Snapshots of the solution of the homogenization advection equation (8.17) with coefficient (8.50) using the first order upwind patch dynamics scheme at certain moments in time. (b) Error with respect to the 'exact' solution of the effective equation (8.18) (top), the finite difference comparison scheme (8.51) (middle), and the finite volume scheme (8.52) (bottom). The total error is dominated by the error of the finite difference scheme, and the error with respect to the finite volume scheme is significantly larger that the error with respect to (8.51).

We choose $H = h + 2 \cdot 10^{-7}$ and $\delta t = 5 \cdot 10^{-9}$ as method parameters, and perform a patch dynamics simulation using a macroscopic mesh size $\Delta x = 2 \cdot 10^{-2}$ and $\Delta t = 5 \cdot 10^{-3}$. We use an upwind initialization and combine this with forward Euler in time.

The simulations show that the patch dynamics scheme is a good approximation to a finite difference approximation of equation (8.18) *in non-conservative form*, see Figure 8.5. It can be shown that the correct comparison scheme would be

$$U_i^{n+1} = U_i^n - \Delta t \left(c^*(x_i) \frac{U_i^n - U_{i-1}^n}{\Delta x} + U_i^n \partial_x c^*(x_i) \right), \quad (8.51)$$

which is not entirely the same as the classical finite volume upwind scheme

$$U_i^{n+1} = U_i^n - \frac{\Delta t}{\Delta x} \left(c^*(x_{i+1/2}) U_i^n - c^*(x_{i-1/2}) U_{i-1}^n \right). \quad (8.52)$$

In particular, scheme (8.51) is not conservative.

8.6 General tooth boundary conditions

Using microscopic simulators of the one-dimensional Burgers' equation, $u_t + u u_x = u_{xx}$, Roberts and Kevrekidis [32, 43] demonstrated the possibility of achieving high order accuracy in the gap-tooth scheme for macroscale dynamics. The particular microsimulator used was a fine-scale discretization of the PDE which executed only in the interior of the teeth (see Figure 8.6). At each microscale time-step during execution, the microsimulator within each tooth requires boundary values which must be continuously updated. The analysis [32, 43] found straightforward rationale for coupling microsimulators with

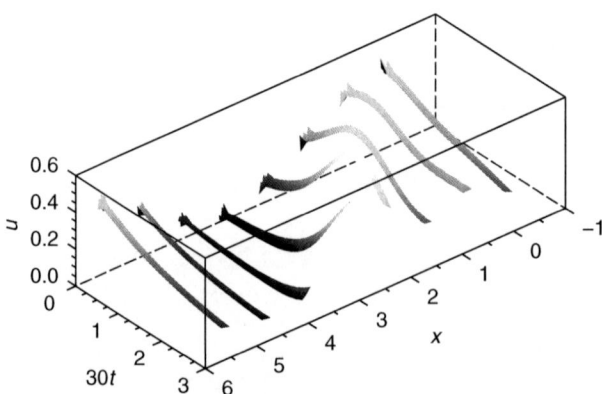

Figure 8.6: Simulation of Burgers' equation using Dirichlet boundary conditions on the teeth (specified u on the edges).

- Neumann boundary conditions of specified *flux* at the edges of their simulation teeth;
- Dirichlet boundary conditions of specified field u at the tooth edges;
- mixed boundary conditions of specified $av_j \pm b\partial_x v_j$ at the tooth edges; or
- nonlocal two-point boundary conditions such as those arising in a microscale discretization of a PDE.

This section reviews the rationale behind spatial coupling of teeth and patches; it does not address the coarse time-stepper of projective time integration. Crucially we find that when the fine scale microstructure is smooth, as in Figure 8.6, then we couple microsimulators by obtaining *whatever values are necessary for the boundaries of the microscopic simulators through classic interpolation of the macroscopic grid values from neighboring teeth*. When the microscale is smooth, the buffers of Section 8.5 are not necessary.

Further research is being done into analytic support for coupling of microsimulators that have detailed fine structure such as the homogenization problems addressed in Sections 8.3–8.5.

This coupling by classic interpolation promotes a strong connection between classic finite difference discretizations of PDEs, classic finite elements, and the methodology of the gap-tooth scheme. The connection is that a PDE encodes the dynamics of a system on an *infinitesimal* patch, whereas a microsimulator encodes the dynamics on a *finite* patch. Roberts and Kevrekidis [32] (§5) show the following desirable properties:

- the approach generates macroscopic discretizations which are consistent with the microscopic dynamics to any specified order in the macroscopic tooth separation Δx;
- the macroscopic model and the microscopic solution field are essentially independent of the size of the teeth h; and
- the macroscopic model and the microscopic solution field are essentially independent of the details of the specific tooth coupling conditions.

To date theoretical support has only been shown for one spatial dimension. Further research, analogous to that of MacKenzie [44], is needed for problems in more spatial dimensions.

8.6.1 Diffusion on Dirichlet teeth

For a definite example [32] (§2), consider gap-tooth simulations of the simple diffusion equation

$$\frac{\partial u}{\partial t} = \frac{\partial^2 u}{\partial x^2}, \quad \text{and } 2\pi\text{-periodic in } x. \tag{8.53}$$

Let $v_j(x,t)$ be the fine-scale, microscopic field in the jth tooth, such as that shown in Figure 8.7, and U_j the jth coarse grid value; that is, the value at the center of each tooth, $U_j := v_j(x_j, t)$ (in previous sections U_j was the average (8.21)). Couple each tooth across space by classic interpolation of the coarse grid values:

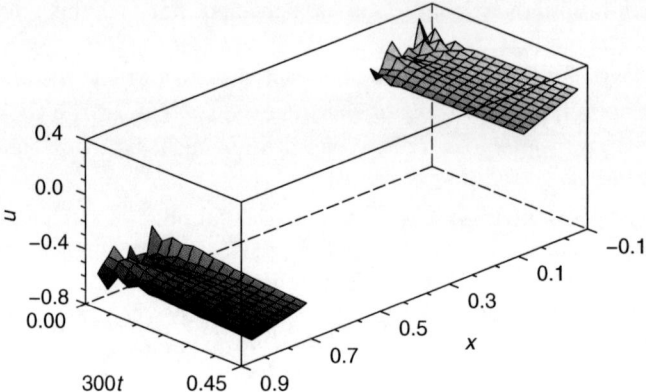

Figure 8.7: View of the initial microscopic evolution within a pair of neighboring teeth with dynamics described by the diffusion PDE (8.53) and also coupled to their neighbors.

on the edge of the jth tooth the Dirichlet-like coupling condition of the fine scale-diffusion is that the field[1]

$$v_j(x_j \pm \tfrac{1}{2}h, t) = \left[1 \pm r\mu\delta + \tfrac{1}{2}r^2\delta^2 \pm \tfrac{1}{6}r(r^2-1)\mu\delta^3 + \tfrac{1}{24}r^2(r^2-1)\delta^4\right]U_j, \tag{8.54}$$

where the fraction $r = h/(2\Delta x)$, and δ and μ are classic finite difference and mean operators, respectively, operating on the coarse grid.

Since the coupling condition (8.54) involves up to fourth order differences, it has the potential, and does achieve, fourth order consistency. Eigenanalysis (Roberts and Kevrekidis [32], Table 2.1) of the microscale diffusion (8.53) on teeth coupled by (8.54) shows that error in simulations is fourth order. Analogous classical interpolation formulae achieve higher order consistency.

8.6.2 Slow manifold support for tooth dynamics

Adapting earlier work [32] (§5), we here explore the dynamics of linear diffusion, (8.53), on small patches coupled by the Dirichlet-like condition (8.54) modified to

$$v_j(x_j \pm \tfrac{1}{2}h, t) = \left\{1 + \gamma r\left[\pm\mu\delta + \tfrac{1}{2}r\delta^2\right] + \gamma^2 \tfrac{1}{6}r(r^2-1)\left[\pm\mu\delta^3 + \tfrac{1}{4}r\delta^4\right]\right\}U_j + \mathcal{O}\gamma^3. \tag{8.55}$$

Why do we introduce the parameter γ? The reason is that, following 'holistic discretisation' [45], when the parameter $\gamma = 0$, the diffusion PDE (8.53) on teeth with coupling (8.55) possesses a slow center manifold parametrized by

---

[1] Here, this coupling condition is centered in the coarse grid. For problems with significant advection we expect to prefer coupling conditions upwind in the coarse grid.

the macroscopic grid values U_j. On this slow manifold the evolution of these macroscopic grid values forms a macroscale model of the diffusion. This model has rigorous theoretical support based upon $\gamma = 0$, and it becomes physically relevant when evaluated at $\gamma = 1$.

We briefly explain how center manifold theory underpins the macroscale model. Setting $\gamma = 0$, the diffusion (8.53) on teeth with coupling (8.55) has spectrum[2] $\lambda = -(n2\pi/h)^2$, $n = 0, 1, 2, \ldots$. Thus, for all initial conditions, all structure within each tooth diffuses exponentially quickly to become constant (the $n = 0$ mode)—but a different constant for each tooth depending upon the initial conditions. (Figure 8.7 shows a similar evolution.) The piecewise constant solutions in each patch form the exponentially attractive slow subspace. Center manifold theory e.g. [46, 47] then asserts that

- the slow subspace is 'bent' to become the slow manifold (with a dimension for each tooth) that exists in some finite neighborhood of $\gamma = 0$, that is, for finite coupling,
- that through its exponential attractiveness (Figure 8.7 shows typical rapidly decaying transients on teeth), the model is relevant in the finite neighborhood,
- and that we may systematically construct the power series approximation to the evolution on the slow manifold.

Thus, from the simple base of piecewise constant fields in each tooth, we construct a description of the field u and its slow evolution as a power series in the coupling parameter γ. The departure of the field u from a constant within each tooth gives the microscopic (subgrid) field inside the jth tooth:

$$v_j = U_j + \gamma[\xi\mu\delta + \tfrac{1}{2}\xi^2\delta^2]U_j + \gamma^2[\tfrac{1}{6}(-\xi + \xi^3)\mu\delta^3 + \tfrac{1}{24}(-\xi^2 + \xi^4)\delta^4]U_j + O\gamma^3, \tag{8.56}$$

where $\xi = (x - x_j)/\Delta x$, as shown for example in the smooth fields of Figure 8.7 that are quickly established. Reassuringly, the microscale subgrid fields (8.56) are independent of the artificial size of the patches; in more general dynamics any size dependence progressively moves to higher orders [32] (§5). The slow evolution of the coarse grid values U_j gives the macroscopic model

$$\dot{U}_j = \frac{1}{\Delta x^2}[\gamma\delta^2 - \tfrac{1}{12}\gamma^2\delta^4]U_j + O\gamma^3. \tag{8.57}$$

We are primarily interested in the physical case of fully coupled teeth for which $\gamma = 1$; then the slow manifold model (8.57) becomes the standard fourth order consistent finite difference approximation to diffusion. Higher order analysis generates higher order consistency.

The various powers of γ in the teeth coupling (8.55) are chosen so that truncation of the expressions to errors $O\gamma^p$ generate a discrete macroscopic

---

[2] The corresponding eigenmodes are, for wavenumber $k = n2\pi/h$, $\sin kx$ and $\sin k|x|$ for all natural numbers n, and $\cos kx$ for even n.

model expressing $\dot{U}_j$ in terms of only $U_{j-p+1}, \ldots, U_{j+p-1}$ (a spatial stencil of width $2p - 1$). Because truncation to errors $O\gamma^p$ results in a model with stencil width $2p - 1$, *such a truncation corresponds to the gap-tooth scheme utilising coupling involving interpolation from only the $2p - 1$ neighboring macroscopic grid values $U_{j-p+1}, \ldots, U_{j+p-1}$.*

In dynamics with significant advection, one could alter the downwind coupling condition (8.55), for example by $\gamma \mapsto \gamma^2$, and hence obtain theoretical support for upwind schemes. This subject requires further research.

The advantage of the center manifold support is that it also applies for nonlinear (e.g. [46, 47, 48]) and/or stochastic systems [49, 50, 51]. Thus, this approach has the potential to not only support macroscale patch models of diffusion, but also very general dynamical systems.

8.7 Conclusions

This chapter overviewed recent progress in the development of patch dynamics, an essential ingredient of the equation-free framework. Patch dynamics is designed to approximate a discretization of an unavailable macroscopic equation using only simulations with a given microscopic model in a number of small regions in the space-time domain (the patches), surrounding the macroscopic grid points. The original formulation by Kevrekidis *et al.* [1] is based on quadratic interpolation and constant gradient boundary conditions, and analyzed for only standard diffusion problems; here, we have discussed how the approach generalizes to other schemes and types of coarse equations, as was done in [31, 32, 41, 43, 52].

We have shown how one can modify the definition of the initial conditions to allow patch dynamics to mimic *any* finite difference scheme (extending the central scheme in the original formulation, much like the heterogeneous multiscale method [34]), and we have investigated to what extent (and at what computational cost) one can avoid the need for specifically designed patch boundary conditions. We have shown that one can surround the patches with buffer regions, where one can impose *arbitrary* boundary conditions. The convergence analysis shows that the buffer size that is required for consistency depends on properties of the macroscopic equation (effective diffusion or advection-coefficient); in general, for advection-dominated problems, much smaller buffer regions are necessary compared to diffusion problems.

Finally, we discussed how buffers are not necessary when the microscale structures are smooth. The center manifold theory support should also apply to nonlinear and stochastic systems in many spatial dimensions; however, this remains to be verified in future research.

8.8 Acknowledgments

GS is a Postdoctoral Fellow of the Research Foundation–Flanders (FWO-Vlaanderen). This work was partially supported by the Research Foundation–Flanders through Research Project 3E060094 and the Belgian Network DYSCO

(Dynamical Systems, Control, and Optimization) (GS); by DARPA, DOE (CPMD) and the ACS/PRF (IGK); and by the Australian Research Council grant DP0774311 (AJR and IGK). The scientific responsibility rests with its authors.

References

[1] Kevrekidis I. G., Gear C. W., Hyman J. M., Kevrekidis P. G., Runborg O., and Theodoropoulos C. (2003). "Equation-free, coarse-grained multiscale computation: enabling microscopic simulators to perform system-level tasks", *Comm. Math. Sciences*, **1**(4), p. 715–762.

[2] Theodoropoulos C., Qian Y. H., and Kevrekidis I. G. (2000). Coarse stability and bifurcation analysis using time-steppers: a reaction-diffusion example. *Proc. Natl Acad. Sci.*, **97**, p. 9840–9845.

[3] Schroff G. M. and Keller H. B. (1993). "Stabilization of unstable procedures: the recursive projection method", *SIAM Journal on Numerical Analysis*, **30**, p. 1099–1120.

[4] Lust K. (1997). *Numerical Bifurcation Analysis of Periodic Solutions of Partial Differential Equations*, Ph. D. thesis, Katholieke Universiteit Leuven.

[5] Lust K., Roose D., Spence A., and Champneys A. R. (1998). "An adaptive newton-picard algorithm with subspace iteration for computing periodic solutions", *SIAM Journal on Scientific Computing*, **19**(4), p. 1188–1209.

[6] Givon D., Kupferman R., and Stuart A. (2004). "Extracting macroscopic dynamics: model problems and algorithms". *Nonlinearity*, **17**, R55–R127.

[7] Gear C. W., Kevrekidis I. G., and Theodoropoulos C. (2002). "'Coarse' integration/bifurcation analysis via microscopic simulators: micro-Galerkin methods", *Computers and Chemical Engineering*, **26**, p. 941–963.

[8] Makeev A. G., Maroudas D., Panagiotopoulos A. Z., and Kevrekidis I. G. (2002b). "Coarse bifurcation analysis of kinetic Monte Carlo simulations: a lattice-gas model with lateral interactions", *J. Chem. Phys.*, **117**(18), p. 8229–8240.

[9] Siettos C. I., Graham M. D., and Kevrekidis I. G. (2003b). "Coarse Brownian dynamics for nematic liquid crystals: bifurcation, projective integration and control via stochastic simulation", *J. Chem. Phys.*, **118**(22), p. 10149–10157. [Can be obtained as cond-mat/0211455 at arxiv.org.]

[10] Van Leemput P., Lust K., and Kevrekidis I. G. (2005). "Coarse-grained numerical bifurcation analysis of lattice boltzmann models", *Physica D*, **210**(1–2), p. 58–76.

[11] Gear C.W., Kaper T. J., Kevrekidis I. G., and Zagaris A. (2005). "Projecting to a slow manifold: Singularly perturbed systems and legacy codes", *SIAM Journal on Applied Dynamical Systems*, **4**(3), p. 711–732.

[12] Gear C. W. and Kevrekidis I. G. (2005). "Constraint-defined manifolds: a legacy code approach to low-dimensional computation", *J. Sci. Comp.*, **25**(1), p. 17–28.

[13] Van Leemput, P., Vanroose, W., and Roose, D. (2007). "Mesoscale analysis of the equation-free constrained runs initialization scheme", *Multiscale Modeling and Simulation*, **6**(4), p. 1234–1255.

[14] Vandekerckhove C., Kevrekidis I. G., and Roose D. (2009). *Journal of Scientific Computing*, **39**(2), 167–188, 2009.

[15] Werder T., Walther J. H., and Koumoutsakos P. (2005). "Hybrid atomistic–continuum method for the simulation of dense fluid flows", *Journal of Computational Physics*, **205**(1), p. 373–390.

[16] Delgado-Buscalioni R. and Coveney P. V. (2003). "USHER: An algorithm for particle insertion in dense fluids", *The Journal of Chemical Physics*, **119**, p. 978.

[17] Frederix Y., Samaey G., Vandekerckhove C., and Roose D., Li T. and Nies E. (2009). *Lifting in Equation-free methods for molecular dynamic simulations of dense fluids*, Discrete and Continuous Dynamical Systems Series B **11**(4), p. 855–874.

[18] Makeev A. G., Maroudas D., and Kevrekidis I. G. (2002a). "Coarse stability and bifurcation analysis using stochastic simulators: kinetic Monte Carlo examples]], *J. Chem. Phys.*, **116**, p. 10083–10091.

[19] Theodoropoulos C., Sankaranarayanan K., Sundaresan S., and Kevrekidis I. G. (2003). "Coarse bifurcation studies of bubble flow Lattice-Boltzmann simulations", *Chem. Eng. Sci.*, **59**, p. 2357–2362. [Can be obtained as nlin.PS/0111040 from arxiv.org.]

[20] Hummer G. and Kevrekidis I. G. (2003). "Coarse molecular dynamics of a peptide fragment: free energy, kinetics and long-time dynamics computations", *J. Chem. Phys.*, **118**(23), p. 10762–10773. [Can be obtained as physics/0212108 at arxiv.org.]

[21] Siettos C. I., Armaou A., Makeev A. G., and Kevrekidis I. G. (2003a). "Microscopic/stochastic timesteppers and coarse control: a kinetic Monte Carlo example", *AIChE J.*, **49**(7), p. 1922–1926. [Can be obtained as nlin.CG/0207017 at arxiv.org.]

[22] Gear C. W. and Kevrekidis I. G. (2003). "Projective methods for stiff differential equations: problems with gaps in their eigenvalue spectrum", *SIAM Journal of Scientific Computation*, **24**(4), p. 1091–1106.

[23] Vandekerckhove C., Roose D., and Lust K. (2007b). "Numerical stability analysis of an acceleration scheme for step size constrained time integrators", *Journal on Computational and Applied Mathematics*, **200**(2), p. 761–777.

[24] Erban R., Kevrekidis I. G., and Othmer H. (2006). "An equation-free computational approach for extracting population-level behavior from individual-based models of biological dispersal", *Physica D*, **215**(1), p. 1–24.

[25] Rico-Martinez R., Gear C. W., and Kevrekidis I. G. (2004). "Coarse projective kmc integration: forward/reverse initial and boundary value problems", *Journal of Computational Physics*, **196**(2), p. 474–489.

[26] Setayeshar S., Gear C. W., Othmer H. G., and Kevrekidis I. G. (2005). "Application of coarse integration to bacterial chemotaxis", *SIAM Multiscale Modeling and Simulation*, **4**(1), p. 307–327.

[27] Kevrekidis I. G. (2000). *Coarse bifurcation studies of alternative microscopic/hybrid simulators.* Plenary lecture, CAST Division, AIChE Annual Meeting, Los Angeles. [Slides can be obtained from http://arnold.princeton.edu/yannis/.]

[28] Gear C. W., Li J., and Kevrekidis I. G. (2003). "The gap-tooth method in particle simulations", *Physics Letters A*, **316**, p. 190–195. [Can be obtained as physics/0303010 at arxiv.org.]

[29] Bensoussan A., Lions J. L., and Papanicolaou G. (1978). *Asymptotic Analysis of Periodic Structures*, Volume 5 of *Studies in Mathematics and its Applications.* North-Holland, Amsterdam.

[30] Cioranescu D. and Donato P. (1999). *An Introduction to Homogenization*, Oxford University Press.

[31] Samaey G., Roose D., and Kevrekidis I. G. (2005). "The gap-tooth scheme for homogenization problems", *SIAM Multiscale Modeling and Simulation*, **4**(1), p. 278–306.

[32] Roberts A. J. and Kevrekidis I. G. (2007). "General tooth boundary conditions for equation free modelling", *SIAM J. Scientific Computing*, **29**(4), p. 1495–1510.

[33] Li J., Kevrekidis P. G., Gear C. W., and Kevrekidis I. G. (2003). "Deciding the nature of the 'coarse equation' through microscopic simulations: the baby-bathwater scheme", *SIAM Multiscale Modeling and Simulation*, **1**(3), p. 391–407.

[34] E W. and Engquist B. (2003). "The heterogeneous multi-scale methods", *Comm. Math. Sci.*, **1**(1), p. 87–132.

[35] Strang G. (1964). "Accurate partial finite difference methods II: Nonlinear problems", *Numerische Mathematik*, **6**, p. 37–46.

[36] Li J., Liao D., and Yip S. (1998). "Imposing field boundary conditions in MD simulation of fluids: optimal particle controller and buffer zone feedback", *Mat. Res. Soc. Symp. Proc.*, **538**, p. 473–478.

[37] Hou T. Y. and Wu X. H. (1997). "A multiscale finite element method for elliptic problems in composite materials and porous media", *Journal of Computational Physics*, **134**, p. 169–189.

[38] Hadjiconstantinou N. G. (1999). "Hybrid atomistic-continuum formulations and the moving contact-line problem", *Journal of Computational Physics*, **154**, p. 245–265.

[39] Abdulle A. and E W. (2003). "Finite difference heterogeneous multi-scale method for homogenization problems", *Journal of Computational Physics*, **191**(1), p. 18–39.

[40] Hindmarsh A. C. (1983). ODEPACK, a systematized collection of ODE solvers. In *Scientific Computing* (eds R. S. *et al.*), p. 55–64. North-Holland, Amsterdam.

[41] Samaey G., Kevrekidis I. G., and Roose D. (2006a). "Patch dynamics with buffers for homogenization problems", *Journal of Computational Physics*, **213**(1), p. 264–287.

[42] Golub G. H. and Loan C. F. Van (1996). *Matrix Computations (3rd ed.)*, Johns Hopkins University Press, Baltimore, MD, USA.

[43] Roberts A. J. and Kevrekidis I. G. (2005). Higher order accuracy in the gap-tooth scheme for large-scale dynamics using microscopic simulators. In *Proc. of 12th Computational Techniques and Applications Conference CTAC-2004* (ed. May R. and Roberts A. J.), Volume 46 of *ANZIAM J.*, p. C637–C657.

[44] MacKenzie T. (2005). *Create Accurate Numerical Models of Complex Spatio-Temporal Dynamical Systems with Holistic Discretisation*. Ph. D. thesis, University of Southern Queensland.

[45] Roberts A. J. (2001). "Holistic discretisation ensures fidelity to Burgers' equation", *Applied Numerical Modelling*, **37**, p. 371–396.

[46] Carr J. (1981). *Applications of Centre Manifold Theory*, Volume 35 of *Applied Math. Sci.* Springer–Verlag.

[47] Kuznetsov Y. A. (1995). *Elements of Applied Bifurcation Theory*, Volume **112** of *Applied Mathematical Sciences*, Springer–Verlag.

[48] Roberts A. J. (2002). "A holistic finite difference approach models linear dynamics consistently", *Mathematics of Computation*, **72**, p. 247–262.

[49] Boxler P. (1991). "How to construct stochastic center manifolds on the level of vector fields", *Lect. Notes in Maths*, **1486**, p. 141–158.

[50] Arnold L. (2003). *Random Dynamical Systems*, Springer Monographs in Mathematics, Springer.

[51] Roberts A. J. (2006). "Resolving the multitude of microscale interactions accurately models stochastic partial differential equations", *LMS J. Computation and Maths*, **9**, p. 193–221.

[52] Samaey G., Roose D., and Kevrekidis I. G. (2006b). Finite difference patch dynamics for advection homogenization problems. In *Model Reduction and Coarse-Graining Approaches for Multiscale Phenomena* (eds Gorban A., Kazantzis N., Kevrekidis I., Öttinger H., and Theodoropoulos C.), p. 205–224. Springer, Berlin–Heidelberg–New York.

9
ON MULTISCALE COMPUTATIONAL MECHANICS WITH TIME-SPACE HOMOGENIZATION

P. Ladevèze, David Néron, Jean-Charles Passieux

9.1 Introduction

Today, in structural mechanics, there is a growing interest in a class of techniques called "multiscale computational approaches", which are capable of analyzing structures in which two or more very different scales can be identified. A typical engineering example is that of a relatively large structure in which local cracking or local buckling occurs [1,2]. Another typical engineering problem is related to the increasing interest in material models described on a scale smaller than that of the macroscopic structural level, with applications ranging from the design of composite materials and structures to manufacturing [3,4]. In such situations, the structure being studied is highly heterogeneous and the local solution involves short-wavelength phenomena in both space and time. As a result, classical finite element codes lead to systems with very large numbers of degrees of freedom and the corresponding calculation costs are generally prohibitive. Therefore, one of today's main challenges is to derive computational strategies capable of solving such engineering problems through true interaction between the two scales in both space and time: the microscale and the macroscale.

This paper focuses on this challenge, with the objective of reducing calculation costs drastically while, at the same time, trying to improve robustness.

The central issue is the transfer of information from one scale to another. A very efficient strategy for linear periodic media consists in applying the homogenization theory initiated by Sanchez-Palencia [5,6]. Further developments and related computational approaches can be found in [7–12]. First, the resolution of the macro problem leads to effective values of the unknowns; then, the micro solution is calculated locally based on the macro solution. The fundamental assumption, besides periodicity, is that the ratio of the characteristic length of the small scale to the characteristic length of the large scale must be small. Boundary zones, in which the material cannot be homogenized, require special treatment. Moreover, this theory is not directly applicable to time-dependent nonlinear problems. Other computational strategies using homogenization techniques based on the Hill-Mandel conditions [13] have also been proposed [14,15] and have similar limitations. Other paradigms for building multiscale computational strategies can be found in [16,17]. All these approaches introduce different scales only in space.

Only relatively few works have been devoted to multi-time-scale computational strategies. What are called multi-time-step methods [18–21] and time-decomposed parallel time integrators [22, 23] deal with different time discretizations and integration schemes. Local enrichment functions were introduced in [24]. In multiphysics problems, coupling between time grids may be envisaged. This type of problem was solved in [25] through the introduction of "micro/macro projectors" between grids. Parareal [26] or PITA [22] approaches belong in this category. However, none of these strategies involves a true time-homogenization technique. Such a technique seems to have been used only for periodic loading histories [27–35].

Our first attempt to meet our challenge was to devise a new micro/macro computational strategy [17] which involved space homogenization over the whole domain while avoiding the drawbacks of classical homogenization theory. This technique was expanded in [36] to include time as well as space thanks to the LATIN Method, which enables one to work globally over the time-space domain [37]. This is an iterative strategy. Here, it will be described in detail for (visco)plastic materials and optional unilateral contact with or without friction, a case already introduced in [17]. More complex types of material behavior could also be taken into account.

The first characteristic of the method resides in the partitioning of the space-time domain. The structure is defined as an assembly of substructures and interfaces. Each component has its own variables and its own equations. The time interval is divided into subintervals, using the discontinuous Galerkin method to handle possible discontinuities. The junction between the macroscale and the microscale takes place only at the interfaces. Each quantity of interest is considered to be the sum of a macro quantity and a micro quantity, where the macro quantities are defined as "mean values" in time and in space, and the associated micro quantities are the complementary parts; this is a choice. An important point is that due to the Saint Venant principle the effects of the micro quantities are localized in space.

The second characteristic of the method is the use of what we call the LATIN method, a nonincremental iterative computational strategy applied over the entire time interval being studied [37]. At each iteration, one must solve a macro problem defined over the entire structure and the entire time interval, along with a family of independent linear problems, each concerning a substructure and its boundary. The latter are "micro" problems in contrast with the "macro" problem which corresponds to the entire structure homogenized in time as well as in space.

The third characteristic of the method concerns the resolution, over the time-space domain, of the numerous micro problems (whose size can be very large) within the cells or substructures. With the LATIN method, a classical approach consists in using radial time-space approximations [37, 38], which reduce calculation and storage costs drastically. Here, a new, more efficient and more robust version is introduced. This technique consists in approximating a function defined in the space-time domain by a sum of products of scalar functions of the time variable by functions of the space variable. As the iterative process goes on, the functions of the space variable constructed in this manner constitute a consistent basis which can be re-used for successive iterations. Moreover, when dealing with similar substructures such as cells in composites, this basis is common to all the substructures.

After reviewing the bases of the multiscale strategy with space and time homogenization, this paper will focus on suitable approximation techniques for the resolution of the micro and macro problems, and particularly on the new radial time-space approximation. Several numerical examples will illustrate the capabilities of the approach presented.

9.2 The reference problem

Under the assumption of small perturbations, let us consider the quasi-static and isothermal evolution of a structure defined in the time-space domain $[0, T] \times \Omega$. This structure is subjected to prescribed body forces $\underline{f}_d$, to traction forces $\underline{F}_d$ over a part $\partial_2 \Omega$ of the boundary, and to prescribed displacements $\underline{U}_d$ over the complementary part $\partial_1 \Omega$ (see Figure 9.1).

The state of the structure is defined by the set of the fields $(\dot{\boldsymbol{\varepsilon}}_p, \dot{\mathbf{X}}, \boldsymbol{\sigma}, \mathbf{Y})$ (using the dot notation $\dot{\Box}$ for the time derivative), in which:

- $\boldsymbol{\varepsilon}_p$ designates the inelastic part of the strain field $\boldsymbol{\varepsilon}$ which corresponds to the displacement field $\underline{U}$, uncoupled into an elastic part $\boldsymbol{\varepsilon}_e$ and an inelastic part $\boldsymbol{\varepsilon}_p = \boldsymbol{\varepsilon} - \boldsymbol{\varepsilon}_e$; $\mathbf{X}$ designates the remaining internal variables;
- $\boldsymbol{\sigma}$ designates the Cauchy stress field and $\mathbf{Y}$ the set of variables which are conjugates of $\mathbf{X}$.

All these quantities are defined over the time-space domain $[0, T] \times \Omega$ and assumed to be sufficiently regular. For the sake of simplicity, only the displacement $\underline{U}$ is assumed to have a nonzero initial value, denoted $\underline{U}_0$.

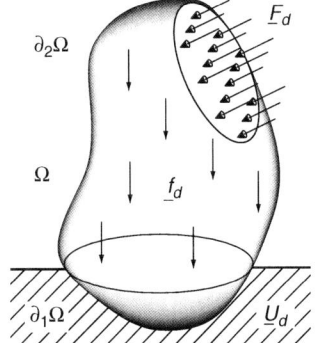
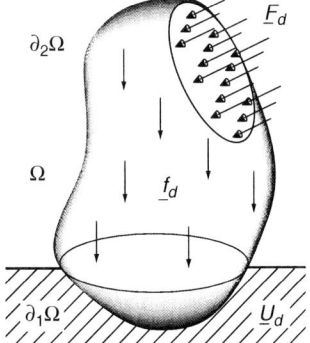

Figure 9.1: The reference problem.

Introducing the following notations for the primal fields:

$$\mathbf{e}_p = \begin{bmatrix} \varepsilon_p \\ -\mathbf{X} \end{bmatrix}, \quad \mathbf{e} = \begin{bmatrix} \varepsilon \\ 0 \end{bmatrix} \text{ and } \mathbf{e}_e = \begin{bmatrix} \varepsilon_e \\ \mathbf{X} \end{bmatrix} \quad \text{so that} \quad \mathbf{e}_p = \mathbf{e} - \mathbf{e}_e \quad (9.1)$$

and for the dual fields:

$$\mathbf{f} = \begin{bmatrix} \boldsymbol{\sigma} \\ \mathbf{Y} \end{bmatrix} \quad (9.2)$$

the mechanical dissipation rate for the entire structure Ω is:

$$\int_\Omega (\dot{\varepsilon}_p : \boldsymbol{\sigma} - \dot{\mathbf{X}} \cdot \mathbf{Y}) d\Omega = \int_\Omega (\dot{\mathbf{e}}_p \circ \mathbf{f}) d\Omega \quad (9.3)$$

where $\cdot$ denotes the contraction adapted to the tensorial nature of $\mathbf{X}$ and $\mathbf{Y}$, and $\circ$ denotes the corresponding operator. Let us introduce the following fundamental "dissipation" bilinear form:

$$\langle \mathbf{s}, \mathbf{s}' \rangle = \int_{[0,T] \times \Omega} \left(1 - \frac{t}{T}\right) (\dot{\mathbf{e}}_p \circ \mathbf{f}' + \dot{\mathbf{e}}'_p \circ \mathbf{f}) d\Omega dt \quad (9.4)$$

along with $\mathbf{E}$ and $\mathbf{F}$, the spaces of the fields $\dot{\mathbf{e}}_p$, and $\mathbf{f}$ which are compatible with (9.4). These spaces enable us to define $\mathbf{S} = \mathbf{E} \times \mathbf{F}$, the space in which the state $\mathbf{s} = (\dot{\mathbf{e}}_p, \mathbf{f})$ of the structure is being sought.

9.2.1 State laws

Following [37], a normal formulation with internal state variables is used to represent the behavior of the material. If ρ denotes the mass density of the material, from the free energy $\rho \Psi(\varepsilon_e, \mathbf{X})$ with the usual uncoupling assumptions, the state law yields:

$$\boldsymbol{\sigma} = \rho \frac{\partial \psi}{\partial \varepsilon_e} = \mathbf{K} \varepsilon_e$$

$$\mathbf{Y} = \rho \frac{\partial \psi}{\partial \mathbf{X}} = \boldsymbol{\Lambda} \mathbf{X} \quad (9.5)$$

where the Hooke's tensor $\mathbf{K}$ and the constant, symmetric, and positive definite tensor $\boldsymbol{\Lambda}$ are material characteristics. These equations can be rewritten in the form:

$$\mathbf{f} = \mathbf{A} \mathbf{e}_e \quad \text{with} \quad \mathbf{A} = \begin{bmatrix} \mathbf{K} & 0 \\ 0 & \boldsymbol{\Lambda} \end{bmatrix} \quad (9.6)$$

where $\mathbf{A}$ is a constant, symmetric, and positive definite operator. Let us note that such an approach is available for most material models [37].

The constitutive equation is given by the positive differential operator $\mathbf{B}$, which is considered to be derived from the dissipation pseudo-potential $\phi^*(\boldsymbol{\sigma}, \mathbf{Y})$:

$$\dot{\mathbf{e}}_p = \begin{bmatrix} \partial_{\boldsymbol{\sigma}} \phi^* \\ \partial_{\mathbf{Y}} \phi^* \end{bmatrix} = \mathbf{B}(\mathbf{f}) \quad \text{with} \quad \mathbf{e}_{p|t=0} = 0 \qquad (9.7)$$

One should note that for the sake of simplicity we are restricting this presentation to the case of a sufficiently smooth pseudo-potential. Should this not be the case, one would modify (9.7) by considering $\partial_\square \phi^*$ to be a subdifferential and replacing the first equality by an inclusion.

For example, if we consider standard viscoplastic behavior with isotropic strain hardening described by the scalar p and kinematic strain hardening described by the second-order tensor $\boldsymbol{\alpha}$, and if the scalar R and the tensor $\boldsymbol{\beta}$ are the conjugate variables of p and $\boldsymbol{\alpha}$, respectively, we have:

$$\begin{aligned} \rho\psi &= \frac{1}{2}\boldsymbol{\varepsilon}_e : \mathbf{K} : \boldsymbol{\varepsilon}_e + \frac{1}{2}c\|\boldsymbol{\alpha}\|^2 + \frac{1}{2}\lambda p^2 \\ \phi^* &= \frac{k}{n+1}\left\langle \|\boldsymbol{\sigma}_D - \boldsymbol{\beta}\| + \frac{a}{2c}\|\boldsymbol{\beta}\|^2 - \ell(R) - R_0 \right\rangle_+^{n+1} \end{aligned} \qquad (9.8)$$

where $\|\boldsymbol{\beta}\| = \sqrt{\boldsymbol{\beta} : \boldsymbol{\beta}}$, $\boldsymbol{\sigma}_D$ is the deviatoric part of Tensor $\boldsymbol{\sigma}$ and $\langle \square \rangle_+$ extracts the positive part of the argument. Scalars k, n, c, λ, a, R_0, and Function ℓ are material characteristics.

9.2.2 Compatibility conditions and equilibrium equations

The compatibility conditions and equilibrium equations are described below and some functional spaces are introduced. We use the notation $\square^*$ to designate the vector space associated with an affine space $\square$.

- The displacement field $\underline{U}$ should match the prescribed displacement $\underline{U}_d$ at Boundary $\partial_1\Omega$ and the initial condition $\underline{U}_0$ at $t = 0$:

$$\underline{U}_{|\partial_1\Omega} = \underline{U}_d \quad \text{and} \quad \underline{U}_{|t=0} = \underline{U}_0 \qquad (9.9)$$

The corresponding space of displacement fields $\underline{U}$ is denoted $\mathcal{U}$.

- The stress field $\boldsymbol{\sigma}$ should be symmetric and in equilibrium with the external prescribed forces $\underline{F}_d$ at $\partial_2\Omega$ and the prescribed body forces $\underline{f}_d$ in Ω. The corresponding variational formulation is:

$$\forall \underline{U}^\star \in \mathcal{U}^\star, \quad -\int_{[0,T]\times\Omega} \boldsymbol{\sigma} : \boldsymbol{\varepsilon}(\underline{\dot{U}}^\star) d\Omega dt$$

$$+ \int_{[0,T]\times\Omega} \underline{f}_d \cdot \underline{\dot{U}}^\star d\Omega dt + \int_{[0,T]\times\partial_2\Omega} \underline{F}_d \cdot \underline{\dot{U}}^\star dS dt = 0 \qquad (9.10)$$

The subspace of $\mathbf{F}$ whose elements $\mathbf{f} = [\boldsymbol{\sigma} \ \mathbf{Y}]^T$ verify the previous condition is denoted $\mathcal{F}$. These fields are said to be "statically admissible".

- The strain rate field $\dot{\varepsilon}$ should derive from the symmetric part of the gradient of a displacement field belonging to Space $\mathcal{U}$. The corresponding variational formulation is:

$$\forall \mathbf{f}^\star \in \mathcal{F}^\star, \quad -\int_{[0,T]\times\Omega} \boldsymbol{\sigma}^\star : \dot{\varepsilon}\,d\Omega dt + \int_{[0,T]\times\partial_1\Omega} \boldsymbol{\sigma}^\star \underline{n} \cdot \dot{\underline{U}}_d dS dt = 0 \quad (9.11)$$

The subspace of $\mathbf{E}$ whose elements $\dot{\mathbf{e}} = [\dot{\varepsilon} - \dot{\mathbf{X}}]^T$ verify the previous condition is denoted $\mathcal{E}$. These fields are said to be "kinematically admissible".

9.2.3 Formulation of the reference problem

The reference problem defined over the time-space domain $[0,T] \times \Omega$ can be formulated as follows:

Find $\mathbf{s_{ref}} = (\dot{\mathbf{e}}_p, \mathbf{f})$ which verifies, with $\mathbf{e} = \mathbf{e}_e + \mathbf{e}_p$,

- the kinematic admissibility $\dot{\mathbf{e}} \in \mathcal{E}$
- the static admissibility $\quad\quad \mathbf{f} \in \mathcal{F}$ \hfill (9.12)
- the state law $\quad\quad\quad\quad\quad\; \mathbf{f} = \mathbf{A}\mathbf{e}_e$
- the evolution law $\quad\quad\quad\; \dot{\mathbf{e}}_p = \mathbf{B}(\mathbf{f}) \quad$ with $\quad \mathbf{e}_{p|t=0} = 0$

which is equivalent to:

Find $\mathbf{s_{ref}} = (\dot{\mathbf{e}}_p, \mathbf{f})$ which verifies

$$(\mathbf{A}^{-1}\dot{\mathbf{f}} + \dot{\mathbf{e}}_p) \in \mathcal{E}, \quad \mathbf{f} \in \mathcal{F}, \quad \dot{\mathbf{e}}_p = \mathbf{B}(\mathbf{f}) \quad \text{with} \quad \mathbf{e}_{p|t=0} = 0 \quad (9.13)$$

9.3 Reformulation of the problem with structure decomposition

Now, the basic idea consists in describing the structure as an assembly of simple components, i.e. substructures and interfaces, each with its own variables and equations (admissibility, equilibrium, and behavior) [37] (see Figure 9.2).

Each substructure Ω_E of Ω is defined by the set of variables $(\dot{\mathbf{e}}_{pE}, \mathbf{f}_E)$ and subjected at its boundary $\partial\Omega_E$ to the action of its environment (the neighboring interfaces), described by a displacement distribution $\underline{W}_E$ and a force distribution $\underline{F}_E$. We will use the subscript $\square_E$ to designate the restriction of variables and operators to Subdomain Ω_E.

Clearly, $\underline{W}_E$ and $\underline{F}_E$ viewed from Substructure Ω_E play the role of prescribed boundary conditions. If these boundary conditions are assumed to be known and compatible, the problem in Subdomain Ω_E consists in finding a solution of an equation similar to (9.13) in which $\underline{W}_E$ participates in the definition of kinematic admissibility and $\underline{F}_E$ in the definition of static admissibility.

Let $\mathbf{s}_E = (\dot{\mathbf{e}}_{pE}, \underline{W}_E, \mathbf{f}_E, \underline{F}_E)$ denote the set of the variables describing the state of Substructure Ω_E and its boundary $\partial\Omega_E$. The mechanical dissipation

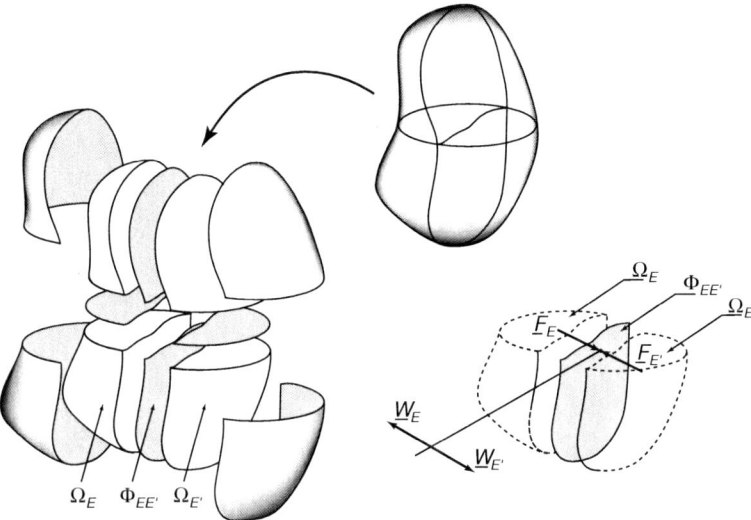

Figure 9.2: Decomposition of the structure into substructures and interfaces.

rate in Substructure Ω_E is:

$$\int_{\Omega_E} (\dot{\mathbf{e}}_{pE} \circ \mathbf{f}_E) d\Omega - \int_{\partial \Omega_E} \underline{\dot{W}}_E \cdot \underline{F}_E dS \qquad (9.14)$$

and we introduce the following fundamental "dissipation" bilinear form:

$$\langle \mathbf{s}_E, \mathbf{s}'_E \rangle_E = \int_{[0,T] \times \Omega_E} \left(1 - \frac{t}{T}\right) (\dot{\mathbf{e}}_{pE} \circ \mathbf{f}'_E + \dot{\mathbf{e}}'_{pE} \circ \mathbf{f}_E) d\Omega dt$$

$$- \int_{[0,T] \times \partial \Omega_E} \left(1 - \frac{t}{T}\right) (\underline{\dot{W}}_E \cdot \underline{F}'_E + \underline{\dot{W}}'_E \cdot \underline{F}_E) dS dt \qquad (9.15)$$

along with $\mathbf{E}_E$, $\mathcal{W}_E$, $\mathbf{F}_E$, and $\mathcal{F}_E$, the spaces of the fields $\dot{\mathbf{e}}_{pE}$, $\underline{\dot{W}}_E$, $\mathbf{f}_E$, and $\underline{F}_E$ which are compatible with (9.15). These spaces enable us to introduce $\mathbf{S}_E = \mathbf{E}_E \times \mathcal{W}_E \times \mathbf{F}_E \times \mathcal{F}_E$, the space within which $\mathbf{s}_E = (\dot{\mathbf{e}}_{pE}, \underline{\dot{W}}_E, \mathbf{f}_E, \underline{F}_E)$ is being sought.

Partitioning a structure into nonoverlapping subdomains is a rather classical idea in mechanics. Another idea consists in considering the interface variables, or at least some of these variables, to be Lagrange multipliers. With this approach, the variables we introduced can be viewed as distributed Lagrange multipliers of both the displacement and force types. One can observe that at each point of an interface one has three displacement-force pairs: one for each substructure on either side of the interface plus one for the interface itself.

One can also note that our description provides a natural framework for dealing with different discretizations in adjacent subdomains (non-matching grids), because the interfaces and subdomains can be meshed independently of one another.

9.3.1 Admissibility conditions for Substructure Ω_E

The following conditions must be verified:

- The displacement field $\underline{U}_E$ should match the interface displacement $\underline{W}_E$ at Boundary $\partial\Omega_E$ and the initial condition $\underline{U}_{E0}$ at $t = 0$:

$$\underline{U}_{E|\partial\Omega_E} = \underline{W}_E \quad \text{and} \quad \underline{U}_{E|t=0} = \underline{U}_{E0} \quad (9.16)$$

The corresponding space of displacement fields $(\underline{U}_E, \underline{W}_E)$ is denoted $\mathcal{U}_E$.

- The stress field σ_E should be symmetric and in equilibrium with the interface forces $\underline{F}_E$ on $\partial\Omega_E$ and the prescribed body forces $\underline{f}_d$ on Ω_E. The corresponding variational formulation is:

$$\forall (\underline{U}_E^\star, \underline{W}_E^\star) \in \mathcal{U}_E^\star, \quad -\int_{[0,T]\times\Omega_E} \sigma_E : \varepsilon(\underline{\dot{U}}_E^\star) d\Omega dt$$

$$+ \int_{[0,T]\times\Omega_E} \underline{f}_d \cdot \underline{\dot{U}}_E^\star d\Omega dt + \int_{[0,T]\times\partial\Omega_E} \underline{F}_E \cdot \underline{\dot{W}}_E^\star dS dt = 0 \quad (9.17)$$

The subspace of $\mathbf{F}_E \times \mathcal{F}_E$ whose elements $\mathbf{f}_E = [\sigma_E \; \mathbf{Y}_E]^T$ verify the previous condition is denoted $\mathcal{F}_E$.

- The strain rate field $\dot{\varepsilon}_E$ should derive from the symmetric part of the gradient of a displacement field belonging to Space $\mathcal{U}_E$. The corresponding variational formulation is:

$$\forall (\mathbf{f}_E^\star, \underline{F}_E^\star) \in \mathcal{F}_E^\star, \quad -\int_{[0,T]\times\Omega_E} \sigma_E^\star : \dot{\varepsilon}_E d\Omega dt + \int_{[0,T]\partial\Omega_E} \underline{F}_E^\star \cdot \underline{\dot{W}}_E dS dt = 0$$
$$(9.18)$$

The subspace of $\mathbf{E}_E \times \mathcal{W}_E$ whose elements $\mathbf{e}_E = [\dot{\varepsilon}_E \; -\dot{\mathbf{X}}_E]^T$ verify the previous condition is denoted $\mathcal{E}_E$.

- Then, the set of variables $\mathbf{s}_E = (\dot{\mathbf{e}}_{pE}, \underline{\dot{W}}_E, \mathbf{f}_E, \underline{F}_E)$ should verify:

$$(\mathbf{A}^{-1}\dot{\mathbf{f}}_E + \dot{\mathbf{e}}_{pE}) \in \mathcal{E}_E \quad \text{and} \quad \mathbf{f}_E \in \mathcal{F}_E \quad (9.19)$$

which defines $\mathbf{A}_{dE}$, the subspace of $\mathbf{S}_E$ whose elements $\mathbf{s}_E$ verify the previous conditions (these fields are said to be "E-admissible"), but also the evolution law:

$$\dot{\mathbf{e}}_{pE} = \mathbf{B}(\mathbf{f}_E) \quad \text{and} \quad \mathbf{e}_{pE|t=0} = 0 \quad (9.20)$$

9.3.2 Interface behavior

The interface concept can be easily extended to the boundary of Ω, $\partial\Omega$, where either the displacements or the forces are prescribed. It suffices to set:

- for a prescribed displacement at $\Phi_{E1} = \partial\Omega_E \cap \partial_1\Omega$: $\quad \underline{W}_E = \underline{U}_d$;
- for a prescribed force at $\Phi_{E2} = \partial\Omega_E \cap \partial_2\Omega$: $\quad \underline{F}_E = \underline{F}_d$.

Let $\mathbf{\Omega}_E$ denote the set of the neighboring substructures of Ω_E and $\Phi_{EE'}$ the interface between Ω_E and $\Omega_{E'} \in \mathbf{\Omega}_E$. This interface is characterized by the restrictions to $\Phi_{EE'}$ of both the displacement field ($\underline{W}_E, \underline{W}_{E'}$) and the force field ($\underline{F}_E, \underline{F}_{E'}$), denoted ($\underline{W}_{EE'}, \underline{W}_{E'E}$) and ($\underline{F}_{EE'}, \underline{F}_{E'E}$), respectively. At Interface $\Phi_{EE'}$, the action-reaction principle:

$$\underline{F}_{EE'} + \underline{F}_{E'E} = \underline{0} \qquad (9.21)$$

holds, along with a constitutive relation of the form:

$$\underline{F}_{EE'|t} = \mathbf{b}_{EE'}\left(\left[\underline{\dot{W}}_{EE'} - \underline{\dot{W}}_{E'E}\right]_{|\tau}, \tau \leqslant t\right) \qquad (9.22)$$

where $\mathbf{b}_{EE'}$ is an operator characterizing the behavior of the interface. For instance, one can have:

- for a perfect connection:

$$\underline{W}_{EE'} = \underline{W}_{E'E} \qquad (9.23)$$

which can be interpreted as $\mathbf{b}_{EE'}$ being a linear stiffness operator with an infinite norm;

- for unilateral contact without friction:

$$\begin{cases} \Pi_{EE'}\,\underline{F}_{EE'} = 0 \\ \underline{n}_{EE'} \cdot (\underline{W}_{EE'} - \underline{W}_{E'E} - \underline{g}_{EE'}) \geqslant 0, \quad \underline{n}_{EE'} \cdot \underline{F}_{EE'} \leqslant 0 \\ \left(\underline{n}_{EE'} \cdot (\underline{W}_{EE'} - \underline{W}_{E'E} - \underline{g}_{EE'})\right)(\underline{n}_{EE'} \cdot \underline{F}_{EE'}) = 0 \end{cases} \qquad (9.24)$$

where $\underline{n}_{EE'}$ is the vector normal to Interface $\Phi_{EE'}$ going from Subdomain Ω_E to Subdomain $\Omega_{E'}$, $\Pi_{EE'}$ is the corresponding orthogonal projector, and $\underline{g}_{EE'}$ is the initial gap between the substructures.

Clearly, in the case of problems with multiple contacts, the philosophy of the method consists in fitting the contact interfaces between the substructures with the material interfaces between the different components of the assembly [37,39]. Each individual component can also be partitioned artificially using a perfect connection interface.

9.3.3 Reformulation of the reference problem

Going back to the reference problem stated at the beginning (9.13), this problem obviously consists in finding the set $\mathbf{s} = (\mathbf{s}_E)_{\Omega_E \subset \Omega}$ in the space $\mathbf{S} = \bigotimes_{\Omega_E \subset \Omega} \mathbf{S}_E$. Let $\mathbf{E}$, $\mathcal{W}$, $\mathbf{F}$, and $\mathcal{F}$ denote the extensions of the previous spaces $\mathbf{E}_E$, $\mathcal{W}_E$, $\mathbf{F}_E$, and $\mathcal{F}_E$ to the entire problem. For the sake of simplicity, we will use the notation $(\dot{\mathbf{e}}_p, \underline{\dot{W}}, \mathbf{f}, \underline{F}) \in \mathbf{E} \times \mathcal{W} \times \mathbf{F} \times \mathcal{F}$ to designate a set $(\dot{\mathbf{e}}_{pE}, \underline{\dot{W}}_E, \mathbf{f}_E, \underline{F}_E)_{\Omega_E \subset \Omega} \in \mathbf{E}_E \times \mathcal{W}_E \times \mathbf{F}_E \times \mathcal{F}_E$.

The decomposed reference problem, defined over the entire time-space domain $[0, T] \times \Omega$, can be formulated as follows:

Find $\mathbf{s}_{\mathbf{ref}} = (\mathbf{s}_E)_{\Omega_E \subset \Omega}$ which verifies, $\forall \Omega_E \subset \Omega$,

- the E-admissibility condition $\mathbf{s}_E \in \mathbf{A}_{\mathbf{d}E}$
- the evolution law $\quad \dot{\mathbf{e}}_{pE} = \mathbf{B}(\mathbf{f}_E) \quad$ with $\quad \mathbf{e}_{pE|t=0} = 0$
- the interface behavior $\quad \forall \Omega_{E'} \in \Omega_E, \ \underline{F}_{EE'} + \underline{F}_{E'E} = \underline{0}\quad$ and

$$\underline{F}_{EE'|t} = \mathbf{b}_{EE'}\left(\left[\underline{\dot{W}}_{EE'} - \underline{\dot{W}}_{E'E}\right]_{|\tau}, \tau \leqslant t\right) \tag{9.25}$$

9.4 Multiscale description in the time-space domain $[0, T] \times \Omega$

9.4.1 A two-scale description of the unknowns

The following idea was initially introduced for multiscale problems in space, then extended to multiscale problems in both time and space in [36]. The approach consists in introducing a two-scale description of the unknowns: these two scales are denoted "macro" and "micro" and concern both space and time. The distinction between the macro level and the micro level is made only at the interfaces.

For the neighboring interfaces of Substructure Ω_E, the unknowns $(\underline{\dot{W}}_E, \underline{F}_E) \in \mathcal{W}_E \times \mathcal{F}_E$ are split into:

$$\underline{\dot{W}}_E = \underline{\dot{W}}_E^M + \underline{\dot{W}}_E^m \quad \text{and} \quad \underline{F}_E = \underline{F}_E^M + \underline{W}_E^m \tag{9.26}$$

where Superscripts $\square^M$ and $\square^m$ designate the macro parts and the micro complements of the fields, respectively. The spaces corresponding to the macro parts are $\mathcal{W}_E^M$ and $\mathcal{F}_E^M$, and the spaces corresponding to the micro parts are $\mathcal{W}_E^m$ and $\mathcal{F}_E^m$. The extensions of these spaces to the entire set of interfaces are $\mathcal{W}^M$, $\mathcal{F}^M$, $\mathcal{W}^m$, and $\mathcal{F}^m$.

Spaces $\mathcal{W}_E^M$ and $\mathcal{F}_E^M$ can be chosen arbitrarily, provided that they are compatible with (9.15) and that $\mathcal{W}_E^M$ includes the trace of the rigid body modes on $\partial \Omega_E$ (which implies that $\mathcal{F}^M$ contains the self-balanced forces). Once these spaces have been chosen, the macro part $\underline{\dot{W}}_E^M$ of Field $\underline{\dot{W}}_E \in \mathcal{W}_E$ is defined by:

$$\forall \underline{F}^\star \in \mathcal{F}_E^M, \quad \int_{[0,T]\times\partial\Omega_E} (\underline{\dot{W}}_E^M - \underline{\dot{W}}_E) \cdot \underline{F}^\star dSdt = 0 \qquad (9.27)$$

and the macro part $\underline{F}_E^M$ of Field $\underline{F}_E \in \mathcal{F}_E$ by:

$$\forall \underline{W}^\star \in \mathcal{W}_E^M, \quad \int_{[0,T]\times\partial\Omega_E} (\underline{F}_E^M - \underline{F}_E) \cdot \underline{\dot{W}}^\star dSdt = 0 \qquad (9.28)$$

Consequently, the micro parts are $\underline{\dot{W}}_E^m = \underline{\dot{W}}_E - \underline{\dot{W}}_E^M$ and $\underline{F}_E^m = \underline{F}_E - \underline{F}_E^M$, and the scales are uncoupled as follows:

$$\int_{[0,T]\times\partial\Omega_E} \underline{\dot{W}}_E \cdot \underline{F}_E dSdt = \int_{[0,T]\times\partial\Omega_E} (\underline{\dot{W}}_E^M \cdot \underline{F}_E^M + \underline{\dot{W}}_E^m \cdot \underline{F}_E^m) dSdt \qquad (9.29)$$

For space, the macroscale is defined by the characteristic length of the interfaces, which is *a priori* much larger than the scale of the spatial discretization. For example, the macro parts are defined as affine functions on each interface $\Phi_{EE'}$.

For time, the macroscale is associated with a coarse partition $\mathcal{T}_h^M = \{0 = t_0^M, \ldots, t_{n^M}^M = T\}$ of the time interval $[0, T]$ being studied. Its characteristic time (i.e. the maximum length of a time step) chosen is much larger than the characteristic time of the initial time discretization $\mathcal{T}_h = \{0 = t_0, \ldots, t_n = T\}$. For example, the macro parts are defined as polynomials of degree p in each macro interval $I_k^M =]t_k^M, t_{k+1}^M[$. Let us note that the choice of functions which are possibly discontinuous implies that one should consider all the equations in the time-discontinuous Galerkin scheme sense [40].

The choices adopted for the definition of the macro quantities are physically sound: these quantities are mean values in time and in space. Fields $\underline{W}_E^M$ and $\underline{F}_E^M$ are written at each space-time point $(\underline{M}, t)$ of $\Phi_{EE'} \times I_k^M$ in the form $\sum_{i,j} \alpha_{ij} \underline{e}_j^M(\underline{M}) f_i^M(t)$, for which a choice of basis functions $\underline{e}_j^M$ and f_j^M is represented in Figures 9.3 and 9.4 in the case of a two-dimensional interface.

9.4.2 Admissibility of the macro quantities

An important feature of the multiscale computational strategy presented here is that the transmission conditions at the interfaces are partially verified *a priori*. The set of the macro forces $\underline{F}^M = (\underline{F}_E^M)_{\Omega_E \subset \Omega}$ is required to verify the transmission conditions systematically, including the boundary conditions:

$$\underline{F}_{EE'}^M + \underline{F}_{E'E}^M = \underline{0} \quad \text{on } \Phi_{EE'}$$
$$\underline{F}_{E2}^M + \underline{F}_d^M = \underline{0} \quad \text{on } \Phi_{E2} \qquad (9.30)$$

The corresponding subspace of $\mathcal{F}^M$ is designated by $\mathcal{F}_{\text{ad}}^M$. We also introduce $\mathcal{W}_{\text{ad}}^M$, the subspace of $\mathcal{W}^M$ whose elements are continuous at the interfaces and equal to the prescribed velocity $\underline{\dot{U}}_d$ on $\partial_1 \Omega$. The subspaces of $\mathcal{W}$ and $\mathcal{F}$ whose elements have their macro parts in $\mathcal{W}_{\text{ad}}^M$ and $\mathcal{F}_{\text{ad}}^M$ are designated by $\mathcal{W}_{\text{ad}}$ and $\mathcal{F}_{\text{ad}}$.

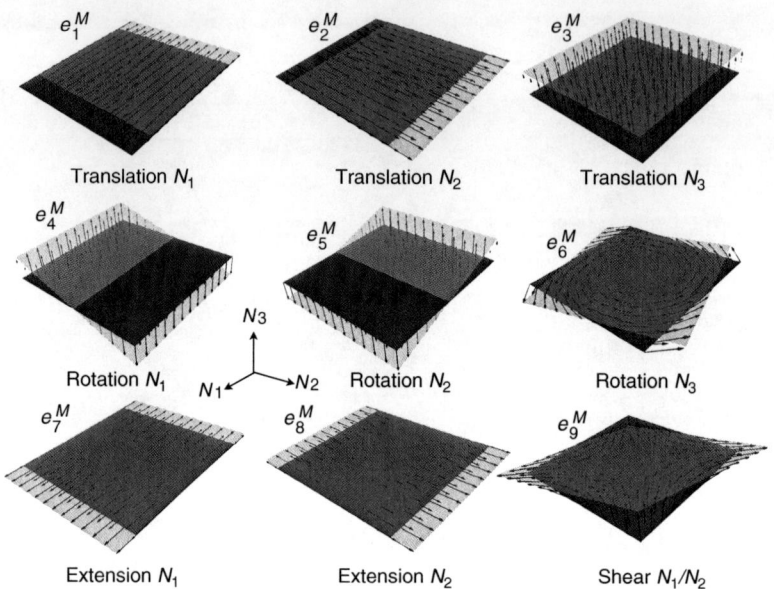

Figure 9.3: Space level: affine basis functions $\{\underline{e}_j^M\}_{j\in\{1,\ldots,9\}}$ for an interface $\Phi_{EE'}$.

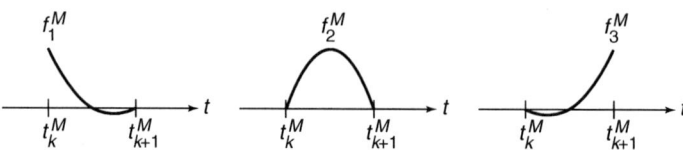

Figure 9.4: Time level: quadratic basis functions $(p=2)$ $\{f_i^M\}_{i\in\{1,\ldots,3\}}$ in Interval I_k^M.

9.5 The multiscale computational strategy

9.5.1 *The driving force of the strategy*

The decomposed reference problem, defined over the time-space domain $[0,T]\times\Omega$, can be formulated as follows:

Find $\mathbf{s_{ref}} = (\mathbf{s}_E)_{\Omega_E\subset\Omega}$ which verifies, $\forall\Omega_E\subset\Omega$,

(a) the E-admissibility condition $\qquad \mathbf{s}_E \in \mathbf{A_{dE}}$
(b) the admissibility of the macro forces $\underline{F} \in \mathcal{F}_{\mathbf{ad}}$
(c) the evolution law $\qquad \dot{\mathbf{e}}_{pE} = \mathbf{B}(\mathbf{f}_E)$ with $\mathbf{e}_{pE|t=0} = 0$
(d) the interface behavior $\qquad \forall\Omega_{E'}\in\mathbf{\Omega}_E,\ \underline{F}_{EE'} + \underline{F}_{E'E} = \underline{0}$ and

$$\underline{F}_{EE'|t} = \mathbf{b}_{EE'}\left(\left[\underline{\dot{W}}_{EE'} - \underline{\dot{W}}_{E'E}\right]_{|\tau}, \tau \leqslant t\right) \tag{9.31}$$

$\cdots \longrightarrow s_n \in A_d \xrightarrow{\text{local stage}} \hat{s}_{n+1/2} \in \Gamma \xrightarrow{\text{linear stage}} s_{n+1} \in A_d \longrightarrow \hat{s}_{n+3/2} \in \Gamma \longrightarrow \cdots \longrightarrow s_{\text{ref}}$

$\underbrace{\hspace{6cm}}_{\text{Iteration } n+1}$

Figure 9.5: Local stage and linear stage at Iteration $n+1$.

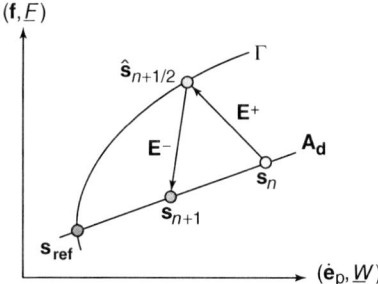

Figure 9.6: One iteration of the LATIN method.

The driving force of the strategy we are about to describe is the LATIN method [37], which is a general, mechanics-based computational strategy for the resolution of time-dependent nonlinear problems which works over the entire time-space domain. It has been successfully applied to a variety of problems: quasi-static and dynamic analysis, post-buckling analysis, analysis of highly heterogeneous systems [17, 39, 41], and multiphysics problems [25].

The first principle of the LATIN method consists in dealing with the difficulties separately by dividing the solutions of the equations into two independent subspaces: the space $\mathbf{A_d}$ of the solutions to the global linear equations (9.31a) and (9.31b) (defined on the level of the whole structure) and the space $\mathbf{\Gamma}$ of the solutions to the local nonlinear equations (9.31c) and (9.31d) (defined on the local level).

The second principle of the method consists in using an iterative scheme to obtain the solution of the problem, which can be interpreted as $s_{\text{ref}} = \mathbf{A_d} \cap \mathbf{\Gamma}$. One iteration consists of two stages, called the "local stage" and the "linear stage". As shown in Figure 9.5, these stages consist in building fields of $\mathbf{\Gamma}$ and $\mathbf{A_d}$ alternatively, an iterative process which, under conditions that will be described later, converges towards the solution s_{ref} of the problem. These stages will be analyzed in the following sections.

Figure 9.6 attempts to give a geometrical interpretation of the method in the space generated by $(\dot{\mathbf{e}}_p, \underline{W})$ and $(\mathbf{f}, \underline{F})$, by showing the sets of equations $\mathbf{A_d}$ and $\mathbf{\Gamma}$, and the "search directions" $\mathbf{E}^+$ and $\mathbf{E}^-$ which are introduced to converge toward the solution.

9.5.2 The local stage at Iteration $n+1$

This stage consists in building $\hat{s}_{n+1/2} = (\hat{s}_E)_{\Omega_E \subset \Omega} \in \mathbf{\Gamma}$ knowing $s_n = (s_E)_{\Omega_E \subset \Omega} \in \mathbf{A_d}$ and using an "ascent" search direction $\mathbf{E}^+$, followed by

$\hat{\mathbf{s}}_{n+1/2} - \mathbf{s}_n = D\mathbf{s}$ (see Figure 9.6). This search direction is defined by:

$$D\mathbf{s} = (D\mathbf{s}_E)_{\Omega_E \subset \Omega} \in \mathbf{E}^+ \iff \forall \Omega_E \subset \Omega, \begin{cases} D\dot{\mathbf{e}}_{pE} + \mathbf{H}D\mathbf{f}_E = 0 \\ D\underline{\dot{W}}_E - \mathbf{h}D\underline{F}_E = 0 \end{cases} \quad (9.32)$$

where $\mathbf{H}$ and $\mathbf{h}$ are symmetric, positive definite operators which are parameters of the method. This search direction can be redefined using a weak formulation for the substructure part:

$$\forall \mathbf{f}^\star \in \mathbf{F}, \quad \sum_{\Omega_E \subset \Omega} \int_{[0,T] \times \Omega_E} (D\dot{\mathbf{e}}_{pE} + \mathbf{H}D\mathbf{f}_E) \circ \mathbf{f}_E^\star d\Omega dt = 0 \quad (9.33)$$

and for the boundary part:

$$\forall \underline{F}^\star \in \mathcal{F}, \quad \sum_{\Omega_E \subset \Omega} \int_{[0,T] \times \partial \Omega_E} (D\underline{\dot{W}}_E - \mathbf{h}D\underline{F}_E) \cdot \underline{F}_E^\star dS dt = 0 \quad (9.34)$$

One can easily show that seeking $\hat{\mathbf{s}}_{n+1/2}$ common to Γ and $\mathbf{E}^+$ leads to the resolution of a set of problems which are local in the space variable (and, very often, also in the time variable), and, therefore, lend themselves to the highest degree of parallelism. This property justifies the term "local" to describe the stage.

9.5.3 The linear stage at Iteration $n + 1$

This stage consists in building $\mathbf{s}_{n+1} = (\mathbf{s}_E)_{\Omega_E \subset \Omega} \in \mathbf{A}_d$ knowing $\hat{\mathbf{s}}_{n+1/2} = (\hat{\mathbf{s}}_E)_{\Omega_E \subset \Omega} \in \Gamma$ and using a "descent" search direction $\mathbf{E}^-$, followed by $\mathbf{s}_{n+1} - \hat{\mathbf{s}}_{n+1/2} = D\mathbf{s}$ (see Figure 9.6). This search direction is defined by:

$$D\mathbf{s} = (D\mathbf{s}_E)_{\Omega_E \subset \Omega} \in \mathbf{E}^- \iff \forall \Omega_E \subset \Omega, \begin{cases} D\dot{\mathbf{e}}_{pE} - \mathbf{H}D\mathbf{f}_E = 0 \\ D\underline{\dot{W}}_E + \mathbf{h}D\underline{F}_E = 0 \end{cases} \quad (9.35)$$

This search direction can be redefined using a weak formulation for the substructure part:

$$\forall \mathbf{f}^\star \in \mathbf{F}, \quad \sum_{\Omega_E \subset \Omega} \int_{[0,T] \times \Omega_E} (D\dot{\mathbf{e}}_{pE} - \mathbf{H}D\mathbf{f}_E) \circ \mathbf{f}_E^\star d\Omega dt = 0 \quad (9.36)$$

and for the boundary part:

$$\forall \underline{F}^\star \in \mathcal{F}_{\mathbf{ad}}, \quad \sum_{\Omega_E \subset \Omega} \int_{[0,T] \times \partial \Omega_E} (D\underline{\dot{W}}_E + \mathbf{h}D\underline{F}_E) \cdot \underline{F}_E^\star dS dt = 0 \quad (9.37)$$

with the adjunction of the condition that the test function $\underline{F}^\star$ belongs to $\mathcal{F}_{\mathbf{ad}}$ instead of $\mathcal{F}$, which enables one to guarantee the admissibility of the macro forces. The last equation is reformulated with the introduction of a Lagrange multiplier $\underline{\dot{\tilde{W}}}^M$ ($\underline{\dot{\tilde{W}}}^M = (\underline{\dot{\tilde{W}}}_E^M)_{\Omega_E \subset \Omega} \in \mathcal{W}_{\text{ad}}^{M\star}$):

$$\forall \underline{F}^\star \in \mathcal{F}, \sum_{\Omega_E \subset \Omega} \left\{ \int_{[0,T] \times \partial \Omega_E} (D\underline{\dot{W}}_E + \mathbf{h} D\underline{F}_E) \cdot \underline{F}_E^\star dS dt \right.$$
$$\left. - \int_{[0,T] \times \partial \Omega_E} \underline{\dot{\tilde{W}}}_E^M \cdot \underline{F}_E^\star dS dt \right\} = 0 \quad (9.38)$$

and the admissibility of the macro forces is expressed by:

$$\forall \underline{\dot{\tilde{W}}}^{M\star} \in \mathcal{W}_{\text{ad}}^{M\star},$$

$$\sum_{\Omega_E \subset \Omega} \left\{ \int_{[0,T] \times \partial \Omega_E} \underline{\dot{\tilde{W}}}_E^{M\star} \cdot \underline{F}_E dS dt - \int_{[0,T] \times \Phi_{E2}} \underline{\dot{\tilde{W}}}_E^{M\star} \cdot \underline{F}_d dS dt \right\} = 0 \quad (9.39)$$

The resolution of the linear stage can be divided into two parts: the resolution of a set of micro problems defined over each time-space substructure $[0,T] \times \Omega_E$, and the resolution of a global macro problem defined over the entire time-space domain $[0,T] \times \Omega$.

9.5.3.1 *The micro problems defined over each $[0,T] \times \Omega_E$ and $[0,T] \times \partial \Omega_E$*
Each micro problem associated with Ω_E is a linear evolution equation:

Find $(\mathbf{s}_E)_{\Omega_E \subset \Omega}$ which verifies, $\forall \Omega_E \subset \Omega$,

- the E-admissibility condition $\mathbf{s}_E \in \mathbf{A}_{\mathbf{d}E}$ \quad (9.40)
- the search direction \quad (9.36, 9.38)

Since (9.38) is local at Boundary $\partial \Omega_E$, the micro problems in the substructures are independent of one another. Since $\mathbf{H}$ and $\mathbf{h}$ are positive definite operators, the micro problem defined over $[0,T] \times \Omega_E$ has a unique solution such that:

$$\mathbf{s}_E = \mathbf{s}_E^{(1)} + \mathbf{s}_E^{(2)}(\underline{\dot{\tilde{W}}}_E^M) \quad (9.41)$$

where $\mathbf{s}_E^{(1)}$ depends on the additional loading and on the previous approximation of the solution $\hat{\mathbf{s}}_E$, and $\mathbf{s}_E^{(2)}$ depends linearly on $\underline{\dot{\tilde{W}}}_E^M$, which is unknown at this stage. In particular, one has:

$$\underline{F}_E^M = \underline{\hat{F}}_{E,d}^M + \mathbf{L}_E \underline{\dot{\tilde{W}}}_E^M \quad (9.42)$$

where $\underline{\hat{F}}_{E,d}$ is due to the additional loading and to the previous approximation to the solution, and $\mathbf{L}_E$ is a linear operator which can be interpreted as a homogenized behavior operator over the time-space substructure $[0,T] \times \Omega_E$. This operator can be calculated by solving a set of micro problems over $[0,T] \times \Omega_E$ in which one takes successively for $\underline{\dot{\tilde{W}}}_E^M$ the macro basis functions of $\mathcal{W}_E^M$.

9.5.3.2 The macro problem defined over $[0, T] \times \Omega$

The macro problem defined over the entire time-space domain $[0, T] \times \Omega$ is:

Find $(\underline{\dot{W}}^M, \underline{F}^M)$ which verifies

- the admissibility of the Lagrange multiplier $\underline{\dot{W}}^M \in \mathcal{W}_{ad}^{M\star}$
- the admissibility of the macro forces $\quad \underline{F} \in \mathcal{F}_{ad}$ (9.43)
- the homogenized behavior (9.42)

Introducing (9.42) into the admissibility condition of the macro forces (9.39), then using the micro-macro uncoupling property (9.29), one has:

$$\forall \underline{\dot{W}}^{M\star} \in \mathcal{W}_{ad}^{M\star}, \sum_{\Omega_E \subset \Omega} \left\{ \int_{[0,T] \times \partial \Omega_E} \underline{\dot{W}}_E^{M\star} \cdot (\hat{\underline{F}}_{E,d}^M + \mathbf{L}_E \underline{\dot{W}}_E^M) dSdt \right.$$

$$\left. - \int_{[0,T] \times \Phi_{E2}} \underline{\dot{W}}_E^{M\star} \cdot \underline{F}_d dSdt \right\} = 0 \quad (9.44)$$

which corresponds to the resolution of a homogenized problem over the whole structure. If the number of macro time-space substructures is large, an approximation technique based on the introduction of a third scale can be used [36].

9.5.3.3 Resolution of the linear stage

The resolution of the linear stage proceeds as follows: first, one solves a series of micro problems, each defined over $[0, T] \times \Omega_E$, in which one takes into account only the data $\hat{\mathbf{s}}_E$ of the previous stage. This leads to $\mathbf{s}_E^{(1)}$. Then, one solves the macro problem defined over $[0, T] \times \Omega$, leading to $\underline{\dot{W}}^M$. Finally, in order to obtain $\mathbf{s}_E^{(2)}$, one solves a second series of micro problems with the Lagrange multiplier as the only data.

Since the macro mesh is defined in time and in space, the micro problems are independent not only from one substructure to another, but also from one macro time interval to another. One should note that the macro quantities are defined at the interfaces only. By treating the medium as a Cosserat material, one can define macro stresses, macro strains... inside a substructure Ω_E. Each cell is assumed to be homogeneous on the macroscale. Thus, macro quantities and conjugate quantities could be derived from the generalized forces and displacements at the interfaces, which would lead to a nonconventional Cosserat-like material.

9.5.4 Choice of the parameters $(\mathbf{H}, \mathbf{h})$ and convergence of the algorithm

Following the proof given in [37], one can prove that the quantity $\frac{1}{2}(\mathbf{s}_{n+1} + \mathbf{s}_n)$ converges towards $\mathbf{s}_{ref}$, the solution of Problem (9.31). The choice of the parameters $(\mathbf{H}, \mathbf{h})$ influences only the convergence of the algorithm, but does not affect the solution.

To ensure the convergence of $\mathbf{s}_n$ and, more generally, to ensure convergence for many types of material behavior, a relaxation technique may be needed. Renaming $\bar{\mathbf{s}}_{n+1}$ the quantity previously denoted $\mathbf{s}_{n+1}$, we redefine $\mathbf{s}_{n+1}$, the approximation generated by the linear stage $n + 1$, as:

$$\mathbf{s}_{n+1} = \mu \bar{\mathbf{s}}_{n+1} + (1 - \mu)\mathbf{s}_n \tag{9.45}$$

where μ is a relaxation parameter usually equal to 0.8.

In the case of linear behavior, one can choose, for example, $\mathbf{H} = \mathbf{B}$ and $\mathbf{h} = \frac{L}{ET}\mathbf{I}$, where E is the Young's modulus of the material, L_E a characteristic length of the interfaces, T the duration of the phenomenon being studied, and $\mathbf{I}$ the identity operator. Other possible choices, especially in the nonlinear case, are discussed in [37].

Since the reference solution $\mathbf{s}_{\mathbf{ref}}$ is the intersection of $\mathbf{\Gamma}$ and $\mathbf{A_d}$, the distance between $\hat{\mathbf{s}}_{n+1/2}$ and $\mathbf{s}_n$ is a good error indicator to verify the convergence of the algorithm [42]. The simplest measure of this distance is:

$$\eta = \frac{\|\hat{\mathbf{s}}_{n+1/2} - \mathbf{s}_n\|}{\frac{1}{2}\|\hat{\mathbf{s}}_{n+1/2} + \mathbf{s}_n\|} \tag{9.46}$$

with:

$$\|\mathbf{s}\|^2 = \frac{1}{2} \sum_{\Omega_E \subset \Omega} \int_{[0,T] \times \Omega_E} \left(1 - \frac{1}{T}\right)(\dot{\mathbf{e}}_{pE} \circ \mathbf{H}^{-1}\dot{\mathbf{e}}_{pE} + \mathbf{f}_E \circ \mathbf{H}\mathbf{f}_E)d\Omega dt \tag{9.47}$$

9.5.5 First example

Let us consider the 3D problem of a composite structure containing cracks (see Figure 9.7(a)). The structure is fixed at the bottom and subjected to forces $\underline{F}_1$,

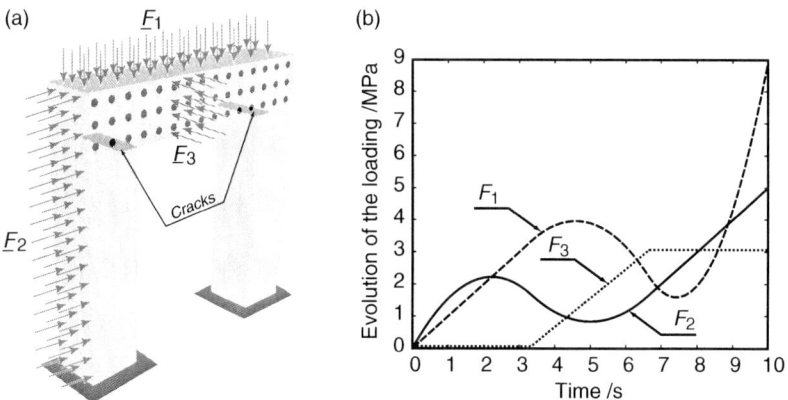

Figure 9.7: Description of the problem. (a) Geometry and loading; (b) Loads $\underline{F}_1$, $\underline{F}_3$, and $\underline{F}_3$

$\underline{F}_3$, and $\underline{F}_3$ (see Figure 9.7(b)). The overall dimensions are $120 \times 120 \times 20$ mm, and the time interval being studied is $T = 10$ s. The cracks are described using unilateral contact with Coulomb friction characterized by Parameter $f = 0.3$.

The structure consists of two types of cells: Type-I cells are homogeneous, made of Type-1 material; Type-II cells consist of a matrix of Type-1 material with inclusions of Type-2 material. Type-1 and Type-2 materials are viscoelastic and their properties are given in Table 9.1. The corresponding constitutive relations are $\dot{\boldsymbol{\varepsilon}}_p = \mathbf{B}_i \boldsymbol{\sigma} = \frac{1}{\eta_i} \mathbf{K}_i^{-1} \boldsymbol{\sigma}$.

The problem was divided into 351 substructures and 1296 interfaces as shown on Figure 9.8, each substructure corresponding to one cell. On the micro level, Type-I and Type-II substructures and interfaces were meshed with 847, 717, and 144 degrees of freedom (DOFs), respectively. The distinction between the macroscale and the microscale was made only at the interfaces and the macro part consisted of a single linear element with only 9 DOFs per interface (see Figure 9.3). With respect to time, the micro level was associated with a refined discretization into 60 intervals using a zero-order discontinuous Galerkin scheme, and the macro level was associated with a coarse discretization into three macro intervals using a second-order discontinuous Galerkin scheme.

Table 9.1. Material properties

Material	Type-1	Type-2
Young's modulus	$E_1 = 50$ GPa	$E_2 = 250$ GPa
Poisson's ratio	$\nu_1 = 0.3$	$\nu_2 = 0.2$
Viscosity parameter	$\eta_1 = 10$ s	$\eta_2 = 1000$ s

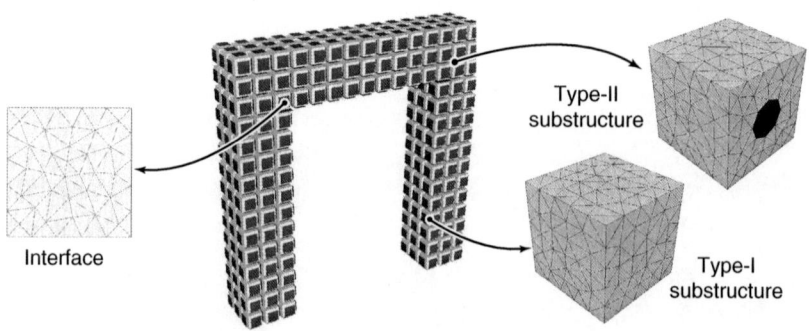

Figure 9.8: Decomposition and microscale discretizations in space.

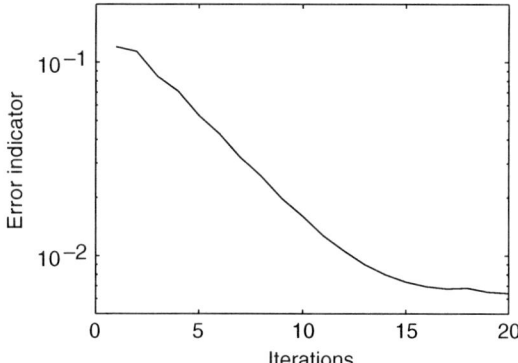

Figure 9.9: Convergence of the method.

Since the constitutive relation is linear, the search direction chosen for the substructures was $\mathbf{H} = \mathbf{B}$. The characteristic length of the interfaces being $L_E = 4$ mm, we chose for all the interfaces the search direction $\mathbf{h} = h\mathbf{I}$, where $h = \frac{L_E}{E_1 \nu_1}$ is a constant scalar.

Figure 9.9 shows the evolution of the error indicator η throughout the iterations. One can observe that the algorithm converges rapidly toward an accurate solution (1% error after 12 iterations). Figure 9.10 shows the approximate Von Mises stress field over the structure (with a zoom near one of the cracks) at the final time $T = 10$ s for Iterations 1, 5, and 20, and after convergence (the reference solution). The evolution over time of the displacement field $\underline{W}$ at Point P_2 is also represented. One can observe that thanks to the resolution of a macro problem the method leads, even on the first iteration, to a rather good approximation of the solution of the problem over both the space and time domains. After a few iterations, the solution becomes even more accurate and the stress and displacement discrepancies tend to zero. After 20 iterations, the difference between the approximate solution and the reference solution is no longer visible.

An example of the micro/macro description of the solution is given in Figure 9.11. Figures 9.11(a) and 9.11(b) show the evolutions of Force $\underline{F}$ and its macro part $\underline{F}^M$, respectively, at time $t = 2/3T$ over a horizontal line L_1 in the heterogeneous part of the structure, and as functions of time at a point P_1 of the previous line. Figures 9.11(a) and 9.11(b) show the same evolutions for Displacement $\underline{W}$ and its macro part $\underline{W}^M$.

One can observe that the macro part of the quantities being considered constitutes a good average approximation of the solution, obtained with only a very small number of basis functions (27 DOFs per interface and per macro interval). The choice of such a basis leads to the resolution at each iteration of a macro problem, with a strong mechanical meaning and with only a few DOFs (in this example, 35,000 DOFs compared to 270,000 DOFs for the assembled reference problem).

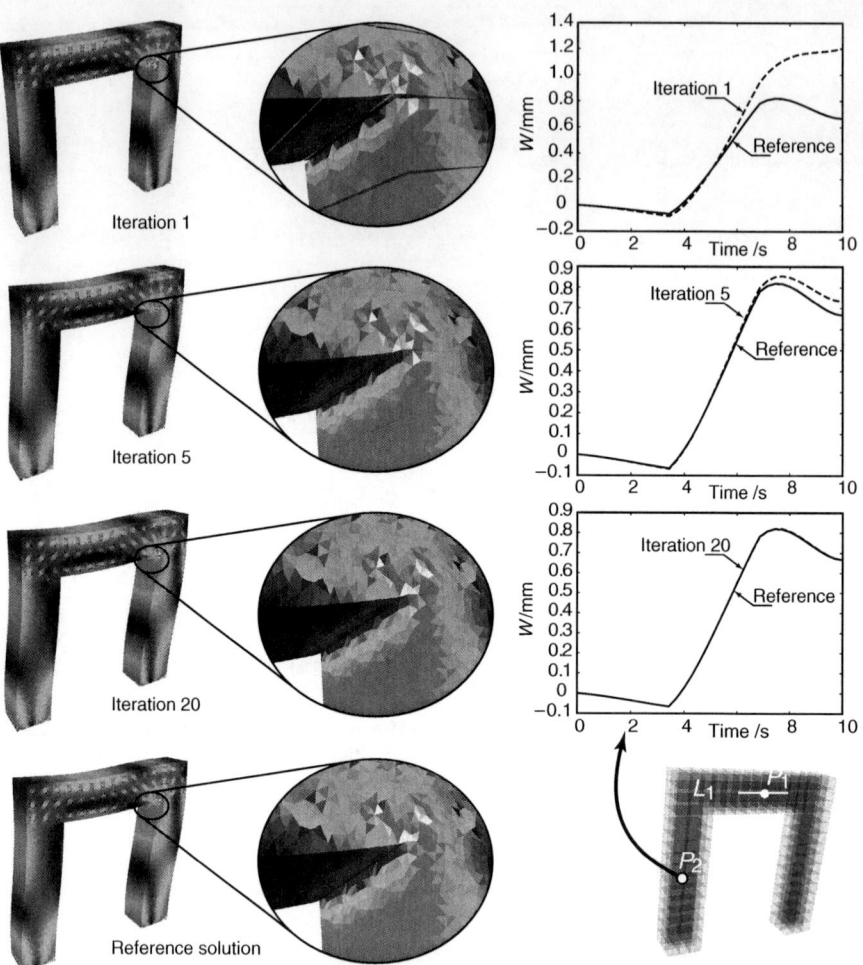

Figure 9.10: Approximate solutions throughout the iterations.

9.6 The radial time-space approximation

The global structure is decomposed into several substructures. Throughout the iterative process, one has to solve for each of these substructures a set of micro problems which represent the equations defined over the corresponding time-space domains. The cost of solving these problems with standard methods can be prohibitive, which led us to the development of what we call the "radial time-space approximation".

The radial time-space approximation was introduced by Ladevèze in 1985 ([43], see also [34, 37]) and is commonly used in the LATIN method. This is the third principle of the LATIN method and it is indeed what makes it so efficient.

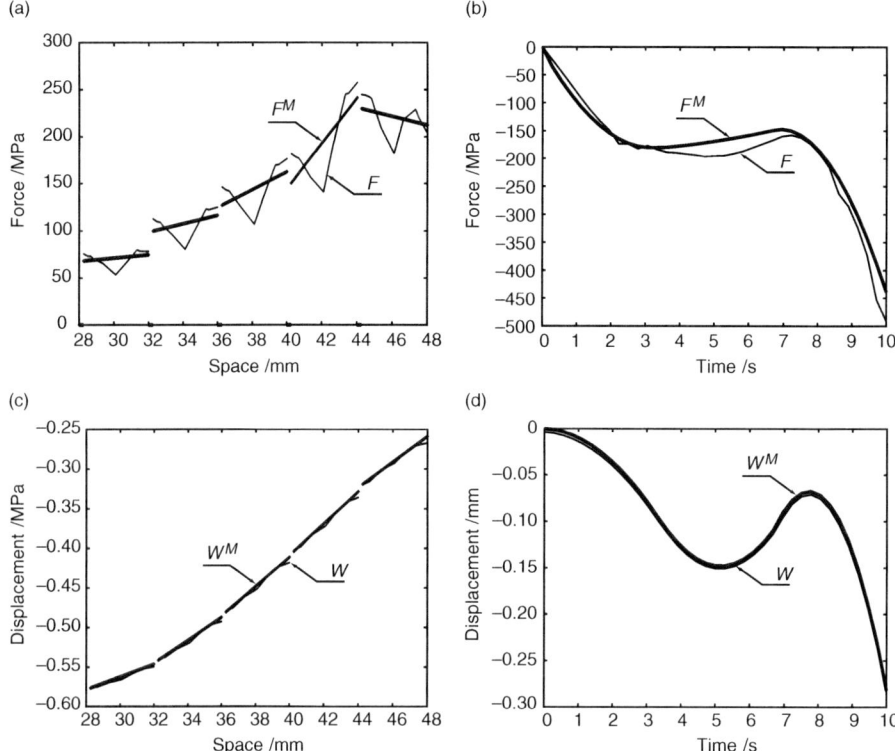

Figure 9.11: Micro/macro description of the solution. (a) $\underline{F}$ and $\underline{F}^M$ at $t = 2/3T$ over Line L_1; (b) $\underline{F}$ and $\underline{F}^M$ at Point P_1 as functions of time; (c) $\underline{W}$ and $\underline{W}^M$ $t = 2/3T$ over Line L_1; (d) $\underline{W}$ and $\underline{W}^M$ at Point P_1 as functions of time.

It was shown in previous works that under the small-displacement assumption this approach reduces the computational cost drastically. The basic idea consists in approximating a function f defined over the time-space domain $[0, T] \times \Omega$ by a finite sum of products of time functions λ_i by space functions Λ_i:

$$\forall (t, M) \in [0, T] \times \Omega, \quad f(t, M) : \sum_i \lambda_i(t) \Lambda_i(\underline{M}) \qquad (9.48)$$

where the products $\lambda_i(t) \Lambda_i(\underline{M})$ are called "radial time-space functions". It is important to note that this is not a spectral decomposition because neither the λ_i nor the Λ_i are known *a priori*.

The starting point of the radial time-space approximation is the radial loading approximation, defined by a single product, which is very well known and commonly used in (visco-)plasticity. This type of approximation could also be seen by replacing the time variable by space variables or stochastic variables. Such developments have been proposed in [44] for radial hyperreduction, in [45, 46] for the resolution of fundamental physics problems, and in [47] for the resolution of stochastic problems. In a certain way, one can say that such approximations belong to the "Proper Orthogonal Decomposition" class of problems [48].

9.6.1 General properties

Let f be a known scalar function defined over the time-space domain $[0,T] \times \Omega$, and let us study the best m^{th}-order time-space approximation of Function f:

$$f_p(t, \underline{M}) = \sum_{i=1}^{p} \lambda_i(t) \Lambda_i(\underline{M}) \qquad (9.49)$$

The following scalar products are introduced:

$$\langle f, g \rangle_{[0,T] \times \Omega} = \int_{[0,T] \times \Omega} fg \, d\Omega dt, \quad \langle f, g \rangle_{[0,T]} = \int_{[0,T]} fg \, dt, \quad \langle f, g \rangle_{\Omega} = \int_{\Omega} fg \, d\Omega \qquad (9.50)$$

It was shown in [37] that the best approximation with respect to the $\|\cdot\|_{[0,T] \times \Omega}$-norm is the result of an eigenvalue problem whose eigenfunctions are the time functions λ_i. This problem can be rewritten as the stationarity of the Rayleigh quotient:

$$R(\lambda) = \frac{\|\langle f, \lambda \rangle_{[0,T]}\|_{\Omega}^2}{\|\lambda\|_{[0,T]}^2} \qquad (9.51)$$

It was also proved in [37] that if $[0,T] \times \Omega$ is the space such that f and $\dot{f}$ belong to $L^2([0,T], L^2(\Omega))$, the eigenvalue problem has a countable sequence of eigensolutions $(\alpha_i^{-1}, \lambda_i)$ where the eigenvalues α_i^{-1} are positive and the eigenfunctions λ_i are orthogonal.

The time functions λ_i having been determined, the corresponding space functions Λ_i are:

$$\Lambda_i = \frac{\langle f, \lambda_i \rangle_{[0,T]}}{\|\lambda_i\|_{[0,T]}^2} \qquad (9.52)$$

The following convergence property is verified:

$$\|f - f_p\|_{[0,T] \times \Omega} \xrightarrow[p \to +\infty]{} 0 \qquad (9.53)$$

and a simple measure of the relative error is:

$$\eta_p = \frac{\|f - f_p\|_{[0,T] \times \Omega}}{\|\frac{1}{2}(f + f_p)\|_{[0,T] \times \Omega}} \qquad (9.54)$$

9.6.2 Illustration

In order to illustrate the relevance of the previous time-space approximation, let us take as an example the case of a randomly-obtained irregular function f defined over a time-space interval $[0,T] \times [0,L]$. Figure 9.12 shows Function f along with its first-, second-, and third-order approximations. The relative error achieved with only three radial functions was less than 1%, which gives an idea of the remarkable accuracy of the proposed time-space approximation.

9.6.3 Practical implementation

Working with the radial time-space description alone constitutes a very convenient framework in which the storage requirement is drastically reduced. Here, we are following [49] to show the potential of this framework.

Let us divide the time interval $[0,T]$ being studied into m subintervals $\{I_i\}_{i=1,\ldots,m}$ of lengths $\{\Delta t_i\}_{i=1,\ldots,m}$, as shown in Figure 9.13. The centers

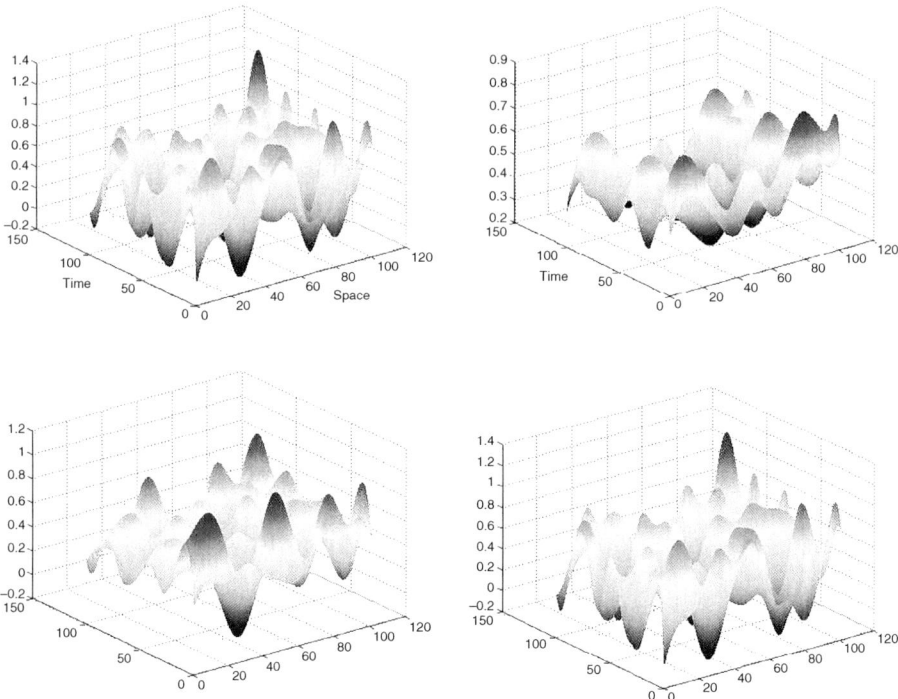

Figure 9.12: Time-space approximations of an irregular function f. (a) Irregular time-space function f; (b) First-order approximation f_1: error $\eta_1 = 3.9\%$; (c) Second-order approximation f_2: error $\eta_2 = 1.5\%$; (d) Third-order approximation f_3: error $\eta_3 = 0.6\%$.

$\{t_i\}_{i=1,\ldots,m}$ of these subintervals are called "reference times" and one has $I_i = [t_i - \Delta t_i/2, t_i + \Delta t_i/2]$.

For space, let us also introduce m' points $\{\underline{M}_j\}_{j=1,\ldots,m'}$ and partition Ω into $\{\Omega_j\}_{j=1,\ldots,m'}$, as shown in Figure 9.14. These points are called "reference points" and the measures of the subdomains are denoted $\{\omega_j\}_{i=j,\ldots,m'}$. In practice, there should be about a few dozen reference points.

The choice of these reference times and points is unrelated to the classical discretizations of the time interval $[0,T]$ and domain Ω. Refined time and space discretizations should still be used for the calculation of the various quantities. What we are doing here describes a field f over the time-space domain $[0,T] \times \Omega$ through:

$$\hat{a}_i^j(t) = \begin{cases} f(t, \underline{M}_j) & \text{if } t \in I_i \\ 0 & \text{otherwise} \end{cases} \quad \text{and} \quad \hat{b}_i^j(\underline{M}) = \begin{cases} f(t_i, \underline{M}) & \text{if } \underline{M} \in \Omega_j \\ 0 & \text{otherwise} \end{cases} \quad (9.55)$$

for $i = 1, \ldots, m$ and $j = 1, \ldots, m'$.

The sets $\{(\hat{a}_i^j, \hat{b}_i^j)\}_{i=1,\ldots,m}^{j=1,\ldots,m'}$ are the generalized components of f. One should note that these quantities verify the following compatibility conditions: for $i = 1, \ldots, m$ and $j = 1, \ldots, m'$,

$$\hat{a}_i^j(t_i) = \hat{b}_i^j(\underline{M}_j) \quad (9.56)$$

The main question is then how to build or rebuild a field from its components. We choose to define Function f from its components by only one product per time-space subdomain $I_i \times \Omega_j$:

$$f(t, \underline{M}) : a_i^j(t) b_i^j(\underline{M}) \quad \forall (t, \underline{M}) \in I_i \times \Omega_j \quad (9.57)$$

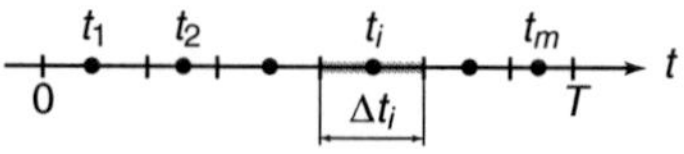

Figure 9.13: The reference times in $[0,T]$.

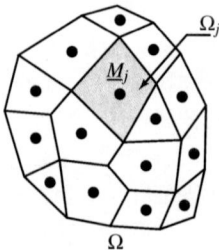

Figure 9.14: The reference points in Ω.

where the sets $\{(a_i^j, b_i^j)\}_{i=1,\ldots,m}^{j=1,\ldots,m'}$ should be defined from the sets $\{(\hat{a}_i^j, \hat{b}_i^j)\}_{i=1,\ldots,m}^{j=1,\ldots,m'}$. However, here, we let the time domain play a special role because there are many more spatial degrees of freedom than time degrees of freedom. Then, Function f is defined by:

$$f(t, \underline{M}) : a_i(t) b_i(\underline{M}) \quad \forall (t, \underline{M}) \in I_i \times \Omega \tag{9.58}$$

Let us introduce the following scalar products:

$$\langle f, g \rangle_{I_i} = \int_{I_i} fg \, dt \quad \text{and} \quad \langle f, g \rangle_{\Omega_j} = \int_{\Omega_j} fg \, d\Omega \tag{9.59}$$

In order to get the sets $\{(a_i, b_i)\}_{i=1,\ldots,m}$, we minimize:

$$J(a_i, b_i) = \sum_{j=1}^{m'} \left[\omega_j \|\hat{a}_i^j(t) - a_i(t) b_i(\underline{M}_j)\|_{I_i}^2 + \Delta t_i \|\hat{b}_i^j(\underline{M}) - a_i(t_i) b_i(\underline{M})\|_{\Omega_j}^2 \right] \tag{9.60}$$

which leads to:

$$a_i(t) = \frac{\sum_{j=1}^{m'} \omega_j \hat{a}_i^j(t) b_i(\underline{M}_j)}{\sum_{j=1}^{m'} \omega_j b_i^2(\underline{M}_j)} \quad \text{and} \quad b_i(\underline{M}) = \frac{\sum_{j=1}^{m'} \hat{b}_i^j(\underline{M})}{m' a_i(t_i)} \tag{9.61}$$

Consequently, $\forall (t, \underline{M}) \in I_i \times \Omega_j$, we obtain:

$$f(t, \underline{M}) : a_i(t) b_i(\underline{M}) = \frac{\sum_{k=1}^{m'} \omega_k \hat{a}_i^k(t) \hat{b}_i^k(\underline{M}_k)}{\sum_{k=1}^{m'} \omega_k \hat{b}_i^k(\underline{M}_k) \hat{b}_i^k(\underline{M}_k)} \hat{b}_i^j(\underline{M}) \tag{9.62}$$

Then, using the compatibility conditions (9.56), we get:

$$f(t, \underline{M}) : a_i(t) b_i(\underline{M}) = \frac{\sum_{k=1}^{m'} \omega_k \hat{a}_i^k(t) \hat{a}_i^k(t_i)}{\sum_{k=1}^{m'} \omega_k \hat{a}_i^k(t_i) \hat{a}_i^k(t_i)} \hat{b}_i^j(\underline{M}) \tag{9.63}$$

9.6.4 Reformulation of the linear stage at Iteration $n+1$

9.6.4.1 Rewriting of a micro problem over $[0, T] \times \Omega_E$

We choose to rewrite the linear stage at Iteration $n+1$ as an incremental correction $\Delta \mathbf{s}$ to the previous approximation $\mathbf{s}_n$, so that the new approximation to the solution is $\mathbf{s}_{n+1} = \mathbf{s}_n + \Delta \mathbf{s}$. If the initial solution $\mathbf{s}_0$ (for example, the solution of a linear elastic calculation) belongs to $\mathbf{A_d}$, then all the corrections are sought in $\mathbf{A_d^\star}$, the space which corresponds to $\mathbf{A_d}$ with homogeneous conditions.

For each $[0, T] \times \Omega_E$, the search direction (9.35) can be interpreted as a linear constitutive relation. Thus, an equivalent formulation consists in minimizing the global constitutive relation error in $\mathbf{A_{dE}^\star}$, which is defined over the time-space substructure $[0, T] \times \Omega_E$. Then, rewriting

$$D\mathbf{s}_E = \mathbf{s}_{E,n+1} - \hat{\mathbf{s}}_{E,n+1/2} = \Delta \mathbf{s}_E - (\hat{\mathbf{s}}_{E,n+1/2} - \mathbf{s}_{E,n}) \tag{9.64}$$

where, at this stage, $(\hat{\mathbf{s}}_{n+1/2} - \mathbf{s}_n)$ is a known quantity, we must solve:

$$\Delta \mathbf{s}_E = \underset{\Delta \mathbf{s}_E \in \mathbf{A}_{dE}^\star}{\text{Arg min}} \; e_{RC,E}^2(\Delta \mathbf{s}_E - (\hat{\mathbf{s}}_{E,n+1/2} - \mathbf{s}_{E,n})) \tag{9.65}$$

where the constitutive relation error is:

$$e_{RC,E}^2(D\mathbf{s}_E) = \|D\dot{\mathbf{e}}_E - \mathbf{H} D\mathbf{f}_E\|_{\mathbf{H},E}^2 + \left\|D\underline{\dot{W}}_E + \mathbf{h} D\underline{F}_E\right\|_{\mathbf{h},E}^2 \tag{9.66}$$

the corresponding norms are:

$$\|\Box\|_{\mathbf{H},E}^2 = \int_{[0,T]\times\Omega_E} (1 - \frac{t}{T})\Box \circ \mathbf{H}^{-1} \Box \, d\Omega dt \tag{9.67}$$

and:

$$\|\Box\|_{\mathbf{h},E}^2 = \int_{[0,T]\times\partial\Omega_E} (1 - \frac{t}{T})\Box \cdot \mathbf{h}^{-1} \Box \, dS dt \tag{9.68}$$

9.6.4.2 Choice of admissible radial time-space functions

The choice of the approximation presented here is an improvement over the version introduced in [36]. The starting point is the introduction as unknowns of the radial time-space approximations of the corrections related to the inelastic strain and to the additional internal variables:

$$\begin{aligned}\Delta \varepsilon_{pE}(t, \underline{M}) &= \sum_{k=1}^{p} a^k(t) \mathbf{E}_{\mathbf{p}}^k(\underline{M}) \\ \Delta \mathbf{X}_E(t, \underline{M}) &= \sum_{k=1}^{p'} b^k(t) \mathbf{D}^k(\underline{M})\end{aligned} \tag{9.69}$$

Using the E-admissibility conditions, one determines the other quantities of interest in terms of the previous unknowns:

$$\begin{aligned}(\Delta \varepsilon_E, \Delta \underline{W}_E)(t, \underline{M}) &= \sum_{k=1}^{p} a^k(t)(\mathbf{E}^k, \underline{Z}^k)(\underline{M}) \\ (\Delta \sigma_E, \Delta \underline{F}_E)(t, \underline{M}) &= \sum_{k=1}^{p} a^k(t)(\mathbf{C}^k, \underline{G}^k)(\underline{M}) \\ \Delta \mathbf{Y}_E(t, \underline{M}) &= \sum_{k=1}^{p'} b^k(t) \mathbf{R}^k(\underline{M})\end{aligned} \tag{9.70}$$

where the space functions are linked by the relations:

$$\mathbf{E}^k = \mathbf{E}_{\mathbf{p}}^k + \mathbf{K}^{-1} \mathbf{C}^k \quad \text{and} \quad \mathbf{R}^k = \Lambda \mathbf{D}^k \tag{9.71}$$

and the space operators are defined through standard finite element approximation over the space domain Ω_E.

Compared to the previous version of the radial loading time-space approximation, we obtain the same quality of approximation with only half the number of time functions.

9.6.4.3 Definition of the best approximation
In order to solve (9.65), the idea is to seek minima alternatively with respect to time (which leads to a system of differential equations) and to space (which leads to a "spatial" problem). Since the construction of the space functions is by far the most expensive step of this process, it is advantageous to store and re-use these functions. Thus, the space functions constructed up to Iteration n are re-used systematically during Iteration $n+1$. Let us note that a reduced basis can be shared by several substructures if these substructures are similar.

9.6.4.4 Practical resolution technique
Let us assume that we are dealing with Iteration $n+1$ and that we have at our disposal a reduced basis made up of the space functions $\{(\mathbf{E}_{\mathbf{p}}^k, \mathbf{D}^k)\}_{k=1,\ldots,m}$ for the approximation of the corrections related to the inelastic strain $\Delta \varepsilon_{\mathrm{p}E}$ and to the additional internal variables $\Delta \mathbf{X}_E$. The space functions related to the other quantities $\Delta \varepsilon_E$, $\Delta \underline{W}_E$, $\Delta \boldsymbol{\sigma}_E$, and $\Delta \underline{F}_E$ are also considered to be known.

Step 1: use of the reduced basis. One introduces the approximation:

$$\Delta \varepsilon_{\mathrm{p}E}(t, \underline{M}) = \sum_{k=1}^{m} a^k(t) \mathbf{E}_{\mathbf{p}}^k(\underline{M})$$
$$\Delta \mathbf{X}_E(t, \underline{M}) = \sum_{k=1}^{m} b^k(t) \mathbf{D}^k(\underline{M})$$
(9.72)

into the constitutive relation error (9.66) where only the time functions are the unknowns. Thus:

$$a^k(0) = b^k(0) = 0, \quad k = 1, \ldots, m \qquad (9.73)$$

These time functions verify a linear differential equation in time with conditions at $t = 0$ and $t = T$, whose solution is obtained classically. This is generally a rather small system. If the value of the constitutive relation error is small enough, one stops the process and selects the approximation obtained. Otherwise, one proceeds to *Step 2*.

Step 2: adding new functions. One adds:

$$\Delta \varepsilon_{\mathrm{p}E}(t, \underline{M}) = \sum_{k=1}^{m+r} a^k(t) \mathbf{E}_{\mathbf{p}}^k(\underline{M})$$
$$\Delta \mathbf{X}_E(t, \underline{M}) = \sum_{k=1}^{m+r} b^k(t) \mathbf{D}^k(\underline{M})$$
(9.74)

where both $\{(a^k, b^k)\}_{k=m+1,\ldots,m+r}$ and $\{(\mathbf{E_p^k}, \mathbf{D}^k)\}_{k=m+1,\ldots,m+r}$ are now unknowns. In practice, one takes $r = 1$. One seeks a minimum alternatively over the time functions and the space functions. These subiterations begin with an initialization of the time functions. In order to do that, one uses the residue written in terms of the reference points and reference times. The minimization with respect to the space functions is standard, with twice the size of a classical finite element calculation. The minimization with respect to the time functions leads to a differential equation with conditions at $t = 0$ and $t = T$, which can be easily solved using a standard technique. In practice, one stops after one or two subiterations. What is important is the complete calculation of the reduced spatial basis.

9.6.5 Numerical example of the resolution of a micro problem

In order to illustrate how this technique is used in the multiscale strategy, let us go back to the example described in Section 9.5.5 and use the radial time-space approximation to represent the unknowns of a problem similar to the micro problem associated with a Type-II substructure in which the loading consists of the distribution of the Lagrange multiplier $\tilde{\underline{W}}_E^M$ alone. For the sake of simplicity, we assume that this loading consists of only a normal force distribution $f(t)$ over the top surface of the substructure (see Figure 9.15).

Figure 9.16 shows the evolution of the constitutive relation error associated with the search direction with respect to the number of functions, using two techniques: the first technique consisted in systematically building new pairs of time/space functions; the second technique consisted in first re-using the basis of space functions previously calculated to update the time functions alone, and only then seeking a new pair of time/space functions. One can see that the accuracy of the approximation is very good because the error was less than 1% using only four radial functions. However, one can observe that the convergence rate of the second technique is higher than that of the first. For example, in order

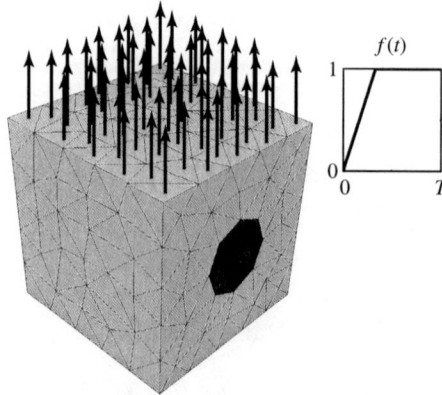

Figure 9.15: Description of the micro problem and its loading.

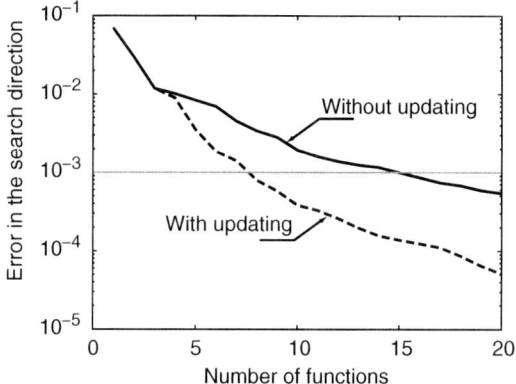

Figure 9.16: Convergence of the approximation.

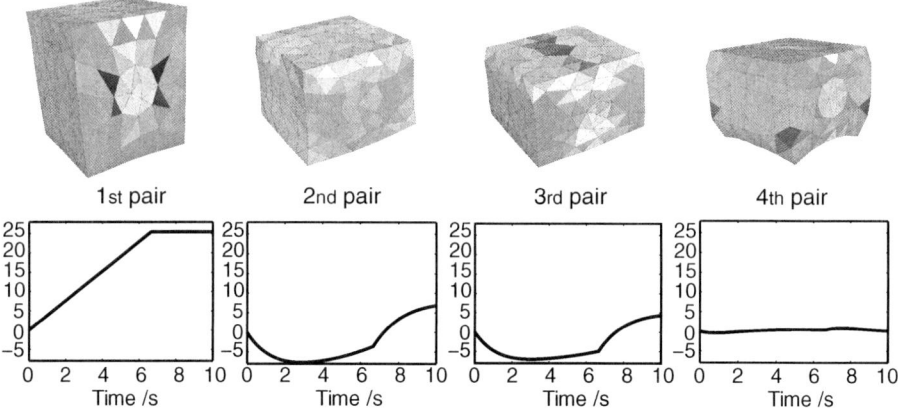

Figure 9.17: The first four radial time-space functions for the problem.

to get less than 0.1% error, one needs to calculate 15 functions if one does not update the time functions, as opposed to only eight functions with the updating procedure. Since the computing cost of the updating stage is much less than that of another space function, it is very important to update the time functions systematically.

Figure 9.17, shows the first four pairs, each constituted by one space function and one time function. The space functions are normalized and, thus, one can observe a decrease in the level of the corresponding time function. Figure 9.18 gives a comparison of the radial time-space approximation and the classical incremental solution in terms of Von Mises stresses over the space and time domains.

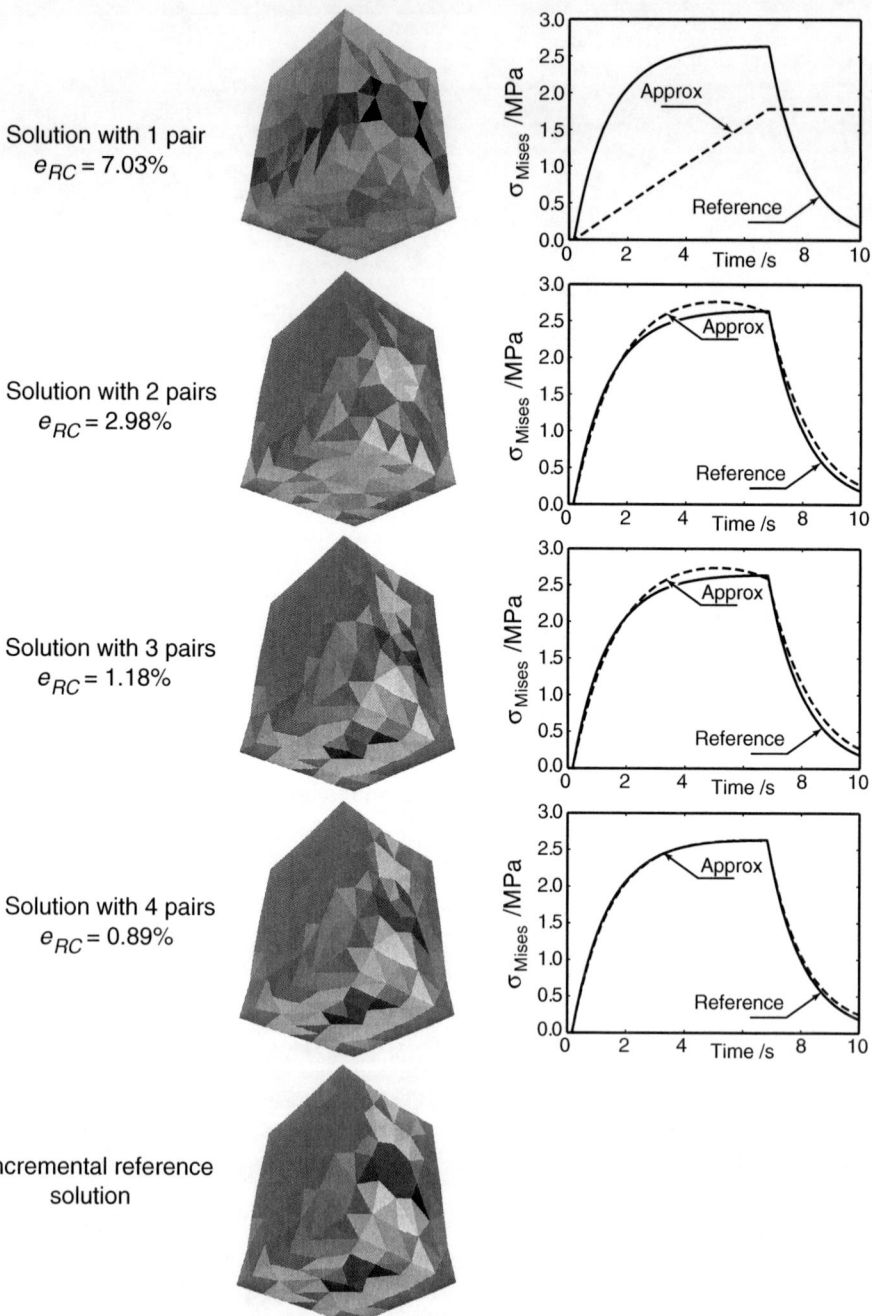

Figure 9.18: Quality of the approximations with 1, 2, 3, and 4 pairs.

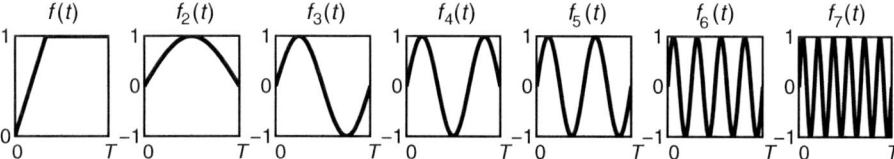

Figure 9.19: The different loading cases for the micro problem.

Table 9.2. Re-use of a space function with the radial time-space approximation

Loading case	$f(t)$	$f_2(t)$	$f_3(t)$	$f_4(t)$	$f_5(t)$	$f_6(t)$	$f_7(t)$
Error e_{CR}	0.179%	0.183%	0.239%	0.291%	0.332%	0.411%	0.434%

A very important point is that the basis of space functions is *a priori* specific to the problem and the loading for which it is defined, but it can be re-used to solve another problem with comparable accuracy. For example, we solved the previous example with six functions and re-used these functions for all the loading cases of Figure 9.19. In order to do that, we carried out a single update stage and evaluated the corresponding error.

Table 9.2 shows that by updating the time functions alone using the same space functions as for a previous problem $f(t)$, one obtains an approximate solution of the new problem $f_i(t)$ with an accuracy comparable to that of the first problem. The robustness of the radial time-space approximation makes it well adapted to multiresolution. Thus, this approximation technique is quite suitable for the multiscale strategy, which involves the resolution of a set of micro problems at each iteration of the LATIN method. We can re-use the same basis for every iteration of these micro problems, and even consider using a common basis for the whole set of substructures.

The coupling of the multiscale time and space aspects of Section 9.4 with the new version of the radial time-space approximation of Section 9.6 is being developed. Nevertheless, examples of the capabilities of the method have already been given in [36, 38].

9.7 Conclusions

The first version of the multiscale computational strategy described here has been applied to several large-scale engineering problems involving multiple scales, such as the prediction of damping in space launcher joints [50] or the simulation of microcracking in composite materials. The radial time-space approximation leads to a drastic reduction in calculation and storage costs, especially with the new version described here, without affecting the robustness and the effectiveness of the method. This, however applies only to quasi-static problems. A first improvement, currently in progress, consists in the introduction of a new algebra, i.e.

a general mathematical framework, in which all functions are described thanks to the radial time-space approximation. Another work in progress is the extension of the scalability proven for multiple space scales to the general case. The final development, still under quasi-static conditions, will deal with the extension to large-displacement problems following the mathematical framework already proposed in [37].

References

[1] Cresta P., Allix O., Rey C., and Guinard S. (2007). "Comparison of multiscale nonlinear strategies for post-buckling analysis", *Computer Methods in Applied Mechanics and Engineering*, **196**(8), p. 1436–1446.

[2] Guidault P.-A., Allix O., Champaney L., and Navarro J.-P. *A micro-macro approach for crack propagation with local enrichment*, in: Proceedings of 7th International Conference on Computational Structures Technology, CST 2004, Lisbon, Portugal, 7–10 September, 2004.

[3] Ladevèze P. (2004). "Multiscale modelling and computational strategies for composites", *International Journal for Numerical Methods in Engineering*, **60**, p. 223–253.

[4] Ladevèze P., Lubineau G., and Violeau D. (2006). "Computational damage micromodel of laminated composites", *International Journal of Fracture*, **137**(1), p. 139–150.

[5] Sanchez-Palencia E. (1974). "Comportement local et macroscopique d'un type de milieux physiques hétérogènes", *International Journal for Engineering Science*, **12**, p. 231–251.

[6] Sanchez-Palencia E. (1980). *Non homogeneous media and vibration theory*, Lecture Notes in Physics, 127, Springer-Verlag.

[7] Feyel F. (2003). "A multilevel finite element method (FE^2) to describe the response of highly nonlinear structures, using generalized continua", *Computer Methods in Applied Mechanics and Engineering*, **192**, p. 3233–3244.

[8] Devries F., Dumontet H., Duvaut G., and Léné F. (1989). "Homogenization and damage for composite structures", *International Journal for Numerical Methods in Engineering*, **27**(2), p. 285–298.

[9] Zohdi T., Oden J., and Rodin G. (1996). "Hierarchical modeling of heterogeneous bodies", *Computer Methods in Applied Mechanics and Engineering*, **138**(1–4), p. 273–298.

[10] Oden J., Vemaganti K., and Moës N. (1999). "Hierarchical modeling of heterogeneous solids", *Computer Methods in Applied Mechanics and Engineering*, **172**(1–4), p. 3–25.

[11] Fish J., Shek K., Pandheeradi M., and S. Shephard M. (1997). "Computational plasticity for composite structures based on mathematical homogenization: Theory and practice", *Computer Methods in Applied Mechanics and Engineering*, **148**(1–2), p. 53–73.

[12] Lefik M. and Schrefler B. (2000). "Modelling of nonstationary heat conduction problems in micro-periodic composites using homogenisation theory with corrective terms", *Archives of Mechanics*, **52**(2), p. 203–223.

[13] Huet C. (1990). "Application of variational concepts to size effects in elastic heterogeneous bodies", *Journal of the Mechanics and Physics of Solids*, **38**(6), p. 813–841.

[14] Kouznetsova V., Geers M., and Brekelmans W. (2002). "Multi-scale constitutive modelling of heterogeneous materials with a gradient-enhanced computational homogenization scheme", *International Journal for Numerical Methods in Engineering*, **54**, p. 1235–1260.

[15] Zohdi T. and Wriggers P. (2005). *Introduction to computational micromechanics*, Springer-Verlag.

[16] Hughes T. (1995). "Multiscale phenomena: Green's function, the dirichlet-to-neumann formulation, subgrid scale models, bubbles and the origin of stabilized methods", *Computer Methods in Applied Mechanics and Engineering*, **127**, p. 387–401.

[17] Ladevèze P., Loiseau O., and Dureisseix D. (2001). "A micro-macro and parallel computational strategy for highly heterogeneous structures", *International Journal for Numerical Methods in Engineering*, **52**(1–2), p. 121–138.

[18] Combescure A. and Gravouil A. (2002). "A numerical scheme to couple subdomains with different time-steps for predominantly linear transient analysis", *Computer Methods in Applied Mechanics and Engineering*, **191**(11–12), p. 1129–1157.

[19] Faucher V. and Combescure A. (2003). "A time and space mortar method for coupling linear model subdomains and non-linear subdomains in explict structural dynamics", *Computer Methods in Applied Mechanics and Engineering*, **192**, p. 509–533.

[20] Gravouil A. and Combescure A. (2003). "Multi-time-step and two-scale domain decomposition method for non-linear structural dynamics", *International Journal for Numerical Methods in Engineering*, **58**, p. 1545–1569.

[21] Gravouil A. and Combescure A. (2001). "Multi-time-step explicit implicit method for non-linear structural dynamics", *International Journal for Numerical Methods in Engineering*, **50**, p. 199–225.

[22] Farhat C. and Chandesris M. (2003). "Time-decomposed parallel time-integrators: theory and feasibility studies for fluid, structure, and fluid-structure applications", *International Journal for Numerical Methods in Engineering*, **58**, p. 1397–1434.

[23] Belytschko T., Smolinski P., and K. Liu W. (1985). "Stability of multi-time step partitioned integrators for first-order finite element systems", *Computer Methods in Applied Mechanics and Engineering*, **49**(3), p. 281–297.

[24] Bottasso C. L. (2002). "Multiscale temporal integration", *Computer Methods in Applied Mechanics and Engineering*, **191**(25–26), p. 2815–2830.

[25] Dureisseix D., Ladevèze P., Néron D., and A. Schrefler B. (2003). "A multi-time-scale strategy for multiphysics problems: application to poroelasticity", *International Journal for Multiscale Computational Engineering*, **1**(4), p. 387–400.

[26] Maday Y. and Turinici G. (2002). "A parareal in time procedure for the control of partial differential equations", *Comptes Rendus Académie des Sciences Paris*, I(335) (Issue 4) p. 387–392.

[27] Maitournam M., Pommier B., and Thomas J.-J. (2002). "Determination de la reponse asymptotique d'une structure anelastique sous chargement thermomecanique cycliquedetermination of the asymptotic response of a structure under cyclic thermomechanical loading", *Comptes Rendus Mecanique*, **330**(10), p. 703–708.

[28] Comte F., Maitournam H., Burry P., and Lan N. T. M. (2006). "A direct method for the solution of evolution problems", *Comptes Rendus Mecanique*, **334**(5), p. 317–322.

[29] Fish J. and Chen W. (2001). "Uniformly valid multiple spatial-temporal scale modeling for wave propagation in heterogeneous media", *Mechanics of Composite Materials and Structures*, **8**, p. 81–99.

[30] Yu Q. and Fish J. (2002). "Multiscale asymptotic homogenization for multiphysics problems with multiple spatial and temporal scales: a coupled thermo-viscoelastic example problem", *International Journal of Solids and Structures*, **39**, p. 6429–6452.

[31] Guennouni T. (1988). "On a computational method for cycling loading: the time homogenization", *Mathematical Modelling and Numerical Analysis* (in french), **22**(3), p. 417–455.

[32] Akel S. and Nguyen Q. S. (1989). *Determination of the limit response in cyclic plasticity*, in: Proceedings of 2nd International Conference on Computational Plasticity, Vol. 639–650, Barcelone, Spain.

[33] Ladevèze P. and Rougée P. (1985). "(Visco) plasticity under cyclic loadings: properties of the homogenized cycle", *Comptes Rendus Académie des Sciences Paris* (in french), II(301) (13), p. 891–894.

[34] Ladevèze P. (1991). New advances in the large time increment method, in: Ladevèze P., Zienkiewicz O. C. (Eds.), *New advances in computational structural mechanics*, Elsevier, p. 3–21.

[35] Cognard J.-Y. and Ladevèze P. (1993). "A large time increment approach for cyclic plasticity", *International Journal of Plasticity*, **9**, p. 114–157.

[36] Ladevèze P. and Nouy A. (2003). "On a multiscale computational strategy with time and space homogenization for structural mechanics", *Computer Methods in Applied Mechanics and Engineering*, **192**, p. 3061–3087.

[37] Ladevèze P. (1999). *Nonlinear computational structural mechanics – new approaches and non-incremental methods of calculation*, Springer, Verlag.

[38] Nouy A. and Ladevèze P. (2004). "Multiscale computational strategy with time and space homogenization: a radial-type approximation technique for solving microproblems", *International Journal for Multiscale Computational Engineering*, **2**(4), p. 557–574.

[39] Boucard P.-A., Ladevèze P., and Lemoussu H. (2000). "A modular approach to 3-D impact computation with frictional contact", *Computer and Structures*, **78**(1–3), p. 45–52.
[40] Eriksson K., Johnson C., and Thomée V. (1985). "Time discretization of parabolic problems by the discontinuous galerkin formulation", *RAIRO Modélisation Mathématique et Analyse Numérique*, **19**, p. 611–643.
[41] Ladevèze P., Nouy A., and Loiseau O. (2002). "A multiscale computational approach for contact problems", *Computer Methods in Applied Mechanics and Engineering*, **191**, p. 4869–4891.
[42] Ladevèze P. and Pelle J.-P. (2004). *Mastering calculations in linear and nonlinear mechanics*, Springer, NY.
[43] Ladevèze P. (1985). New algorithms: mechanical framework and development (in french), Tech. Rep. 57, LMT-Cachan.
[44] Ryckelynck D., Hermanns L., Chinesta F., and Alarcon E. (2005). "An efficient 'a priori' model reduction for boundary element models", *Engineering Analysis with Boundary Elements*, **29**(8), p. 796–801.
[45] Ammar A., Mokdad B., Chinesta F., and Keunings R. (2006). "A new family of solvers for some classes of multidimensional partial differential equations encountered in kinetic theory modelling of complex fluids", *Journal of Non-Newtonian Fluid Mechanics*, **139**(3), p. 153–176.
[46] Ammar A., Mokdad B., Chinesta F., and Keunings R. (2007). "A new family of solvers for some classes of multidimensional partial differential equations encountered in kinetic theory modelling of complex fluids: Part ii: Transient simulation using space-time separated representations", *Journal of Non-Newtonian Fluid Mechanics*, **144**(2–3), p. 98–121.
[47] Nouy A. (2007). "A generalized spectral decomposition technique to solve a class of linear stochastic partial differential equations", *Computer Methods in Applied Mechanics and Engineering*, **196**(45–48), p. 4521–4537.
[48] Chatterjee A. (2000). "An introduction to the proper orthogonal decomposition", *Current Science*, **78**(7), p. 808–817.
[49] Ladevèze P. (1997). A computational technique for the integrals over the time-space domain in connection with the LATIN method, Tech. Rep. 193, LMT-Cachan .
[50] Caignot A., Ladevèze P., Néron D., Le Loch S., Le Gallo V., Ma K., and Romeuf T. (2005). Prediction of damping in space lauch vehicles using a virtual testing strategy, in: Proc. of 6th International Symposium on Launcher Technologies.
[51] Brezzi F. and Marini L. (1993). A three-field domain decomposition method, in: Proceedings of the 6th International Conference on Domain Decomposition Methods, p. 27–34.
[52] Lions J.-L., Maday Y., and Turinici G. (2001). "Résolution d'edp par un schéma en temps 'pararéel'", *Comptes-Rendus de l'Académie des Sciences*, I(332), p. 661–668.

[53] Feyel F. and Chaboche J.-L. (2000). "FE^2 multiscale approach for modelling the elastoviscoplastic behaviour of long fibre sic/ti composite materials", *Computer Methods in Applied Mechanics and Engineering*, **183**(3–4), p. 309–330.
[54] Ladevèze P., Néron D., and Gosselet P. (2007). "On a mixed and multiscale domain decomposition method", *Computer Methods in Applied Mechanics and Engineering*, **196**, p. 1525–1540.
[55] Séries L., Feyel F., and Roux F.-X. (2003). A domain decomposition method with two Lagrange multipliers, in: Proceedings of the 16th congrès français de mécanique, in french.
[56] Gosselet P. and Rey C. (2002). On a selective reuse of krylov subspaces in newton-krylov approaches for nonlinear elasticity, in: Proceedings of the 14th conference on domain decomposition methods, p. 419–426.
[57] Gosselet P. and Rey C. (2005). "Non-overlapping domain decomposition methods in structural mechanics", *Archives of Computational Methods in Engineering* submitted.
[58] Rebel G., Park K., and Felippa C. (2002). "A contact formulation based on localized lagrange multipliers: formulation and application to two dimensional problems", *International Journal for Numerical Methods in Engineering*, **54**, p. 263–297.
[59] Le Tallec P., De Roeck Y.-H., and Vidrascu M. (1991). "Domain-decomposition methods for large linearly elliptic three dimensional problems", *Journal of Computational and Applied Mathematics*, **34**, p. 93–117.
[60] Farhat C., Lesoinne M., Le Tallec P., Pierson K., and Rixen D. (2001). "FETI-DP: a dual-primal unified FETI method – part i: a faster alternative to the two-level FETI method", *International Journal for Numerical Methods in Engineering*, **50**(7), p. 1523–1544.

PART IV

ADAPTIVITY, ERROR ESTIMATION AND UNCERTAINTY QUANTIFICATION

10

ESTIMATION AND CONTROL OF MODELING ERROR: A GENERAL APPROACH TO MULTISCALE MODELING

J.T. Oden, S. Prudhomme, P.T. Bauman, and L. Chamoin

10.1 Problem setting

The following points provide the setting for the theory and methodologies described in this chapter.

- We wish to develop reliable and predictive computer simulations of the behavior of very large and complex molecular and atomistic systems. Such systems are encountered with increasing frequency in nanomanufacturing, in the design of advanced materials, in semiconductor manufacturing, in the analysis of biological systems, in drug design, and in numerous bio-medical applications. The simulation of the behavior of such systems is a critical challenge facing advancements in many areas of science and engineering.

- Detailed computational models of such large systems may involve hundreds of millions of unknowns. There may also be uncertainties in the parameters defining the models of such systems, so the unknowns are generally random variables and the models are stochastic. Such problems are well outside the capabilities of the largest and fastest computers that exist today or are likely to exist for many decades.

- In all computer simulations of physical systems, there are generally specific features of the solution that are of primary interest, the so-called *quantities of interest* (QoIs) or target outputs. It is generally assumed that in many cases these quantities are largely dependent on local fine-scale features of the model, and that at distances remote from those at which the quantities of interest are defined, behavior is determined by coarser scales representing averages in some sense of the fine-scale behavior. Thus, multiscale models are needed to reduce the size of the problem to one in which only the phenomena at specific scales affect the accuracy of the quantities of interest. This is really the only reason to consider multiscale modeling: to include in the computational model only the scales, and correspondingly the numbers of unknowns, needed to deliver quantities of interest with sufficient accuracy. But how does one know what level of fine-scale or coarse-scale information is needed in a model to obtain approximations of the QoIs with sufficient accuracy?

- The only way to resolve this last question is to develop methods of estimating the error produced in quantities of interest by averaging or filtering out fine-scale efffects. Such *a posteriori* error estimates clearly pertain to the relative error between the full fine-scale (and generally intractable) model and other approximations of coarser scale. Coarse-scale models with dramatically fewer numbers of degrees of freedom compared to the base fine-scale model are obtained by various coarsening methods and the terms coarse-graining, upscaling, homogenization, dimensional reduction, etc., are used to describe such processes. Many of such methods are *ad hoc* and do not attempt to estimate and control errors due to coarsening.

- We are interested here in a general class of methods for multiscale modeling that derive from the general setting just described:
    1. They assume the existence of a well-posed fine-scale base model of molecular systems that is generally intractable but of sufficient detail and sophistication to capture all events of interest with acceptable accuracy;
    2. Specific quantities of interest are identified that are representable as functionals of the fine-scale solution;
    3. Various averaging techniques are used to produce coarser-scale models in particular subdomains of the solutions of the fine-scale model, and, hence, hybrid models of multiple scales may be produced, and these models are tractable;
    4. The coarse-grained hybrid models are "solved" and approximations of the quantities of interest are computed using the (incorrect) solutions of the hybrid models;
    5. *A posteriori* estimates of error in the hybrid coarse-scale QoIs compared to the actual fine-scale values are computed; if they are small (compared to a preset tolerance level), the analysis is terminated and the hybrid model is accepted as a sufficiently accurate approximation of the fine-scale model; if the error is large, the hybrid model contains insufficient fine-scale information and it must be refined to reduce the error in the QoIs;
    6. The hybrid model is adapted by the addition of fine-scale features in appropriate subdomains in a way that systematically reduces the estimated error in the QoIs until error tolerances are met; the process terminates when a near-optimal multiscale model is generated which yields acceptable values of the QoIs.

    These six steps characterize the *Goals Algorithm* for adaptive modeling. We discuss averaging (coarse-graining) techniques, error estimation, the adaptive strategy, and the all-important coupling algorithms that provide interfaces between the fine and coarse-scale regions of the model in subsequent sections.

We will focus on a specific class of molecular models: lattice-based models of polymer systems in static equilibrium (zero temperature). These types of molecular statics problems are encountered in nanomanufacturing of semiconductor devices using imprint lithography, an application addressed in the dissertation of Bauman [1] and in related publications (e.g. [2, 3, 4, 5, 6]). The general methodology described is also applicable to problems of molecular dynamics, as shown in [5], but the polymer equilibrium problem has features of particular interest in multiscale modeling: the calculation of the structure and properties of the molecular system itself through the chemical process of polymerization. We also discuss this aspect of the modeling process.

10.2 The general theory of modeling error estimation

The idea of replacing a general mathematical model of physical events (the fine-scale *base model*) with a surrogate with coarser features and potentially fewer degrees of freedom (coarse-graining, etc.) can be understood in a quite general abstract setting. We wish to find the vectors $\mathbf{u}$ in some topological vector space of trial functions U such that

$$A(\beta)\mathbf{u} = \mathbf{F} \qquad (10.1)$$

where $A : U \longrightarrow V'$ is a nonlinear operator, possibly dependent on a set of parameters β, and $\mathbf{F}$, are data that belong to topological dual V' of a space of test vectors V. The boundary and initial conditions pertinent to the model are embedded in the definitions of U and $\mathbf{F}$. We assume that solutions $\mathbf{u} = \mathbf{u}(\beta)$ exist for each $\mathbf{F}$ and for all β in some subset of parameter space in which (10.1) remains well posed, but that the solutions may not be unique, and, of course, that they depend upon the data β. In general, the parameters are random variables and, therefore, $\mathbf{u}$ is a random variable. If it were possible to solve (10.1) for $\mathbf{u}(\beta)$ (which, in general, it is not), we are primarily interested in calculating particular quantities of interest (QoIs) that can be represented as functionals on U:

$$\boxed{\begin{array}{c} \text{Quantity of Interest} \quad Q(\mathbf{u}(\beta)) \\ Q : U \longrightarrow \mathbb{R} \\ (\mathbf{u}(\beta) \text{ is a solution of (10.1)}) \end{array}} \qquad (10.2)$$

For example, (10.1) may represent a large model of the dynamics of a complex molecular system, with $\mathbf{u}$ the set of trajectories of individual molecules over a specified time interval, and $Q(\mathbf{u})$ could represent the spatial and temporal average of the motion of a small subset of molecules over a time interval. The general problem is (10.1) and (10.2), of course, can represent virtually any class of problems in mathematical physics, including problems in continuum mechanics.

Since (10.1) is generally unsolvable, we replace it with an approximate model, here perhaps one with coarser-scale features, but one that we presume can be solved in some sense:

$$A_0(\beta_0)\mathbf{u}_0 = \mathbf{F} \qquad (10.3)$$

This model is a surrogate to (10.1), involving an operator $A_0 : U \longrightarrow V'$ which approximates A in some sense, and a set of parameters β_0. We hope that the solution $\mathbf{u}_0$ is "close" to $\mathbf{u}(\beta)$ in some sense. Obviously, this "closeness" makes sense only in regard to quantities of interest. Thus, upon solving (10.3), we compute $Q(\mathbf{u}_0)$. The modeling error is then

$$\varepsilon_0 = Q(\mathbf{u}(\beta)) - Q(\mathbf{u}_0(\beta_0)) \qquad (10.4)$$

How well the surrogate model approximates the base model is determined by the magnitude of ε_0 and whether or not it is small enough for the application at hand. The error ε_0 is thus also a guide to coarse-graining and dimensional reduction; it quantifies exactly the effect of any particular choice of local averaging, coarse-graining, etc., on values of the principal targets of the simulation: the QoIs.

10.2.1 The error estimate

Remarkably, if the data of the base problem are known (but not the solution $\mathbf{u}(\beta)$) the error ε_0 can be estimated provided one computes the solution of an adjoint problem corresponding to the particular quantity of interest Q. The basic steps are as follows:

- Since $A(\beta)\mathbf{u} - \mathbf{F}$ lies in the dual space V', we may view (10.1) as the condition that the residual functional $R : V \longrightarrow \mathbb{R}$ vanishes:

$$R(\beta, \mathbf{u}; \mathbf{v}) = \langle A(\beta)\mathbf{u} - \mathbf{F}, \mathbf{v}\rangle = 0 \quad \forall \mathbf{v} \in V \qquad (10.5)$$

Here $\langle \cdot, \cdot \rangle$ denotes duality pairing on $V' \times V$. We employ here the convention that functionals are linear in arguments to the right of the semi-colon, but possibly nonlinear functions of those entries to the left of the semi-colon. Hence, R is a linear functional of the test vector $\mathbf{v}$.

- The *adjoint problem* consists of finding $\mathbf{p} \in V$ such that

$$A'(\beta, \mathbf{u})^T \mathbf{p} = Q'(\beta, \mathbf{u}) \quad \text{in } V' \qquad (10.6)$$

where $A'(\beta, \mathbf{u})$ is the *linear* operator defined by

$$\lim_{\theta \to 0} \frac{1}{\theta} \langle A(\beta)(\mathbf{u} + \theta \mathbf{p}) - A(\beta)\mathbf{u} - \theta A'(\beta, \mathbf{u})\mathbf{p}, \mathbf{v}\rangle = 0 \quad \forall \mathbf{v} \in V$$

and $Q'(\beta, \mathbf{u})$ is the linear functional,

$$\langle Q'(\beta, \mathbf{u}), \mathbf{v}\rangle = \lim_{\theta \to 0} \frac{1}{\theta}\left(Q(\beta, \mathbf{u} + \theta \mathbf{v}) - Q(\beta, \mathbf{u})\right)$$

The solution $\mathbf{p}$ of the adjoint problem may be thought of as the *generalized Green function* corresponding to the particular choice Q of the QoI. It is also interpreted as the Lagrange multiplier associated with minimizing $Q(\mathbf{v})$ subject to the constraint (10.1).

- With $R(\beta, \mathbf{u}; \cdot)$ given by (10.5) and $\mathbf{p}$ the solution of (10.6) for given Q, it is shown in [4] that

$$\varepsilon_0 = R(\beta, \mathbf{u}_0; \mathbf{p}) + \Delta \qquad (10.7)$$

where Δ is a remainder depending on terms of quadratic and higher order in the error ε_0. Our assumption is that if the surrogate model (10.3) is close enough to (10.1), Δ is negligible and

$$\varepsilon_0 \approx R(\beta, \mathbf{u}_0; \mathbf{p}) \qquad (10.8)$$

Our goal is to derive a family of algorithms that generate a sequence of surrogates and a sequence of approximations of the adjoint and the residual that reduces the modeling error ε_0 to a tolerable level.

We remark that the functionals in (10.5) and (10.6) define semilinear and linear forms,

$$\begin{aligned}
B(\beta, \mathbf{u}; \mathbf{v}) &= \langle A(\beta)\mathbf{u}, \mathbf{v}\rangle \\
F(\mathbf{v}) &= \langle F, \mathbf{v}\rangle \\
B'(\beta, \mathbf{u}; \mathbf{v}, \mathbf{p}) &= \langle A'(\beta, \mathbf{u})^T \mathbf{p}, \mathbf{v}\rangle \\
Q'(\beta, \mathbf{u}; \mathbf{v}) &= \langle Q'(\mathbf{u}(\beta)), \mathbf{v}\rangle
\end{aligned} \qquad (10.9)$$

so that the primal and adjoint problems can be written as

$$\begin{aligned}
&\text{Find } (\mathbf{u}, \mathbf{p}) \in U \times U \\
&B(\beta, \mathbf{u}; \mathbf{v}) = F(\mathbf{v}) \qquad \forall \mathbf{v} \in V \\
&B'(\beta, \mathbf{u}; \mathbf{v}, \mathbf{p}) = Q'(\mathbf{u}(\beta), \mathbf{v}) \qquad \forall \mathbf{v} \in V
\end{aligned} \qquad (10.10)$$

The rates of change (Gâteaux derivatives) of these forms due to variations in the parameters β can also be computed

$$B_\beta(\beta, \mathbf{u}; \mathbf{v}, w) = \lim_{\theta \to 0} \theta^{-1}\big(B(\beta + \theta w, \mathbf{u}(\beta + \theta w); \mathbf{v}) - B(\beta, \mathbf{u}; \mathbf{v})\big)$$

$$Q_\beta(\beta, \mathbf{u}; \mathbf{v}, w) = \lim_{\theta \to 0} \theta^{-1}\big(Q(\beta + \theta w, \mathbf{u}(\beta + \theta w); \mathbf{v}) - Q(\beta, \mathbf{u}; \mathbf{v})\big)$$

10.3 A large-scale molecular statics model

While the framework described up to this point is quite general, we shall focus on a class of problems in nanomanufacturing that exhibit many of the features and challenges of multiscale modeling at the molecular level. The target application is Step and Flash Imprint Lithography, a process of stamping polymer etch barriers at room temperature to produce high-precision features of semiconductor

290   *Estimation and control of modeling error*

devices. Full details of this process are discussed in [1]. For our present purposes, it suffices to assert that the process involves creating a polymer through a chemical process that fills a quartz template designed to mold the surface features of devices with dimensions on the order of 40-70 nanometers or smaller. We wish to model the creation of the polymer and its densification (static deformation to an equilibrium configuration) using a lattice representation of the polymer molecules and chains, ignoring thermal effects. While such lattice models of polymer statics are often used (e.g. [7]), much more complex models could also be used to define the base model. Our adaptive modeling procedure also remains valid for such systems.

Figures 10.1 and 10.2 show the color-coded molecule locations in a lattice and the network of polymer chains arising from one realization of the polymerization process. For semiconductor units of dimensions $70 \times 200 \times 1000$ nanometers, these models can involve up to 10^7 unknown site displacement components.

10.4 The family of six algorithms governing multiscale modeling of polymer densification

Six major algorithms are developed and implemented to model the molecular statics of lattice-type models of polymer systems. These are listed below in the order of their implementation.

10.4.1 *Polymerization: Kinetic Monte Carlo method*

A liquid solution of specific monomers, cross-link molecules, reactants, each with given initial volume fractions, is subjected to ultraviolet light, which initiates a chemical reaction. As a result, a network of polymer chains and cross-links is created, the chains representing long molecules with monomer links connected

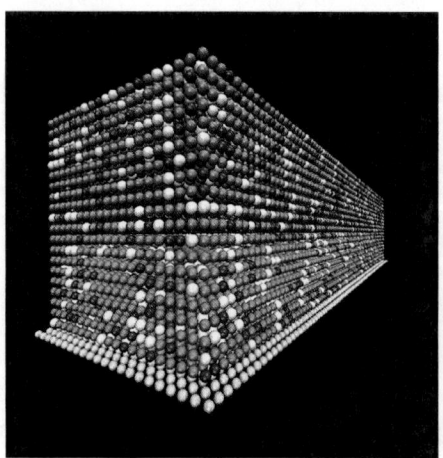 

Figure 10.1: Lattice model of one realization of the polymerization of a system of two monomers, with cross-links, reactants, and voids. (See Plate 7).

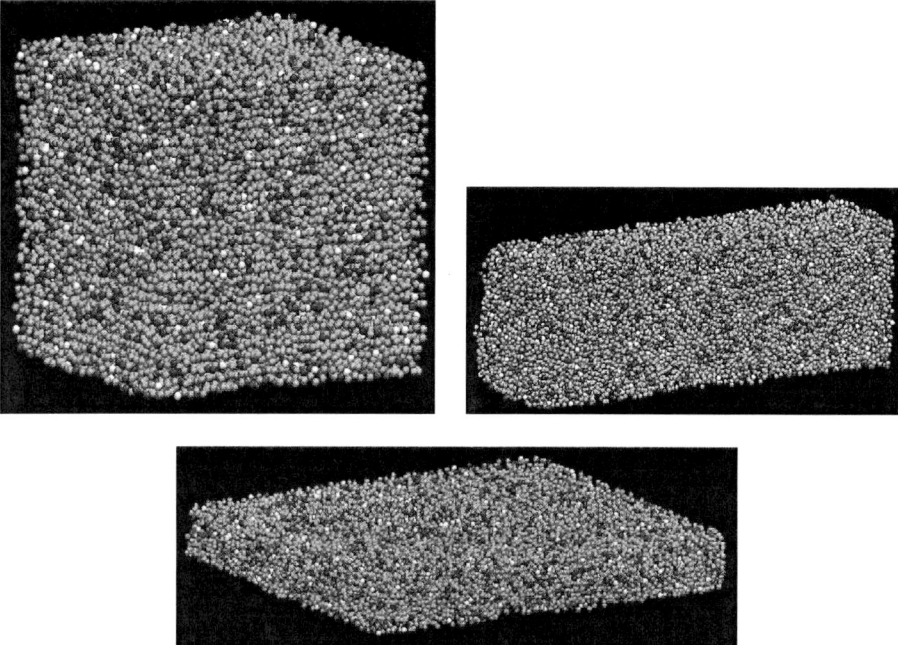

Figure 10.2: Virtual experiments on a computer-generated Representative Volume Element (RVE) of a polymer: relaxation (top left), uniaxial stretch (top right), biaxial stretch (bottom). (See Plate 8).

by covalent bonds. Other bonds, such as Van der Waals bonds, may also be created across chains. The process is called *polymerization*. Possible conformations of molecules and the configuration of the polymer structure are determined using a biased Monte Carlo algorithm in which molecular constituents are randomly distributed over a 3D lattice in proportion to the initial volume fraction. A reactant or molecule with free electrons existing in one cell is allowed to react with those of neighboring cells which have the highest probability of reaction as determined by the Arrhenius law,

$$R = Ce^{-kE_aT}$$

k being the reaction rate and E_a the activation energy. The probability of a reaction is proportional to the rate k:

$$p \propto k$$

and the rates are experimentally determined [8]. This aspect of the algorithm is similar to the classical Metropolis method [9]. At the conclusion of n steps of this process, one realization of the polymer, such as that shown in Figure 10.1, is obtained. Complete details of this algorithm are given in [1].

10.4.2 Molecular potentials

With a given molecular distribution over the polymer determined by Algorithm 1, molecular potentials are selected to represent the intermolecular forces. In this work, covalent bonds on polymer chains are represented by harmonic potentials,

$$V(\mathbf{r}) = \frac{k}{2}(\mathbf{r} - \mathbf{r}_0)^2$$

weak Van Der Waals bonds are represented by 6-12 Lennard-Jones potentials, etc. For simplicity, only nearest neighbors are accounted for here, but in principal, long-range effects could easily be included at the cost of greater computational complexity.

10.4.3 Densification algorithm: inexact Newton-Raphson with trust region

It is only at this stage of the analysis that we actually arrive at the possible base model of equilibrium configurations of the polymer structure, and this for only one realization of the polymer. For this realization, we are faced with the problem of solving very large nonlinear algebraic systems of the form (10.10) with

$$\left. \begin{aligned} B(\beta, \mathbf{u}; \mathbf{v}) &= \sum_{i=1}^{N} \sum_{k=1}^{N_i} \frac{\partial E_{ik}(\beta; \mathbf{u})}{\partial \mathbf{u}_i} \cdot \mathbf{v}_i \\ F(\mathbf{v}) &= \sum_{i=1}^{N} \mathbf{f}_i \cdot \mathbf{v}_i \end{aligned} \right\} \qquad (10.11)$$

where $\mathbf{u}$ is a vector of lattice site displacements $\mathbf{u} = (\mathbf{u}_1, \mathbf{u}_2, \ldots, \mathbf{u}_N)$, $\mathbf{v}$ is a set of N test vectors, $E_{ik}(\beta; \mathbf{u})$ is the energy associated with site $\mathbf{x}_i$ and N_i neighbors $\mathbf{x}_k$, for a particular polymer realization, and $\mathbf{f}_i$ is the applied force at site $\mathbf{x}_i$. We solve the nonlinear system,

$$\sum_{k=1}^{N_i} \left(\frac{\partial E_{ik}(\beta; \mathbf{u})}{\partial \mathbf{u}_i} - \mathbf{f}_i \right) = 0 \quad 1 \leq i \leq N \qquad (10.12)$$

by the inexact Newton trust region method or we solve, for each β, the equivalent optimization problem,

$$\mathbf{u} = \arg \min_{\mathbf{v}} E(\mathbf{v}) \qquad (10.13)$$

where $E(\mathbf{v}) = \sum_{i=1}^{N} \sum_{k=1}^{N_i} E_{ik}(\beta; \mathbf{v}) - \sum_{i=1}^{N} \mathbf{f}_i \cdot \mathbf{v}_i$ using the TAO/PETSc codes [10, 11].

10.4.4 Algorithms for computing surrogate models

To reduce the enormous size of the systems of equations to a manageable level, we construct a coarse-scale continuum model of the polymer using virtual experiments on Representative Volume Elements (RVEs) and we construct an interface region between the discrete polymer domain and the continuum.

10.4.4.1 Virtual experiments

The polymerization routine is used to generate a sequence of realizations of a cubic RVE of the polymer. Owing to the choice of monomers, we know that rate effects are minimal, and can be neglected, and that the imprint process is performed at constant room temperature. No biases in the polymerization can lead to macroscopic heterogeneities or anisotropies over many realizations. Thus, we assume that at the macroscale level, the corresponding continuum is characterized as an isotropic, homogeneous, compressible, hyperelastic material with a stored energy functional of the form,

$$W = \hat{W}(I_1, I_2, I_3) = \alpha(I_1 - 3) + \beta(I_2 - 3) + \gamma(J - 1)^2 \\ - (2\alpha + 4\beta) \ln J \tag{10.14}$$

where α, β, γ are material constants, I_1, I_2 are the first two principal invariants of the deformation tensor $\mathbf{C} = \mathbf{F}^T\mathbf{F}$, and $J = \det \mathbf{F}$, $\mathbf{F}$ being the deformation gradient. For each realization, we subject the RVE to independent homogeneous deformation (stretch, lateral compression, etc.) to determine histograms of the material constants α, β, γ as indicated in Figure 10.2. We increase the dimensions of the RVE for each realization until essentially no changes in the experimentally-determined parameters occur.

10.4.4.2 The interface

We construct an overlap domain that provides an interface between regions where the molecular model exists and the continuum model generated in step 10.4.4.1 above. This is accomplished using the discrete-continuum Arlequin method [12].

10.4.4.3 Residual force calculation

At the conclusion of the polymerization step, residual forces (and "strains") exist in the molecular model that force it to comply to the regular geometry shape of the lattice relative to a cartesian coordinate system. We account for these residual forces using an algorithm developed by Bauman [1]. We consider again a cubic RVE, compute the forces on the faces required to maintain the cubic shape, and apply equal-and-opposite forces to make the RVE forces free of external forces. The densification of the cube then takes it into a configuration with no forces on the molecules on the original RVE faces. Virtual experiments are implemented using these relaxed configurations. Details of this important process are given in [1].

10.4.5 The adjoint problem

We identify one or more quantities of interest (QoIs). For example, the slump of a patch Ω of a polymer device after it is removed from a template can be characterized by the functional

$$Q(\mathbf{u}) = \frac{1}{|\Omega|} \sum_{i \in I} \mathbf{u}_i \cdot \mathbf{k} \tag{10.15}$$

where $|\Omega|$ is the area of the surface, I an index set of lattice sites in Ω, and $\mathbf{k}$ a unit exterior vector normal to the exterior face of Ω. The adjoint problem corresponding to (10.15) is

$$B'(\beta, \mathbf{u}; \mathbf{v}, \mathbf{p}) = Q(\mathbf{v}) \quad \forall \mathbf{v} \in V \qquad (10.16)$$

where now

$$B'(\beta, \mathbf{u}; \mathbf{v}, \mathbf{p}) = \sum_{i=1}^{N} \sum_{k=1}^{N_i} \frac{\partial^2 E_{ik}(\beta; \mathbf{u})}{\partial \mathbf{u}_i \partial \mathbf{u}_k} \mathbf{v}_i \cdot \mathbf{p}_k \qquad (10.17)$$

This is the adjoint problem corresponding to the full fine-scale base problem (10.12). In general, it cannot be solved because the solution $\mathbf{u}$ to the primal problem is not known. Instead, we may solve a surrogate problem for an approximation $\hat{\mathbf{p}}$ of $\mathbf{p}$:

$$B'(\beta, \hat{\mathbf{u}}; \mathbf{v}, \hat{\mathbf{p}}) = Q(\mathbf{v}) \quad \forall \mathbf{v} \in V \qquad (10.18)$$

Here $\hat{\mathbf{u}}$ is a solution of one of a sequence of approximations of $\mathbf{u}$ generated using the goals algorithm.

10.4.6 The goals algorithm

For simplicity, consider the two-dimensional polymer lattice domain D shown in Figure 10.3 and suppose that the quantity of interest involves the motion of lattice sites confined to a subdomain Ω, as shown. We construct a larger open domain Θ_1 containing Ω in which all of the fine-scale degrees of freedom are contained. The interface region Γ_1 is created using the Arlequin method that connects the particle domain Θ_1 with the continuum model domain $D_1 = D - \Theta_1$. The deformation of the continuum domain D_1 is modeled using the finite element method so that in the hybrid model, the solution is discrete and can be compared with $\mathbf{u}$ in the trial space U. Let $\hat{\mathbf{u}}^{(1)}$ be the solution of this hybrid model. We compute the residual

$$R^{(1)} = R^{(1)}(\hat{\beta}, \hat{\mathbf{u}}^{(1)}, \mathbf{p}^{(1)}) \approx \varepsilon_0^{(1)} \qquad (10.19)$$

where $\mathbf{p}^{(1)}$ is the solution of a surrogate adjoint problem to be described momentarily. The subregion Θ_1 is a member of a set of mutually disjoint sets forming a partition of D:

$$\bar{D} = \bigcup_{k=1}^{N} \bar{\Theta}_k \quad , \quad \Theta_k \cap \Theta_j = 0, k \neq j$$

The total residual can be represented as the sum

$$R^{(1)} = \sum_{k=1}^{N} R_k^{(1)}$$

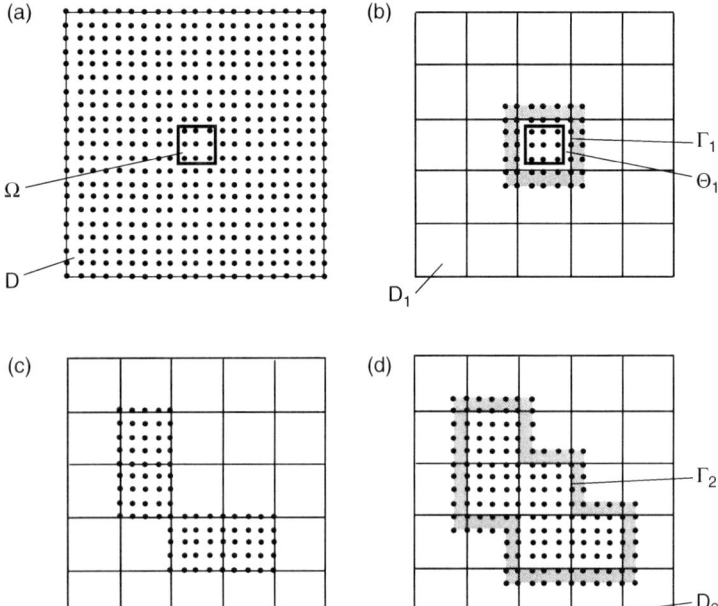

Figure 10.3: The Goals Algorithm: (a) a subdomain Ω associated with a quantity of interest in a fine-scale molecular model with domain D; (b) a hybrid model consisting of a partition of D with Θ_1 containing Ω, and interface region Γ_1 and a discretized continuum model D_1; (c) subdomains with residual contributions exceeding or equal to αR_{max}; and (d) an adapted (second) hybrid model in the sequence of models.

where $R_k^{(1)}$ is the portion of the residual contributed by subdomain Θ_k. We choose $\alpha \in (0,1)$ and invoke the condition, refine every subdomain Θ_k such that

$$R_k^{(1)} \geq \alpha R_{max}^{(1)} \quad ; \quad R_{max}^{(1)} = \max_k R_k^{(1)}$$

By "refine" we mean introducing all of the fine-scale unknowns (the molecular displacements) associated with that subdomain. The union of Θ_1 and all Θ_k satisfying $R_k^{(1)} \geq \alpha R_{max}^{(1)}$ creates a new hybrid model, with particles in $\Theta_1 \cup \Theta_k$, and with a new Arlequin interface Γ_2. A new residual $R^{(2)}$ is computed. This process is repeated until a preset tolerance $|R^{(s)}| < \gamma_{tol}$ is met, or the process is terminated after a preset number of iterates.

In computing the residuals, the adjoint vectors $\mathbf{p}^{(k)}$ must generally be computed by solving an adjoint problem on a finer-scale model than that used to calculate the hybrid solutions. More details on that important feature of these methods is given in [1].

10.5 Representative results

10.5.1 Verification of error estimator and adaptive strategy

We reproduce here results of [1] and [3] designed to test the effectivity of the *a posteriori* error estimate for a large molecular system and to test the performance of the Goals Algorithm for a simplified case in which the solution of the base problem is actually known. As a model problem we consider a hypothetical case of a cubic $22 \times 22 \times 22$ lattice of molecules, fixed at its base, and undergoing a volume change, a 30% shrinkage during the densification of the polymer system. All covalent bonds are modeled using the harmonic potential with equal spring constants k. The resulting system has 31,944 degrees of freedom. As a quantity of interest, we choose the slump of the polymer as represented by the average vertical displacement of a subset of molecules in a small region Ω centered at the top five lattice layers, as shown in Figure 10.4. The average vertical displacement of the 5×5 patch of molecules at the center of the upper face of the cube will define the slump. A base solution $\mathbf{u}$ of the set of displacement vectors of the sites is computed using the Newton algorithm described earlier, and adjoint solution $\mathbf{p}$ corresponding to this particular quantity of interest is also computed. The intensity of the adjoint solution vectors is shown color-coded in Figure 10.5, the largest non-zero values in red near the quantity of interest subdomain Ω, and quickly dropping off to near zero at sites farther removed from the QoI.

We next produce a sequence of surrogate models. The lattice domain is partitioned into 105 cubic subdomains, with seven divisions in the vertical direction (normal to the fixed base) and 5×5 divisions in the planes perpendicular to the axis. These define the partition $\{\Theta_j\}$ used in the Goals Algorithm. As a first surrogate, we choose a subdomain, including Ω, of $13 \times 13 \times 7$ particles (molecules), outside of which the hyperelastic continuum, modeled as a compressible Mooney-Rivlin material described earlier, is approximated using only $5 \times 5 \times 7$ trilinear finite elements. A color-coded depiction of this partition is shown in Figure 10.6, the pink denoting the fully discrete domain Ω, the red the overlap domain where the Arlequin method imposes the connection of the fine-scale molecular model,

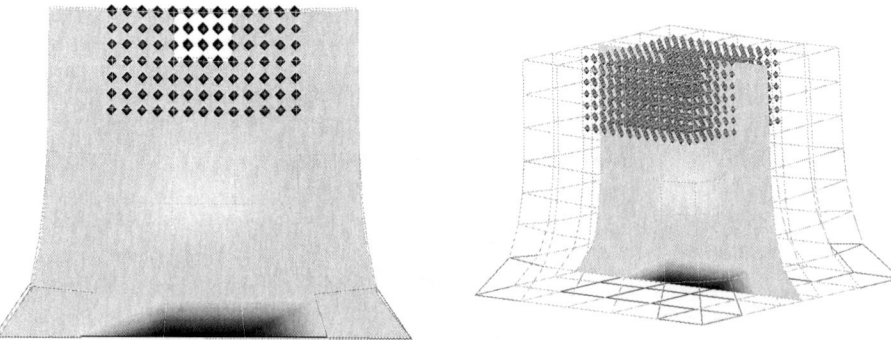

Figure 10.4: The Arlequin problem. (See Plate 9).

Plate 1: Schematic description of unstructured coarsening for upscaling of transport equation. A random color is assigned to each coarse grid block. See Page 6.

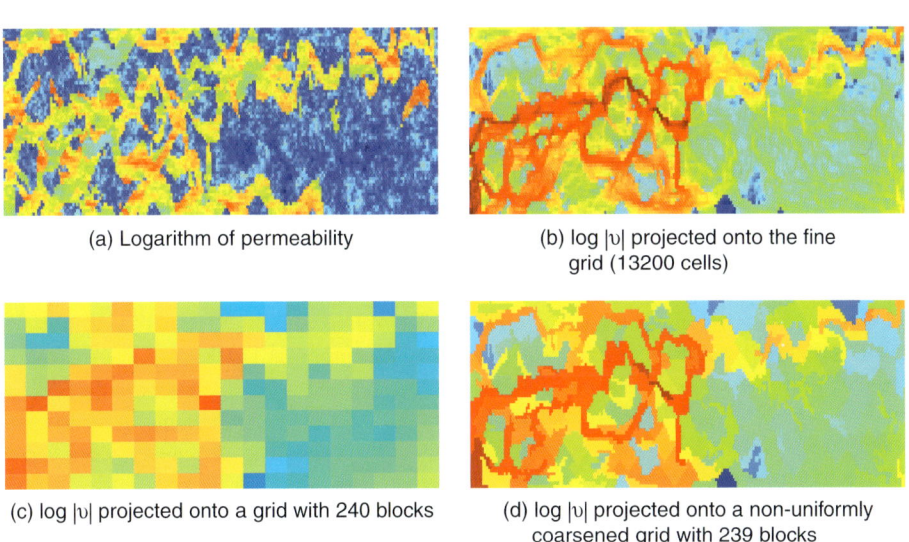

(a) Logarithm of permeability

(b) log |υ| projected onto the fine grid (13200 cells)

(c) log |υ| projected onto a grid with 240 blocks

(d) log |υ| projected onto a non-uniformly coarsened grid with 239 blocks

Plate 2: Illustration of the non-uniform coarsening algorithms ability to generate grids that resolve flow patterns and produce accurate production estimates. Results from paper [2]. See Page 11.

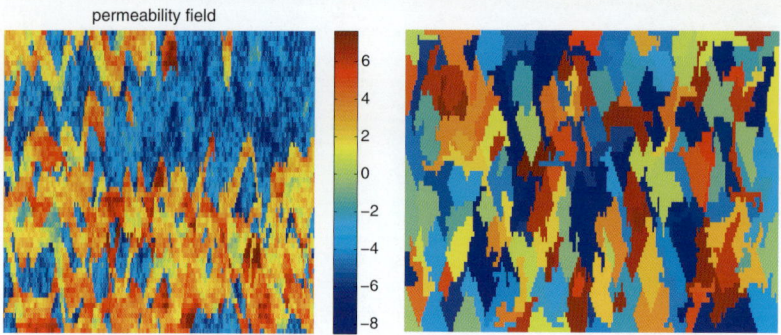

Plate 3: 60×220 permeability field and the coarse grid with 180 blocks. A random color is assigned to each coarse grid block. See Page 12.

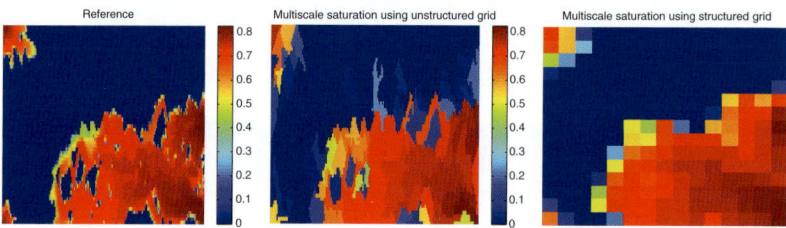

Plate 4: Saturation comparisons. See Page 13.

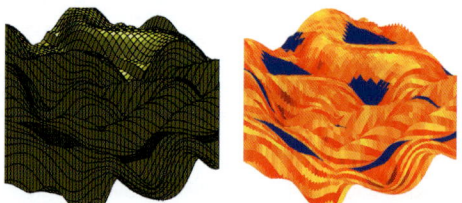

Plate 5: A corner-point model with vertical pillars and 100 layers. To the right is a plot of the permeability field on a logarithmic scale. The model is generated with SBED^{TM}, and is courtesy of Alf B. Rustad at STATOIL. See Page 15.

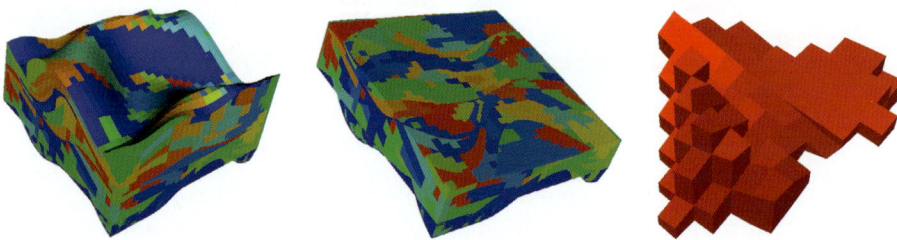

Plate 6: Left: schematic description of unstructured coarsening (each coarse grid block is assigned a random color). Middle: a horizontal slice of unstructured coarsening presented on the left. Right: a coarse grid block (enlarged). See Page 16.

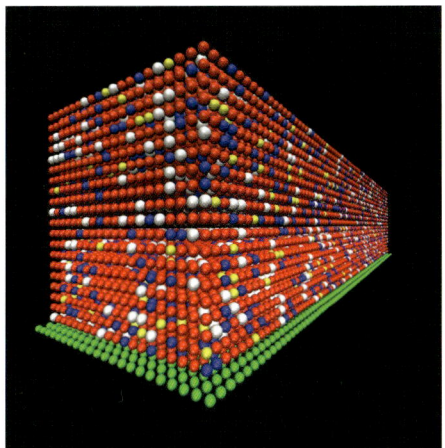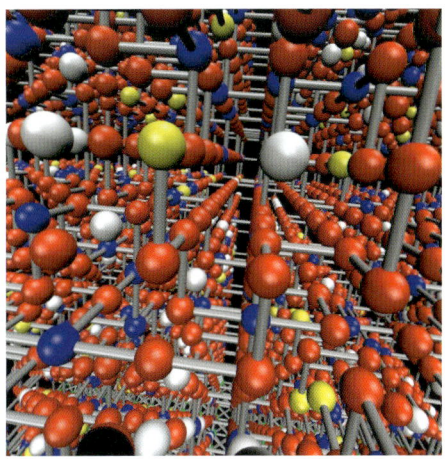

Plate 7: Lattice model of one realization of the polymerization of a system of two monomers, with cross-links, reactants, and voids. See Page 290.

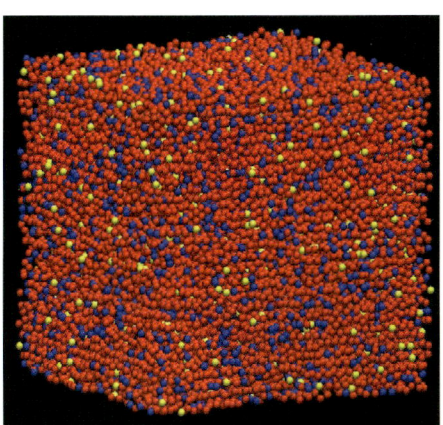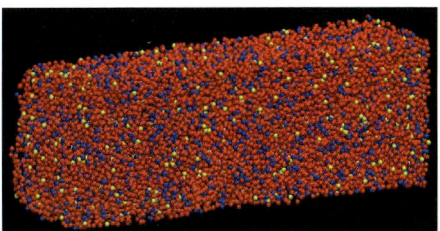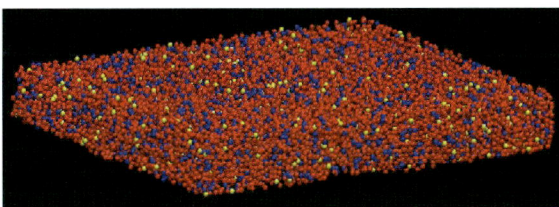

Plate 8: Virtual experiments on a computer-generated Representative Volume Element (RVE) of a polymer: relaxation (top left), uniaxial stretch (top right), biaxial stretch (bottom). See Page 291.

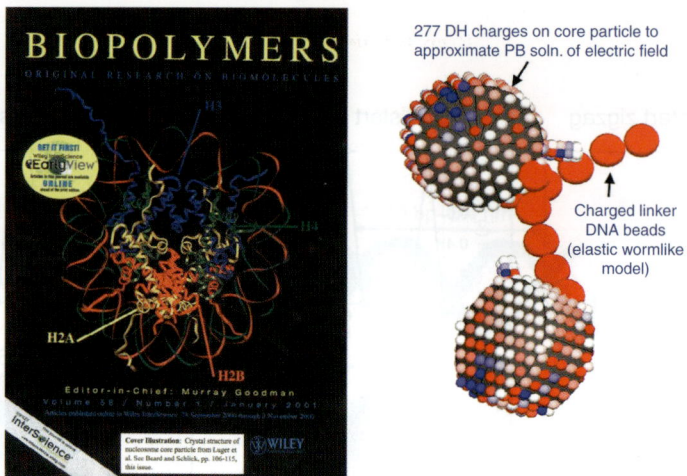

Plate 16: First-generation chromatin model. The nucleosome core (left) is represented by charged bodies denoting the nucleosomes with linker DNA represented as connecting beads (right). See Page 519.

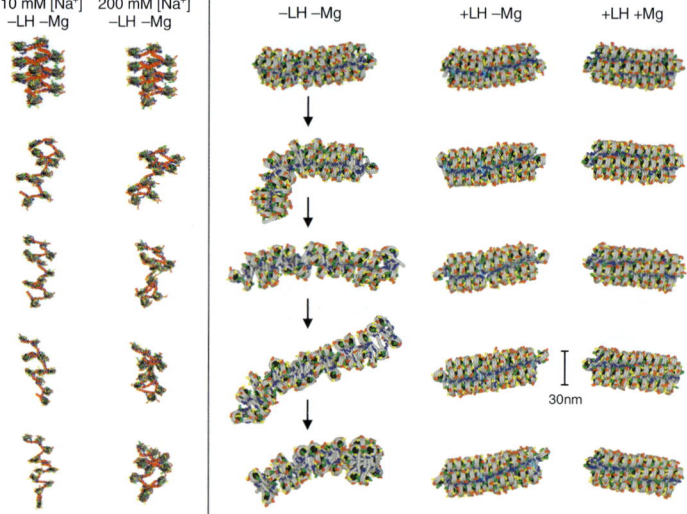

Plate 17: Nucleosome arrays with flexible histone tails simulated using the mesoscale chromatin model by Monte Carlo. The tails are colored as: H2A—yellow, H2B—red, H3—blue, and H4—green. The leftmost two columns illustrate behavior of 12-nucleosome arrays under low (10 mM) and high (200 mM) univalent salt concentrations. At low salt, the negatively-charged linker DNAs (red) repel one another and expand the array, while at high salt, the chromatin fiber condenses. The other columns show the behavior of 48-nucleosome arrays at high (200 mM) monovalent salt concentrations with linker DNA colored grey and nucleosome cores colored black at three compositions: without linker histones without magnesium (–LH–Mg); with linker histone H1 (cyan) without magnesium(+LH–Mg) ; and with linker histone H1with magnesium (+LH+Mg). Note the compaction introduced by linker histone as well as divalent ions. See Page 526.

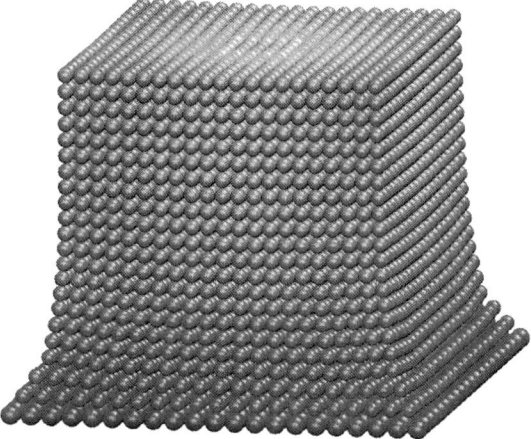

Figure 10.5: Solution of the adjoint problem. (See Plate 10).

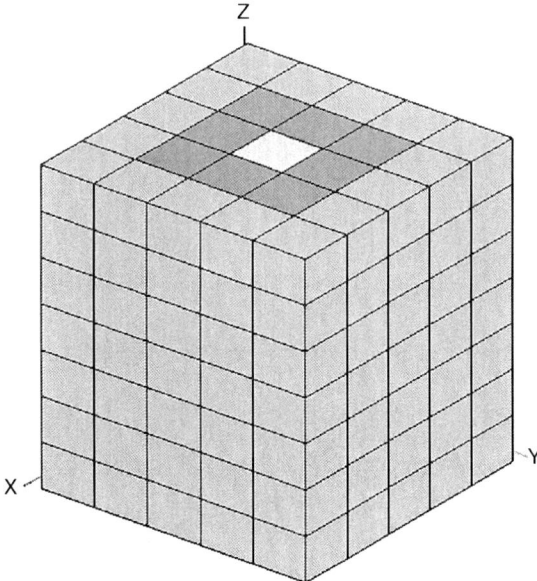

Figure 10.6: Partition of the domain. (See Plate 11).

and the FEM-discretization of the hyperelastic continuum, which is green in the figure. The projection of the lattice site displacements within a continuum element to the current configuration of the surrogate lattice is denoted Π, as indicated in Figure 10.7. Thus, if $\mathbf{u}_0$ is the solution of the discrete surrogate model, representing a set of displacement vectors at sites in the reduced model,

$\Pi\mathbf{u}_0$ is the projection of this set of vectors on the lattice sites, enabling one to compare solutions of the surrogate to that of the base model.

A sequence of five model adaptations were used in the Goals Algorithm. Estimated errors were computed using the estimate,

$$\varepsilon_0 = R(\Pi\mathbf{u}_0, \beta; \mathbf{p})$$

with $\mathbf{p}$ the exact adjoint solution. The process reduced the error in the quantity of interest from near 10% to less than 5% in five steps (Table 10.1). The effectivity indices for the estimated error in each surrogate were very good, averaging over 90%. It is estimated that for this example, the remainder Δ in the estimate (10.7) was around 8% of the total error. The sequence of surrogates generated by the Goals Algorithm with changing particle subdomains, Arlequin overlap domains, and discretized continuum elements, are shown in Figure 10.8.

There are many important details in implementing these types of adaptive modeling algorithms that cannot be discussed in the limited space available here. These include such effects as refining the mesh approximating the continuum, varying the size of the Arlequin overlap domain, approximating the residual

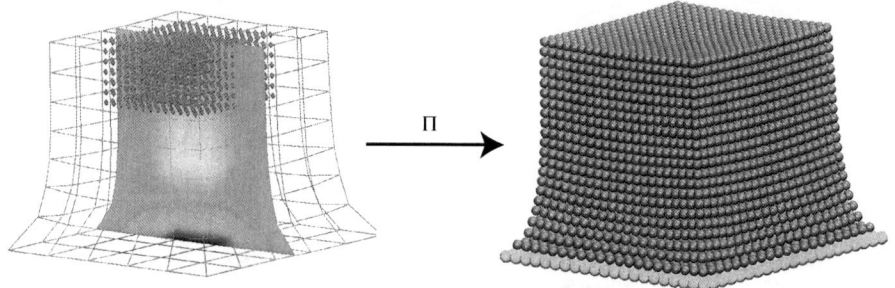

Figure 10.7: Projection to the current configuration of the surrogate lattice. (See Plate 12).

Table 10.1. Results obtained with the Goals Algorithm

Adaptive Step	Error	Effectivity Index
0	9.77%	94.2%
1	8.24%	93.8%
2	7.71%	93.3%
3	6.75%	91.9%
4	5.08%	88.0%
5	4.74%	92.1%

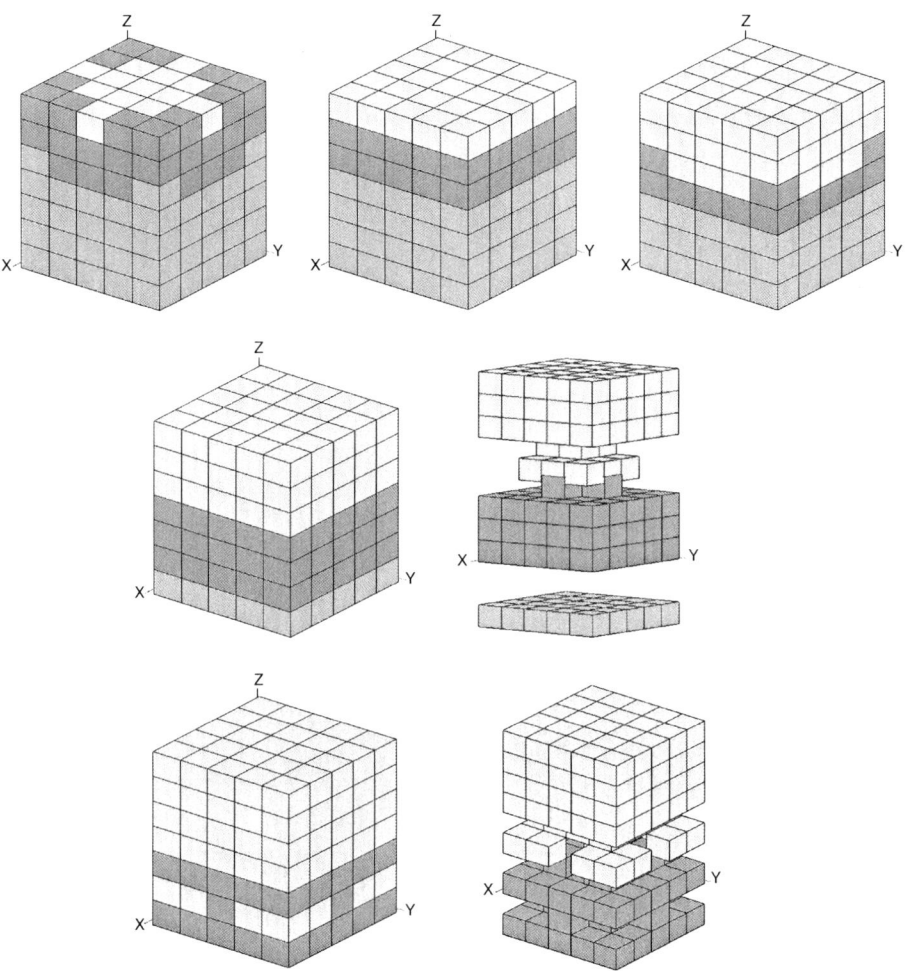

Figure 10.8: Sequence of surrogates generated by the Goals Algorithm. (See Plate 13).

functional by evaluating it only over a subset of the partition domains, various approximations of the adjoint solution, and issues with developing scalable parallel code. These issues are discussed more fully in [1, 3, 13].

10.6 Extensions

We close this chapter with brief discussions of extensions of the method to important classes of problems in multiscale modeling. One area is the extension of goals algorithms to stochastic systems and the quantification of uncertainties in quantities of interest. The idea of quantifying uncertainty in errors in quantities of interest is discussed in [14]. The difficult problem of extending the Arlequin-type methods for modeling the interface of discrete and continuum models to problems with random parameters and molecular structure is discussed in connection with model 1D problems in [15].

The framework developed earlier in this chapter can be extended to time-dependent problems in molecular dynamics and, as an important step in producing predictive simulations, to problems of calibration, inverse analysis, and optimal control. The mathematical formulation follows again arguments typical in optimal control theory (e.g. [16]). One such extension involves introducing the following generalizations of (10.9) and (10.10):

$$B(\beta, \mathbf{u}; \mathbf{v}) = \int_0^T \mathbf{v} \cdot \big(M\ddot{\mathbf{u}} - A(\beta, \mathbf{u})\big)\mathrm{d}t + \mathbf{v}(0) \cdot M\dot{\mathbf{u}}(0) - \dot{\mathbf{v}}(0) \cdot M\mathbf{u}(0)$$

$$F(\beta, \mathbf{v}) = \mathbf{v}(0) \cdot M\psi_0 - \dot{\mathbf{v}}(0) \cdot M\varphi_0$$

$$B'(\beta, \mathbf{u}; \mathbf{v}, \mathbf{p}) = \int_0^T \mathbf{v} \cdot \big(M\ddot{\mathbf{p}} - A'(\beta, \mathbf{u})^T \mathbf{p}\big)\mathrm{d}t + \dot{\mathbf{v}}(T) \cdot M\mathbf{p}(T) - \mathbf{v}(T) \cdot M\dot{\mathbf{p}}(T)$$

$$\Gamma(\beta, \mathbf{v}) = Q(\beta, \mathbf{v}) + \frac{\mu}{2}\|\varphi_0 - \varphi\|_{\mathcal{X}}^2 + \frac{\lambda}{2}\|\psi_0 - \psi\|_{\mathcal{Y}}^2 + \frac{1}{2}\|C\mathbf{v} - \gamma_{\mathrm{obs}}\|_{\mathcal{Z}}^2$$

$$(10.20)$$

Here M is the symmetric mass matrix for the entire molecular system and superimposed dots (˙), (¨) indicate time derivatives. Thus, for instance,

$$\mathbf{v} \cdot M\ddot{\mathbf{u}} = \sum_{i=1}^N \sum_{j=1}^N \mathbf{v}_i \cdot M_{ij} \frac{\mathrm{d}^2 \mathbf{u}_j}{\mathrm{d}t^2}$$

The initial velocity and displacement vectors are ψ_0 and φ_0 and are typically unknown. In the expression for Γ in (10.20), μ and λ are real positive parameters, and $\mathcal{X}$, $\mathcal{Y}$, and $\mathcal{Z}$ are Hilbert spaces of initial displacement fields, velocity fields, and time-dependent vectors γ_{obs} of values of experimental observations of the system. Thus, $C : U \to \mathcal{Z}$ is a *calibration operator*, mapping vectors in the trial space U into the observation space $\mathcal{Z}$. The functional $\Gamma : \mathbb{P} \times U \to \mathbb{R}$ is the cost functional providing control of the quantity of interest Q, itself a functional on U, the initial data, and the calibration data. The data vector $\beta \in \mathbb{R}^m$ of model parameters, primarily the parameters appearing in the molecular energy potentials and initial data, belongs to a manifold $\mathbb{P}$ of real, constant-in-time parameters, and discrete-valued functions defined on the initial configurations of the system. The mass matrix M_{ij} also may be random as different monomers, and hence different molecular masses, can occupy site i for different realizations of the polymer structure. Thus β can be written,

$$\beta = (\omega, \varphi_0, \psi_0)$$

where ω is a vector of model data parameters, such as those appearing in the molecular potentials, and φ and ψ are the initial displacement and velocity vectors.

We are concerned with the optimal control problem,

$$\boxed{\begin{aligned}&\text{Find } (\beta, \mathbf{u}) \in \mathbb{P} \times U \text{ such that}\\ &\Gamma(\beta, \mathbf{u}) = \inf_{\mathcal{W}} \Gamma(\gamma, \mathbf{v})\\ &\mathcal{W} = \{(\gamma, \mathbf{v}) \in \mathbb{P} \times U : B(\gamma, \mathbf{v}; \mathbf{w}) = F(\gamma; \mathbf{w}) \quad \forall \mathbf{w} \in V\}\end{aligned}} \quad (10.21)$$

The solution to (10.21) is characterized by the following system:

- The forward-primal problem:

$$\begin{aligned}\mathbf{u} &= \mathbf{u}(\beta) \in U\\ B(\beta, \mathbf{u}; \mathbf{v}) &= F(\beta; \mathbf{v}) \quad \forall \mathbf{v} \in V\end{aligned} \quad (10.22)$$

- The backward or adjoint problem:

$$\begin{aligned}\mathbf{p} &= \mathbf{p}(\beta) \in U\\ B'(\beta, \mathbf{u}; \mathbf{v}, \mathbf{p}) &= Q'(\beta, \mathbf{u}; \mathbf{v}) + (C\mathbf{u} - \gamma_{\text{obs}}, C\mathbf{v})_{\mathcal{Z}} \quad \forall \mathbf{v} \in V\end{aligned} \quad (10.23)$$

- The sensitivity of the cost functional

$$\Gamma_\beta(\beta, \mathbf{u}, \delta\beta) + F_\beta(\beta, \mathbf{p}, \delta\beta) - B_\beta(\beta, \mathbf{u}; \mathbf{p}, \delta\beta) = 0 \quad \forall \delta\beta \in \mathbb{P} \quad (10.24)$$

The system (10.22)–(10.24) is equivalent to the dynamical problem,

$$\begin{aligned}M\frac{d^2\mathbf{u}}{dt^2} &= A(\beta, \mathbf{u})\\ \mathbf{u}(0) &= \varphi; \quad \dot{\mathbf{u}}(0) = \psi\end{aligned} \quad (10.25)$$

$$\begin{aligned}M\frac{d^2\mathbf{p}}{dt^2} &= A'(\beta, \mathbf{u})^T \mathbf{p}(t) + \mathbf{q}'(\beta, \mathbf{u}) + C^T(C\mathbf{u} - \gamma_{\text{obs}})\\ \mathbf{p}(T) &= 0; \quad \dot{\mathbf{p}}(T) = 0\end{aligned} \quad (10.26)$$

$$\begin{aligned}&-B_\omega(\beta, \mathbf{u}; \mathbf{p}) + Q_\omega(\beta, \mathbf{u}) = 0\\ &-B_{\varphi_0}(\beta, \mathbf{u}; \mathbf{p}) + Q_{\varphi_0}(\beta, \mathbf{u}) + \mu(\varphi_0 - \varphi) - M\dot{\mathbf{p}}(0) = 0\\ &-B_{\psi_0}(\beta, \mathbf{u}; \mathbf{p}) + Q_{\psi_0}(\beta, \mathbf{u}) + \lambda(\psi_0 - \psi) + M\mathbf{p}(0) = 0\end{aligned} \quad (10.27)$$

where now

$$Q'(\beta, \mathbf{u}; \mathbf{v}) = \int_0^T \mathbf{q}'(\beta, \mathbf{u}) \cdot \mathbf{v} \, dt$$

$$A(\beta, \mathbf{u}) = \sum_{i=1}^N \sum_{j=1}^{N_i} \frac{\partial E_{ij}(\beta, \mathbf{u})}{\partial u_i} \tag{10.28}$$

$$A'(\beta, \mathbf{u})^T \mathbf{p} = \sum_{i=1}^N \sum_{j=1}^{N_i} \sum_{k=1}^N \frac{\partial^2 E_{ij}(\beta, \mathbf{u})}{\partial u_k \partial u_i} p_k$$

To compute sensitivities of the solution $\mathbf{u}$ to changes in the parameters β, we consider a path $S(\beta)$ in U, parameterized by β, along which the forward problem is solved; i.e. $\forall \beta \in S(\beta)$, we have $B(\beta, \mathbf{u}; \mathbf{v}) = F(\beta, \mathbf{v})$ $\forall \mathbf{v} \in V$. The change in $\mathbf{u}$ due to the change in β is denoted $\mathbf{u}'(\beta)$. Thus, along $S(\beta)$, the change in $B(\beta, \mathbf{u}; \mathbf{v}) - F(\beta, \mathbf{v})$ is zero:

$$B_\beta(\beta, \mathbf{u}; \mathbf{v}, \delta\beta) + B'(\beta, \mathbf{u}; \mathbf{u}', \mathbf{v}) - F_\beta(\beta, \mathbf{v}, \delta\beta) = 0 \quad \forall \mathbf{v} \in V$$

The backward (adjoint) problem along $S(\beta)$ satisfies

$$B'(\beta, \mathbf{u}; \mathbf{u}', \mathbf{p}) = \Gamma'(\beta, \mathbf{u}; \mathbf{u}')$$

where $\Gamma'(\beta, \mathbf{u}; \mathbf{u}')$ is given by the right-hand side of (10.23) with $\mathbf{v} = \mathbf{u}'$. Thus, the change in the objective functional Γ along $S(\beta)$ is

$$D_\beta \Gamma = \partial_\beta \Gamma_{|(\mathbf{u}, \varphi, \psi)} - B_\beta(\beta, \mathbf{u}; \mathbf{p}, \delta\beta) + F_\beta(\beta, \mathbf{p}, \delta\beta)$$

These functions define the sensitivity of the solution to the parameters β.

Let $(\mathbf{u}_0, \mathbf{p}_0)$ denote a solution pair to any surrogate model approximating the system (10.22)–(10.24) with a parameter set β_0. Then arguments similar to those leading to (10.8) lead to the *a posteriori* error estimate,

$$\Gamma(\beta, \mathbf{u}) - \Gamma(\beta_0, \mathbf{u}_0) \approx \mathcal{R}(\beta_0, \mathbf{u}_0; \mathbf{p}) \tag{10.29}$$

where $\mathcal{R}(\beta_0, \mathbf{u}_0; \mathbf{p})$ is the time-dependent residual:

$$\mathcal{R}(\beta_0, \mathbf{u}_0; \mathbf{p}) = F(\beta_0; \mathbf{p}) - B(\beta_0, \mathbf{u}_0; \mathbf{p}) + F_\beta(\beta_0; \mathbf{p}, \beta - \beta_0)$$

$$- B_\beta(\beta_0, \mathbf{u}_0; \mathbf{p}, \beta - \beta_0) \tag{10.30}$$

The difference $\beta - \beta_0$ can, in principle, be estimated using the inverse analysis and calibration process alluded to earlier. Further refinements of the estimate (10.29) are believed to be possible, but await further research. This modeling error estimate can be readily used as a basis for adaptive modeling via the goals algorithm described earlier.

10.7 Concluding comments

An important goal of contemporary computational science is quantifiable predictability, the systematic prediction of the behavior of physical systems with quantifiable metrics of confidence and uncertainty. What is remarkable is that the core of the success of predictive methods is the selection and ultimately the control of the models used as the bases of simulation. It is hoped that the methodologies described here and their applications to large-scale problems provide some bases for advancing this important subject.

The idea of estimating and controlling errors induced in computational models by invoking assumptions is an extremely powerful concept and can bring a new level of sophistication and rigor to the analyses of the most complex problems in multiscale modeling. Many issues remain to be resolved in applications of these ideas to large-scale problems, and these will involve advances in goals-type algorithms used to implement adaptive modeling strategies. Extensions to stochastic models and the use of frameworks similar to those discussed in Section 10.6 will make feasible the development of methods for uncertainty quantifications in large multiscale problems.

10.8 Acknowledgments

The support of this work by the Department of Energy under contract DE-FG02-05ER25701 is gratefully acknowledged.

References

[1] Bauman P. T. (2008). *Adaptive Multiscale Modeling of Polymeric Materials Using Goal-Oriented Error Estimation, Arlequin Coupling, and Goals Algorithms.* PhD dissertation, The University of Texas at Austin.

[2] Bauman P. T., Ben Dhia H., Elkhodja N., Oden J. T., and Prudhomme S. (2008). "On the application of the Arlequin method to the coupling of particle and continuum models", *Computational Mechanics*, **42**(4), p. 511–530.

[3] Bauman P. T., Oden J. T., and Prudhomme S. (2008). "Adaptive multiscale modeling of polymeric materials: Arlequin coupling and goals algorithm", *Computer Methods in Applied Mechanics and Engineering*, **198**(5–8), p. 799–818.

[4] Oden J. T. and Prudhomme S. (2002). "Estimation of modeling error in computational mechanics", *Journal of Computational Physics*, **182**, p. 496–515.

[5] Oden J. T., Prudhomme S., Romkes A., and Bauman P. T. (2006). "Multiscale modeling of physical phenomena: adaptive control of models", *SIAM Journal of Scientific Computing*, **28**(6), p. 2359–2389.

[6] Prudhomme S., Ben Dhia H., Bauman P. T., Elkhodja N., and Oden J. T. (2008). "Computational analysis of modeling error for the coupling of particle and continuum models by the Arlequin method", *Computer Methods in Applied Mechanics and Engineering*, **197**(41–42), p. 3399–3409.

[7] Vanderzande C. (1998). *Lattice Models of Polymers*, Cambridge University Press.
[8] Long B. K., Keitz B. K., and Willson C. G. (2007). "Materials for step and flash imprint lithography (S-FIL)", *Journal of Materials Chemistry*, **17**, p. 3575–3580.
[9] Metropolis N., Rosenbluth A. W., Rosenbluth M. N., Teller M., and Teller E. (1953). "Equation of state calculations by very fast computing machines". *Journal of Chemical Physics*, **21**, p. 1087.
[10] Balay S., Buschelman K., Gropp W. D., Kaushik D., Knepley M. G., McInnes L. C., Smith B. F., and Zhang H. (2001). PETSc Web page: http://www.mcs.anl.gov/petsc.
[11] Benson S., McInnes L. C., Moré J., Munson T., and Sarich J. (2007). *TAO User Manual (Revision 1.9)*. Mathematics and Computer Science Division, Argonne National Laboratory, ANL/MCS-TM-242. Web page: http://www.mcs.anl.gov/tao.
[12] Ben Dhia H. and Rateau G. (2005). "The Arlequin method as a flexible engineering design tool". *International Journal for Numerical Methods in Engineering*, **62**(11), p. 1442–1462.
[13] Prudhomme S., Chamoin L., Ben Dhia H., and Bauman P. T. (2009). "An adaptive strategy for the control of modeling error in two-dimensional atomic-to-continuum coupling simulations", *Computer Methods in Applied Mechanics and Engineering*, (in press).
[14] Oden J. T., Babuska I., Nobile F., Feng Y., and Tempone R. (2005). "Theory and methodology for estimation and control of errors due to modeling, approximation, and uncertainty", *Computer Methods in Applied Mechanics and Engineering*, **194**(42–44), p. 4506–4527.
[15] Chamoin L., Oden J. T., and Prudhomme S. (2008). "A stochastic coupling method for atomic-to-continuum Monte-Carlo simulations", *Computer Methods in Applied Mechanics and Engineering*, **197**(43–44), p. 3530–3546.
[16] Le Dimet F. X. and Shutyaev V. (2005). "On deterministic error analysis in variational data assimilation", *Nonlinear Processes in Geophysics*, **12**, p. 481–490.

11

ERROR ESTIMATES FOR MULTISCALE OPERATOR DECOMPOSITION FOR MULTIPHYSICS MODELS

D. Estep

11.1 Introduction

Multiphysics, multiscale models that couple different physical processes acting across a large range of scales are encountered in virtually all scientific and engineering applications. Such systems present significant challenges in terms of computing accurate solutions and for estimating the error in information computed from numerical solutions. In this chapter, we discuss the problem of computing accurate error estimates for one of the most common, and powerful, numerical approaches for multiphysics, multiscale problems.

Examples of multiphysics models

Without any attempt to be complete, we describe three examples of multiphysics models that illustrate some different ways in which physical processes may be coupled.

Example 1. A thermal actuator

A thermal actuator is a MEMS (microelectronic mechanical switch) device. A contact rests on thin braces composed of a conducting material. When a current is passed through the braces, they heat up and consequently expand to close the contact, see Figure 11.1. The actuator is modeled by a system of three coupled

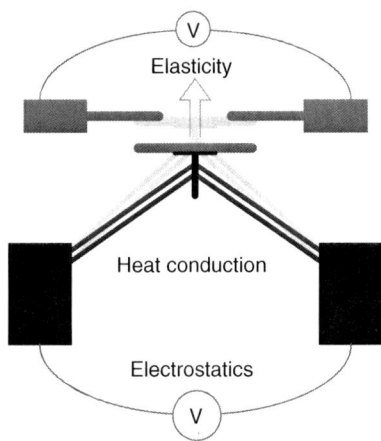

Figure 11.1: Sketch of a thermal actuator.

equations, each representing a distinct physical process acting on its own scale. The first is an electrostatic current equation

$$\nabla \cdot (\sigma \nabla u_1) = 0, \qquad (11.1)$$

governing potential u_1 (where the current is $J = -\sigma \nabla u_1$), the second is a steady-state energy equation

$$\nabla \cdot (\kappa(u_2) \nabla u_2) = \sigma(\nabla u_1 \cdot \nabla u_1), \qquad (11.2)$$

for the governing temperature u_2, and a linear elasticity equation giving the steady-state displacement u_3,

$$\nabla \cdot \left(\lambda \operatorname{tr}(E) I + 2\mu E - \beta(u_2 - u_{2,ref}) I\right) = 0, \quad E = \left(\nabla u_3 + \nabla u_3^\top\right)/2. \qquad (11.3)$$

This is an example of "parameter passing", in which the solution of one component is used to compute the parameters and/or data for another component. Note that the electric potential u_1 can be calculated independently of u_2 and u_3. The temperature u_2 can be calculated once the electric potential u_1 is known, while the calculation of displacement u_3 requires prior knowledge of u_2, and therefore of u_1.

Example 2.

The Brusselator problem First introduced by Prigogine and Lefever [1] as a model of chemical dynamics, the Brusselator problem consists of a coupled set of equations,

$$\begin{cases} \frac{\partial u_1}{\partial t} - k_1 \frac{\partial^2 u_1}{\partial x^2} = \alpha - (\beta + 1) u_1 + u_1^2 u_2, & x \in (0,1), t > 0, \\ \frac{\partial u_2}{\partial t} - k_2 \frac{\partial^2 u_2}{\partial x^2} = \beta u_1 - u_1^2 u_2, & x \in (0,1), t > 0, \\ u_1(0,t) = u_1(1,t) = \alpha, \ u_2(0,t) = u_2(1,t) = \beta/\alpha, & t > 0, \\ u_1(x,0) = u_{1,0}(x), \ u_2(x,0) = u_{2,0}(x), & x \in (0,1), \end{cases} \qquad (11.4)$$

where u_1 and u_2 are the concentrations of species 1 and 2, respectively. Solutions of the Brusselator problem exhibit a wide range of behavior depending on parameter values.

Reaction-diffusion equations are an example of a problem that combines different physics—in this case, reaction and diffusion—in one equation. The generic picture for a reaction-diffusion equation is a relatively fast, destabilizing reaction component interacting with a relatively slow, stabilizing diffusion component. Thus, the physical components have both different scales and different stability properties.

Example 3.

Conjugate heat transfer between a fluid and solid object We consider the flow of a heat-conducting Newtonian fluid past a solid cylinder as shown in Figure 11.2.

The model consists of the heat equation in the solid and the equations governing the conservation of momentum, mass, and energy in the fluid, where we apply the Boussinesq approximation to the momentum equations in the fluid. The temperature field is advected by the fluid and couples back to the momentum equations through the buoyancy term.

Let Ω_F and Ω_S be polygonal domains in $\mathbb{R}^2$ with boundaries $\partial \Omega_F$ and $\partial \Omega_S$ intersecting along an interface $\Gamma_I = \partial \Omega_S \cap \partial \Omega_F$. The complete coupled problem is

$$\begin{cases} -\mu \Delta u + \rho_0 (u \cdot \nabla) u + \nabla p + \rho_0 \beta T_F g = \rho_0 (1 + \beta T_0) g, & x \in \Omega_F, \\ -\nabla \cdot u = 0, & x \in \Omega_F, \\ -k_F \Delta T_F + \rho_0 c_p (u \cdot \nabla T_F) = Q_F, & x \in \Omega_F, \\ \begin{cases} T_S = T_F, \\ k_F (n \cdot \nabla T_F) = k_S (n \cdot \nabla T_S), \end{cases} & x \in \Gamma_I, \\ -k_S \Delta T_S = Q_S, & x \in \Omega_S, \end{cases} \quad (11.5)$$

where ρ_0 and T_0 are reference values for the density and temperature, respectively, μ is the molecular viscosity, β is the coefficient of thermal expansion, c_p is the specific heat, k_F and k_S are the thermal conductivities of the fluid and solid, respectively, Q_F and Q_S are source terms, and n is the unit normal vector directed into the fluid. Note that u is a vector.

We define $\Gamma_{u,D}$ and $\Gamma_{u,N}$ to be the boundaries on which we apply Dirichlet and Neumann conditions for the velocity field respectively, and set

$$\begin{cases} u = g_{u,D}, & x \in \Gamma_{u,D}, \\ \mu \dfrac{\partial u}{\partial n} = g_{u,N}, & x \in \Gamma_{u,N}. \end{cases}$$

Similarly, we define $\Gamma_{T_F,D}$, $\Gamma_{T_F,N}$, $\Gamma_{T_S,D}$, and $\Gamma_{T_S,N}$ to be the boundaries on which we impose Dirichlet and Neumann conditions for the temperature fields

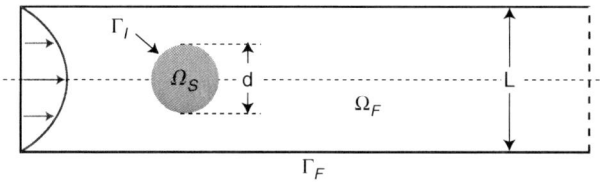

Figure 11.2: Computational domain for flow past a cylinder.

in the fluid and the solid respectively, and set

$$\begin{cases} T_F = g_{T_F, D}, & x \in \Gamma_{T_F, D}, \\ k_F(n \cdot \nabla T_F) = g_{T_F, N}, & x \in \Gamma_{T_F, N}, \\ T_S = g_{T_S, D}, & x \in \Gamma_{T_S, D}, \\ k_S(n \cdot \nabla T_S) = g_{T_S, N}, & x \in \Gamma_{T_S, N}. \end{cases}$$

This presents a class of problems where different physics in different physical domains are coupled through interactions across a common boundary.

11.1.1 Challenges and goals of multiscale, multiphysics models

Multiscale, multiphysics models are characterized by intimate interactions between different physics across a wide range of scales. This poses challenges for solving such problems, e.g.:

Accurate and efficient computation Computing information that depends on solution behavior occurring at very different scales is problematic. It is rarely possible to simply use a discretization sufficiently fine to resolve the finest scale behavior.

Complex stability A multiphysics model generally offers a complex stability picture that results from a fusion of the stability properties of different physics. For example, consider a reacting fluid that combines fluid flow with the dynamical properties of reaction-diffusion equations.

Linking different physics across scales Understanding the significance of linkages between physical components and how those affect model output is another complicated issue. In many situations, the output of one physical component must be transformed and/or scaled to obtain information relevant to the other components.

Another complication is the range of applications of multiphysics models. These include:

Model prediction Perhaps the chief goal of mathematical modeling is to predict the behavior of the modeled system outside of the range of physical observation.

Sensitivity analysis The reliability of model predictions depends on analyzing the effects of uncertainties and variation in the physical properties of the model on its output.

Parameter optimization In design problems, the goal is to determine optimal parameter values with respect to producing a desired observation or consequence.

Such applications require computation of solutions corresponding to a wide range of data and parameters. We expect the solution behavior to vary significantly and the ability to obtain accurate numerical solutions therefore to vary as well. This raises a critical need for quantification and control of numerical error.

The solution and application of multiphysics, multiscale models invoke two computational goals:

- Compute specific information from multiscale, multiphysics models accurately and efficiently
- Accurately quantify the error and uncertainty in any computed information.

The context is important:

It is often difficult or impossible to obtain solutions of multiscale, multiphysics models that are uniformly accurate throughout space and/or time.

Thus, we are interested in computing accurate error estimates for solutions that are relatively inaccurate. This is an important consideration, given that much of classical error analysis is derived under conditions that amount to assuming that the numerical solution is in the "asymptotic range of convergence", meaning that the solution is sufficiently accurate that the rate of convergence can be observed by uniform refinement of the discretization. It is rarely possible to reach this level of discretization in a multiphysics, multiscale problem.

11.1.2 Multiscale, multidiscretization operator decomposition

Multiscale, multidiscretization operator decomposition is a widely used technique for solving multiphysics, multiscale models. The general approach is to decompose the multiphysics and/or multiscale problem into components involving simpler physics over a relatively limited range of scales, and then to seek the solution of the entire system through some sort of iterative procedure involving numerical solutions of the individual components. We illustrate in Figure 11.3. In general, different components are solved with different numerical methods as well as with different scale discretizations. This approach is appealing because there is generally a good understanding of how to solve a broad spectrum of single physics problems accurately and efficiently, and because it provides an alternative to accommodating multiple scales in one discretization.

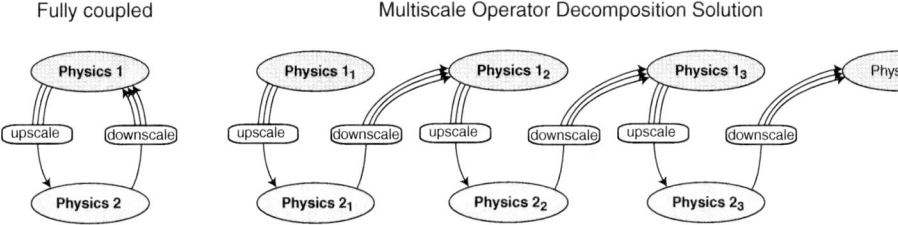

Figure 11.3: Left: Illustration of a multiscale, multiphysics model. Right: Illustration of a multiscale operator decomposition solution.

Example 4.

A classic example of multiscale operator decomposition is operator splitting for a reaction-diffusion equation [46, 49, 50, 51, 53],

$$\begin{cases} \dfrac{\partial u}{\partial t} = \epsilon \Delta u + f(u), & x \in \Omega, 0 < t, \\ \text{suitable boundary conditions}, & x \in \partial\Omega, 0 < t, \\ u(\cdot, 0) = u_0(\cdot) \end{cases} \quad (11.6)$$

where $\Omega \subset \mathbb{R}^d$ is a spatial domain and f is a smooth function. Accuracy considerations dictate the use of relatively small steps to integrate a fast reaction component. On the other hand, stability considerations over moderate to long time intervals suggests the use of implicit, dissipative numerical methods for integrating diffusion problems. Such methods are expensive to use per step, but relatively large steps can be used on a purely dissipative problem. If the reaction and diffusion components are integrated together, then the small steps required for accurate resolution of the reaction lead to an expensive computation.

In a multiscale operator splitting approach, the reaction and diffusion components are integrated independently inside each time interval of a discretization of time and "synchronized" in some fashion only at the nodes of the interval. The reaction component is often integrated by using significantly smaller sub-steps (e.g. 10^{-5} smaller is not uncommon) than those used to integrate the diffusion component, which can lead to tremendous computational savings.

Employing the method of lines, if we discretize in space using a continuous, piecewise linear finite element method with M elements, see Section 11.4.1, we obtain the initial value problem: find $y \in \mathbb{R}^M$ such that

$$\begin{cases} \dot{y} = Ay(t) + F(y(t)), & 0 < t \leq T, \\ y(0) = y_0, \end{cases} \quad (11.7)$$

where A is an $l \times l$ constant matrix representing a "diffusion component" and $F(y) = (F_1(y), F_2(y), \ldots, F_l(y))^\top$ is a vector of nonlinear functions representing a "reaction component".

We first discretize $[0, T]$ into $0 = t_0 < t_1 < t_2 < \cdots < t_N = T$ with diffusion time steps $\{\Delta t_n\}_{n=1}^N$, $\Delta t_n = t_n - t_{n-1}$, and $\Delta t = \max_{1 \leq n \leq N}(\Delta t_n)$. We define a piecewise continuous approximate solution

$$\tilde{y}(t) = \frac{t_n - t}{\Delta t_n} \tilde{y}_{n-1} + \frac{t - t_{n-1}}{\Delta t_n} \tilde{y}_n, \quad t_{n-1} \leq t \leq t_n, \quad (11.8)$$

with nodal values $\{\tilde{y}_n\}$ obtained from the following procedure:

Algorithm 11.1.8 Abstract Operator Splitting for Reaction-Diffusion Equations

Set $\tilde{y}_0 = y_0$
for $n = 1, \ldots, N$ **do**
    Compute $y^r(t_n^-)$ satisfying the reaction component

$$\begin{cases} \dot{y}^r = f(y^r(t)), & t_{n-1} < t \leq t_n, \\ y^r(t_{n-1}^+) = \tilde{y}_{n-1} \end{cases} \quad (11.9)$$

    Compute $y^d(t_n^-)$ satisfying the diffusion component

$$\begin{cases} \dot{y}^d = A y^d(t), & t_{n-1} < t \leq t_n, \\ y^d(t_{n-1}^+) = y^r(t_n^-) \end{cases} \quad (11.10)$$

    Set $\tilde{y}_n = y^d(t_n^-)$
end for

With a little thought, we recognize that this algorithm has the potential to be a multiscale solution procedure since we can now resolve the solution of each component on independent scales. That is one benefit of using operator decomposition. Unfortunately, this decomposition has unforseen effects on both accuracy and stability. The reason is that we have discretized the instantaneous interaction between the reaction and diffusion components.

Example 5.
In [2], we consider a problem in which the reaction component exhibits finite time blow up when undamped by the diffusion component. The problem is

$$\begin{cases} \dot{y} + \lambda y = y^2, & t > 0, \\ y(0) = y_0 \in \mathbb{R}, \end{cases} \quad (11.11)$$

which has the exact solution

$$y(t) = \frac{\lambda y_0}{y_0 - (y_0 - \lambda) e^{\lambda t}}, \quad (11.12)$$

when $\lambda \neq 0$. The exact solution exists for all time and tends to zero as $t \to \infty$ when $\lambda > y_0$. On the other hand, there is finite time blow up, e.g. $y \to \infty$ at a finite time, if $\lambda < y_0$.

Applying the operator splitting to (11.11), the solutions of the two components and the operator splitting solution are,

$$y^r(t) = \frac{y_{n-1}^{d-}}{1 - y_{n-1}^{d-}(t - t_{n-1})}, \quad y^d(t) = e^{-\lambda(t - t_{n-1})} y_n^{r-}, \quad \tilde{y}_n = \frac{e^{-\lambda \Delta t_n} \tilde{y}_{n-1}}{1 - \Delta t_n \, \tilde{y}_{n-1}},$$

when the reaction component is defined. We see that decoupling the smoothing effect provided by instantaneous interaction with the diffusion component means that the reaction component can blow up in finite time. This has an effect on numerical solution.

We consider the time steps introduced above, $\{\Delta t_n\}_{n=1}^N$, to be diffusion time steps. For each diffusion step, we choose a (small) time step $\Delta s_n = \Delta t_n / M_n$ with $\Delta s = \max_{1 \leq n \leq N} \Delta s_n$, and the nodes $t_{n-1} = s_{0,n} < s_{1,n} < \cdots < s_{M_n,n} = t_n$ (see Figure 11.4). We associate the time intervals $I_n = [t_{n-1}, t_n]$ and $I_{m,n} = [s_{m-1,n}, s_{m,n}]$ with these discretizations. Without going into details, we solve the components (11.9), (11.10) using the forward and backward Euler method respectively,

$$Y^{r-}_{m,n} = Y^{r-}_{m-1,n} + f(Y^{r-}_{m-1,n})\Delta s_n, \quad Y^{d-}_n = Y^{d-}_{n-1} + AY^{d-}_n \Delta t_n.$$

See Section 11.4.5 for details on discretization of evolution problems. We compute a piecewise linear discrete approximation $\tilde{Y}$ using the nodal values of Y^d.

On the left side of Figure 11.5, we plot the true solution and the nodal values of the approximation $\tilde{Y}$ computed with $N = 50$ diffusion steps and $M = 1$ reaction step per diffusion step. The approximation is reasonably accurate. Next,

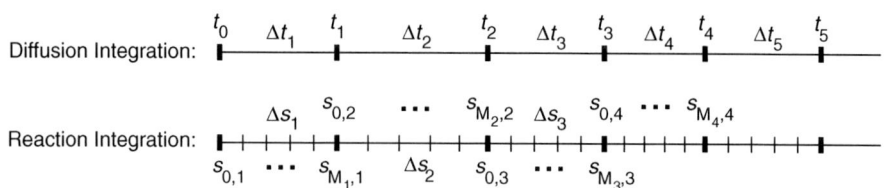

Figure 11.4: Discretization of time used for multiscale operator splitting.

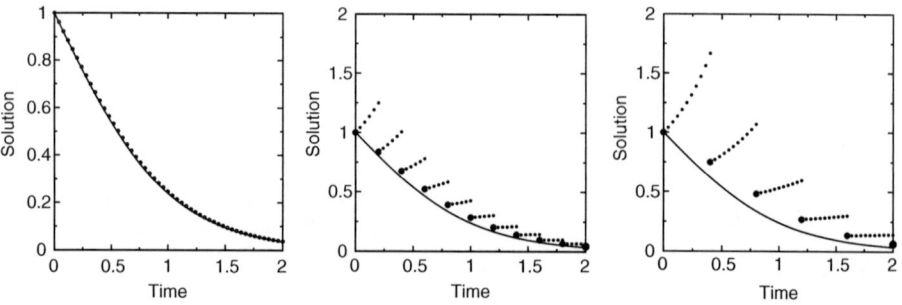

Figure 11.5: Plots of the approximation $\tilde{Y}$ and the true solution. Left: $N = 50$, $M = 1$. Middle: $N = 10$, $M = 5$. Right: $N = 5$, $M = 10$. The nodal values of $\tilde{Y}$ are denoted by the larger points while the smaller points denote node values of the reaction component Y^r.

we show the results when the diffusion step is increased by choosing $N = 10$ and, in order to maintain the same resolution of the reaction component, we correspondingly increase to $M = 5$ reaction steps per diffusion step. The node values of $\tilde{y}$ are relatively close to those of y. The subsequent nodal values of the reaction component solution y^r inside each step move away from the true solution. This large departure is somewhat counteracted by application of the diffusion operator. The reaction components exhibit significant growth inside each diffusion step, which severely affects accuracy.

If we increase the diffusion step by taking $N = 5$ and maintain resolution in the reaction component by taking $M = 10$, the approximation becomes even less accurate. If we increase the diffusion step further, then the reaction component actually blows up inside a diffusion step.

We emphasize that the error in this example is a consequence of a kind of instability introduced by multiscale operator decomposition. We will see below that multiscale operator decomposition commonly affects both accuracy and stability in a wide variety of problems.

11.2 The key is stability. But what is stability... and stability of what?

Stability is likely one of the most shopworn terms in mathematics, given so many meanings so as to cause a high probability of miscommunication in any mixed crowd. Nonetheless, stability is the key to quantifying the effects of error and uncertainty on the output of a computed model solution.

Generally, it is impossible to give a definitive definition of stability in the context of a multiphysics model. A computational mathematician might think in terms of numerical stability, with instability characterized by oscillations on the scale of the discretization. A mathematician in partial differential equations might be thinking in terms of well-posedness, i.e. continuous dependence on data. A physicist might be worried about preserving the conservation of important quantities like mass and energy. A dynamicist might think in terms of the stability properties of stationary solutions and attracting manifolds.

Indeed, all of these views of stability, and likely others, are important in the right contexts. In fact, the only definitive thing to be said about stability is that it is very unlikely that just one view of stability will suffice when solving a multiphysics, multiscale problem.

11.2.1 *Pointwise stability of the Lorenz problem*

To illustrate the complexity of stability, we consider the infamous Lorenz problem,

$$\begin{cases} \dot{u}_1 = -10u_1 + 10u_2, \\ \dot{u}_2 = 28u_1 - u_2 - u_1 u_3, \quad 0 < t, \\ \dot{u}_3 = -\frac{8}{3}u_3 + u_1 u_2. \end{cases} \quad (11.13)$$

The Lorenz equations were derived by Lorenz [3] as a gross simplification of a weather model to explain why weather predictions become inaccurate after a few days. We have chosen parameter values believed to lead to chaotic behavior. In Figure 11.6 we plot a solution. All solutions rapidly approach the "strange attractor" where they subsequently remain. The dynamical behavior is always the same in qualitative terms. There are two nonzero steady-state solutions and a generic solution is either "orbiting" one of these solutions or transitioning between orbits. The orbits spiral away from the steady-state solution at the center until a point when a solution is sufficiently far away from the fixed point, whereupon it moves to orbit around the other fixed point. In a crude way, solutions behave in a very predictable fashion.

Chaos is often described as "sensitivity to initial conditions", which means that solutions that begin close by to each other eventually move apart. In Figure 11.7, we plot a second solution to the Lorenz problem that begins near the solution plotted in Figure 11.6 along with the distance between the solutions. The two solutions start close together and actually remain close until around time 17.5, when there is a rapid increase in their separation. For a brief period between 18 and 21, the separation remains fairly constant, and then it begins to increase again, with another very rapid increase around 24. All solutions must remain in a compact region around the origin, so at some point the distance between the solutions reaches the order of size of the compact region and cannot grow further.

We conclude that two solutions that are pointwise close at some time eventually diverge pointwise at a later time. The chaotic nature of the Lorenz problem means that it is difficult to predict the pattern of orbits around the fixed points

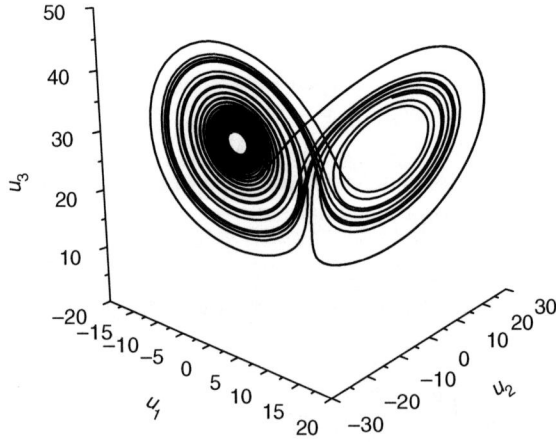

Figure 11.6: Solution of the Lorenz problem (11.13) corresponding to initial condition $(-9.408, -9.096, 28.581)$.

with any accuracy very far into the future. On the other hand, nearby solutions may remain close for quite some time and may even become closer before eventually diverging.

The source of chaotic behavior in the Lorenz problem is actually rather complex [4]. However, one important factor is relatively easy to explain. Following [4], in Figure 11.8, we show a plot looking straight down the vertical axis at parts of many solutions. The solutions shown in the lower left corner are orbiting around one of the nonzero fixed points or, if they are in the "outer" orbit,

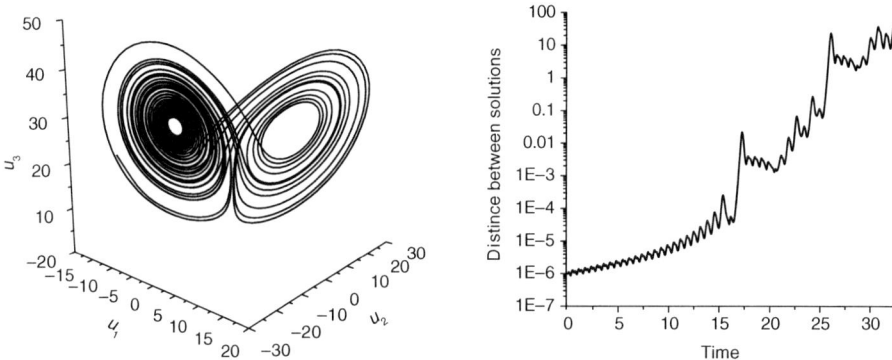

Figure 11.7: A second solution of the Lorenz problem (11.13) corresponding to initial condition $(-9.408, -9.096001, 28.581)$ and the pointwise difference between the two solutions.

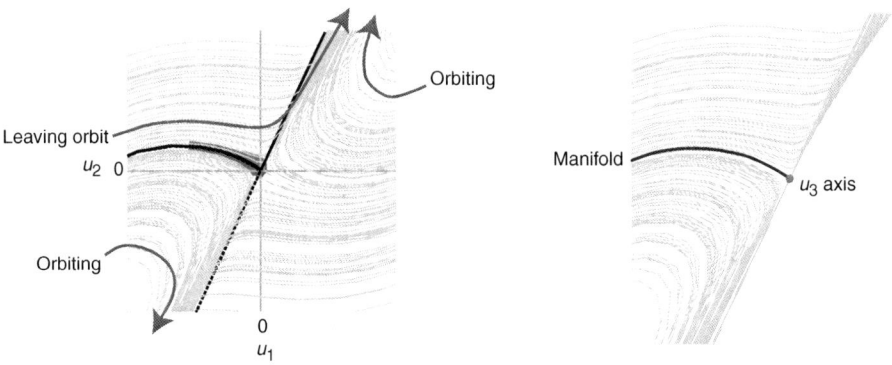

Figure 11.8: Left: Looking straight down the u_3 (vertical) axis at many solutions of the Lorenz equations (11.13). Solutions passing through a neighborhood of a separatrix coming out of the u_3 axis, shaded in the figure, are very sensitive to small perturbations while they are in that neighborhood. Right: A plot of the separatrix.

moving to the neighborhood of the other fixed point. Likewise, the solutions plotted in the upper right corner are either orbiting around a fixed point or moving to a neighborhood of the other fixed point.

We note that there are two solutions in the lower left-hand region, that are very close until they approach the vertical u_3 axis but then rapidly move apart after that. In fact, there is a separatrix, or manifold, coming out of the u_3 axis that separates solutions taking another orbit around a fixed point from those that transition to the other fixed point. Solutions on either side of this manifold move apart rapidly. Thus, we see how small perturbations can lead to rapid separation. Any solution that passes near the neighborhood of the separatrix, shaded in the figure, become very sensitive to small perturbations during the short time the solution remains in the neighborhood. Eventually, all solutions pass nearby the separatrix and thus become sensitive to perturbation. Away from the neighborhood of the separatrix, the distance between solutions may grow or shrink slowly, e.g. at a polynomial rate. This explains the pattern of separation seen in the plot of distance between two solutions in Figure 11.7.

Not surprisingly, chaotic behavior affects numerical solutions as well. In Figure 11.9, we show the effects of varying step sizes on pointwise accuracy. We plot the difference between the numerical solutions on the left in Figure 11.10. The pointwise numerical error clearly follows an increasing trend, but it does not increase monotonically. In fact, the pointwise error actually decreases during some short periods of time, see the plot on the right in Figure 11.10.

11.2.2 *Classic* a priori *stability analysis*

But what about the classic *a priori* stability analysis that is taught in courses in differential equations and numerical analysis (*a priori* means that it is carried out before any solutions are computed)? Do these classic notions of stability present

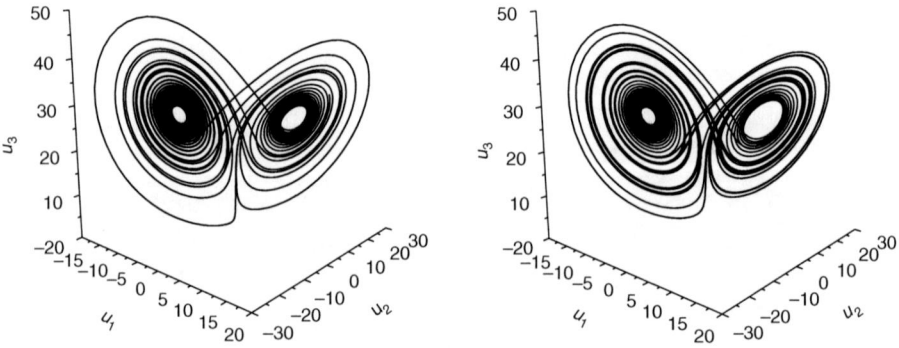

Figure 11.9: Two numerical approximations of the Lorenz solution shown in Figure 11.6 are shown; the solution on the left is accurate while the solution on the right is computed with larger step sizes. The distance between the solutions is shown on the right.

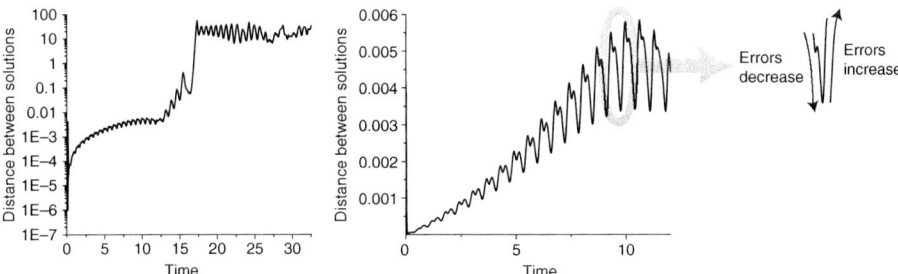

Figure 11.10: Left: Plot of the pointwise difference between the numerical solutions of Lorenz shown in Figure 11.9. Right: A blow up of the difference for $0 \leq t \leq 12$.

a reasonable picture for a particular solution? In fact, the answer is decidedly no in most cases. The classic view of stability for linear problems tends to be black and white; a particular solution is either stable or unstable with respect to perturbations. We see that this point of view fails for solutions of simple nonlinear problems, e.g. the Lorenz problem.

Example 6.

It is easy to illustrate the shortcomings of *a priori* stability analysis using error analysis of the numerical solution of a matrix system of equations. Consider a numerical solution $Y \approx y$ of a matrix system

$$\mathbf{A}y = b, \tag{11.14}$$

computed using Gaussian elimination. The computable residual of Y is $R = \mathbf{A}Y - b$ and the classic relative error bound [5] is

$$\frac{\|Y - y\|}{\|y\|} \leq C\kappa(\mathbf{A})\frac{\|R\|}{\|b\|}, \tag{11.15}$$

where the condition number $\kappa(\mathbf{A}) = \|\mathbf{A}\| \, \|\mathbf{A}^{-1}\|$ is a measure of the sensitivity of the solution of (11.14) to perturbations in the data, i.e. the stability properties of the inverse operator of $\mathbf{A}$. In particular,

$$\kappa(\mathbf{A}) = \frac{1}{distance\ from\ \mathbf{A}\ to\ \{singular\ matrices\}}.$$

(To be precise, we have to specify norms, but that level of detail is not important here.)

We now solve (11.14) using Gaussian elimination, where $\mathbf{A}$ is a random 800×800 matrix, where the coefficients in the random matrix are uniformly distributed $U(-1, 1)$ on $(-1, 1)$. The goal is to determine the first component y_1 of the solution. The condition number of $\mathbf{A}$ is 6.7×10^4. Straightforward computation yields

$$\text{actual error in the quantity of interest} \approx 1.0 \times 10^{-15},$$
$$\text{traditional error bound for the error} \approx 3.5 \times 10^{-5}.$$

We see that the traditional error bound is orders of magnitude larger than the actual error and is essentially useless as far as estimating the error in the particular computed information.

We remark that the bound (11.15) is a specific example of a general "meta-theorem", which reads

11.2.3 Theorem

$$\|\textit{effect of perturbation on the output of an operator}\|$$
$$\leq \|\textit{measure of stability of the operator}\| \times \|\textit{size of the perturbation}\|. \tag{11.16}$$

In the linear algebra example, we have

$$AY = b + R,$$

hence we can think of the numerical solution Y as solving the linear system (11.14) with perturbed data $b + R$.

The pessimism of a classic *a priori* stability bound is not surprising given a little reflection. After all, the goal of such a bound is to account for the largest possible error in a large class of solutions corresponding to a large set of data, not produce an accurate error estimate for particular information computed from a particular solution. The power of an *a priori* error bound is that it characterizes the general behavior of the numerical method.

The situation for nonlinear problems is worse. For example, in nonlinear evolution problems, the classic stability analysis uses a Gronwall argument to obtain a bound in the form,

$$\text{effect of perturbation at time } t \leq C\,e^{Lt} \times \text{size of perturbation,}$$

where C and L are constants with L typically large (L is on the order of 100 in the Lorenz example). Such bounds are nondescriptive past a very short initial transient, e.g. even for the chaotic Lorenz problem. The factor Ce^{Lt} plays the role of a condition number (for an absolute error estimate).

11.2.4 Stability for stationary problems

There is a long tradition of conducting careful, precise analysis of stability for evolutionary problems and, in particular, distinguishing different types of stability properties. It is perhaps fair to say that stability for stationary problems tends to be treated more crudely, at least for elliptic problems. This is unfortunate because stability is just as complex an issue for stationary problems as evolutionary problems.

Example 7.
To illustrate this claim, we consider an elliptic problem

$$\begin{cases} L(u) = -\nabla \cdot ((.001 + |\tanh(10(y+1)|)\nabla u) \\ \qquad - \alpha \times \begin{pmatrix} 50(x - 1.5) \\ 50 \end{pmatrix} \cdot \nabla u = f(x,y) = 10, \quad (x,y) \in \Omega, \\ u(x,y) = 0, \qquad\qquad\qquad\qquad\qquad\qquad\qquad (x,y) \in \partial\Omega, \end{cases} \quad (11.17)$$

where $\alpha = 0, 1$ and $\Omega = [-2, 2] \times [-2, 2]$. This diffusion parameter is $\mathbf{O}(1)$ except for a narrow region around the line $y = -1$, where it dips rapidly to .001. When $\alpha = 0$ there is no convection and when $\alpha = 1$, there is strong convection. We plot a solution with no convection in Figure 11.11.

We now consider the effect of perturbing the data f to $f + \rho$ by a very "pointed" function

$$\rho(x,y) = 100 \times e^{(-10 \times ((x+1)^2 + (y-.5)^2))},$$

which is nearly zero outside of a small neighborhood of $(-1, .5)$. Because the problem (11.17) is linear, we can compute the effects directly. Namely with $L(u) = f$ and U denoting the perturbed solution $L(U) = f + \rho$, we can compute the perturbation to the solution, $w = U - u$, directly as the solution of

$$L(w) = L(U - u) = L(U) - L(u) = \rho.$$

In Figure 11.11, we plot the perturbation w to the solutions for $\alpha = 0$ and $\alpha = 1$.

We observe that the effect of the perturbation decreases dramatically to zero sufficiently far from the region where the perturbation is nonzero, e.g. close to the

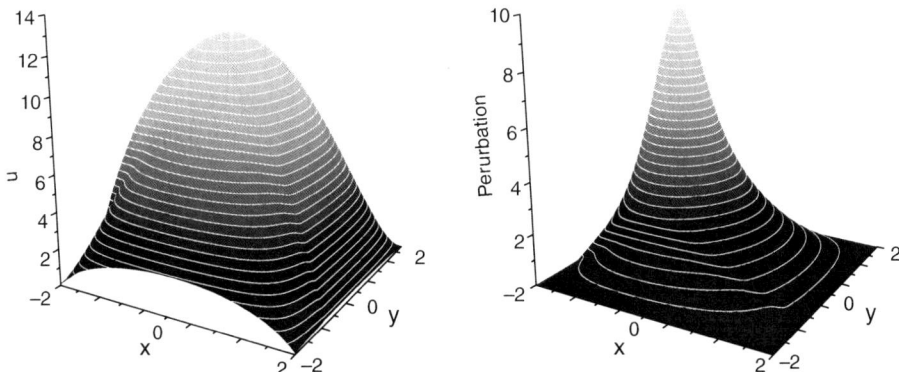

Figure 11.11: Left: A plot of the solution of (11.17) with $\alpha = 0$. Right: A plot of the effect of the perturbation ρ when $\alpha = 0$.

boundaries of $x = 2$ and $y = -2$, see Figure 11.11. This kind of "decay of influence" is characteristic of the Greens function associated with Poisson's problem. The situation for more general elliptic problems is much more complex however.

For example, convection has a strong effect on the way in which the perturbation disturbs the solution. In Figure 11.12, we plot the solution with $\alpha = 1$. Now, some of the effects of the perturbation are felt right across the domain, even very near the boundary at $y = -2$. Note also that the perturbation does not decay uniformly in all directions.

In general, the "decay of influence" of the effects of a localized perturbation is an important stability property of elliptic problems. It has strong consequences for devising efficient adaptive mesh refinement for example. It can also be exploited to devise new approaches to domain decomposition, see [6]. However, the decay is very problem-dependent and exploiting it fully requires numerical solution of the adjoint problem in general.

We can contrast these ideas with the classic analysis of elliptic stability, which typically yields a result of the form

$$\|w\|_* \leq C\|\rho\|_{**},$$

for some appropriate norms $\|\ \|_*, \|\ \|_{**}$, where ρ belongs in some reasonable space of functions and C is some constant independent of the choice of particular ρ in this space. We see that such a result does not describe the decay of influence in the example above. As with the linear algebra example, an *a priori* analysis of stability tends to be much too pessimistic when applied to a particular solution.

11.2.5 *The meaning of stability depends on the information to be computed*

In the examples above, we concentrated on the stability of pointwise values. But, we must broaden the point of view here. For example, a little reflection suggests

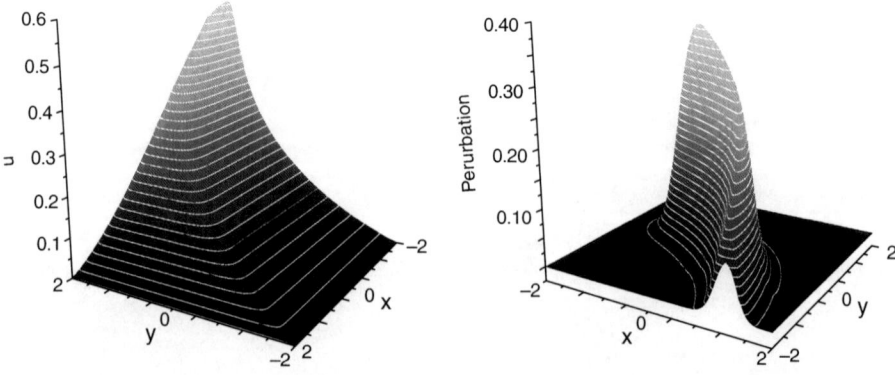

Figure 11.12: Left: A plot of the solution of (11.17) with $\alpha = 1$. Right: A plot of the effect of the perturbation ρ when $\alpha = 1$. Note how the influence "curves" because of the singular perturbation in the diffusion.

that worrying about the pointwise behavior of Lorenz solutions is not very well motivated in terms of physical modeling. In whatever sense the Lorenz problem presents a model of weather, it is certainly not a pointwise representation of weather! Rather, it is the qualitative behavior of the Lorenz problem that is meant to represent some characteristic of weather patterns. It is more reasonable to consider a quantity of interest computed from solutions that better represents qualitative behavior of all solutions.

This is an important observation because it turns out that the effect of perturbations on a solution depends strongly on the information being computed from the solution.

Example 8.

To illustrate this, we consider the average of the instantaneous distance from a solution of the Lorenz problem to the origin, see Figure 11.13. The motivation is that all solutions must remain in a neighborhood of the origin. We compare results for numerical solutions with a coarse time step .001 and fine time step .0001, and see that the distances are completely different after a moderate time.

We give values for the average instantaneous distance along with the variance over three intervals in Table 11.1. The accuracy of the numerical solution appears to have little effect on the average distance.

In order to verify this observation, we compare these results with the average distance to the origin computed from an ensemble of 100 accurate solutions, each computed using time step .0001 for 15 time units in Table 11.2. Initial values for these solutions were drawn at random from values of the long-time solution after $T = 50$, insuring the initial values are distributed appropriately on the strange attractor. Again, there is close agreement in values.

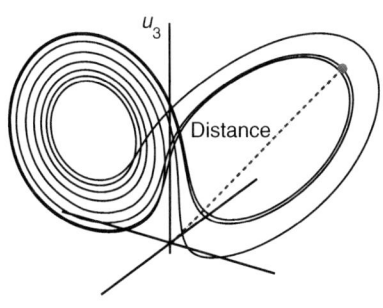
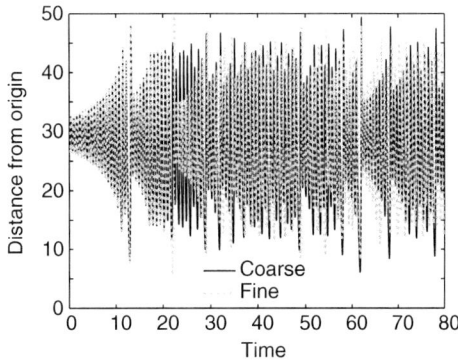

Figure 11.13: The quantity of interest is the average of the instantaneous distance from the solution to the origin. On the right, we plot the average instantaneous distance for solutions with time steps .001 and .0001, respectively. The distances agree during an initial period and are completely different after that.

Example 9.

A characteristic of Lorenz solutions that is linked to chaotic behavior is the pattern of orbits around the nonzero fixed points. We compute the average number of orbits around a particular fixed point made by a solution before it moves to orbit the other fixed point, see Figure 11.14. We compare the probability density of orbits for an ensemble average over many short-time, accurate solutions to that for a long-time solution.

Table 11.1. Average instantaneous distance of numerical Lorenz solutions to the origin computed with time steps .001 and .0001

End Time	Coarse Solution		Fine Solution	
	Ave	Var	Ave	Var
20	27.6	52.0	27.6	51.9
80	26.5	79.5	26.5	79.2
320	26.3	83.7	26.3	83.0

Table 11.2. Average distance of numerical Lorenz solutions to the origin computed with over a long time and using an ensemble average

End Time	Coarse Solution		Fine Solution		Ensemble Average	
	Ave	Var	Ave	Var	Ave	Var
320	26.3	83.7	26.3	83.0	26.3	83.7

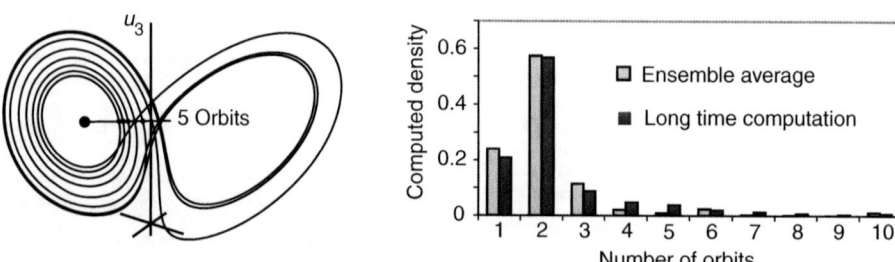

Figure 11.14: The quantity of interest is the average number of orbits around a nonzero fixed point. On the right, we plot the probability density for the orbits computed from a long-time solution and an ensemble average of short-time accurate solutions.

The conclusion is that solutions of the chaotic Lorenz problem are sensitive pointwise to perturbations and errors, yet there are quantities of interest that can be computed from the solutions that are relatively insensitive to perturbation and error. An analysis to obtain accurate estimates of the effects of perturbation and error must be conducted relative to the information that is to be computed from solutions.

11.3 The tools for quantifying stability properties: functionals, duality, and adjoint operators

In the previous Section, we saw that stability plays a critical role in determining the effects of perturbation and error. We saw that stability is often a complex issue with many facets that are not easily determined. We also saw that classic *a priori* analyses may be too crude to be used to quantify variation and error arising in particular information computed from particular solutions of particular problems with any reasonable accuracy.

All of these observations provide motivation to find another approach to determine the effects of stability. The approach we describe in this chapter is *a posteriori*, which means that the stability of particular information of a particular solution is determined after the information is computed. *A posteriori* and *a priori* analyses are fundamentally different. For example, an *a priori* error analysis of a numerical method describes the general accuracy properties for a wide class of solutions, yet generally overestimates the error in any particular solution to a significant extent. An *a posteriori* error analysis provides an estimate of the error in particular computed information, but the estimate changes when the solution changes and consequently it is generally difficult to draw any conclusions about convergence of the method. Both *a priori* and *a posteriori* analyses play key roles in analyzing the effects of uncertainty and error.

We use duality and adjoint operators to quantify stability *a posteriori*. We combine these with variational analysis to produce accurate estimates of the effects of perturbation and error. These tools have a long history of application in model sensitivity analysis and optimization, dating back to Lagrange. We present a very brief overview here, see the references [7, 8, 9, 10, 11, 12, 13, 14, 15] for more details. The application of these tools to *a posteriori* error estimation has a more recent history, see the references in [16, 17, 18, 19, 20, 21, 22, 23].

11.3.1 *Functionals and computing information*

We focus on computing a particular piece of information, or a quantity of interest, from a solution of a model. We use linear functionals, which are a special kind of linear map, to do this. A continuous linear functional ℓ is a continuous linear map from a vector space X to the reals $\mathbb{R}$.

Example 10.

Let v in $\mathbb{R}^n$ be fixed. The map $\ell(x) = v \cdot x = (x, v)$ is a linear functional on $\mathbb{R}^n$.

Example 11.

Consider $C([a,b])$. Both $\ell(f) = \int_a^b f(x)\,dx$ and $\ell(f) = f(y)$ for $a \le y \le b$ are linear functionals.

Example 12.

There are important nonlinear functionals, e.g. norms.

It is useful to think of a linear functional as providing a "one-dimensional snapshot" of a vector.

Example 13.

In Example 10, consider $v = e_i$, the ith standard basis function. Then $\ell(x) = x_i$ where $x = (x_1, \ldots, x_n)$.

Example 14.

If δ_y denotes the delta function at a point y in a region Ω. This gives a linear functional on sufficiently smooth, real valued functions via

$$\ell(u) = u(y) = \int_\Omega \delta_y(x) u(x)\,dx.$$

Example 15.

The expected value $E(X)$ of a random variable X is a linear functional.

Example 16.

The Fourier coefficients of a continuous function f on $[0, 2\pi]$,

$$c_j = \int_0^{2\pi} f(x) e^{-ijx}\,dx$$

Using linear functionals means settling for a set of snapshots rather than the entire solution. Presumably, it is easier to compute accurate snapshots than solutions that are accurate everywhere. In many situations, we settle for an "incomplete" set of samples.

Example 17.

We are often happy with a small set of moments of a random variable.

Example 18.

In applications of Fourier series, we typically use a finite sum truncation of the infinite series. We require increasing amounts of information, e.g. values, of a function in order to compute increasingly higher order Fourier coefficients.

We define the dual space to be the collection of "reasonable" snapshots. More precisely, if X is a normed vector space with norm $\|\ \|$, the space of continuous linear functionals on X is called the dual space of, or on, or to X, and is denoted by X^*. The dual space is a vector space. We can define the dual norm for $y \in X^*$ by

$$\|y\|_{X^*} = \sup_{\substack{x \in X \\ \|x\|_X = 1}} |y(x)| = \sup_{\substack{x \in X \\ x \neq 0}} \frac{|y(x)|}{\|x\|}.$$

Example 19.
When $X = \mathbb{R}^n$ with the usual dot product $(\,,\,)$ and norm $\|\ \| = \|\ \|$, we saw that every vector v in $\mathbb{R}^n$ is associated with a linear functional $\ell_v(\cdot) = (\cdot, v)$. This functional is continuous since $|(x, v)| \leq \|v\| \|x\|$ (The "C" in the definition is $\|v\|$). A classic result in linear algebra is that **all** linear functionals on $\mathbb{R}^n$ have this form, i.e., we can make the identification $(\mathbb{R}^n)^* \equiv \mathbb{R}^n$.

Example 20.
Recall Hölder's inequality for $f \in L^p(\Omega)$ and $g \in L^q(\Omega)$ with $\frac{1}{p} + \frac{1}{q} = 1$ for $1 \leq p, q \leq \infty$ is

$$\|fg\|_{L^1(\Omega)} \leq \|f\|_{L^p(\Omega)} \|g\|_{L^q(\Omega)}.$$

This implies that each g in $L^q(\Omega)$ is associated with a bounded linear functional on $L^p(\Omega)$ when $\frac{1}{p} + \frac{1}{q} = 1$ and $1 \leq p, q \leq \infty$ by

$$\ell(f) = \int_\Omega g(x) f(x)\, dx.$$

An important, and difficult, result is that we can "identify" $(L^p)^*$ with L^q when $1 < p, q < \infty$, The cases $p = 1, q = \infty$ and $p = \infty, q = 1$ are trickier. The case L^2 is special in that we can identify $(L^2)^*$ with L^2.

There is a useful notation for the value of a functional. If x is in X and y is in X^*, we define the bracket of x and y as

$$y(x) = \langle x, y \rangle.$$

It is not surprising that norms on X and its dual X^* are closely related. An important inequality is:

11.3.2 Theorem

The generalized Cauchy inequality is

$$|\langle x, y \rangle| \leq \|x\|_X \|y\|_{X^*}, \quad x \in X, y \in X^*.$$

Combining this with the Hahn-Banach theorem yields a "weak" representation of the norm on X,

11.3.3 Theorem

If X is a Banach space, then

$$\|x\|_X = \sup_{\substack{y \in X^* \\ y \neq 0}} \frac{|y(x)|}{\|y\|_{X^*}} = \sup_{\substack{y \in X^* \\ \|y\|_{X^*}=1}} |y(x)|$$

for all x in X.

This says that we can determine the size of a vector in X by examining a sufficient number of "snapshots".

Example 21.

In Example 19, we saw that $\mathbb{R}^n$ with the standard Euclidean norm can be identified with its dual space. Likewise, L^2 can be identified with its dual space. Both of these spaces are Hilbert spaces.

Remarkably, Example 19 generalizes to infinite dimensions. If X is a Hilbert space with inner product (x, y), then each $y \in X$ determines a linear functional $\ell_y(x) = \langle x, y \rangle = (x, y)$ for x in X. This functional is continuous by Cauchy's inequality, which says that $|(x, y)| \leq \|x\| \, \|y\|$.

It turns out that is the only kind of continuous linear functional on a Hilbert space.

11.3.4 Theorem

Riesz Representation for Hilbert Spaces For every bounded linear functional ℓ on a Hilbert space X, there is a unique element y in X such that

$$\ell(x) = (x, y) \text{ for all } x \in X, \text{ and } \|y\|_{X^*} = \sup_{\substack{x \in X \\ x \neq 0}} \frac{|\ell(x)|}{\|x\|}.$$

This means that the dual space to a Hilbert space X can be identified with X. Abusing notation, it is common to replace the bracket notation and the generalized Cauchy inequality by the inner product and the "real" Cauchy inequality without comment.

11.3.5 The adjoint operator

To motivate the definition of the adjoint operator, let X and Y be two Banach spaces, $L: X \to Y$ be a continuous linear map, and consider the problem of computing a functional value

$$\ell(L(x))$$

for some input $x \in X$. Some important questions are
- Given that we only want a functional value of the solution, can we find a way to compute the functional value efficiently?
- What is the error in the functional value if approximations are involved?

- Given a functional value, what can we say about x?
- Given a collection of functional values, what can we say about L?

We can address these questions using the adjoint operator. Suppose L is a continuous linear transformation. For each $y^* \in Y^*$,

$$y^* \circ L(x) = y^*(L(x)) = \langle Lx, y^* \rangle$$

assigns a number to each $x \in X$, hence defines a functional $\ell(x)$. The functional $\ell(x)$ is clearly linear. It is also continuous since

$$|\ell(x)| = |y^*(L(x))| \le \|y^*\|_{Y^*} \|L(x)\|_Y \le \|y^*\|_{Y^*} \|L\| \, \|x\|_X = C\|x\|_X,$$

where $C = \|y^*\|_{Y^*} \|L\|$. By the definition of the dual space, there is an $x^* \in X^*$ such that $y^*(L(x)) = x^*(x)$ for all $x \in X$. x^* is unique.

We have defined $\ell = x^*$ implicitly. Given x, we first apply the operator L, then compute a functional of the result. This may seem a little strange. We put it into context of a classic example.

Example 22.
Consider the elliptic problem

$$\begin{cases} -\Delta u = f, & y \in \Omega, \\ u = 0, & y \in \partial\Omega, \end{cases} \tag{11.18}$$

where we wish to evaluate $u(y_0)$ for some $y_0 \in \Omega$. In this example, the data f plays the role of x above in the definition of ℓ. Note that we do **not** evaluate $f(y_0)$. Instead, we have to solve (11.18), where $u = L(f)$ is determined by the solution operator L of the Dirichlet problem. We then apply the functional

$$u(y_0) = (u, \delta_{y_0}) = (L(f), \delta_{y_0}).$$

We now apply this implicit definition to each $y^* \in Y^*$. For each y^*, assign a unique $x^* \in X^*$ and in this way define a linear transformation $L^* : Y^* \to X^*$, called the *adjoint* or *dual operator* to L.

Example 23.
Continuing Example 22, we pose the adjoint problem

$$\begin{cases} -\Delta \phi = \delta_{y_0}, & y \in \Omega, \\ \phi = 0, & y \in \partial\Omega, \end{cases}$$

and denote the solution $\phi = L^*(\delta_{y_0})$. We have

$$(u, \delta_{y_0}) = (L(f), \delta_{y_0}) = (f, L^*(\delta_{y_0})) = (f, \phi).$$

This is just the method of Greens function.

Note that we have defined the adjoint transformation via computing snapshots using elements in the dual space. We can write these relations as

$$y^*(L(x)) = L^*y^*(x)$$

or using the bracket notation,

$$\langle L(x), y^* \rangle = \langle x, L^*(y^*) \rangle \quad x \in X,\ y^* \in Y^*. \tag{11.19}$$

Equation (11.19) is called the bilinear identity.

Example 24.

Let $X = \mathbb{R}^m$ and $Y = \mathbb{R}^n$, where we take the standard inner product and norm. By the Riesz Representation theorem, the bilinear identity for $L \in \mathcal{L}(\mathbb{R}^m, \mathbb{R}^n)$ reads

$$(Lx, y) = (x, L^*y), \quad x \in \mathbb{R}^m,\ y \in \mathbb{R}^n.$$

We know that L is represented by a unique $n \times m$ matrix $\mathbf{A}$ so that if $y = L(x)$,

$$\mathbf{A} = \begin{pmatrix} a_{11} & \cdots & a_{1m} \\ \vdots & & \vdots \\ a_{n1} & \cdots & a_{nm} \end{pmatrix},\quad y = \begin{pmatrix} y_1 \\ \vdots \\ y_n \end{pmatrix},\quad x = \begin{pmatrix} x_1 \\ \vdots \\ x_m \end{pmatrix}$$

then

$$y_i = \sum_{j=1}^m a_{ij} x_j, \quad 1 \le i \le n.$$

For a linear functional $y^* = (y_1^*, \ldots, y_n^*)^\top \in Y^*$, we have

$$L^* y^*(x) = y^*(L(x)) = (y_1^*, \ldots, y_n^*) \begin{pmatrix} \sum_{j=1}^m a_{1j} x_j \\ \vdots \\ \sum_{j=1}^m a_{nj} x_j \end{pmatrix} = \sum_{j=1}^m \left(\sum_{i=1}^n y_i^* a_{ij} \right) x_j$$

Therefore, $L^*(y^*)$ is given by the inner product with $\tilde{y} = (\tilde{y}_1, \ldots, \tilde{y}_m)^\top$ where

$$\tilde{y}_j = \sum_{i=1}^n y_i^* a_{ij}.$$

This implies the matrix $\mathbf{A}^*$ of L^* is

$$\mathbf{A}^* = \begin{pmatrix} a_{11}^* & \cdots & a_{1n}^* \\ \vdots & & \vdots \\ a_{m1}^* & \cdots & a_{mn}^* \end{pmatrix} = \begin{pmatrix} a_{11} & a_{21} & \cdots & a_{n1} \\ \vdots & & & \vdots \\ a_{1m} & a_{2m} & \cdots & a_{nm} \end{pmatrix} = A^\top.$$

We can write the bilinear identity as
$$y^\top \mathbf{A} x = x^\top \mathbf{A}^\top y$$
using the fact that $(x, y) = (y, x)$.

We give more examples below.

We conclude with some basic facts:

11.3.6 Theorem

Let X and Y be normed vector spaces and $L : X \to Y$. Then, L^* is a continuous linear operator and $\|L^*\| = \|L\|$. Also, $0^* = 0$. If $M : X \to Y$ is another continuous linear operator, then $(L + M)^* = L^* + M^*$ and $(\alpha L)^* = \alpha L^*$ for all scalars α.

If Z is another normed vector space and $N : Y \to Z$, then $NL : X \to Z$ is a continuous linear operator and $(NL)^* = L^* N^*$.

11.3.7 Four good reasons to use adjoints

We can provide some immediate motivations to introduce the concepts of duality adjoint operators, which indeed turn out to be fundamentally important for the analysis of operators.

Reason # 1 X^* often has good properties that X may lack.

11.3.8 Theorem

If X is a normed vector space over $\mathbb{R}$, then X^* is a Banach space, i.e. Cauchy sequences in X converge to a limit in X, whether or not X is a Banach space.

Reason # 2 There is a close connection between the stability properties of an operator and its adjoint.

11.3.9 Theorem

The singular values of a matrix $\mathbf{L}$ are the eigenvalues of the square, symmetric transformations $\mathbf{L}^* \mathbf{L}$ or $\mathbf{L} \mathbf{L}^*$.

This connects the condition number of a matrix $\mathbf{L}$ to $\mathbf{L}^*$.

Reason # 3 If L is a linear transformation between normed vector spaces, the solvability of $L(y) = b$ is closely related to the solvability of $L^*(\phi) = \psi$.

11.3.10 Theorem

Let X and Y be normed linear spaces and $L : X \to Y$ a continuous linear transformation. A necessary condition that $L(x) = y$ has a solution is that $y^*(y) = 0$ for all continuous functionals y^* such that $L^* y^* = 0$. This is a sufficient condition if the range of L is closed in Y.

Example 25.

Suppose that $L : \mathbb{R}^m \to \mathbb{R}^n$ is associated with the $n \times m$ matrix $\mathbf{A}$, i.e., $L(x) = \mathbf{A}x$. The necessary and sufficient condition for the solvability of $\mathbf{A}x = b$ is that b is orthogonal to all linearly independent solutions of $\mathbf{A}^\top y = 0$.

In general, all kinds of information about the solvability and deficiency of the linear system $\mathbf{A}x = b$ can be determined by considering $\mathbf{A}^*\mathbf{A}$. In the overdetermined and underdetermined cases, it yields a "natural" definition of a solution or gives conditions for a solution to exist, see [8].

Reason # 4 Suppose we wish to compute a functional $\langle \cdot, \psi \rangle$ of the solution y of an inverse problem for a linear operator A and data b,

$$Ay = b.$$

We define the adjoint (inverse) problem

$$A^*\phi = \psi.$$

Then we obtain a representation of the solution,

$$\langle y, \psi \rangle = \langle y, A^*\phi \rangle = \langle Ay, \phi \rangle = \langle b, \phi \rangle.$$

Such an error representation is very useful in practice. For example, we can compute the effect of many perturbations in the data very efficiently by computing one adjoint solution and taking inner products with the perturbations.

This formal argument is essentially the entire foundation for error estimation and uncertainty quantification described in this chapter. The reader may notice that it generalizes the method of Greens functions. We discuss this further below.

11.3.11 Adjoint operators for linear differential equations

We briefly discuss the computation of adjoints to differential equations. On a simple level, given a differential operator L on a domain Ω, we seek to **evaluate the bilinear identity**,

$$\langle Lu, v^* \rangle - \langle u, L^*v^* \rangle = 0, \quad \text{all } u \in X, \, v^* \in Y^*. \tag{11.20}$$

But, there are a lot of details needed to make this a computational process, e.g. what does it mean to compute $\langle \, , \, \rangle$!

In the common situation in which we consider functions in a Hilbert space like $L^2(\Omega)$, we attempt to replace $\langle \, , \, \rangle$ by the inner project $(\, , \,)$,

$$\langle Lu, v^* \rangle - \langle u, L^*v^* \rangle = (Lu, v) - (u, L^*v).$$

Even using the familiar $L^2(\Omega)$ inner product, however, computing an adjoint can be tricky. On one hand, the process for computing an adjoint is simple to state: multiply the differential equation by a test function, integrate over the entire space-time domain (which amounts to taking the L^2 inner product of the differential equation and the test function), and keep integrating by parts until all derivatives fall on the test function. The differential operator that ends up being applied to the test function "is" the adjoint operator. On the other hand, details lead to all kinds of technical difficulties. A general abstract theory is difficult to present, see [24].

First of all, the definition of the adjoint of a given forward operator depends heavily on the spaces involved with the maps. On the face, the process described above only works for functions sufficiently smooth that all the integration by parts are defined. Technically, we compute the adjoint for smooth functions and then pass to a limit (a "density" argument) to the full spaces on which the operators are defined.

Second of all, integration by parts leaves behind integrals over boundaries and these have to be accounted for when defining the adjoint operator. The reason is simply that a differential operator is generally under determined and we add boundary and initial conditions in order to get an invertible operator. Clearly, the boundary and initial conditions therefore must affect the definition of the adjoint operator.

To simplify life, we compute the adjoint in two stages. We first assume that the functions involved are smooth and have compact support inside Ω, i.e. the functions and all their derivatives vanish at the boundary. In this way, we carry out the integration by parts while ignoring boundary terms. Given a differential operator L on a domain Ω, the formal adjoint L^* is the differential operator that satisfies

$$(Lu, v) = (u, L^*v)$$

for all sufficiently smooth u and v with compact support in Ω.

Example 26.
For

$$Lu(x) = -\frac{d}{dx}\left(a(x)\frac{d}{dx}u(x)\right) + \frac{d}{dx}(b(x)u(x))$$

on $[0,1]$. Integration by parts neglecting boundary terms gives the formal adjoint

$$L^*v = -\frac{d}{dx}\left(a(x)\frac{d}{dx}v(x)\right) - b(x)\frac{d}{dx}(v(x)).$$

Example 27.
A general linear second order differential operator L in $\Omega \subset \mathbb{R}^n$ can be written

$$L(u) = \sum_{i=1}^{n}\sum_{j=1}^{n} a_{ij}\frac{\partial^2 u}{\partial x_i \partial x_j} + \sum_{i=1}^{n} b_i \frac{\partial u}{\partial x_i} + cu,$$

where $\{a_{ij}\}$, $\{b_i\}$, and c are functions of $x_1, x_2, \ldots, x_n$. Then,

$$L^*(u) = \sum_{i=1}^{n}\sum_{j=1}^{n} \frac{\partial^2 (a_{ij}v)}{\partial x_i \partial x_j} - \sum_{i=1}^{n} \frac{\partial (b_i v)}{\partial x_i} + cv.$$

It can be verified directly that

$$vL(u) - uL^*(v) = \sum_{i=1}^{n} \frac{\partial p_i}{\partial x_i},$$

where

$$p_i = \sum_{j=1}^{n} \left(a_{ij} v \frac{\partial u}{\partial x_j} - u \frac{\partial (a_{ij} v)}{\partial x_j} \right) + b_i uv.$$

The expression on the right is a divergence expression and the divergence theorem yields

$$\int_\Omega (vL(u) - uL^*(v))\, dx = \int_{\partial \Omega} p \cdot n\, ds = 0,$$

where $p = (p_1, \ldots, p_n)$ and n is the outward normal in $\partial \Omega$.

Example 28.

Let L be a differential operator of order $2p$ of the form

$$Lu = \sum_{|\alpha|, |\beta| \leq p} (-1)^{|\alpha|} D^\alpha \left(a_{\alpha\beta}(x) D^\beta u \right),$$

then

$$L^* v = \sum_{|\alpha|, |\beta| \leq p} (-1)^{|\alpha|} D^\alpha \left(a_{\beta\alpha}(x) D^\beta v \right),$$

and L is elliptic if and only if L^* is elliptic. Some special cases.

$$\text{grad}^* = -\text{div}$$
$$\text{div}^* = -\text{grad}$$
$$\text{curl}^* = \text{curl}$$

and if

$$Lu = \sum_{|\alpha| \leq p} a_\alpha(x) D^\alpha u$$

then

$$L^* v = \sum_{|\alpha| \leq p} (-1)^{|\alpha|} D^\alpha (a_\alpha(x) v(x)).$$

Ignoring initial as well as boundary conditions, evolution problems are treated similarly in the sense that we integrate by parts over space and time. There is an important difference however because time has a direction and the time variable for the adjoint problem runs "backwards."

Example 29.
If we have a parabolic problem
$$Lu = u_t - \nabla \cdot (a\nabla u) + bu, \quad x \in \Omega, \ 0 < t \leq T,$$
then
$$L^*v = -v_t - \nabla \cdot (a\nabla v) + bv, \quad x \in \Omega, \ T > t \geq 0.$$
The adjoint problem is also parabolic, and not an "ill-posed" or "backwards" parabolic problem as suggested by the "−" in front of the time derivative term. This is easily seen by making the substitution $t \to s = T - t$, so that
$$L^*v = v_s - \nabla \cdot (a(T-s)\nabla v) + b(T-s)v, \quad x \in \Omega, \ 0 < s \leq T.$$
We find it convenient to use this change of variables when solving the adjoint problem in practice.

In the second stage of computing the adjoint, we remove the assumption that the functions involved in evaluating the bilinear identity have compact support. The integrations by parts that produces the formal adjoint yield additional terms involving integrals of the functions and their derivatives over the boundary of Ω. We then choose boundary conditions for the adjoint problem depending on what we want to happen with the boundary terms from evaluating the bilinear identity.

For example, the standard approach is to pose the **minimal** boundary conditions on the adjoint problem necessary to make the boundary terms that appear when evaluating the bilinear identity vanish. These are called the adjoint boundary conditions. This definition is rather vague, but it can be made completely precise, see [24].

Note that for the purpose of defining the adjoint boundary conditions, the form of the boundary conditions imposed on the original operator L are important, but the values given for these conditions are not. If the boundary conditions for L are not homogeneous, we make them so for the purpose of determining the adjoint. It follows that some of the boundary terms that appear when evaluating the bilinear identity vanish because of the homogeneous boundary conditions imposed on L and the adjoint boundary conditions insure that the rest vanish.

Example 30.
Consider Newton's equation of motion $s''(t) = f(t)$, normalized with mass 1. If we assume $s(0) = s'(0) = 0$, and $0 < t < 1$, then we have
$$s''v - sv'' = \frac{d}{dt}(vs' - sv')$$
and
$$\int_0^1 (s''v - sv'') \, dt = (vs' - sv')\big|_0^1. \tag{11.21}$$

Now the boundary conditions imply the contributions at $t = 0$ vanish, while at $t = 1$ we have

$$v(1)s'(1) - v'(1)s(1).$$

To insure this vanishes, we must have $v(1) = v'(1) = 0$. (We cannot specify $s(1)$ or $s'(1)$ of course.) These are the adjoint boundary conditions.

Example 31.
Since

$$\int_\Omega (u\Delta v - v\Delta u)\, dx = \int_{\partial\Omega} \left(u\frac{\partial v}{\partial n} - v\frac{\partial u}{\partial n} \right) ds,$$

the Dirichlet and Neumann boundary value problems for the Laplacian are their own adjoints.

Example 32.
Let $\Omega \subset \mathbb{R}^2$ be bounded with a smooth boundary and let $s =$ arclength along the boundary. Consider

$$\begin{cases} -\Delta u = f, & x \in \Omega, \\ \dfrac{\partial u}{\partial n} + \dfrac{\partial u}{\partial s} = 0, & x \in \partial\Omega. \end{cases}$$

Since

$$\int_\Omega (u\Delta v - v\Delta u)\, dx = \int_{\partial\Omega} \left(u\left(\frac{\partial v}{\partial n} - \frac{\partial v}{\partial s}\right) - v\left(\frac{\partial u}{\partial n} + \frac{\partial u}{\partial s}\right) \right) ds,$$

the adjoint problem is

$$\begin{cases} -\Delta v = g, & x \in \Omega, \\ \dfrac{\partial v}{\partial n} - \dfrac{\partial v}{\partial s} = 0, & x \in \partial\Omega. \end{cases}$$

11.4 *A posteriori* error analysis using adjoints

We now apply functionals, adjoint operators, and variational analysis to the problem of estimating the error of a finite element solution of a partial differential equation. The analysis rests on the observation in Reason # 4 above and we begin by extending that argument to differential equations. Given a domain Ω, which could be a time interval, a space domain, or a space-time domain, we consider a problem of the form

$$\begin{cases} Lu = f, & \text{on } \Omega, \\ \text{bound. cond. and init. val.}, & \text{on } \partial\Omega, \end{cases} \tag{11.22}$$

where L is a linear differential operator and we specify the correct boundary and/or initial conditions so that (11.22) has a unique solution. We assume that the goal of solving (11.22) is to compute a quantity of interest given as a linear functional $\ell(u) = (u, \psi)$ for some ψ. The generalized Greens function for (11.22) corresponding to ψ satisfies

$$\begin{cases} L^*\phi = \psi, & \text{on } \Omega, \\ \text{adjoint bound. cond. and init. val.}, & \text{on } \partial\Omega, \end{cases} \quad (11.23)$$

where L^* is the formal adjoint of L. There are minor variations of this definition if we pose the data ψ on the boundary of Ω rather than the interior (i.e., as boundary or initial data), see [25]. We obtain the basic representation formula,

$$(u, \psi) = (u, L^*\phi) = (Lu, \phi) = (f, \phi).$$

We use this argument to derive an *a posteriori* error estimate.

Example 33.

We begin by returning to Example 6, and estimating the error $e = X - x$ in the numerical solution X of a linear system of equations

$$Ax = b.$$

We derive an estimate of the error in a quantity of interest given by a linear functional (e, ψ), where ψ is an given vector. We introduce the generalized Greens vector solving the adjoint problem

$$A^\top \phi = \psi.$$

Arguing as above,

$$(e, \psi) = (e, A^\top \phi) = (Ae, \phi) = (R, \phi),$$

where $R = AX - b$. We obtain a representation of the error as an inner product of the computable residual and the solution of the adjoint problem. In practice, we approximate ϕ and obtain a computable estimate.

We can also derive a bound

$$|(e, \psi)| \leq \|\phi\| \, \|R\|. \quad (11.24)$$

Returning to the specific example in Example 6, we find

estimate of the error in the quantity of interest $\approx 1.0 \times 10^{-15}$,

a posteriori error bound for the quantity of interest $\approx 5.4 \times 10^{-14}$,

traditional error bound for the error $\approx 3.5 \times 10^{-5}$.

The *a posteriori* estimate is very accurate. The *a posteriori* bound overestimates the error since any cancellation in the inner product (R, ϕ) is lost, but it is still much better than the traditional condition number bound.

The adjoint quantity $\|\phi\|$ is called the stability factor. It is related to the condition number of A, since

$$\left| \left(\frac{e}{\|x\|}, \psi \right) \right| \leq \operatorname{cond}_\psi(A) \frac{\|R\|}{\|b\|},$$

where

$$\operatorname{cond}_\psi(A) = \|\phi\| \, \|A\| = \|A^{-\top}\psi\| \, \|A\|$$

is a kind of "weak" condition number of A with respect to the targeted quantity of interest. If we take the supremum of $\operatorname{cond}_\psi(A)$ over all possible ψ with norm 1, we obtain the standard condition number of A. Hence, the stability factor obtained from the generalized Greens function is a measure of the sensitivity of particular information computed from a numerical solution of the problem to computational errors.

11.4.1 *Discretization of elliptic problems*

We first consider a general second order linear elliptic boundary value problem for a scalar unknown,

$$\begin{cases} Lu = f, & x \in \Omega, \\ u = 0, & x \in \partial\Omega, \end{cases} \quad (11.25)$$

where

$$L(D, x)u = -\nabla \cdot a(x)\nabla u + b(x) \cdot \nabla u + c(x)u(x), \quad (11.26)$$

with $u : \mathbb{R}^n \to \mathbb{R}$, a is a $n \times n$ matrix function of x, b is a n-vector function of x, and c is a function of x. We assume that $\Omega \subset \mathbb{R}^n$, $n = 2, 3$, is a smooth or polygonal domain; $a = (a_{ij})$, where $a_{i,j}$ are continuous in $\overline{\Omega}$ for $1 \leq i, j \leq n$ and there is a $a_0 > 0$ such that $v^\top a v \geq a_0$ for all $v \in \mathbb{R}^n \setminus \{0\}$ and $x \in \Omega$; $b = (b_i)$ where b_i is continuous in $\overline{\Omega}$; and finally c and f are continuous in $\overline{\Omega}$.

We discretize (11.25) by applying a finite element method to the associated variational formulation:

Find $u \in H_0^1(\Omega)$ such that

$$A(u, v) = (a\nabla u, \nabla v) + (b \cdot \nabla u, v) + (cu, v) = (f, v) \text{ for all } v \in H_0^1(\Omega), \quad (11.27)$$

where $H_0^1(\Omega)$ is the subset of functions in $H^1(\Omega)$ that are zero on $\partial\Omega$ and $H^1(\Omega)$ consists of functions that together with their first derivatives are square integrable on Ω.

To construct a finite element discretization, we form a piecewise polygonal approximation of $\partial\Omega$ whose nodes lie on $\partial\Omega$ and which is contained inside Ω. This forms the boundary of a convex polygonal domain Ω_h. We let $\mathcal{T}_h$ denote a simplex triangulation of Ω_h that is locally quasi-uniform. We let h_K denote the length of the longest edge of $K \in \mathcal{T}_h$ and define the piecewise constant mesh function h by $h(x) = h_K$ for $x \in K$. We also use h to denote $\max_K h_K$. We choose a finite element solution from the space V_h of functions that are continuous on Ω, piecewise linear on Ω_h with respect to $\mathcal{T}_h$, zero on the boundary $\partial\Omega_h$, and finally extended to be zero in the region $\Omega \setminus \Omega_h$. With this construction, we have $V_h \subset H_0^1(\Omega)$, and for smooth functions, the error of interpolation into V_h is $\mathbf{O}(h^2)$ in $\|\ \|$, but not better. The finite element method is:

$$\text{Compute } U \in V_h \text{ such that } A(U,v) = (f,v) \text{ for all } v \in V_h. \quad (11.28)$$

In these notes, we take for granted the usual *a priori* convergence results for finite element methods and concentrate on the *a posteriori* analysis used to produce computational error estimates. In particular, by standard results, we know that U exists and converges to u as $h \to 0$.

11.4.2 *A posteriori analysis for elliptic problems*

The goal of the *a posteriori* error analysis is to estimate the error in a quantity of interest (u, ψ) computed from the finite element solution U. To do this, we use a generalized Greens function ϕ solving the adjoint problem corresponding to ψ,

Find $\phi \in H_0^1(\Omega)$ such that

$$A^*(v, \phi) = (\nabla v, a\nabla \phi) - (v, \text{div}(b\phi)) + (v, c\phi) = (v, \psi) \text{ for all } v \in H_0^1(\Omega). \quad (11.29)$$

This is just the weak form of the adjoint problem $L^*(D,x)\phi = \psi$. Extending the analysis above,

$$(e, \psi) = (\nabla e, a\nabla \phi) - (e, \text{div}(b\phi)) + (e, c\phi)$$
$$= (a\nabla e, \nabla \phi) + (b \cdot \nabla e, \phi) + (ce, \phi)$$
$$= (a\nabla u, \nabla \phi) + (b \cdot \nabla u, \phi) + (cu, \phi) - (a\nabla U, \nabla \phi) - (b \cdot \nabla U, \phi) - (cU, \phi)$$
$$= (f, \phi) - (a\nabla U, \nabla \phi) - (b \cdot \nabla U, \phi) - (cU, \phi).$$

Letting $\pi_h \phi$ denote an approximation of ϕ in V_h, using Galerkin orthogonality (11.28), we conclude

11.4.3 *Theorem*

The error in the quantity of interest computed from the finite element solution (11.28) satisfies the error representation,

$$(e, \psi) = (f, \phi - \pi_h \phi) - (a\nabla U, \nabla(\phi - \pi_h \phi)) - (b \cdot \nabla U, \phi - \pi_h \phi) - (cU, \phi - \pi_h \phi), \quad (11.30)$$

where the generalized Greens function ϕ satisfies the adjoint problem (11.29) corresponding to data ψ.

The most accurate a posteriori error estimates are obtained by using (11.30) directly as opposed to making further estimates. To use the estimate, we approximate ϕ using a finite element method. Since $\phi - \pi_h \phi \sim \sum_{|\alpha|=2} D^\alpha \phi$ where ϕ is smooth, we use a higher order finite element than that used to solve the original boundary value problem. For example, good results are obtained using the space V_h^2 of continuous, piecewise quadratic functions with respect to $\mathcal{T}_h$. The approximate generalized Greens function is

Compute $\Phi \in V_h^2$ such that
$$A^*(v, \Phi) = (\nabla v, a\nabla \Phi) - (v, \mathrm{div}\,(b\Phi)) + (v, c\Phi) = (v, \psi) \text{ for all } v \in V_h^2. \quad (11.31)$$

The approximate error representation is
$$(e, \psi) \approx (f, \Phi - \pi_h \Phi) - (a\nabla U, \nabla(\Phi - \pi_h \Phi)) - (b \cdot \nabla U, \Phi - \pi_h \Phi)$$
$$- (cU, \Phi - \pi_h \Phi). \quad (11.32)$$

Example 34.

In [36], we estimate the error in the average value of the solution of
$$\begin{cases} -\Delta u = 200 \sin(10\pi x) \sin(10\pi y), & (x, y) \in \Omega = [0, 1] \times [0, 1], \\ u = 0, & (x, y) \in \partial\Omega \end{cases}$$

The solution is $u = \sin(10\pi x) \sin(10\pi y)$, see Figure 11.15.

In Figure 11.15, we show a plot of error/estimate ratios for various degrees of accuracy. Ideally, we would get a ratio of 1. In practice, the accuracy of the estimate is affected by the numerical error in the adjoint and the errors arising from quadrature applied to the integrals in the representation (11.32). At the

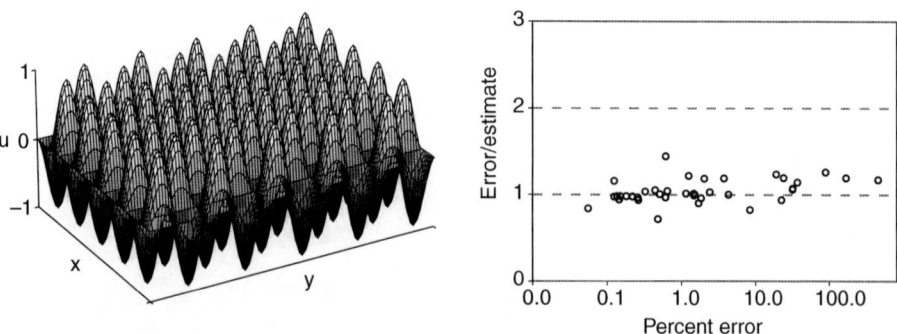

Figure 11.15: The solution $u = \sin(10\pi x) \sin(10\pi y)$ and a plot of error/estimate ratios for various mesh sizes.

inaccurate end, we are using meshes with 5×5 to 10×10 elements. We emphasize that the computed numerical solution bears almost no resemblance to the true solution at those discretization levels, yet the estimate is reasonably accurate.

Note that in practice, the error to estimate ratio tends to vary quite a bit, even in the best of circumstances. The accuracy of the estimate is affected by numerical considerations including the accuracy of the computed adjoint solution and the use of quadrature to evaluate the integrals yielding the error estimate. In nonlinear problems, it turns out that there is a linearization error that may be significant.

11.4.4 Adjoint analysis for nonlinear problems

So far in the discussion of *a posteriori* analysis, we have treated linear problems. This is no coincidence because the notion of an adjoint to an operator explicitly depends on the linearity of the operator, which means in particular that the operator is independent of the input. This in turn implies that the bilinear identity serves to determine a unique adjoint operator. On reflection, this fact is apparent in the computations in Example 24.

Above, we derived an *a posteriori* error estimate for a linear problem by modifying the representation formula involving the generalized Greens vector/function. Subtracting the representation formulae for a solution and an approximation leads to a representation of the error. Now we apply the adjoint analysis technique directly to the problem of deriving an error estimate for an approximate solution of a nonlinear problem.

We assume that $F : X \to Y$ is a nonlinear map between Banach spaces X, Y with a convex domain $\mathcal{D}(F)$. Convexity is a typical assumption, with one important consequence being that mean value theorems hold. We let $u \in \mathcal{D}(F)$ solve the nonlinear problem

$$F(u) = b, \tag{11.33}$$

for some data $b \in Y$ in the range of F. We let $U \approx u$ be an approximate solution, where $U \in \mathcal{D}(F)$. The nonlinear residual of U is

$$R(U) = F(U) - b.$$

With $e = U - u$, we have

$$F(U) - F(u) = R(U). \tag{11.34}$$

Now we write $U = u + e$ and define the operator

$$\mathcal{E}(e) = \mathcal{E}(e; u) = F(u+e) - F(u),$$

where $\mathcal{E}(0) = 0$. The fundamental observation is that if F is smooth, e.g. Frechet differentiable, then $\mathcal{E}(e) \approx F'(u)e$ behaves linearly in e to first order when e is small. Hence, it makes sense to try to define an adjoint to $\mathcal{E}(e)$. The domain of $\mathcal{E}$ is

$$\mathcal{D}(\mathcal{E}) = \{v \in X | u + v \in \mathcal{D}(F)\}.$$

To be technically precise, we assume that $\mathcal{D}(\mathcal{E})$ is a dense vector subspace of X and independent of e.

We now define the adjoint operator $\mathcal{E}^*$ through

$$\langle \mathcal{E}(v), w \rangle = \langle v, \mathcal{E}(v)^* w \rangle, \quad \text{for all sufficiently small } v \in \mathcal{D}(\mathcal{E}) \text{ and all } w \in \mathcal{D}(\mathcal{E}^*). \tag{11.35}$$

(Note the notation is somewhat confusing, since $\mathcal{E}(v)$ is an operator applied to v while $\mathcal{E}(v)^*$ is an operator.) This gives the basic representation formula that is useful for error estimation. In the Sobolev space setting, we realize this as

$$(\mathcal{E}(v), w) = (v, \mathcal{E}(v)^* w), \quad \text{for all sufficiently small } v \in \mathcal{D}(\mathcal{E}) \text{ and all } w \in \mathcal{D}(\mathcal{E}^*).$$

Example 35.

Consider the map $F : \mathbb{R}^2 \to \mathbb{R}^2$ given by

$$F(u) = \begin{pmatrix} u_1^2 + 3u_2 \\ u_1 e^{u_2} \end{pmatrix}.$$

It is easy to compute

$$\mathcal{E}(\varepsilon) = F(u + \varepsilon) - F(u) = \begin{pmatrix} 2u_1 \varepsilon_1 + \varepsilon_2^2 + 3\varepsilon_2 \\ u_1 e^{u_2}\left(e^{\varepsilon_2} - 1\right) + \varepsilon_1 e^{u_2 + \varepsilon_2} \end{pmatrix}.$$

We can write

$$\mathcal{E}(v) = \begin{pmatrix} 2u_1 + v_1 & 3 \\ e^{u_2+v_2} & u_1 e^{u_2}\left(\dfrac{e^{v_2} - 1}{v_2}\right) \end{pmatrix} \begin{pmatrix} v_1 \\ v_2 \end{pmatrix}.$$

Evaluating (11.35) yields

$$\mathcal{E}(v)^* = \begin{pmatrix} 2u_1 + v_1 & e^{u_2+v_2} \\ 3 & u_1 e^{u_2}\left(\dfrac{e^{v_2} - 1}{v_2}\right) \end{pmatrix}.$$

In the limit of small v, we recognize that $\mathcal{E}(v)^* \approx (F'(u))^*$, where $F'(u)$ is the Jacobian of F at u,

$$F'(u) = \begin{pmatrix} 2u_1 & 3 \\ e^{u_2} & u_1 e^{u_2} \end{pmatrix}.$$

This example suggests one systematic way to define an adjoint operator for error analysis. When F is Frechet differentiable, we use the Integral Mean Value Theorem to write

$$\mathcal{E}(e) = F(u+e) - F(u) = \left(\int_0^1 F'(u+se)\,ds\right) e$$

$$= \left(\int_0^1 F'(su + (1-s)U)\,ds\right) e. \qquad (11.36)$$

If we define the "average" Jacobian

$$\overline{F'} = \int_0^1 F'(u+se)\,ds = \int_0^1 F'(su + (1-s)U)\,ds,$$

then we can use $\mathcal{E}(e)^* = (\overline{F'})^*$ as an adjoint in the analysis.

In much of the literature on *a posteriori* error analysis for nonlinear problems, the standard way to define an adjoint operator is to use the Integral Mean Value Theorem approach. Note that in practice, u is typically unknown so $\overline{F'}$ is not computable. Typically, we simply linearize around U, i.e. replace $\overline{F'} \to F'(U)$. This may be an issue if u and U are sufficiently far apart that they are associated with significantly different adjoint operators.

Example 36.
We continue Example 35 by examining the invertibility of $\mathcal{E}(v)^*$. We can use column operations to obtain the triangular matrix

$$\mathcal{E}(v)^* \to \begin{pmatrix} 0 & 1 \\ 3 - (2u_1 + v_1)u_1\left(\frac{1-\exp(-v_2)}{v_2}\right) & u_1\left(\frac{1-\exp(-v_2)}{v_2}\right) \end{pmatrix}$$

Thus, the invertibility of $\mathcal{E}(v)^*$ is determined by the distance of $3 - (2u_1 + v_1)u_1\left(\frac{1-\exp(-v_2)}{v_2}\right)$ to 0.

Expanding this quadratic function in u_1, we find that the roots are equal to $\pm\sqrt{3/2}$ when $v_1 = v_2 = 0$ and nearby for small v_1, v_2. If we bound u_1 away from these critical values,

$$\left| u_1^2 - \frac{3}{2} \right| \geq c^2 > 0,$$

for some constant c, we find that

$$\left| 3 - (2u_1 + v_1)u_1\left(\frac{1-\exp(-v_2)}{v_2}\right) \right| \geq 2c^2 - |u_1|\mathbf{O}(|v_1|) - |u_1|^2\mathbf{O}(|v_2|)$$

We conclude that there is a constant $\tilde{c}$ such that $\mathcal{E}(v)^*$ is uniformly invertible for

$$|v_1| \leq \frac{\tilde{c}}{|u_1|}, \quad |v_2| \leq \frac{\tilde{c}}{|u_1|^2}.$$

Hence, linearization around two points with nearby values of u_1 produces adjoint operators with nearly the same stability (invertibility) properties.

On the other hand, if $|u_1| \approx \sqrt{3/2}$ then there are critical values of v_1 and v_2 which make the operator $\mathcal{E}(v)^*$ non-invertible. In this case, linearization around two points that are near $u_1 \approx \pm\sqrt{3/2}$ may yield adjoint operators with substantially different stability properties.

We also note that (11.35) does not define a unique adjoint operator in general.

Example 37.

Suppose that $\mathcal{E}(e)$ can be written as $\mathcal{E}(e) = A(e)e$, where $A(e)$ is a linear operator with $\mathcal{D}(\mathcal{E}) \subset \mathcal{D}(A)$. For a fixed $e \in \mathcal{D}(\mathcal{E})$, we can define the adjoint of A satisfying $(A(e)w, v) = (w, A^*(e)v)$ for all $w \in \mathcal{D}(A)$, $v \in \mathcal{D}(A^*)$ as usual. Substituting $w = e$ shows this defines an adjoint of $\mathcal{E}$ as well. If there are several such linear operators A, then there are generally several different possible adjoints.

Following [7], for $(t, x) \in \Omega = (0, 1) \times (0, 1)$, we let $X = X^* = Y = Y^* = L^2$ equal to the space of periodic functions in t and x with period 1. Consider the Burgers equation

$$u_t + uu_x + au = f, \tag{11.37}$$

where $a > 0$ is constant and f is a periodic function, and we apply periodic boundary conditions on Ω. Straightforward computation yields

$$\mathcal{E}(e) = \frac{\partial e}{\partial t} + (u + e)\frac{\partial e}{\partial x} + (u_x + a)e.$$

We have $\mathcal{E}(e) = A_1(e)e$, where

$$A_1(e)v = \frac{\partial v}{\partial t} + (u + e)\frac{\partial v}{\partial x} + (u_x + a)v$$

and the adjoint is

$$A_1(e)^* w = -\frac{\partial w}{\partial t} - \frac{\partial((u + e)w)}{\partial x} + (u_x + a)w.$$

We also have $\mathcal{E}(e) = A_2(e)e$, where

$$A_2(e)v = \frac{\partial v}{\partial t} + u\frac{\partial v}{\partial x} + (u_x + e_x + a)v$$

and

$$A_2(e)^* w = -\frac{\partial w}{\partial t} - \frac{\partial(uw)}{\partial x} + (u_x + e_x + a)w.$$

Returning to the original problem, once the adjoint is defined, we can derive a representation formula for the error. In the Sobolev space setting, we note that

$$(R(U), w) = (F(U) - F(u), w) = (\mathcal{E}(e), w) = (e, \mathcal{E}(e)^* w).$$

To estimate the error in a quantity of interest (e, ψ), we let the generalized Greens function ϕ solve

$$\mathcal{E}(e)^* \phi = \psi,$$

and we obtain

$$(e, \psi) = (R(U), \phi).$$

11.4.5 *Discretization of evolution problems*

We consider a reaction-diffusion equation for the solution u on an interval $[0, T]$,

$$\begin{cases} \dot{u} - \nabla \cdot (\epsilon(x,t) \nabla u) = f(u, x, t), & (x,t) \in \Omega \times (0, T], \\ u(x,t) = 0, & (x,t) \in \partial\Omega \times (0, T], \\ u(x,0) = u_0(x), & x \in \Omega, \end{cases} \quad (11.38)$$

where Ω is a convex polygonal domain in $\mathbb{R}^d$ with boundary $\partial\Omega$, $\dot{u}$ denotes the partial derivative of u with respect to time, and there is a constant $\epsilon > 0$ such that

$$\epsilon(x,t) \geq \epsilon, \quad x \in \Omega, \ t > 0.$$

We also assume that ϵ and f have smooth second derivatives and for simplicity, we write $f(u, x, t) = f(u)$. Everything in this paper extends directly to problems with different boundary conditions, convection, nonlinear diffusion coefficients, and systems of equations, see [18].

We describe two finite element space-time discretizations of (11.38) called the continuous and discontinuous Galerkin methods, see [18, 26, 27, 28]. We can represent many standard finite element in space-finite difference scheme in time methods as one of these two methods with an appropriate choice of quadrature for evaluating the integrals defining the finite element approximation. We partition $[0, T]$ as $0 = t_0 < t_1 < t_2 < \cdots < t_n < \cdots < t_N = T$, denoting each time interval by $I_n = (t_{n-1}, t_n]$ and time step by $k_n = t_n - t_{n-1}$. We use k to denote the piecewise constant function that is k_n on I_n. We discretize Ω using a set of elements $\mathcal{T}$ as described in Section 11.4.1. We describe the notation when the space mesh is the same for all time steps. In general, we can employ different meshes for each time step.

The approximations are polynomials in time and piecewise polynomials in space on each space-time "slab" $S_n = \Omega \times I_n$. In space, we let $V \subset H_0^1(\Omega)$ denote the space of piecewise linear continuous functions defined on $\mathcal{T}$, where each function is zero on $\partial\Omega$. Then on each slab, we define

$$W_n^q = \left\{ w(x,t) : w(x,t) = \sum_{j=0}^q t^j v_j(x), \ v_j \in V, \ (x,t) \in S_n \right\}.$$

Finally, we let W^q denote the space of functions defined on the space-time domain $\Omega \times [0, T]$ such that $v|_{S_n} \in W_n^q$ for $n \geq 1$. Note that functions in W^q may be discontinuous across the discrete time levels and we denote the jump across t_n by $[w]_n = w_n^+ - w_n^-$ where $w_n^\pm = \lim_{s \to t_n^\pm} w(s)$.

We use a projection operator into V, $Pv \in V$, e.g. the L^2 projection satisfying $(Pv, w) = (v, w)$ for all $w \in V$, where $(\cdot, \cdot)$ denotes the $L_2(\Omega)$ inner product. We use the $\| \ \|$ for the L_2 norm. We also use a projection operator into the piecewise polynomial functions in time, denoted by $\pi_n : L^2(I_n) \to \mathcal{P}^q(I_n)$, where $\mathcal{P}^q(I_n)$ is the space of polynomials of degree q or less defined on I_n. The global projection operator π is defined by setting $\pi = \pi_n$ on S_n.

The continuous Galerkin cG(q) approximation $U \in W^q$ satisfies $U_0^- = P_0 u_0$ and

$$\begin{cases} \int_{t_{n-1}}^{t_n} \left((\dot{U}, v) + (\epsilon \nabla U, \nabla v) \right) dt = \int_{t_{n-1}}^{t_n} (f(U), v) \, dt \\ \qquad\qquad\qquad\qquad\qquad\qquad\qquad\qquad \text{for all } v \in W_n^{q-1}, \quad 1 \leq n \leq N, \\ U_{n-1}^+ = U_{n-1}^-. \end{cases} \tag{11.39}$$

Note that U is continuous across time nodes.

The discontinuous Galerkin dG(q) approximation $U \in W^q$ satisfies $U_0^- = Pu_0$ and

$$\int_{t_{n-1}}^{t_n} \left((\dot{U}, v) + (\epsilon \nabla U, \nabla v) \right) dt + ([U]_{n-1}, v^+) = \int_{t_{n-1}}^{t_n} (f(U), v) \, dt$$
$$\text{for all } v \in W_n^q, \quad 1 \leq n \leq N. \tag{11.40}$$

Note that the true solution satisfies both (11.39) and (11.40).

Example 38.

To illustrate, we discretize the scalar problem

$$\begin{cases} \dot{u} - \Delta u = f(u), & (x,t) \in \Omega \times \mathbb{R}^+, \\ u(x,t) = 0, & (x,t) \in \partial\Omega \times \mathbb{R}^+, \\ u(x,0) = u_0(x), & x \in \Omega, \end{cases} \tag{11.41}$$

using the dG(0) method. Since U is constant in time on each time interval, we let $\vec{U}_n^-$ denote the M vector of nodal values with respect to the nodal basis $\{\eta_i\}_{i=1}^M$ for V. We let $\mathbf{B} : (B)_{ij} = (\eta_i, \eta_j)$ for $1 \leq i, j \leq M$ denote the mass matrix and $\mathbf{A} : (A)_{ij} = (\nabla \eta_i, \nabla \eta_j)$ denote the stiffness matrix. Then U_n satisfies

$$(\mathbf{B} + k_n \mathbf{A}) \vec{U}_n^- - \vec{F}(\vec{U}_n^-) k_n = \mathbf{B} \vec{U}_{n-1}^-, \quad 1 \leq n \leq N,$$

where $(\vec{F}(U_n^-))_i = (f(U_n^-), \eta_i)$.

As mentioned, with an appropriate use of quadrature to evaluate the integrals in the variational formulation, these Galerkin methods yield standard difference schemes. We write these standard numerical methods as space-time finite element methods in order to make use of adjoints and variational analysis.

Example 39.
In the example above, if the lumped mass quadrature is used to evaluate the coefficients of **B**, then the resulting set of equations for the dG(0) approximation is the same as the equations for the nodal values of the backward Euler-second order centered difference scheme for (11.41).

The dG(0) method is related to the backward Euler method, the cG(1) method is related to the Crank-Nicolson scheme, and the dG(1) method is related to the third order sub-diagonal Padé difference scheme, see [26, 27, 28, 29, 30, 31, 32, 33, 34].

Under general assumptions, the cG(q) and dG(q) have order of accuracy $q+1$ in time and 2 in space at any point. In addition, they enjoy a superconvergence property in time at time nodes. The dG(q) method has order of accuracy $2q+1$ in time and the cG(q) method has order $2q$ in time at time nodes for sufficiently smooth solutions.

11.4.6 *Analysis for discretizations of evolution problems*

We begin the *a posteriori* analysis by defining a suitable adjoint problem for error analysis. The adjoint problem is a linear parabolic problem with coefficients obtained by linearization around an average of the true and approximate solutions.

$$\bar{f} = \bar{f}(u, U) = \int_0^1 \frac{\partial f}{\partial u}(us + U(1-s)) \, ds. \tag{11.42}$$

The regularity of u and U typically implies that $\bar{f}$ is piecewise continuous with respect to t and a continuous, H^1 function in space.

Written out pointwise for convenience, the adjoint problem to (11.38) for the generalized Greens function associated to the data ψ, which determines the quantity of interest,

$$\int_0^T (u, \psi) \, dt, \tag{11.43}$$

is

$$\begin{cases} -\dot{\phi} - \nabla \cdot (\epsilon \nabla \phi) - \bar{f}\phi = \psi, & (x,t) \in \Omega \times (T, 0], \\ \phi(x,t) = 0, & (x,t) \in \partial\Omega \times (T, 0], \\ \phi(x,T) = 0, & x \in \Omega, \end{cases} \tag{11.44}$$

Using this definition, for the dG method we have

$$\int_0^T (e, \psi) \, dt = \int_0^T (e, -\dot{\phi} - \nabla \cdot (\epsilon \nabla \phi) - \bar{f}\phi) \, dt$$

$$= \sum_{n=1}^N \int_{I_n} (e, -\dot{\phi} - \nabla \cdot (\epsilon \nabla \phi) - \bar{f}\phi) \, dt.$$

We integrate by parts in time for

$$\int_{I_n} (e, -\dot\phi)\, dt = -(e_n^-, \phi_n) + (e_{n-1}^+, \phi_{n-1}) + \int_{I_n} (\dot e, \phi)\, dt.$$

Likewise,

$$\int_{I_n} (e, -\nabla \cdot (\epsilon\nabla\phi))\, dt = \int_{I_n} (\epsilon\nabla e, \nabla\phi)\, dt.$$

Finally,

$$\int_{I_n} (e, \overline{f}\phi)\, dt = \int_{I_n} (\overline{f}e, \phi)\, dt = \int_{I_n} (f(U) - f(u), \phi)\, dt.$$

Next we realize that the true solution satisfies the weak formulation

$$\int_{I_n} \big((\dot u, \phi) + (\epsilon\nabla u, \nabla\phi) - (f(u), \phi)\big)\, dt = 0,$$

hence,

$$\int_{I_n} \big((\dot e, \phi) + (\epsilon\nabla e, \nabla\phi) - (f(U) - f(u), \phi)\big)\, dt$$
$$= \int_{I_n} \big((\dot U, \phi) + (\epsilon\nabla U, \nabla\phi) - (f(U), \phi)\big)\, dt,$$

The sum of terms arising from the integration by parts in time simplifies

$$\sum_{n=1}^{N} \big(-(e_n^-, \phi_n) + (e_{n-1}^+, \phi_{n-1})\big) = (e_0^+, \phi_0) + \sum_{n=1}^{N-1} (e_n^+ - e_n^-, \phi_n) - (e_N^-, \phi_N),$$

and then simplifies further upon realizing that $u_n^- = u_n^+$ and $\phi_N = 0$. Using the definition (11.40) for the dG method, we obtain

11.4.7 Theorem

$$\int_0^T (e, \psi)\, dt = ((I - P)u_0, \phi(0)) + \sum_{j=1}^{N} ([U]_{j-1}, (\pi P\phi - \phi)_{j-1}^+)$$
$$+ \int_0^T \big((\dot U, \pi P\phi - \phi) + (\epsilon(U)\nabla U, \nabla(\pi P\phi - \phi)) - (f(U), \pi P\phi - \phi)\big)\, dt.$$
(11.45)

The initial error is $e^-(0) = (I - P)u_0$.

If instead we desire to estimate $(u(T), \psi)$, for a function ψ, then the adjoint problem is

$$\begin{cases} -\dot{\phi} - \nabla \cdot (\epsilon \nabla \phi) - \bar{f}\phi = 0, & (x,t) \in \Omega \times (T, 0], \\ \phi(x, t) = 0, & (x,t) \in \partial\Omega \times (T, 0], \\ \phi(x, T) = \psi, & x \in \Omega. \end{cases} \quad (11.46)$$

The resulting estimate is

11.4.8 Theorem

$$(e(T), \psi) = ((I - P)u_0, \phi(0)) + \sum_{j=1}^{N} \left([U]_{j-1}, (\pi P \phi - \phi)_{j-1}^+ \right)$$

$$+ \int_0^T \left((\dot{U}, \pi P \phi - \phi) + (\epsilon(U) \nabla U, \nabla(\pi P \phi - \phi)) \right.$$

$$\left. - (f(U), \pi P \phi - \phi) \right) dt. \quad (11.47)$$

A similar argument for the cG method, say for the global quantity of interest (11.43), yields

11.4.9 Theorem

$$\int_0^T (e, \psi) \, dt = ((I - P)u_0, \phi(0))$$

$$+ \int_0^T \left((\dot{U}, \pi P \phi - \phi) + (\epsilon(U) \nabla U, \nabla(\pi P \phi - \phi)) \right.$$

$$\left. - (f(U), \pi P \phi - \phi) \right) dt. \quad (11.48)$$

In practice, we compute a numerical solution of a linear adjoint problem obtained from (11.44). Typically, we linearize around the computed approximate solution and solve using a higher order method in space and time. Without specifying the details, we denote the approximate adjoint solution by Φ. Focussing on the dG method, where application to the cG method is obvious, the approximate *a posteriori* error estimate then reads

$$\left| \int_0^T (e, \psi) \, dt \right| \approx E(U) = E(U; \psi)$$

$$= \left| ((I - P)u_0, \Phi(0)) + \sum_{j=1}^{N} \left([U]_{j-1}, (\pi P \Phi - \Phi)_{j-1}^+ \right) \right.$$

$$\left. + \int_0^T \left((\dot{U}, \pi P \Phi - \Phi) + (\epsilon(U) \nabla U, \nabla(\pi P \Phi - \Phi)) - (f(U), \pi P \Phi - \Phi) \right) dt \right|.$$

$$(11.49)$$

Example 40.

In [35], we consider the accuracy of the *a posteriori* error estimate applied to the chaotic Lorenz problem

$$\begin{cases} \dot{u}_1 = -10u_1 + 10u_2, \\ \dot{u}_2 = 28u_1 - u_2 - u_1 u_3, \\ \dot{u}_3 = -\frac{8}{3}u_3 + u_1 u_2, \\ u_1(0) = -6.9742, u_2(0) = -7.008, u_3(0) = 25.1377. \end{cases} \quad 0 < t,$$

The terms in (11.49) describing space discretization simply drop out in this case, and we compute the resulting estimate.

In Figure 11.16, we show the accuracy of the *a posteriori* error estimate for pointwise values of each component at many times. Similar accuracy is obtained for other functionals, e.g. average error.

We illustrate the idea that the solution of the adjoint problem provides a kind of condition number for the computed solution. Following [4], in Figure 11.17, we show that the adjoint solution grows very rapidly when the solution passes through the tiny region near separatix. On the other hand, the residual error of the solution remains small in this region. In this case, using only the residual, or indeed the "local error", fails completely to indicate that the error of the solution increases rapidly in a neighborhood of the separatix.

Example 41.

In [18, 36, 37], we compute the *a posteriori* error estimate for the well known bistable (Allen-Cahn) problem $\dot{u} - \epsilon \Delta u = u - u^3$ posed with Neumann boundary conditions. This is used to model the motion of domain walls in a ferromagnetic

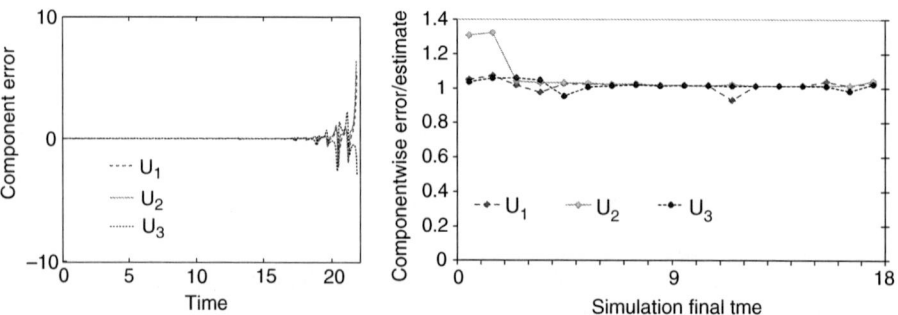

Figure 11.16: Left: We plot an approximate error in each component of a numerical solution of the Lorenz problem computed by taking the difference between solutions with estimated error .001 and .0001 at many times. Right: we plot the pointwise error/estimate ratios for each component versus time at many time points.

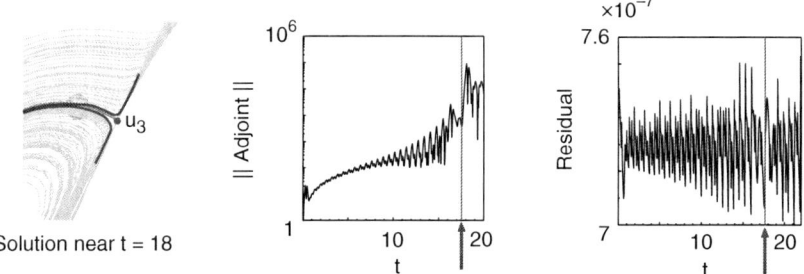

Figure 11.17: Left: We plot the accurate and inaccurate numerical solutions during the time that the inaccurate solution becomes 100% inaccurate along with the separatix. The inaccurate solution steps on the wrong side of the separatix. Middle: We plot the norm of the adjoint solutions corresponding to pointwise error. The adjoint solution grows exponentially rapidly only when the solution passes near the separatix. Right: We plot the residual for the inaccurate solution, which remains small even when the error becomes large.

material. The problem has two attracting steady-state solutions 1 and -1. Generic solutions eventually converge to one of these two steady-state solutions, but the evolution towards a steady state can take some time because of the interesting competition between competing stable processes, e.g. the diffusion that tends to drive a solution towards zero and the reaction that tends to drive a solution towards 1 or -1.

In one dimension, generic solutions form a pattern of layers between the values of -1 and 1, then solutions undergo long periods of metastability during which the motion of the layers "horizontally" is extremely slow punctuated by rapid transients in which the solution moves to another metastable state or the final stable state. We show the evolution of numerical approximation of a metastable solution with two metastable periods $[0, 44]$ and $[44, 144]$ in Figure 11.18. The time scale of metastable periods increases exponentially in $1/\sqrt{\epsilon}$ as the diffusion coefficient ϵ decreases.

We plot the L^2 space norms of the residuals versus time for numerical solutions computed on uniform discretizations. The time residuals reflect an initial transient and then the two transients concluding the metastable periods. The space residual simply becomes smaller as the layers disappear. Finally, we plot the L^2 space norm of the adjoint weights corresponding to the time and space parts of the error estimate for the quantities of interest equal to pointwise values of the solution at a set of uniform time points. The weights grow in advance of the transients concluding the metastable periods but immediately decrease to 1 or smaller right after the transient, indicating that accumulated errors have damped. Thus, while the effects of errors grow during metastable periods, the overall error accumulation remains bounded, implying that accurate long-time solutions can be computed provided the meshes are sufficiently refined.

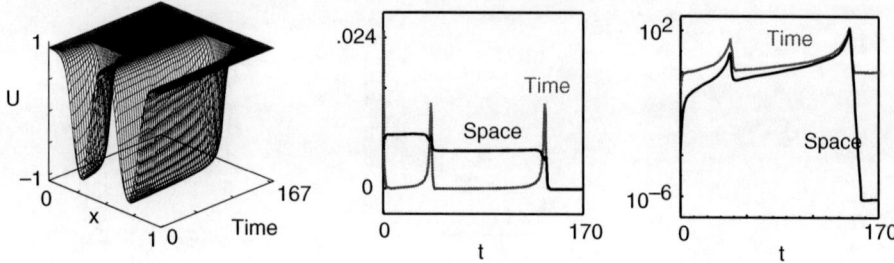

Figure 11.18: Left: The evolution of a metastable solution of the bistable problem with $\epsilon = .0009$. Middle: Evolution of the time and space residuals on a uniform discretization. Right: The evolution of the absolute adjoint weights for pointwise errors for time and space.

In two dimensions, the dynamics of the problem are much different because the evolution is governed by "motion by mean curvature", meaning that the normal velocity of a transition layer is proportional to the sum of the principle curvatures of the layer. Consequently, the time scale for the evolution increases only at an algebraic rate, κ/ϵ, where κ is the mean curvature, as the diffusion coefficient ϵ decreases. We solve the bistable problem using initial data consisting of two "mesas" corresponding to the two wells in the solution shown in Figure 11.18 using $\epsilon = .00003$ so that the evolution occurs over the same time scale. We shows four snapshots of the solution in Figure 11.19. The time evolution of the adjoint weights shows a pattern of growth and decay as for metastable solutions in one dimension. However, solutions in two dimensions are much less sensitive to perturbations than solutions in one dimension, and the adjoint weights are much smaller overall.

11.4.10 General comments on a posteriori analysis

We can abstract the four steps for *a posteriori* error analysis as:

1. Identify functionals that yield the quantities of interest
2. Define appropriate adjoint problems for the quantities of interest
3. Derive a computable residual for each source of error
4. Derive an error representation using suitable adjoint weights for each residual.

We also note that in general we have to account for all sources of error in the analysis. Typical sources include:

- space and time discretization (approximation of the solution space)
- use of quadrature to compute integrals in a variational formulation (approximation of the differential operator)
- solution error in solving any linear and nonlinear systems of equations
- model error

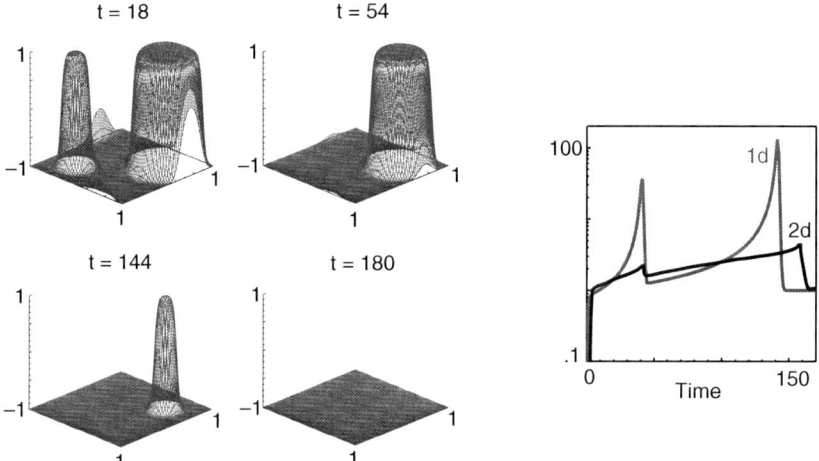

Figure 11.19: Left: Four snapshots of a bistable solution in two dimensions. Right: The evolution of the absolute adjoint weights for pointwise errors for time and space along with the weights for a solution in one dimension.

- data and parameter error
- operator decomposition

We have not discussed most of these sources in this note. However, it is important to realize that different sources of error typically accumulate and propagate at different rates, and so must be accounted for individually in any analysis.

11.5 *A posteriori* error estimates and adaptive mesh refinement

Computing accurate error estimates provides the tantalizing idea of optimizing discretizations. We briefly discuss the use of *a posteriori* error estimates for guiding adaptive mesh refinement.

A typical goal of adaptive error control is to generate a mesh with a relatively small number of elements such that for a given tolerance TOL and data ψ,

$$\text{error in the quantity of interest } = |(e,\psi)| \lessapprox \text{TOL}. \quad (11.50)$$

We note that (11.50) cannot be verified in practice because the error is unknown, so we use an error estimate and try to construct a mesh to achieve

$$a \text{ posteriori estimate of the error in the quantity of interest} \lessapprox \text{TOL}. \quad (11.51)$$

The general idea is to write the estimate as a sum of "element contributions" that indicate the contribution from discretization on each element to the total error. We identify the elements that contribute most and then refine those elements.

However, this simple description belies a number of theoretical and practical difficulties.

11.5.1 Adaptive mesh refinement in space

We first consider adaptive mesh refinement for a stationary problem. In the case of an elliptic problem, we use the estimate (11.32) to implement (11.51). To do so, we rewrite (11.32) as a sum of (signed) element contributions,

$$(e, \psi) \approx \sum_{K \in \mathcal{T}_h} \int_K \left((f - b \cdot \nabla U - cU)(\Phi - \pi_h \Phi) - a \nabla U \cdot \nabla (\Phi - \pi_h \Phi) \right) dx. \tag{11.52}$$

Thus, using (11.52), (11.51) gives the goal of satisfying the following condition: The mesh acceptance criterion is

$$\left| \sum_{K \in \mathcal{T}_h} \int_K \left((f - b \cdot \nabla U - cU)(\Phi - \pi_h \Phi) - a \nabla U \cdot \nabla (\Phi - \pi_h \Phi) \right) dx \right| \leq \text{TOL}. \tag{11.53}$$

If the current approximation satisfies (11.53), then the solution is deemed acceptable and the refinement process is stopped.

The difficulties start when (11.53) is not satisfied. We have to decide how to "enrich" the discretization, e.g., refine the mesh or increase the order of the element functions, in order to improve the accuracy. The problem is that generally there is a great deal of cancellation among the contributions from each element. For example, consider that large positive contributions from one subregion might cancel the large negative contributions from another region so that the sum of the contributions from the two regions together is small, see Example 42 below. In fact, we make the certainly controversial claim that

There is currently no theory or practical method for accommodating cancellation of errors in an adaptive error control in a way that truly optimizes efficiency.

The standard approach is to formulate the discretization enrichment problem as a constrained optimization problem after replacing the error estimate by an error bound consisting of a sum over elements of positive quantities. For example, we obtain a bound from (11.52) by inserting norms in some way, e.g., we use

$$|(e, \psi)| \leq \sum_{K \in \mathcal{T}_h} \int_K \left| (f - b \cdot \nabla U - cU)(\Phi - \pi_h \Phi) - a \nabla U \cdot \nabla (\Phi - \pi_h \Phi) \right| dx. \tag{11.54}$$

Thus, if (11.53) is **not** satisfied, then the mesh is refined in order to achieve the more conservative criterion,

$$\sum_{K \in \mathcal{T}_h} \int_K \left| (f - b \cdot \nabla U - cU)(\Phi - \pi_h \Phi) - a \nabla U \cdot \nabla (\Phi - \pi_h \Phi) \right| dx \lesssim \text{TOL}. \tag{11.55}$$

The adaptive error control problem is the constrained minimization problem of finding a mesh with a minimal number of degrees of freedom on which the approximation satisfies (11.55). Using the fact that the bound in (11.55) is a sum of positive terms, and assuming the solution is asymptotically accurate, a calculus of variations argument yields the **Principle of Equidistribution**, which states that the solution of this constrained optimization problem is achieved when the elements contributions are all approximately equal. An adaptive mesh algorithm is a procedure for solving the constrained minimization problem associated with (11.55). If the Principle of Equidistribution is used, then the algorithm seeks to choose meshes so that the element contributions are approximately equal.

Depending on the argument, two possible element acceptance criteria for the element indicators are

$$\max_K \left| (f - b \cdot \nabla U - cU)(\Phi - \pi_h \Phi) - a \nabla U \cdot \nabla (\Phi - \pi_h \Phi) \right| \lesssim \frac{\text{TOL}}{|\Omega|}, \quad (11.56)$$

or

$$\int_K \left| (f - b \cdot \nabla U - cU)(\Phi - \pi_h \Phi) - a \nabla U \cdot \nabla (\Phi - \pi_h \Phi) \right| dx \lesssim \frac{\text{TOL}}{M}, \quad (11.57)$$

where M is the number of elements in $\mathcal{T}_h$. Elements that fail one of these tests are marked for refinement.

Computing a mesh using these criteria is usually performed by a "compute-estimate-mark-refine" adaptive algorithm that begins with a coarse mesh and then refines those elements on which (11.56) respectively (11.57) fail successively. See [6, 16, 17, 19, 21, 38] for more details.

The problem with any claims of "optimal" mesh selection is that generically the bound (11.54) is typically orders of magnitude larger than the estimate (11.52).

Example 42.

We illustrate the issue of the effect of cancellation of errors on the choice of an optimal adapted mesh with a simple computation. Assume that we solve an elliptic problem on a square domain using bilinear elements on a mesh consisting of rectangles, and the *a posteriori* estimate yields the signed element contributions shown on the left in Figure 11.20. The total *a posteriori* estimate is

$$.1 + 3 \times -.0333 + 17 \times .001 + 4 \times .01 = .0571.$$

Note that the large element contribution in the lower left corner is nearly canceled by the contributions of its three neighbors, $.01 + 3 \times -.0333 = .0001$, so that the region ends up contributing relatively little to the error. If we refine in the upper right-hand corner by subdividing each square into four smaller elements as indicated (and assume the element contributions decrease by a factor of $2^2 = 4$ without any change in sign), the new estimate becomes

$$.1 + 3 \times -.0333 + 17 \times .001 + 16 \times .01 \times \tfrac{1}{4} \times \tfrac{1}{4} = .0271.$$

Note that while we change four elements into 16 smaller elements, the element contribution in each goes down by a factor of four while the area of each smaller element is four times smaller than its parent.

On the other hand, if we use the absolute element contributions to guide refinement, then the elements in the lower left-hand corner are refined as shown on the left in Figure 11.20, and the new estimate becomes

$$4 \times .1 \times \tfrac{1}{4} \times \tfrac{1}{4} + 3 \times \left(4 \times -.0333 \times \tfrac{1}{4} \times \tfrac{1}{4}\right) + 17 \times .001$$
$$+ 16 \times .01 \times \tfrac{1}{4} \times \tfrac{1}{4} = .057025.$$

There is almost no improvement in overall accuracy in the quantity of interest.

Regardless of the issue of dealing with cancellation of errors efficiently, there is still a crucial difference between adaptive mesh refinement based on adjoint-weighted residual estimates and traditional "error indicators" that often amount to using only residuals or "local errors." In the adjoint-based approach, the element residuals are scaled by an adjoint weight, which reflects how much error in that element affects the error in the quantity of interest. This has a significant effect of mesh refinement patterns in general.

Example 43.

In [6], we apply these ideas to the adaptive solution of

$$\begin{cases} -\nabla \cdot \left((.05 + \tanh(10(x-5)^2 + 10(y-1)^2))\nabla u\right) \\ \quad + \begin{pmatrix} -100 \\ 0 \end{pmatrix} \cdot \nabla u = 1, \quad (x,y) \in \Omega = [0,10] \times [0,2], \\ u = 0, \quad (x,y) \in \partial\Omega \end{cases} \quad (11.58)$$

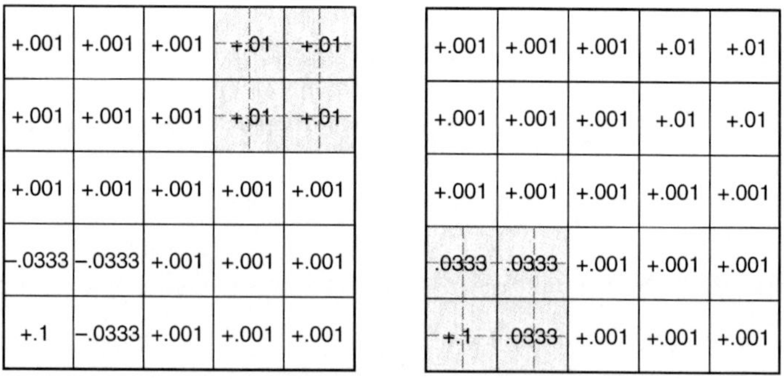

+.001	+.001	+.001	+.01	+.01
+.001	+.001	+.001	+.01	+.01
+.001	+.001	+.001	+.001	+.001
−.0333	−.0333	+.001	+.001	+.001
+.1	−.0333	+.001	+.001	+.001

+.001	+.001	+.001	+.01	+.01
+.001	+.001	+.001	+.01	+.01
+.001	+.001	+.001	+.001	+.001
.0333	.0333	+.001	+.001	+.001
+.1	.0333	+.001	+.001	+.001

Figure 11.20: On the left, we display (simulated) signed element contributions. We shade four elements chosen for refinement using nonstandard criteria. On the right, we display the corresponding absolute element contributions and shade four elements marked for refinement by a standard adaptive algorithm.

In Figure 11.21 we show the mesh required to obtain a numerical solution whose average value is accurate to within 4%. The adaptive pattern is obtained by refining from a coarse uniform mesh using (11.57). Convection causes a nonuniform pattern of refinement.

In the first computation, the quantity of interest is given by a function ψ that is constant over the entire domain. In the next computation, we take the quantity of interest to be the average value in a square in one corner of the domain. We now require much fewer elements to achieve the desired accuracy. The pattern of refinement shows the effects of the adjoint solution, see Figures 11.22 and 11.25. In particular, the adjoint solution decreases rapidly to zero towards the side of the domain opposite to the quantity of interest region and there is less dense mesh refinement along that side. The influence of regions far "upstream" is also diminished.

In the next computation, we take the quantity of interest to be the average value in a square in the middle of the domain. Again, the pattern of refinement shows the effects of the adjoint solution, see Figures 11.23 and 11.25. In

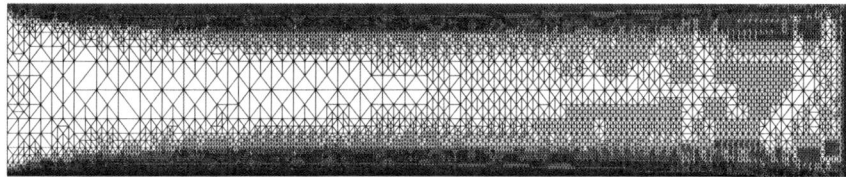

Figure 11.21: The mesh used to solve (11.58) with an error of 4% in the average value requires 24,4000 elements.

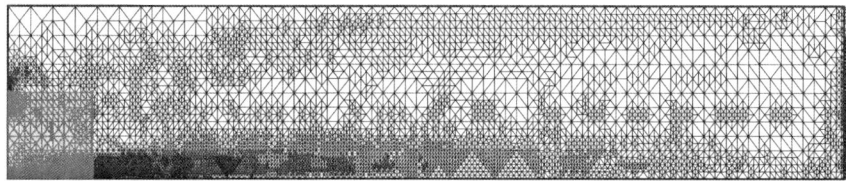

Figure 11.22: The quantity of interest is the average value in the shaded square. The final mesh requires 7300 elements.

Figure 11.23: The quantity of interest is the average value in the shaded square. The final mesh requires 7300 elements.

particular, the adaptive mesh refinement makes no attempt to resolve the boundary layer at the "outflow" boundary as the accuracy there has no effect on the accuracy of the quantity of interest.

In the final computation, we take the quantity of interest to be the average value in a square at the far end of the domain. Again, the pattern of refinement shows the effects of the adjoint solution, see Figures 11.24 and 11.25.

In Figure 11.25, we plot the solutions of the adjoint problems corresponding to the quantities of interest equal to average values in squares at the opposite ends of the domain.

11.5.2 *Adaptive mesh refinement for evolutionary problems*

Traditionally, different approaches for adaptive mesh algorithms are used to handle spatial meshes and time discretization. Influenced by the long history of "local error control", the traditional time algorithm achieves an equidistribution of element contributions by insuring that the contribution from each time interval is smaller than, but approximately equal to, a "local error tolerance" (LTOL) before proceeding to the next time step. Often, (LTOL) is input directly without any attempt to relate it to the desired tolerance (TOL) on the error. Given a true global error estimate, however, and the asymptotic accuracy of the integration scheme, there are various heuristic arguments for determining LTOL in terms of TOL.

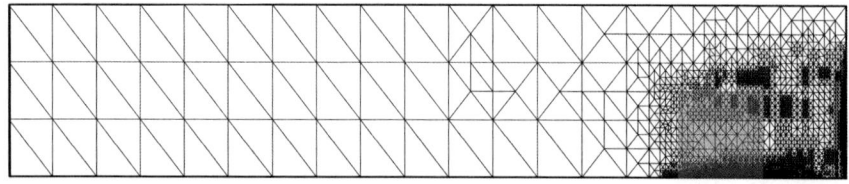

Figure 11.24: The quantity of interest is the average value in the shaded square. The final mesh requires 3500 elements.

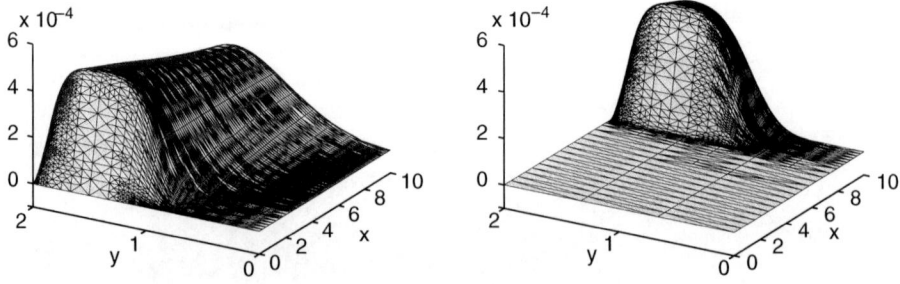

Figure 11.25: The adjoint solutions for the computations in Figures 11.22 and 11.23.

For an evolutionary partial differential equation, space and time mesh refinement strategies have to be combined somehow. In the case of a parabolic problem, we distinguish time and space contributions in (11.49) by "splitting" the projections on the adjoint solution,

$$E(U) = \Big|((I-P)u_0, \Phi(0)) + \sum_{j=1}^{N}([U]_{j-1}, (P\Phi - \Phi)^+_{j-1})$$

$$+ \int_0^T \left((\dot{U}, P\Phi - \Phi) + (\epsilon(U)\nabla U, \nabla(P\Phi - \Phi)) - (f(U), P\Phi - \Phi)\right) dt$$

$$+ \sum_{j=1}^{N}([U]_{j-1}, ((\pi - 1)P\Phi)^+_{j-1})$$

$$+ \int_0^T \left((\dot{U}, (\pi - 1)P\Phi) + (\epsilon(U)\nabla U, \nabla((\pi - 1)P\Phi))\right.$$

$$- (f(U), (\pi - 1)P\Phi)\Big) dt \Big|. \tag{11.59}$$

We define bounds on the time and space contributions,

$$\text{Æ}_t(U) = \sum_{j=1}^{N} \sum_{K \in \mathcal{T}_h} \left| \int_K ([U]_{j-1})(((\pi - 1)P\Phi)^+_{j-1}) \, dx \right|$$

$$+ \sum_{n=1}^{N} \sum_{K \in \mathcal{T}_h} \left| \int_{t_{n-1}}^{t_n} \int_K \left(\dot{U}((\pi - 1)P\Phi) + (\epsilon(U)\nabla U) \cdot (\nabla(\pi - 1)P\Phi) \right. \right.$$

$$- f(U)((\pi - 1)P\Phi) \Big) \, dx \, dt \Big|, \tag{11.60}$$

$$\text{Æ}_x(U) = \sum_{K \in \mathcal{T}_h} \sum_{j=1}^{N} \sum_{K \in \mathcal{T}_h} \left| \int_K ([U]_{j-1})((P\Phi - \Phi)^+_{j-1}) \, dx \right|$$

$$+ \sum_{n=1}^{N} \sum_{K \in \mathcal{T}_h} \left| \int_{t_{n-1}}^{t_n} \int_K \left(\dot{U}(P\Phi - \Phi) + (\epsilon(U)\nabla U) \cdot (\nabla(P\Phi - \Phi)) \right. \right.$$

$$- f(U)(P\Phi - \Phi) \Big) \, dx \, dt \Big|. \tag{11.61}$$

Note that the space discretization may affect the time contribution and likewise, the time discretization may affect the space contribution.

We may now split the adaptive mesh problem into two sub-problems, refining the space and time steps in order to achieve

$$Æ_x(U) \lessapprox \frac{\text{TOL}}{2} \quad \text{and} \quad Æ_t(U) \lessapprox \frac{\text{TOL}}{2}. \tag{11.62}$$

On a given time interval, this requires an iteration during which both the space mesh and time steps are refined.

11.6 Multiscale operator decomposition

We now turn to the main goal of this chapter, which is to describe how the techniques of *a posteriori* error analysis can be extended to multiscale operator decomposition solutions of multiphysics, multiscale problems. Recall that the general approach is to decompose the multiphysics problem into components involving simpler physics over a relatively limited range of scales, and then to seek the solution of the entire system through some sort of iterative procedure involving solutions of the individual components.

While the particulars of the analysis vary considerably with the problem, there are several key ideas underlying a general approach to treat operator decomposition multiscale methods, including:

- We identify auxiliary quantities of interest associated with information passed between physical components and solve auxiliary adjoint problems to estimate the error in those quantities.
- We deal with scale differences by introducing projections between discrete spaces used for component solutions and estimate the effects of those projections.
- The standard linearization argument used to define an adjoint operator associated with error analysis for a nonlinear problem may fail, requiring another approach to define adjoint operators.
- In this regard, the adjoint operator associated with a multiscale operator decomposition solution method is often different than the adjoint associated with the original problem, and the difference may have a significant impact on the stability of the method.
- In practice, solving the adjoint associated with the original fully-coupled problem may present the same kinds of multiphysics, multiscale challenges posed by the original problem, so attention must be paid to the solution of the adjoint problem.

We explain these ideas in the context of three examples.

11.6.1 *Multiscale decomposition of triangular systems of elliptic problems*

Following [39], we can capture the essential features of the thermal actuator model described in Example 1 using a two component "one-way" coupled system of the form

$$\begin{cases} -\nabla \cdot a_1 \nabla u_1 + b_1 \cdot \nabla u_1 + c_1 u_1 = f_1(x), & x \in \Omega, \\ -\nabla \cdot a_2 \nabla u_2 + b_2 \cdot \nabla u_2 + c_2 u_2 = f_2(x, u_1, Du_1), & x \in \Omega, \\ u_1 = u_2 = 0, & x \in \partial\Omega, \end{cases} \quad (11.63)$$

where a_i, b_i, c_i, f_i are smooth functions, with $a_1, a_2 \geq \alpha > 0$ on a bounded domain Ω in $\mathbb{R}^N$ with boundary $\partial\Omega$, and α is a constant. Note that the problems are coupled through f_2. The "lower triangular" form of this system means that we can either solve it as a coupled system or we can solve the first equation and then use the solution to generate the parameters for the second problem. The latter approach fits the idea of a multiscale, operator decomposition discretization.

The weak form of the first component of (11.63) reads: find $u_1 \in \tilde{W}_2^1(\Omega)$ satisfying

$$\mathcal{A}_1(u_1, v_1) = (f_1, v_1), \quad \text{for all } v_1 \in H_0^1(\Omega), \quad (11.64)$$

where

$$\mathcal{A}_1(u_1, v_1) \equiv (a_1 \nabla u_1, \nabla v_1) + (b_1(x) \cdot \nabla u_1, v_1) + (c_1 u_1, v_1)$$

is a bilinear form on Ω and $H_0^1(\Omega)$ is the subspace of functions in $H^1(\Omega)$ that are zero on $\partial\Omega$. Likewise, the weak formulation of the second component of (11.63) reads: find $u_2 \in H_0^1(\Omega)$ satisfying

$$\mathcal{A}_2(u_2, v_2) = (f_2(x, u_1, Du_1), v_2), \quad \text{for all } v_2 \in H_0^1(\Omega), \quad (11.65)$$

where

$$\mathcal{A}_2(u_2, v_2) \equiv (a_2 \nabla u_2, \nabla v_2) + (b_2(x) \cdot \nabla u_2, v_2) + (c_2 u_2, v_2)$$

is another bilinear form on Ω.

We introduce the finite element space $\mathcal{S}_{h,1}(\Omega) \subset H_0^1(\Omega)$, corresponding to a discretization $\mathcal{T}_{h,1}$ of Ω for the first component, and another finite element space $\mathcal{S}_{h,2}(\Omega)$, on a different mesh $\mathcal{T}_{h,2}$, for the second component. Using different finite element spaces for different components in a system of equations raises a serious practical difficulty. Namely, evaluating integrals defining finite element approximate solutions involving functions from different spaces is problematic. In practice, quadrature formulae are used to approximate the integrals defining a finite element function. This raises a potential difficulty because quadrature formulae work best when the integrands are smooth, whereas the standard finite element functions are **only** continuous. We avoid potential difficulties by writing any integrals as a sum of integrals over elements,

$$\int_\Omega \text{integrand}\, dx = \sum_{K \in \mathcal{T}_h} \int_K \text{integrand}\, dx,$$

and applying quadrature formulae on each element on which the finite element functions are smooth. However, in the case of a system in which the components are solved in different finite element spaces, it is not so straightforward to apply quadrature formulae to evaluate integrals. A function in one finite element space may only be continuous on an element associated with another finite element space. To avoid this problem, we introduce projections $\Pi_{i \to j}$ from $\mathcal{S}_{h,i}$ to $\mathcal{S}_{h,j}$, e.g. interpolants or an L^2 orthogonal projection. We apply these projections before applying quadrature formulae.

Algorithm 11.6.2 Multiscale Operator Decomposition for Triangular Systems of Elliptic Equations

Construct discretizations $\mathcal{T}_{h,1}, \mathcal{T}_{h,2}$ and corresponding spaces $\mathcal{S}_{h,1}, \mathcal{S}_{h,2}$. Compute $U_1 \in \mathcal{S}_{h,1}(\Omega)$ satisfying

$$\mathcal{A}_1(U_1, v_1) = (f_1, v_1), \quad \text{for all } v_1 \in \mathcal{S}_{h,1}(\Omega). \tag{11.66}$$

Compute $U_2 \in \mathcal{S}_{h,2}(\Omega)$ satisfying

$$\mathcal{A}_2(U_2, v_2) = (f_2(x, \Pi_{1\to 2}U_1, \Pi_{1\to 2}DU_1), v_2), \quad \text{for all } v_2 \in \mathcal{S}_{h,2}(\Omega). \tag{11.67}$$

We observe that any errors made in the solution of the first component affect the solution of the second component. This turns out to be a crucial fact for *a posteriori* error analysis.

Example 44.

In [39], we solve a system

$$\begin{cases} -\Delta u_1 = \sin(4\pi x)\sin(\pi y), & x \in \Omega \\ -\Delta u_2 = b \cdot \nabla u_1 = 0, & x \in \Omega, \\ u_1 = u_2 = 0, & x \in \partial\Omega, \end{cases} \quad b = \frac{2}{\pi}\begin{pmatrix} 25\sin(4\pi x) \\ \sin(\pi x) \end{pmatrix} \tag{11.68}$$

using a standard piecewise linear, continuous finite element method, where $\Omega = [0,1] \times [0,1]$, in order to compute the quantity of interest

$$u_2(.25, .25).$$

We solve for u_1 first and then solve for u_2 using independent meshes and show the solutions in Figure 11.26.

Using uniform meshes, evaluating the standard *a posteriori* error estimate for the second component problem, ignoring any effect arising from error in the solution of the first component, yields an estimate of component solution error to be ≈.0042. However, the true error is ≈.0048 and there is discrepancy of ≈.0006 (≈13%) in the estimate. This is a consequence of ignoring the transfer error.

If we adapt the mesh for the solution of the second component based on the *a posteriori* error estimate of the error in that component while neglecting the effects of the decomposition, see Figure 11.27, the discrepancy becomes alarmingly worse. For example, we can refine the mesh until the estimate of the error in the second component is ≈.0001. But, we find that the true error is ≈.2244!

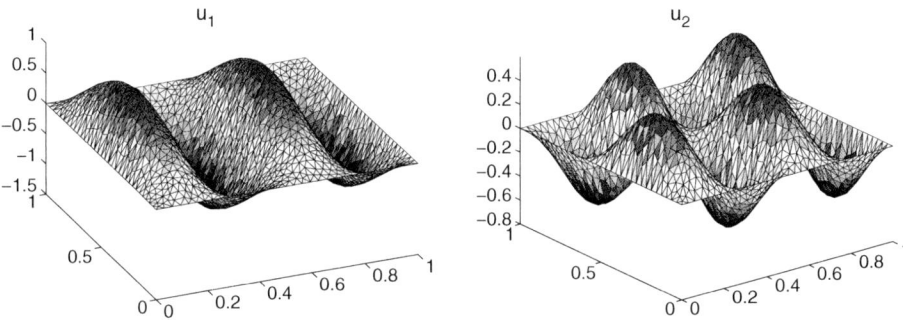

Figure 11.26: Solutions of the component problems of (11.68) computed on uniform meshes.

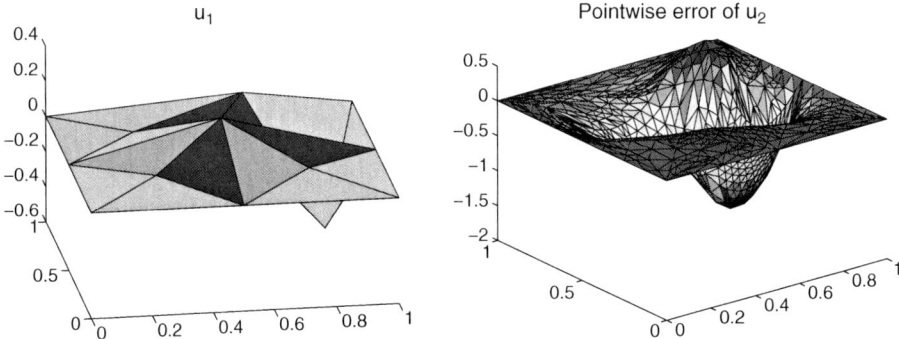

Figure 11.27: Results obtained after refining the mesh for the second component so that the *a posteriori* error estimate of the error only in the second component is less than .001. The mesh for the first component remains coarse; consequently the error in the first component becomes relatively much larger.

A linear algebra example

We can describe some of the essential features of the analysis using a block lower triangular linear system of equations. We consider the system

$$\begin{pmatrix} \mathbf{A}_{11} & 0 \\ \mathbf{A}_{21} & \mathbf{A}_{22} \end{pmatrix} \begin{pmatrix} u_1 \\ u_2 \end{pmatrix} = \begin{pmatrix} b_1 \\ b_2 \end{pmatrix} = b, \quad (11.69)$$

with approximate solution

$$U = \begin{pmatrix} U_1 \\ U_2 \end{pmatrix} \approx \begin{pmatrix} u_1 \\ u_2 \end{pmatrix} = u.$$

We estimate the error in a primary quantity of interest involving only u_2,

$$(\psi^{(1)}, u) = (\psi_2^{(1)}, u_2) \text{ where } \psi = \begin{pmatrix} 0 \\ \psi_2^{(1)} \end{pmatrix}.$$

We require a superscript $^{(1)}$ since we later define an auxiliary quantity of interest. The lower triangular structure of the system matrix yields residuals

$$R_1 = b_1 - \mathbf{A}_{11} U_1,$$
$$R_2 = (b_2 - \mathbf{A}_{21} U_1) - \mathbf{A}_{22} U_2.$$

Note that the residual R_2 of the approximate solution of the second component depends upon the solution of the first component, and any attempt to decrease this residual requires a consideration of the accuracy of U_1. The adjoint problem to (11.69) is

$$\begin{pmatrix} \mathbf{A}_{11}^T & \mathbf{A}_{21}^T \\ 0 & \mathbf{A}_{22}^T \end{pmatrix} \begin{pmatrix} \phi_1^{(1)} \\ \phi_2^{(1)} \end{pmatrix} = \begin{pmatrix} 0 \\ \psi_2^{(1)} \end{pmatrix},$$

and the resulting error representation is

$$\begin{aligned}
(\psi^{(1)}, e) = (\psi_2^{(1)}, e_2) &= (\mathbf{A}_{22}^T \phi_2^{(1)}, e_2) \\
&= (\phi_2^{(1)}, \mathbf{A}_{22} u_2) - (\phi_2^{(1)}, \mathbf{A}_{22} U_2) \\
&= (\phi_2^{(1)}, b_2 - \mathbf{A}_{21} u_1) - (\phi_2^{(1)}, \mathbf{A}_{22} U_2) \quad (11.70) \\
&= (\phi_2^{(1)}, b_2 - \mathbf{A}_{21} U_1 - \mathbf{A}_{22} U_2) + (\phi_2^{(1)}, \mathbf{A}_{21} e_1) \\
&= (\phi_2^{(1)}, R_2) + (\phi_2^{(1)}, \mathbf{A}_{21} e_1).
\end{aligned}$$

The first term of the error representation requires only U_2 and $\phi_2^{(1)}$. Since the adjoint system is upper triangular and

$$\phi_2^{(1)} = \left(\mathbf{A}_{22}^T\right)^{-1} \psi_2^{(1)}$$

is independent of the first component, the calculation of $(\phi_2^{(1)}, R_2)$ remains within the "single physics paradigm", that is, we solve the adjoint problem using individual component solves rather than forming and solving a global problem. The second term $\left(\phi_2^{(1)}, \mathbf{A}_{21} e_1\right)$ represents the effect of errors in U_1 on the solution U_2. At first glance this term is uncomputable, but we note that it is a linear functional of e_1 since

$$\left(\phi_2^{(1)}, \mathbf{A}_{21} e_1\right) = \left(\mathbf{A}_{21}^\top \phi_2^{(1)}, e_1\right).$$

We therefore form the adjoint problem for the transfer error,

$$\begin{pmatrix} \mathbf{A}_{11}^\top & \mathbf{A}_{21}^\top \\ 0 & \mathbf{A}_{22}^\top \end{pmatrix} \begin{pmatrix} \phi_1^{(2)} \\ \phi_2^{(2)} \end{pmatrix} = \begin{pmatrix} \psi_1^{(2)} \\ 0 \end{pmatrix} = \begin{pmatrix} \mathbf{A}_{21}^\top \phi_2^{(1)} \\ 0 \end{pmatrix}.$$

The upper triangular block structure of $\mathbf{A}^\top$ immediately yields $\phi_2^{(2)} = 0$. As noted earlier, error estimates of u_1 should be independent of u_2. Thus, $\mathbf{A}_{11}^\top \phi_1^{(2)} = \psi_1^{(2)} = \mathbf{A}_{21}^\top \phi_2^{(1)}$, so that once again we can solve for $\phi^{(2)}$ in the "single physics paradigm." Given $\phi^{(2)}$ we obtain the auxiliary error representation

$$\left(\psi^{(2)}, e\right) = \left(\psi_1^{(2)}, e_1\right) = \left(\mathbf{A}_{21}^\top \phi_2^{(1)}, e_1\right) = \left(\mathbf{A}_{11}^\top \phi_1^{(2)}, e_1\right) = \left(\phi_1^{(2)}, R_1\right). \quad (11.71)$$

Combining the first term of (11.70) with (11.71) yields the complete error representation

$$\left(\psi^{(1)}, e\right) = \left(\phi_2^{(1)}, R_2\right) + \left(\phi_1^{(2)}, R_1\right)$$

which is a sum of the inner products of "single physics" residuals and adjoint solutions computed using the "single physics" paradigm.

Example 45.

We consider a 101×101 system with

$$\begin{aligned} \mathbf{A}_{11} &= I + .2 \times \text{random matrix}, \\ \mathbf{A}_{21} &= \text{random matrix}, \\ \mathbf{A}_{22} &= I + .1 \times \text{random matrix}, \end{aligned}$$

where the coefficients in the random matrices are $U(-1, 1)$. The right-hand side is also a random vector with $U(-1, 1)$ coefficients. We solve the linear systems using the Gauss-Seidel iteration with varying numbers of iterations, so that we have control over the accuracy of the solutions. The quantity of interest is the average of the coefficients of u_2.

In the first computation, we solve the first component $\mathbf{A}_{11} U_1 = b_1$ using 20 iterations. The error in the resulting solution is $\approx .031$. When we solve the second component $\mathbf{A}_{22} U_2 = b_2 - \mathbf{A}_{21} U_1$ using 40 or more iterations, we find an error in the quantity of interest of $\approx 3.9 \times 10^{-4}$. We cannot improve the accuracy of the second component regardless of how many iterations we use beyond 40. Solving the adjoint problems, we find

$$(\psi^{(1)}, e) \approx 3.86 \times 10^{-4},$$
$$(\phi_2^{(1)}, R_2) \approx 3.1 \times 10^{-8},$$
$$(\phi_1^{(2)}, R_1) \approx 3.86 \times 10^{-4}.$$

The error in the quantity of interest is almost entirely due to the error in the solution of the first component.

In the second computation, we solve the first component $\mathbf{A}_{11}U_1 = b_1$ using 200 iterations. The error in the resulting solution is $\approx 10^{-6}$. When we solve the second component $\mathbf{A}_{22}U_2 = b_2 - A_{21}U_1$ using 40 or more iterations, we find an error in the quantity of interest of $\approx 10^{-7}$. This confirms that the error in the solution of the second component in the first computation is dominated by the error in the solution of the first component.

Description of the a posteriori analysis

We seek the error e_2 in a quantity of interest given by a functional of u_2, noting that that a quantity of interest involving u_1 can be computed without solving for u_2. Since we introduce some additional, auxiliary quantities of interest, we denote the primary quantity of interest by $(\psi_2^{(1)}, e_2)$. We use the adjoint operators

$$\mathcal{A}_1^*(\phi_1, v_1) = (a_1 \nabla \phi_1, \nabla v_1) - (\text{div}(b_1 \phi_1), v_1) + (c_1 \phi_1, v_1)$$
$$\mathcal{A}_2^*(\phi_2, v_2) = (a_2 \nabla \phi_2, \nabla v_2) - (\text{div}(b_2 \phi_2), v_2) + (c_2 \phi_2, v_2).$$

We also use the linearization

$$L f_2(u_1)(u_1 - U_1) = \int_0^1 \frac{\partial f_2}{\partial u_1}(u_1 s + U_1(1 - s))\, ds.$$

Noting that the solution of the first adjoint component is not needed to compute the quantity of interest $(\psi, e) = (\psi_2^{(1)}, e_2)$, we define the primary adjoint problem to be

$$\mathcal{A}_2^*(\phi_2^{(1)}, v_2) = (\psi_2^{(1)}, v_2), \quad \text{for all } v_2 \in \tilde{W}_2^1(\Omega).$$

The standard argument yields the error representation,

$$(\psi, e) = (\psi_2^{(1)}, e_2) = (f_2(x, u_1, Du_1), \phi_2^{(1)}) - \mathcal{A}_2(U_2, \phi_2^{(1)}). \qquad (11.72)$$

To simplify notation, we denote the weak residual of each solution component by

$$\mathcal{R}_i(U_i, \chi; \nu) = (f_i(\nu), \chi) - \mathcal{A}_i(U_i, \chi), \quad i = 1, 2,$$

so (11.72) becomes
$$(\psi, e) = \mathcal{R}_2(U_2, \phi_2^{(1)}; u_1),$$

At this point, it is not clear that (11.72) is computationally useful since the residual on the right-hand side of (11.72) involves the unknown true solution u_1. One consequence is that we cannot immediately use Galerkin orthogonality by inserting a projection of $\phi^{(1)}$ into the representation, since Galerkin orthogonality for U_2 holds for residual $\mathcal{R}_2(U_2, \phi_2^{(1)}; U_1)$ not $\mathcal{R}_2(U_2, \phi_2^{(1)}; u_1)$.

To deal with this, we add and subtract $(f_2(x, U_1, DU_1), \phi_2^{(1)})$ in (11.72) and, assuming the same meshes are used for U_1 and U_2, use Galerkin orthogonality to obtain

$$(\psi, e) = \mathcal{R}_2\big(U_2, (I - \Pi_2)\phi_2^{(1)}; U_1\big) + \big(f_2(x, u_1, Du_1) - f_2(x, U_1, DU_1), \phi_2^{(1)}\big), \tag{11.73}$$

where Π_2 is a projection into the finite element space for U_2. The first term on the right of (11.73) is the standard *a posteriori* error expression for the second component while the remaining difference represents the transfer error that arises from using an approximation of u_1 in defining the coefficients in the equation for u_2. The goal now is to estimate this transfer error and its effect on the quantity of interest.

We recognize that the transfer error is a (nominally nonlinear) functional of the error in u_1, defining an auxiliary quantity of interest. We approximate it by a linear functional,

$$\big(f_2(x, u_1, Du_1) - f_2(x, U_1, DU_1), \phi_2^{(1)}\big) \approx \big(Df_2(U_1) \times e_1, \phi_2^{(1)}\big) = \big(e_1, \psi_1^{(2)}\big).$$

We define the corresponding transfer error adjoint problem

$$\mathcal{A}_1^*(\phi_1^{(2)}, v_1) = \big(\psi_1^{(2)}, v_1\big) \quad \text{for all } v_1 \in \tilde{W}_2^1(\Omega), \tag{11.74}$$

noting that as for the primary problem, we do not have to solve the second component of the full adjoint problem. The transfer error representation formula is given by

$$\big(\psi_1^{(2)}, e_1\big) = \big(f_1, (I - \Pi_1)\phi_1^{(2)}\big) - \mathcal{A}_1\big(U_1, (I - \Pi_1)\phi_1^{(2)}\big),$$

where Π_1 is a projection into the finite element space of U_1. We obtain the error representation,

$$(\psi, e) = \mathcal{R}_2\big(U_2, (I - \Pi_2)\phi_2^{(1)}; U_1\big) + \mathcal{R}_1\big(U_1, (I - \Pi_1)\phi_1^{(2)}\big). \tag{11.75}$$

In the final step, we account for the error induced by using a multiscale discretization, i.e. different meshes for U_1 and U_2. Example 44 shows that this can have a significant effect on overall accuracy.

One issue is that we use

$$f_2(x, \Pi_{1\to 2}U_1, \Pi_{1\to 2}DU_1)$$

in the equations defining the finite element approximation. Correspondingly, we alter the definition of the residual

$$\mathcal{R}_2(U_2, \chi; \nu) = (f_2(\Pi_{1\to 2}\nu), \chi) - \mathcal{A}_2(U_2, \chi).$$

In addition, we use projections to treat any integral involving functions that are defined on both discretizations, i.e. functions of U_i and ϕ_i, $i = 1, 2$. After decomposing the original estimate to account for all the projections, the new error representation formula for the transfer error becomes

$$\left(Df_2(U_1) \times e_1, \Pi_{2\to 1}\phi_2^{(1)}\right) + \left(Df_2(U_1) \times e_1, (I - \Pi_{2\to 1})\phi_2^{(1)}\right)$$

which is the error contribution arising from the transfer as well as an additional term that is large when the approximation spaces are significantly different.

The data $\psi^{(2)}$ defining the transfer error adjoint is now

$$\left(f_2(u_1) - f_2(U_1), \phi_2^{(1)}\right) \approx \left(Df_2(U_1) \times e_1, \Pi_{2\to 1}\phi_2^{(1)}\right) = \left(\psi_1^{(2)}, e_1\right).$$

The additional term $\left(Df_2(U_1) \times e_1, (I - \Pi_{2\to 1})\phi_2^{(1)}\right)$ is a linear functional, so we define an additional auxiliary quantity of interest

$$\left(\psi_1^{(3)}, e_1\right) = \left(Df_2(U_1) \times e_1, (I - \Pi_{2\to 1})\phi_2^{(1)}\right)$$

and the corresponding adjoint problem

$$\mathcal{A}_1^*(\phi_1^{(3)}, v_1) = \left(\psi_1^{(3)}, v_1\right) \quad \text{for all } v_1 \in \tilde{W}_2^1(\Omega). \tag{11.76}$$

The final error representation is therefore [39].

11.6.2 Theorem

$$(\psi, e) = \mathcal{R}_2(U_2, (I - \Pi_2)\phi_2^{(1)}; U_1) + \mathcal{R}_1(U_1, (I - \Pi_1)(\phi_1^{(2)} + \phi_1^{(3)}))$$

$$+ \left(\Pi_{1\to 2}f_2(U_1) - f_2(\Pi_{1\to 2}U_1), \phi_2^{(1)}\right) + \left((I - \Pi_{1\to 2})f_2(U_1), \phi_2^{(1)}\right).$$
$$\tag{11.77}$$

We emphasize that evaluating the integrals in (11.77) is far from trivial. We have used Monte-Carlo techniques with good results, see [39].

Example 46.

In Example 44, we estimate the contributions to the error reported in that computation using the relevant portions of (11.77). To produce the adaptive mesh results shown in Figure 11.27, we construct the adapted mesh using equidistribution based on a bound derived from the first term in (11.77), i.e. neglecting the terms that estimate the transfer error.

Instead, we consider the system (11.68) for the quantity of interest equal to the average value of U_2. We begin with the same initial coarse meshes as in Figure 11.27, but add the transfer error expression to the mesh refinement criterion. Adapting the mesh so that the total error in the quantity of interest for U_2 has error estimates less than 10^{-4} yields the meshes shown in Figure 11.28. We see that the first component solve requires significantly more refinement than the second component.

Example 47.

This example shows that differences in mesh discretization scale between the two components can contribute significantly to the error. We again solve (11.68) for the quantity of interest equal to $U_2(.15, 15)$. We begin with identical coarse meshes for the two components, but refine only the mesh for U_2. We solve the primary adjoint problem as well as the two auxiliary adjoint problems and show the results in Table 11.3.

11.6.3 Multiscale decomposition of reaction-diffusion problems

We follow the presentation in [40]. In the introduction (Section 11.1), we presented operator splitting Algorithm 11.1.8 for reaction-diffusion problems as a classic example of multiscale operator decomposition. Upon discretizing a reaction-diffusion equation (11.4.5) in space using a standard piecewise linear, continuous finite element method as described in Section 11.4.1 we obtain a (high dimensional) initial value problem of the form (11.7). We can then apply

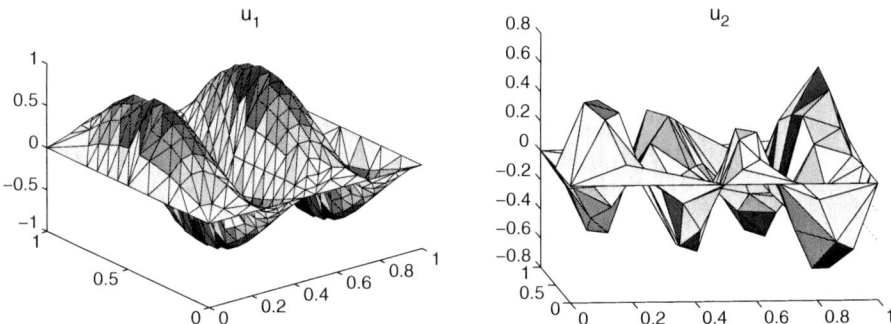

Figure 11.28: The adapted meshes resulting from the full estimate that accounts for "primary" and "transfer" errors. The transfer error dominates and drives the adaptive refinement.

Table 11.3. Error contributions when there is a scale difference in the meshes

Primary Error	Transfer Error	Error from Scale Differences
0.000713	0.0905	0.0325

the operator splitting Algorithm 11.1.8. Finally, we discretize the component problems of the operator splitting method using the either the dG or cG methods on the independent time discretizations $\{t_n\}$, $\{s_{m,n}\}$ described in Figure 11.4.

For example, if we use the dG methods for both components, then the finite element approximate solutions are sought in a piecewise polynomial space for the diffusion and reaction components respectively,

$$\mathcal{V}^{(q_d)} = \left\{ U : U|_{I_n} \in \mathcal{P}^{(q_d)}(I_n), 1 \leq n \leq N \right\},$$

$$\mathcal{V}^{(q_r)}(I_n) = \left\{ U : U|_{I_{m,n}} \in \mathcal{P}^{(q_r)}(I_{m,n}), 1 \leq m \leq M_n \right\},$$

for $n = 1, \ldots, N$, and $I_n = [t_{n-1}, t_n]$ and $I_{m,n} = [s_{m-1,n}, s_{m,n}]$. $\mathcal{P}^{(q_d)}(I_n)$ denotes the space of polynomials in $\mathbb{R}^l$ of degree q_d on I_n. A similar definition holds for $\mathcal{P}^{(q_r)}(I_{m,n})$. We let $U_n^{+,-}$ denote the left- and right-hand limits of U at t_n, and $[U]_n = U_n^+ - U_n^-$ the jump value of U at t_n.

Let $\tilde{Y}(t)$ be the piecewise continuous finite element approximation of the operator splitting with

$$\tilde{Y}(t) = \frac{t_n - t}{\Delta t_n} \tilde{Y}_{n-1} + \frac{t - t_{n-1}}{\Delta t_n} \tilde{Y}_n, \quad t_{n-1} \leq t \leq t_n.$$

The nodal values $\tilde{Y}_n$ are obtained from the following procedure:

Algorithm 11.6.9 Multiscale Operator Splitting for Reaction-Diffusion Equations

---

Set $\tilde{Y}_0 = y_0$
for $n = 1, \ldots, N$ **do**
    Set $Y^r{}_{0,n}^- = \tilde{Y}_{n-1}$
    for $m = 1, \ldots, M_n$ **do**
        Compute $Y^r|_{I_{m,n}} \in \mathcal{P}^{(q_r)}(I_{m,n})$ satisfying

$$\int_{I_{m,n}} \left(\dot{Y}^r, W \right) dt + \left([Y^r]_{m-1,n}, W_{m-1}^+ \right) = \int_{I_{m,n}} (F(Y^r), W) \, dt \quad \text{for all } W \in \mathcal{P}^{(q_r)}(I_{m,n}) \tag{11.78}$$

    end for
    Set $Y^d{}_{n-1}^- = Y^r{}_{M_n,n}^-$
    Compute $Y^d|_{I_n} \in \mathcal{P}^{(q_d)}(I_n)$ satisfying

$$\int_{I_n} (\dot{Y}^d, V) dt + ([Y^d]_{n-1}, V_{n-1}^+) = \int_{I_n} (AY^d, V) dt \quad \text{for all } V \in \mathcal{P}^{(q_d)}(I_n) \tag{11.79}$$

    Set $\tilde{y}_n = y^d(t_n^-)$
end for

Adapting standard convergence analysis techniques, we can show that if f is Lipschitz continuous, then for $q_d = 0, 1$ and $q_r = 0, 1$, there exist constants C_1, C_2, C_3 such that,

$$|y_N - \tilde{Y}_N| \leq C_1 \Delta t + C_2 \Delta t^{q_d+1} + C_3 \Delta s^{q_r+1}.$$

In Example 5, we present an example in which multiscale operator decomposition affects the stability, and hence accuracy, of the solution. Such affects can take myriad forms.

Example 48.

In [40], we illustrate the instability of operator splitting applied to the Brusselator problem (11.4). We apply a standard first order splitting scheme to a space discretization of the Brusselator model with 500 discrete points with $\alpha = .6$, $\beta = 2$, $k_1 = k_2 = .025$ consisting of the cG(1) method for the diffusion with time step of .2, and dG(0) method for the reaction with time step of .004. On the left of Figure 11.29, we show a numerical solution that exhibits nonphysical oscillations that developed after some time. On the right, we show plots of the error versus time steps at different times. There is a critical time step above which the instability develops. Moreover, changing the space discretization does not improve the accuracy. In fact, using a finer spatial discretization for a constant time-step size leads to significantly more error in the long-time solution, see [52].

The goal is to derive an accurate *a posteriori* estimate of the error in a specified quantity of interest computed from a multiscale operator splitting approximate solution of (11.7). The estimate must account for the stability effects arising from operator splitting. In the analysis, we distinguish the effects of operator splitting from the effects of numerical discretization of the components. The

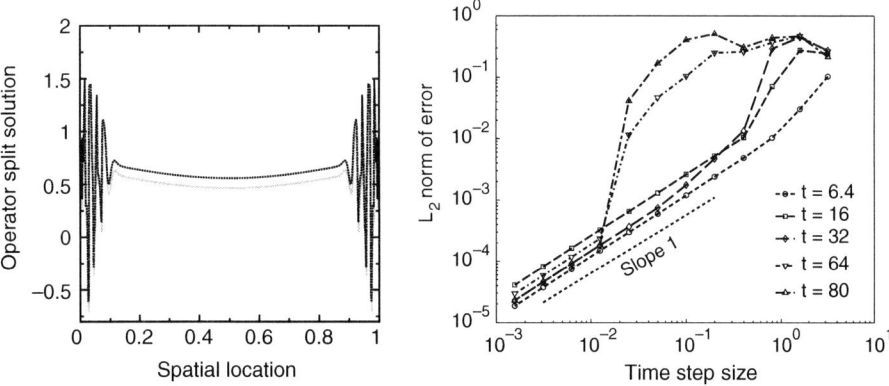

Figure 11.29: The left-hand plot illustrates typical instability that can arise from multiscale operator splitting applied to the Brusselator problem. Solution is shown at time 80. On the right, we show plots of the error in the L_2 norm versus time-step size at different times.

operator splitting discretization Algorithm 11.6.9 is a consistent discretization of the formal "analytic" operator splitting Algorithm11.1.8 and the numerical error arising in each component can be treated with the standard *a posteriori* analysis discussed previously. Estimating the error arising from the operator splitting itself requires a new approach.

A main technical issue is the definition of a suitable adjoint problem because the standard approach used for nonlinear problems described in Section 11.4.4 fails. Indeed, the adjoint operator corresponding to the solution operator for an operator decomposition discretization is typically different from the adjoint operator associated with the true solution operator. Because an adjoint problem carries the global stability information about the quantity of interest, accounting for the differences between adjoint problems associated with the original problem and a numerical discretization are critical for obtaining accurate error estimates.

In the estimate described below, this difference takes the form of "residuals" between certain adjoint operators associated with the fully-coupled problem and an analytic operator split version. A practical difficulty with such a result is that solving the adjoint for the fully-coupled problem poses the same multiphysics challenges as solving the original forward problem. We therefore develop a new hybrid *a priori–a posteriori* estimate that combines a computable leading order expression obtained using *a posteriori* arguments with a provably higher order bound obtained using *a priori* convergence result.

A linear algebra example

We illustrate the ideas of the analysis in the context of solving a linear system

$$\mathbf{M}y = b, \qquad (11.80)$$

where $\mathbf{M}$ is a $n \times n$ matrix of the form

$$\mathbf{M} = \mathbf{I} - \epsilon(\mathbf{A} + \mathbf{B}),$$

with $\mathbf{I}$ denoting the identity matrix, $\mathbf{A}$ and $\mathbf{B}$ denoting $n \times n$ matrices, and ϵ denotes a small parameter. By standard results, $\mathbf{M}$ is invertible for all sufficiently small ϵ.

The analog of an operator splitting for (11.80) is the problem

$$\mathbf{N}\tilde{y} = b, \qquad (11.81)$$

where

$$\mathbf{N} = (\mathbf{I} - \epsilon\mathbf{A})(\mathbf{I} - \epsilon\mathbf{B})$$

is also invertible for ϵ small. This follows from the observation

$$\mathbf{M}^{-1} = \bigl(\mathbf{I} - \epsilon(\mathbf{A} + \mathbf{B})\bigr)^{-1}, \quad \mathbf{N}^{-1} = \bigl(\mathbf{I} - \epsilon\mathbf{B}\bigr)^{-1}\bigl(\mathbf{I} - \epsilon\mathbf{A}\bigr)^{-1},$$

so that inverting $\mathbf{N}$ involves inverting operators involving only $\mathbf{A}$ and $\mathbf{B}$ individually. Using the Neumann series to represent the inverse operators leads to the estimate,

$$\begin{aligned}\mathbf{M}^{-1} - \mathbf{N}^{-1} &= \mathbf{I} + \epsilon(\mathbf{A}+\mathbf{B}) + \epsilon^2(\mathbf{A}+\mathbf{B})^2 + \epsilon^3(\mathbf{A}+\mathbf{B})^3 + \cdots \\ &\quad - \left(\mathbf{I} + \epsilon\mathbf{B} + \epsilon^2\mathbf{B}^2 + \epsilon^3\mathbf{B}^3 + \cdots\right) \times \left(\mathbf{I} + \epsilon\mathbf{A} + \epsilon^2\mathbf{A}^2 + \epsilon^3\mathbf{A}^3 + \cdots\right) \\ &= \epsilon^2 \mathbf{B}\mathbf{A} + O(\epsilon^3). \end{aligned} \tag{11.82}$$

We consider the problem of solving (11.80) to compute the quantity of interest (y, ψ). We have the associated adjoint problems

$$\mathbf{M}^* \phi = \psi, \quad \mathbf{N}^* \tilde{\phi} = \psi.$$

If $\tilde{Y} \approx \tilde{y}$ is a numerical solution, the standard *a posteriori* analysis in Example 33 gives

$$(\tilde{e}, \psi) = (\tilde{Y} - \tilde{y}, \psi) = (\tilde{\phi}, \tilde{R}), \quad \tilde{R} = \mathbf{N}\tilde{Y} - b.$$

Since we assume that problems involving N are solvable, this is a computable estimate. However, the error we wish to estimate is $(\tilde{Y} - y, \psi)$. We write this as

$$(\tilde{Y} - y, \psi) = (\tilde{Y} - \tilde{y}, \psi) + (\tilde{y} - y, \psi) = (\tilde{\phi}, \tilde{R}) + (\tilde{y} - y, \psi).$$

The second term on the right is the error arising from the operator splitting. Since these problems are linear, we can use the Greens function formulas

$$(y, \psi) = (\phi, b), \quad (\tilde{y}, \psi) = (\tilde{\phi}, b),$$

to conclude

$$(\tilde{Y} - y, \psi) = (\tilde{\phi}, \tilde{R}) + (\tilde{\phi} - \phi, b). \tag{11.83}$$

Example 49.

We let $\mathbf{A}$ and $\mathbf{B}$ be random 500×500 matrices, where the coefficients in the random matrices are $U(-1, 1)$, normalized in the 2-norm and $\epsilon = .01$. We set y to be a random vector, with $U(-1, 1)$ coefficients, and set $b = My$, which insures that y is a solution within machine precision. Finally, we set $\psi = (1, 0, \ldots)^\top$. We compute $\tilde{Y}$ using Gaussian elimination. We find

$$(\tilde{Y} - y, \psi) \approx 5.376 \times 10^{-5}$$

$$(\tilde{\phi}, \tilde{R}) \approx 1.221 \times 10^{-15}$$

$$(\tilde{\phi} - \phi, b) \approx 5.376 \times 10^{-5}.$$

This means that nearly all of the error is captured by the effect of operator splitting on the adjoint solution.

As noted above, (11.83) is problematic because it requires the solution of the "true" adjoint problem, which is unavailable in the operator splitting paradigm.

Description of the hybrid a posteriori–a priori error analysis

We now describe an error estimate for a multiscale operator decomposition solution of (11.7) that is composed of a leading expression which is *a posteriori* and an *a priori* expression which is provably higher order. See [40] for the full analysis.

We begin with the decomposition

$$\tilde{Y} - y = (\tilde{Y} - \tilde{y}) + (\tilde{y} - y), \tag{11.84}$$

where y solves (11.7), $\tilde{y}$ is computed via the abstract operator splitting Algorithm 11.1.8, and $\tilde{Y}$ is computed via the numerical operator splitting Algorithm 11.6.9.

The first expression on the right of (11.84) is the error of $\tilde{Y}$ as a solution of the operator split problem. Note that $\tilde{Y}$ is a consistent numerical solution for the analytic operator split problem and the expression for its error can be estimated using the standard *a posteriori* error analysis. We let ϑ^d define the adjoint solution associated with the diffusion component (11.79) satisfying

$$\begin{cases} -\dot{\vartheta}^d = A^\top \vartheta^d(t), & t_n > t \geq t_{n-1}, \\ \vartheta^d(t_n^-) = \psi_n. \end{cases}$$

Furthermore, we let ϑ^r define the adjoint solution associated with the reaction component (11.78) satisfying

$$\begin{cases} -\dot{\vartheta}^r = (\hat{F}'(y^r, Y^r))^\top \vartheta^r(t), & s_{m,n} > t \geq s_{m-1,n}, \\ \vartheta^r(s_{m,n}) = \psi^r_{m,n}, \end{cases}$$

for $m = M_n, \ldots, 1$, with $\psi^r_{M_n,n} = \vartheta^{d+}_{n-1}$ and $\psi^r_{m,n} = \vartheta^r_{m,n}$ for $m < M_n$. Thus, ϑ^r is continuous across the internal reaction time nodes $s_{m,n}$, $m = 1, \ldots, M_n - 1$. Here,

$$\hat{F}'(y^r, Y^r) = \int_0^1 F'(sy^r + (1-s)Y^r) \, ds.$$

The second expression on the right of (11.84) is an abstract error of operator splitting. Following the analysis for the linear algebra example, we use analogs of the classic representation formula involving the Greens function of a linear elliptic problem to construct an estimate. The nonlinearity complicates the analysis, however, because we have to use linearization to define unique adjoint problems, which raises the issue of choosing a trajectory around which to linearize. We cannot use the standard approach of linearizing the error representation described in

Section 11.4.4 because of the operator splitting. Instead, we assume that both the original problem and the operator split version have a common solution and we linearize each problem in a neighborhood of this common solution. For example, we assume that $y = 0$ is a steady-state solution of both problems, which can be achieved by assuming that

$$\text{Homogeneity Assumption:} \quad F(0) = 0,$$

and we linearize in a region around 0. In terms of applications to reaction-diffusion problems, there are mathematical reasons for making the homogeneity assumption and it is satisfied in a great many applications. However, we can modify the analysis to allow for linearization around any known common solution, see [40].

To motivate this definition, we derive an abstract Greens function representation. On time interval (t_{n-1}, t_n), we consider the linearized problem,

$$\begin{cases} \dot{y} = A\,y(t) + \overline{F'(y)}\,y(t), & t_{n-1} < t \le t_n, \\ y(t_{n-1}) = y_{n-1}, \end{cases}$$

where

$$\overline{F'(y)} = \int_0^1 F'(sy)\,ds.$$

We note that $\overline{F'(y)}y = F(y)$ because $F(0) = 0$. The generalized Greens function φ satisfies the adjoint problem

$$\begin{cases} -\dot{\varphi} = A^\top \varphi(t) + \overline{F'(y)}^\top \varphi(t), & t_n > t \ge t_{n-1}, \\ \varphi(t_n) = \psi_n, \end{cases} \quad (11.85)$$

where ψ_n determines the quantity of interest $(y(t_n), \psi_n)$, and $A^\top$ and $\overline{F'(y)}^\top$ denote the transpose of A and $\overline{F'(y)}$, respectively. We choose $\psi_n = \varphi(t_n^+)$, which couples the local adjoint problems (11.85) to form a global adjoint problem. This definition yields a simple representation of the solution value over one time step

$$(y_n, \psi_n) = (y_{n-1}, \varphi_{n-1}),\ n = 1, 2, \ldots, N \implies (y_N, \psi_N) = (y_0, \varphi_N). \quad (11.86)$$

We use analogs for (11.86) for solutions of each component in the operator splitting discretization. For $n = 1, \ldots, N$, we define the three adjoint problems. The diffusion problem is simpler because it is linear,

$$\begin{cases} -\dot{\varphi}^d = A^\top \varphi^d(t), & t_n > t \ge t_{n-1}, \\ \varphi^d(t_n^-) = \psi_n^d. \end{cases} \quad (11.87)$$

It is convenient to let Φ_n^d denote the solution operator, so $\varphi^d(t_{n-1}) = \Phi_n^d \psi_n^d$. We require two adjoint problems to treat the reaction component. The difference between the problems is the function around which they linearized,

$$\begin{cases} -\dot{\varphi}_1^r = \overline{F'(\tilde{Y})}^\top \varphi_1^r(t), & t_n > t \geq t_{n-1}, \\ \varphi_1^r(t_n^-) = \psi_n^r. \end{cases} \tag{11.88}$$

$$\begin{cases} -\dot{\varphi}_2^r = \overline{F'(Y^r)}^\top \varphi_2^r(t), & t_n > t \geq t_{n-1}, \\ \varphi_2^r(t_n^-) = \psi_n^r. \end{cases} \tag{11.89}$$

If $\Phi_n^r(z)$ denotes the solution operator for the problem linearized around a function z, then we have $\varphi_1^r(t_{n-1}) = \Phi_n^r(\tilde{Y})\psi_n^r$ and $\varphi_2^r(t_{n-1}) = \Phi_n^r(Y^r)\psi_n^r$. We can now prove [40].

11.6.4 Theorem

A hybrid a posteriori–a priori *error estimate for the multiscale operator splitting dG finite element method is*

$$(\tilde{Y}_N - y_N, \psi_N) = \sum_{n=1}^{N} \sum_{m=1}^{M_n} \left(\int_{I_{m,n}} (\dot{Y}^r - F(Y^r), \vartheta^r - \Pi\vartheta^r) \, dt \right.$$

$$\left. + ([Y^r]_{m-1,n}, \vartheta^{r+}_{m-1,n} - \Pi\vartheta^{r+}_{m-1,n}) \right)$$

$$+ \sum_{n=1}^{N} \left(\int_{I_n} (\dot{Y}^d - AY^d, \vartheta^d - \Pi\vartheta^d) \, dt \right.$$

$$\left. + ([Y^d]_{n-1}, \vartheta^{d+}_{n-1} - \Pi\vartheta^{d+}_{n-1}) \right)$$

$$+ \sum_{n=1}^{N} (\tilde{Y}_{n-1}, (E_1 + E_2)\psi_n) + \mathbf{O}(\Delta t^{q_d+2}) + \mathbf{O}(\Delta t \, \Delta s^{q_r+1}),$$

where

$$E_1 = \frac{1}{2}\Delta t_n \left(A^\top \mathcal{F}(\tilde{Y}) - \mathcal{F}(\tilde{Y}) A^\top \right), \quad \mathcal{F}(\tilde{Y}) = \int_{I_n} \overline{F'(\tilde{Y})} \, dt,$$

$$E_2 = \left(\Phi_n^r(\tilde{Y}) - \Phi_n^r(Y^r) \right) \Phi_n^d.$$

The first expression on the right is the error introduced by the numerical solution of the reaction component. Likewise, the second expression on the right is the error introduced by the numerical solution of the diffusion component. The third expression on the right measures the effects of operator splitting. The expression E_1 is a leading order estimate for the effects of operator splitting while E_2 accounts for issues arising from the differences in linearizing around the global computed solution as opposed to the solution of the reaction component, which affects the formulation of the adjoint problems. Both of these quantities are scaled by the solution itself, so that these effects become negligible when the solution approaches zero. Finally, the remaining terms represent bounds on terms that are not computable but are higher order. In practice, we neglect those terms when computing an estimate.

Using the estimate requires the solution of five adjoint problems. But we avoid the need to solve an adjoint problem corresponding to linearization around the true solution by deriving the hybrid estimate.

Numerical examples
We describe some examples in [40].

Example 50.
The first example is partial differential equation version of Example 5,

$$\begin{cases} \frac{\partial u}{\partial t} - 0.05 \frac{\partial^2 u}{\partial x^2} = u^2, & x \in (0,1), t > 0, \\ u(0,t) = u(1,t) = 0, & t > 0, \\ u(x,0) = 4x(1-x), & x \in (0,1). \end{cases}$$

The solution of the reaction component exhibits finite time blow up when undamped by the diffusion component. This is perhaps the most extreme form of instability. For this computation, we use 20 spatial finite elements. Table 11.4 shows the ratio of the error to the estimate computed at the final time $T = 1$. In this computation, we keep the reaction time step constant and vary the diffusion time step and number of reaction time steps. We see that the estimate is very accurate for a range of time steps.

Table 11.4. Operator splitting error estimate for the blow up problem at $T = 1$, reaction time step $= 10^{-3}$

Δt	M	Exact Err (%)	Error/Estimate
10^{-1}	100	11.07	1.0286
10^{-2}	10	1.35	1.0067
10^{-3}	1	0.45	1.0020

Example 51.

We next consider the Brusselator problem (11.4) with $\alpha = 2$, $\beta = 5.45$, $k_1 = 0.008$, $k_2 = 0.004$ and initial conditions $u_1(x,0) = \alpha + 0.1\sin(\pi x)$ and $u_2(x,0) = \beta/\alpha + 0.1\sin(\pi x)$, which yields an oscillatory solution. In this case, the reaction is very mildly unstable, with at most polynomial rate accumulation of perturbations as time passes. We use a 32 node spatial finite element discretization, resulting in a differential equation system with dimension 62. We note that in original form, the reaction terms do not satisfy the requirement $F(0) = 0$ so we linearize around the steady-state solution c with $c_i = \alpha$ for $i = 1, \ldots, N_e - 1$, and $c_i = \beta/\alpha$ for $i = N_e, \ldots, 2N_e - 2$, so that $F(c) = 0$.

Figure 11.30 compares the errors computed using $\Delta t = 0.01$ and $M = 10$ reaction time steps to the hybrid *a posteriori* error estimates. We show results for $[0, 2]$, when the solution is still in a transient stage, and at $T = 40$ when the solution has become periodic. All the results show that the exact and estimated errors are in remarkable agreement.

11.6.5 Multiscale decomposition of a fluid–solid conjugate heat transfer problem

Following [40], we next consider the multiscale decomposition solution of the heat transfer problem described in Example 3. The weak formulation of (11.5) consists of computing $u \in V_F$, $p \in L_0^2(\Omega_F)$, $T_F \in W_F$, and $T_S \in W_S$ such that

$$\begin{cases} a_1(u,v) + c_1(u,u,v) + b(v,p) + d(T_F,v) = (f,v), \\ b(u,q) = 0, \\ a_2(T_F, w_F) + c_2(u, T_F, w_F) + a_3(T_S, w_S) = (Q_F, w_F) + (Q_S, w_S), \end{cases} \quad (11.90)$$

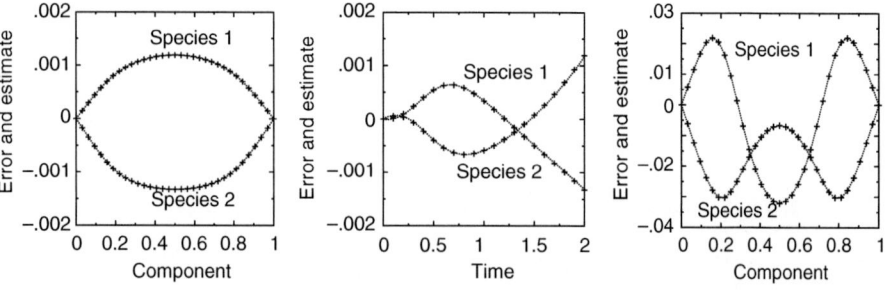

Figure 11.30: Brusselator results. Left: Comparison of errors against the spatial location at $T = 2$. Middle: Time history of errors at the midpoint location on $[0, 2]$. Right: Comparison of errors against the spatial location at $T = 40$. The dotted line is the exact error and the (+) is the estimated error.

for all $v \in V_{F,0}$, $q \in L_0^2(\Omega_F)$, $w_F \in W_{F,0}$ and $w_S \in W_{S,0}$, where

$$a_1(u,v) = \int_{\Omega_F} \mu(\nabla u : \nabla v)\, dx, \qquad a_2(T_F, w_F) = \int_{\Omega_F} k_F(\nabla T_F \cdot \nabla w_F)\, dx,$$

$$a_3(T_S, w_S) = \int_{\Omega_S} k_S(\nabla T_S \cdot \nabla w_S)\, dx, \qquad b(v,q) = -\int_{\Omega_F} (\nabla \cdot v) q\, dx,$$

$$c_1(u,v,z) = \int_{\Omega_F} \rho_0 (u \cdot \nabla) v \cdot z\, dx, \qquad c_2(u,T,w) = \int_{\Omega_F} \rho_0 c_p (u \cdot \nabla T)\, w\, dx,$$

$$d(T,v) = \int_{\Omega_F} \rho_0 \beta T g \cdot v\, dx, \qquad f = \rho_0 (1 + \beta T_0) g,$$

and

$$V_F = \{v \in H^1(\Omega_F) \,|\, v = g_{u,D} \text{ on } \Gamma_{u,D}\}, \qquad V_{F,0} = \{v \in V_F \,|\, v = 0 \text{ on } \Gamma_{u,D}\},$$
$$W_F = \{w \in H^1(\Omega_F) \,|\, w = g_{T_F,D} \text{ on } \Gamma_{T_F,D}\}, \qquad W_{F,0} = \{w \in W_F \,|\, w = 0 \text{ on } \Gamma_{T_F,D}\},$$
$$W_S = \{w \in H^1(\Omega_S) \,|\, w = g_{T_S,D} \text{ on } \Gamma_{T_S,D}\}, \qquad W_{S,0} = \{w \in W_S \,|\, w = 0 \text{ on } \Gamma_{T_S,D}\}.$$

To discretize, we construct independent, locally quasi-uniform triangulations $\mathcal{T}_{F,h}$ and $\mathcal{T}_{S,h}$ of Ω_F and Ω_S, respectively. We use the piecewise polynomial spaces

$$V_F^h = \{v \in V_F \,|\, v \text{ continuous on } \Omega_F,\ v_i \in P^2(K) \text{ for all } K \in \mathcal{T}_{F,h}\},$$
$$Z^h = \{z \in Z \,|\, z \text{ continuous on } \Omega_F,\ z \in P^1(K) \text{ for all } K \in \mathcal{T}_{F,h}\},$$
$$W_F^h = \{w \in W_F \,|\, w \text{ continuous on } \Omega_F,\ w \in P^2(K) \text{ for all } K \in \mathcal{T}_{F,h}\},$$
$$W_S^h = \{w \in W_S \,|\, w \text{ continuous on } \Omega_S,\ w \in P^2(K) \text{ for all } K \in \mathcal{T}_{S,h}\},$$

and the associated subspaces

$$V_{F,0}^h = \{v \in V^h \,|\, v = 0 \text{ on } \Gamma_{u,D}\},$$
$$W_{F,0}^h = \{w \in W_F^h \,|\, w = 0 \text{ on } \Gamma_{T_F,D}\},$$
$$W_{S,0}^h = \{w \in W_S^h \,|\, w = 0 \text{ on } \Gamma_{T_S,D} \text{ and } w = 0 \text{ on } \Gamma_I\},$$

where $P^q(K)$ denotes the space of polynomials of degree q on an element K. Note that $W_{S,0}^h$ is different than $W_{F,0}^h$ in an important way since $\Gamma_{T_S,D}$ does not include Γ_I. We let π_V, π_{W_F}, π_{W_S}, and π_Z be projections into V_F^h, W_F^h, W_S^h, and Z^h, respectively. We also use π_{W_F} and π_{W_S} to denote projections into W_F^h and W_S^h, respectively along the interface Γ_I.

Algorithm 11.6.16 Multiscale Decomposition Method for Conjugate Heat Transfer

k = 0
while ($\|T_S^{\{k\}} - \pi_S T_F^{\{k\}}\|_{\Gamma_I} > TOL$) **do**
 k = k+1
 Compute $T_{S,h}^{\{k\}} \in W_S^h$ such that $T_{S,h}^{\{k\}} = \pi_{W_S} T_{F,h}^{\{k-1\}}$ along the interface Γ_I and

$$a_3(T_{S,h}^{\{k\}}, w) = (Q_S, w), \quad \forall w \in W_{S,0}^h, \tag{11.91}$$

Compute $u_h^{\{k\}} \in V_F^h$, $p_h^{\{k\}} \in Z^h$ and $T_{F,h}^{\{k\}} \in W_F^h$ such that

$$\begin{cases} a_1(u_h^{\{k\}}, v) + c_1(u_h^{\{k\}}, u_h^{\{k\}}, v) + b(v, p_h^{\{k\}}) + d(T_{F,h}^{\{k\}}, v) = (f, v), \forall v \in V_{F,0}^h, \\ b(u_h^{\{k\}}, q) = 0, \forall q \in Z^h, \\ a_2(T_{F,h}^{\{k\}}, w) + c_2(u_h^{\{k\}}, T_{F,h}^{\{k\}}, w) = (Q_F, w) - (k_S(n \cdot \nabla T_{S,h}^{\{k\}}), w)_{\Gamma_I}, \\ \hspace{8cm} \forall w \in W_{F,0}^h. \end{cases} \tag{11.92}$$

end while

To compute a stable solution of the fluid equations, we choose V_F^h and Z^h to be the Taylor-Hood finite element pair satisfying the discrete inf-sup condition

$$\inf_{q \in Z^h} \sup_{v \in V_F^h} \frac{b(v,q)}{\|v\|_1 \cdot \|q\|_0} \geq \beta > 0. \tag{11.93}$$

We also note that the convergence of the iteration defined by this Algorithm depends on the values of k_S and k_F along the interface and the geometry of each region. Often, a "relaxation" is used to help improve convergence properties. We choose $\alpha \in [0, 1)$ and impose the relaxed Dirichlet interface values

$$T_F^{\{k\}} = \alpha T_F^{\{k-1\}} + (1-\alpha)\pi_F T_S^{\{k-1\}}.$$

This affects the analysis, but we do not discuss that here, see [40, 41].

Description of an a posteriori error analysis
We define the adjoint using the standard linearization approach. We define the errors

$$e_u = u - u_h^{\{k\}}, \quad e_p = p - p_h^{\{k\}}, \quad e_{T_F} = T_F - T_{F,h}^{\{k\}} \text{ and } e_{T_S} = T_S - T_{S,h}^{\{k\}}.$$

The adjoint problem for the quantity of interest

$$(\psi, e) = (\psi_u, e_u) + (\psi_p, e_p) + (\psi_{T_F}, e_{T_F}) + (\psi_{T_S}, e_{T_S})$$

for the coupled problem (11.5) is

$$\begin{cases} -\mu\Delta\phi + \bar{c}_1^*(\phi) + \nabla z + \bar{c}_{2u}^*(\theta_F) = \psi_u, & x \in \Omega_F, \\ -\nabla \cdot \phi = \psi_p, & x \in \Omega_F, \\ -k_F\Delta\theta_F + \bar{c}_{2T}^*(\theta_F) + \rho_0\beta(g \cdot \phi) = \psi_{T_F}, & x \in \Omega_F, \\ \begin{cases} \theta_F = \theta_S, \\ k_F(n \cdot \nabla\theta_F) = k_S(n \cdot \nabla\theta_S), \end{cases} & x \in \Gamma_I, \\ -k_S\Delta\theta_S = \psi_{T_S}, & x \in \Omega_S, \end{cases} \quad (11.94)$$

with adjoint boundary conditions

$$\begin{cases} \phi = 0, & x \in \Gamma_{u,D}, \\ \mu\dfrac{\partial \phi}{\partial n} = 0, & x \in \Gamma_{u,N}, \\ \theta_F = 0, & x \in \Gamma_{T_F,D}, \\ k_F(n \cdot \nabla\theta_F) = 0, & x \in \Gamma_{T_F,N}, \\ \theta_S = 0, & x \in \Gamma_{T_S,D}, \\ k_S(n \cdot \nabla\theta_S) = 0, & x \in \Gamma_{T_S,N}. \end{cases} \quad (11.95)$$

Here, we have used the linearizations

$$\bar{c}_1^*(\phi) = \tfrac{1}{2}\rho_0\nabla(u + u_h) \cdot \phi - \tfrac{1}{2}\rho_0(u + u_h) \cdot \nabla\phi - \tfrac{1}{2}\rho_0(\nabla \cdot (u + u_h))\phi,$$
$$\bar{c}_{2u}^*(\theta) = \tfrac{1}{2}\rho_0 c_p \nabla(T + T_h)\theta,$$
$$\bar{c}_{2T}^*(\theta) = -\tfrac{1}{2}\rho_0 c_p (u + u_h) \cdot \nabla\theta - \tfrac{1}{2}\rho_0 c_p (\nabla \cdot (u + u_h))\theta.$$

We solve (11.94) numerically using an iterative operator decomposition approach as for the forward problem. These iterations are completely independent of the forward iterations. In [40, 41], we derive estimates that only require adjoint solutions of the two component problems.

To write out the *a posteriori* error representation, we introduce an additional projection $\pi_{W_S}^0 : H^2 \to W_{S,0}^h$ defined such that for any node x_i

$$\pi_{W_S}^0 \theta_S(x_i) = \begin{cases} \pi_{W_S}\theta_S(x_i), & x_i \notin \Gamma_I, \\ 0, & x_i \in \Gamma_I, \end{cases}$$

along with $\pi_\partial \theta_S = \pi_{W_S}\theta_S - \pi_{W_S}^0\theta_S$. The role of these projections is made clear in the context of improving accuracy, see [40] and remarks below.

We can now prove [40],

11.6.6 Theorem

The errors satisfy

$$(\psi, e) = (f, \phi - \pi_V \phi) - a_1(u_h^{\{k\}}, \phi - \pi_V \phi_1) - c_1(u_h^{\{k\}}, u_h^{\{k\}}, \phi - \pi_V \phi)$$
$$- b(\phi - \pi_V \phi, p_h) - d(T_{F,h}^{\{k\}}, \phi - \pi_V \phi) - b(u_h^{\{k\}}, z - \pi_Z z) \quad (11.96)$$

$$+ (Q_F, \theta_F - \pi_{W_F} \theta_F) - a_2(T_{F,h}^{\{k\}}, \theta_F - \pi_{W_F} \theta_F)$$
$$- c_2(u_h^{\{k\}}, T_{F,h}^{\{k\}}, \theta_F - \pi_{W_F} \theta_F) + (Q_S, \theta_S - \pi_{W_S} \theta_S)$$
$$- a_3(T_{S,h}^{\{k\}}, \theta_S - \pi_{W_S} \theta_S) \quad (11.97)$$

$$+ \left(T_{S,h}^{\{k\}} - \pi_S T_{F,h}^{\{k\}}, k_S(n \cdot \nabla \theta_S)\right)_{\Gamma_I} + \left(\pi_S T_{F,h}^{\{k\}} - T_{F,h}^{\{k\}}, k_S(n \cdot \nabla \theta_S)\right)_{\Gamma_I} \quad (11.98)$$

$$+ \left(k_S(n \cdot \nabla T_{S,h}^{\{k\}}), \pi_{W_F} \theta_F\right)_{\Gamma_I} + (Q_S, \pi_{W_S} \theta_S) - a_3(T_{S,h}^{\{k\}}, \pi_{W_S} \theta_S). \quad (11.99)$$

The contributions to the error are

- Equations (11.96)–(11.97) represent the contribution of the discretization error arising from each component solve.
- Equation (11.98) represents the contribution from the iteration.
- The first term in (11.99) represents the contribution of the transfer error while the remaining terms represent the contribution arising from projections between two different discretizations.

Example 52.

We consider an example from [40]. For the flow past a cylinder shown in Figure 11.2, we solve the steady non-dimensionalized Boussinesq equations in the fluid domain and the non-dimensional heat equation in the solid domain. To simulate the flow of water past a cylinder made from stainless steel, we set the dimensionless constants $Pr = 6.6$ and $k_R = 30$, and choose the inflow velocity and the temperature gradient so that,

$$Re = 75, \quad Pe = 495, \quad Fr = 0.001, \quad Ra = 50.$$

The temperature gradient is imposed by setting different temperatures along the top and bottom boundaries, with a linear temperature gradient on the inflow boundary, and an adiabatic condition on the outflow boundary.

We show results for two quantities of interest. The first is the temperature in a small region in the wake, located approximately one channel width downstream of the center of the cylinder and one quarter of a channel width below the upper wall. The second is temperature at a small region in the center of the cylinder. In each case, we derive an *a posteriori* bound by the usual methods, and base adaptivity on an element tolerance of 1×10^{-8}.

We show the final adaptive meshes for the flow in Figure 11.31 and for the solid in Figure 11.32. For the first quantity of interest, the flow mesh is most refined near the region of interest and upstream of the region of interest, locating more elements between the cylinder and the top wall than the cylinder and the bottom wall since the flow advecting heat to the region of interest passes above rather than below the cylinder. The solution downstream of the region of interest can be computed with less accuracy as is recognized by the coarser mesh. For the

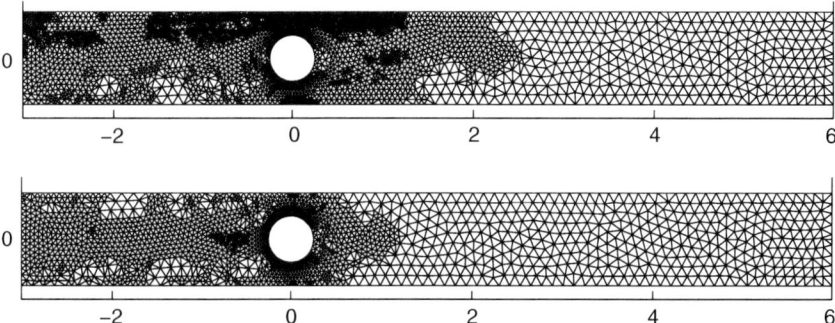

Figure 11.31: Upper: Final adaptive mesh in the fluid when the quantity of interest is the temperature in a small region in the wake above the cylinder. Lower: Final adaptive mesh in the fluid when the quantity of interest is the temperature in a small region in the center of the solid.

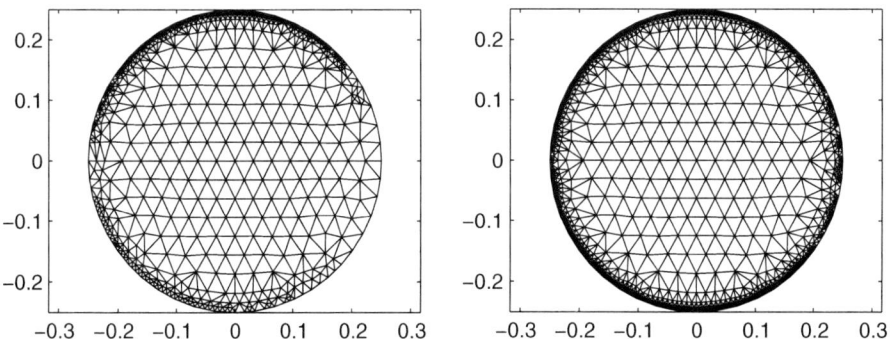

Figure 11.32: Left: Final adaptive mesh in the solid when the quantity of interest is the temperature in a small region in the wake above the cylinder. Right: Final adaptive mesh in the solid when the quantity of interest is the temperature in a small region in the center of the solid.

solid, the mesh is highly refined along the top in order to increase the accuracy of the normal derivative that is computed in the solid and used as a boundary condition in the fluid computation. Evidently, the normal derivatives elsewhere on the interface have less of an influence on the first quantity of interest. For the second quantity of interest, the mesh is highly refined upstream of the cylinder. We note that the refinement downstream of the cylinder corresponds closely to the recirculation region, and the mesh refinement is slightly asymmetric about the midplane of the channel due to the asymmetric initial mesh. The mesh in the solid is refined uniformly near the boundary, reflecting the fact that the error in the finite element flux makes a significant contribution to the error in the quantity of interest.

Loss of order and flux correction
The meshes shown in Figures 11.31 and 11.32 are highly refined near the interface. This reflects the fact that there is significant error in the numerical flux passed between the components. It turns out that this pollutes the entire computation, so that overall the method loses an entire order of accuracy.

Example 53.

We apply Algorithm 11.6.16 to the steady flow of a Newtonian fluid in a two-dimensional channel connected along one boundary to a solid which is heated from below as shown in Figure 11.33.

The Reynolds number (based on the channel width and the flux averaged inlet velocity) is $Re = 2.5$ and the thermal conductivities are $k_F = 0.9$ and $k_S = 1 + 0.5\sin(2\pi x)\sin(2\pi y)$, which are chosen so that the solution is smooth, but nontrivial. The temperature fields are displayed in Figure 11.33.

We solve the problem iteratively and, to approximate the error, we compute a reference solution with a higher order method on the same mesh. In Figure 11.34, we compare the L^2 errors in the temperature fields over $\Omega_S \cup \Omega_F$ on a series of meshes that align along the interface Γ_I.

We see that the solution converges at a second-order rate, rather than the optimal third-order rate. This loss of order is a consequence of the operator decomposition as the computed boundary flux obtained from the finite element

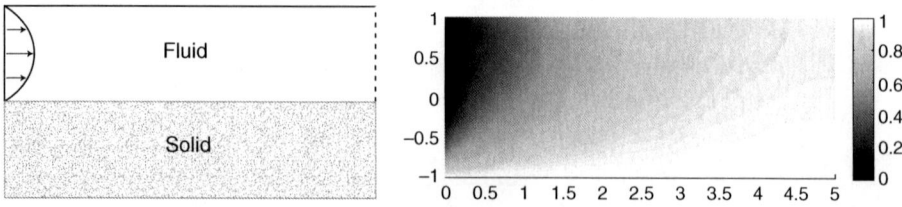

Figure 11.33: Left: Computational domain for a motivational example. Right: Temperature fields within the fluid and the solid.

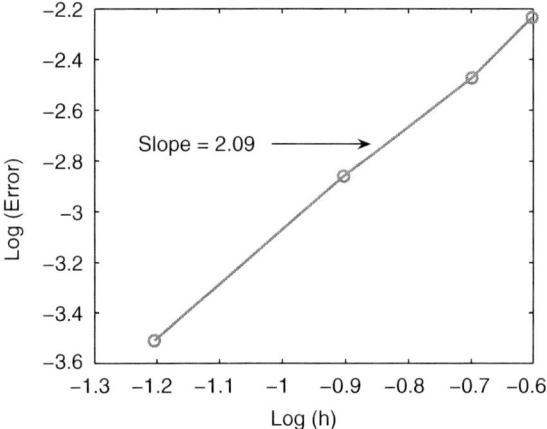

Figure 11.34: Comparison of the mesh size, h, versus the L^2 error in the temperature field when the finite element flux is passed.

solution is one order less accurate than the solution itself and this error pollutes the rest of the computation.

One way to compensate for the loss of order is by refining the mesh locally near the interface. Another way is to compute the particular information, in this case the flux on the interface, more accurately. It turns out that we can adapt a postprocessing technique called flux correction developed originally by Wheeler [42] and Carey [43, 44] to recover boundary flux values with increased accuracy.

We denote the set of elements in $\tau_{S,h}$ that intersect the interface boundary by

$$\tau_{S,h}^{\Gamma_I} = \{K \in \tau_{S,h} \mid \overline{K} \cap \Gamma_I \neq \emptyset\},$$

and consider the corresponding finite element space

$$\Sigma_h = \{v \in P^2(K) \text{ with } K \in \tau_{S,h}^{\Gamma_I},\ v(\eta_i) = 0 \text{ if } \eta_i \notin \Gamma_I\},$$

where $\{\eta_i\}$ denotes the nodes of element K. The degrees of freedom correspond to the nodes on the boundary. We compute $\sigma^{\{k\}} \in \Sigma_h$ satisfying

$$-\left(\sigma^{\{k\}}, v\right)_{\Gamma_I} = (Q_S, v) - a_3(T_{S,h}^{\{k\}}, v), \quad \text{for all } v \in \Sigma_h,$$

where $T_{S,h}^{\{k\}}$ solves (11.91). Since the dimension of the problem scales with the number of nodes on a boundary, it is relatively inexpensive to solve.

The modified Algorithm is given in Algorithm 11.6.20.

Algorithm 11.6.20 Multiscale Decomposition Method for Conjugate Heat Transfer with Flux Correction

k = 0
while ($\|T_S^{\{k\}} - \pi_S T_F^{\{k\}}\|_{\Gamma_I} > TOL$) **do**
    k = k+1
    Compute $T_{S,h}^{\{k\}} \in W_S^h$ such that $T_{S,h}^{\{k\}} = \pi_{W_S} T_{F,h}^{\{k-1\}}$ along the interface Γ_I and

$$a_3(T_{S,h}^{\{k\}}, w) = (Q_S, w), \quad \forall w \in W_{S,0}^h, \tag{11.100}$$

Compute $\sigma^{\{k\}} \in \Sigma_h$ solving

$$-\left(\sigma^{\{k\}}, v\right)_{\Gamma_I} = (Q_S, v) - a_3(T_{S,h}^{\{k\}}, v), \ \forall v \in \Sigma_h, \tag{11.101}$$

Compute $u_h^{\{k\}} \in V_F^h$, $p_h^{\{k\}} \in Z^h$ and $T_{F,h}^{\{k\}} \in W_F^h$ such that

$$\begin{cases} a_1(u_h^{\{k\}}, v) + c_1(u_h^{\{k\}}, u_h^{\{k\}}, v) + b(v, p_h^{\{k\}}) + d(T_{F,h}^{\{k\}}, v) = (f, v), \ \forall v \in V_{F,0}^h, \\ b(u_h^{\{k\}}, q) = 0, \ \forall q \in Z^h, \\ a_2(T_{F,h}^{\{k\}}, w) + c_2(u_h^{\{k\}}, T_{F,h}^{\{k\}}, w) = (Q_F, w) - \left(\sigma^{\{k\}}, w\right)_{\Gamma_I}, \ \forall w \in W_{F,0}^h. \end{cases} \tag{11.102}$$

end while

It turns out that using the recovered boundary flux leads to a cancelation of the "transfer error" term in the error representation formula, which is the source of the loss of order. The new theorem reads [40],

11.6.7 Theorem

The errors satisfy

$$(\psi, e) = (f, \phi - \pi_V \phi) - a_1(u_h^{\{k\}}, \phi - \pi_V \phi_1) - c_1(u_h^{\{k\}}, u_h^{\{k\}}, \phi - \pi_V \phi)$$
$$- b(\phi - \pi_V \phi, p_h) - d(T_{F,h}^{\{k\}}, \phi - \pi_V \phi) - b(u_h^{\{k\}}, z - \pi_Z z) \tag{11.103}$$
$$+ (Q_F, \theta_F - \pi_{W_F} \theta_F) - a_2(T_{F,h}^{\{k\}}, \theta_F - \pi_{W_F} \theta_F)$$
$$- c_2(u_h^{\{k\}}, T_{F,h}^{\{k\}}, \theta_F - \pi_{W_F} \theta_F) + (Q_S, \theta_S - \pi_{W_S} \theta_S)$$
$$- a_3(T_{S,h}^{\{k\}}, \theta_S - \pi_{W_S} \theta_S) \tag{11.104}$$

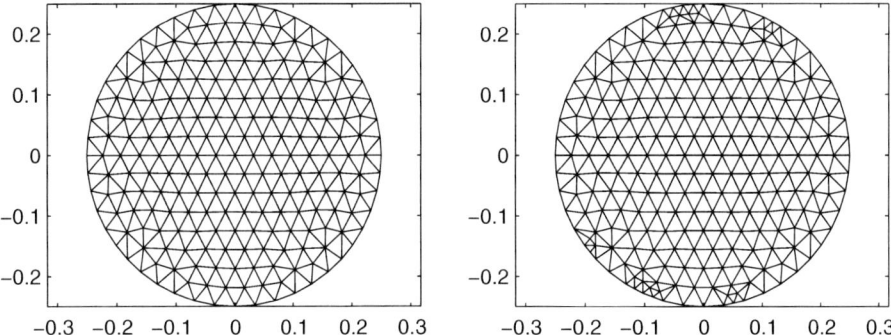

Figure 11.35: Left: Final adaptive mesh in the solid when the quantity of interest is the temperature in a small region in the wake above the cylinder. Right: Final adaptive mesh in the solid when the quantity of interest is the temperature in a small region in the center of the solid.

$$+ \left(T_{S,h}^{\{k\}} - \pi_S T_{F,h}^{\{k\}}, k_S(n \cdot \nabla \theta_S)\right)_{\Gamma_I} + \left(\pi_S T_{F,h}^{\{k\}} - T_{F,h}^{\{k\}}, k_S(n \cdot \nabla \theta_S)\right)_{\Gamma_I} \tag{11.105}$$

$$+ \left(\sigma^{\{k\}}, \pi_{W_F}\theta_F - \pi_{W_S}\theta_S\right)_{\Gamma_I}. \tag{11.106}$$

Note the difference in (11.106) compared to (11.99); now there is only a projection error arising from a change of scale, without any transfer error expression.

We can prove that using the recovered flux recovers the expected cubic order of convergence, see [40].

Example 54.

The recovered accuracy is easily demonstrated by considering the adapted meshes produced by (11.103)–(11.106). We repeat the computations in Example 52 using the modified error bound with the recovered flux derived from (11.103)–(11.106) to guide adaptive mesh refinement. We show the final adaptive meshes for the solid in Figure 11.35. There is no mesh refinement near the boundaries, indicating that the flux error is no longer dominant.

11.7 The effect of iteration

In the presentation above, we have minimized the effects arising from the solution of nonlinear and/or fully-coupled systems by carefully choosing the models and results that are discussed. Referring back to Figure 11.3, we generally expect multiscale operator decomposition to require a number of iterations between the physical components. This raises additional issues that need to be addressed, e.g.:

- The convergence of the iteration is always paramount. Note that the convergence is strongly affected by the fact that we are repeatedly upscaling and downscaling information. Indeed, this may even affect the definition of convergence, e.g. when coupling stochastic models to continuum models.
- When iteration is required, then we are passing information, along with error, between iteration levels as well as physical components, and this requires defining additional auxiliary quantities of interest and corresponding adjoint operators.
- The "single physics paradigm" means that in practice we may have access only to adjoint operators associated to the components, but not to the entire system. This has strong consequences on which contributions to the error may be estimated.

These issues are discussed in [40, 41, 42, 45].

11.8 Conclusion

Multiphysics, multiscale models present significant challenges in terms of computing accurate solutions and for estimating the error in information computed from numerical solutions. In this chapter, we discuss the problem of computing accurate error estimates for one of the most common, and powerful, numerical approaches for multiphysics, multiscale problems called multiscale operator decomposition. This is a widely used technique for solving multiphysics, multiscale models. The general approach is to decompose the multiphysics and/or multiscale problem into components involving simpler physics over a relatively limited range of scales, and then to seek the solution of the entire system through some sort of iterative procedure involving numerical solutions of the individual components. In general, different components are solved with different numerical methods as well as with different scale discretizations. This approach is appealing because there is generally a good understanding of how to solve a broad spectrum of single physics problems accurately and efficiently, and because it provides an alternative to accommodating multiple scales in one discretization.

In the first part of this chapter, we describe the ingredients of adjoint-based *a posteriori* error analysis. We stress the need to accurately quantify stability of particular information to be computed from a model and the role of the adjoint problem for this purpose.

Turning to specific examples of multiscale, multiphysics models, we illustrate the general observation that the stability properties of such models are exceedingly complex. This heightens the importance of obtaining accurate information about stability.

We then describe how the techniques of *a posteriori* error analysis can be extended to multiscale operator decomposition solutions of multiphysics, multiscale problems. While the particulars of the analysis vary considerably with the problem, there are several key ideas underlying a general approach to treat operator decomposition multiscale methods, including:

- We identify auxiliary quantities of interest associated with information passed between physical components and solve auxiliary adjoint problems to estimate the error in those quantities.
- We deal with scale differences by introducing projections between discrete spaces used for component solutions and estimate the effects of those projections.
- The standard linearization argument used to define an adjoint operator associated with error analysis for a nonlinear problem may fail, requiring another approach to define adjoint operators.
- In this regard, the adjoint operator associated with a multiscale operator decomposition solution method is often different from the adjoint associated with the original problem, and the difference may have a significant impact on the stability of the method.
- In practice, solving the adjoint associated with the original fully-coupled problem may present the same kinds of multiphysics, multiscale challenges posed by the original problem, so attention must be paid to the solution of the adjoint problem.

We explain these ideas in the context of three specific examples.

References

[1] Prigogine I. and Lefever R. (1968). "Symmetry breaking instabilities in dissipative systems", *J. Chem. Phys*, **48**(4), p. 1695–1700.

[2] Estep D., Tavener S., and Wildey T. (2008b). "A posteriori error estimation and adaptive mesh refinement for a multi-discretization operator decomposition approach to fluid-solid heat transfer", *JCP*. submitted.

[3] Lorenz E. N. (1963). "Deterministic non-periodic flows", *J. Atmos. Sci.*, **20**, p. 130–141.

[4] Estep D. and Johnson C. (1998). "The computability of the Lorenz system", *Math. Models Meth. Appl. Sci.*, **8**, p. 1277–1305.

[5] Higham N. J. (2002). *Accuracy and Stability of Numerical Algorithms*, SIAM, Philadelphia.

[6] Estep D., Holst M., and Larson M. (2005). "Generalized Green's functions and the effective domain of influence", *SIAM J. Sci. Comput.*, **26**, p. 1314–1339.

[7] Marchuk G. I., Agoshkov V. I., and Shutyaev V. P. (1996). *Adjoint Equations and Perturbation Algorithms in Nonlinear Problems*, CRC Press, Boca Raton, FL.

[8] Lanczos C. (1997). *Linear Differential Operators*, Dover Publications.

[9] Cacuci D. (1997). *Sensitivity and Uncertainty Analysis: Theory*, Volume I. Chapman & Hall/CRC.

[10] Marchuk G. I. (1995). *Adjoint Equations and Analysis of Complex Systems*, Kluwer.
[11] Atkinson K. and Han W. (2001). *Theoretical Numerical Analysis: A Functional Analysis Framework*. Springer.
[12] Aubin J. (2000). *Applied Functional Analysis*, John Wiley & Sons, Inc.
[13] Cheney W. (2000). *Analysis for Applied Mathematics*, John Wiley & Sons, Inc.
[14] Folland G. (1999). *Real Analysis*, John Wiley & Sons, Inc.
[15] Schechter M. (2002). *Real Analysis*, American Mathematical Society.
[16] Eriksson K., Estep D., Hansbo P., and Johnson C. (1995). *Introduction to adaptive methods for differential equations*. In *Acta Numerica, 1995*, p. 105–158. Cambridge Univ. Press, Cambridge.
[17] Eriksson K., Estep D., Hansbo P., and Johnson C. (1996). *Computational Differential Equations*, Cambridge University Press, Cambridge.
[18] Estep D., Larson M. G., and Williams R. D. (2000). "Estimating the error of numerical solutions of systems of reaction-diffusion equations", *Mem. Amer. Math. Soc.*, **146**(696), p. viii+109.
[19] Becker R. and Rannacher R. (2001). "An optimal control approach to a posteriori error estimation in finite element methods", *Acta Numerica*, **10**, p. 1–102.
[20] Giles M. and Süli E. (2002). "Adjoint methods for PDEs: A posteriori error analysis and postprocessing by duality", *Acta Numerica*, **11**, p. 145–236.
[21] Bangerth, Wolfgang and Rannacher, Rolf (2003). *Adaptive Finite Element Methods for Differential Equations*, Birkhauser, Verlag.
[22] Paraschivoiu M., Peraire J., and Patera A. T. (1997). "A posteriori finite element bounds for linear functional output of elliptic partial differential equations", *Comput. Methods Appl. Mech. Engrg.*, **150**, p. 289–312.
[23] Barth T. J. (2004). *A-Posteriori Error Estimation and Mesh Adaptivity for Finite Volume and Finite Element Methods*, Volume 41 *of Lecture Notes in Computational Science and Engineering*, Springer, New York.
[24] Lions J. and Magenes E. (1972). *Non-Homogeneous Boundary Value Problems and Applications*, Volume 1, Springer-Verlag, New York.
[25] Wildey T., Estep D., and Tavener S. (2008). "A *posteriori* error estimation of approximate boundary fluxes", *Commun. Num. Meth. Engin.*, **24**, p. 421–434.
[26] Estep D. (1995). "A posteriori error bounds and global error control for approximation of ordinary differential equations", *SIAM J. Numer. Anal.*, **32**(1), p. 1–48.
[27] Estep D. and French D. (1994). "Global error control for the continuous Galerkin finite element method for ordinary differential equations", *RAIRO Modél. Math. Anal. Numér.*, **28**, p. 815–852.
[28] Estep D. and Stuart A. M. (2002). "The dynamical behavior of the discontinuous Galerkin method and related difference schemes", *Math. Comp.*, **71**(239), p. 1075–1103 (electronic).

[29] Jamet P. (1978). "Galerkin-type approximations which are discontinuous in time for parabolic equations in a variable domain", *SIAM J. Numer. Anal.*, **15**, p. 912–928.

[30] Delfour M. and Dubeau F. (1986). "Discontinuous polynomial approximations in the theory of one-step, hybrid and multistep methods for nonlinear ordinary differential equations", *Math. Comp.*, **47**, p. 169–189.

[31] Delfour M., Hager W., and Trochu F. (1981). "Discontinuous Galerkin methods for ordinary differential equations", *Math. Comp.*, **36**, p. 455–472.

[32] Thomée V. (1980). *Galerkin Finite Element Methods for Parabolic Problems*, Springer-Verlag, New York.

[33] Eriksson K., Johnson C., and Thomée V. (1985). "Time discretization of parabolic problems by the Discontinuous Galerkin method", *RAIRO Modél. Math. Anal. Numér.*, **19**, p. 611–643.

[34] Estep D. and Larsson S. (1993). "The discontinuous Galerkin method for semilinear parabolic problems", *RAIRO Modél. Math. Anal. Numér.*, **27**, p. 35–54.

[35] Sandelin, J. (2006). *Blogal Estimate and Control of Model, Numerical, and Parameter Error*, Ph.D. thesis, Department of Mathematics, Colorado State University, Fort Collins, CO 80523.

[36] Estep, D., Holst, M., and Mikulencak, D. (2002). "Accounting for stability: a posteriori error estimates based on residuals and variational analysis", *Comm. Num. Meth. Engin.*, **18**, p. 15–30.

[37] Estep, D. and Williams, R. (1996). "Accurate parallel integration of large sparse systems of differential equations", *Math. Models Meth. Appl. Sci.*, **6**, p. 535–568.

[38] Carey V., Estep D., Larson M., and Tavener S. (2008b). Blockwise adaptivity for time dependent problems based on coarse scale adjoint solutions, *SIAM J. Sci. Comput.*

[39] Carey V., Estep D., and Tavener S. (2009). "A posteriori analysis and adaptive error control for operator decomposition methods for coupled elliptic systems I: One way coupled systems", *SINUM*, **47**, p. 740–761.

[40] Estep D., Ginting V., Shadid J., and Tavener S. (2008b). "An a posteriori-a priori analysis of multiscale operator splitting", *SIAM J. Num. Analysis*, **46**, p. 1116–1146.

[41] Estep D., Ginting V., and Tavener S. (2008a). "A posteriori analysis of multiscale iterative operator decomposition for coupled ODEs", *SINUM*, submitted.

[42] Wheeler M. F. (1974). "A Galerkin procedure for estimating the flux for two-point boundary-value problems using continuous picewise-polynomial spaces", *Numer. Math*, **2**, p. 99–109.

[43] Carey G. F., Chow S. S., and Seager M. K. (1985). "Approximate boundary-flux calculations", *Comp. Meth. in Applied Mech. and Engr.*, **50**, p. 107–120.

[44] Carey G. F. (1982). "Derivative calculation from finite element solutions", *Comp. Meth. Applied Mech. Engr.*, **35**, p. 1–14.

[45] Carey V., Estep D., and Tavener S. (2008a). "A posteriori analysis and adaptive error control for operator decomposition methods for elliptic systems II: Fully coupled systems", *In preparation*.

[46] Dawson C. N. and Wheeler M. F. (1992). Time-splitting methods for advection-diffusion-reaction equations arising in contaminant transport. In *ICIAM* 91 (Washington, DC, 1991), p. 71–82. *SIAM*, Philadelphia, PA.

[47] Estep D. (2004). *A short course on duality, adjoint operators, Green's functions, and a posteriori error analysis*. Sandia National Laboratories, Albuquerque, New Mexico. Notes can be downloaded from http://math.colostate.edu/~estep.

[48] Estep D., Tavener S., and Wildey T. (2008a). "A posteriori analysis and improved accuracy for an operator decomposition solution of a conjugate heat transfer problem", *SINUM*, **46**, p. 2068–2089.

[49] Hundsdorfer W. and Verwer J. (2003). *Numerical solution of time-dependent advection-diffusion-reaction equations*, Volume 33 of *Springer Series in Computational Mathematics*. Springer-Verlag, Berlin.

[50] Marchuk G. I. (1970). On the theory of the splitting-up method. In *In: Proceedings of the Second Symposium on Numerical Solution of Partial Differential Equations, SVNSPADE*, p. 469–500.

[51] Marchuk G. I. (1990). Splitting and alternating direction methods. In *Handbook of Numerical Analysis*, Volume I, p. 197–462. North-Holland, New York.

[52] Ropp D. L. and Shadid J. N. (2005). "Stability of operator splitting methods for systems with indefinite operators: reaction-diffusion systems", *J. Comput. Phys.*, **203**(2), p. 449–466.

[53] Strang G. (1968). "On the construction and comparison of difference schemes", *SIAM J. Numer. Anal.*, **5**, p. 506–517.

PART V

MULTISCALE SOFTWARE

12

COMPONENT SOFTWARE FOR MULTISCALE SIMULATION

M.S. Shephard, M.A. Nuggehally, B. FranzDale, C.R. Picu, J. Fish, O. Klaas, and M.W. Beall

12.1 Introduction

Increasingly simulation processes must consider multiple physical scales where different classes of models are used to represent phenomena taking direct consideration of the interactions across scales. The complexity of the development of a compatible set of multiscale models and methods has led researchers to develop implementations in which all methods, numerical algorithms, and associated data structures are tightly coupled. Although a viable approach for initial research, it is a suboptimal for the continued development of future multiscale simulation methods. What is required are methods for the representation and implementation of multiscale simulation models that support the effective combination of various single-scale models and methods through scale linking models and methods. Such approaches effectively support the ability to address the multiscale modeling needs for a broad range of applications. Considering the thousands of person years of effort that has gone into the development, verification, and validation of single-scale simulation software, an approach that can effectively use such simulation software as components is the only practical approach for the widespread application of multiscale simulation.

The development of a set of tools to support multiscale simulation must explicitly account for the variability of the models at different scales and the methods of communicating information between them accounting for scale transformations. Since the best combination of models, methods, and numerical discretizations is not known *a priori* it is critical that multiscale simulation tools support the application of adaptive control techniques based on *a-posteriori* measures for the adaptive selection of scales and models, and the adaptive construction of optimal model discretization and control of scale transformations of the discretized models on multiple scales.

This chapter defines a set of developing interoperable components designed for the effective combination with existing single components to support the development of multiscale simulations. The next section defines an abstraction of multiscale simulation that leads naturally to the definition of a set of interoperable components and associated structures to support multiscale simulations, which are discussed in Section 12.3. Section 12.4 overviews a set of component tools added to support the definition and multiscale modeling of crystaline materials. Section 12.5 overviews an adaptive concurrent atomistic/continuum multiscale simulation capability developed using the methods, structures, and operators described in this paper.

12.2 Abstraction of adaptive multiscale simulation

The behavior of systems is determined by phenomena operating on a range of scales that may include electronic, atomic, micro-structural, and system-wide effects. Distinct mathematical models are used to account for the behavior of interest at the various scales. The divisions between models on the various scales is largely delineated by a breakdown of the assumptions associated with a particular model. Analyses performed at a selected scale typically rely on available experimental data to calibrate "constitutive laws" at that scale. This validity of this approach becomes questionable when applied to problems where scale interactions are critical. As the other chapters in this volume have clearly indicated, efforts to explicitly account for scale linking have been, and continue to be, developed. These methods can be categorized into two broad classes. The first is hierarchic models in which the fine scale is modeled and its gross response is infused into the coarser scale. The second is concurrent methods in which the scales are simultaneously resolved.

12.2.1 Current multiscale simulation implementations

Efforts to date on the development of multiscale modeling technologies have been focused primarily on specific combinations of methods to demonstrate the capability of a specific scale-linking technology. Among the most mature of these developments are the implementations of the quasicontinuum methods for which one is available from the web [44]. The quasicontinuum method [20, 28, 56] starts from consideration of lattice level atomic modeling in which standard molecular mechanics potentials are applied to the atoms in critical regions. In noncritical regions entire groups of atoms are approximated in terms of representative atoms with specific assumptions applied on the variation of fields. Adaptive versions of these procedures have been developed building on finite element error estimation and refinement procedures. Although the quasicontinuum method employs molecular mechanics, the approximations applied require careful consideration be given to the transfer of information between the regions resolved down to the lattice level and those where only the representative atoms are used.

Other scale-linking procedures consider the construction of operators to link continuum PDE fields to discrete atomic fields. In several cases the scale-linking operators consider the discretized PDE form (e.g., element mesh) when constructing these operators (i.e., see [9, 11, 60]). Others define the operators at the equation level. The heterogeneous multiscale methods [14] define compression operators to relate discrete to continuum and reconstruction operators to relate continuum to discrete scales. The equation-free multiscale method [19] links statistically averaged fine-scale realizations to the coarse scale. The concurrent domain decomposition methods employ a blending process between the two models that is easily carried into weak forms appropriate for solution by generalized discretization methods [15, 33].

Errors introduced in the representation of physical phenomena depend on the mathematical model selected, the physical scale on which the mathematical model is applied, the domain of analysis, and the boundary/initial conditions [3]. Methods to control the errors associated with discrete solutions of PDEs are known for several equations [1, 5]. These methods have been extended to estimate errors in quantities of interest [37, 43], and to estimate modeling errors [35]. Recently, these methods have been further extended and applied to atomistic/continuum multiscale methods [36, 42]. Another approach to the adaptive control of models, including those acting at different scales, is the use of model breakdown indicators [11, 33].

To this point there are no generalized structures and tools to support the effective implementation of a full range of scale-linking methods. Their development must address specific steps and operations that include:

- Specifying the scales and models to be applied over the various portions of the space/time domain.
- Generating the discretization (when needed) and computational methods that will be used to solve each model over its portion of the space/time domain.
- Supporting the transfer of information between the models and scales over interacting portions of the domain including the transformation operations required.

Ideally the structures and tools developed will support the construction of interoperable procedures that can interact with existing and developing single-scale analysis tools and scale-linking methodologies.

There is a wide variety of software to support continuum and discrete analysis methods. On the continuum side, general purpose finite difference, finite element, or finite volume procedures solve discretized version of PDEs. A large number of commercial software packages are available for solid mechanics, fluid mechanics, electromagnetics, etc. Open-source software for many of these areas is also available from the web and within text books. The construction of the discretized domains used in these analyses is supported by a number of commercial and open-source programs for mesh generation, including ones that operate from domain definitions defined in commercial CAD systems [8, 51].

A large number of programs are available for discrete level modeling [24, 29, 61, 21]. The NSF supported Network for Computational Nanotechnology [31] is collecting nanotechnology modeling software with a current emphasis on microelectronic applications.

12.2.2 Multiscale simulation abstraction

The abstraction of multiscale simulation processes must carefully consider the hierarchy of information needed and transformations required to go from a physical description of the problem, through the application of appropriate mathematical models, to the construction of the computer models used to solve those mathematical descriptions.

The highest level definition of the problem must be invariant with respect to all of the models and methods applied in a simulation. For example, if multiple continuum analyses with different meshes are used, then defining the boundary conditions, material properties, etc. on a mesh is not appropriate since this specification is mesh-dependent and typically incomplete. Mesh-independent specification of this information is properly supported as tensors acting on the geometry that represents the domain [34]. These same methodologies can be extended to account for the transformation of information between scales.

Defining and generalizing multiscale simulations begins with consideration of single-scale simulations. Even with the amount of effort that has gone into single-scale simulation software, the vast majority does not interact with a general abstraction of the simulation. This is one of the reasons that, even though adaptive computation procedures have been known for years, they are still not in common use. Figure 12.1 shows an abstraction of a single-scale simulation into four levels in which the information on each level is placed into one of three basic information groups. The three basic information groups are the domain definition, the models that govern the behavior of interest over the domain, and the fields that describe how the parameters associated with the behavioral models are defined and act over the domains of interest. The four levels define the quantification of the problem information beginning with the problem specification level that represents a generalized statement of the simulation problem. Each of the lower levels represents the result of a transformation of the information from the previous level until a computational system appropriate for solution on a digital computer is constructed. Each step of transformation changes the form of the information into the one needed for the next step. The operations involved with the transformations can be complex and involve the introduction of various levels of approximation. The effective identification and adaptive control of these approximations is central to reliable simulation processes.

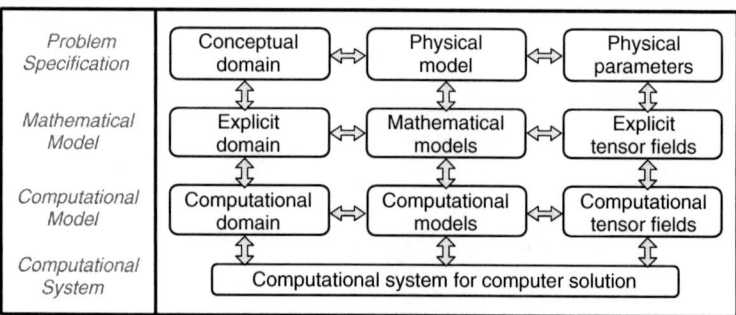

Figure 12.1: Abstraction of Single Scale Simulation.

Current software supports only specific portions of such an abstracted simulation process. CAE systems for solving continuum PDE equations and programs that solve discrete atomic models operate at only the computational model and computational systems level, with the operations performed being the transformation of the computational model into a computational system. In the case of continuum problems current CAD systems provide effective support of explicit (mathematical) domain definitions. There is limited software for the support of the remainder mathematical model information and its transformation to the computational model information. Although a number of CAD systems indicate the support of higher level models such as feature models, available engineering simulation software requires extensions to support the remainder of the problem specification level and its transformations to mathematical model information.

The abstraction to a multiscale simulation (Figure 12.2) extends the single-scale simulation process to account for scale interactions. The key additions to the single-scale simulations are the multiscale problem specification and the scale-linking functions that operate between the single-scale problems. The linking functions that perform the transformation of information between scales interact with the model, domain, and tensor field information in a manner as depicted in Figure 12.3. The model relationships indicate the mathematical relationships governing the transformation of the equation parameters between the two scales. This information governs the process of transforming the tensor fields between

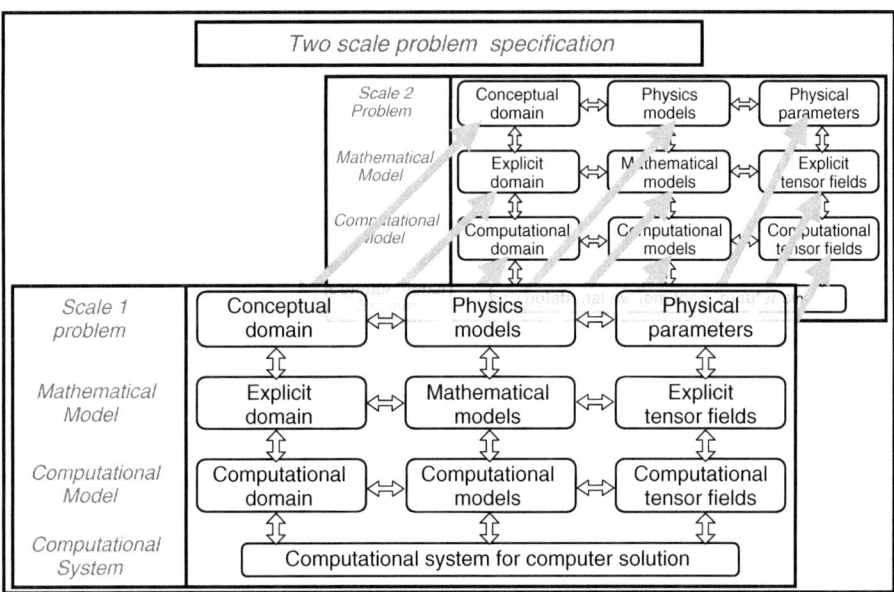

Figure 12.2: Example of a two-scale multiscale simulation.

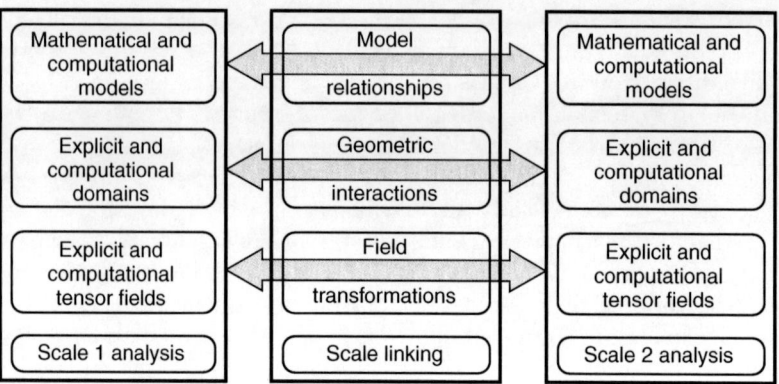

Figure 12.3: Scale-linking that transforms information between scales.

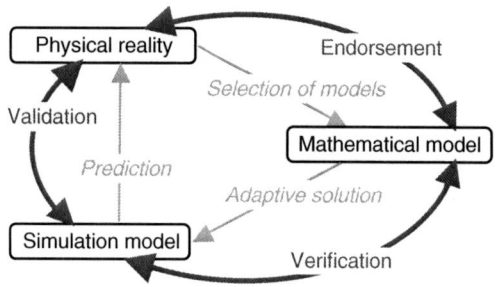

Figure 12.4: Verification and Validation (fashioned after [2]).

scales. In cases where atomistic and continuum methods are linked, the tensor field transformations must deal with technical issues such as relating quantities like atomistic forces and continuum stresses, defining quantities not defined on the fine scale (e.g., temperature), filtering unneeded high frequency components as moving up in scale, and accounting for statistical variation as scales are traversed. Since the computational tensor fields are typically constructed through a discretization process, the transformation functions must also account for that process.

The extensive use of simulation is leading to increased attention to verification and validation (V&V) [4, 45]. Efforts on V&V include professional society consideration, such as AIAA and ASME [2]. As indicated in Figure 12.4, the four-level abstraction proposed for multiscale simulation processes maps directly into the V&V processes being defined with problem specification mapping to physical reality, mathematical model to mathematical model, and computational model and computational system to the simulation model.

12.3 Multiscale simulation components and structures

In the context of multiscale simulation, interoperable components will be associated with the transformation and flow of information associated with the arrows connecting the boxes in Figures 12.1 through 12.3 when those associations are between different software functions that do not operate on a single set of structures. Specifically, the interoperable components will take information of a specific type, either domain, model, or field, and either (i) perform operations to associate it to one of the other information types on the same level, (ii) execute a transformation needed to convert the information between levels in a simulation, or (iii) execute a transformation needed to convert the information between two different scales. The ability to construct flexible multiscale components that can effectively interoperate in the construction of multiscale simulations for multiple applications requires a general application programming interface (API) that effectively captures the transformation processes and information transfers involved in a multiscale simulation.

Critical to the development of interoperable multiscale simulation components is a set of structures that can house the data associated with each of the three information groups and can support the associations between the information groups and the transformations applied during the execution of a multiscale simulation. The structures being developed for this build on a set of component-based methods developed for solving continuum PDEs [7, 8, 52], and more recent procedures being developed that include scale-linking for multiscale analysis [33, 53]. The methodologies are being used for the development of both commercial [54] and open-source [53] components for the construction of adaptive simulation procedures. The key structures being used and/or extended for component-based multiscale simulation include:

- An abstract model that is used to house the information associated with all of the problem specification as well as information used in the control of the mathematical and computational models.
- Topological and voxel representations that control both the explicit and computational domain definitions and their associations.
- Tensor fields for the definition and grouping of tensors for both the explicit and computational tensor fields.

To begin to qualify the functions such an API must support, the three subsections that follow will consider the contents of the three basic information groups of models, domains, and fields, and the types of transformations and information transfer they must support in the four levels of a multiscale simulation process.

12.3.1 Models

The problem specification level indicates the overall physical principles governing the problem at each scale considered. It must also indicate which scales might need to be considered, and physical relationships between the field parameters and domains on those scales. The physical principals are assigned to the appropriate parts of the domain, and the interactions between parts, including scale-linking, are associated with interactions between parts. The model becomes functional at the next level where the sets of mathematical models representing the governing physical principles are associated with the parts and the interactions between them. Since the mathematical models may be adapted as part of the simulation process, it must be possible to relate the alternative mathematical models for different behaviors, and information about when they are appropriate to apply, in the relationships between the physical models and mathematical models.

The transformation of the mathematical model to the computational model is focused on providing the specific inputs needed to construct the computational system, which is ultimately sets of algebraic equations, to be solved. In the case of discrete systems like molecular dynamics the potentials are applied to discrete entities directly yielding algebraic relationships. Since the unmodified application of general potentials typically leads to unacceptably large systems, there are often approximations made at this level. These approximations can be as simple as the use of a cutoff function to limit the distance of interactions considered, or may be a more complex set of coarse-graining operations. Maintaining a knowledge of the approximations made here is needed so that adaptive procedures to improve those approximations can be applied when needed.

In the case of continuum mathematical models the transformation of the mathematical model to the computational model typically employs a formal discretization process. The most commonly applied discretization methods are based on a double discretization process in which the domain is decomposed into a geometric grid or mesh and the equation parameters, the computational tensor fields, are approximated over the entities of the mesh by piecewise distribution functions times yet to be determined multipliers, referred to as degrees of freedom (DOF). The result of this process is a set of mesh entity level algebraic systems that are assembled into the computational system that is then solved to determine the values of the DOF. The process of executing the computational model defines specific relationships between entities in the computational domain and the computational tensor fields. Three common methods that employ different combinations of interactions between the mesh entities and the DOF, and the distributions that define the computational tensor fields, are finite difference, finite volume, and finite element methods [53]

12.3.1.1 Abstract model
The function of the abstract model is to support the definition of the conceptual domain, the physical model, the physical parameters, and the relationships between them at the problem specification level. To do this there must be entities within the abstract model representing each physical part involved in the problem specification. Within a single-scale analysis, the abstract model must

support the specification of general relationships between the artifacts such that the relationship of physical models to the conceptual domain entities and physical parameters maintained used to construct the viewpoints implied by alternative transformations needed to traverse to the other levels in the simulation process. In the case of multiscale simulation this also includes supporting the interactions of models acting on different scales and their transformation processes.

To date the concept of an abstract model is being developed to support simulation-based design systems [52]. In this case, structures and methods have been developed to support the morphing as needed during design evolution into viewpoint-specific forms that account for the interactions conceptual domain components and a given set of single-scale physical models and physical parameters within a single-scale analysis. A workable form of such a representation is a graph structure similar to those used to define assembly and feature models in CAD systems [17] with extensions to support hierarchal decompositions and multiple viewpoints [17, 18, 32] that account for model and scale interactions [52].

Key to an operational abstract model is an abstract model manager that is responsible for supporting the definition of entities and their interactions. In the case of the entities within the conceptual domain, recent extensions have been made to support high level geometric reasoning and interactions [54] at a level of entity definition above that in the typical boundary representation of a CAD system. These methods can also be used to support the definition of relationships between domain entity representations on multiple physical scales. Currently, information on the physical model relationships via the links between abstract model entities and physical parameter information is attributed to either the entities or the links. The abstract model within the Simmetrix Simulation Applications Suite [54] is being used to support the needs of the problem specification for multiscale simulations; this model, and its ability to maintain associations with information at both the mathematical and computational model level, can support all the needs just indicated. Although a workable approach to support the development of specific multiscale simulations, extensions are desired to improve the ability to conveniently define physical model representations, associate them for conceptual domain entities, define the physical parameters associated with those models in a general way, and support the interactions and transformations between physical models including accounting for scale-linking.

12.3.2 Domains

The definition of the problem domain must indicate the spatial and temporal extents used in the definition and execution of a multiscale simulation. The current discussion in focused on the spatial domain. There are a number of general forms used to represent spatial domains. To meet the needs of multiscale simulation the domain representations must support (i) the transformation of the conceptual domain into the explicit and computational domains, (ii) the association of the physical parameters and fields with the domain models, (iii) any domain interrogation required during the simulation, and (iv) the geometric interactions between related domains used in a multiscale simulation.

The conceptual domain definition is a set of parts with relationship between the parts defined by the combination of spatial or model interactions. The explicit domain definitions are a function of the type of mathematical model being used over the domain. For example, continuum domain definitions are needed in the case of PDEs while a discrete set of atomic positions is needed in MD.

The two primary sources for domain definitions used with continuum equations are CAD models and image data. CAD systems employ some form of boundary representation. Image data are defined using a volumetric form such as voxels or octrees. Except in cases of directly using the image data in the construction of the computational domain, it is generally accepted that boundary representations are well suited for continuum-level simulations. Common to all boundary representations is the use of the abstraction of topological entities and their adjacencies to represent the domain entities of different dimensions.

The most common form of continuum computational domains are meshes which are piecewise decompositions of the domain. To support a full set of operations needed for reliable multiscale analysis the mesh must maintain an association with the continuum domain representation, which can be effectively done by relating appropriate sets of topological entities between the explicit and computational domain [7].

The domain definitions for the discrete models are the positions of the entities the potentials are written to relate. The overall domain can be a representative volume, a pre-defined geometric domain, or an adaptively-defined overlay that has portions of its boundary interior to a higher level domain.

The parameters and transformations used to define the atomic positions are a strong function of the type of material being defined. In the case of crystalline materials, the initial atom positions are defined by the set of lattice vectors. The definition of the geometric configuration of a set of crystals is a nontrivial process and various grain-growth procedures can be applied [47, 48]. In the case of polymeric materials the atomic positions must be defined by their spatial position along a molecular chain where there are strong bonds between neighboring atoms in the chain. Statistically-based geometric constructs can be used to define these material-dependent chains in the simulation box.

One of the methods used to bridge scales is to combine sets of atoms such that they can be defined in terms of a smaller number of discrete points. One such approach well suited to lattice-type atomic structures is the quasi-continuum method where the movements of atoms over simple shapes (triangles, tetrahedra) are described to vary linearly [20, 28]. In the case of polymeric chains all the atoms along a chain may be represented by a small number of beads placed along the chain [25, 38].

The three general forms of domain interactions in multiscale simulations are:

1. Disjoint domains that share information over a common boundary.
2. Overlapping domains where the higher-scale domain overlaps all or part of the finer scale-domain and the information is shared over the overlap region.
3. Telescoping domains where the finite, but very small with respect to the higher-scale domain, fine-scale domain only passes information to a point in the higher-scale domain.

In each case the operations used to transfer parameters between the scales must be consistent with the form of domain interaction. Effective methods to relate domains across multiple scales include a combination of topological and spatial methods. The topologically-based procedures will relate boundary representation entities across the scales. Spatially-based methods (e.g., octree) will be used when topological relations are not convenient or possible (e.g., when the fine-scale model is only a set of atom positions).

The representations of the explicit and computational spatial domains play a central role in both quantifying the domain definition and supporting the relationships between models and fields. The abstraction of topological entities and their adjacencies as defined in a general nonmanifold geometric model boundary representation [57] can (i) effectively meet the needs of all continuum-level domain representations, effectively represent meso-scale domains (e.g., grain boundaries), (iii) support reducing the volume of data needed when major portions of the atomistic scale data are regular, (iv) support the various relationships to other information, and (v) the interactions with the conceptual domain representation within the abstract model.

Often the methods used to construct the continuum domains directly employ a boundary representation. Examples are commercial CAD systems and polygonal representations of crystal structures. The definition of geometric domains within CAD systems are built from modeling kernels like ACIS and Parasolid [39, 55]. The geometric modeling kernels support an application programming interface (API) that supports a full range of geometry definition, modification, and interrogation operations, most of which are keyed in terms of the entities and adjacencies of vertices, edges, loops, faces, shells, and regions that define the boundary representation.

The use of topological entities and their adjacencies is also ideal for the definition of continuum level computational domains. In addition to providing a clear mechanism for maintaining the relationship of the entities in the computational domain to the mathematical domain, this approach has been effective in support of a full range of mesh-based operations such as automatic mesh generation [8, 51], automated adaptive mesh modification [13, 22], and variable order p-version analysis procedures [22]. Since the individual entities in a mesh are less topologically complex than those of general geometric models, more storage effective mesh boundary representations are possible [7, 49].

In some applications the initial explicit domain definition comes in the form of values associated with the cells, referred to as voxels in 3-D, of a regular grid. One example of this type is image date where there is an intensity value associated with each voxel. Based on knowledge of the system being imaged and the imaging process, those intensity values are converted to the identification of one or more materials associated with the pixel. A second example of this type is where the explicit domain definition is constructed using a grid-based algorithm such as those used to simulate grain growth [46], where each voxel has a material identification given to it. In the case where the initial explicit domain definition is a voxel form there are two options by which to proceed. The first is to use a computational domain definition that can be generated for the voxel data such as a standard octree structure [23]. The second approach is to apply algorithms that can convert the voxel data into a boundary representation for which a mesh, and associated mesh boundary representation, can be constructed [58].

12.3.3 *Fields*

At the problem specification level the overall physical parameters involved with the models at the various scales are indicated. In a multiscale simulation this information must indicate the relationships between parameters of related scales. The physical parameters used in the mathematical equations are tensor quantities defined over various portions of the domain that can be general functions of the independent variables of space and time as well as other dependent variables.

The definition, storage, and transformation of the tensor fields must support

- the structure of various order tensor quantities and their basic transformation (e.g., representing the tensor in alternative coordinate systems),
- the application of the complex sets of transformations seen in multiscale simulations that can alter the form of tensor and its representation,
- the relationship of those tensors to the entities in the explicit and computational domain definitions,
- the specification of how the tensors vary over the appropriate domain entities, and
- the ability to combine different groups of both explicit and computational tensor fields together to support the needs of the current multiscale simulation.

The tensor fields that are known *a priori* are defined by given distribution functions over domain entities. For example, in the case of solving PDEs over continuum domains, the distribution of the given input tensors is effectively related to topological entities of regions, faces, edges, and vertices [34] (e.g., constitutive matrix for a region, a distributed traction over a surface, etc.).

Those tensor fields that are to be determined as part of the computation are ultimately related to the appropriate explicit model entity. However, their actual definition is in terms of distribution functions associated with entities in the computational model that represent that model entity. For example, in the case of a continuum PDE solved using finite elements, the computational field is defined over the mesh entities that represent the appropriate explicit model entities. Over each of those mesh entities the tensor field is discretized into a set of shape functions times degrees of freedom. A single tensor field can be used by a number of different analysis routines that interact, and the field may be associated with multiple computational models having alternative relationships between them. In addition, different distributions can be used by a field to discretize its associated tensor.

The basic information indicating the relationships of fields across scales is defined at the problem specification level. However, the execution of these processes for those tensor fields that are defined as computational tensors must properly account for the processes involved in the construction of that tensor field over the computational domain. In the case of continuum methods this means properly accounting for the distribution functions. In the case of discrete-level models, that will include accounting for the influence of any coarse-graining process. To better maintain control on the consistency of scale-linking processes, the methods used require the qualification of the scale-linking functions at each of the levels in the process. Although this may appear to be a burden, it is needed to avoid potential errors that could be made in directly linking across scales at the computational level, even though that is where the operations are ultimately applied.

12.4 Component tools for crystaline materials

There are two critical scales below the macro scale that must be considered in the multiscale analysis of crystaline materials. The first defines the grain structure and the second is the atomistic position within that grain structure. To support the application of multiscale simulations for these classes of materials, a set of component-level tools was developed. These tools include (i) the definition and storage of grain structures, (ii) the lattice-level definition and efficient storage of atomistic information on the grains, and (iii) a fast relaxation procedure to define more energy-favorable positions for atoms near the boundary of grains that were initially positioned only based on a lattice definition.

12.4.1 Defining the grain structure

Crystaline materials are made up of sets of grains that are of a scale much finer than can be explicitly represented in models of the entire domain of real parts. At the same time the crystals exist at a scale substantially larger that the atomic scale. Of specific importance to multiscale simulations is the fact that a substantial amount of the behavior at scales finer than the full part are associated with the boundaries between grains. Thus, the ability to address this scale explicitly within multiscale simulations is critical for these classes of materials. One class of fine-scale models applies continuum models at the grain level with the grain boundaries are material interfaces. Another class considers the atomistic scale where specific advantage is taken of the fact that away from the crystal interfaces, the atomistic configurations are very close to an ideal lattice structure. In both cases, it is necessary to define mesoscale models of the crystal structure when and where it is needed.

The natural choice for this representation is as a nonmanifold geometric boundary representation in which each crystal region is bounded by a shell consisting of faces, edges, and vertices that de-mark the two-, one-, and zero-dimensional interfaces between various crystals. The compact representation of atoms in a grain structure and the fast relaxation procedure discussed in the next two subsections are examples of procedures that require a boundary representation of the grain structure.

For simple grain structures, the capabilities of a solid modeling system can be used to defined the domains. Extensions can be added to a graphical user interface to speed the process for the construction of faceted crystal domains. For more realistic domains this process is tedious and thus, algorithmic means are desired to define this information. Such procedures have been found to provide realistic microstructures for a variety of crystaline materials of interest [46, 47, 48]. Some of the methods for the algorithmic construction of such grain structures employ a process that anneals a set of multimaterial voxels into a grain structure defined by the grouping of voxels of the same material that neighbor each other. Sometimes, the next step in the simulation process may be able to work directly from the voxel data. Mesh generation procedures that employ octree structures are one such possibility that have been extended to do this [54]. Similar methods can be used to extract a more formal boundary representation for use by a more extensive range of algorithms. An example of such an algorithmic approach is a multimaterial extension [59] of marching cubes.

The left-hand image in Figure 12.5 shows an example of a grain structure generated using the methods presented in [47, 46] while the right-hand image shows a mesh automatically generated within that grain structure [54].

12.4.2 *Defining and storing the atomistic information*

Given the computational resources required to represent millions or billions of atoms, it is desirable to store as few atoms as possible explicitly in memory. In particular, we would like the time and memory performance of multiscale simulation to surpass that of its smallest scale in both the big-O sense and in an absolute sense. For that reason, explicitly representing every atom for complete domains would defeat the purpose of the multiscale approach. However, for a multiscale model to be free to adapt, it must be able to determine and change the location of any atom at any time. These two constrains require that the majority of atoms be stored implicitly. We can do this for entire domains by making use of the polycrystalline structure of the materials in which we are interested. This section discusses an appropriate structure for sparse storage of atoms in polycrystals. Next, we discuss one artifact that can arise with such a structure, and provide an efficient way to alleviate this artifact.

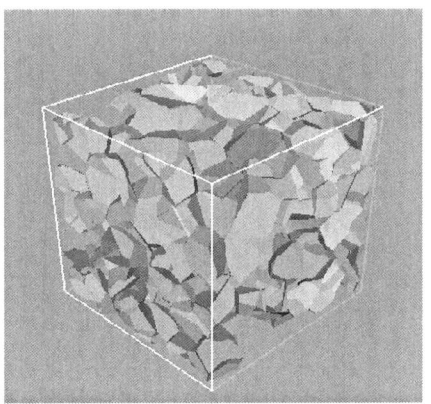 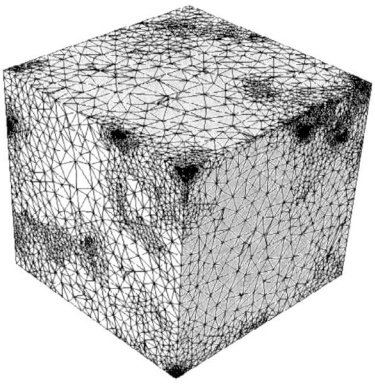

Figure 12.5: An algorithmically-defined grain structure (left) and a finite element mesh generated in that grain structure (right).

Consider the task of representing an infinite perfect crystal. A perfect crystal comprises atoms that fill space in a regular pattern. A small number of atoms defined in a unit cell can be repeated to fill space. The unit cell of a face-centered cubic crystal can be represented by three basis vectors, $\mathbf{a}$, $\mathbf{b}$, and $\mathbf{c}$, defining the geometry of the unit cell. The positions of the four atoms in the unit cell are defined by the lattice vectors, $\mathbf{L}_0$ through $\mathbf{L}_3$ [50].

With this construction, we can compute the position of every atom in a perfect crystal given only lattice and basis vectors by tessellating the cell throughout space. Furthermore, we can uniquely label every atom in the crystal with four integers, $(i, j, k; \ell)$. The first three integers, (i, j, k) indicate which unit cell contains the atom, and ℓ is a subscript indicating which atom it is in that cell. So atom $(i, j, k; \ell)$ is located at

$$\mathbf{x}(i, j, k, \ell) = i\,\mathbf{a} + j\,\mathbf{b} + k\,\mathbf{c} + \mathbf{L}_\ell. \tag{12.1}$$

For the following discussion, we distinguish between "lattice position", meaning the $(i, j, k; \ell)$ position of an atom in a lattice, and "spatial position", meaning the (x, y, z) position of the atom in 3D space. The "perfect" atomistic layout of a crystal can be represented in code as a structure consisting of the three basis vectors and a collection of lattice vectors. This represents the positions of an infinite number of atoms. Although it represents the positions of an infinite number of atoms, it uses a small, constant amount of memory and requires a small constant amount of time to determine the position of any one atom.

By providing the ability to combine base lattice definitions through basic vectors, transformations, and superposition of multiple-base lattice definition, the lattice structures of much more complex multiple element atom combinations is easily supported.

Because accessing an atom position is a constant-time operation, we can quickly go from atom to nearby atom within the lattice. For example, we can find all atoms in a neighborhood of the atom $(0,0,0;0)$ by checking the other atoms in the $(0,0,0;\ell)$ unit cell, then those in unit cell $(-1,-1,-1;\ell)$, $(0,-1,-1;\ell)$, etc.—27 unit cells in all. This is a constant-time operation, assuming a limited number of atoms per unit cell. This proves useful when using this structure to compute forces between atoms.

The above is an effective way to describe a perfect lattice, but it does not allow for an imperfect lattice, in which some atoms have moved. We can represent an imperfect lattice by building on this, explicitly storing only those atoms that have been displaced [10]. By using a hash table with an effective hash function to store the displacement of displaced atoms, we can move atoms while maintaining constant access time and while using memory proportional to the number of displaced atoms.

We can represent a finite crystal grain as an infinite crystal along with a polygonal representation of the grain's geometry, masking most of the infinite lattice. This describes a single grain using memory only for the grain geometry and for the atoms that have been displaced. A very simple way to represent a polycrystal in software, then, is as an array of crystal grains. This is sufficient to represent all of the atoms in a polycrystal but makes it difficult to find the atoms on the opposite side of an interface between grains. By also storing grain adjacency information explicitly, we can determine the atoms neighboring each atom, even when that neighbor is on the opposite side of a crystal boundary, in constant time.

12.4.3 A fast atomistic relaxation scheme

The data structure described above is realistic overall, but is problematic at grain boundaries. There, atoms can become arbitrarily close together because an atom can be arbitrarily close to the surface of its own grain, as shown in Figure 12.6. We could address this problem by relaxing all atoms, but that would involve moving all atoms, a cost of time and memory proportional to the total number of atoms in the problem. We observe that only atoms near grain boundaries are moved significantly from their ideal lattice positions. With this in mind, we can approximate the relaxed state by moving only those atoms that are near grain boundaries. Because we have an explicit representation of the grains, we know where the grain boundaries are located and so can quickly find only those atoms near grain boundaries. By contrast, if we had all n_atoms atoms listed explicitly without any grain-structure information, it would be an $O(n_\text{atoms})$ operation simply to find the problematic atoms.

In an effort to address the grain-boundary issues, we apply a simple energy-minimization algorithm to only those atoms at the surface of the grains. By leaving other atoms fixed, we are able to approximately minimize the energy of the system in time proportional to the number of atoms near grain boundaries

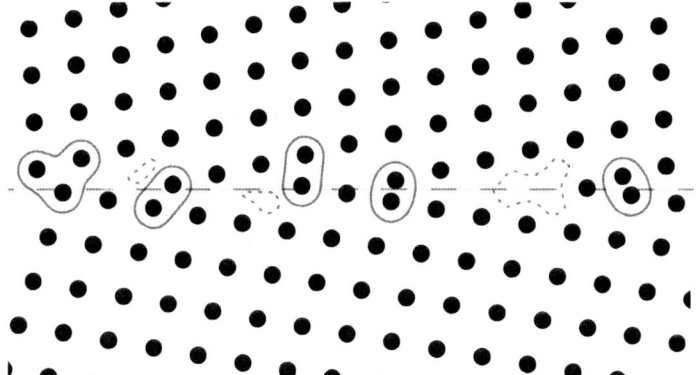

Figure 12.6: Two perfect crystals interfacing at an angle. This is not a relaxed state; there are high-energy atoms (circled solid) as well as vacancies (dashed lines), but we can approximate a minimum energy solution by moving only those atoms near the boundary (shaded area).

rather than to the total number of atoms. We consider the assumptions that underlie this algorithm, many of which are common assumptions for molecular dynamics. We can visit all n_{boundary} boundary atoms in $O(n_{\text{boundary}})$ time. For each atom in turn, we apply steepest descent to the energy of the atom as a function of its position by computing the force on each boundary atom due to its neighbors.

Our implementation of this algorithm performs as expected, scaling smoothly to millions of atoms where full relaxation scales linearly with the number of atoms. Figure 12.7 shows the time required to relax the energy of a polycrystal consisting of two cube-shaped grains relaxed using this algorithm and using LAMMPS to perform a full relaxation. We changed the size of the grains and watched the time required to minimize the energy scale. As shown, the test computer began to run out of memory with less than three million atoms. While our algorithm does not achieve the exact minimum-energy configuration, it is able to quickly move atoms that are unrealistically close together, and as a result, to remove roughly half of the energy that would be removed by full relaxation [16].

As described in [16], we applied the algorithm to a more complicated polycrystal consisting of 44 grains of copper representing a volume 1200 Å on a side and containing over 146 million atoms. Again, because our algorithm is sublinear, we were able to relax the atoms at an effective rate of 3871 atoms per processor second on a single computer as opposed to 244 atoms per processor second for full relaxation. Furthermore, full relaxation required nearly 80 gigabytes of memory and so needed a 96-processor cluster, whereas our relaxation ran on a single computer.

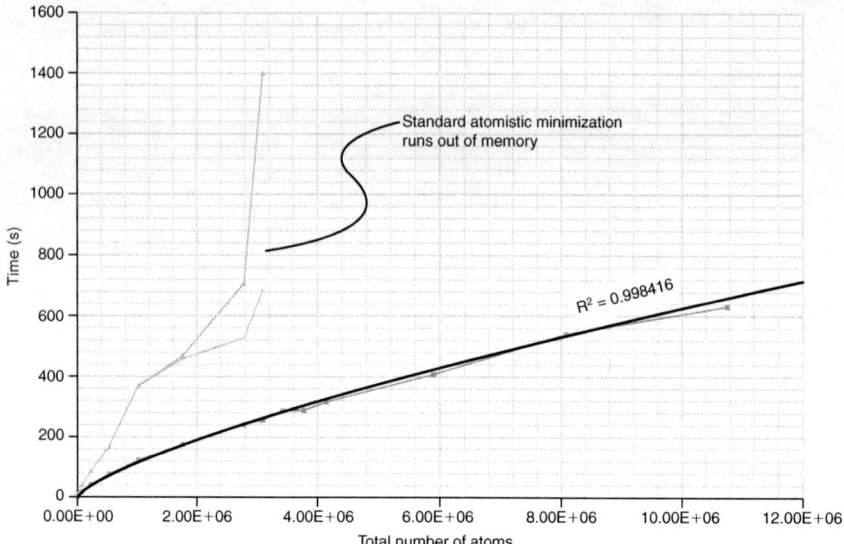

Figure 12.7: Real time and user time for one iteration of minimization of a model consisting of two cube-shaped crystal grains for LAMMPS (steep curves) and for our algorithm (lower curve). The real and user times only diverge significantly for the blue lines as the full atomistic simulation begins to run out of memory. The dark solid curve is a least-squares power fit of time $= n_{\text{atoms}}^{0.739} \cdot 0.00419\,\text{s}$ ($R^2 = 0.998416$).

12.5 A concurrent atomistic/continuum adaptive multiscale simulation tool

12.5.1 Concurrent atomistic/continuum multiscale formulation

The nucleation of defects and their interactions at the atomistic level provide a fundamental understanding of the mechanical response of materials at the nanoscale that influences the macroscale response. Crystalline material behavior is elastic in the absence of dislocation nucleation and motion. However, continuum models fail to capture inhomogeneity due to the nucleation of defects unless mixed continuum-discrete models are used [27]. To address such problems a model hierarchy consisting of an atomistic model and two continuum models is adopted [15, 33]. The finest model in the hierarchy is the atomistic model with the Embedded Atom Method (EAM) [12]-based interatomic potential. Outside the regions with defects a homogeneous elastic deformation prior to the nucleation of defects can be closely captured by a nonlinear elastic continuum model. In the critical continuum regions the third order elastic constants are found to be essential for a proper treatment of the finite deformation of crystals [30]. In the least critical regions a linear elastic continuum model is satisfactory.

In the concurrent model problem formulation the problem domain Ω shown in Figure 12.8 is composed of subdomains in which various models from the

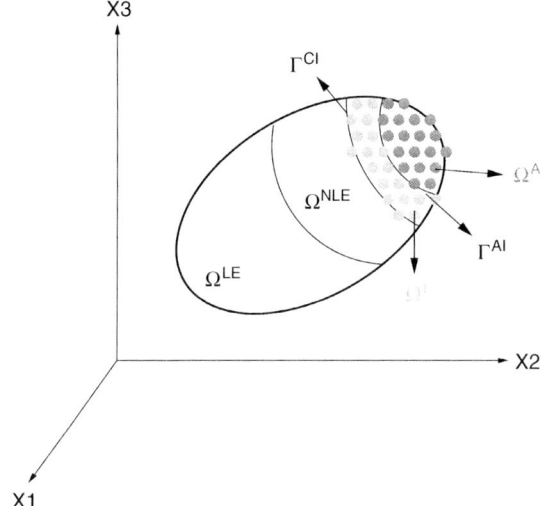

Figure 12.8: Hybrid concurrent model domain.

hierarchy are used. The linear elastic description is used in Ω^{LE}, the nonlinear description is defined for Ω^{NLE}, the EAM atomistic description is defined for Ω^A, and Ω^I is the interphase subdomain which has a blended description of the nonlinear elastic continuum and the atomistic models. $\Omega^C = \Omega^{LE} \cup \Omega^{NLE}$ is the continuum subdomain. The details of the problem formulation are given in [15]. The governing equations are standard equilibrium equations in the continuum region, the equilibrium of forces for each atom in the atomistic region and a blend of the continuum stress, and the atomistic force in the interphase region. Key points are given here.

The weak form of the equilibrium equation is stated as: Given $b_i : \Omega^C \cup \Omega^I \to \Re$, $b_{i\alpha} : \Omega^A \cup \Omega^I \to \Re$, $g_i : \Gamma_{g_i} \to \Re$, $h_i : \Gamma_{h_i} \to \Re$, Find displacements $u_i^C(\mathbf{x}) \in \mathcal{U}_i^C$ and $u_{i\alpha}^A \in \mathcal{U}_i^A$ such that $\forall w_i^C \in \mathcal{W}_i^C$ and $\forall w_{i\alpha}^A \in \mathcal{W}_i^A$

$$-\int_\Omega w_{i,j}^C \Theta^C \sigma_{ij} \, d\Omega + \int_{\Gamma_{h_i}} w_i^C \Theta^C h_i \, d\Gamma + \int_\Omega w_i^C \Theta^C b_i \, d\Omega$$

$$+ \int_\Omega \sum_\alpha^n w_{i\alpha}^A \left[\sum_\beta^{neig^\alpha} \left(\Theta_{\alpha\beta}^A f_{i\alpha\beta} \right) + \Theta_\alpha^A b_{i\alpha} \right] \delta(\mathbf{x} - \mathbf{x}_\alpha) \, d\Omega = 0 \qquad (12.2)$$

and a weak displacement compatibility Λ is satisfied as follows:

$$\Lambda \left(u_i^C(\mathbf{x}_\alpha) - u_{i\alpha}^A \right) = \sum_\alpha^{n^I} \lambda_{i\beta} \left\{ u_i^C(\mathbf{x}_\alpha) - u_{i\alpha}^A \right\} = 0 \ \ \forall \lambda_{i\beta} \in \mathcal{R}^{n^I} \qquad (12.3)$$

$\mathcal{U}_i^C$ and $\mathcal{W}_i^C$ are continuum function spaces defined as:

$$\mathcal{U}_i^C = \{u_i^C | u_i^C \in H^1, \Lambda\left(u_i^C(\mathbf{x}_\alpha) - u_{i\alpha}^A\right) = 0 \text{ on } \Omega^I, u_i^C = g_i \text{ on } \Gamma_{g_i}\} \quad (12.4)$$

$$\mathcal{W}_i^C = \{w_i^C | w_i^C \in H^1, \Lambda\left(w_i^C(\mathbf{x}_\alpha) - w_{i\alpha}^A\right) = 0 \text{ on } \Omega^I, w_i^C = 0 \text{ on } \Gamma_{g_i}\} \quad (12.5)$$

$\mathcal{U}_i^A$ and $\mathcal{W}_i^A$ belong to the discrete phase space of the atomistic system given by

$$\mathcal{U}_i^A = \{u_{i\alpha}^A | u_{i\alpha}^A \in \Re^n, \Lambda\left(u_i^C(\mathbf{x}_\alpha) - u_{i\alpha}^A\right) = 0 \text{ on } \Omega^I\} \quad (12.6)$$

$$\mathcal{W}_i^A = \{w_{i\alpha}^A | w_{i\alpha}^A \in \Re^n, \Lambda\left(w_i^C(\mathbf{x}_\alpha) - w_{i\alpha}^A\right) = 0 \text{ on } \Omega^I\} \quad (12.7)$$

$\Re^{n^I}$ in (12.3) is the phase space of atoms n^I in the interphase Ω^I; b_i and $b_{i\alpha}$ are the body forces for the continuum and atomistics, respectively; g_i and h_i are the essential and natural boundary conditions on essential boundary Γ_{g_i} and natural boundary Γ_{h_i}, respectively, where $\Gamma_{g_i} \cup \Gamma_{h_i} = \Gamma$, the domain boundary, and $\Gamma_{g_i} \cap \Gamma_{h_i} = \emptyset$. The internal force $f_{i\alpha\beta}$ acting on atom α due to the atom β is given by

$$f_{i\alpha\beta} = \frac{\partial \Phi_\beta}{\partial d_{i\alpha}} \quad (12.8)$$

where $d_{i\alpha}$ is the displacement of the atom α. Blend functions Θ^A and Θ^C are defined as:

$$\Theta_\alpha^A = 1 - \Theta^C(\mathbf{x}_\alpha)$$

$$\Theta_{\alpha\beta}^A = 1 - \frac{1}{2}\{\Theta^C(\mathbf{x}_\alpha) + \Theta^C(\mathbf{x}_\beta)\} \quad (12.9)$$

The continuum blend function $\Theta^C(\mathbf{x}) = 0$ on Ω^A, $\Theta^C(\mathbf{x}) = 1$ on Ω^C and on Ω^I $0 < \Theta^C(\mathbf{x}) < 1$, which is evaluated based on the proximity of the point $\mathbf{x} \in \Omega^I$ to the boundaries Γ^{CI} and Γ^{AI}. By defining $s \in [0, 1]$ as a normalized distance in the physical domain from Γ^{CI} to Γ^{AI}, $\Theta(s)$ can be approximated to be a function of the scalar parameter s; $\delta(\mathbf{x} - \mathbf{x}_\alpha)$ is the Dirac delta function at $\mathbf{x}_\alpha$.

The finite element discretization of the problem domain Ω is denoted by Ω^h. The discrete compatibility equation is constructed by choosing $\lambda_{i\beta}$ in (12.3) using piecewise constant shape functions defined to be constant over the finite element domains $\Omega^e \in \Omega^{h^I}$ as:

$$\lambda_{i\beta} = \sum_{\Omega^e \in \Omega^{h^I}} N^e \zeta_{i\beta}^e \quad (12.10)$$

where

$$N^e = \begin{cases} 1 \text{ on } \Omega^e \\ 0 \text{ elsewhere} \end{cases} \quad (12.11)$$

Substituting (12.10) and (12.11) into the (12.3) we obtain

$$\sum_{\Omega^e \in \Omega^{h^I}} \sum_{\alpha}^{n^e} \zeta_{i\beta}^e \left\{ u_i^C(\mathbf{x}_\alpha) - u_{i\alpha}^A \right\} = 0 \qquad (12.12)$$

Requiring the arbitrariness of $\zeta_{i\beta}^e$ yields the following discrete compatibility equation for every element Ω^e, n^e being the number of atoms in an element Ω^e:

$$\Lambda^h \left\{ u_i^h(\mathbf{x}_\alpha) - u_{i\alpha}^A \right\} = \sum_{\alpha}^{n^e} \left\{ N_B(\mathbf{x}_\alpha) d_{iB}^C - u_{i\alpha}^A \right\} = 0 \quad \forall \Omega^e \qquad (12.13)$$

Equation (12.13) yields a number of constraint equations equal to the number of spatial dimension for each finite element $\Omega^e \in \Omega^{h^I}$. From (12.13) the DOF, of one atom in Ω^e can be expressed in terms of the DOF of the finite element Ω^e and the DOF of the remaining atoms in the element. Note that at least one atom has to be positioned with an element Ω^e in the interphase.

The continuum displacement and test functions defined over $\Omega^C \cup \Omega^I$ are discretized using C^0 continuous finite element shape functions. The discretized displacement is denoted by $u_i^h \in \mathcal{U}_i^h$ and the discretized test function is denoted by $w_i^h \in \mathcal{W}_i^h$. The spaces $\mathcal{U}_i^h$ and $\mathcal{W}_i^h$ are given by

$$\mathcal{U}_i^h = \left\{ u_i^h | u_i^h = N_B d_{iB}^C, \Lambda^h \left(u_i^h(\mathbf{x}_\alpha) - u_{i\alpha}^A \right) = 0 \text{ on } \Omega^{h^I}, u_i^h = g_i \text{ on } \Gamma_{g_i}^h \right\} \qquad (12.14)$$

$$\mathcal{W}_i^h = \left\{ w_i^h | w_i^h = N_B a_{iB}^C, \Lambda^h \left(w_i^h(\mathbf{x}_\alpha) - w_{i\alpha}^A \right) = 0 \text{ on } \Omega^{h^I}, w_i^h = 0 \text{ on } \Gamma_{g_i}^h \right\} \qquad (12.15)$$

where N_B are the finite element shape functions associated with the finite element nodes B, d_{iB}^C are the nodal DOF, and a_{iB}^C are the nodal multipliers corresponding to test functions. Summation convention over repeated index B is employed. The discretized system of equations shown below is obtained by using (12.14) and (12.15) in the equilibrium equation (12.2)

$$-\int_{\Omega^h} N_{A,j} \Theta^C \sigma_{ij} \, d\Omega + \int_{\Gamma_{h_i}^h} N_A \Theta^C h_i \, d\Gamma + \int_{\Omega^h} N_A \Theta^C b_i \, d\Omega$$

$$+ \sum_{\alpha}^{n^m} \left\{ \sum_{\beta}^{neig^\alpha} (\Theta_{\alpha\beta}^A f_{i\alpha\beta}) + \Theta_\alpha^A b_{i\alpha} \right\} = 0 \qquad (12.16)$$

where n^m is the number of independent atoms. Equation (12.16) is a nonlinear system of equations in continuum DOF and independent atomistic DOF. Assuming no follower forces and no body forces, (12.16) can be written in terms of residuals as

$$r_{kP} = -\int_{\Omega^h} N_{A,j}\Theta^C \sigma_{ij}\, d\Omega + \sum_\alpha^{n^m} \left\{ \sum_\beta^{neig^\alpha} \Theta^A_{\alpha\beta} f_{i\alpha\beta} \right\} = 0 \qquad (12.17)$$

σ_{ij} is expressed in terms of the continuum DOFs according to appropriate constitutive equations and $f_{i\alpha\beta}$ is expressed in terms of the independent atomistic DOFs [15]. The nonlinear system of (12.17) can then be solved either by the newton method or by conjugate gradient minimization of the residuals. Several options for the blend function $\Theta^C(\mathbf{x})$ are discussed in [15].

12.5.2 Analysis components for multiscale simulation tool

12.5.2.1 Atomistic component

LAMMPS (Large-scale Atomic/Molecular Massively Parallel Simulator) is a molecular dynamics package developed at Sandia National Laboratories. It solves molecular dynamics and statics problems and scales well to parallel computer clusters with over 60,000 processors [40]. We chose LAMMPS for this project for those reasons and because it already included a library interface, it scales well in parallel, it was written with solid-mechanics applications in mind, and it is written in C++ [41]. LAMMPS required some modification to work well as a multiscale component, but this included only a few key features to provide access to LAMMPS's internal state. By carefully making a few small additions, we were able to effectively use LAMMPS as a component within a multiscale solver without weighing down LAMMPS with any changes that pertain explicitly to multiscale simulation. Rather, each change expanded LAMMPS's flexibility in general.

One important change was to replace the file operations with functions. This allowed us to easily request the locations of atoms, for example, without requiring all atom positions to be accessed and without going to disk. This also has advantages in parallel in that initializing LAMMPS with atoms went from an $O(n_\text{atoms} + n_\text{CPU})$ operation with a large constant to an $O(n_\text{atoms}/n_\text{CPU})$ operation with a small constant. We no longer require any temporary files and all of our interaction with LAMMPS that occurs at each iteration is now $O(n_\text{atoms}+n_\text{CPU})$. For more details, see [16].

The changes made to LAMMPS were few and self-contained. A separate multiscale version of LAMMPS was not created; rather, the developers of LAMMPS improved its API, adding new features that can be used for nonmultiscale applications. As a case study, the changes we made demonstrate the effectiveness of exposing the domain, model, and fields of a simulation.

12.5.2.2 Continuum component

The continuum-scale component started from an object-oriented framework, named Trellis [6], for automated adaptive continuum simulations. By maintaining a strong separation between the mathematical, computational, and numerical descriptions of the problem being solved, Trellis supports the continuum finite element component and provides an overall structure for the control of the multiscale simulation processes.

To more clearly demonstrate extending Trellis into the multiscale component tools kit (MCTK) consider the domain interactions. MTCK uses an API based on topological entities and their adjacencies for both the explicit and computational domains within the adaptive concurrent multiscale analysis process. At the mathematical model level the domain is subdivided into a linear and nonlinear regions, an atomistic region and an interface region between the atomistic and continuum regions where both models exist. The detailed specification of these regions is actually by the assignment of groups of finite elements for which the various models are applied (in a discretized manner in the case of the continuum models). Thus, the actual implementation of the adaptive model selection is based on an element-level assignment that, since the mesh maintains a topological hierarchy, makes it easy to define and properly control the boundaries between the different model regions and to apply specific functions over them such as the blending of the atomistic and continuum models over the elements in the interface region [15]. Since the mesh maintains its association to the CAD solid model of the domain, and the mesh modification procedures interact directly with that model, the mesh's representation can be adapted for any of the models in such a way that as the mesh is refined, and its approximation to the geometry can be automatically improved.

Within the continuum components of MCTK there is a strong separation of the modules that execute the various transformations associated with converting a computational model into the computational system. For example, the element shape functions are defined independently of the integral terms defined by the selected weak form, which is strongly separated from the numerical integration schemes used to evaluate the resulting discretized integrals. This maximizes the ability to re-use codes and encourages the application of alternative methods (e.g., hierarchic vs. low order shape functions, alternative Petrov-Galerkin forms, etc.)

12.5.3 Adaptive solution procedure

In order to use this hierarchy of models effectively in a concurrent multiscale problem setup, we need an automated technique to identify subdomains of the problem domain to properly apply different computational models. Two error indicators are adopted to facilitate the adaptive refinement and adaptive coarsening of the models in the hierarchy [33]. The relative error in the energy computed by the two continuum models over an element Ω^e of a finite element discretization is used as an indicator between the nonlinear model and the linear model. A stress gradient-based dislocation nucleation criterion [26] is adopted as the atomistic indicator. See reference [33] for more information on these error indicators. The sequence of steps in the adaptive procedure is as follows:

1. Start the load step τ of a quasistatically-loaded problem with a coarse model (note: scale the load to a value where more than just a linear elastic model is needed to solve the problem).
2. Increment the load step $\tau = \tau + 1$.
3. Solve the problem with the given concurrent model problem formulation.
4. Compute appropriate error indicators for the model hierarchy chosen.
5. Select or deselect the models from the hierarchy for subdomains of the problem domain according to the error indicators and corresponding tolerance values.
6. If models change go to step 3, else continue with step 2 till the end of load steps.

12.5.4 An example result

The example of a block of material with four spherical nanovoids is subjected to hydrostatic tensile load is considered. When the nanovoids are close to each other they begin to interact faster and tend to coalesce. Each void is $\sim 35\,\text{Å}$ in diameter and the block is $\sim 625\,\text{Å}$ cube. The load is applied quasistatically through Dirichlet boundary conditions in increments of $0.4665\,\text{Å}$ of each face of the cube. The simulation is run adaptively as indicated above. The left portion of Figure 12.9 shows a concurrent model with different model subdomains during an adaptive simulation, while the right portion shows the high energy atoms that form stacking fault tetrahedra around voids.

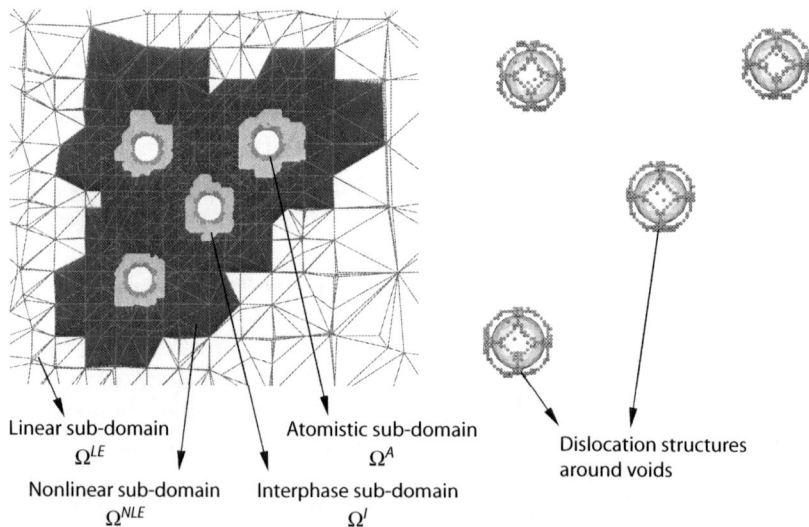

Linear sub-domain Ω^{LE}
Nonlinear sub-domain Ω^{NLE}
Atomistic sub-domain Ω^{A}
Interphase sub-domain Ω^{I}
Dislocation structures around voids

Figure 12.9: The concurrent model and the dislocation structures around voids–load step 20 (trace of strain = 0.0945).

Figure 12.10: Advanced dislocation structure around voids—load step 28 (trace of strain = 0.126).

The results shown in Figure 12.9 indicate the formation of dislocation loops around the voids in the presence of hydrostatic load, their growth and reaction to form Lomer-Cottrell junctions and formation of stacking fault tetrahedra around the voids. At larger loads these start interacting with each other, new dislocations nucleate from the voids, and the overall dislocation configuration becomes much more complicated as shown in Figure 12.10.

12.6 Acknowledgments

We acknowledge the support of NSF for grants 0303902 for "NIRT: Modeling and Simulation Framework at the Nanoscale: Application to Process Simulation, Nanodevices and Nanostructured Composites" and 0310596 for "Multiscale Systems Engineering for Nanocomposites". We also acknowledge the support of the DOE program "A Mathematical Analysis of Atomistic to Continuum Coupling Methods".

References

[1] Ainsworth M. and Oden J. T. (2000). *A Posteriori Error Estimation in Finite Element Analysis*, John Wiley & Sons, Inc.
[2] ASME (2006). *Guide for Verification and Validation in Computational Solid Mechanics*, ASME, New York, NY, V&V 10 – 2006.
[3] Babuska I., ed. (1989). "Adaptive Mathematical Modeling", *SIAM*.

[4] Babuska I. and Oden J. T. (2004). "Verification and validation in computational engineering and science: basic concepts", *Comp. Meth. Appl. Mech. Engng*, **193**, p. 457–466.

[5] Babuska I. and Strouboulis T. (2001). *The Reliability of the FE Method*, Oxford Press.

[6] Beall M. W. and Shephard M. S. (1999). "An object-oriented framework for reliable numerical simulations", *Engineering with Computers*, **15**(1) p. 61–72.

[7] Beall M. W. and Shephard M. S. (1997). "A general topology-based mesh data structure", *International Journal for Numerical Methods in Engineering*, **40**, p. 1573–1596.

[8] Beall M. W., Walsh J., and Shephard M. S. (2004). "A comparison of techniques for geometry access related to mesh generation", *Engineering with Computers*, **20**, p. 210–221.

[9] Belytschko T. and Xiao S. P. (2003). "Coupling methods for continuum model with molecular model", *International Journal for Multiscale Computational Engineering*, **1**(1) p. 115–126.

[10] Cormen T. H. (2001). *Introduction to Algorithms*, MIT Press.

[11] Datta D. K., Picu C. R., and Shephard M. S. (2004). "Composite grid atomistic continuum method: An adaptive approach to bridge continuum with atomistic analysis", *International Journal for Multiscale Computational Engineering*, **2**, p. 401–420.

[12] Daw M. S. and Baskes M. I. (1984). "Embedded-atom method: Derivation and application to impurities, surfaces, and other defects in metals", *Physical Review B*, **29**(12), p. 6443–6453.

[13] de Cougny H. L. and Shephard M. S. (1999). "Parallel refinement and coarsening of tetrahedral meshes", *International Journal for Numerical Methods in Engineering*, **46**, p. 1101–1125.

[14] E W. and Engquist B. (2003). "The heterogeneous multiscale methods", *Communications in Mathematical Sciences*, **1**(1), p. 87–133.

[15] Fish J., Nuggehally M. A., Shephard M. S., Picu C. R., Badia S., Parks M. L., and Gunzburger M. (2007). "Concurrent atc coupling based on a blend of the continuum stress and the atomistic force", *Computer Methods in Applied Mechanics and Engineering*, **196**, p. 4548–4560.

[16] FrantzDale B. (2007). *Software Design for Multiscale Engineering Simulation*, Master's thesis, Mechanical Engineering, Rensselaer Polytechnic Institute, 110 8th Street, Troy, NY 12180.

[17] Hoffmann C. M. and Joan-Arinyo R. (1998). "Cad and the product master model", *Computer-Aided Design*, **30**(11), p. 905–918.

[18] Hoffmann C. M. and Joan-Arinyo R. (2000). "Distributed maintenance of multiple project views", *Computer-Aided Design*, **32**, p. 421–431.

[19] Kevrekidis I. G., Gear C. W., Hyman J. M., Kevrekidis P. G., Runborg O., and Theodoropoulos C. (2003). "Equation-free multiscale computation:

enabling microscopic simulators to perform system-level tasks", *Communications in Mathematical Sciences*, **1**(4), p. 715–762.
[20] Knap J. and Ortiz M. (2001). "An analysis of the quasicontinuum method", *J. Mechanics and Physics of Solids*, **49**, p. 1899–1923.
[21] LAMMPS home page, http://lammps.sandia.gov/.
[22] Li X., Shephard M. S., and Beall M. W. (2003). "Accounting for curved domains in mesh adaptation", *International Journal for Numerical Methods in Engineering*, **58**, p. 247–276.
[23] Lorensen W. E. and Cline H. E. (1987). "Marching cubes: A high resolution 3d surface construction algorithm", *Computer Graphics (Proceeding of SIGGRAPH '87)*, **21**(4), p. 163–169.
[24] Mathmol home page, http://www.nyu.edu/pages/mathmol/software.htm.
[25] Mavrantzas V. G., Boone T. D., Zervopoulou E., and Theodorou D. N. (1999). "End-bridging monte carlo: An ultrafast algorithm for atomistic simulation of condensed phases of long polymer chains", *Macromolecules*, **32**, p. 5072–5096.
[26] Miller R. E. and Acharya A. (2004). "A stress-gradient based criterion for dislocation nucleation in crystals", *Journal of the Mechanics and Physics of Solids*, **52**, p. 1507–1525.
[27] Miller R. E., Shilkrot L. E., and Curtin W. A. (2004). "A coupled atomistics and discrete dislocation plasticity simulation of nanoindentation into single crystal thin films", *Acta Materialia*, **52**, p. 271–284.
[28] Miller R. E. and Tadmor E. B. (2002). "The quasicontinuum method: Overview, applications and current directions", *Journal of Computer-Aided Materials Design*, **9**(3), p. 203–239.
[29] MolecularModeling home page, http://www.msi.umn.edu/user_support/software/molecularmodeling.html.
[30] Musgrave M. J. P. (1970). *Crystal Acoustics. Introduction to the study of elastic waves and vibrations in crystals*, Holden-Day.
[31] Network for Computational Nanotechnology home page, http://ncn.purdue.edu/.
[32] Noort A., Hoek G. F. M., and Bronsvoort W. F. (2002). "Integrated part and assembly modeling", *Computer-Aided Design*, **34**, p. 899–912.
[33] Nuggehally M. A., Shephard M. S., Picu C. R., and Fish J. (2007). "Adaptive model selection procedure for concurrent multiscale problems", Accepted.
[34] O'Bara R. M., Beall M. W., and Shephard M. S. (2002). "Attribute management system for engineering analysis", *Engineering with Computers*, **18**, p. 339–351.
[35] Oden J. T., Babuska I., Nobile, Y. F., Feng, and Tempone R. (2005). "Theory and methology for estimation and control of errors due to modeling, approximation and uncertainty", *Comp. Meth. Appl. Mech. Engng.*, **194**, p. 195–204.

[36] Oden J. T., Bauman P. T., and Prudhomme S. (2005). "On the extension of goal-oriented error estimation and hierarchical modeling to discrete lattice models", *Comp. Meth. Appl. Mech. Engng.*, **194**(34–35), p. 3668–3688.

[37] Oden J. T. and Prudhomme S. (2001). "Goal-oriented error estimation and adaptivity for the finite element method", *Comp. Meth. Appl. Mech. Engng.*, **41**(5–6), p. 735–756.

[38] Padding J. T. and Briels W. J. (2002). "Time and length scales of polymer melts studied by coarse-grained molecular dynamics simulations", *J. Chem. Phys.*, **117**(2), p. 925.

[39] Parasolid home page, http://www.ugs.com/products/open/parasolid/.

[40] Plimpton S., LAMMPS benchmarks, http://lammps.sandia.gov/bench.html.

[41] Plimpton S., LAMMPS documentation, http://lammps.sandia.gov/doc/Manual.html.

[42] Prudhomme S., Bauman P. T., and Oden J. T. (2006). "Error control for molecular statics problems", *Int. J. Multiscale Computational Engineering*, **4**(56), p. 647–662.

[43] Prudhomme S. and Oden J. T. (1999). "On goal-oriented error estimation for elliptic problems: Application to the control of pointwise errors", *Comp. Meth. in Appl. Mech. and Eng.*, **176**, p. 313–331.

[44] Quasicontinuum home page, http://www.qcmethod.com/.

[45] Roache P. J. (1998). *Verification and Validation in Computational Science and Engineering*, Hermosa Publisher, Albuquerque, NM.

[46] Rollett A. D., Brahme A. P., and Roberts C. G. (2007) "An overview of accomplishments and challenges in recrystallization and grain growth", *Materials Science Forum*, **558–559**, p. 33–42.

[47] Rollett A. D., Saylor D. M., Friday J., El-Dasher, Brahme A., Lee S.-B., Cornwell C., and Noack R. (2004). "Modeling polychrystalline microstructures in 3-d", *CP712, Materials Processing and Design: Modeling, Simulation, and Applications, NUMIFORM 2004*, p. 71–77.

[48] Saylor D. M., Fridy J., El-Dasher B. S., Jung K.-Y., and Rollett A. D. (2004). "Statistically representative three-dimensional microstructures based on orthogonal observation sections", *Metallurgical and Materials Transactions*, **35A**(7), p. 1969–1979.

[49] Seol E. S. and Shephard M. S. (2006). "Efficient distributed mesh data structure for parallel automated adaptive analysis", *Engineering with Computers*, **22**(3–4), p. 197–213.

[50] Shames I. H. and Cozzarelli F. A. (1992). *Elastic and inelastic stress analysis*, Prentice-Hall.

[51] Shephard M. S. (2000). "Meshing environment for geometry-based analysis", *International Journal for Numerical Methods in Engineering*, **47**, p. 169–190.

[52] Shephard M. S., Beall M. W., O'Bara R. M., and Webster B. E. (2004). "Toward simulation-based design", *Finite Elements in Analysis and Design*, **40**(12), p. 1575–1598.

[53] Shephard M. S., Seol E. S., and FrantzDale B. (2007). *Toward a multi-model hierarchy to support multiscale simulation*. In Fishwick P. A., editor, *CRC Handbook of Dynamic System Modeling*, p. 12.1–12.18. Chapman and Hall, Boca Raton.
[54] Simmetrix home page, http://www.simmetrix.com/.
[55] Spatial Technologies home page, http://www.spatial.com/components/acis.
[56] Tadmor E. B., Ortiz M., and Phillips R. (1996). "Quasicontinuum analysis of defects in solids", *Philosophical Magazine A*, **73**(6), p. 1529–1563.
[57] Weiler K. J. (1988). "The radial-edge structure: a topological representation for non-manifold geometric boundary representations", *Geometric Modeling for CAD Applications*, p. 3–36.
[58] Wilson N., Wang K., Dutton R., Taylor C. A., and Taylor C. A. (2001). *A software framework for creating patient specific geometric models from medical imaging data for simulation based medical planning of vascular surgery*, Lecture Notes in Computer Science, **2208**, p. 449–456.
[59] Wu Z., John J. M., and Sullivan M. (2003). "Multiple material marching cubes algorithm", *International Journal for Numerical Methods in Engineering*, **58**, p. 189–207.
[60] Xiao S. P. and Belytschko T. (2004) "A bridging domain method for coupling continua with molecular dynamics", *Computer Methods in Applied Mechanics and Engineering*, **193**, p. 1645–1669.
[61] Zeus home page, http://zeus.polsl.gliwice.pl/~nikodem/linux4chemistry.html.

PART VI

SELECTED MULTISCALE APPLICATIONS

13

FINITE TEMPERATURE MULTISCALE METHODS FOR SILICON NEMS

Z. Tang and N. R. Aluru

13.1 Introduction

Rapid advances in nanotechnology have led to the fabrication of nanoscale mechanical structures with applications in nanoelectromechanical systems (NEMS) [1–3]. The design, optimization and fabrication of NEMS for various applications can be accelerated by developing accurate physical theories and computational design tools that describe the motion and operation of nanostructures [4, 5]. There are two major challenges in the physical modeling and computational analysis of nanostructures. First, when the characteristic length of NEMS scales down to several tens of nanometers, nanoscale effects, such as quantum effects, material defects and surface effects become significant. Classical theories based on continuum assumptions or the computational design tools that have been developed for micro and macro systems may not be directly applicable for nanosystems because of the small scales encountered in NEMS. Second, although the characteristic length of NEMS is often a few nanometers, the entire system could still be of the order of micrometers. Therefore, typical NEMS can still contain millions of atoms. In this case, atomistic simulation methods such as *ab initio* calculations, molecular dynamics (MD), and Monte Carlo (MC) simulations, that can be employed for accurate analysis of systems comprising several hundreds of atoms, are computationally impractical for design and optimization of practical NEMS.

To achieve the goal of accurately capturing the atomistic physics and yet retaining the efficiency of continuum models, multiscale modeling, and simulation techniques which connect and integrate the atomistic and continuum theories have recently attracted considerable research interest [6–21]. Broadly defined, there are three multiscale modeling strategies: direct coupling, top-down, and bottom-up approaches. Direct coupling methods [6–10] typically decompose the physical domain into atomistic, continuum and interface regions. Atomistic and continuum calculations are performed separately and the interface regions are used to exchange information between the atomistic and continuum regions. Top-down approaches, such as the quasicontinuum (QC) method [11–13], the bridging scale method [14], and the heterogeneous multiscale method [15], solve the continuum equations by extracting constitutive laws from the underlying atomistic descriptions. In contrast, bottom-up methods such as the coarse-grained molecular dynamics [16] and multigrid bridging approaches [17] coarse-grain the atoms of the system into macroatoms and the fundamental equations defined on the atoms are coarse-grained into equivalent macroscale equations. Of these multiscale modeling techniques, the QC approach is attractive due to its simplicity and generality. The QC method [11–13] was originally formulated using a continuum finite element framework and restricted to zero temperature. To accurately predict the mechanical behavior of nanosystems, it is necessary to take into account the effect of finite temperature. Recently, the QC approach has been extended to deal with finite temperature solid systems [18–21]. In [18] and [19], a QC Monte Carlo (QCMC) method and a QC free energy minimization (QCFEM) method were proposed to study equilibrium properties of defects at finite temperature. In these methods, a local quasiharmonic approximation [25] of the interatomic potential was used to compute the entropic energy contribution from the thermal vibration of the atoms. The entropic energy contribution was then added to the zero temperature QC energy to construct the effective energy and the free energy in the QCMC and QCFEM methods, respectively. In [20], a finite temperature QC method was proposed to investigate the thermal and mechanical properties of single-wall carbon nanotubes, where the local quasiharmonic approximation of the Brenner's potential was employed to compute the Helmholtz free energy density of carbon atoms. In [21], the QC concepts are employed to develop a coarse-grained alternative to molecular dynamics. The thermal vibrational part of the coarse-grained potential energy is obtained from the local quasiharmonic approximation.

Here, within the top-down framework, we provide an alternative approach to extend the QC method to perform mechanical analysis of nanostructures at finite temperature. In particular, we formulate a finite temperature QC method to calculate the mechanical response of nanostructures subjected to externally applied forces. We solve the continuum elasticity governing equations, but extract the material constitutive laws using an underlying atomistic description where the atoms are described by empirical interatomic potential. At finite temperature, for an isothermal system, the constitutive relations are computed by using the Helmholtz free energy density of the representative atoms. The static part of the Helmholtz free energy density is obtained directly from interatomic potential while the vibrational part (or the finite temperature part) is calculated by using the quantum-mechanical lattice dynamics. Here, we employ three quasiharmonic models, namely the real space quasiharmonic (QHM) model, the local quasiharmonic (LQHM) model, and the reciprocal space (or k-space) quasiharmonic (QHMK) model, to compute the vibrational Helmholtz free energy density. Next, the temperature and strain effects on the phonon density of states (PDOS), phonon Grüneisen parameters, and the elastic properties are calculated. By comparing the results obtained from the three quasiharmonic models with experimental and MD results, we observe that: (1) the QHM model predicts the material properties accurately; however, it is extremely inefficient when the system contains more than several hundred atoms; (2) the LQHM model is simple and efficient, but it can be inaccurate in predicting elastic constants, especially when the material is under strain; (3) the QHMK model can predict the material properties accurately and efficiently. Based on these observations, we propose to employ the QHMK model along with a semi-local approximation of the vibrational Helmholtz free energy density to extract the material properties. We then compute the mechanical response of silicon nanostructures subjected to various external loads by using the proposed finite temperature QC method. It is shown that, for silicon nanostructures larger than a few nanometers, the QHMK model predicts the mechanical response of the nanostructure accurately over a large temperature range, while the LQHM model can be inaccurate.

13.2 Finite temperature QC method

The key idea in the QC approach is to adopt the framework of continuum mechanics, but to extract the material properties and the constitutive relations from the atomistic description of the underlying local environment. Figure 13.1 illustrates the basic idea in the QC approach by using a simple example: a silicon structure subjected to a uniaxial external force. In the QC approach, the structure is described in two levels: the continuum level and the atomistic level.

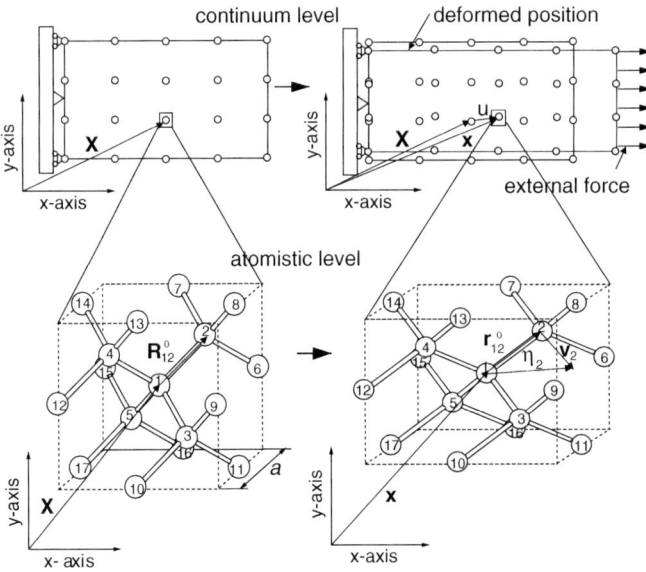

Figure 13.1: The two-level paradigm of the QC approach: at the continuum level, the structure is represented by a set of discrete nodes and each node corresponds to a large number of silicon atoms at the atomistic level. Note that although only 17 atoms are shown for the continuum node in the figure, the actual number of atoms represented by the node can be much larger. When the structure deforms due to an external force, the equilibrium position of the atoms of the underlying crystal lattice changes to balance the external force.

In the continuum level, the structure is represented by a set of discrete nodes or elements and the deformation of the structure is determined by the laws of continuum mechanics. In the atomistic level, the crystalline silicon structure consists of atoms which are connected by covalent bonds and each atom is tetrahedrally bonded to four neighboring silicon atoms, as shown in Figure 13.1. When the structure is subjected to an external force, the configuration of the silicon atoms changes to balance the external force. The change in position of the atoms in the crystal lattice manifests as deformation at the continuum level. In the QC approach, the material volume surrounding each continuum node corresponds to a large number of atoms in the atomistic scale as shown in Figure 13.1 and the covalent bonding of the atoms is described by atomistic models, e.g. empirical interatomic potentials and tight-binding descriptions [12, 26]. At the continuum level, the governing equations can be solved by using a variety of numerical methods. The deformation gradient at each continuum node is used in the underlying crystal lattice (in accordance with the Cauchy-Born rule [27]) to determine the deformed configuration of the atoms. The material constitutive relations of the continuum nodes are then extracted from the underlying crystal lattice. Based on the above description, the QC approach can be classified as a "top-down" multiscale strategy.

In the original QC method [11–13], where no temperature is considered (i.e. $T = 0$ K), the mechanical response of the atoms and, consequently, the constitutive laws of the material at the continuum nodes are solely determined by the interatomic potential energy of the atoms. In the case of finite temperature, the crystal structure is a thermodynamic system and the mechanical response of the system depends on the thermodynamic quantities such as the internal energy (for the adiabatic system) or the Helmholtz free energy (for the isothermal system) [28]. The constitutive relations for the continuum nodes are obtained from the internal energy or the Helmholtz free energy density. The internal energy or the Helmholtz free energy is extracted from the atomic lattice by using the theory of quantum-mechanical lattice dynamics. Several lattice dynamics models based on the quasiharmonic approximation of the interatomic potential are discussed in this chapter.

Before proceeding to the details on the finite temperature QC approach, we introduce the notation used here. As shown in Figure 13.1, at the continuum level, the initial and the deformed positions of a continuum node are denoted by $\mathbf{X}$ and $\mathbf{x}$, respectively. The displacement of the node is denoted by $\mathbf{u}$ with the relation $\mathbf{x} = \mathbf{X} + \mathbf{u}$. At the atomistic level, as shown in Figure 13.1, the initial and the deformed *equilibrium* position of the center atom 1, which is the representative atom corresponding to the continuum node, are denoted by $\mathbf{X}$ and $\mathbf{x}$, respectively. The vectors between the equilibrium positions of atom 1 and atom 2 in the initial and the deformed configurations are denoted by $\mathbf{R}_{12}^0$ and $\mathbf{r}_{12}^0$, respectively. At finite temperature, the atom fluctuates around its equilibrium position. The thermal vibrational displacements of atom 1 and atom 2 are denoted by $\mathbf{v}_1$ and $\mathbf{v}_2$, respectively. Note that, for clarity, the thermal vibrational displacements are only shown in the deformed configuration in Figure 13.1. The instantaneous positions of atom 1 and atom 2, denoted by $\mathbf{x}_1$ and $\mathbf{x}_2$, in the deformed configuration are given by

$$\mathbf{x}_1 = \mathbf{x} + \mathbf{v}_1, \tag{13.1}$$

$$\mathbf{x}_2 = \mathbf{x} + \mathbf{r}_{12}^0 + \mathbf{v}_2. \tag{13.2}$$

Denoting the vectors from the equilibrium position of atom 1 (in the deformed configuration) to the instantaneous position and equilibrium position of an atom α ($\alpha = 1, 2, \ldots$) by $\boldsymbol{\eta}_\alpha$ and $\boldsymbol{\eta}_\alpha^0$, respectively, we obtain

$$\boldsymbol{\eta}_\alpha = \mathbf{x}_\alpha - \mathbf{x}, \tag{13.3}$$

$$\boldsymbol{\eta}_\alpha^0 = \mathbf{r}_{1\alpha}^0. \tag{13.4}$$

Note that $\boldsymbol{\eta}_1^0 = \mathbf{r}_{11}^0 = 0$. At any instant, the distance between atom α and atom β, denoted by $r_{\alpha\beta}$, is given by

$$r_{\alpha\beta} = |\mathbf{r}_{\alpha\beta}| = |\mathbf{x}_\beta - \mathbf{x}_\alpha| = |\boldsymbol{\eta}_\beta - \boldsymbol{\eta}_\alpha|. \tag{13.5}$$

where $\mathbf{r}_{\alpha\beta}$ is the vector between the instantaneous positions of atoms α and β. At finite temperature, the vector, $\mathbf{R}_{\alpha\beta}^0$, between the equilibrium positions of atoms α and β in the initial configuration is given by

$$\mathbf{R}_{\alpha\beta}^0 = \frac{a\overline{\mathbf{R}}_{\alpha\beta}}{\overline{a}}, \tag{13.6}$$

where a is the lattice constant at a given temperature, $\overline{a}$ is the 0 K static lattice constant (i.e. neglecting thermal fluctuation), and $\overline{\mathbf{R}}_{\alpha\beta}$ is the vector between atoms α and β in the initial configuration at 0 K.

13.2.1 Continuum level description

The continuum elastostatic governing equations for an arbitrary domain Ω are given by [29]

$$\nabla \cdot (\mathbf{FS}) + \mathbf{B} = \mathbf{0} \quad \text{in } \Omega, \tag{13.7}$$

$$\mathbf{u} = \overline{\mathbf{G}} \quad \text{on } \Gamma_g, \tag{13.8}$$

$$\mathbf{P} \cdot \mathbf{N} = \overline{\mathbf{T}} \quad \text{on } \Gamma_h, \tag{13.9}$$

where $\mathbf{u}$ is the displacement vector of a continuum node from the initial configuration $\mathbf{X}$ to the deformed configuration $\mathbf{x}$ and $\mathbf{F}$ is the deformation gradient, which is given by $\mathbf{F} = \mathbf{I} + \partial \mathbf{u}/\partial \mathbf{X}$ where $\mathbf{I}$ is the identity tensor, $\mathbf{S}$ is the second Piola-Kirchhoff stress tensor, $\mathbf{N}$ is the unit outward normal vector in the initial configuration, $\mathbf{B}$ is the body force vector per unit undeformed volume, $\overline{\mathbf{G}}$ is the prescribed displacement vector on the boundary portion Γ_g, $\bar{\mathbf{T}}$ is the surface traction vector per unit undeformed area on the boundary Γ_h, and $\mathbf{P}$ is the first Piola-Kirchhoff stress tensor given by $\mathbf{P} = \mathbf{FS}$. In continuum mechanics, the second Piola-Kirchhoff stress $\mathbf{S}$ can be described by

$$\mathbf{S} = \frac{dW}{d\mathbf{E}}, \tag{13.10}$$

where W is the strain energy density function and $\mathbf{E}$ is the Green-Lagrange strain tensor, which is given by $\mathbf{E} = \frac{1}{2}(\mathbf{F}^T\mathbf{F} - \mathbf{I})$. Note that (13.10) is the general form of the constitutive law describing the material response to external forces. Depending on the form of the strain energy density function, different classes of materials can be defined. In classical continuum mechanics, analytical formulations of the strain energy density function have been developed for linear and nonlinear elastic, hyperelastic, and other types of materials [29]. Given a strain energy density function, the elastic constants can be obtained as

$$\mathbf{C}_{ijkl} = \frac{d\mathbf{S}_{ij}}{d\mathbf{E}_{kl}} = \frac{d^2W}{d\mathbf{E}_{ij}d\mathbf{E}_{kl}}, \quad i,j,k,l = 1,2,3. \tag{13.11}$$

The divergence of the first Piola-Kirchhoff stress tensor in the governing equation, (13.7), is thus given by

$$\nabla \cdot \mathbf{P} = \frac{\partial \mathbf{P}_{ij}}{\partial \mathbf{X}_j} = \frac{\partial \mathbf{F}_{ik}}{\partial \mathbf{X}_j}\mathbf{S}_{kj} + \mathbf{F}_{ik}\frac{\partial \mathbf{S}_{kj}}{\partial \mathbf{X}_j} = \frac{\partial \mathbf{F}_{ik}}{\partial \mathbf{X}_j}\mathbf{S}_{kj} + \mathbf{F}_{ik}\mathbf{C}_{kjlm}\frac{\partial \mathbf{E}_{lm}}{\partial \mathbf{X}_j},$$

$$i,j,k,l,m = 1,2,3. \tag{13.12}$$

Equations (13.7–13.12) provide a description of the elastostatic behavior of solid structures at the continuum level. For nanostructures, however, the use of continuum constitutive laws can be questionable, since both the form and the material properties in the constitutive laws can vary spatially due to material inhomogeneities and device geometry. To overcome this limitation, in the QC approach, the stress-strain relations, (13.10), are obtained directly from the underlying crystal lattice, i.e., by using an atomistic level description of the structure.

13.2.2 *Atomistic level description*

We use the Tersoff potential model [24] for microscopic description of silicon in this work. The Tersoff potential has been used extensively to investigate the energetics and elastic properties of crystalline silicon [30]. In the Tersoff empirical potential model, the total lattice potential energy is expressed as the sum of local many-body interactions. The total potential energy U of the system is given by

$$U = \sum_{\alpha} U_\alpha = \frac{1}{2}\sum_{\alpha \neq \beta} V_{\alpha\beta}, \tag{13.13}$$

where α and β are the atom indices, U_α is the potential energy of atom α, $V_{\alpha\beta}$ is the bond energy between atoms α and β, and is given by

$$V_{\alpha\beta} = f_C(r_{\alpha\beta}) \left[A e^{-\lambda_1 r_{\alpha\beta}} - b_{\alpha\beta} B e^{-\lambda_2 r_{\alpha\beta}} \right], \tag{13.14}$$

where $r_{\alpha\beta}$ is the distance between atoms α and β, A is the coefficient of the repulsive term, B is the coefficient of the attractive term, λ_1 and λ_2 are constants, f_C is the cut-off function which has the form

$$f_C(r) = \begin{cases} 1 & r < R - D \\ \frac{1}{2} - \frac{1}{2}\sin\left[\frac{\pi}{2}(r-R)/D\right] & R - D \leq r \leq R + D \\ 0 & r > R + D \end{cases}, \tag{13.15}$$

where R and D are constants, the function $b_{\alpha\beta}$ is a measure of the bond order which describes the dependence of the bond-formation energy on the local atomic arrangement due to the presence of other neighboring atoms, and is given by

$$b_{\alpha\beta} = \left(1 + \mu^n \zeta_{\alpha\beta}^n\right)^{-\frac{1}{2n}}, \tag{13.16}$$

$$\zeta_{\alpha\beta} = \sum_{\gamma \neq \alpha, \beta} f_C(r_{\alpha\gamma}) g(\cos\theta_{\alpha\beta\gamma}) e^{\lambda_3^3 (r_{\alpha\beta} - r_{\alpha\gamma})^3}, \tag{13.17}$$

$$g(\cos\theta) = 1 + c^2/d^2 - c^2/\left[d^2 + (h - \cos\theta)^2\right], \tag{13.18}$$

where γ denotes an atom other than α and β, $\zeta_{\alpha\beta}$ is called the effective coordination number, $\theta_{\alpha\beta\gamma}$ is the bond angle between the bonds $\alpha\beta$ and $\alpha\gamma$, and $g(\cos\theta)$ is the stabilization function of the tetrahedral structure. The remaining variables are constants. While several different parameter sets have been presented for the Tersoff model, the parameter values from [24] are used in this work.

13.2.3 QC method at classical zero temperature

Here, in the continuum level description of a silicon nanostructure, the structure is represented by a set of discrete continuum nodes and a continuum mechanical description of the structure is given by quantities such as the stress, **S**, and the strain, **E**, which are functions of the deformation gradient **F**. From a microscopic point of view, each continuum node is associated with an underlying crystal lattice consisting of silicon atoms bonded in the form of a diamond structure. When the macroscale structure deforms, the underlying lattice deforms accordingly. The interactions among the atoms in the underlying lattice are described by the Tersoff model, which takes into account the actual atom configuration at a given location and provides a description of the material behavior. To connect the macroscopic level description of the continuum nodes with the atomistic description of the underlying crystal lattice, we employ the hypotheses of Cauchy-Born rule [27], which states that the crystal lattice surrounding a continuum node is homogeneously distorted according to the deformation gradient at the continuum node. In addition, there exist additional inner displacements for the two interpenetrating FCC Bravais lattices (denoted by B_1 and B_2) of silicon crystal. For example, atoms 1 and 6–17 shown in Figure 13.1 belong to one Bravais lattice and atoms 2–5 belong to the other Bravais lattice. Here, the inner displacements associated with B_1 and B_2 are denoted by $\boldsymbol{\xi}_1$ and $\boldsymbol{\xi}_2$, respectively. The Cauchy-Born rule can be expressed as

$$\mathbf{r}^0_{\alpha\beta} = \mathbf{F}\mathbf{R}^0_{\alpha\beta} + \boldsymbol{\xi}_j - \boldsymbol{\xi}_i, \quad \alpha \in B_i; \quad \beta \in B_j; \quad i,j = 1,2, \qquad (13.19)$$

where $\boldsymbol{\xi}_i$, $i = 1,2$, are the additional inner displacements of the two Bravais lattices which can be determined by the energy minimization for a given deformation gradient $\mathbf{F}$. Since the inner displacements $\boldsymbol{\xi}_1$ and $\boldsymbol{\xi}_2$ are relative displacements between the two Bravais lattices, in order to rule out rigid-body translations we fix the lattice by setting $\boldsymbol{\xi}_2 = 0$. Therefore, $\boldsymbol{\xi}_2$ can be simply discarded. To simplify the notation, $\boldsymbol{\xi}_1$ is denoted as $\boldsymbol{\xi}$ in the rest of the chapter. The Cauchy-Born rule can then be rewritten as

$$\mathbf{r}^0_{\alpha\beta} = \mathbf{F}\mathbf{R}^0_{\alpha\beta} + \boldsymbol{\xi}, \quad \text{if } \alpha \in B_2 \text{ and } \beta \in B_1, \qquad (13.20)$$

$$\mathbf{r}^0_{\alpha\beta} = \mathbf{F}\mathbf{R}^0_{\alpha\beta} - \boldsymbol{\xi}, \quad \text{if } \alpha \in B_1 \text{ and } \beta \in B_2, \qquad (13.21)$$

$$\mathbf{r}^0_{\alpha\beta} = \mathbf{F}\mathbf{R}^0_{\alpha\beta}, \quad \text{if } \alpha,\beta \in \text{same Bravais lattice}. \qquad (13.22)$$

When the temperature is 0 K and the system is considered as a classical system, there is no thermal fluctuation of the atoms (i.e., $\mathbf{v}_i = 0$, $i = 1,2,\ldots$, as shown in Figure 13.1) and the strain energy density function in (13.10) is simply the potential energy per unit volume of the silicon lattice. Due to the symmetric properties of the diamond structure silicon lattice, and as the Tersoff potential only includes the nearest neighbor interactions, the calculation of the strain energy density function and its derivatives can be limited to within a 5-atom unit cell (see Fig. 1 in [31] for the 5-atom unit cell) where, without losing generality, the four outer atoms are assumed to belong to the first FCC Bravais lattice, B_1, and the center atom belongs to the second FCC Bravais lattice, B_2. The strain energy density, thus, can be rewritten as

$$W = W_U = \frac{U_1}{V_A} = \frac{1}{2V_A}\sum_{\beta=2}^{5} V_{1\beta}, \qquad (13.23)$$

where W_U is the potential energy density, U_1 is the potential energy of the center atom which is numbered as 1, and V_A is the volume of an atom in the initial configuration, which is given by $a^3/8$, where a is the silicon lattice constant. Note that $a = \bar{a} = 5.432$ Å for Tersoff silicon at classical 0 K [30]. In (13.14–13.18), $V_{1\beta}$ is a function of the distances between the atoms at their instantaneous positions.

At zero temperature, since $\mathbf{r}_{\alpha\beta} = \mathbf{r}^0_{\alpha\beta}$, (13.14–13.18) can be combined as

$$V_{1\beta} = f_C(r^0_{1\beta}) \left\{ Ae^{-\lambda_1 r^0_{1\beta}} - Be^{-\lambda_2 r^0_{1\beta}} \right.$$

$$\left. \times \left[1 + \mu^n \left(\sum_{\gamma \neq 1,\beta} f_C(r^0_{1\gamma}) g(\cos\theta^0_{1\beta\gamma}) e^{\lambda_3^3 (r^0_{1\beta} - r^0_{1\gamma})^3} \right)^n \right]^{-\frac{1}{2n}} \right\},$$
(13.24)

where $\cos\theta^0_{1\beta\gamma} = ((r^0_{1\beta})^2 + (r^0_{1\gamma})^2 - (r^0_{\beta\gamma})^2)/(2r^0_{1\beta} r^0_{1\gamma})$. By using the Cauchy-Born rule given in (13.20–13.22), one obtains

$$r^0_{1\beta} = |\mathbf{r}^0_{1\beta}| = |\mathbf{F}\mathbf{R}^0_{1\beta} + \boldsymbol{\xi}|, \quad \beta = 2, \ldots, 5, \tag{13.25}$$

and

$$r^0_{\beta\gamma} = |\mathbf{r}^0_{\beta\gamma}| = |\mathbf{F}\mathbf{R}^0_{\beta\gamma}|, \quad \beta, \gamma = 2, \ldots, 5; \quad \gamma \neq \beta. \tag{13.26}$$

Note that the inner displacements for atoms β and γ cancel out in (13.26) since the two atoms belong to the same Bravais lattice. From (13.24–13.26), we note that $V_{1\beta}$ is a function of the atom positions in the initial configuration, the deformation gradient $\mathbf{F}$ and the inner displacement $\boldsymbol{\xi}$. The second Piola-Kirchhoff stress can then be rewritten as

$$\mathbf{S} = \frac{1}{2V_A} \frac{d}{d\mathbf{E}} \left(\sum_{\beta=2}^{5} V_{1\beta} \right) = \frac{1}{2V_A} \sum_{\beta=2}^{5} \left(\frac{\partial V_{1\beta}}{\partial \mathbf{E}} + \frac{\partial V_{1\beta}}{\partial \boldsymbol{\xi}} \frac{\partial \boldsymbol{\xi}}{\partial \mathbf{E}} \right). \tag{13.27}$$

For a given deformation gradient $\mathbf{F}$, the inner displacement $\boldsymbol{\xi}$ can be determined by minimizing the strain energy density function, i.e.,

$$\frac{1}{2V_A} \left(\sum_{\beta=2}^{5} \frac{\partial V_{1\beta}}{\partial \boldsymbol{\xi}} \right)_{\mathbf{F}} = 0. \tag{13.28}$$

Substituting (13.28) into (13.27), one obtains

$$\mathbf{S} = \frac{1}{2V_A} \sum_{\beta=2}^{5} \left(\frac{\partial V_{1\beta}}{\partial \mathbf{E}} \right) = \frac{\mathbf{F}^{-1}}{2V_A} \sum_{\beta=2}^{5} \left(\frac{\partial V_{1\beta}}{\partial \mathbf{F}} \right). \tag{13.29}$$

Substituting (13.24) into (13.29) and using the chain rule, the second Piola-Kirchhoff stress can be further rewritten as

$$\mathbf{S} = \frac{\mathbf{F}^{-1}}{2V_A} \sum_{\beta=2}^{5} \left[\frac{\partial V_{1\beta}}{\partial r^0_{1\beta}} \frac{\partial r^0_{1\beta}}{\partial \mathbf{F}} + \sum_{\substack{\gamma=2 \\ \gamma \neq \beta}}^{5} \left(\frac{\partial V_{1\beta}}{\partial r^0_{1\gamma}} \frac{\partial r^0_{1\gamma}}{\partial \mathbf{F}} + \frac{\partial V_{1\beta}}{\partial \cos\theta^0_{1\beta\gamma}} \frac{\partial \cos\theta^0_{1\beta\gamma}}{\partial \mathbf{F}} \right) \right].$$
(13.30)

The derivatives of the bond lengths $r_{1\beta}^0$, $r_{1\gamma}^0$ and the cosine of the bond angle $\cos\theta_{\alpha\beta\gamma}^0$ with respect to the deformation gradient $\mathbf{F}$ and the inner displacement $\boldsymbol{\xi}$ are given in Appendix A in [32]. The elastic constants can be computed from

$$C_{ijkl} = \frac{\partial^2 W_U}{\partial \mathbf{E}_{ij} \partial \mathbf{E}_{kl}} - \frac{\partial^2 W_U}{\partial \mathbf{E}_{ij} \partial \boldsymbol{\xi}_m} \left(\frac{\partial^2 W_U}{\partial \boldsymbol{\xi}_m \partial \boldsymbol{\xi}_n} \right)^{-1} \frac{\partial^2 W_U}{\partial \boldsymbol{\xi}_n \partial \mathbf{E}_{kl}}, \quad i,j,k,l,m,n = 1,2,3, \tag{13.31}$$

where the first term on the right hand side of (13.31) is the homogeneous part, which describes the elastic response when all the atoms are displaced homogeneously upon the application of the strain, and the second term on the right hand side of (13.31) is the inhomogeneous part associated with the inner displacement $\boldsymbol{\xi}$ between the two FCC Bravais lattices under a uniform strain. As shown in Appendix B in [32], (13.31) can be rewritten as

$$C_{ijkl} = \mathbf{F}_{in}^{-1} \frac{\partial^2 W_U}{\partial \mathbf{F}_{nj} \partial \mathbf{F}_{mk}} \mathbf{F}_{lm}^{-1} - \mathbf{F}_{in}^{-1} \mathbf{F}_{km}^{-1} \frac{\partial W_U}{\partial \mathbf{F}_{mj}} \mathbf{F}_{ln}^{-1}$$

$$- \mathbf{F}_{ip}^{-1} \frac{\partial^2 W_U}{\partial \mathbf{F}_{pj} \partial \boldsymbol{\xi}_m} \left(\frac{\partial^2 W_U}{\partial \boldsymbol{\xi}_m \partial \boldsymbol{\xi}_n} \right)^{-1} \mathbf{F}_{kq}^{-1} \frac{\partial^2 W_U}{\partial \mathbf{F}_{ql} \partial \boldsymbol{\xi}_n},$$

$$i,j,k,l,m,n,p,q = 1,2,3, \tag{13.32}$$

where the elastic constants are expressed in terms of $\mathbf{F}$ and $\boldsymbol{\xi}$ and the derivatives of the strain energy density with respect to the deformation gradient $\mathbf{F}$ and the inner displacement $\boldsymbol{\xi}$ are obtained in a straightforward manner by using the results in Appendix A in [32]. After all the strain and stress tensors and their derivatives are calculated, the governing equations (13.7)–(13.9) can be solved either by constructing a weak form (as is typically done in the finite element method) or by collocating the governing equations at the continuum nodes (as is typically done in a finite difference method or in a collocation meshless method [33]).

13.2.4 *Finite temperature formulation*

In the finite temperature case, the silicon structure is considered as a thermodynamic system. For an isothermal system, the second Piola-Kirchhoff stress tensor at a constant temperature, T, is defined by [28]

$$\mathbf{S} = \left(\frac{dW_A}{d\mathbf{E}} \right)_T, \tag{13.33}$$

where the strain energy density function, W, in (13.10) is replaced by the Helmholtz free energy density function W_A. The Helmholtz free energy is calculated by using the theory of lattice dynamics with a quasiharmonic approximation of the interatomic potential. In a harmonic approximation, the Tersoff potential function is written in a quadratic form by neglecting the higher-order (>2) terms in its Taylor's series expansion. The total potential energy of a system of N atoms at any instantaneous position can thus be rewritten as

$$U(\boldsymbol{\eta}_1,\ldots,\boldsymbol{\eta}_N) = U(\boldsymbol{\eta}_1^0,\ldots,\boldsymbol{\eta}_N^0)$$
$$+ \frac{1}{2} \sum_{\beta,\gamma=1}^{N} \sum_{j,k=1}^{3} \left. \frac{\partial^2 U(\boldsymbol{\eta}_1,\ldots,\boldsymbol{\eta}_N)}{\partial \eta_{\beta j} \partial \eta_{\gamma k}} \right|_{\boldsymbol{\eta}_1,\ldots,\boldsymbol{\eta}_N = \boldsymbol{\eta}_1^0,\ldots,\boldsymbol{\eta}_N^0} v_{\beta j} v_{\gamma k}, \tag{13.34}$$

where $\eta_{\beta j}$ and $\eta_{\gamma k}$ are the jth and kth component of the relative position of atoms β and γ, respectively, and $\mathbf{v}_{\beta j}$ and $\mathbf{v}_{\gamma k}$ are the jth and kth component of the thermal vibrational displacement of atoms β and γ, respectively. Denoting $\boldsymbol{\eta} = (\boldsymbol{\eta}_1,\ldots,\boldsymbol{\eta}_N)$ and $\boldsymbol{\eta}^0 = (\boldsymbol{\eta}_1^0,\ldots,\boldsymbol{\eta}_N^0)$, (13.34) can be rewritten in a matrix form as

$$U(\boldsymbol{\eta}) = U(\boldsymbol{\eta}^0) + \tfrac{1}{2} \mathbf{v}^T \boldsymbol{\Phi} \mathbf{v}, \tag{13.35}$$

where $\mathbf{v} = (\mathbf{v}_1,\ldots,\mathbf{v}_N)^T$ and $\boldsymbol{\Phi}$ is the $3N \times 3N$ force constant matrix given by

$$\Phi_{3\beta+j-3, 3\gamma+k-3} = \left. \frac{\partial^2 U(\boldsymbol{\eta})}{\partial \eta_{\beta j} \partial \eta_{\gamma k}} \right|_{\boldsymbol{\eta}=\boldsymbol{\eta}^0}, \quad \beta,\gamma = 1,\ldots,N; \; j,k = 1,2,3. \tag{13.36}$$

Note that $\boldsymbol{\eta}^0$ is a function of lattice constant a, the deformation gradient $\mathbf{F}$, and the inner displacement $\boldsymbol{\xi}$, as shown in (13.4, 13.6, 13.20–13.22). In a quasi-harmonic approximation, the lattice constant is a function of temperature. For a crystal lattice, the normal vibrational frequencies can be computed from the force constant matrix by using the theory of lattice dynamics. Once the vibrational frequencies are obtained, the Helmholtz free energy A can be readily computed as [27]

$$A = U^0 + \frac{1}{2} \sum_n \hbar \omega_n + k_B T \sum_n \ln\left(1 - e^{-\frac{\hbar \omega_n}{k_B T}}\right), \tag{13.37}$$

where $U^0 \equiv U(\boldsymbol{\eta}^0)$ denotes the total potential energy evaluated at the equilibrium position of the system, $\hbar$ is the reduced Planck's constant, k_B is the Boltzmann constant, and ω_n is the vibrational frequency of the nth normal mode of the crystal lattice. Note that on the right hand side of (13.37), the first term is the static potential energy, the second term is the quantum-mechanical zero point energy, and the sum of the second and the third terms is the vibrational Helmholtz free energy. The Helmholtz free energy density for an atom α is given by

$$W_A = \frac{1}{V_A}\left\{U_\alpha^0 + \frac{1}{N}\sum_n \left[\frac{1}{2}\hbar\omega_n + k_B T \ln\left(1 - e^{-\frac{\hbar\omega_n}{k_B T}}\right)\right]\right\}, \tag{13.38}$$

where V_A is the volume of atom α in the initial configuration, and $U_\alpha^0 \equiv U_\alpha(\boldsymbol{\eta}^0)$ denotes the potential energy of atom α evaluated using the equilibrium position of the system. Note that $U^0 = \sum_{\alpha=1}^{N} U_\alpha^0$.

In the quasiharmonic approximation, the elements of the force constant matrix given in (13.36) are a function of the lattice constant a, the deformation gradient $\mathbf{F}$, and the inner displacement $\boldsymbol{\xi}$. Therefore, the vibrational frequencies, ω_n, and the Helmholtz free energy density function, W_A, are also functions of a, $\mathbf{F}$, and $\boldsymbol{\xi}$. For a given temperature, T, the lattice constant, a, is first determined on the unstrained silicon crystal by

$$\left(\frac{\partial A}{\partial a}\right)_T = 0. \tag{13.39}$$

More details on the determination of the lattice constant can be found in [31]. As discussed in Section 13.2.3, the inner displacement $\boldsymbol{\xi}$ can be determined by minimizing W_A for a given deformation gradient $\mathbf{F}$, i.e.,

$$\left(\frac{\partial W_A}{\partial \boldsymbol{\xi}}\right)_{\mathbf{F}} = 0. \tag{13.40}$$

After $\boldsymbol{\xi}$ is determined, the second Piola-Kirchhoff stress is then calculated as

$$\mathbf{S} = \left(\frac{\partial W_A}{\partial \mathbf{E}} + \frac{\partial W_A}{\partial \boldsymbol{\xi}} \frac{\partial \boldsymbol{\xi}}{\partial \mathbf{E}}\right) = \frac{\partial W_A}{\partial \mathbf{E}} = \mathbf{F}^{-1} \frac{\partial W_A}{\partial \mathbf{F}}. \tag{13.41}$$

Substituting (13.38) into (13.41), we have

$$\mathbf{S} = \frac{\mathbf{F}^{-1}}{V_A} \left[\frac{\partial U_\alpha^0}{\partial \mathbf{F}} + \frac{\hbar}{N} \sum_n \left(\frac{1}{2} + \frac{1}{e^{\frac{\hbar \omega_n}{k_B T}} - 1}\right) \frac{\partial \omega_n}{\partial \mathbf{F}}\right]. \tag{13.42}$$

By defining the generalized phonon Grüneisen parameter (GPGP) [28],

$$\gamma^{(n)}(\mathbf{F}) = -\mathbf{F}^{-1} \frac{d\ln\omega_n}{d\mathbf{F}}, \tag{13.43}$$

$\mathbf{S}$ can also be rewritten as

$$\mathbf{S} = \frac{1}{V_A} \left[\mathbf{F}^{-1} \frac{\partial U_\alpha^0}{\partial \mathbf{F}} - \frac{\hbar}{N} \sum_n \left(\frac{1}{2} + \frac{1}{e^{\frac{\hbar \omega_n}{k_B T}} - 1}\right) \omega_n \gamma^{(n)}\right]. \tag{13.44}$$

The isothermal elastic constants are now given by (see Appendix B in [32]),

$$\mathbf{C}_{ijkl} = \mathbf{F}_{in}^{-1} \frac{\partial^2 W_A}{\partial \mathbf{F}_{nj} \partial \mathbf{F}_{mk}} \mathbf{F}_{lm}^{-1} - \mathbf{F}_{in}^{-1} \mathbf{F}_{km}^{-1} \frac{\partial W_A}{\partial \mathbf{F}_{mj}} \mathbf{F}_{ln}^{-1}$$

$$- \mathbf{F}_{ip}^{-1} \frac{\partial^2 W_A}{\partial \mathbf{F}_{pj} \partial \boldsymbol{\xi}_m} \left(\frac{\partial^2 W_A}{\partial \boldsymbol{\xi}_m \partial \boldsymbol{\xi}_n}\right)^{-1} \mathbf{F}_{kq}^{-1} \frac{\partial^2 W_A}{\partial \mathbf{F}_{ql} \partial \boldsymbol{\xi}_n},$$

$$i, j, k, l, m, n, p, q = 1, 2, 3. \tag{13.45}$$

The procedure for computing the derivatives of the vibrational frequencies with respect to the deformation gradient and the inner displacement, i.e., $\partial\omega_n/\partial\mathbf{F}$, and $\partial\omega_n/\partial\boldsymbol{\xi}$, is given in Appendix C in [32]. Once the isothermal elastic constants are obtained, the adiabatic elastic constants $\bar{\mathbf{C}}_{ijkl}$ can be computed by [28]

$$\bar{\mathbf{C}}_{ijkl} = \mathbf{C}_{ijkl} - \left(\frac{\partial^2 W_A}{\partial T^2}\right)^{-1} \frac{\partial^2 W_A}{\partial T \partial \mathbf{E}_{ij}} \frac{\partial^2 W_A}{\partial T \partial \mathbf{E}_{kl}}. \tag{13.46}$$

In the general procedure illustrated in (13.33–13.46), a key step is to calculate the vibrational frequencies ω_n from the force constant matrix. Several quasiharmonic models can be used to calculate the vibrational frequencies [31], including the QHM model [27], the LQHM model [25], and the QHMK model [34]. In the following sections, we describe how the vibrational frequencies are obtained with each of the models and provide expressions for the second Piola-Kirchhoff stress and other quantities that are specific to each model.

13.2.4.1 Real space quasiharmonic (QHM) model

In the QHM model, the vibrational frequencies are given by $\omega_j = \sqrt{\lambda_j/M}$, where M is the mass of the silicon atom, and λ_j ($j = 1, 2, \ldots, 3N$) are the eigenvalues computed from the $3N \times 3N$ force constant matrix $\boldsymbol{\Phi}$ which is defined in (13.36). The Helmholtz free energy for an N-atom crystal lattice is given by [28]

$$A = \sum_{\alpha=1}^{N} U_\alpha^0 + \frac{1}{2}\sum_{j=1}^{3N} \hbar\omega_j + k_B T \sum_{j=1}^{3N} \ln\left(1 - e^{-\frac{\hbar\omega_j}{k_B T}}\right). \tag{13.47}$$

The Helmholtz free energy density for an atom α is given by

$$W_A = \frac{1}{V_A}\left\{U_\alpha^0 + \frac{1}{N}\sum_{j=1}^{3N}\left[\frac{\hbar\omega_j}{2} + k_B T \ln\left(1 - e^{-\frac{\hbar\omega_j}{k_B T}}\right)\right]\right\}. \tag{13.48}$$

The second Piola-Kirchhoff stress is calculated from (13.42) as

$$\mathbf{S} = \frac{\mathbf{F}^{-1}}{V_A}\left[\frac{\partial U_\alpha^0}{\partial \mathbf{F}} + \frac{\hbar}{N}\sum_{j=1}^{3N}\left(\frac{1}{2} + \frac{1}{e^{\frac{\hbar\omega_j}{k_B T}} - 1}\right)\frac{\partial \omega_j}{\partial \mathbf{F}}\right]. \tag{13.49}$$

The elastic constants can be calculated by using (13.45, 13.46).

The QHM model directly computes all the normal modes and the vibrational frequencies of the system. While the surface effects and/or defects in a nanostructure can be readily captured in the QHM model, the number of atoms that can be considered in this approach is limited due to the fact that the eigenvalues and their derivatives of a $3N \times 3N$ system must be computed. Typically, the number of atoms N that can be considered is of the order of several hundreds. For nanostructures with a characteristic length larger than a few nanometers the QHM approach described by (13.48, 13.49) can be inefficient.

13.2.4.2 Local quasiharmonic (LQHM) model

The LQHM model has been proposed as a simple and a reasonably accurate model to compute the Helmholtz free energy of solid systems [25]. In the local quasiharmonic approximation, the coupling between the vibrations of different atoms is neglected and the atoms of the system are considered as independent harmonic oscillators. In other words, for a homogeneous system such as a perfect crystal lattice, all the atoms have the same vibrational frequencies. The force constant matrix (13.36) can be decomposed into N 3×3 local force constant matrices. The 3×3 local force constant matrix $\mathbf{\Phi}(\alpha)$ for an atom α is given by

$$\Phi_{j,k}(\alpha) = \frac{\partial^2 U_{\text{local}}(\alpha)}{\partial \eta_{\alpha j} \partial \eta_{\alpha k}}\bigg|_{\eta=\eta^o}, \quad j,k = 1,2,3, \qquad (13.50)$$

where $U_{\text{local}}(\alpha)$ is the local potential energy of atom α, which contains contributions from the first and the second nearest neighbors of center atom α. In the LQHM model, the center atom α vibrates about its equilibrium position while the surrounding atoms are considered fixed. As shown by the 17-atom cluster in Figure 13.1, the instantaneous position of the center atom ($\alpha = 1$) affects the potential energy of atoms 1-5, i.e., the potential energy U_β, $\beta = 1, \ldots, 5$, is a function of the center atom position. Therefore, $U_{\text{local}}(\alpha)$ can be calculated within a cell that includes the first and the second nearest neighbors of the center atom (i.e., 17 atoms as shown in Fig. 13.1). $U_{\text{local}}(\alpha)$ is given by

$$U_{\text{local}}(\alpha) = \sum_{\beta=1}^{5} U_\beta(\boldsymbol{\eta}_\sigma), \quad \alpha = 1; \ \sigma \in \{1, \ldots, 17\}, \qquad (13.51)$$

where $\boldsymbol{\eta}_\sigma$ (σ takes values of β and its neighbors) denotes the position of atom β and its four neighbor atoms, and U_β is the potential energy of atom β. By diagonalizing the local force constant matrix $\mathbf{\Phi}(\alpha)$, the three vibrational frequencies $\omega_{\alpha j}$ can be determined, i.e., $\omega_{\alpha j} = \sqrt{\lambda_{\alpha j}/M}$, where $\lambda_{\alpha j}$ (j=1, 2, 3) are the eigenvalues of the 3 × 3 force constant matrix $\mathbf{\Phi}(\alpha)$. The Helmholtz free energy for an atom α is given by [25]

$$A(\alpha) = U_\alpha^0 + \frac{1}{2}\sum_{j=1}^{3} \hbar\omega_{\alpha j} + k_B T \sum_{j=1}^{3} \ln\left(1 - e^{-\frac{\hbar\omega_{\alpha j}}{k_B T}}\right). \qquad (13.52)$$

The Helmholtz free energy for an N-atom crystal lattice is given by $A = \sum_{\alpha=1}^{N} A(\alpha)$. The Helmholtz free energy density function for an atom α is then given by $W_A = A(\alpha)/V_A$, and the second Piola-Kirchhoff stress for an atom α is given by

$$\mathbf{S} = \frac{\mathbf{F}^{-1}}{V_A}\left[\frac{\partial U_\alpha^0}{\partial \mathbf{F}} + \hbar \sum_{j=1}^{3}\left(\frac{1}{2} + \frac{1}{e^{\frac{\hbar\omega_{\alpha j}}{k_B T}} - 1}\right)\frac{\partial \omega_{\alpha j}}{\partial \mathbf{F}}\right]. \qquad (13.53)$$

The elastic constants can again be computed by using (13.45, 13.46).

The LQHM model is computationally attractive, as it reduces the degrees of freedom by neglecting the correlations between the vibrations of different atoms, and has been used [18–21] to extend the QC method at classical 0 K to finite temperature cases. As shown in [31] and as pointed out in Section 13.2.5.3, the LQHM model can be inaccurate in describing the thermal and elastic properties as it neglects the vibrational coupling of the atoms. For this reason, nonlocal quasiharmonic models are necessary to accurately calculate the material properties.

13.2.4.3 k-space Quasiharmonic (QHMK) Model

For a perfect crystal lattice with homogeneous deformation, the Helmholtz free energy can be computed efficiently in the reciprocal space by using the Bloch's theorem [35] with the Born-von Karman boundary condition. In the reciprocal representation, the force constant matrix defined by (13.36) can be reduced to a 6×6 dynamical matrix by using Fourier transformation as (see [31] for details)

$$\mathbf{D}(\mathbf{k}) = \frac{1}{M} \begin{bmatrix} \sum_{\beta=1}^{N} \Phi_{j,k}^{11}(\alpha,\beta) e^{i\mathbf{k}\cdot\mathbf{R}_{\beta\alpha}^{0}} & \sum_{\beta=1}^{N} \Phi_{j,k}^{12}(\alpha,\beta) e^{i\mathbf{k}\cdot(\mathbf{R}_{\beta\alpha}^{0}-\mathbf{F}^{-1}\boldsymbol{\xi})} \\ \sum_{\beta=1}^{N} \Phi_{j,k}^{21}(\alpha,\beta) e^{i\mathbf{k}\cdot(\mathbf{R}_{\beta\alpha}^{0}+\mathbf{F}^{-1}\boldsymbol{\xi})} & \sum_{\beta=1}^{N} \Phi_{j,k}^{22}(\alpha,\beta) e^{i\mathbf{k}\cdot\mathbf{R}_{\beta\alpha}^{0}} \end{bmatrix},$$

$$\alpha = 1; \quad j, k = 1, 2, 3, \tag{13.54}$$

where $\mathbf{k}$ is the wave vector and, for a given N atom silicon lattice, $\frac{N}{2}$ $\mathbf{k}$ points can be chosen to lie in the first Brillouin zone (FBZ) due to the periodicity of the reciprocal lattice [34], and $\Phi_{j,k}^{pq}(\alpha,\beta)$ is given by

$$\Phi_{j,k}^{pq}(\alpha,\beta) = \left.\frac{\partial^2 U(\boldsymbol{\eta})}{\partial \eta_{\alpha j} \partial \eta_{\beta k}}\right|_{\boldsymbol{\eta}=\boldsymbol{\eta}^0, \alpha \in B_p, \beta \in B_q},$$

$$\alpha = 1; \quad \beta = 1, \ldots, N; \quad p, q = 1, 2; \quad j, k = 1, 2, 3. \tag{13.55}$$

In the above equations, α is the center atom selected to compute the dynamical matrix, and B_p and B_q denote the pth and the qth Bravais lattices, respectively. In the calculation of the dynamical matrix, for the center atom α, atom β loops over all the atoms in the crystal lattice. However, as the Tersoff potential only includes the nearest neighbor interactions, it can be shown that $\Phi_{j,k}^{pq}(\alpha,\beta)$ has nonzero values only if the atom β is within two layers of atoms surrounding the center atom α. Therefore, the force constant matrix can be obtained by the calculation within a cell that includes the second nearest neighbors (total 17 atoms for diamond structure crystal silicon) of a center atom α, as shown by the 17-atom cluster in Figure 13.1. More details on the calculation of the dynamical matrix for Tersoff silicon can be found in [31].

The vibrational frequencies can be calculated by $\omega_s(\mathbf{k}) = \sqrt{\lambda_s(\mathbf{k})}$ where $\lambda_s(\mathbf{k})$ are the eigenvalues of the 6×6 dynamical matrix $\mathbf{D}$ and s is the index of the polarization for silicon crystal. In the reciprocal representation, the Helmholtz free energy for an N-atom crystal lattice is given by [31, 34]

$$A = \sum_{\alpha=1}^{N} U_\alpha^0 + \frac{1}{2} \sum_{\mathbf{k}} \sum_{s=1}^{6} \hbar\omega_s(\mathbf{k}) + k_B T \sum_{\mathbf{k}} \sum_{s=1}^{6} \ln\left(1 - e^{-\frac{\hbar\omega_s(\mathbf{k})}{k_B T}}\right). \quad (13.56)$$

For a bulk silicon crystal, $\mathbf{k}$ can be taken as a continuous variable and $\sum_{\mathbf{k}}$ in (13.56) can be replaced by an integral. Therefore, the Helmholtz free energy density for an atom α is given by

$$W_A = \frac{U_\alpha^0}{V_A} + \frac{1}{2V_A V_B}\left[\frac{1}{2}\int_{\mathbf{k}} \sum_{s=1}^{6} \hbar\omega_s(\mathbf{k}) d\mathbf{k} + k_B T \int_{\mathbf{k}} \sum_{s=1}^{6} \ln\left(1 - e^{-\frac{\hbar\omega_s(\mathbf{k})}{k_B T}}\right) d\mathbf{k}\right], \quad (13.57)$$

where V_B is the volume of the first Brillouin zone of the reciprocal lattice and the factor 2 is due to the two Bravais lattices of a silicon crystal. The integration domain is chosen to be one quadrant of the first Brillouin zone, which is decomposed into nine tetrahedrons as shown in Figure 13.2. The integration is carried out by using Gaussian quadrature for each tetrahedron. The first term and the second term on the right hand side of (13.57) are the static and the vibrational component of the Helmholtz free energy density, respectively. After the Helmholtz free energy density is obtained, the second Piola-Kirchhoff stress

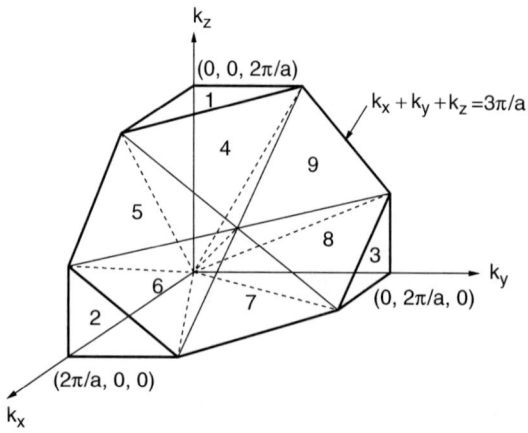

Figure 13.2: A quadrant of the first Brillouin zone is decomposed into nine tetrahedrons.

is given by

$$\mathbf{S} = \frac{\mathbf{F}^{-1}}{V_A} \left[\frac{\partial U_\alpha^0}{\partial \mathbf{F}} + \frac{\hbar}{2V_B} \sum_{s=1}^{6} \int_{\mathbf{k}} \left(\frac{1}{2} + \frac{1}{e^{\frac{\hbar \omega_s(\mathbf{k})}{k_B T}} - 1} \right) \frac{\partial \omega_s(\mathbf{k})}{\partial \mathbf{F}} d\mathbf{k} \right]. \quad (13.58)$$

The elastic constants can again be obtained by using (13.45, 13.46).

The QHMK model is mathematically equivalent to the QHM model with the Born-von Karman boundary condition (see [31] for more details). In the QHMK model, the vibrational component of the Helmholtz free energy density (the second term on the right hand side of (13.57)) for the atom corresponding to the continuum node, $\mathbf{X}$, is computed by using a bulk (*nonlocal*) silicon crystal lattice. The *nonlocal* silicon crystal lattice is, however, assumed to be subjected to a homogeneous deformation given by the *local* deformation gradient $\mathbf{F}(\mathbf{X})$. We refer to this approximation as the "semilocal" approximation to compute the vibrational free energy density. For large crystal structures where the surface effects are negligible and each continuum node represents a large number of atoms, the semilocal approximation can be an accurate and an efficient way to compute the vibrational Helmholtz free energy of the structure [31].

13.2.5 *Results and discussion*

13.2.5.1 *Lattice constants*

Lattice constant or thermal expansion as a function of temperature is a fundamental property of materials. It is well known that the classical harmonic approximation predicts no thermal expansion; therefore, the lattice constant does not change when temperature varies. Thermal expansion is indeed due to the anharmonic characteristics of the interatomic potential. As discussed in previous sections, the quasiharmonic approximation, which accounts for the dependence of the phonon frequencies on the temperature, is a simple extension of the classical harmonic approximation. We compute the zero pressure lattice parameter $a(T)$ at various temperatures (0–1500 K) by using the three quasiharmonic models. The lattice parameter $a(T)$ is obtained when the Helmholtz free energy is minimized with respect to $a(T)$. Figure 13.3 shows the comparison of the computed lattice parameter with the MD results. The values computed from the QHMK model are in good agreement with the results from the MD simulations. However, the LQHM approach overestimates, by more than 50%, the variation of the lattice parameter with temperature. The lattice parameters obtained from the QHM model with 64, 216, and 512 atoms are also in good agreement with the MD data. However, the computational cost of the QHM model is much higher than that of the QHMK approach. In addition, anharmonic effects become significant at high temperature as evidenced by the deviation between the QHMK approach results and the MD results at a temperature above 1000 K.

13.2.5.2 *Strain and temperature effects on PDOS and Grüneisen parameters*

The PDOS represents the number of vibrational modes per unit phonon frequency per atom. In this chapter, for the calculation of PDOS, we adopt the

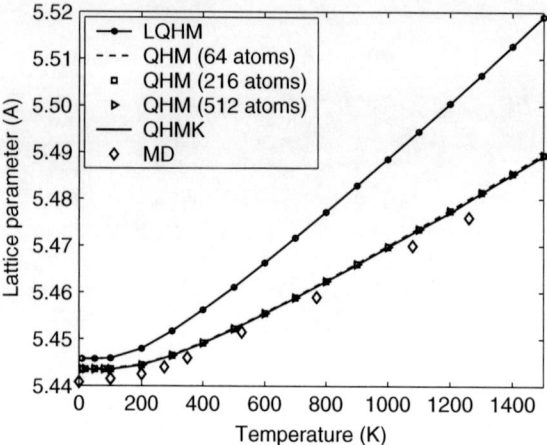

Figure 13.3: Variation of the lattice parameter with temperature obtained from the LQHM approach, the QHM approach with 64, 216, and 512 atoms, the QHMK approach, and MD (MD results are from [40]).

direct sampling method, which generates a large number of uniformly distributed **k**-points in the FBZ and approximates the PDOS by a normalized histogram [39]. In our calculation, $100 \times 100 \times 100/2$ points uniformly distributed in the first quadrant of the FBZ are used. To investigate the strain effect on the PDOS at finite temperature, the lattice constant is first determined by solving (13.39). Next, for a given deformation gradient **F**, the inner displacement $\boldsymbol{\xi}$ is determined by solving (13.40). The phonon frequencies can then be obtained from the eigenvalues of the dynamical matrix. The effect of strain on the PDOS at $T=1000$ K is shown in Figure 13.4. We observe that when a uniaxial deformation is applied, a shift of optical phonons and splitting of their degeneracies occurs, while the acoustic phonons do not vary much, as shown in Figure 13.4(a). Figure 13.4(b) shows that a shear deformation primarily causes a splitting of the degeneracies, especially for the optical phonons.

The GPGPs $\gamma_{ij}^{(n)}$ defined by (13.43) measure the variation of the phonon frequency ω_n with deformation at a constant temperature. These parameters can be used directly in the calculation of the stress tensor as shown in (13.44). We compute the generalized Grüneisen parameters as a function of the deformation gradient **F** at $T = 1000$ K for stretch and shear tests, by using the QHMK method, as shown in Figures 13.5–13.10. The volumetric phonon Grüneisen parameter (PGP) for unstrained silicon crystal, which is defined as $\gamma_V^{(n)} = -d(\ln \omega_n)/d(\ln V)$ where V is the lattice volume, can be calculated from the GPGPs by $\gamma_V^{(n)} = (\gamma_{11}^{(n)} + \gamma_{22}^{(n)} + \gamma_{33}^{(n)})/3$. The volumetric PGP for unstrained Tersoff silicon at 0 K (classical) has already been calculated in [40]. In this chapter, we investigate the temperature effect on the PGP for unstrained Tersoff

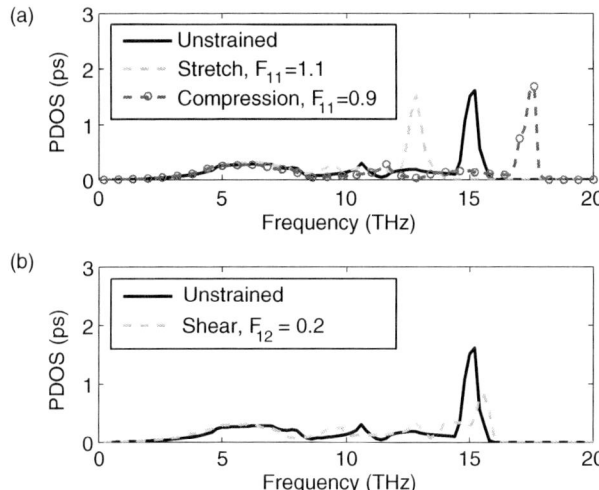

Figure 13.4: (a) PDOS for stretch and compression conditions; (b) PDOS under a shear condition. $T=1000$ K.

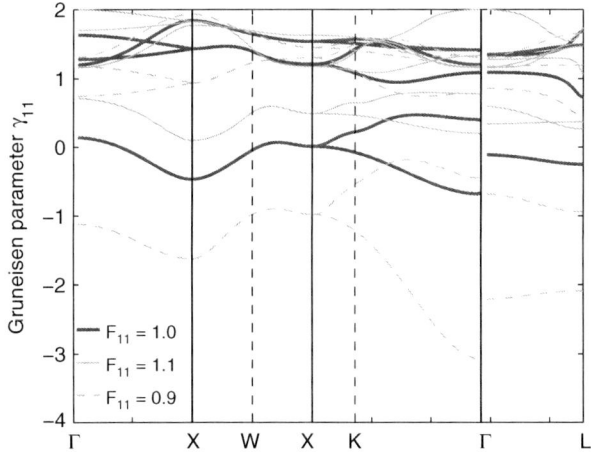

Figure 13.5: The generalized Grüneisen parameter $\gamma_{11}^{(\mathbf{k}s)}$ for Tersoff silicon at 1000 K under deformation from tension to compression ($\mathbf{F}_{11} = 1.1, 1.0, 0.9$, $\mathbf{F}_{22} = \mathbf{F}_{33} = 1$ and $\mathbf{F}_{ij} = 0$ for $i \neq j$).

silicon at finite temperature. First, the lattice constant is determined by using (13.39). In the QHMK method, the phonon dispersion relations $\omega = \omega_s(\mathbf{k})$ are obtained by numerically diagonalizing (13.54) for different wave vectors $\mathbf{k}$, which are usually chosen along several special directions in the FBZ. By differentiating the phonon dispersion relations with respect to the deformation

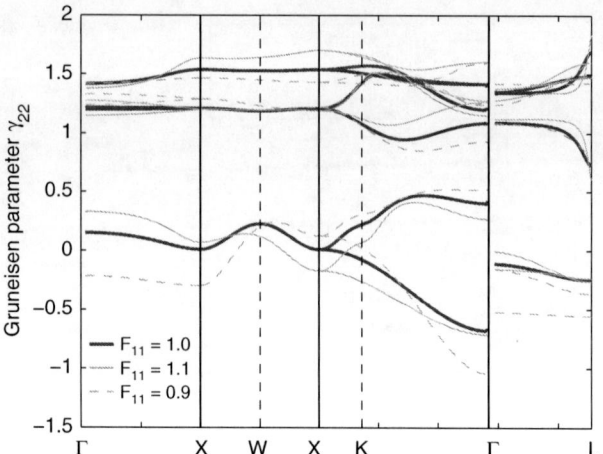

Figure 13.6: The generalized Grüneisen parameter $\gamma_{22}^{(\mathbf{k}s)}$ for Tersoff silicon at 1000 K under deformation from tension to compression ($\mathbf{F}_{11} = 1.1, 1.0, 0.9$, $\mathbf{F}_{22} = \mathbf{F}_{33} = 1$ and $\mathbf{F}_{ij} = 0$ for $i \neq j$).

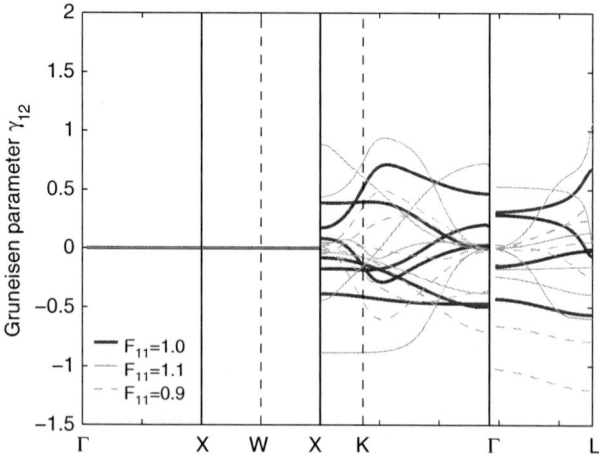

Figure 13.7: The generalized Grüneisen parameter $\gamma_{12}^{(\mathbf{k}s)}$ for Tersoff silicon at 1000 K under deformation from tension to compression ($\mathbf{F}_{11} = 1.1, 1.0, 0.9$, $\mathbf{F}_{22} = \mathbf{F}_{33} = 1$ and $\mathbf{F}_{ij} = 0$ for $i \neq j$).

gradient $\mathbf{F}$, the GPGPs as a function of the wave vector $\mathbf{k}$ are obtained by using $\gamma_{ij}^{(\mathbf{k}s)} = -\mathbf{F}_{im}^{-1} d(\ln \omega_s(\mathbf{k}, \mathbf{F}))/d\mathbf{F}_{mj}$, $i, j, m = 1, 2, 3$. The volumetric PGPs, γ_V, obtained from the QHMK method at two different temperatures are shown in Figure 13.11. We observe that the Grüneisen parameters for both the acoustic

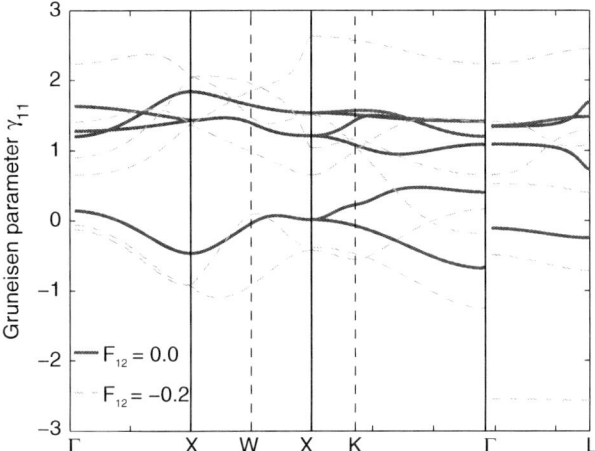

Figure 13.8: The generalized Grüneisen parameter $\gamma_{11}^{(\mathbf{k}s)}$ for Tersoff silicon at 1000 K under shear ($\mathbf{F}_{12} = -0.2$, $\mathbf{F}_{ii} = 1.0$, and $\mathbf{F}_{ij} = 0$ for $i \neq j$ excluding $\mathbf{F}_{12}$).

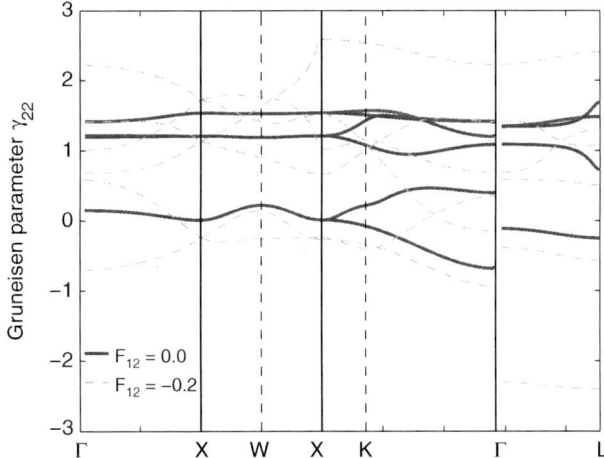

Figure 13.9: The generalized Grüneisen parameter $\gamma_{22}^{(\mathbf{k}s)}$ for Tersoff silicon at 1000 K under shear ($\mathbf{F}_{12} = -0.2$, $\mathbf{F}_{ii} = 1.0$, and $\mathbf{F}_{ij} = 0$ for $i \neq j$ excluding $\mathbf{F}_{12}$).

modes and the optical modes increase as the temperature increases and the shifts are due to the thermal expansion. In the LQHM method, the three phonon frequencies degenerate into one value for the unstrained silicon, and the volumetric PGPs are calculated as 1.295 at $T = 0$ K and 1.317 at $T = 1000$ K.

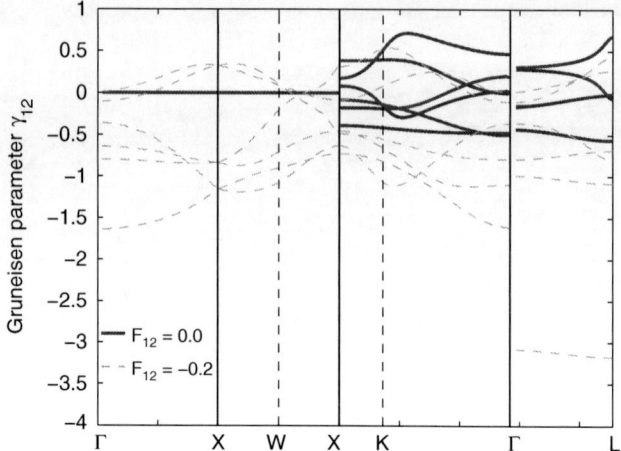

Figure 13.10: The generalized Grüneisen parameter $\gamma_{12}^{(\mathbf{k}s)}$ for Tersoff silicon at 1000 K under shear ($\mathbf{F}_{12} = -0.2$, $\mathbf{F}_{ii} = 1.0$, and $\mathbf{F}_{ij} = 0$ for $i \neq j$ excluding $\mathbf{F}_{12}$).

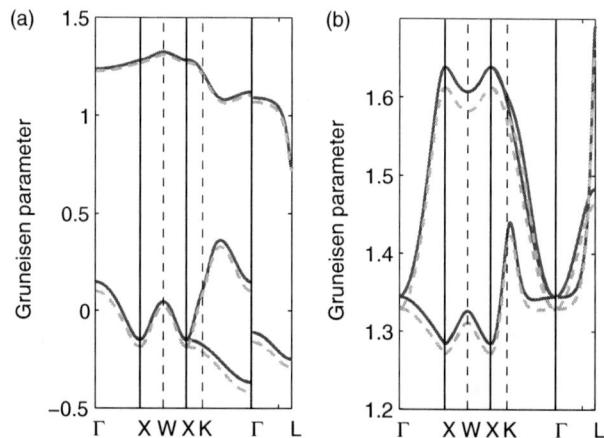

Figure 13.11: The volumetric PGP, γ_V, for unstrained Tersoff silicon: (a) The acoustic modes at 0 K (the dashed lines) and 1000 K (the solid line); (b) The optical modes at 0 K (the dashed lines) and 1000 K (the solid lines).

13.2.5.3 Bulk elastic constants

The elastic constants are the key parameters in the constitutive laws describing a material. We have determined the isothermal and adiabatic elastic constants of Tersoff silicon for several cases.

First, when quantum-mechanical effects are neglected at 0 K, i.e., when the zero point energy (the second term on the right-hand side of (13.56)) is neglected, the characteristic energy is the lattice potential energy. In this case, the elastic constants are simply the second derivatives of the potential energy density function with respect to the strain and the inner displacement as given by (13.31). Since the unstrained silicon crystal lattice is cubic symmetric [41], the pairs ij and kl in $\mathbf{C}_{ijkl}$ can be replaced by a single index p in the Viogt notation: $ij = 11 \rightarrow p = 1$, $ij = 22 \rightarrow p = 2$, and $ij = 12$ or $21 \rightarrow p = 4$. For example, $\mathbf{C}_{1212}$ is replaced by $\mathbf{C}_{44}$. The elastic constants computed at classical 0 K for the second Tersoff (T2) model [42] are, $C_{11} = 1.218$ Mbar, $C_{12} = 0.859$ Mbar, $C_{44} = 0.103$ Mbar, and $C_{44}^0 = 0.924$ Mbar, where C_{44}^0 is C_{44} computed by neglecting the inner displacement $\boldsymbol{\xi}$. For the third Tersoff (T3) model [24], $C_{11} = 1.428$ Mbar, $C_{12} = 0.756$ Mbar, $C_{44} = 0.690$ Mbar, and $C_{44}^0 = 1.189$ Mbar. These results match the calculations performed in [30]. The Cauchy discrepancy, which is defined as $(C_{12} - C_{44})$, for the T2 model is 0.756 Mbar, for the T3 model is 0.066 Mbar, and from experiments is -0.16 Mbar.

Second, the adiabatic elastic constants at finite temperature and zero strain are calculated by using the QHMK, LQHM, and QHM models. We first calculate the isothermal elastic constants and the adiabatic elastic constants are then obtained by using (13.46). We use 4000 **k**-points for the QHMK calculations. For various temperatures, the calculated elastic constants and the comparison with published data are summarized in Table 13.1. In the calculations using the

Table 13.1. Adiabatic elastic constants (in Mbars) for Tersoff silicon at various temperatures

T (K)	QHMK			LQHM			QHM			MD/MC simulations		
	C_{11}	C_{12}	C_{44}	C_{11}	C_{12}	C_{44}	C_{11}	C_{12}	C_{44}	C_{11}	C_{12}	C_{44}
100	1.392	0.740	0.672	1.392	0.738	0.673						
180.7	1.386	0.738	0.669	1.387	0.734	0.671	1.386	0.738	0.669	1.388[a]	0.742[a]	0.673 ±0.138[a]
300	1.372	0.732	0.662	1.373	0.725	0.667						
500	1.345	0.719	0.648	1.346	0.708	0.656	1.345	0.719	0.648			
700	1.315	0.706	0.633	1.317	0.691	0.644						
844.6	1.293	0.696	0.622	1.295	0.678	0.635	1.293	0.696	0.622	1.307 ±0.002[b]	0.708 ±0.002[b]	0.620 ±0.01[b]
900	1.285	0.692	0.618	1.287	0.673	0.631						
1000	1.270	0.685	0.611	1.272	0.664	0.625	1.270	0.685	0.611			
1100	1.255	0.678	0.603	1.256	0.655	0.618						
1300	1.225	0.663	0.588	1.225	0.636	0.605						
1460.6	1.200	0.652	0.575	1.200	0.621	0.594	1.200	0.652	0.575	1.228[b]	0.681[b]	0.592 ±0.01[b]
1500	1.194	0.649	0.572	1.194	0.617	0.591						

[a] Data from [43].

[b] Data from [44].

QHM model, 5×5×5 unit cells (1000 atoms) with periodic boundary conditions are used. We observe that the results from the QHMK model match well with those from atomistic simulations over a large temperature range while those from the LQHM model deviate from atomistic simulations (especially the results for C_{12}). The QHM model results are identical to the results obtained from the QHMK model with $1000/2 = 500$ k-points. The maximum deviation between the QHM model results with 1000 atoms and the QHMK model results with 4000 k-points (shown in Table 13.1) is less than 5×10^{-4} Mbar. However, the computational cost of the QHM model is several orders of magnitude higher that that of the QHMK model. For this reason, the QHM model is not used for the rest of the calculations shown in this chapter.

Third, we investigated the strain effect on the isothermal elastic constants at both 0 K and 1000 K. Both QHMK and LQHM models are employed to calculate the elastic constants. In Figure 13.12, the component $\mathbf{F}_{11}$ of the deformation gradient is varied from 0.85 to 1.15 to investigate the effect of compression ($\mathbf{F}_{11} < 1$) and tension ($\mathbf{F}_{11} > 1$) along the x-direction. The elastic constants decrease as the material is stretched and increase as the material is compressed. Note that, the cubic symmetry of the unstrained silicon crystal lattice breaks down due to the strain effect and this results in $\mathbf{C}_{1111} \neq \mathbf{C}_{2222}$, as shown in Figure 13.12. Figure 13.13 shows the variation of the elastic constants with $\mathbf{F}_{12}$, which represents a shear deformation along the y-direction. In this case, due to the introduction of shear strain, the elastic constants corresponding to the shear

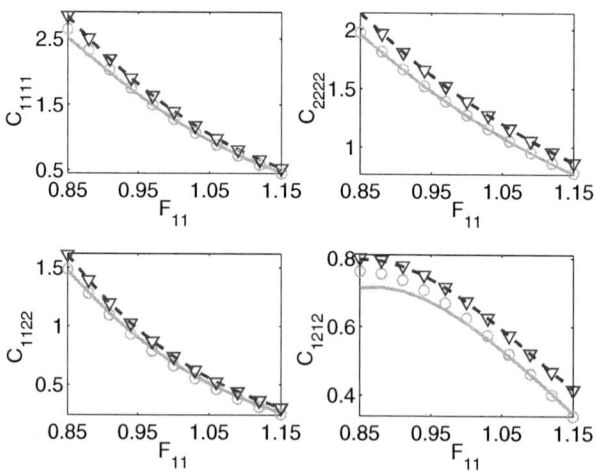

Figure 13.12: Elastic constants (in Mbars) when the lattice is under compression ($\mathbf{F}_{11} < 1$) and tension ($\mathbf{F}_{11} > 1$). The dashed lines are the QHMK results for $T = 0\,\text{K}$, the triangles are the LQHM results for $T = 0\,\text{K}$, the solid lines are the QHMK results for T=1000 K, and the circles are the LQHM results for $T = 1000\,\text{K}$.

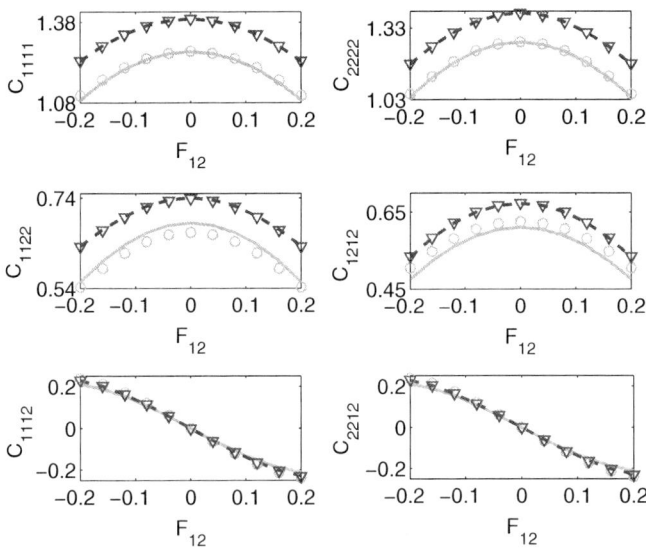

Figure 13.13: Elastic constants (in Mbars) as a function of shear deformation. The dashed lines are the QHMK results for $T = 0\,\text{K}$, the triangles are the LQHM results for $T = 0\,\text{K}$, the solid lines are the QHMK results for $T = 1000\,\text{K}$, and the circles are the LQHM results for $T = 1000\,\text{K}$.

strain and stress relationships become nonzero, i.e., $\mathbf{C}_{1112}, \mathbf{C}_{2212} \neq 0$. Since $\mathbf{E}_{12} = \mathbf{E}_{21}$ under any deformation, the following symmetric properties remain: $\mathbf{C}_{1112} = \mathbf{C}_{1121}$, $\mathbf{C}_{2212} = \mathbf{C}_{2221}$, and $\mathbf{C}_{1212} = \mathbf{C}_{1221} = \mathbf{C}_{2112} = \mathbf{C}_{2121}$. As shown in Figures 13.12 and 13.13, the LQHM model results can deviate by as much as 15% compared to the QHMK model results at 1000 K. The LQHM model results can deviate significantly (>30%) when the silicon crystal is under a general strain condition at 1000 K as shown in Figure 13.14. From these results, we can conclude that the elastic constants from the LQHM model can be inaccurate for Tersoff silicon, especially when the material is under strain at high temperature.

13.2.5.4 *Mechanical behavior of nanostructures under external loads*

We have described the multiscale QC approach for zero and finite temperature conditions. In this section, we employ the multiscale QC approach to perform mechanical analysis of silicon nanostructures under external loads at zero and finite temperatures. Both the LQHM and the QHMK models are employed in all the calculations for finite temperature, and all the calculations for 0 K are carried out with quantum-mechanical effects included, i.e., the zero point energy is calculated as shown in (13.52) for the LQHM model and (13.56) for the QHMK model. Shown in Figure 13.15 are results for both compression and stretch tests on a silicon beam with a geometry of 10 (length) × 20 (width) × 10 (thickness) unit

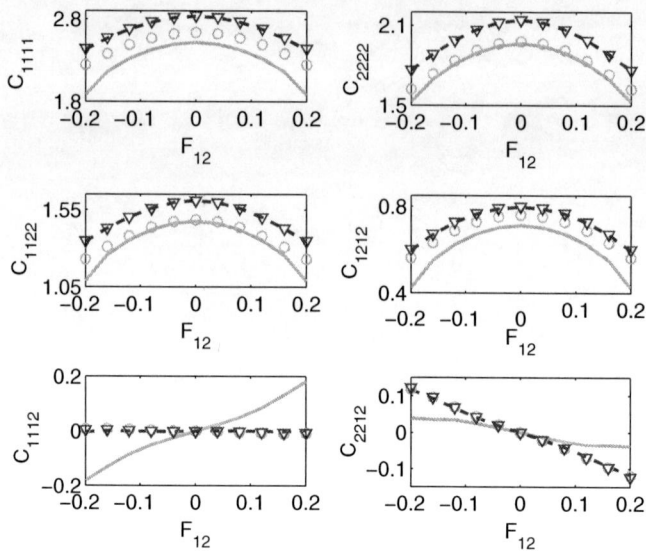

Figure 13.14: Elastic constants (in Mbars) at 1000 K under a combined deformation of $\mathbf{F}_{11} = 0.85$ and $\mathbf{F}_{12}$ varies from -0.2 to 0.2. The dashed lines are the QHMK results for $T = 0\,\text{K}$, the triangles are the LQHM results for $T = 0\,\text{K}$, the solid lines are the QHMK results for $T = 1000\,\text{K}$, and the circles are the LQHM results for $T = 1000\,\text{K}$.

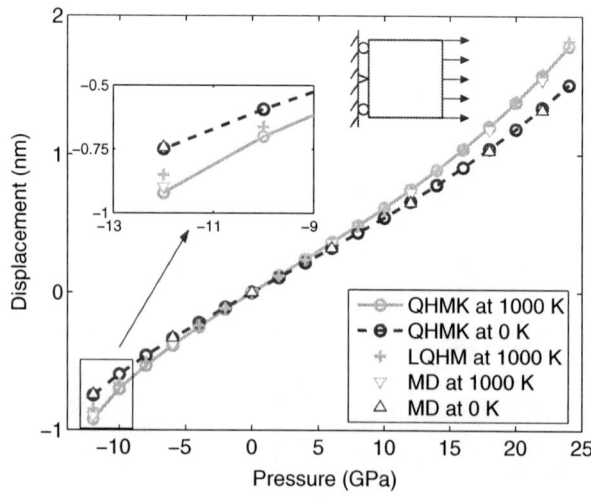

Figure 13.15: The peak displacement as a function of the applied compression pressure(<0) and tensile pressure (>0) for a silicon beam at 0 K and 1000 K.

cells, i.e., $5.432 \times 10.864 \times 5.432$ nm^3 (17,037 atoms in MD simulations) at 0 K and 1000 K. The MD results are also presented in Figure 13.15 for comparison. The temperature in MD is scaled by using Eq. (59) in [32]. The results in Figure 13.15 show the significance of temperature on the displacement of the structure. In addition, we observe that in the compression region (pressure < 0) the LQHM model underestimates the displacement of the structure, due to its overestimation of the elastic constants when the crystal structure is under compression (see Figure 13.12); while in the tensile region the results from the LQHM model are close to the QHMK model results, since both models give similar elastic constants under tension. The temperature effect is captured by the finite temperature QC approach and the accuracy of the method is verified by the MD results. Figures 13.16 and 13.17 show bending tests at 0 K and 1000 K on a silicon beam with a geometry of $40 \times 20 \times 10$ unit cells, i.e., $21.728 \times 10.864 \times 5.432$ nm^3 (66,867 atoms in MD simulations), with fixed-fixed and cantilever boundary conditions, respectively. When the temperature is high (1000 K in this case) the results from the QC method start to deviate from the MD results due to the anharmonic effects of the interatomic potential. Figure 13.18 shows results for a silicon beam with a geometry of $10 \times 20 \times 10$ unit cells, i.e., $5.432 \times 10.864 \times 5.432\,nm^3$ (17,037 atoms in MD simulations), subjected to a general loading condition. The deformation of the beam resembles the deformation condition under which the elastic constants shown in Figure 13.14 are computed. In this case, the results obtained from the QHMK model agree well with the MD results, while the LQHM results are in significant error compared with the MD results. The error can be explained by the elastic constants shown in Figure 13.14, where most of

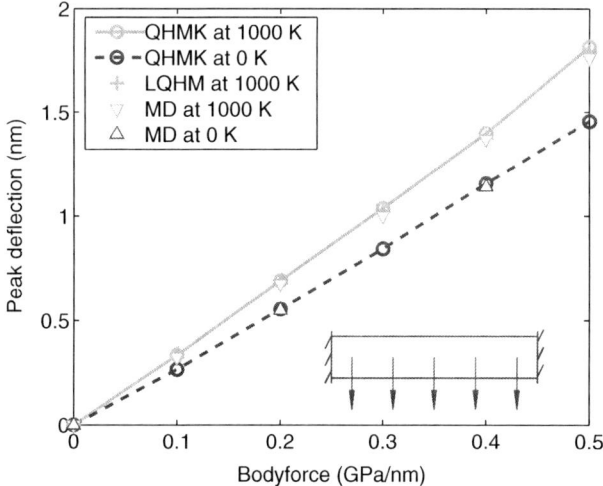

Figure 13.16: The peak displacement as a function of the applied body-force (acts vertically downwards) for a fixed-fixed silicon beam at 0 K and 1000 K.

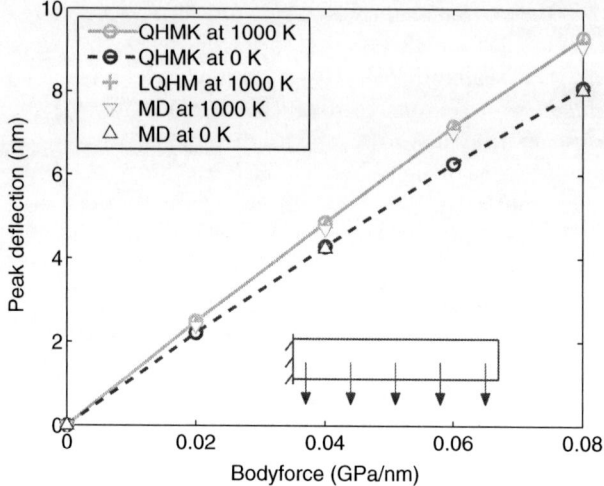

Figure 13.17: The peak displacement as a function of the applied body-force (acts vertically downwards) for a cantilever silicon beam at 0 K and 1000 K.

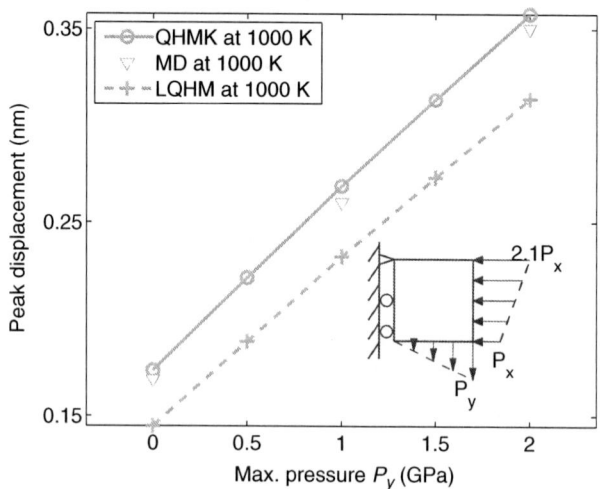

Figure 13.18: The peak displacement of a silicon beam subjected to a general loading at 1000 K. $P_x = 5$ GPa and P_y vary from 0 to 2 GPa.

the elastic constants predicted by the LQHM model are significantly higher than those computed by using the QHMK model.

When the characteristic length of the nanostructure is large, the surface effects, which arise due to the different atomic configuration of the surface atoms from that of the interior atoms, on the mechanical behavior of the nanostructure are typically negligible. However, when the characteristic length of the nanostructure is less than a few nanometers, the surface effects can become significant. The surface atomic configuration depends on the material as well as on the surrounding environment, which can be quite complicated for silicon nanostructures. In general, the Cauchy-Born rule may not hold for surface atoms. Accurate treatment of the surface atoms by the quasicontinuum method at finite temperatures is still an open research topic [19, 45]. In this work, we adopt a simple surface model, where the deformed configuration of the surface atoms is still assumed

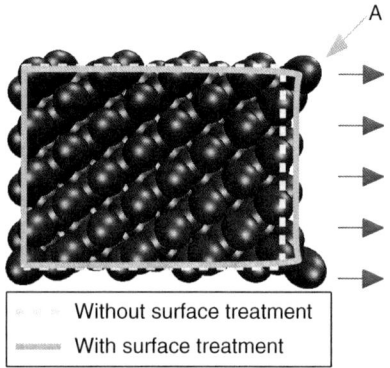

Figure 13.19: Deformed shape of a silicon nanostructure subjected to an axial pressure. Comparison of the MD and the QC approaches (874 K).

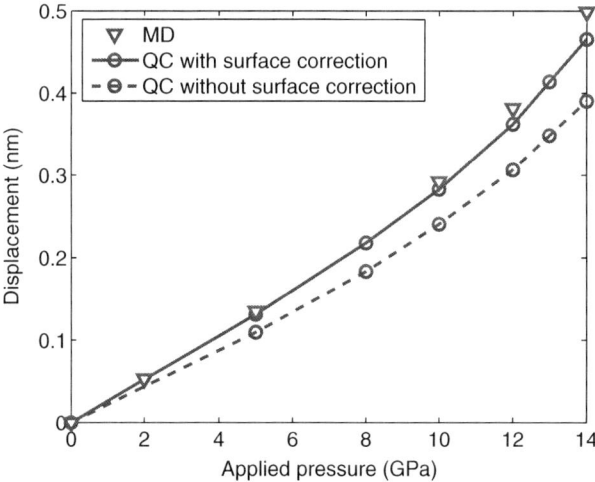

Figure 13.20: The displacement of point A shown in Figure 13.19 as a function of the applied pressure.

to follow the Cauchy-Born rule and the elastic properties at the surface are calculated by averaging over the first and the second layer of atoms. Although the surface model employed in this chapter is simple, more advanced surface models can be readily incorporated in the finite temperature QC framework. Figure 13.19 shows a 4×10×4 unit cell, i.e., $2.173 \times 5.432 \times 2.173 \, nm^3$ (1487 atoms in MD simulations) silicon structure with one end clamped and the other end subjected to a tensile pressure. The deformed shape of the nanostructure obtained from MD and the QC approaches (using the QHMK model) is compared. To investigate the significance of the surface effect, we also computed the deformed shape of the nanostructure without taking into account the surface effects, i.e., the surface atoms are treated the same as the interior atoms. As shown in Figure 13.19, the primary local deformation features due to the surface effects are captured by the QC approach. Figure 13.20 shows the displacement as a function of the applied pressure for the corner point on the right surface of the structure shown in Figure 13.19. The computed results compared well with the MD results.

13.3 Local phonon density of states approach

In this section, we present another approach within the finite temperature quasicontinuum method framework, to investigate thermodynamic and mechanical properties of an inhomogeneous system at finite temperature. Thermodynamic properties for finite temperature solid systems under isothermal conditions are characterized by the Helmholtz free energy density. The static part of the Helmholtz free energy is obtained directly from the interatomic potential, while the vibrational part is calculated by using the theory of local phonon density of states (LPDOS). The LPDOS is calculated efficiently from the on-site phonon Green's function by using a recursion technique based on a continued fraction representation. The Cauchy-Born hypothesis is employed to compute the mechanical properties. By considering ideal Si{001}, (2 × 1) reconstructed Si{001} and monolayer-hydrogen passivated (2 × 1) reconstructed Si{001} surfaces of a silicon nanowire, we calculate the local phonon structure, local thermodynamic, and mechanical properties at finite temperature and observe that the surface effects on the local thermal and mechanical properties are localized to within one or two atomic layers of the silicon nanowire.

13.3.1 Theory

13.3.1.1 Lattice dynamics

The most commonly used method to compute the thermodynamic properties of crystals is the lattice dynamics theory based on the quasiharmonic approximation. [34] Considering an N-atom system, the total potential energy is first expanded using a Taylor's series expansion. In a quasiharmonic approximation, the higher-order (>2) terms are neglected and the total potential energy can thus be written in a quadratic form,

$$U(\mathbf{x}) = U(\mathbf{x}^0) + \frac{1}{2} \sum_{\alpha,\beta=1}^{N} \sum_{j,k=1}^{3} \left. \frac{\partial^2 U(\mathbf{x})}{\partial x_{\alpha j} \partial x_{\beta k}} \right|_{\mathbf{x}=\mathbf{x}^0} (x_{\alpha j} - x_{\alpha j}^0)(x_{\beta k} - x_{\beta k}^0), \quad (13.59)$$

where **x** denotes the instantaneous position of all the atoms in the system, i.e., $\mathbf{x} = (\mathbf{x}_1, \mathbf{x}_2, \ldots, \mathbf{x}_N)$, $\mathbf{x}^0$ denotes the equilibrium position of all the atoms in the system, i.e., $\mathbf{x}^0 = (\mathbf{x}_1^0, \mathbf{x}_2^0, \ldots, \mathbf{x}_N^0)$, $\mathbf{x}_\alpha$ and $\mathbf{x}_\alpha^0$ are the instantaneous and equilibrium position vectors of atom α, $\alpha = 1, 2, \ldots, N$, respectively, $x_{\alpha j}$ and $x_{\beta k}$ are the components of the position vectors $\mathbf{x}_\alpha$ and $\mathbf{x}_\beta$ along the jth and kth directions, respectively, and $x_{\alpha j}^0$ and $x_{\beta k}^0$ are the components of the position vectors $\mathbf{x}_\alpha^0$ and $\mathbf{x}_\beta^0$ along the jth and kth directions, respectively. The elements of the $3N \times 3N$ force constant matrix, Φ, are defined as

$$\Phi_{3\alpha-3+j, 3\beta-3+k} = \left. \frac{\partial^2 U(\mathbf{x})}{\partial x_{\alpha j} \partial x_{\beta k}} \right|_{\mathbf{x}=\mathbf{x}^0}, \quad \alpha, \beta = 1, 2, \ldots, N, \ j, k = 1, 2, 3. \tag{13.60}$$

Furthermore, a mass-weighted force constant matrix can be defined by

$$\hat{\Phi}_{3\alpha-3+j, 3\beta-3+k} = \frac{1}{\sqrt{M_\alpha M_\beta}} \Phi_{3\alpha-3+j, 3\beta-3+k}, \quad \alpha, \beta = 1, 2, \ldots, N, \ j, k = 1, 2, 3, \tag{13.61}$$

where M_α and M_β are the masses of atoms α and β, respectively. In the theory of lattice dynamics, the phonon frequencies can be calculated from the eigenvalues of the mass-weighted force constant matrix, i.e., $\omega_l = \sqrt{\lambda^{(l)}}$, $l = 1, 2, \ldots, 3N$, where ω_l is the lth phonon frequency, and $\lambda^{(l)}$ is the lth eigenvalue of $\hat{\Phi}$. Once the phonon frequencies are known, the Helmholtz free energy, A, for an N-atom system is given by [27]

$$A = U(\mathbf{x}^0) + k_B T \sum_{l=1}^{3N} \ln \left[2\sinh\left(\frac{\hbar \omega_l}{2 k_B T}\right) \right], \tag{13.62}$$

where the first term, $U(\mathbf{x}^0)$, on the right-hand side, is the static potential energy, the second term is the vibrational energy, k_B is the Boltzmann constant, T is the temperature, and $\hbar$ is the reduced Planck's constant.

13.3.1.2 LPDOS and local thermodynamic properties

The Helmholtz free energy, A, computed by (13.62), is a global property of the system. To compute the local thermodynamic properties of a system, a useful and an important quantity is the LPDOS, [47, 48] which is defined as

$$n(\omega, \mathbf{x}_\alpha) = \sum_{j=1}^{3} \sum_{l=1}^{3N} \delta(\omega - \omega_l) |\psi_l(x_{\alpha j})|^2, \quad \alpha = 1, 2, \ldots, N, \tag{13.63}$$

where $\psi_l(x_{\alpha j})$ is the element of the eigenvector corresponding to ω_l at position $x_{\alpha j}$, and ω is the phonon frequency. Using the definition of LPDOS, the Helmholtz free energy, A, in (13.62) can be rewritten as [47]:

$$A_\alpha = U_\alpha + k_B T \int_0^{\omega_{\max}} \ln\left(2\sinh\frac{\hbar \omega}{2 k_B T}\right) n(\omega, \mathbf{x}_\alpha) d\omega, \tag{13.64}$$

where A_α and U_α are the Helmholtz free energy and the static potential energy of atom α, respectively, and $\omega_{\max}$ is the maximum phonon frequency. The total Helmholtz free energy $A = \sum_{\alpha=1}^{N} A_\alpha$. Other thermodynamic quantities can be easily derived once the Helmholtz free energy is known [28]. For example, the local internal energy of atom α is given by, $E_\alpha = A_\alpha - T(\partial A_\alpha/\partial T)$, the local heat capacity of an atom α is given by, $C_{v\alpha} = -T(\partial^2 A_\alpha/\partial T^2)$, and the local entropy of an atom α is given by, $S_\alpha = (E_\alpha - A_\alpha)/T$.

13.3.1.3 Phonon GF and the recursion method

The calculation of LPDOS by (13.63) requires that all the phonon frequencies, ω_l, $l = 1, 2, \ldots, 3N$, are known. The calculation of all the phonon frequencies of the system can be quite expensive. An alternative approach is to calculate the LPDOS by using the phonon GF. In this section, the LPDOS is first related to the phonon GF [49]. Then an efficient numerical scheme, the recursion method [50–52], is used to calculate the phonon GF and LPDOS.

The general inter-site phonon GF is given by [49]

$$G_{\alpha j, \beta k}(\omega^2) = \left[(\omega^2 \mathbf{I} - \hat{\Phi})^{-1}\right]_{3\alpha-3+j, 3\beta-3+k}$$

$$= \sum_{l=1}^{3N} \frac{\psi_l(x_{\alpha j})\psi_l^\dagger(x_{\beta k})}{\omega^2 - \omega_l^2}, \quad \alpha, \beta = 1, 2, \ldots, N; \ j, k = 1, 2, 3, \quad (13.65)$$

where the combination αj denotes the row index $3\alpha - 3 + j$, and the combination βk denotes the column index $3\beta - 3 + k$ of the $3N \times 3N$ matrix $G(\omega^2)$, ω is the frequency, $\mathbf{I}$ is the identity matrix, and $\psi_l^\dagger(x_{\beta k})$ is the complex conjugate of $\psi_l(x_{\beta k})$. When $\omega = \omega_l$, the phonon GF in (13.65) is defined by a limiting procedure [49], i.e.,

$$\lim_{\epsilon \to 0^+} G_{\alpha j, \beta k}(\omega^2 + i\epsilon) = \lim_{\epsilon \to 0^+} \sum_{l=1}^{3N} \frac{\psi_l(x_{\alpha j})\psi_l^\dagger(x_{\beta k})}{\omega^2 - \omega_l^2 + i\epsilon}. \quad (13.66)$$

By using the identities for the δ function,

$$\delta(\omega^2 - \omega_l^2) = -\frac{1}{\pi} \lim_{\epsilon \to 0^+} \operatorname{Im} \frac{1}{\omega^2 - \omega_l^2 + i\epsilon}, \quad (13.67)$$

$$\delta(\omega^2 - \omega_l^2) = \frac{1}{2\omega} \delta(\omega - \omega_l), \quad (13.68)$$

the diagonal entries of the phonon GF matrix, $G(\omega^2)$, can be written as

$$2\omega \left[-\frac{1}{\pi} \lim_{\epsilon \to 0^+} \operatorname{Im} G_{\alpha j, \alpha j}(\omega^2 + i\epsilon)\right] = \sum_{l=1}^{3N} \delta(\omega - \omega_l)|\psi_l(x_{\alpha j})|^2. \quad (13.69)$$

By comparing (13.63) and (13.69), the LPDOS for an atom α can be calculated from the on-site phonon GFs (on-site phonon GFs refer to the diagonal entries of the $3N \times 3N$ phonon GF matrix $G(\omega^2)$) by [49]

$$n(\omega, \mathbf{x}_\alpha) = 2\omega \left[-\frac{1}{\pi} \lim_{\epsilon \to 0^+} \text{Im} \sum_{j=1}^{3} G_{\alpha j, \alpha j}(\omega^2 + i\epsilon) \right], \quad (13.70)$$

where $\text{Im} \sum_{j=1}^{3} G_{\alpha j, \alpha j}$ represents the imaginary part of the summation of the on-site phonon GFs for atom α.

To compute the LPDOS by using (13.70), only the diagonal entries of the phonon GF matrix are needed. The direct matrix inversion in (13.65) (i.e., $G(\omega^2) = (\omega^2 \mathbf{I} - \hat{\Phi})^{-1}$) to compute the diagonal entries of $G(\omega^2)$ is numerically very expensive. If the mass-weighted force constant matrix $\hat{\Phi}$ is diagonal, the diagonal entries of the phonon GF matrix can be easily obtained. However, the process of diagonalizing the mass-weighted force constant matrix is also computationally expensive when the dimension of $\hat{\Phi}$ is large. As an alternative method, if the mass-weighted force constant matrix can be transformed into a tridiagonal form, then the diagonal entries of the phonon GF can be easily obtained. Note that, if the mass-weighted force constant matrix $\hat{\Phi}$ is symmetric, the tridiagonalized force constant matrix, Φ^{TD}, obtained by using orthogonal transformation matrices, will also be symmetric. Expressing Φ^{TD} as,

$$\Phi^{TD} = \begin{bmatrix} a_1 & b_2 & & & \\ b_2 & a_2 & b_3 & & \\ & \ddots & \ddots & \ddots & \\ & & b_{3N-1} & a_{3N-1} & b_{3N} \\ & & & b_{3N} & a_{3N} \end{bmatrix}, \quad (13.71)$$

the modified GF matrix using Φ^{TD} is given by,

$$\tilde{G}(Z) = (Z\mathbf{I} - \Phi^{TD})^{-1}, \quad (13.72)$$

where $Z = \omega^2 + i\epsilon$. By using the symmetric and tridiagonal properties of Φ^{TD}, the entries of $\tilde{G}(Z)$ can be represented as continued fractions [53]. For example, the first diagonal entry of $\tilde{G}(Z)$ is given by

$$\tilde{G}_{11}(Z) = \cfrac{1}{Z - a_1 - \cfrac{b_2^2}{Z - a_2 - \cfrac{b_3^2}{\ddots \cfrac{b_{3N}^2}{Z - a_{3N}}}}}. \quad (13.73)$$

Denoting L to be the orthogonal transformation matrix used to construct Φ^{TD}, the relation between $G(Z)$ and $\tilde{G}(Z)$ is given by

$$G(Z) = L\tilde{G}(Z)L^T. \qquad (13.74)$$

The calculation of all the diagonal and the off-diagonal elements of the tridiagonalized force constant matrix Φ^{TD} can still be expensive, i.e., $\mathcal{O}(N^2)$ computational cost. [54] The convergence property of the coefficients, a_l, $l = 1,\ldots,3N$, and b_l, $l = 2,\ldots,3N$, allows that one can compute the first n-level coefficients (n being far less than $3N$) and approximate all the high-order coefficients as a_∞ and b_∞. Based on this approximation, the continued fraction in (13.73) is calculated exactly up to n-levels and the rest of the continued fraction is approximated by an infinite level continued fraction, i.e., $\tilde{G}_{11}(Z)$ can be rewritten as

$$\tilde{G}_{11}(Z) = \cfrac{1}{Z - a_1 - \cfrac{b_2^2}{Z - a_2 - \cfrac{b_3^2}{\ddots \cfrac{b_n^2}{Z - a_n - b_{n+1}^2 t(Z)}}}}, \qquad (13.75)$$

where $t(Z)$ is defined as the square root terminator (SRT) function [55] and is given by

$$t(Z) = \frac{1}{b_\infty}\left[\eta(Z) - i\sqrt{1 - \eta^2(Z)}\right], \qquad (13.76)$$

where i is the complex number and $\eta(Z) = \dfrac{Z - a_\infty}{2b_\infty}$. The terminating coefficients, a_∞ and b_∞, need to be computed appropriately as the higher-order coefficients represent less local and more distant structural information. Although the higher order coefficients can be predicted by an extrapolation technique [56], in our work, we choose a_∞ and b_∞ by averaging over the first n-levels, i.e., $a_\infty = \frac{1}{n}\sum_{l=1}^{n} a_l$, $b_\infty = \frac{1}{n}\sum_{l=2}^{n+1} b_l$.

The tridiagonalization of the mass-weighted force constant matrix can be realized by using the Lanczos algorithm [57]. In the numerical scheme, for a given atom index α and a direction index j (i.e., to compute $G_{\alpha j,\alpha j}$), the starting Lanczos state is set as

$$\phi_1 = (\ldots,0,1,0,\ldots)^T \qquad (13.77)$$

where the only nonzero entry is located in the $(3\alpha-3+j)$th position. We also set $b_1 = 0$, $\phi_0 = \mathbf{0}$, and we can compute a_1 as $a_1 = \phi_1^T \hat{\Phi} \phi_1$. The Lanczos recursion relation for the mass-weighted force constant matrix $\hat{\Phi}$ applied to a sequence of vectors,

$$\tilde{\phi}_{l+1} = \left(\hat{\Phi} - a_l \mathbf{I}\right)\phi_l - b_l \phi_{l-1}, \quad l = 1,2,\ldots,n-1, \qquad (13.78)$$

generates the recursion coefficients (RCs), $a_{l+1} = \tilde{\phi}_{l+1}^T \hat{\Phi} \tilde{\phi}_{l+1}$, $b_{l+1} = \sqrt{\tilde{\phi}_{l+1}^T \tilde{\phi}_{l+1}}$, and the set of normalized vectors, $\phi_{l+1} = \tilde{\phi}_{l+1}/b_{l+1}$, $l = 1, 2, \ldots, n-1$. From the Lanczos algorithm, the orthogonal transformation matrix, L, is computed as $L = [\phi_1 \phi_2 \cdots \phi_n \cdots]$. Because of the construction of ϕ_1, it is easy to show that

$$G_{\alpha j, \alpha j}(Z) = \tilde{G}_{11}(Z). \tag{13.79}$$

The above process can be repeated to compute all the diagonal entries of G, i.e., for a given starting Lanczos state ϕ_1, the first n-levels of the RCs, a_l, $l = 1, \ldots, n$, and b_l, $l = 2, \ldots, n$, are first calculated by the Lanczos algorithm. Then, a_∞, b_∞, and the SRT function, $t(Z)$, are computed. The first diagonal entry of the modified GF matrix, $\tilde{G}_{11}(Z)$, can be calculated by (13.75), which is identical to the desired diagonal entry of the original phonon GF matrix, $G_{\alpha j, \alpha j}(Z)$. Since only the first n (which is far less than N and does not depend on N) recursion levels are calculated, the recursion technique is an $\mathcal{O}(N)$ method.

13.3.1.4 Local mechanical properties

Using the local Helmholtz free energy defined in (13.64), the local constitutive relation for mechanical analysis of nanostructures is given by,

$$\mathbf{S}_{ij}(\mathbf{x}_\alpha) = \partial W_A^\alpha / \partial \mathbf{E}_{ij}, \quad i, j = 1, 2, 3, \tag{13.80}$$

where $\mathbf{S}_{ij}(\mathbf{x}_\alpha)$ is the second Piola-Kirchhoff stress tensor at position $\mathbf{x}_\alpha$, W_A^α is the Helmholtz free energy density of atom α which is defined by $W_A^\alpha = A_\alpha/\Omega_\alpha$, Ω_α is the volume of atom α in the initial configuration, and $\mathbf{E}$ is the Green-Lagrange strain tensor. The elastic constants for each atom position are given by [28],

$$\mathbf{C}_{ijkl}(\mathbf{x}_\alpha) = \frac{\partial^2 W_A^\alpha}{\partial \mathbf{E}_{ij} \partial \mathbf{E}_{kl}} - \frac{\partial^2 W_A^\alpha}{\partial \mathbf{E}_{ij} \partial \xi_m} \left(\frac{\partial^2 W_A^\alpha}{\partial \xi_m \partial \xi_n} \right)^{-1} \frac{\partial^2 W_A^\alpha}{\partial \xi_n \partial \mathbf{E}_{kl}},$$
$$i, j, k, l, m, n = 1, 2, 3, \tag{13.81}$$

where $\mathbf{C}_{ijkl}(\mathbf{x}_\alpha)$ is the elastic constant tensor for atom α, and ξ are the additional inner displacements for a complex crystal structure such as diamond silicon. Due to the cubic symmetry of unstrained silicon, the pairs ij and kl in $\mathbf{C}_{ijkl}$ can be replaced by a single index p in the Voigt notation: $ij = 11 \to p = 1$, $ij = 22 \to p = 2$, and $ij = 12$ or $ij = 21 \to p = 4$. Thus the three independent elastic constants for unstrained silicon with cubic symmetry are $\mathbf{C}_{11}$, $\mathbf{C}_{12}$, and $\mathbf{C}_{44}$. The derivatives of the Helmholtz free energy density with respect to the deformation parameters (i.e., $\mathbf{E}$ and ξ) are given by

$$\frac{\partial W_A^\alpha}{\partial \mathbf{v}} = \frac{1}{\Omega_\alpha} \left[\frac{\partial U_\alpha}{\partial \mathbf{v}} + k_B T \int_0^{\omega_{max}} \ln\left(2\sinh\frac{\hbar\omega}{2k_B T} \right) \frac{\partial n(\omega, \mathbf{x}_\alpha)}{\partial \mathbf{v}} d\omega \right], \tag{13.82}$$

where $\mathbf{v}$ can be replaced by $\mathbf{E}$ or ξ. The derivatives of the LPDOS can be obtained from the derivatives of the GF,

$$\frac{\partial n(\omega, \mathbf{x}_\alpha)}{\partial \mathbf{v}} = 2\omega \left[-\frac{1}{\pi} \lim_{\epsilon \to 0^+} \text{Im} \sum_{j=1}^{3} \frac{\partial G_{\alpha j, \alpha j}(\omega^2 + i\epsilon)}{\partial \mathbf{v}} \right]. \qquad (13.83)$$

In the recursion technique, since the GF is expressed as a continued fraction and only the RCs, a_l and b_l, are a function of the deformation parameters, explicit expressions for the derivatives of the GF can be easily derived by using the chain rule. The calculation of these derivatives is presented in the Appendix in [58].

13.3.2 Semilocal model

In the QHMG-n method, as outlined above, for a given atom position α and direction j, the first n-levels of recursion coefficients are calculated. Then the on-site phonon Green's function is calculated by using (13.73). Finally, the Helmholtz free energy of atom α can be calculated by using (13.64 and 13.70). The accuracy of the Helmholtz free energy computed by the above approach depends on the number of recursion coefficients employed. The inclusion of the more levels of recursion coefficients in the calculation of the on-site phonon Green's function leads to a more accurate calculation of the Helmholtz free energy of the atom. In an earlier work [58], it has been shown that $n = 20$ (i.e. QHMG-20) can typically provide accurate results for thermodynamic and mechanical properties of silicon nanostructures.

To understand the significance of the number of recursion levels on the calculation of LPDOS, consider the results shown in Figure 13.21, where the QHMG-n method with $n = 1, 2, 5$ and 20 is compared with the QHMK method for bulk silicon. The QHMK method, which has been discussed in detail in [31] and [32], has been shown to accurately capture the LPDOS of bulk silicon. When $n = 1$, the LPDOS obtained from QHMG-1 (see Figure 13.21(a)) is very different from that computed by the QHMK method. As shown in Figure 13.21(b), QHMG-2 performs better than QHMG-1 as it can capture the acoustic and the optical phonon peaks. The accuracy further increases with QHMG-5 as shown in Figure 13.21(c). Even though QHMG-5 captures the lower and the higher frequency spectrum, it does not accurately capture the middle frequencies in the range of around 10 THz. When the number of levels is increased to 20, as shown in Figure 13.21(d), the LPDOS obtained from QHMG-20 is very similar to the LPDOS computed by using the QHMK method.

The results presented above indicate that typically, a value of at least $n = 20$ is needed to accurately compute the thermal and mechanical properties of silicon nanostructures. The implementation of QHMG-20 can, however, be quite expensive. For example, when $n = 1$, in order to compute the recursion coefficients a_1 and b_2 accurately for bulk silicon, we need to consider the nearest two layers of atoms around an atom. When $n = 2$, the nearest four layers of atoms need to be taken into account. Similarly, for $n = 20$, the nearest 40 layers of atoms

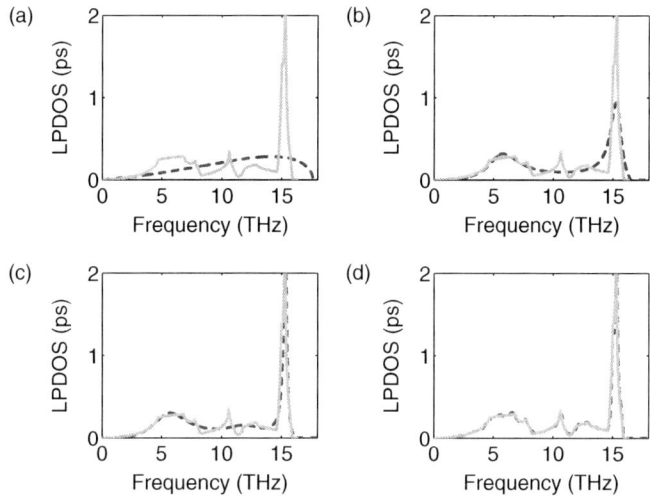

Figure 13.21: Comparison of LPDOS obtained with QHMG-n and QHMK methods: (a) QHMG-1, (b) QHMG-2, (c) QHMG-5, and (d) QHMG-20. The dashed lines are LPDOS computed by QHMG-n method. The solid lines are LPDOS obtained with the QHMK method. $T = 300\,\text{K}$.

need to be considered to accurately compute the first 20 levels of recursion coefficients. As the number of layers increases, the number of atoms that need to be considered increases rapidly and the size of the force constant matrix becomes very large and computationally intractable. As a result, for large n, QHMG-n can be expensive for the calculation of the thermal and mechanical properties of nanostructures.

As discussed above, QHMG-n with a large n (typically at least a value of $n = 20$) can be an accurate method to compute the spatial variation of thermal and mechanical properties of silicon nanostructures. However, QHMG-n with a large n can be quite expensive. As pointed out in the introduction (Section 13.1), other alternative methods also have limitations. For example, LQHM [25], which considers only the vibration of the atom of interest, is a local approach which is fast but the accuracy can be questionable [31, 32]. QHM, a global approach, considers the vibrations of all the atoms. This approach can accurately predict the spatial variation of thermal and mechanical properties but can be an expensive approach for a large number of atoms in the system. QHMK is a fast and an accurate approach for bulk systems but the extension of the approach to compute spatial variation of thermal and mechanical properties, especially under deformation, is not trivial. Here, we develop a semilocal QHMG method, which combines—to a large extent—the accuracy of the QHMG-n approach and the efficiency of the LQHM approach. The key features of the semilocal QHMG approach are (i) it considers two layers of atoms around an atom (it

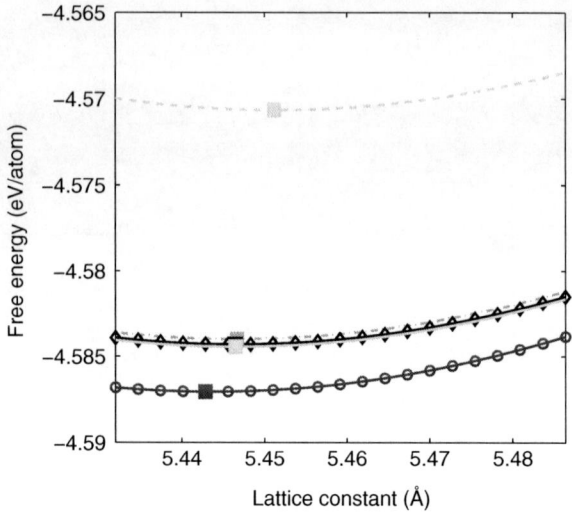

Figure 13.22: Variation of the Helmholtz free energy with lattice constant at 300 K. Comparison between LQHM (dashed line), QHMK (solid line), and QHMG-n (QHMG-1: solid line with circle symbol, QHMG-2: dash dot line and QHMG-20: solid line with diamond symbol). The filled square symbol on each curve indicates the lattice constant corresponding to the minimum Helmholtz free energy.

includes two greater numbers of atoms compared to the LQHM method, but much fewer compared to the QHMG-20 approach—hence, the name semilocal) and (ii) it computes the asymptotic recursion coefficients, a_∞ and b_∞, from the force constant matrix obtained by the LQHM method instead of the average value of the first n-levels of the recursion coefficients as was done in the QHMG-n approach.

To understand why two layers of atoms can be a reasonable approximation for nearby interactions, consider the calculation of the Helmholtz free energy as a function of the lattice constant for bulk Tersoff silicon at 300 K. Figure 13.22 shows the variation of the Helmholtz free energy with lattice constant by using different levels of recursion coefficients. Results from LQHM and QHMK are also shown in the figure for comparison. The filled square symbol on each curve represents the minimum Helmholtz free energy, and the corresponding value on the x-axis is the lattice constant. As the results in the figure indicate, QHMG-1 is in significant error compared to the QHMK results, while QHMG-2 provides a good approximation to both the free energy and the lattice constant. QHMG-2, however, uses four layers of atoms around the atom of interest to accurately compute the first two levels of recursion coefficients. If we use only two layers of atoms, instead of four layers of atoms, the first two levels of recursion coefficients for various temperatures are shown in Figure 13.23. The recursion coefficients,

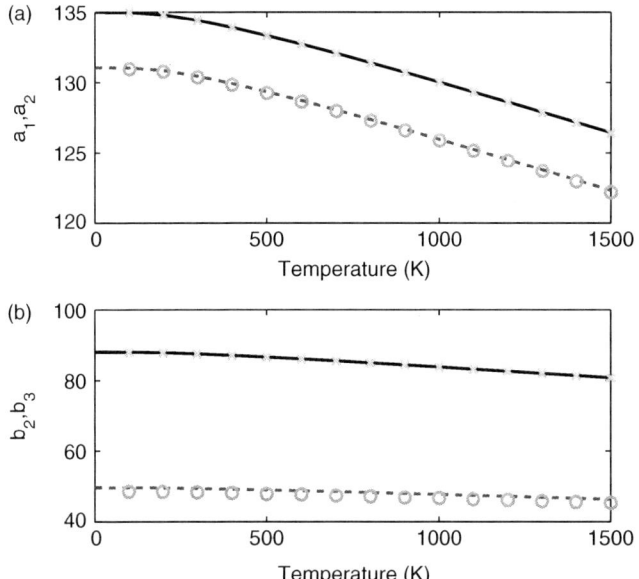

Figure 13.23: Comparison of the first two levels of recursion coefficients between QHMG-2 and semilocal QHMG method. Computed by four layers of atoms, a_1 and a_2 are shown as the solid line and dashed line in the (a); b_2 and b_3 are shown as the solid line and dashed line in the (b). Computed by two layers of atoms, a_1 and a_2 are shown as the cross symbol and circle symbol in the (a); b_2 and b_3 are shown as the cross symbol and the circle symbol in the (b).

a_1, b_2, and a_2, obtained with two layers of atoms and four layers of atoms are identical. The recursion coefficient, b_3, obtained with two layers of atoms deviates slightly when compared to its value obtained with four layers of atoms. This is due to the fact that the interactions with the atoms in the third and the fourth layers are neglected. Since the effect of the atoms in the third and the fourth layers is negligible on a_1, b_2, and a_2 and small on b_3, a semilocal approximation with two layers of atoms can be a reasonable approximation to determine the thermodynamic properties.

In the rest of this section, first, we will present an efficient way to compute a_∞ and b_∞ using the semilocal approximation, and then the moment method is introduced to calculate the first two levels of recursion coefficients by using the nearest two layers of atoms.

a_∞ and b_∞ are the asymptotic values of the recursion coefficients a_n and b_n. For a large system (i.e., if N is large), a_∞ can be approximated by the trace of the tridiagonalized force constant matrix $\mathbf{\Phi}^{TD}$ divided by $3N$ [56]. Since the transformation matrix $\mathbf{L}$ used to compute $\mathbf{\Phi}^{TD}$, i.e., $\mathbf{\Phi}^{TD} = \mathbf{L}^T \mathbf{\Phi} \mathbf{L}$, is an orthonormal matrix [58], the trace of $\mathbf{\Phi}$ is preserved, i.e., trace($\mathbf{\Phi}$) = trace($\mathbf{\Phi}^{TD}$). Thus, a_∞

can also be defined as trace($\mathbf{\Phi}$)/3N. Furthermore, by using the local quasiharmonic approximation [25], without changing the trace of $\mathbf{\Phi}$, the force constant matrix $\mathbf{\Phi}$ can be decomposed into N 3 × 3 matrices, $\mathbf{\Phi}(\alpha)$, $\alpha = 1, 2, \ldots, N$, by neglecting all the interactions between atoms. The local force constant matrix $\mathbf{\Phi}(\alpha)$, is given by

$$\Phi_{j,k}(\alpha) = \left.\frac{\partial^2 U_{\text{local}}(\alpha)}{\partial \mathbf{x}_{\alpha j} \partial \mathbf{x}_{\alpha k}}\right|_{\mathbf{x}=\mathbf{x}^0}, \quad ij, k = 1, 2, 3, \tag{13.84}$$

where $U_{\text{local}}(\alpha)$ is the local potential energy of atom α, which contains contributions from the first and second nearest neighbors. If the system is homogeneous, the local force constant matrix of each atom, $\mathbf{\Phi}(\alpha)$, is identical. Thus, we have the relation

$$a_\infty \approx \frac{\text{trace}(\mathbf{\Phi}^{TD})}{3N} = \frac{\text{trace}(\mathbf{\Phi})}{3N} = \frac{\sum_{\alpha=1}^{N} \text{trace}(\mathbf{\Phi}(\alpha))}{3N} \approx \frac{\text{trace}\mathbf{\Phi}(\alpha)}{3}. \tag{13.85}$$

Furthermore, for a perfect silicon lattice structure, a_∞ and b_∞ are related by the expression [59, 60]

$$b_\infty = \frac{a_\infty}{2}. \tag{13.86}$$

Once a_∞ is obtained from (13.85), b_∞ is obtained by using (13.86).

In the QHMG-n method, the force constant matrix is tridiagonalized by using the Lanczos algorithm to compute the recursion coefficients. An alternative, and a more flexible approach, to compute the recursion coefficients is by using the moments of the density of states. The moments approach has the advantage that it allows an interpretation of the recursion coefficients in terms of the atomic structure.

The r-th moment of the phonon density of states $n(\omega)$, denoted as μ_r, is given by [47]:

$$\mu_r = \int_{-\infty}^{\infty} (m\omega^2)^r n(\omega) d\omega = \sum_i \lambda_i^r = \text{trace}(\mathbf{\Phi}^r) = \sum_{\alpha=1}^{N} \sum_{j=1}^{3} \psi^T(\mathbf{x}_{\alpha j}) \mathbf{\Phi}^r \psi(\mathbf{x}_{\alpha j}), \tag{13.87}$$

where λ_i is the eigenvalue of the force constant matrix $\mathbf{\Phi}$, $\psi(\mathbf{x}_{\alpha j}) = [\cdots, 0, 1, 0, \cdots]^T$ is the $3N \times 1$ orthogonal vector with the only nonzero entry being the $(3\alpha + j - 3)$-th position corresponding to atom α in the j-th direction, and r is the order of the moment. The corresponding r-th moment of LPDOS of atom α in the j-direction is given by:

$$\mu_r(\mathbf{x}_{\alpha j}) = \psi^T(\mathbf{x}_{\alpha j}) \mathbf{\Phi}^r \psi(\mathbf{x}_{\alpha j}) = \sum_{abc\ldots k} \underbrace{\Phi_{la}\Phi_{ab}\Phi_{bc}\ldots\Phi_{kl}}_{r}, \tag{13.88}$$

where $l = 3\alpha + j - 3$ and the last step follows from simple matrix multiplication. Since the matrix elements Φ_{ij} are zero except for the atoms with interactions, the only nonzero contributions to the above equation come from the "chains" of atoms with interactions. These must be closed chains starting and finishing at atom α in the j-direction. Thus, $\mu_r(\mathbf{x}_{\alpha j})$ can be calculated by counting the number of such chains of length r [50]. For the force constant matrix obtained by using the Tersoff potential, the first moment represents a hop on a single atom, the second to hops of the nearest two layers of neighbors and back, and so on. Based on these hopping properties, two more layers of atoms need to be considered when we increase the order of the moments by two.

The connection between moments and recursion coefficients can be seen by expressing the moments in terms of the recursion coefficients [55]. The relation between the moments with an irreducible chain [50] and the recursion coefficients is as follows:

$$\bar{\mu}_0(\mathbf{x}_{\alpha j}) = 1,$$
$$\bar{\mu}_1(\mathbf{x}_{\alpha j}) = a_1,$$
$$\bar{\mu}_2(\mathbf{x}_{\alpha j}) = b_2^2, \qquad (13.89)$$
$$\bar{\mu}_3(\mathbf{x}_{\alpha j}) = a_2 b_2^2,$$
$$\bar{\mu}_4(\mathbf{x}_{\alpha j}) = a_2^2 b_2^2 + b_2^4 + b_2^2 b_3^2.$$

As an example, for the 17-atom cluster centered at atom α, we have a 51×51 force constant matrix

$$\Phi = \begin{bmatrix} \mathbf{A} & \mathbf{B}^T \\ \mathbf{B} & \mathbf{D} \end{bmatrix}, \qquad (13.90)$$

where $\mathbf{A} = \Phi_{1,1}$ is a scalar, $\mathbf{B}$ is a 50×1 vector, and $\mathbf{D}$ is a 50×50 matrix. By using (13.90), for atom α in the x-direction ($j = 1$), the irreducible chain moments in (13.89) can be rewritten as

$$\bar{\mu}_0(\mathbf{x}_{\alpha 1}) = 1,$$
$$\bar{\mu}_1(\mathbf{x}_{\alpha 1}) = \mathbf{A},$$
$$\bar{\mu}_2(\mathbf{x}_{\alpha 1}) = \mathbf{B}^T \mathbf{B}, \qquad (13.91)$$
$$\bar{\mu}_3(\mathbf{x}_{\alpha 1}) = \mathbf{B}^T \mathbf{D} \mathbf{B},$$
$$\bar{\mu}_4(\mathbf{x}_{\alpha 1}) \approx (\mathbf{B}^T \mathbf{B})^2 + \mathbf{B}^T \mathbf{D} \mathbf{D} \mathbf{B}.$$

By using Eqs. (13.89, 13.91), the first two levels of recursion coefficients can be computed. Since only the nearest two layers of neighbor atoms are considered in obtaining the force constant matrix Φ, the recursion constant b_3 has a small error (when compared to b_3 computed by considering four layers of atoms), but the recursion constants a_1, b_2, and a_3 are accurate as shown in Figure 13.23.

13.3.3 Silicon surface models

In this work, we refer to ideal surfaces as the Si{001} surfaces (with dangling bonds). In addition to ideal surfaces we also consider reconstructed and hydrogen passivated reconstructed surfaces. The reconstruction of the Si{001} surfaces has previously been systematically examined using both classical potentials and *ab initio* calculations, and most calculations agree on the essentials of the Si{001} (2×1) reconstruction with the dimer bond along the <110> direction [61–63]. The periodicity of (2×1) is explained by the time average of the thermal flip-flop motion of asymmetric dimmers on the Si{001} surfaces [64, 65]. Moreover, for a hydrogen passivation of one monolayer, the surface retains a (2×1) reconstruction with hydrogen atoms terminating the dangling bonds of silicon. The ideal Si{001} surfaces, the (2×1) reconstructed Si{001} surfaces, and the monolayer-hydrogen passivated (2×1) reconstructed Si{001} surfaces are studied in this work. These three configurations of the silicon surfaces are shown in Figure 13.24. For both bulk silicon and the Si{001} surfaces, the Tersoff interatomic potential model [24] is used to approximate the Si-Si covalent bond interactions. For Si-H interactions, the empirical interatomic potential proposed in [68] is adopted in this work, since this extended version of the Tersoff potential has been tested successfully for its accuracy in decribing the Si-H system in solid form [69, 70]. Note that the

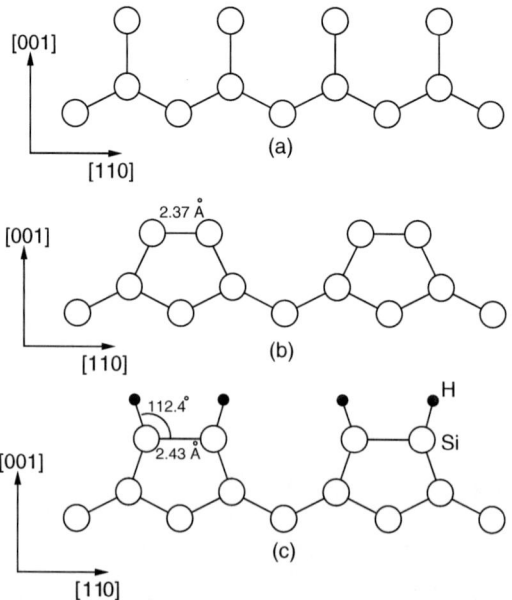

Figure 13.24: The different silicon surfaces. (a) The ideal Si(001) surface, (b) the Si(001) surface with a (2×1) reconstruction, and (c) the (2×1) reconstructed Si(001) surface with monolayer-hydrogen passivation.

Tersoff potential gives a Si-Si dimer bond length of 2.37 Å. When a monolayer of hydrogen atoms are added on the (2×1) reconstructed Si{001} surfaces, the Si-Si dimer bond length increases to 2.43 Å, as shown in Figure 13.24. For the real Si{001} surfaces, there exist steps, which belong to a generic defect class, line defects [66, 67]. In this chapter, we do not consider the existance of the steps on the silicon surfaces, and neglect the effect of the steps on the phonon structures, thermodynamic, and mechanical properties of silicon nanowires.

13.3.4 Results and discussion

13.3.4.1 Thermal and mechanical properties of bulk silicon

In addition to the calculation of the lattice constant, Helmholtz free energy, and LPDOS as discussed above, we have also used the semilocal QHMG method to compute the internal energy, entropy, heat capacity, and elastic constants (C_{11}, C_{12}, and C_{44}) of bulk silicon at 300 K. These results are summarized in Table 13.2. The predictions from other methods such as LQHM, QHM, QHMK, and QHMG-20 are also shown in Table 13.2 for comparison. The results from the semilocal QHMG model are in good agreement with the results obtained from QHM, QHMK, and QHMG-20, while the results from LQHM can have significant error. For example, the relative error in entropy between LQHM and QHMK models is 24.7606%, but the relative error in entropy between semilocal QHMG and QHMK models is only 1.5647%. Table 13.2 also summarizes the computational cost of each method. The computational cost of the semilocal QHMG method is comparable to the computational cost of the LQHM method

Table 13.2. Comparison between different models at $T = 300$ K

	LQHM	QHM	QHMK	QHMG-20	Semilocal QHMG
space	real	real	reciprocal	real	real
model	local	nonlocal	nonlocal	nonlocal	local
vibration correlations	no	yes	yes	yes	yes
dimension of the force constant matrix	3×3	$3N \times 3N$	6×6 for silicon	$3N \times 3N$	51×51 for Tersoff silicon
Lattice Parameter (Å)	1.00353	1.00267	1.00267	1.00267	1.00270
Free Energy (eV/atom)	−4.571	−4.580	−4.584	−4.584	−4.583
Internal Energy (eV/atom)	−4.531	−4.531	−4.531	−4.531	−4.531
Entropy (J/Mol-K)	12.887	15.833	17.128	17.090	16.860
Heat Capacity (J/Mol-K)	18.903	19.142	19.2308	19.201	19.148
C_{11}	1.373	1.377	1.372	1.372	1.372
C_{12}	0.726	0.733	0.732	0.732	0.731
C_{44}	0.667	0.664	0.662	0.662	0.660
CPU time (second)	0.01	0.2 ($N = 64$)	0.1	4.97	0.03

while QHM, QHMK, and QHMG-20 are clearly more expensive. From the results presented here, it can be seen that the semilocal QHMG method achieves the computational efficiency of the LQHM method and the accuracy of the QHM, QHMK, and QHMG-20 methods.

13.3.4.2 LPDOS of bulk silicon and nanoscale silicon structures

In the quasiharmonic approximation, the elements of the mass-weighted force constant matrix $\hat{\Phi}$ are a function of the lattice constant a, the strain tensor $\mathbf{E}$, and the inner displacement ξ. Therefore, the LPDOS, $n(\omega, \mathbf{x}_\alpha)$ and the Helmholtz free energy are also a function of a, $\mathbf{E}$, and ξ. For a given temperature, T, the lattice constant, a, is first determined on the unstrained bulk silicon crystal by minimizing the Helmholtz free energy, i.e., by solving the equation $(\partial A/\partial a)_T = 0$. The expression for the derivative of the Helmholtz free energy with respect to lattice constant a can be easily obtained from (13.82–13.83) and the expressions given in the Appendix in [58] (by replacing $\mathbf{v}$ with a in the expressions in the Appendix in [58]). For the calculation of the mass-weighted force constant matrix, we use the analytical expressions, i.e., the second derivatives of the Tersoff potential for Si-Si interactions and Tersoff-type potential for Si-H interactions with respect to the atomic displacements. The analytical expressions for the derivatives of the Tersoff potential are given in [32]. The expressions for the derivatives of the Tersoff type potential for Si-H interactions are derived by following the same steps and are not provided here for the sake of brevity.

Figure 13.25 shows the computed lattice constant as a function of temperature by using the quasiharmonic phonon GF approach with different recursion levels (QHMG-n, where n represents the recursion levels). The results from QHMK, LQHM, and MD simulations are also included in the figure for comparison. Once the lattice constant a at the given temperature is computed, all the phonon structures can be easily obtained. For bulk silicon at 300 K, Figure 13.26 shows a comparison of LPDOS between QHMK and QHMG-n by using Tersoff

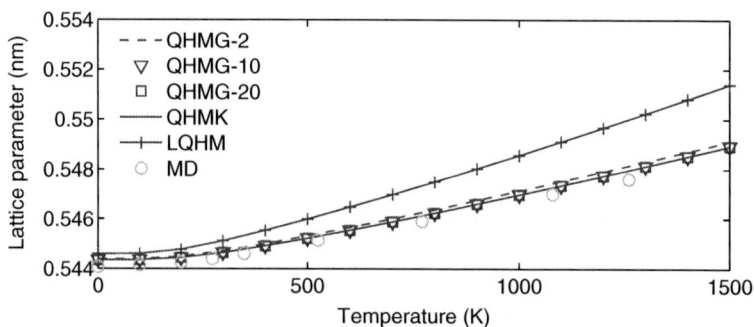

Figure 13.25: The lattice constant for Tersoff silicon at different temperatures. Results from QHMG-n are compared with those from QHMK, LQHM, and molecular dynamics (MD) simulations (MD data are from [40]).

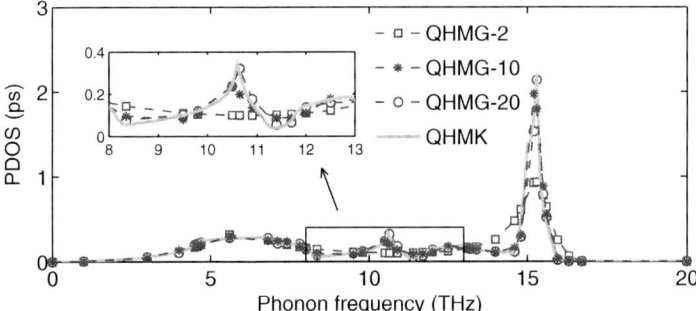

Figure 13.26: LPDOS for bulk Tersoff silicon: comparison between QHMK and QHMG-n at 300 K. Note that the LPDOS from LQHM is a δ function located at $\omega = 11$ THz.

interatomic potential. Note that the LPDOS for the LQHM model is a delta function located at ω=11 THz. From the results on lattice constant and LPDOS for bulk silicon, we find that the calculation with 20 recursion levels results in a very good approximation.

Next, we calculate the LPDOS for a silicon nanowire consisting of 100 × 5 × 5 unit cells, as shown in Figure 13.27. We adopt three surface configurations as discussed in Section 13.3.3 to end the nanowire. For the preparation of the silicon nanowire with different surface configurations, we first generate the silicon diamond structure by using the lattice constant of the Tersoff silicon, which is 5.432 Å. Then the (2×1) reconstructed surfaces and the hydrogen terminated surfaces are formed by using the parameters given in [68, 70]. Then we use MD simulations to relax the structure (no external loads applied) and obtain the equilibrium geometry of the nanowire. For the nanowire with ideal surfaces, the interested atom positions are exactly as shown in Figure 13.27, and the surface atoms (1 and 12) have only two bonds. For the nanowire with (2×1) reconstructed surfaces, the positions of atoms 1 and 12 are slightly different, and atoms 1 and 12 have three bonds where the new bonds are generated from the surface reconstruction. For the nanowire with (2×1) reconstructed surfaces with hydrogen passivation, the positions of atoms 1 and 12 are also slightly different and atoms 1 and 12 have four bonds. Figure 13.28 gives the LPDOS for different atom positions in the nanowire (see Figure 13.27 for the locations of the atoms) at 300 K. From Figure 13.28(a), we find that, for the surface atoms, the acoustic phonons have lower frequencies. This is because the atoms on the surface have fewer bonds (compared to bulk) and this leads to the softening of the phonon frequency. Moreover, from Figure 13.28(a), we find that the acoustic phonons are shifted to the right (towards higher frequencies) for reconstructed surfaces. The reason is that the surface reconstruction and hydrogen passivation generate more bonds for the surface atoms. However, for the reconstructed surfaces with hydrogen passivation, even though the surface silicon atoms also have four bonds

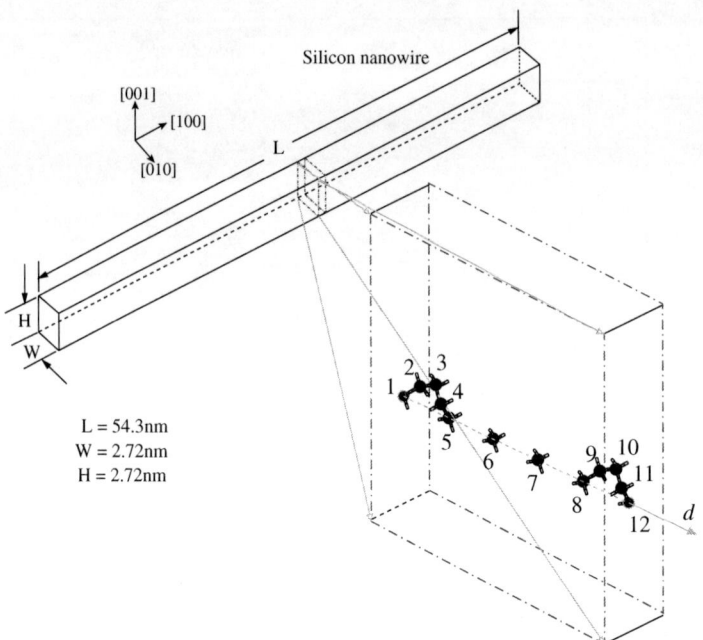

Figure 13.27: The silicon nanowire (100×5×5 unit cells). The interested atoms (atoms 1 to 12) are shown. For the nanowire with ideal surfaces, the interested atom positions are exactly as shown in the figure. For the nanowire with (2×1) reconstructed surfaces and (2×1) reconstructed surfaces with hydrogen passivation, the positions of atoms 1 and 12 are slightly different.

(the same number as for bulk silicon atoms), their LPDOS are still quite different from the LPDOS of bulk silicon atoms. This is because Si-H interactions are weaker than the Si-Si interactions. A new peak is observed at about 57 THz for the surface silicon atom in the nanowire with hydrogen passivation on the surfaces as shown in Figure 13.28(a). This high frequency is due to the Si-H bond stretching vibrations (e.g., the stretching vibrational frequency for SiH molecule is 61 THz [68]). From Figures 13.28(b) and 13.28(c), we observe that for the atom positions inside the surface, the LPDOS are quite similar to that of the bulk. The highest peak (at about 57 THz) disppears for both atoms 2 and 5. The phonons in the range of 15–18 THz are shifted towards higher frequencies for atom 2, due to the effect of the Si-H bond on the surface. This effect, however, is not seen for atom 5, where the LPDOS is almost the same as that of the bulk, for all three silicon nanowires.

13.3.4.3 Local thermal properties

After the LPDOS are calculated, the Helmholtz free energy can be obtained from (13.64). All other thermodynamic properties can then be obtained easily.

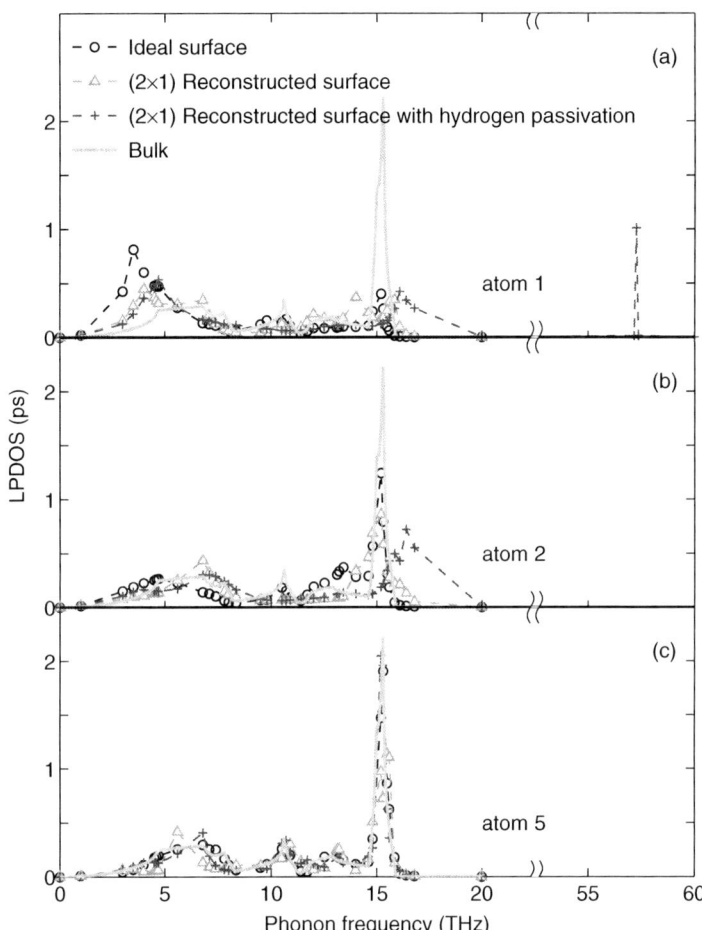

Figure 13.28: The LPDOS calculated by QHMG-20 for different atom positions in the silicon nanowire shown in Figure 13.27 with ideal surface (circles with dashed line), (2×1) reconstruction (triangles with dashed line), and (2×1) reconstruction with hydrogen passivation (plus with dashed line) at 300 K. The QHMK result for bulk silicon (solid line) is also given for comparison.

Figure 13.29 shows the local thermodynamic properties (Helmholtz free energy, internal energy, vibrational energy, kinetic energy, entropy, and heat capacity) as a function of the atom positions.

From Figures 13.29(a) and 13.29(b), we find that the free energy and the internal energy for surface atoms are higher than those for interior atoms, for silicon nanowires with ideal Si{001} surfaces and with (2×1) surface reconstructions. For the silicon nanowire with (2×1) surface reconstruction and hydrogen

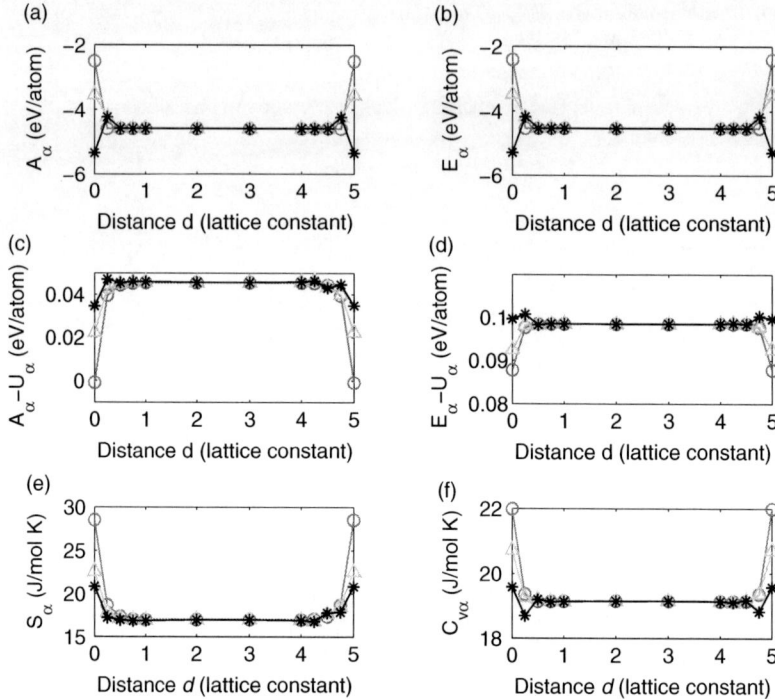

Figure 13.29: Variation of thermodynamic properties ((a) Helmholtz free energy, (b) internal energy, (c) vibrational energy, (d) kinetic energy, (e) entropy, (f) heat capacity) with different atom positions (denoted by the distance from the rear surface, d) computed by QHMG-20 for the silicon nanowire shown in Figure 13.27 with three surface configurations: ideal surface (circles with solid line), (2×1) reconstruction (triangles with solid line), and (2×1) reconstruction with hydrogen passivation (stars with solid line). T = 300 K.

passivation, the free energy and internal energy for the surface silicon atoms are lower than those for the interior atoms. This is because the static energy is the dominant contribution to the Helmholtz free energy and internal energy, and the presence of hydrogen atoms on the (2×1) reconstructed Si{001} surfaces lowers the static potential energy as the bond energy for Si-H is lower than the Si-Si bond energy (e.g., the bond energy for SiH molecule is −3.1 eV, which is lower than the bond energy for Si_2 molecule, −2.66 eV [68]).

Figures 13.29(c) and 13.29(d) indicate that the vibrational energy and the kinetic energy of the surface atoms for silicon nanowire with ideal Si{001} surfaces are lower than those for the interior atoms. The decrease in the vibrational and kinetic energies of the surface atoms when compared to the interior atoms is due to the phonon softness on the surface, as described in Section 13.3.4.2. Since the vibrational energy and the kinetic energy are a measure of the vibrational movement in the system, low frequency phonons result in a lower vibrational energy and kinetic energy for the surface atoms. Moreover, by comparing the results for all the three surfaces in Figures 13.29(c) and 13.29(d), we find that the surface reconstruction first increases the vibrational and kinetic energy, and the introduction of hydrogen atoms on the reconstructed surface further increases the vibrational and the kinetic energies. The reason is, as described in Section 13.3.4.2, due to the surface reconstruction and hydrogen passivation, new Si-Si and Si-H bonds are formed, and the low frequency phonons are shifted towards higher frequencies. Since the high frequency phonons have higher energy, the shift of the low frequency phonons results in the increase of the vibrational and kinetic energies for the atoms near the surfaces.

Figure 13.29(e) shows that the entropy of the surface atoms is higher compared to the interior atoms. This is because the entropy is a measure of the disorder of the system, and the surface atoms increase the disorder of the system. Since the surface reconstruction forms new Si-Si bonds on the surfaces, the disorder of the system is decreased and the entropy is thus decreased. The introduction of hydrogen atoms on the reconstructed surfaces forms new Si-H bonds and the disorder of the system is further decreased. Therefore, the silicon nanowire with (2×1) surface reconstruction and hydrogen passivation has the lowest entropy for those atom positions near the surfaces, as shown in Figure 13.29(e). From Figure 13.29(f), we find that the heat capacity of the surface atoms is also higher compared to the interior atoms for the silicon nanowire with ideal Si{001} surfaces, and the surface reconstruction and hydrogen passivation decrease the heat capacities.

13.3.4.4 *Mechanical properties*

The local elastic properties for the different atom positions in the silicon nanowire are calculated by (13.81). Figure 13.30 shows the variation of the elastic

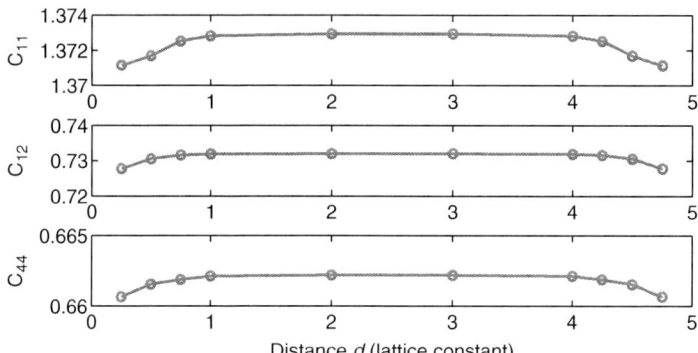

Figure 13.30: Variation of elastic constants (in Mbars) with different atom positions (denoted by the distance from the rear surface, d) for the silicon nanowire with ideal surfaces shown in Figure 13.27. T = 300 K.

properties with atom positions at room temperature (T = 300 K) for the silicon nanowire with ideal Si{001} surfaces. We find that the elastic constants for the atoms near the surface are smaller than those farther away from the surface. The variation of the elastic constants for interior atoms is small ($\Delta C_{11} < 0.5\%$, $\Delta C_{12} < 0.7\%$, and $\Delta C_{44} < 0.3\%$). Thus, the effect of the surface on the elastic constants of the interior atoms is negligible. The calculation of the elastic constants for the surface atoms (i.e., atoms 1 and 12) is a nontrivial task as the Cauchy-Born rule is not applicable for the surface atoms. To investigate the surface effect on the elastic properties, we calculate the average elastic constants for a silicon nanowire with a length of 44 nm and various square crosssectional areas. The averaged elastic constants are calculated by differentiating the total Helmholtz free energy with respect to $\mathbf{E}$ and ξ. Figure 13.31(a) shows the elastic constants as a function of the nanowire crosssectional area at T = 300 K. Note that the cubic symmetry of the bulk silicon crystal lattice breaks down due to the finite size along [010] direction and this results in $\mathbf{C}_{1111} \neq \mathbf{C}_{2222}$, as shown in Figure 13.31(a). The Young's modulus along [100] direction can be obtained by $E_{[100]} = \mathbf{C}_{1111} - 2\mathbf{C}_{1122}^2/(\mathbf{C}_{2222} + \mathbf{C}_{1122})$. Figure 13.31(b) shows the Young's modulus along [100] direction as a function of the nanowire crosssectional area at T = 300 K. For comparison, the Young's modulus from stretch tests with MD simulations at constant temperature [5] is also shown in Figure 13.31(b). From these results we can conclude that, when the crosssectional area is larger than $10 \sim 15\,\mathrm{nm}^2$, the Young's modulus of the silicon nanowire is the same as that of the bulk silicon. For the silicon nanowire with ideal Si{001} surfaces, the deformation of the atoms in the surface region can be approximated as homogeneous and the calculation of the averaged elastic constant is possible. However, for the

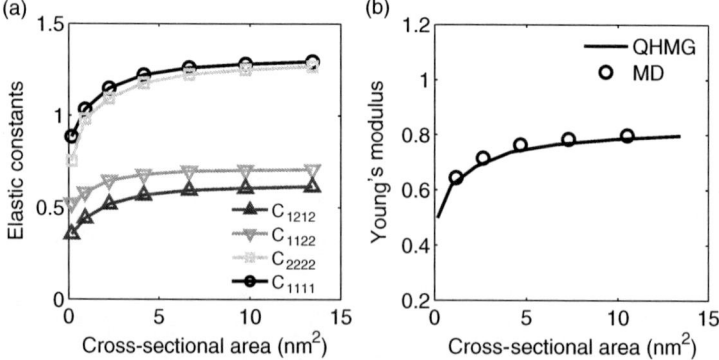

Figure 13.31: (a) Variation of elastic constants (in Mbars) with cross-sectional area for the silicon nanowire with ideal surfaces. (b) Variation of Young's modulus (in Mbars) along [100] direction with cross-sectional area. The length of the silicon nanowire (with ideal surfaces) is chosen as 44 nm, which is large enough to eliminate the surface effect in the length direction. T = 300 K.

silicon nanowires with surface reconstruction and hydrogen passivation, surface region has a curvature. The use of the standard Cauchy-Born rule for atoms in the curved surfaces can be inaccurate. Recently, the standard Cauchy-Born rule has been extended for nanostructures with curvatures e.g. carbon nanotubes [71,72]. The application of the extended Cauchy-Born rule for surface reconstructions in silicon is not a trivial task and we leave this topic for future work.

13.4 Conclusion

The original QC method has been extended to treat the solid structures at finite temperature. For isothermal systems at finite temperature, the constitutive relations are computed by using the Helmholtz free energy density. The static part of the Helmholtz free energy density is obtained directly from the interatomic potential while the vibrational part is calculated by using the real space quasiharmonic model, the local quasiharmonic model, and the **k**-space quasiharmonic model. The calculated results indicate that even though the real space quasiharmonic model predicts the material properties accurately, it is inefficient when the system contains more than several hundreds of atoms. The local quasiharmonic model is simple and efficient, but it can be inaccurate to predict elastic constants of Tersoff silicon, especially when the material is under strain. The **k**-space quasiharmonic model can predict the material properties accurately and efficiently. Using the finite temperature QC method, we have investigated the effect of temperature and strain on the PDOS and phonon Grüneisen parameters, and calculated the elastic constants and the mechanical response of silicon nanostructures under external loads at various temperatures. The results indicate that for silicon nanostructures larger than a few nanometers in critical dimension, the **k**-space quasiharmonic model predicts the mechanical response of the nanostructure accurately over a large temperature range, while the local quasiharmonic model can be inaccurate for Tersoff silicon.

Furthermore, the local thermodynamic and mechanical properties of silicon nanostructures have been investigated by using the local phonon GF method, and the constitutive relation which plays a key role in finite temperature multiscale analysis of nanostructures has also been established. For silicon nanowires with ideal Si{001} surfaces, with (2×1) surface reconstruction, and with surface reconstruction and hydrogen passivation, the phonon structures and local thermal properties have been calculated. For the silicon nanowire with ideal surfaces, the local elastic constants and the averaged elastic properties are computed and the averaged elastic constants are compared with molecular dynamics simulation results. The calculations on the local thermal and mechanical properties show that the surface effects for silicon nanostructures are quite localized.

To simplify the QHMG method, a semilocal QHMG method is also presented to compute the thermodynamic and mechanical properties of both bulk crystalline silicon and silicon nanostructures described by the Tersoff interatomic potential. The semilocal QHMG method considers two layers of atoms to compute the local phonon density of states. It shows that the semilocal QHMG method can approach the efficiency and simplicity of the local quasiharmonic method, and at the same time, predict the thermodynamic and mechanical properties of both bulk silicon and silicon nanostructures more accurately compared to the local quasiharmonic method.

References

[1] Craighead H. G. (2000). "Nanoelectromechanical systems", *Science*, **290**, p. 1532.
[2] Yang Y. T., Ekinci K. L., Huang X. M. H., Schiavone L. M., Roukes M. L., Zorman C. A., and Mehregany M. (2001). "Monocrystalline silicon carbide nanoelectromechanical systems", *Appl. Phys. Lett.*, **78**, p. 162.
[3] Pescini L., Lorenz H., and Blick R. H., (2003). "Mechanical gating of coupled nanoelectromechanical resonators operating at radio frequency", *Appl. Phys. Lett.* **82**, p. 352.
[4] Dequesnes M., Rotkin S., and Aluru N. R., (2002). "Calculation of pull-in voltages for carbon-nanotube-based nanoelectromechanical switches", *Nanotechnology*, **13**, p. 120.
[5] Tang Z., Xu Y., Li G., and Aluru N. R. (2005). "Physical models for coupled electromechanical analysis of silicon nanoelectromechanical systems", *J. Appl. Phys.*, **97**, p. 114304.
[6] Broughton J. Q., Abraham F. F., Bernstein N., and Kaxiras E. (1999). "Concurrent coupling of length scales: methodology and application", *Phys. Rev. B*, **60**, p. 2391.
[7] Atkas O. and Aluru N. R. (2002). "A combined continuum/DSMC technique for multiscale analysis of microfluidic filters", *J. Comp. Phys.*, **178**, p. 342.
[8] Dequesnes M., Tang Z., and Aluru N. R. (2004). "Static and dynamic analysis of carbon nanotube based switches", *J. Eng. Mat. Tech.*, **126**, p. 230.
[9] Deymier P. A. and Vasseur J. O. (2002). "Concurrent multiscale model of an atomic crystal coupled with elastic continua", *Phys. Rev. B*, **66**, p. 134106.
[10] Belytschko T. and Xiao S. P. (2003). "Coupling methods for continuum model with molecular model", *Int. J. Multiscale Comp. Eng.*, **1**, p. 115.
[11] Tadmor E. B., Ortiz M., and Philips R. (1996). "Quasicontinuum analysis of defects in solids", *Philos. Mag. A*, **73**, p. 1529.
[12] Tadmor E. B., Smith G. S., Bernstein N., and Kaxiras E. (1999). "Mixed finite element and atomistic formulation for complex crystals", *Phys. Rev. B*, **59**, p. 235.
[13] Shenoy V. B., Miller R., Tadmor E. B., Rodney D., Phillips R., and Ortiz M. (1999). "An adaptive finite element approach to atomic-scale mechanics – the quasicontinuum method", *J. Mech. Phys. Solids*, **47**, p. 611.
[14] Wagner G. J. and Liu W. K. (2003). "Coupling of atomistic and continuum simulations using a bridging scale decomposition", *J. Comp. Phys.*, **190**, p. 249.

[15] Weinan E., Engquist B., and Huang Z. (2003). "Heterogeneous multiscale method: a general methodology for multiscale modeling", *Phys. Rev. B*, **67**, p. 092101.
[16] Rudd R. E. and Broughton J. Q. (1998). "Coarse-grained molecular dynamics and the atomic limit of finite elements", *Phys. Rev. B*, **58**, p. 5893.
[17] Fish J. and Chen W. (2004). "Discrete-to-continuum bridging based on multigrid principles", *Comput. Methods Appl. Mech. Engrg.*, **193**, p. 1693.
[18] Shenoy V., Shenoy V., and Phillips R. (1999). *Mater. Res. Soc. Sym. Proc.*, **538**, p. 465.
[19] Miller R. E. and Tadmor E. B. (2002). "The quasicontinuum method: overview, applications and current directions", *J. Comput. Aid. Mater. Des.*, **9**, p. 203.
[20] Jiang H., Huang Y., and Hwang K. C. (2005). "A finite-temperature continuum theory based on interatomic potentials", *J. Engr. Mater. Tech.*, **127**, p. 408.
[21] Dupuy L. M., Tadmor E. B., Miller R. E., and Phillips R. (2005). "Finite-temperature quasicontinuum: molecular dynamics without all the atoms", *Phys. Rev. Lett.*, **95**, p. 060202.
[22] Bøggild P., Hansen T. M., Tanasa C., and Grey F. (2001). "Fabrication and actuation of customized nanotweezers with a 25nm gap", *Nanotechnology*, **12**, p. 331.
[23] Cleland A. N. and Roukes M. L. (1996). "Fabrication of high frequency nanometer scale mechanical resonators from bulk Si crystals", *Appl. Phys. Lett.*, **69**(18), p. 28.
[24] Tersoff J. (1988). "Empirical interatomic potential for Silicon with improved elastic properties", *Phys. Rev. B*, **38**, p. 9902.
[25] LeSar R., Najafabadi R., and Srolovitz D. J. (1989). "Finite-temperature defect properties from free-energy minimization", *Phys. Rev. Lett.*, **63**, p. 624.
[26] Harrison W. A. (1980). *Electronic Structure and the Properties of Solids*, Freeman, San Fransisco.
[27] Born M. and Huang K. (1954). *Dynamical theory of crystal lattices*, Oxford, Clarendon.
[28] Wallace D. C. (1972). *Thermodynamics of Crystals*, John Wiley & Sons.
[29] Malvern L. E. (1969). *Introduction to the Mechanics of Continuum Medium*, Prentice Hall, Englewood Cliffs, NJ.
[30] Balamane H., Halicioglu T., and Tiller W. A. (1992). "Comparative study of siliocn empirical interatomic potentials", *Phys. Rev. B*, **46**, p. 2250.
[31] Zhao H., Tang Z., Li G., and Aluru N. R. (2006). "Quasiharmonic models for the calculation of theremodynamic properties of crystalline silicon under strain", *J. Appl. Phys.*, **99**, p. 064314.
[32] Tang Z., Zhao H., Li G., and Aluru N. R. (2006). "Finite-temperature quasicontinuum method for multiscale analysis of siliocn nanostructures", *Phys. Rev. B*, **74**, p. 064110.

[33] Aluru N. R. and Li G. (2001). "Finite cloud method: a true meshless technique based on a fixed reproducing kernel approximation", *Int. J. Num. Meth. Eng.*, **50**, p. 2373.
[34] Ashcroft N. W. and Mermin N. D. (1976). *Solid State Physics*, Harcourt, Inc.
[35] Bloch F. (1928). "über die quantenmechanik der electronen in kristalligittern", *Z. Phys.*, **52**, p. 555.
[36] Hoover W. G. (1985). "Canonical dynamics: equilibrium phase-space distributions", *Phys. Rev. A*, **31**, p. 1695.
[37] Allen M. P. and Tildesley D. J. (1987). *Computer Simulation of Liquid*, Oxford, Clarendon.
[38] Wang C. Z., Chan C. T., and Ho K. M. (1990). "Tight-binding molecular-dynamics study of phonon anharmonic effects in silicon and diamond", *Phys. Rev. B*, **42**, p. 11276.
[39] Maradudin A. A., Montroll E. W., Weiss G. H., and Ipatova I. P. (1971). *Theory of Lattice Dynamics in the Harmonic Approximation*, Academic Press.
[40] Porter L. J., Yip S., Yamaguchi M., Kaburaki H., and Tang M. (1997). "Empirical bond-order potential description of thermodynamic properties of crystalline silicon", *J. Appl. Phys.*, **81**, p. 96.
[41] Boresi A. P. and Lynn P. P. (1974). *Elasticity in Engineering Mechanics*, Prentice-Hall, Inc.
[42] Tersoff J. (1988). "New empirical approach for the structure and energy of covalent systems", *Phys. Rev. B*, **37**, p. 6991.
[43] Izumi S. and Sakai S. (2004). "Internal displacement and elastic properties of the silicon Tersoff model", *JSME International Journal*, Serial A, **47**, p. 54.
[44] Karimi M., Yates H., Ray J. R., Kaplan T., and Mostoller M. (1998). "Elastic constants of silicon using Monte Carlo simulations", *Phys. Rev. B*, **58**, p. 6019.
[45] Shenoy V. B., Miller R., Tadmor E. B., Phillips R., and Ortiz M. (1999). "Quasicontinuum models of interfacial structure and deformation", *Phys. Rev. Lett.*, **80**, p. 742.
[46] Feynman R. P. (1939). "Forces in molecules", *Phys. Rev.*, **56**, p. 340.
[47] Maradudin A. A., Montroll E. W., Weiss G. H., and Ipatova I. P. (1971). *Theory of lattice dynamics in the harmonic approximation*, Academic Press.
[48] Li J. and Yip S. (1997). "Order-N method to calculate the local density of states", *Phys. Rev. B*, **56**, p. 3524.
[49] Economou E. N. (1983). *Green's functions in quantum physics*, Springer-Verlag.
[50] Haydock R., Heine V., and Kelly M. J. (1972). "Electronic structure based on the local atomic environment for tight-binding bands", *J. Phys. C: Solid State Phys.*, **5**, p. 2845.
[51] Meek P. E. (1976). "Vibrational spectra and topological structure of tetrahedrally bonded amorphous semiconductors", *Phil. Mag.*, **33**, p. 897.

[52] Kelly M. J. (1980). *Solid State Physics*, **35**, p. 295, New York, Academic.
[53] Wall H. S. (1948). *Analytical theory of continued fractions*, Van Nostrand-Reinhold, Princeton, New Jersey.
[54] Jennings A. and McKeown J. J. (1992). *Matrix computation*, John Wiley & Sons.
[55] Horsfield A. P., Bratkovsky A. M., Fearn M., Pettifor D. G., and Aoki M. (1996). "Bond-order potentials: theory and implementation", *Phys. Rev. B*, **53**, p. 12694.
[56] Pettifor D. G. and Weaire D. L. (1985). *The recursion method and its applications*, Springer-Verlag.
[57] Lanczos C. (1950). "An iteration method for the solution of the eigenvalue problem of linear differential and integral operators", *J. Res. Natl. Bur. Stand.*, **45**, p. 255.
[58] Tang Z. and Aluru N. R. (2006). "Calculation of thermodynamic and mechanical properties of silicon nanostructures using the local phonon density of states", *Phys. Rev. B*, **74**(23), p. 235441.
[59] Magnus A. (1985). *The Recursion Method and its Applications* (edited by Pettifor D. G., Weaire D. L., Springer, Berlin).
[60] Zheng Z. and Lee J. (1986). "Recursion method for electron and phonon spectra of Si with stacking faults", **19**, p. 6739–6750.
[61] Tromp R. M., Hamers R. J., and Demuth J. E. (1985). "Si(001) dimer structure observed with scanning tunneling microscopy", *Phys. Rev. Lett.*, **55**, p. 1303.
[62] Ihara S., Ho S. L., Uda T., and Hirao M. (1990). "Ab initio molecular-dynamics study of defects on the reconstructed Si(001) surface", *Phys. Rev. Lett.*, **65**, p. 1909.
[63] Batra I. P. (1990). "Atomic structure of the Si(001)-(2 × 1) surface", *Phys. Rev. B*, **41**, p. 5048.
[64] Dabrowski J. and Mussig H. J. (2000). *Silicon surfaces and formation of interfaces*, World Scientific, Singapore.
[65] Shirasawa T., Mizuno S., and Tochihara H. (2005). "Electron-beam-induced disordering of the Si(001)-$c(4 \times 2)$ surface structure", *Phys. Rev. Lett.*, **94**, p. 195502.
[66] Tromp R. M. and Reuter M. C. (1993). "Step morphologies on small-miscut Si(001) surfaces", *Phys. Rev. B*, **47**, p. 7598.
[67] Zhong L., Hojo A., Aiba Y., and Chaki K. (1996). "Atomic steps on a silicon (001) surface tilted toward an arbitrary direction", *Appl. Phys. Lett.*, **68**, p. 1823.
[68] Murty M. V. R. and Atwater H. A. (1995). "Empirical interatomic potential for Si-H interactions", *Phys. Rev. B*, **51**, p. 4889.
[69] Hawa T. and Zachariah M. R. (2004). "Molecular dynamics study of particle-particle collisions between hydrogen-passivated silicon nanoparticles", *Phys. Rev. B*, **69**, p. 035417.

[70] Hansen U. and Vogl P. (1998). "Hydrogen passivation of silicon surfaces: a classical molecular dynamics study", *Phys. Rev. B*, **57**, p. 13295.
[71] Arroyo M. and Belytschko T. (2004). "Finite crystal elasticity of carbon nanotubes based on the exponential Cauchy-Born rule", *Phys. Rev. B*, **69**, p. 115415.
[72] Chandraseker K., Mukherjee S., and Mukherjee Y. X. (2006). "Modifications to the CauchyCBorn rule: applications in the deformation of single-walled carbon nanotubes", *Int. J. Solids and Structures*, **43**, p. 7128.

14
MULTISCALE MATERIALS
Sidney Yip

14.1 Materials modeling and simulation (computational materials)

The way a scientist looks at the materials world is changing dramatically. Advances in the synthesis of nanostructures and in high-resolution microscopy are allowing us to create and probe assemblies of atoms and molecules at a level that was unimagined only a short time ago. Another factor is the advent of large-scale computation, once a rare and sophisticated resource accessible only to a privileged few. In this environment *multiscale materials modeling* is emerging to be a multidisciplinary research area encompassing a very wide range of physical structures and phenomena.

There are certain problems in the fundamental description of matter which previously were regarded as intractable but now are amenable to simulation and analysis. The *ab initio* calculation of solid-state properties using electronic-structure methods and the direct estimation of free energies based on statistical mechanical formulations are two such examples. Because materials modeling draws from essentially all the disciplines in physical science and engineering, it can have very broad impact in cross fertilization between traditionally different communities. One might say that *computational materials*, the intersection between computational science and materials research, is just as vital as computational physics or chemistry; it offers a robust framework for focused scientific studies and exchanges, from the introduction of new university curricula to the formation of centers for collaborative research among academia, corporate, and government laboratories [1]. The common appeal to all members of this new community is the challenge and opportunity to advance fundamental understanding of our materials world and at the same time impact technologies critical to society [2].

The foundation of computational materials is multiscale modeling and simulation where conceptual models and simulation techniques are linked across the micro-to-macro length and time scales for the purpose of analyzing and eventually controlling the outcome of specific materials processes. Invariably these phenomena are highly nonlinear, inhomogeneous, or nonequilibrium. In this paradigm, electronic structure would be treated by quantum mechanical calculations, atomistic processes by molecular dynamics or Monte Carlo simulations, mesoscale microstructure evolution by methods such as finite-element, dislocation dynamics, or kinetic Monte Carlo, and continuum behavior by field equations central to continuum elasticity and computational fluid dynamics. By combining these different methods, one can deal with complex problems in an integrative manner which is not possible when the methods are used individually.

> "*Modeling* is the physicalization of a concept,
> *simulation* is its computational realization."

This is a highly simplified way to describe the two processes that give computational materials its unique character. Since there appears to be no consensus on what each term means by itself and in what sense one complements the other, we suggest here an all-purpose definition that is at least brief and general. By concept we have in mind an idea, an idealization, or a picture of a system (or scenario of a process) which embodies an aspect of functionality.

14.1.1 Characteristic length/time scales

Many physical phenomena have significant manifestations on more than one level of length or time scale. For example, wave propagation and attenuation in a fluid can be described at the continuum level by using the equations of fluid dynamics, while the determination of shear viscosity and thermal conductivity is best treated at the level of molecular dynamics. While each level has its own set of relevant phenomena, an even more powerful description would result if the microscopic treatment of transport could be integrated into the calculation of macroscopic flows. Generally speaking, one can identify four distinct length (and corresponding time) scales where materials phenomenon are typically analyzed. As illustrated in Figure 14.1 the four regions may be referred to

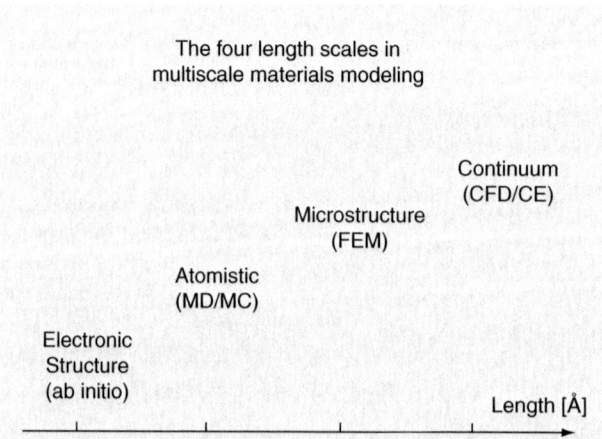

Figure 14.1: Length scales in materials modeling showing that many applications take place on the micron scale and higher, while our basic understanding and predictive ability lie at the microscopic levels.

as electronic structure, atomistic, microstructure, and continuum [3]. Imagine a piece of material, say a crystalline solid. The smallest length scale of interest is about a few angstroms (10^{-8} cm). On this scale one deals directly with the electrons in the system which are governed by the Schrödinger equation of quantum mechanics. Because the techniques for solving the Schrödinger equation are very computationally intensive, they can be applied only to small simulation systems, at present no more than about 300 atoms. However, these calculations are most rigorous from the theoretical standpoint; they are especially valuable for developing and validating more approximate but computationally more tractable descriptions.

The scale at the next level, spanning from tens to about a thousand angstroms, is called atomistic. Here, discrete particle simulation techniques, principally molecular dynamics and Monte Carlo, are well developed, requiring the specification of an empirical classical interatomic potential function with parameters fitted to experimental data and electronic-structure calculations. The most important advantage of atomistic simulation is that one can now study a system of large number of atoms, at present as many as 10^7. On the other hand, by ignoring the electrons atomistic simulations are not as reliable as *ab initio* calculations.

Above the atomistic level the relevant length scale is a micron (10^4 Å). Whether this level should be called microscale or mesoscale is a matter for which convention has not been clearly established. The simulation technique commonly in use is finite-element calculations. Because many useful properties of materials are governed by the microstructure in the system, this is perhaps the most critical level for materials performance and design. However, the information required to carry out such calculations, for example, the stiffness matrix, or any material-specific physical parameters, has to be provided from either experiment or calculations at the atomistic or *ab initio* level. To a large extent, the same can be said for the continuum level, namely, the parameters needed to perform the calculations have to be supplied externally. There are definite benefits when simulation techniques at different scales can be linked. Continuum or finite-element methods are often most practical for design calculations. They require parameters or properties which cannot be generated within the methods themselves. Also they cannot provide the atomic-level insights needed for design. For these reasons continuum and finite element calculations should be coupled to atomistic and *ab initio* methods. It is only when methods at different scales are effectively integrated that one can expect materials modeling to give fundamental insight as well as reliable predictions across the scales. The efficient bridging of the scales in Figure 14.1 is a significant challenge in the further development of multiscale modeling.

14.1.2 *Intellectual merits*

Computational materials has the potential for development of new materials and the technological improvement of materials already in use. The creation of new materials structures and devices with remarkable physical, chemical, electronic, optical, and magnetic properties, through the manipulation of metallic, ceramic, semiconductor, supramolecular, and polymeric materials heralds a whole new era of molecular design. The principle of materials design is rooted in the correlation of molecular structure with physical properties to formulate predictive models of microstructure evolution. Through these models one has the means to investigate the mechanisms underlying critical materials behavior and to systematically arrive at improved designs. Structure-property correlation through simulation can be superior to relying completely on experimental data because one has complete information on the evolving microstructure, as well as complete control over the initial and boundary conditions. A great deal of progress is currently being made not only in first-principles quantum mechanical calculations of electronic structure and atomistic simulations individually, but also in connecting these two powerful techniques of probing physical phenomena in materials.

The benefits of new materials development are evident in the developments of nanoscience and technology. The linking of electronic structure methods, necessary for dealing with novel nanostructures and functional properties, with atomistic and mesoscale techniques ensures the different phases of materials innovation, from design to testing to performance and lifetime evaluation, all can be examined and optimized through simulation. The power of multiscale computation also can be seen in a number of high-profile applications involving the behavior of known materials in extreme environments. The mechanical behavior of plastic deformation in metals at high pressure and high strain rate is a problem relevant to national security which has occupied the attention of a sizable community of researchers for several years [4]. The challenge is to conduct multiscale simulation, linking calculation of the core of a dislocation using electronic-structure methods to modeling dislocation mobility using molecular dynamics simulation, to developing constitutive relations for continuum-level codes using dislocation dynamics simulation, under the requirements of high pressure, greater than 10 Mbar, and high strain rate, greater than 10^8.

Even without the extreme conditions of high pressure and intense radiation fields, the field of materials innovation is rich with challenges in which a better understanding of the underlying microstructure of the material can reap enormous benefits. Such a list could include the formulation of a molecular model of cement, the most widely used man-made substance in the world, that could be used in designing enhanced creep-resistant and environmentally-durable structures. Similar advances in materials performance, each with its own specification of desired performance improvements, are being actively investigated for the development of catalysts for fuel cells in electric vehicles, and for oilfield exploration, where instrumentation and digital management of hydrocarbon reservoirs are critical issues. Indeed everywhere one looks there are problems important to society that require optimizing the functional properties of materials through the control of their microstructure. When one considers why computational materials should have long-term impact on materials innovation, three attributes stand out.

14.1.2.1 *Exceptional bandwidth*
The conceptual basis of materials modeling and simulation covers all of physical science without regard to what belongs to physics versus chemistry versus engineering. This means the scientific bandwidth of computational materials is as broad as all of science and engineering.

14.1.2.2 *Removing empiricism*
A virtue of multiscale modeling is that results pertaining to both model and simulation are conceptually and operationally quantifiable. This means that empirical assumptions can be systematically replaced by science-based descriptions. Quantifiability implies any part of the modeling and simulation can be scrutinized and upgraded in a controlled manner, allowing a complex phenomenon to be probed part by part.

14.1.2.3 *Visual insights*
The numerical output from a simulation are data on the degrees of freedom characterizing the model. Their availability allows not only direct animation, but also visualization of analyzed properties which are not accessible to experimental observation. In microscopy, for example, one has structural information but usually without the energetics, whereas by simulation one can have both. The same may be said of deformation mechanisms and reaction pathways.

We have claimed that the conceptualization of a problem (modeling) and the computational solution of this problem (simulation), is the foundation of computational materials. This coupled endeavor is unique in several respects. It allows practically any complex system to be analyzed with predictive capability by invoking the multiscale paradigm—linking unit-process models at lower-length (or time) scales where fundamental principles have been established to calculations at the system level. It allows the understanding and visualization of cause-effect through simulations where initial and boundary conditions are prescribed specifically to gain insight. Furthermore, it can complement experiment and theory by providing the details that cannot be measured nor described through equations. When these conceptual advantages in modeling are coupled to unprecedented computing power through simulation, one has a vital and enduring scientific approach destined to play a central role in solving the formidable problems of our society. Yet, to translate these ideals into successful applications requires specific case studies.

In the next two sections, we will discuss a number of case studies of deformation behavior and defect mobility to illustrate the physical insights that have been gained in atomistic simulations. These may be considered "unit process" problems which form the basis for a bottom-up approach to understanding materials properties and performance in various applications. To face the challenge of improving materials that are already in use, it is necessary to focus only on a particular functional behavior of the material in question, an approach which is similar in spirit to solving an inverse problem. This is briefly considered in the concluding section of this chapter where we give an outlook on applying modeling and simulation to real materials.

14.2 Atomistic measures of strength and deformation

Understanding materials strength at the atomic level has been a goal in all disciplines of science and engineering. The theoretical basis for describing the mechanical stability of a crystal lattice lies in the formulation of stability conditions which specify the critical level of external stress that the system can withstand. Lattice stability is not only one of the most central issues in elasticity, it is also fundamental in any analysis of structural transitions in solids, such as polymorphism, amorphization, fracture, or melting. In these notes our goal is to discuss the role of elastic stability criteria at finite strain in elucidating the competing mechanisms underlying a variety of structural instabilities, and the physical insights that may be gained by probing stress- and temperature-induced structural responses through atomistic simulations.

14.2.1 Limits to strength: homogeneous deformation

M. Born first showed that expanding the internal energy of a crystal in a power series in the strain and requiring positivity leads to conditions on the elastic constants if structural stability of the lattice is to be maintained [5, 6]. This concept of ideal strength of perfect crystals has been examined by Hill [7] and Hill and Milstein [55], as well as used in various applications [8]. That Born's results are valid only when the solid is under zero external stress has been explicitly pointed out in a later derivation by Wang *et al.* [9] invoking the formulation of a Gibbs integral. Further discussions were given by Zhou and Joos [10] and by Morris and Krenn [11], the latter emphasizing the equivalence to a thermodynamic formulation given by Gibbs' original formulation [12]. A consequence of these investigations is the clarification that theoretical strength can vary with the symmetry and magnitude of the applied load, rather than being an intrinsic property of the material system only. In this respect the study of theoretical limits to material strength using atomistic models, including first-principles calculations [13], promises to yield new insights into mechanisms of structural instability.

Consider a perfect lattice undergoing homogeneous deformation under an applied stress τ, where the system configuration changes from X to Y = JX, with J being the deformation gradient or the Jacobian matrix. The associated Lagrangian strain tensor is

$$\eta = (1/2)(J^T J - 1) \tag{14.1}$$

Let the change in the Helmholtz free energy be expressed by an expansion in η to second order,

$$\begin{aligned}\Delta F &= F(X, \eta) - F(X, 0) \\ &= V(X)[t(X)\eta + (1/2)C(X)\eta\eta]\end{aligned} \tag{14.2}$$

where V is the volume, t the conjugate stress which is also known as the thermodynamic tension or the second (symmetric) Piola-Kirkhoff stress, and C the fourth-order elastic constant tensor. For the work done by an applied stress τ, which is commonly called the Cauchy or true stress, we imagine a virtual move near Y along a path where $J \to J + \partial J$ which results in an incremental work

$$\Delta W = \oint_S \tau_{ij} n_j \delta u_, dS$$
$$= V(Y) \frac{\tau_{ij}}{2} \left(\frac{\partial u_i}{\partial Y_j} + \frac{\partial u_j}{\partial Y_i} \right)$$
$$= V(Y) Tr(J^{-1} \tau J^{-T} \delta \eta) \qquad (14.3)$$

The work done over a deformation path ℓ, $\Delta W(\ell)$, is the integral of δW, given by (14.3), over the path. To examine the lattice stability at configuration X, we now consider the difference between the increase in Helmholtz free energy and the work done by the external stress,

$$\Delta G(Y, \ell) = \Delta F(X, \eta) - \Delta W(\ell)$$
$$= \int_\ell g(Y) d\eta \qquad (14.4)$$

where

$$g(Y) = \frac{\partial F}{\partial \eta} - V(Y) J^{-1} \tau J^{-t} \qquad (14.5)$$

One may also interpret ΔG in the spirit of a virtual work argument. If the work done by the applied stress exceeds that which is absorbed as the free energy increase, then an excess amount of energy would be available to cause the displacement to increase and the lattice would become unstable.

We regard ΔG as a Gibbs integral in analogy with the Gibbs free energy. Notice that in general ΔG depends on the deformation path through the external work contribution. This means that strictly speaking it is not a true thermodynamic potential on which one can perform the usual stability analysis. Nevertheless, $-g(Y)$ can be treated as a force field in deformation space for the purpose of carrying out a stability analysis [9]. Suppose the lattice, initially at equilibrium at X under stress τ, is perturbed to configuration Y with corresponding strain η. A first-order expansion of g(Y) gives

$$g_{ij}(\eta) = V(Y) B_{ijkl} \eta_{kl} + \cdots \qquad (14.6)$$

where, by using $V(Y) = V(X) \det|J|$, one obtains

$$B_{ijkl} = C_{ijkl} - \left[\frac{\partial (\det|J| J_{im}^{-1} \tau_{mn} J_{nj}^{-1})}{\partial \eta_{kl}} \right]_{\eta=0, J=I}$$
$$= C_{ijkl} + \Lambda_{ijkl}(\tau) \qquad (14.7)$$

with

$$\Lambda_{ijkl}(\tau) = (1/2)[\delta_{ik}\tau_{jl} + \delta_{jk}\tau_{il} + \delta_{il}\tau_{jk} + \delta_{jl}\tau_{ik} - 2\delta_{kl}\tau_{ij}] \quad (14.8)$$

δ_{ij} being the Kronecker delta symbol for indices i and j. The physical implication of (14.6) is that in deformation space the shape of the force field around the origin is described by B. The stability condition is then the requirement that all the eigenvalues of B be positive, or

$$\mathbf{det}|A| > \mathbf{0} \quad (14.9)$$

where $A = (\mathbf{1/2})(B^T + B)$, with B being in general asymmetric [9]. In cases where the deformation gradient J is constrained to be symmetric, as in certain atomistic simulations at constant stress, one can argue that the condition det $|B| > 0$ is quite robust [9]. Thus, lattice stability is governed by the fourth-rank tensor B, a quantity which has been called the elastic stiffness coefficient [14]. It differs from the conventional elastic constant by the tensor Λ which is a linear function of the applied stress. The foregoing derivation shows clearly the effect of external work which was not taken into account in Born's treatment. In the limit of vanishing applied stress one recovers the stability criteria given by Born [5].

In the present discussion we will consider only cubic lattices under hydrostatic loading in which case the stability criteria take on a particularly simple form,

$$\begin{aligned} K &= (\mathbf{1/3})(C_{\mathbf{11}} + 2C_{\mathbf{12}} + P) > 0 \\ G' &= (\mathbf{1/2})(C_{\mathbf{11}} + C_{\mathbf{12}} - 2P) > 0 \\ G &= C_{\mathbf{44}} - P > 0 \end{aligned} \quad (14.10)$$

where C_{ij} are the elastic constants at current pressure P, P > 0 (< 0) for compression (tension), K is seen to be the isothermal bulk modulus, G' and G the tetragonal and rhombohedral shear moduli, respectively. The theoretical strength is that value of P for which one of the three conditions in (14.10) is first violated. A simple demonstration showing that the external load must appear in the stability criteria is to subject a crystal to hydrostatic tension by direct atomistic simulation using a reasonable interatomic potential. In this case one finds the instability mode is the vanishing of K, whereas the Born criteria, (14.10) with P set equal to zero, would predict the vanishing of G' [15].

It is worth mentioning that the six components of the eigenmodes of deformation corresponding to the three zero eigenvalues of det(B) are (1,1,1,0,0,0) $\delta\eta$, $(\delta\eta_{xx}, \delta\eta_{yy}, \delta\eta_{zz}, \mathbf{0}, \mathbf{0}, \mathbf{0})$ with $\delta\eta_{xx} + \delta\eta_{yy} + \delta\eta_{zz} = \mathbf{0}$ in the order indicated in (14.10) [9]. The deformation when the bulk modulus vanishes (spinodal instability) preserves the cubic symmetry, while for the tetragonal shear instability the cubic symmetry must be broken.

The connection between stability criteria and theoretical strength is rather straightforward. For a given applied stress $\underline{\tau}$ one can imagine evaluating the current elastic constants to obtain the stiffness coefficients B. Then by increasing the magnitude of $\underline{\underline{\tau}}$ one will reach a point where one of the eigenvalues of the matrix A (cf. (14.3)) vanishes. This critical stress at which the system becomes structurally unstable is then a measure of theoretical strength of the solid. In view of this, one has a direct approach to strength determination through atomistic simulation of the structural instability under a prescribed loading. If the simulation is performed by molecular dynamics, temperature effects can be taken into account naturally by following the particle trajectories at the temperature of interest.

Under a uniform load the deformation of a single crystal is homogeneous up to the point of structural instability. For a cubic lattice under an applied hydrostatic stress, the load-dependent stability conditions are particularly simple, being of the form

$$B = (C_{11} + 2C_{12} + P)/3 > 0, \quad G' = (C_{11} - C_{12} - 2P)/2 > 0,$$
$$G = C_{44} - P > 0, \quad (14.11)$$

where P is positive (negative) for compression (tension), and the elastic constants C_{ij} are to be evaluated at the current state. While this result is known for some time [16, 17, 18], direct verification against atomistic simulations showing that the criteria do accurately describe the critical value of P (P_c) at which the homogeneous lattice becomes unstable has been relatively recent [9, 15, 19, 20, 21, 22, 60]. One may therefore regard P_c as a definition of theoretical or ideal tensile (compressive) strength of the lattice.

Turning now to molecular dynamics simulations we show in Figure 14.2 the stress-strain response for a single crystal of Ar under uniaxial tension at 35.9 K. At every step of fixed strain, the system is relaxed and the virial stress evaluated. One sees the expected linear elastic response at small strain up to about 0.05; thereafter the response is nonlinear but still elastic up to a critical strain of 0.1 and corresponding stress of 130 MPa. Applying a small increment strain beyond this point causes a dramatic stress reduction (relief) at point (b). Inspection of the atomic configurations at the indicated points shows the following. At point (a) several point-defect like inhomogeneities have been formed; most probably one or more will act as nucleation sites for a larger defect which causes the strain energy to be abruptly released. At the cusp, point (b), one can clearly discern an elementary slip on an entire (111) plane, the process being so sudden that it is difficult to capture the intermediate configurations. Figuratively speaking, we suspect that a dislocation loop is spontaneously created on the (111) plane which expands at a high speed to join with other loops or inhomogeneities until it annihilates with itself on the opposite side of the periodic border of the simulation cell, leaving a stacking fault. As one increases the strain the lattice loads up again until another slip occurs. At (c) one finds that a different slip system is activated.

14.2.2 Soft modes

One may regard the stability criteria, (14.5), as manifestation in the long wavelength limit of the general condition for vibrational stability of a lattice. The

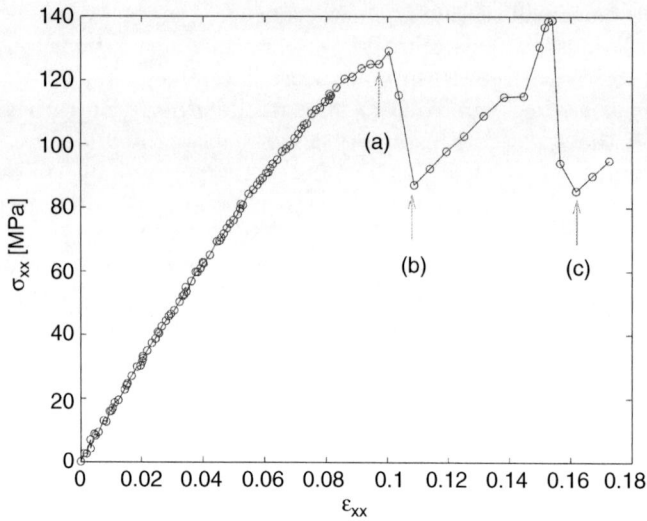

Figure 14.2: Atomistic stress-strain response of a single crystal of Ar under uniaxial tensile deformation at constant strain at a reduced temperature of 0.3 (35.9 K). Simulated data are indicated as circles and a solid line is drawn to guide the eye.

vanishing of elastic constants then corresponds to the phenomenon of soft phonon modes in lattice dynamics. Indeed one finds that under sufficient deformation such soft modes do occur in a homogeneously-strained lattice. To see the lattice dynamical manifestation of this condition, we apply molecular dynamics to relax a single crystal sample with periodic boundary condition at essentially zero temperature for a specified deformation at constant strain. The resulting atomic configurations are then used to construct and diagonalize the dynamical matrix. Figure 14.3 shows two sets of dispersion curves for the Lennard-Jones interatomic potential describing Ar which has fcc structure, one for the crystal at equilibrium (for reference) and the other when the lattice is deformed under a uniaxial tensile strain of 0.138, which is close to the critical value [23]. One can see in the latter a Γ'-point soft mode in the (011) direction. Similar results for deformation under shear or hydrostatic tensile strain would show soft modes Γ'-point in the (111) direction and Γ-point in the (100) direction, respectively. All these are acoustic zone-center modes, and therefore they would correspond to elastic instabilities. For a more complicated lattice such as SiC in the zinc blende structure, one would find that soft modes can also occur at the zone center [23]. The overall implication here is that lattice vibrational analysis of a deformed crystal offers the most general measure of structural instability, and this again demonstrates that strength is not an intrinsic property of the material, rather it depends on the mode of deformation.

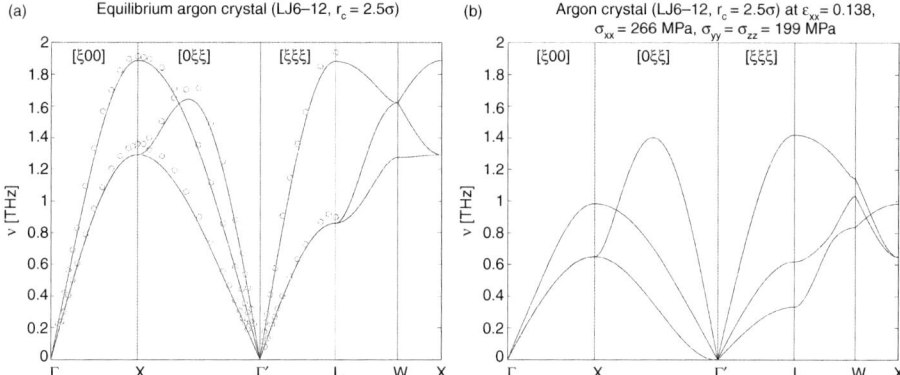

Figure 14.3: Phonon dispersion curves of a single crystal of Ar as described by the Lennard-Jones potential (lines); (a) comparison of results for equilibrium condition with experimental data (circles); (b) results for uniaxial tension deformation at strain of 0.138 (corresponding stresses of 266 MPa and 119 MPa along the tensile and transverse direction) [23].

14.2.3 *Microstructural effects*

Figure 14.2 is a typical stress-strain response on which one can conduct very detailed analysis of the deformation using the atomic configurations available from the simulation. This atomic-level version of structure-property correlation can be even more insightful than the conventional macroscopic counterpart simply because in simulation the microstructure can be as well characterized as one desires. As an illustration we repeat the deformation simulation using as initial structures other atomic configurations which have some distinctive microstructural features. We have performed such studies on cubic SiC (3C or beta phase), which has zinc-blends structure, using an empirical bond-order potential [56] and comparing the results for a single crystal, and prepared amorphous and nanocrystalline structures [23]. Figure 14.4 shows the stress-strain response to hydrostatic tension at 300 K. At every step of incremental strain, the system is relaxed and the virial stress evaluated. Three samples are studied, all with periodic boundary conditions, a single crystal (3C), an amorphous system that is an enlargement of a smaller configuration produced by electronic-structure calculations [25], and a nanocrystal composed of four distinct grains with random orientations (7810 atoms). As in Figure 14.1, the single-crystal sample shows in Figure 14.4 the expected linear elastic response at small strain up to about 0.03; thereafter the response is nonlinear but still elastic up to a critical strain of 0.155 and corresponding stress of 38 GPa. Applying a small increment strain beyond this point causes a dramatic change with the internal stress suddenly reduced by a factor of four. Inspection of the atomic configurations (not shown) reveals the nucleation of an elliptical microcrack in the lattice along the direction

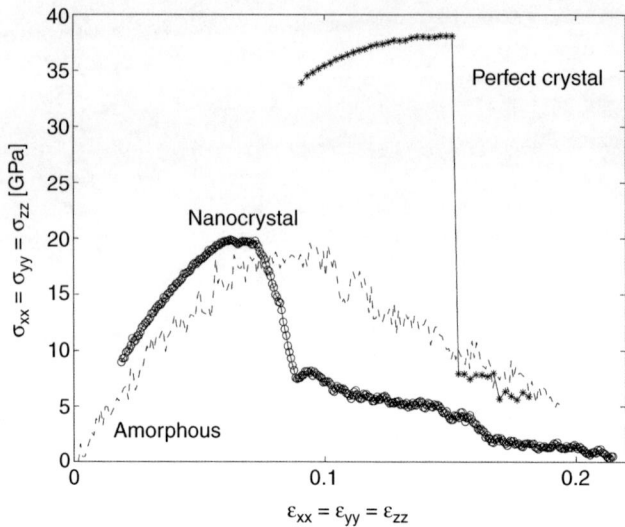

Figure 14.4: Variation of virial stress at constant strain from MD simulations of SiC (3C) under hydrostatic tension at 300 K in perfect crystal, amorphous, and nanocrystalline phases.

of maximum tension. With further strain increments the specimen deforms by strain localization around the crack with essentially no change in the system stress.

The responses of the amorphous and nanocrystal SiC differ significantly from that of the single crystal. The former shows a broad peak, at about half the critical strain and stress, suggesting a much more gradual structural transition. Indeed, the deformed atomic configuration reveals channel-like decohesion at a strain of 0.096 and stress 22 GPa. Another feature of the amorphous sample is that the response to other modes of deformation, uniaxial tension and shear, is much more isotropic relative to the single crystal, which is perhaps understandable with bonding in SiC being quite strongly covalent and therefore directionally dependent. For the nanocrystal, the critical strain and stress are similar to the amorphous phase, except that the instability effect is much more pronounced, qualitatively like that of the single crystal. The atomic configuration shows rather clearly the failure process to be intergranular decohesion. These observations allow us to correlate the qualitative behavior of the stress-strain responses with a gross feature of the system microstructure, namely, the local disorder (or free volume). This feature is of course completely absent in the single crystal, well distributed in the amorphous phase, and localized at the grain boundaries in the nanocrystal. The disorder can act as a nucleation site for structural instability, thereby causing a reduction of the critical stress and strain for failure. Once a site is activated, it will tend to link up with neighboring activated sites, thus giving rise to different behavior between the amorphous and nanocrystal samples.

14.2.4 *Instability in nano-indentation*

We continue with a second case study—the observation of dislocation nucleation in nano-indentation experiments. A single crystal, subjected to a local external stress through the action of the indenter, deforms in a manner shown in Figure 14.5 [24]. The variation of the compressive stress with depth of indentation, typically seen in such measurements, shows a continuous increase of depth with load as expected in elastic deformation, with intermittent *discontinuous* jumps which suggest the loss of structural stability in a local region. These have been regarded as onsets of plastic deformation, in the form of dislocation nucleation and multiplication events.

For *inhomogeneous* deformation such as the case of nano-indentation, one expects local defects to be nucleated at certain sites in the system (the weak spots) when the system is driven across a saddle point. A continuum-level description of homogeneous nucleation was first explored by R. Hill [25] in the concept of discontinuity of "acceleration waves". Later J. R. Rice [26] treated shear localization in much the same spirit and derived a formal criterion

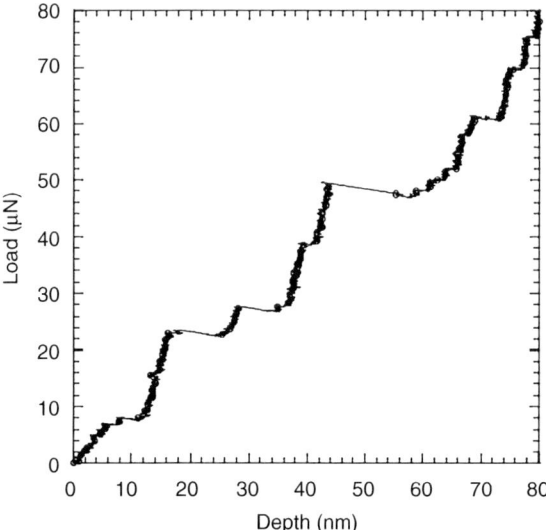

Figure 14.5: Typical load–displacement curve in a nano-indentation measurement (load control) [24]. Continuous increase of indentation depth with indenter loading shows elastic deformation interspersed with horizontal jumps. These "bursts" are interpreted as the onset of dislocation activity, each jump corresponding to a local instability somewhere in the system (not necessarily right under the indenter).

characterized by a tensor **L** playing the same role as the stiffness tensor **B**. This formalism can be taken to the discrete-particle level to obtain a spatially-dependent nucleation criterion for practical implementation [27, 28]. Consider a representative volume element (RVE) undergoing homogeneous deformation at finite strain to a current configuration **x**. Expanding the free energy F to second order in incremental displacement **u(x)**, one obtains

$$\Delta F = \frac{1}{2} \int_{V(x)} D_{ijkl} u_{i,j}(\mathbf{x}) u_{k,l}(\mathbf{x}) dV \qquad (14.12)$$

where $D_{ijkl} = C_{ijkl} + \tau_{jl}\delta_{ik}, \tau_{ij}$ being the internal (Cauchy) stress, and $u_{i,j} \equiv \partial u_i(\mathbf{x})/\partial x_j$. In (14.12) C is the elastic constant tensor and **u(x)** the strain at the current state of stress. By resolving the displacement as a plane wave, $u_i(\mathbf{x}) = w_i e^{i\mathbf{k}\cdot\mathbf{x}}$, one arrives at the stability condition for the RVE,

$$\Lambda(\mathbf{w}, \mathbf{k}) = (C_{ijkl} w_i w_k + \tau_{jl}) k_j k_l > 0 \qquad (14.13)$$

The structure of (14.13) is analogous to that of the stiffness tensor **B**, the presence of the stress term represents the work done by the external load [3]. Whereas **B** determines the overall crystal stability in homogeneous deformation, Λ is in contrast a site-dependent quantity, with its sign indicating the concavity of F. The significance of (14.13) is that if a pair of **w**, **k** exists such that Λ vanishes or becomes negative, then homogeneity of the RVE cannot be maintained and a defect singularity will form internally. In other words, the inequality can be used to interrogate the elastic stability of the RVE by minimizing Λ with respect to the polarization vector **w** and the wave vector **k**. The minimum value of Λ, $\Lambda_{\min}$, acts as a measure of the *local micro-stiffness*; wherever $\Lambda_{\min}$ vanishes, an instability is predicted at that spatial position. Equation (14.13) is an energy-based criterion applicable to finite-strain deformation, with the minimization of Λ being the process where the local environment is sampled from point to point. Notice that the Helmholtz form of the free energy is used rather than the Gibbs form. This is because we are applying the plane wave resolution effectively in an RVE, the region containing the weak spot, with periodic boundary condition, so no external work is involved [29].

The usefulness of (14.13) has been demonstrated in analysis of MD simulation results on nano-indentation [27, 28, 30]. The first few predictions of instability according to the Λ criterion were found to correspond closely with jumps in the simulated load-displacement curve, as shown in Figure 14.6. Because the simulations were performed at increments of fixed strain (displacement control mode), discontinuities appear as vertical jumps in contrast to Figure 14.5. We refer interested readers to the original publications for further discussions of atomic-level understanding of dislocation nucleation and propagation in nano-indentation.

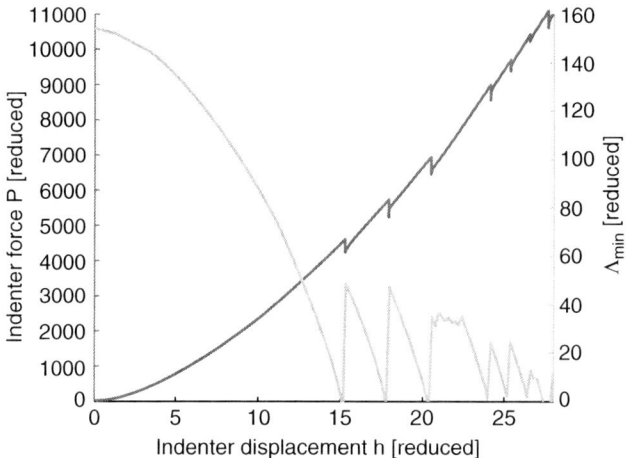

Figure 14.6: Load-displacement response in an MD simulation of nano-indentation (load control) on a slab of single crystal of Cu (rising curve), showing several vertical jumps corresponding to the dislocation bursts of Figure 14.1 [27]. The decreasing curve gives the variation of $\Lambda_{\min}$ with indenter depth. According to (14.13), each vanishing of $\Lambda_{\min}$ signals a structural instability.

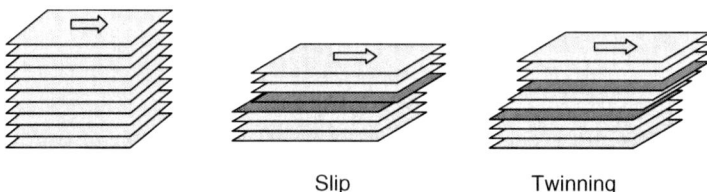

Figure 14.7: Schematic of a stack of undeformed crystal planes (left) about to undergo shear localization. A dislocation is nucleated by a *single* relative displacement (middle), whereas twinning requires *two or more* consecutive relative displacements (right, three displacements).

14.2.5 Instability in shear deformations—dislocation slip versus twinning

Slip and twinning of crystal planes are competing processes by which a crystal can plastically accommodate large shear strains. They are distinguished by the number of planes which undergo sliding, as indicated schematically in Figure 14.7. In the lattice response to shear; slip refers to *only one* relative displacement between two adjacent layers, while for twinning a stack of *three or more layers* must undergo uniform shear. When a crystal is being sheared uniformly, all the planes initially respond elastically. This would continue until symmetry is broken, when the system spontaneously transforms into a crystal containing a defect (loss of homogeneity). In practice considerations of factors such as the crystal structure and the material in question, the planes on which deformation is taking place, the temperature and the shear rate allow one to predict on the basis of conventional wisdom whether slip or twinning is favored. In the absence of such knowledge, one can resort to simulation for observing the atomistic details concerning system response near a saddle point.

Molecular dynamics (MD) simulations have been performed to observe the spontaneous nucleation of a deformation twin in a bcc crystal [31, 32]. A simulation cell containing 500,000 atoms with periodic boundary conditions is chosen to have the X (horizontal), Y (normal to plane of paper), and Z (vertical) axes oriented as shown in Figure 14.8. Shear is applied at 3×10^6 s^{-1} on the xy plane in the X (twinning) direction, extremely fast by laboratory standards but quite slow by MD standards. At 10 K we observe homogeneous nucleation of a deformation twin at a critical shear stress of 12.2 GPa (7.8% strain). Once nucleation sets in, a sharp decrease in strain energy and shear stress is observed. The full 3-D configuration of the twinned crystal has too many degrees of freedom to be efficiently processed in any kind of detailed analysis. Given that the twinned region is well localized, an overwhelming majority of the particles are not relevant for the characterization of defect configuration and energy. We therefore introduce a

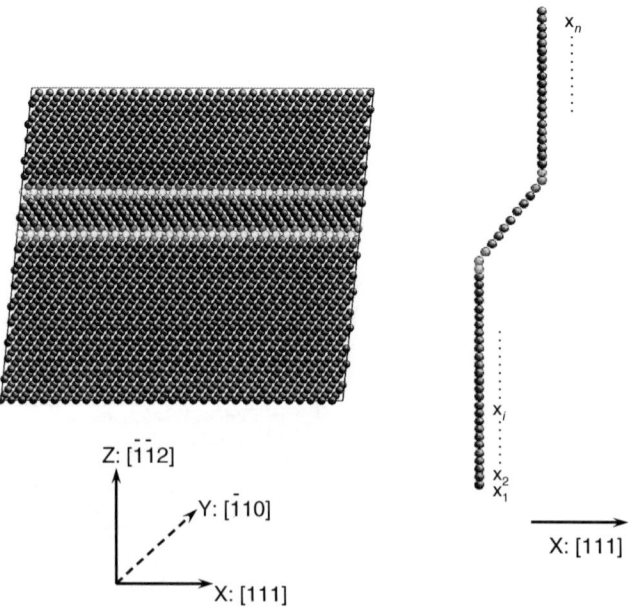

Figure 14.8: A 3-D crystal in shear along the X axis containing a twinned region of seven planes, bounded by two twin boundaries (left). Schematic of the 1-D chain model with nine twinned layers (in this illustration) [31, 32].

1-D chain model to single out those coordinates essential to define the twin structure. As indicated in Figure 14.8, we regard the twin as a one-dimensional chain of "defect atoms" specified by a set of coordinates x_i, measured on the X-axis in the twinning direction. Each "defect atom" i represents a plane of physical atoms, the layer being perpendicular to the chain direction, with coordinates (x_i, y_i, z_i). In this model the only degrees of freedom are the *relative displacements* in the twinning direction *between adjacent layers*, that is, $\Delta x_i = x_{i+1} - x_i$. For a twin consisting of N relative displacements of 'defect atoms', there are N *primary* degrees of freedom, Δx_i, i = 1, ..., N, specifying the defect. All the other degrees of freedom, the relative displacements in the Y and Z directions will be considered frozen and taken out of the analysis. The system energy is therefore only a function of N variables, $E = E(0, \ldots 0, \Delta x_j, \ldots, \Delta x_{j+N}, 0, \ldots 0)$, with the displaced planes starting at position j and ending at j+N.

Using the chain model, with the energy functional $E(\Delta x_1, \ldots, \Delta x_N)$ evaluated numerically treating all the degrees of freedom except $\Delta x_1, \ldots, \Delta x_N$ in the 3-D crystal as either frozen or under constrained relaxation, one has a means to examine the energetics of the twin defect for an arbitrary number of layers undergoing relative displacements. The simplest case is the one-component chain involving a rigid translation of the upper half of the lattice relative to the lower half. The one-dimensional energy $E(0, \ldots, \Delta x_1, \ldots 0)$, allowing relaxation in the other two directions, is conventionally known as the γ-surface, a quantity commonly used to characterize lattice deformation in shear. The two-component chain is the one of present interest, as it is able to describe either a one-layer slip or a two-layer twin. The two-dimensional energy surface is shown in Figure 14.9, where a minimum is now seen around the displacements (b/3, b/3) [32]. The significance is that under positive shear the present system can either twin or slip. The energy barrier for twinning is found to be 0.672eV with the saddle point at (0.36b, 0.16b), while for slip the barrier is 0.736eV with the saddle point at (0.5b, 0.09b). Under negative shear, only slip is allowed, at a barrier of 0.808eV. This is the kind of atomic-level energy landscape that enables one to analyze the competition between dislocation slip and deformation twin. Another useful way to examine the energy surface is in the form of a contour plot, also shown in Figure 14.9. In this one can trace out the minimum-energy path for the two deformations, using any of the reaction pathway techniques in the literature. The particular result given here was obtained by the method of nudged elastic band [33]. The path is seen to connect the initial configuration at the perfect lattice energy minimum (0,0) with the two possible final-state configurations, an energy minimum corresponding to a two-layer twin at (b/3, b/3), and another minimum corresponding to a dislocation slip at (b, 0). The two paths bifurcate at (0.29b, 0.03b) before either of the saddle points is encountered. The system can either twin or slip after the bifurcation point; however, since the twinning path has a lower energy barrier than the slip path, 0.672eV to 0.736eV, twinning is expected to be favored.

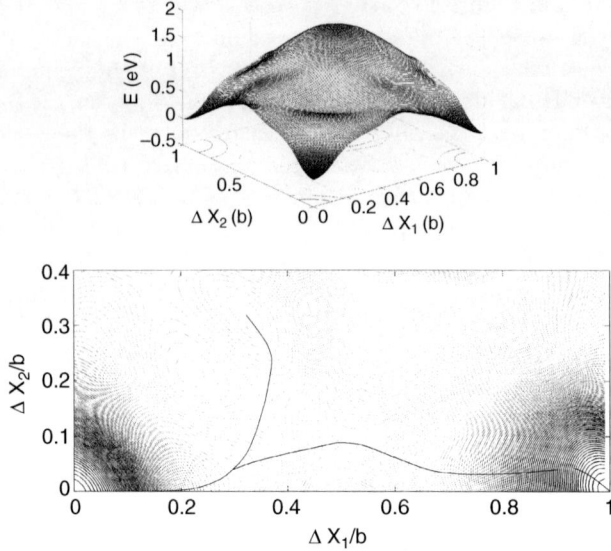

Figure 14.9: Strain energy surface (upper figure) and energy contour plot (lower figure) for the $(\bar{1}\bar{1}2)[111]$ deformation, with Y and Z relaxations, in a two-layer analysis [31]. An energy minimum at relative displacements of b/3 for sliding between adjacent layers confirms the existence of a twin defect in the present model.

14.2.6 *Strain localization*

In our MD simulation the process which revealed the formation of a twin involved the application of a uniform shear strain to a perfect crystal in incremental steps. Initially the system responded uniformly, the deformation being elastic and reversible; if the strain were released the system would return precisely to its initial configuration. When the applied strain reaches a critical level, the system undergoes a structural transition whereby all the strain in the system is concentrated in the immediate surrounding of the defect, while the other parts of the crystal return to a state of zero strain. The initial and final configurations of such a transition are given in Figure 14.10. To visualize the rather complex sequence of atomic rearrangement that must take place, we introduced a simple tiling device to enable us to see the mechanism of the localization process [31]. The tiling procedure consists of imposing a modulating perturbation wave, a sinusoidal wave of negligible amplitude across the system, as shown in Figure 14.10. The perturbation is much too weak to have any effect on the dynamical evolution of the system, while the distortion of the wave during localization helps to reveal the local atomic shuffling details that otherwise would be difficult to extract from all the particle displacements in the system. The idea is analogous

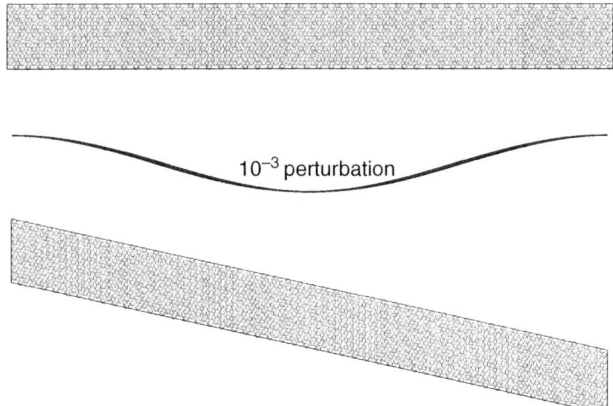

Figure 14.10: Nucleation of a deformation twin (shear localization), single crystal undeformed (upper) and in final sheared state (lower). Modulating perturbation wave (middle) follows the localization process by virtue of its distortion.

to painting a line across a wall before it starts to collapse and then following how the line breaks up as different parts of the wall start to crumble. By examining a video of the breakup of the perturbation wave, one is able to describe the sequence of structural transformations occurring in the nucleation of the deformation twin as a four-stage scenario, which is depicted in Figure 14.11. In the initial stage the linear wave grows in amplitude as the applied strain increases. When the amplitude reaches a certain level nonlinearities set in and the wave starts to distort (steepen). As the distortion approaches the critical level the wave front becomes increasingly sharp and collapses into a shock. In the final stage of localization the wave settles into a profile indicating the presence of a twin.

There is a lattice deformation involving shear localization that is much more extensively investigated than nano-indentation and affine shear. This is the problem of crack tip behavior in a ductile solid. A recent simulation study using reaction pathway sampling has shown that a sharp crack in an fcc lattice, such as Cu, will emit a dislocation loop under critical mode I (uniaxial tension) loading [34]. When the same method is applied to a brittle material, such as Si, a different result is found; the crack front advances by a series of bond breaking and reformation [35]. Thus, the deformation response of a solid can be very sensitive to the nature of the chemical bonding. What happens when a relatively brittle material, a semiconductor or a ceramic, is subjected to nano-indentation or affine shear? Preliminary results indicate that the system can undergo local disordering, suggesting yet another competing mechanism of local response to critical stress or strain. Clarification of this kind of phenomenon is work for the future. What hopefully is clear from the present discussion is that informed atomistic simulations will continue to provide a wealth of structural, energetic, and

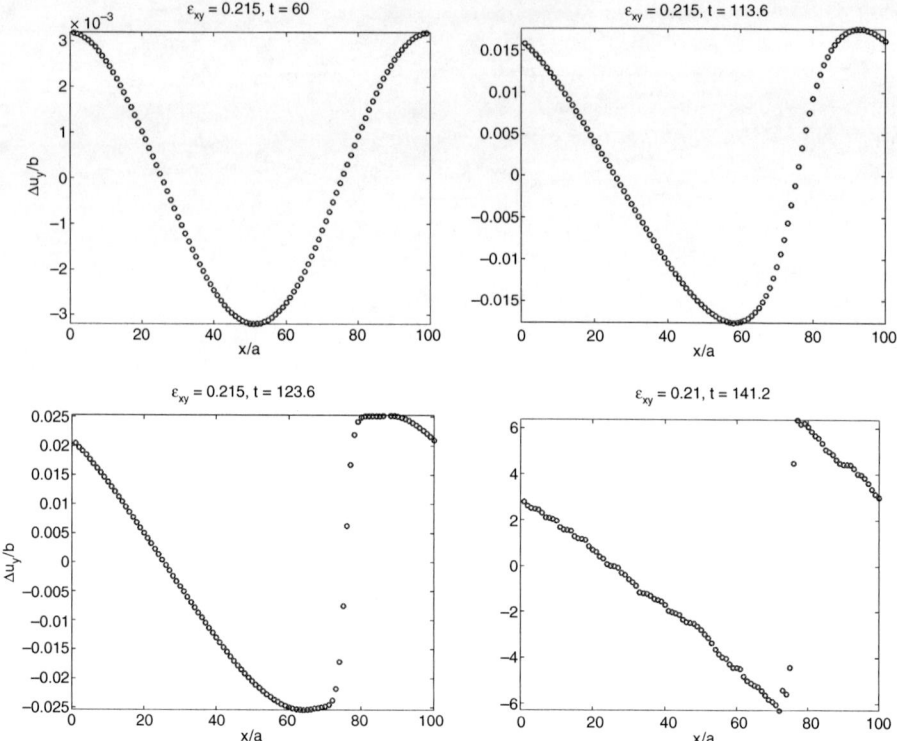

Figure 14.11: Evolution of the modulating perturbation wave during nucleation of a deformation twin. The wave distortions can be classified in four stages—growth of linear wave, onset of nonlinearity, increasing wave steepening toward a singularity, and final state of strain localization. Wave forms are those of the perturbation wave during the localization event [32].

dynamical information about the collective behavior of simple condensed matter, provoking in this way further questions for the statistical-physics community.

14.3 Atomistic measure of defect mobility

14.3.1 Single dislocation glide in a metal

Most mechanical properties of crystalline materials are affected, to a greater or lesser extent, by the presence of dislocations. Dynamic properties of dislocations, in terms of intrinsic properties of the dislocation and the effects of temperature and the driving force, play an important role in crystal plasticity [36, 37]. At zero temperature continuous glide motion occurs only if the applied stress σ exceeds a critical value σ_c^o given by the maximum glide resistance. At finite temperatures macroscopic continuous glide can occur with the help of fluctuations at any stress up to a value $\sigma_c \leq \sigma_c^o$. In bcc metals dislocation velocity can fall into four regions [38, 39]. In the first region, $v/c_s \leq 10^{-5}$, c_s being sound speed in the solid, the motion is controlled by the thermal release of dislocation from some equilibrium positions in the obstacle resistance profile. This is the thermally-activated region where the motion is jerky and the stress dependence is highly nonlinear. Also the velocity increases with temperature, the motion is sensitive to imperfections. The second region, $10^{-5} < v/c_s < 10^{-2}$, is a continuation of the first, the motion being sensitive to all contributions to the glide resistance. In the third region where $v/c_s \geq 10^{-2}$, the drag resistance is predominant. Here $\sigma > \sigma_c$, the dependence of $v(\sigma)$ is linear, the velocity characteristically decreases with temperature and depends weakly on the imperfection concentrations. In the fourth region at $v \sim c_s$, relativistic effects can be observed.

We will employ a periodic simulation cell that contains an edge dislocation dipole, prepared as indicated in Figure 14.12 [40]. We first remove the atoms on two identical half planes, then we displace the atoms immediately to the left and right of the gap by $[111]/3$ and allow the system to relax to zero stress. This gives a dislocation dipole configuration with Burgers vector b = a/2[111] and glide plane $(1\bar{2}1)$. Simulation is carried out in the $NT\sigma$ ensemble, with an overall shear stress imposed on the simulation cell through the Parrinello-Rahman procedure [41].

To extract the dislocation velocity from the atomic trajectories we need to locate the dislocation core position at each time step. Since the dislocation line is approximately parallel to the Z-axis, we divide the simulation cell into slices along this direction, with each slice containing one layer of atoms. For each slice we first identify the two rows of atoms immediately above and below the slip

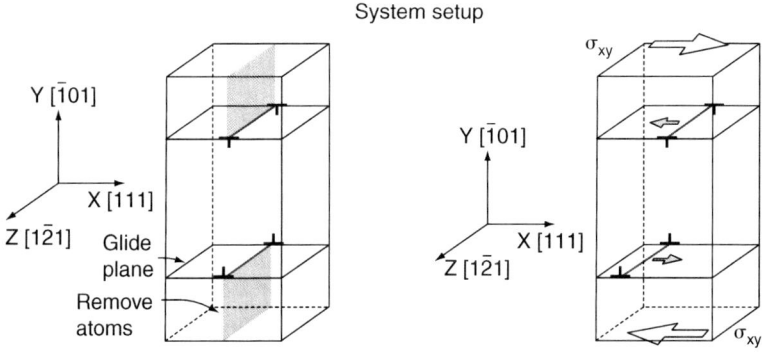

Figure 14.12: Periodic simulation cell containing an edge dislocation dipole under applied shear stress. The two dislocations will glide on the $(1\bar{2}1)$ slip plane.

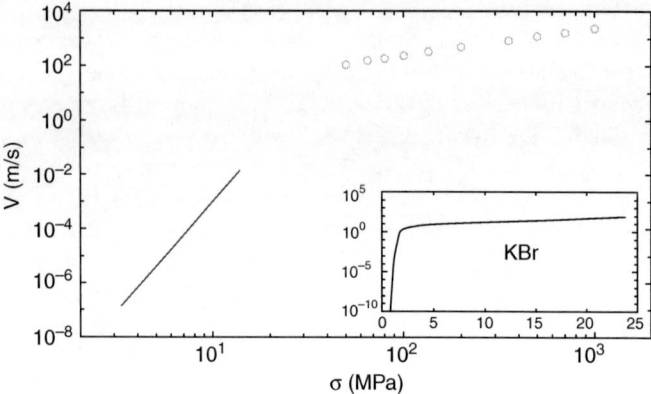

Figure 14.13: Stress dependence of edge dislocation velocity in Mo at T = 77 K. Simulation results shown as open circles and experimental results denoted by a solid line. Inset shows the change in dislocation velocities (edge and screw) as a function of shear stress in calcium-doped KBr at room room temperature where the stress exponent m decreases from m = 40 to m = 1 [42].

plane. Then we calculate the disregistry between the two rows as a function of the X-coordinate, keeping in mind that this function is periodic along X. The X-position of maximum disregistry is taken to be the dislocation core position. By averaging along the dislocation line we determine the average line position at each time step, and from this information we deduce the dislocation velocity. The dislocation velocity results obtained from a series of MD simulations on bcc Mo over a temperature range of 2150 K, and stresses 5 MPa to 5 GPa are displayed in Figures 14.13 and 14.14 concerning stress and temperature dependence, respectively.

Figure 14.13 shows that in the stress range we have studied, the velocity varies essentially linearly with the stress which is a distinctive feature of the third region discussed above. This behavior is almost to be expected in that the MD method is known to be restricted to microscopic distances and times. Consequently, one can only probe the high-velocity, high-stress region, while experiments are generally confined to the low-velocity, low-stress region. In order to reach stresses comparable to the experimental range, one needs to resort to acceleration techniques that allow simulations over a considerably longer time period.

The dislocation drag coefficient B, $B = b\sigma/v$, is shown in Figure 14.13, in normalized form, where θ_D is the Debye temperature. We see that the drag coefficient increases with increasing temperature, which means the velocity decreases with increasing temperature. This is fundamentally different from the behavior in the thermal activation region. The implication is that the underlying mechanism for the damping of edge dislocations is phonon drag. Although there are no experimental data on Mo to validate our interpretation, experiments on other

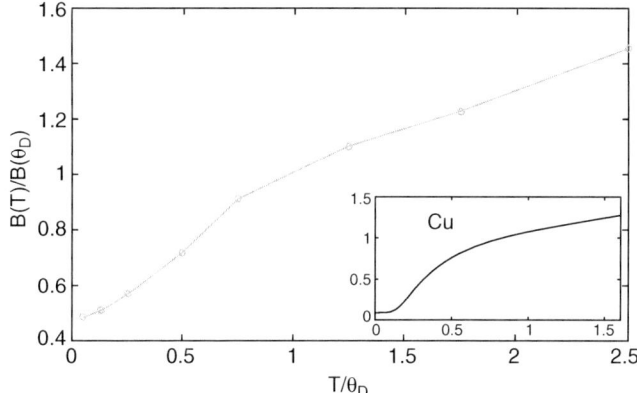

Figure 14.14: Temperature dependence of drag coefficient for edge dislocation glide in Mo. Simulation results are denoted by open circles. Inset shows experimental results for copper [43].

materials do show a positive temperature dependence as illustrated in the inset figure. Theoretical calculations yield similar results [44]. Another feature of the temperature dependence results is a plateau in the limit of low temperature, a behavior that is also suggested by the experiments.

14.3.2 Dislocation mobility in silicon: kink mechanism

We describe a kinetic MC method of modeling the glide mobility of a single dislocation based on the sequence of elementary processes of double-kink nucleation, kink migration, and kink annihilation. The method is applied to study the coupling effects of the dissociated partial dislocation in Si, a system well known to have a high secondary Peierls barrier [45]. As illustrated in Figure 14.15, the potential energy landscape has undulations along two directions, the primary and secondary barrier heights are u_1 and u_2, respectively. The kink mechanism for a dislocation line, lying initially along one of the potential valleys, to move to the adjacent valley is to first nucleate a double-kink (blue line) over the primary barrier, and then let this double-kink expand across the secondary barrier so the dislocation line effectively slides over the primary barrier. The entire process therefore involves two activation events, nucleation of a double-kink of certain width and propagation of the left and right kinks (see Figure 14.15), each with a characteristic energy that will have to be specified.

The kinetic MC method is a way to track the cumulative effects of nucleation anywhere along a dislocation line, and once a double-kink is nucleated it can grow by kink migration, as illustrated schematically in Figure 14.16. In this case we have a dissociated dislocation described by a leading and a trailing partial dislocation bounding an area that is a stacking fault. The direction of motion is upward in the figure.

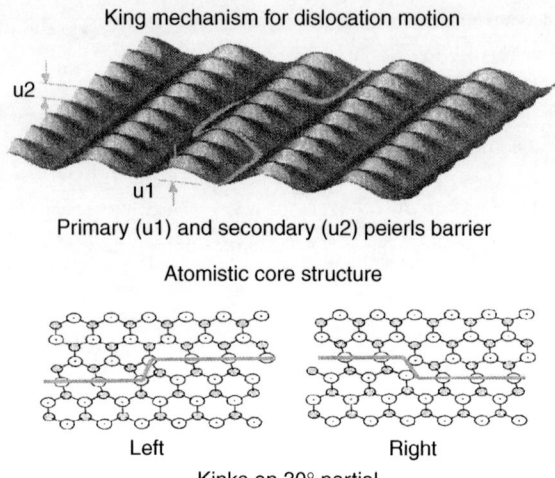

Figure 14.15: Schematic of Peierls barriers for double-kink mechanism of dislocation motion (upper) and atomic configurations of left and right kinks on the 30° partial dislocation (lower).

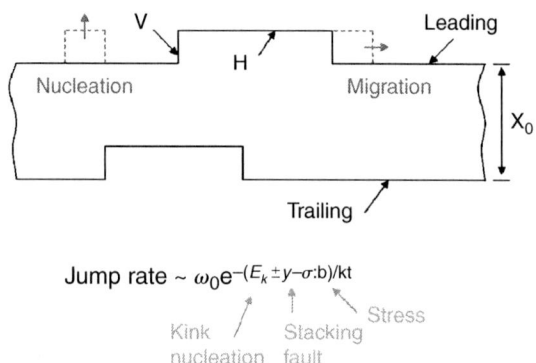

Figure 14.16: Schematic of dissociated dislocation (leading and trailing partials bounding a region of stacking fault). Along the leading partial nucleation and migration events are sampled according to a distribution which considers in addition to nucleation (or migration) activation the effects of the stacking fault energy and Peach-Koehler stresses [45].

The simulation is designed to produce the overall dislocation movement averaged over a large number of individual events, requiring for input the kink formation and migration energies which in principle can be computed by electronic structure or atomistic calculations. The model we describe deals with a screw dislocation dissociated into two 30° partials in the (111) plane. Each partial is represented by a piecewise straight line composed of alternating horizontal (H) and vertical (V) segments. The length of H segments can be any multiple of the Burgers vector b, while the V segments all have the same length, the kink height h. The stacking fault bounded by the two parallel partials has a width measured in multiples of h. The simulation cell is oriented with the partials running horizontally so that a periodic boundary condition can be applied in this direction.

In each simulation step, a stochastic sampling is performed to determine which event will take place next. The rates of these elementary events are calculated based on the energetics of the corresponding kink mechanisms. For example, for a kink pair nucleation three energy terms are considered, a formation energy, an energy bias favoring the reduction of the stacking-fault area, and the elastic interaction between a given segment and all the other segments and applied stress (so-called Peach-Koehler interaction). The estimate of the nucleation rate is what is shown in Figure 14.16. A similar expression also exists for kink migration. The formation and migration energies used in the simulation are given in the table in Figure 14.17. Figure 14.7 shows the dislocation velocities obtained from the kMC simulations using two sets of atomistic input. While the results are seen to differ by some four orders of magnitude, they do bracket the experiments.

One could be cautiously optimistic about this comparison, considering the roughness of the calculation (kink multiplicity not considered and using rather approximate values for the frequency factor and entropy). Of the materials parameters that entered into the model, the kink energies reamin uncertain despite efforts to obtain more accurate results using tight-binding and density

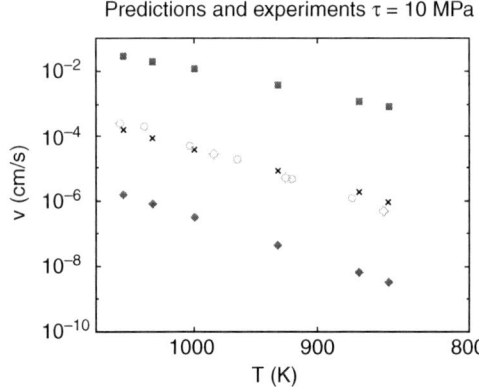

Figure 14.17: Temperature variation and magnitude of velocity of screw dislocation in Si, kMC results using two sets of kink nucleation and migration energies are denoted by square (EDIP potential) and closed diamond (tight binding), while experimental data are shown as open diamonds [45].

functional theory methods. When more reliable atomistic results become available, we believe this approach can lead to significant understanding of the effects of applied stress on dislocation mobility in a system like Si.

14.3.3 Crack front extension in metal and semiconductor

Is it possible to study how a sharp crack evolves in a crystal lattice without actually driving the system to the point of instability? By this we mean determining the pathway of a crack front motion while the lattice resistance against such displacement is still finite. Despite a large number of molecular dynamics simulations on crack tip propagation (e.g., see [57, 58]), this particular issue has not been examined. Most studies to date have been carried out in an essentially 2-D setting, with periodic boundary condition imposed along the direction of the crack front. In such simulations the crack tip is sufficiently constrained that the natural response of the crack front cannot be investigated. Besides the size constraint on the crack front, there is also the problem that in direct MD simulation one frequently drives the system to instability, resulting in abrupt crack-tip displacements which make it difficult to characterize the slow crack growth by thermal activation. We discuss here an approach capable of probing crack front evolution without subjecting the system to critical loading. This involves using reaction pathway sampling to probe the minimum energy path (MEP) [33] for the crack front to advance by one atomic lattice spacing, while the imposed load on the system is below the critical threshold. We have applied this method to characterize the atomistic configurations and energetics of crack extension in a metal (Cu) [34] and a semiconductor (Si) [35]. Thus we can compare the results of the two studies to show how ductility or brittleness of the crystal lattice can manifest in the mechanics of crack front deformation at the nanoscale.

Consider a 3-D atomically sharp crack front which is initially straight, as shown in Figure 14.18(a). Suppose we begin to apply a mode-I load in incremental steps. Initially the crack would not move spontaneously because the driving force is not sufficient to overcome the intrinsic lattice resistance. What does this mean? Imagine a final configuration, a replica of the initial configuration with the crack front translated by an atomic lattice spacing in the direction of the crack advancement. At low loads, e.g., K'_1 as shown in Figure 14.18(b), the initial configuration (open circle) has a lower energy than the final configuration (closed circle). They are separated by an energy barrier which represents the intrinsic resistance of the lattice. As the loading increases, the crack will be driven toward the final configuration; one can regard the overall energy landscape as being tilted toward the final configuration with a corresponding reduction in the activation barrier (see the saddle-point states (shaded circle) in Figure 14.18(b). As the load increases further the biasing becomes stronger. So long as the barrier remains finite the crack will not move out of its initial configuration without additional activation, such as thermal fluctuations. When the loading reaches the point where the lattice-resistance barrier disappears altogether, the crack is then unstable at the initial configuration; it will move without any thermal activation.

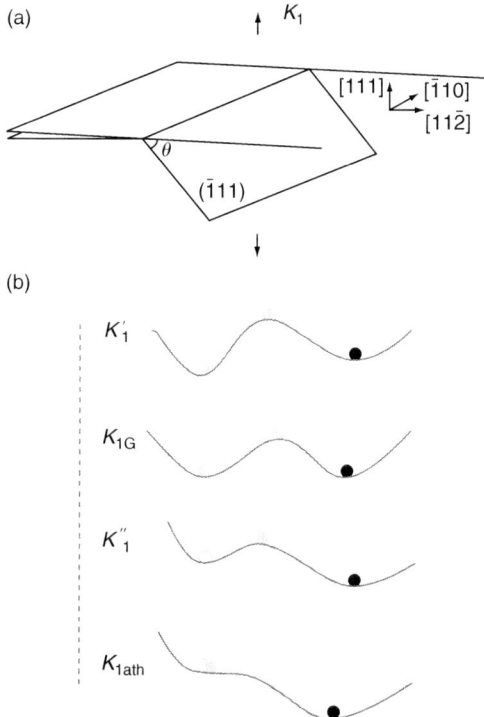

Figure 14.18: (a) Schematic of a 3-D atomically sharp crack front under a mode-I (uniaxial tensile) load K_1; (b) energy landscape of the crack system at different loads ($K_1' < K_{1G} < K_1'' < K_{1ath}$). Open circle represents the initial state of a straight crack front under an applied load, closed circle denotes the final state after the crack front has uniformly advanced by one atomic spacing (under the same load as the initial state), and shaded circle denotes the saddle-point [46].

This is the athermal load threshold, denoted by K_{1ath} in Figure 14.18(b). In our simulation, we study the situation where the applied load is below this threshold, thereby avoiding the problem of a fast moving crack that is usually overdriven.

The cracks in Cu and Si that we will compare are both semi-infinite cracks in a single crystal, with the crack front lying on a (111) plane and running along the $(\bar{1}10)$ direction. The simulation cells consist of a cracked cylinder cut from the crack tip, with a radius of 80 Å. The atoms located within 5 Å of the outer surface are fixed according to a prescribed boundary condition, while all the remaining atoms are free to move. To probe the detailed deformation of the crack front, the simulation cell along the cylinder is taken to be as long as computationally feasible, 24 (Cu) and two(Si) unit cells. A periodic boundary condition is imposed along this direction. With this setup, the number of atoms in the system are 10,392 (Cu) and 7720 (Si). For interatomic potentials we use a many-body potential of the embedded atom method type for Cu [47] for which the unstable stacking energy has been fitted to the value of 158 mJ/m^2, given by an *ab initio* calculation, and a well known three-body potential model for Si proposed by Stillinger and Weber [48].

Prior to applying reaction pathway sampling, we first determine the athermal energy release rate, denoted by G_{emit} (corresponding to K_{Iath} in Figure 14.18(b)). This is the critical value at which the activation energy barrier for dislocation nucleation vanishes, or equivalently a straight dislocation is emitted without thermal fluctuations [49, 50]. As detailed in [34], the athermal load is determined to be $K_{Iath} = 0.508$ Mpa$\sqrt{m}$ (or $G_{emit} = 1.629$ J/m^2, based on the Stroh solution [59]) for the nucleation of a Shockley partial dislocation across the inclined ($\bar{1}10$) slip plane. So long as the applied load is below K_{Iemit} the crack front will remain stable. It is in such a state that we will probe the reaction pathway for dislocation using the method of nudged elastic band (NEB) [33]. The quantity we wish to determine is the minimum energy path (MEP) for the emission of a partial dislocation loop from an initially straight crack front. MEP is a series of atomic configurations connecting the initial and final states. Here the initial configuration is a crack front as prescribed by the Stroh solution which is then relaxed by energy minimization, while the final configuration is a fully formed straight Shockley partial dislocation, a state obtained by taking the simulation cell containing a pre-existing Shockley partial dislocation and reducing to the same load as the initial state. (The pre-existing partial was generated by loading the simulation cell at a level above the threshold G_{emit}.) To find the MEP, 15 intermediate replicas of the system which connect the initial and final states are constructed. We choose intermediate replicas containing embryonic loops that result from the relaxation of a straight crack front, allowing for the nucleation of a curved dislocation. The relaxation of each replica is considered converged when the potential force vertical to the path is less than a prescribed value, 0.005 eV/A in our case.

The sequence of replicas defines a reaction coordinate in the following sense. Each replica in the sequence is a specific configuration in 3N configuration hyperspace, where N is the number of movable atoms in the simulation. For each replica we calculate the hyperspace arc length ℓ between the initial state $\mathbf{x}_i^{3N}$ and the state of the replica $\mathbf{x}^{3N}$. The normalized reaction coordinate s is defined to be $s \equiv \ell/\ell_f$, where ℓ_f denotes the hyperspace arc length between the initial and final states. The relaxed energy of any replica is a local minimum within a 3N-1 hyperplane vertical to the path. By definition, MEP is a path that begins at $\Delta E = 0 (s = 0)$, where ΔE is the relaxed energy measured relative to the energy of the initial state. Along the path (reaction coordinate s) ΔE will vary. The state with the highest energy on this path is the saddle point, and the activation energy barrier is the energy difference between the saddle point and the initial state. Figure 14.19 shows the MEP for the nucleation of a dislocation loop from the crack front in Cu, loaded at G = 0.75 G_{emit}. Notice that at this loading

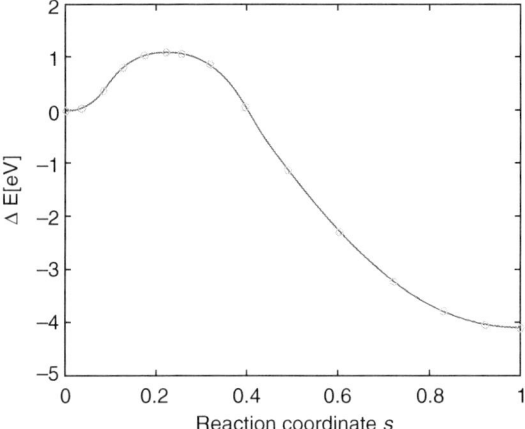

Figure 14.19: MEP of dislocation loop emission in Cu at a load of G = $0.75G_{emit}$ [34].

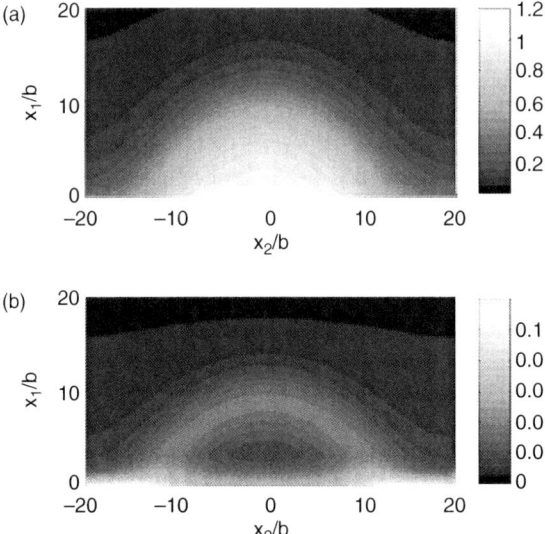

Figure 14.20: Shear (a) and opening (b) displacement contours across the slip plane at the saddle-point state [46].

the final state is strongly favored over the initial state. Figure 14.19 also shows clearly the presence of a lattice resistance barrier at this particular loading.

To visualize the variation of atomic configurations along the MEP, we turn to displacement distributions between atomic pairs across the slip plane.

Figure 14.20(a) shows a contour plot of the shear displacement distribution along the crack front at the saddle-point state. One can see clearly the shape of a dislocation loop bowing out; the profile of b/2 shear displacement is a reasonable representation of the locus of dislocation core. Also this is an indication that the enclosed portion of the crack front has been swept by the Shockley partial dislocation loop. Besides shear displacement, normal or opening displacement in the direction [$\overline{1}$11] is also of interest. The corresponding distribution is shown in Figure 14.20(b). One sees that large displacements are not at the center of the crack front. In Figures 14.20(a) and 14.20(b) we have a detailed visualization of the crack front evolution in three dimensions. The largest displacements are indeed along the crack front, but they are not of the same character. The atoms move in a shear mode in the central region, and in an opening mode on the two sides. For a similar discussion of crack advancement in Si, the reader can consult [46].

14.4 Outlook for multiscale materials

It is safe to predict that the computational materials research community has an opportunity to play an increasingly important role in developing the scientific models and simulation tools to solve very complex materials problems faced by our society, for example, problems concerning the properties and behavior of materials in extreme environments. Besides the challenges in understanding the kind of unit "process" problems that we have discussed, there will be a need to focus on larger-scale ("overarching") problems that can drive an entire team of investigators with complementary capabilities and interest. Virtual, microscale models will be needed to probe specific functional behavior of a real material on the mesoscale. In essence this is but another realization of linking nanoscale mechanics and physics with mesoscale and macroscale performance, with a specific identification of the behavior of interest. Problems of this type, which are being considered only recently, come from application areas which can be very different from each other, such as understanding at the molecular level the setting kinetics of cement [51, 52], analyzing the fatigue resistance of bearing steel [53] using atomistic methods, and probing the oxidation resistance of an ultra-high temperature ceramic by combining molecular simulation with experiments [54]. Yet in terms of multiscale modeling and simulation they are all characterized by similar fundamental complexities, those associated with the interplay between chemical reactivity and the mechanics of deformation and transport. Without the focus on a specific functional behavior, these materials systems are much too complex for multiscale modeling and simulation at the present time. The approach we are contemplating is analogous to the notion of the inverse problem. An optimistic outlook is that any breakthrough in solving one problem would have widespread impact on the evolution of the entire field.

14.5 Acknowledgment

The research described here was carried out over a number of years with collaborators Ju Li, Ting Zhu, Jingpeng Chang, Wei Cai, Jinghan Wang, Shigenobu Ogata, Krystyn Van Vliet, Dieter Wolf, and Subra Suresh.

References

[1] Yip S. (2003). "Synergistic Science", *Nature Mater.*, **2**, p. 3.
[2] National Science Foundation Report, *Revolutionizing Engineering Science through Simulation*, Report of the National Science Foundation Blue Ribbon Panel on Simulation-Based Engineering Science, February 2006.
[3] Yip S., ed. (2005). *Handbook of Materials Modeling*, Springer, New York.
[4] Chandler E. and Diaz de la Rubia, T. guest eds. (2002). special issue of *J. Computer-Aided Materials Design*, vol. **9**, p. 2.
[5] Born M. (1940). *Cambridge Philos. Soc.*, **36**, p. 160.
[6] Born M. and Huang K. (1956). *Dynamical theory of Crystal Lattices*, Clarendon, Oxford.
[7] Hill R. (1975). *Math. Proc. Camb. Phil. Soc.*, **77**, p. 225.
[8] Kelly A. and Macmillan N. H. (1986). *Strong Solids*, 3rd ed, Clarendon, Oxford.
[9] Wang J., Li J., Yip S., Phillpot S., and Wolf D. (1995). *Phys. Rev. B*, **52**, p. 12627.
[10] Zhou Z. and Joos B. (1996). *Phys. Rev. B*, **54**, p. 3841.
[11] Morris J. W. and Krenn C. R. (2000). *Philos Mag. A*, **80**, p. 2827.
[12] Gibbs J. W. (1993). In *The Scientific Papers of J. Willard Gibbs, Vol. 1: Thermodynamics*, Ox Bow Press, Woodbridge, Conn. p. 55.
[13] Morris J. W., Krenn C. R., Roundy D., and Cohen M. L. (2000). In *Phase Transformations and Evolution in Materials*, Turchi P. E. and Gonis A., eds., TMS, Warrendale, p. 187.
[14] Wallace D. C. (1972). *Thermodynamics of Crystals*, Wiley, New York.
[15] Wang J., Yip S., Phillpot S., and Wolf D. (1993). *Phys. Rev. Lett.*, **71**, p. 4182.
[16] Barrons T. H. K. and Klein M. L. (1965). *Proc. Phys. Soc*, **85**, p. 523.
[17] Hoover W. G., Holt A. C., and Squire D. R. (1969). *Physica*, **44**, p. 437.
[18] Basinski Z. S., Duesbery M. S., Pogany A. P., Taylor R., and Varshni Y. P. (1970). *Can. J. Phys.*, **48**, p. 1480.
[19] Mizushima K., Yip S., and Kaxiras E. (1994). *Phys. Rev. B*, **50**, p. 14952.
[20] Tang M. and Yip S. (1994). *J. Appl. Phys.*, **76**, p. 2716.
[21] Li J. and Yip S. (2002). *Computer Modelling in Engineering and Sciences*, **3**, p. 219.
[22] Tang M. and Yip S. (1995). *Phys. Rev. Lett.*, **75**, p. 2738.
[23] Li J. (2000). Ph.D. Thesis, MIT.
[24] Gouldstone A., Koh H.-J., Zeng K. Y., Giannakopoulos A. E., and Suresh S. (2000). *Acta Mater*, **48**, p. 2277.
[25] Hill R. (1962). *J. Mech. Phys. Solids*, **10**, p. 1.
[26] Rice J. R. (1976). In *Theoretical and Applied Mechanics*, Koiter W. T., ed., North-Holland, Amsterdam, vol **1**, p. 207.
[27] Li J., Van Vliet K. J., Zhu T., Yip S., and Suresh S. (2002). *Nature*, **418**, p. 307.
[28] Zhu T., Li J., Van Vliet K. J., Suresh S., Ogata S., and Yip S. (2004). *J. Mech. Phys Solids*, **52**, p. 691.

[29] Li J., Zhu T., Yip S., Van Vliet K. J., and Suresh S. (2004) "Elastic Criterion for Dislocation Nucleation", *Mat. Sci. & Eng*, **A365**, p. 25.
[30] Van Vliet K. J., Li J., Zhu T., Yip S., and Suresh S. (2003). *Phys. Rev.* **B67**, p. 104105.
[31] Chang J., (2003). Ph.D. Thesis, MIT.
[32] Chang J., Zhu T., Li J., Lin X., Qian X., and Yip S. (2004). In *Mesoscopic Dynamics of Fracture Process and Materials Strength*, Kitagawa H. and Shibutani Y. eds., Kluwer Academic, Dordrecht, p. 223.
[33] Jonsson H., Mills G., and Jacobsen K. W. (1998). In *Classical and Quantum Dynamics in Condensed Phase Simulations*, Ed., Berne B. J., Ciccotti G., Coker D. F. eds., Plenum Press, New York, p. 385.
[34] Zhu T., Li J., and Yip S. (2004a). *Phys. Rev. Lett.*, **93**, p. 020503.
[35] Zhu T., Li J., and Yip S. (2004b) *Phys. Rev. Lett.*, **93**, p. 205504.
[36] Hirth J. P. and Lothe J. (1982). *Theory of Dislocations*, Wiley, New York.
[37] Nadgornyi E. (1988). *Dislocation Dynamics and Mechanical Properties of Crystals*, Progress in Materials Science, vol. **31**.
[38] Klahn D., Mukherjee A. K., and Dorn J. E. (1970). *Strength of Metals and Alloys*, Proceedings of the Second International Conference, American Society of Metals, Cleveland, OH, p. 951.
[39] Neuhauser H. (1979). In Haasen P., *et al.* eds., *Strength of Metals and Alloys*, Proceedings of the Fifth International Conference, Pergamon Press, Oxford, p. 1531.
[40] Chang J., Bulatov V. V., and Yip S. (1999). *J. Computer-Aided Mater. Design*, **6**, p. 165.
[41] Parrinello M. and Rahman A. (1981). *J. Appl. Phys.*, **52**, p. 7182.
[42] Pariiskii V. B., Lubenets S. V., and Startsev V. I. (1966). *Soviet Phys. Solid State*, **8**, p. 976.
[43] Jasby K. M. and Vreeland T. (1973). *Phys. Rev. B*, **8**, p. 3537.
[44] Alshits V. I. and Indenbom V. L. (1986). *Dislocation in Solids*, **7**, p. 43.
[45] Cai W., Bulatov V. V., Justo J. F., Argon A. S., and Yip S. (2000). *Phys. Rev. Lett.*, **15**, p. 3346.
[46] Zhu T., Li J., and Yip S. (2005). *J. Appl. Mech.*, **72**, p. 1.
[47] Mishin Y., Mehl J., Papaconstantopoulos D. A., Voter A. F., Kress J. D. (2001). *Phys. Rev. B*, **63**, p. 224106.
[48] Stillinger F. H. and Weber T. A. (1985). *Phys. Rev. B*, **31**, p. 5262.
[49] Rice J. R. (1992). *J. Mech. Phys. Solids*, **40**, p. 239.
[50] Xu G., Argon A. S., and Ortiz M. (1997). *Philos. Mag. A*, **75**, p. 341.
[51] Lootens D., Hebraud P., Lecolier E., and van Damme H. (2004). *Oil & Gas Sci. Tech – Rev.IFP*, **59**, p. 31.
[52] Pellenq R. J.-M. and van Damme H. (2004). *MRS Bull*, **29**, p. 319.
[53] Forst C. J., Slycke J., Van Vliet K. J., and Yip S. (2006). *Phys. Rev. Lett.*, **96**, p. 175501.
[54] Bongiorno A., Forst C. J., Kalia R. K., Li J., Marschall J., Nakano A., Opeka M. M., Talmy I. G., Vashishsta P., and Yip S. (2006). *MRS Bull.*, **31**, p. 410.

[55] Hill R. and Milstein F. (1977). *Phys. Rev. B*, **15**, p. 3087.
[56] Tersoff J. (1989). *Phys. Rev. B*, **39**, p. 5566.
[57] Cleri F., Wang J., and Yip S. (1995). *J. Appl. Phys.*, **77**, p. 1449.
[58] Buehler M. J., Abraham F. F., and Gao H. (2003). *Nature*, **426**, p. 141.
[59] Stroh A. N. (1958). *Philos. Mag.*, **7**, p. 625.
[60] Wang J., Li J., Yip S., Wolf D., and Phillpot S. (1997). *Physica A*, **240**, p. 396.

15

FROM MACROSCOPIC TO MESOSCOPIC MODELS OF CHROMATIN FOLDING

Tamar Schlick

15.1 Introduction

One of the current challenges in scientific computing is model development that entails bridging the resolution among different spatial and temporal scales. In biological applications, a wide range of spatial scales defines systems, from the quantum particles to atomic, molecular, cellular, organ, system, and genome entities. Temporal scales range from sub-femtosecond for electronic motion to billions of years of evolutionary changes. As our computing power and algorithms have improved, problems of greater scientific significance can be addressed with enhanced confidence and accuracy. However, developing appropriate molecular models and simulation algorithms to answer specific biological questions that require bridging all-atom details with the macroscopic view of activity on the cellular level remains an *ad hoc* endeavor which requires as much art as science.

As a special volume of SIAM's journal on *Multiscale Modeling and Simulation* illustrated [1], multiscale biology is being developed by many varied techniques and applied to a variety of problems, such as involving protein and RNA three-dimensional structures, DNA supercoiling, ribosomal motions, DNA packaging in viruses, heart muscle motion, RNA translation, or fruit fly circadian rhythm. For these applications, techniques involve hierarchical methods that transform fast, low-resolution to slower, higher-resolution models; dynamics propagation using projection of standard molecular dynamics to longer timescales using master-equation methods; rigid-body dynamics; elastic or normal-mode models; and coarse-grained studies of slow, large-scale motions and features using differential equations for global properties or statistical methods.

This chapter describes the evolution of our group's models from macroscopic to mesoscopic scales developed to study chromatic folding; the focus is on presenting an overview of the models and results rather than simulation details which can be found in the individual papers. The chromatin fiber is the protein/DNA complex in eukaryotes that stores the genetic material [2]. With recent discoveries that point to a "second code" in DNA, a nucleosome-positioning code [3], understanding chromatin structure and dynamics becomes central, since transcription regulation and hence, the most fundamental biological processes, depend on chromatin architecture. This is because such fundamental cellular processes that are DNA-template directed require direct access to the

DNA material, which must be unraveled from its compact folded state with the cellular protein matrix [2].

Following a brief introduction into the biology of chromatin, we describe development of physical models, energy functions, and simulation algorithms (Monte Carlo, Brownian Dynamics) for chromatin that gradually incorporated greater molecular complexity. We then mention how the models were validated against experiments and describe the biological findings concerning low-salt/high-salt unfolding/folding dynamics and the structural aspects of the chromatin fiber as a function of the ionic environment (ion charge and type) and the presence of linker histones, as well as the stabilizing role of the histone tails. We conclude by mentioning future applications and required model developments.

15.2   Chromatin structure

Superimposed upon the canonical right-handed B-DNA helix is a left-handed superhelical (or supercoiled) structure that facilitates fundamental template-directed biological process like transcription and replication. In higher organisms, this supercoiled DNA is wrapped around proteins, much like a long yarn around many spools (Fig. 15.1, taken from [4]). These protein anchors consist of a core of eight histone proteins—two copies each of H2A, H2B, H3, and H4—whose flexible positively-charged tails extrude from the cylindrical-like core to help shield the negatively-charged DNA chain. These proteins, along with the wrapped DNA, form the basic building block of chromatin, the nucleosome [5]. Additional, linker-histone proteins are needed for compacting the chromatin fiber into condensed states [6, 7, 8, 9, 10, 11]. The diameter of the chromatin fiber—30 nm—is widely quoted as the dimension at physiological ionic environments, but its detailed structure is unknown. However, this dimension represents only a DNA compaction factor of about 40 (Fig. 15.1). Very condensed states at certain stages of the cell cycle likely involve fiber/fiber interactions and further compaction by structural proteins [12, 13, 14, 15]. Thus, to pack the genomic material into the cell whose diameter of ∼5 µm is smaller by more than five orders of magnitude than the linear length of the DNA stored, a severe folding problem must be solved. This multiscale "DNA folding problem" is a challenge to both experimentalists and modelers because the resolution of the respective techniques is limited [16].

X-ray crystallography has produced very detailed structures of the nucleosome building block (∼150 bp of DNA wrapped around the histone octamer) [17, 5], including the recent tetranucleosome [18], while electron microscopy can provide macroscopic fiber views, and dynamic techniques can measure polymer properties like diffusion properties or sedimentation coefficients [19]. Modeling on the level of the nucleosome is already too challenging for all-atom approaches, while macroscopic elastic models for long supercoiled DNA (e.g., [20, 21]) are inappropriate for the molecular complexity of the fiber. Hence modeling fiber structure and motion requires specialized models that treat the system's electrostatics as accurately as possible—since these features are thought to be crucial for

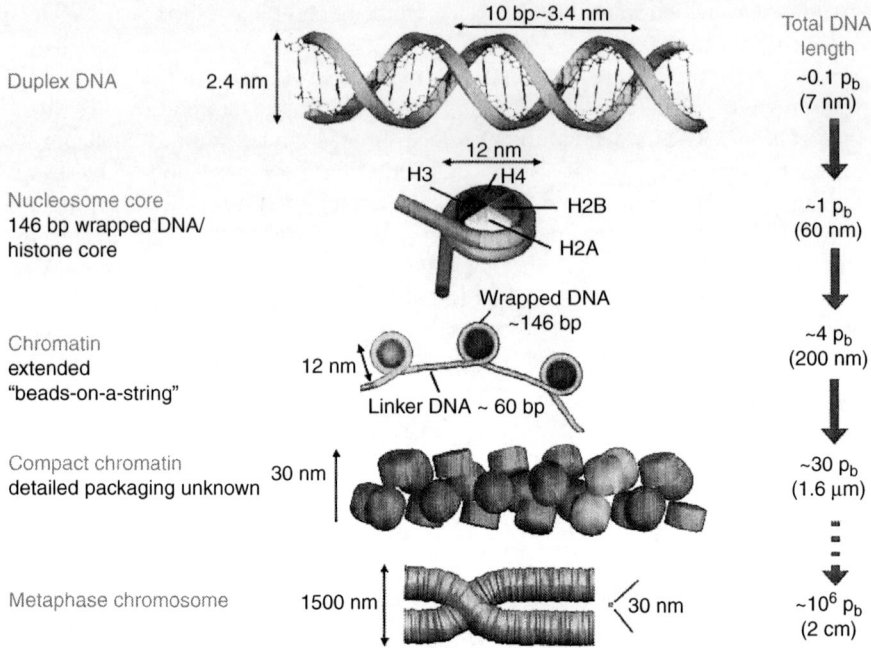

Figure 15.1: Eukaryotic DNA in the cell (from [4]). In eukaryotes, the DNA is organized in the chromatin fiber, a complex made of DNA wrapped around core of eight proteins, two copies each of H2A, H2B, H3, H4. The chromatin building block is the nucleosome. At low salt, the fiber forms "beads on a string"-like models and, at physiological salt concentrations, it compacts into the 30 nm fiber, whose detailed structure is unknown. Much more condensed fiber structures are formed during transcription-silent phases of the cell cycle. In each level of organization shown, the pink unit represents the system from the prior level, to emphasize the compaction ratio involved. Note also that at lengths much less than the DNA persistence length (which is ∼150 bp), the DNA is relatively straight but, at much longer lengths, the DNA is very flexible. (See Plate 14).

chromatin organization—while approximating other components so as to make possible studies of nucleosome arrays (or oligonucleosomes) with sufficient configurational sampling (e.g., 100 million configurations) or simulation times of milliseconds and longer to be biologically relevant. In recent years, a variety of computational models have been developed (e.g., [22, 23, 24, 25, 26, 27, 28]).

A key question for investigation is what is the precise organization of the chromatin fiber at various physiological conditions (e.g., ionic concentrations). In particular, how do these structures depend on the presence of linker histones and variations in the linker DNA length? What is the role of the histone tails in fiber compaction? And what are the energetics of folding/unfolding?

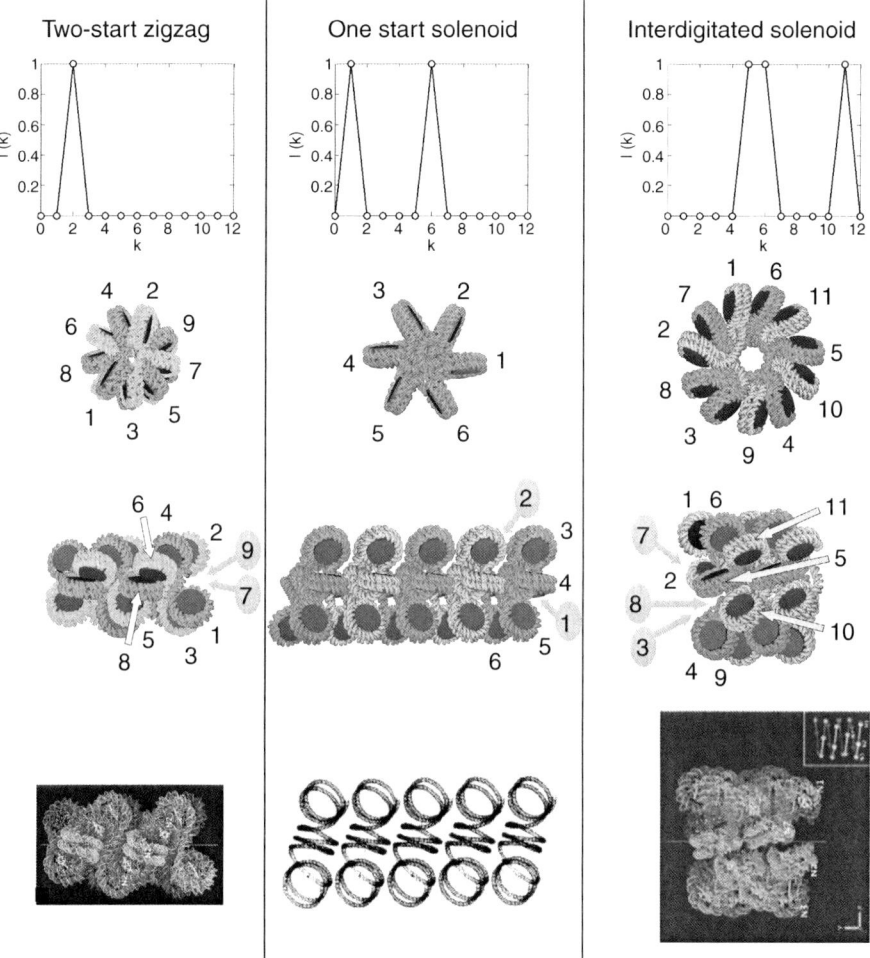

Figure 15.2: Three hypothesized chromatin arrays are shown from the top and side with black nucleosome cores. In the classical zigzag model [18], DNA is shown in orange and blue, and nucleosomes i interact with i ± 2. In the classical solenoid model [30], DNA is shown in pink and blue, and nucleosomes i interact with i ± 1 and i ± 6. In the interdigitated solenoid model [9], DNA is shown in pink and blue, and nucleosomes i interact on the flat sides with i ± 5, i ± 6, and on the narrow edges at i ± 11. Yellow arrows point to visible nucleosomes, and grey arrows point behind the chromatin to hidden nucleosomes. Images at the bottom were reprinted with permission from: (left) Macmillan Publishers Ltd. (Fig. 15.3 of T. Schlach et al., Nature **436**(7047): 138-141, 2005, license 1943250688673); (middle) Elsevier (Fig. 15.5 of J.D. McGhee et al., Cell **33**(3): 831-841, 1983, license 1982084); and (right) National Academy of Sciences (Fig. 15.3 of P.J. Robinson et al., PNAS USA **103**(17): 6506-6511, 2006, copyright 2006 National Academy of Sciences USA). (See Plate 15).

Already, many experimental and theoretical studies have proposed various structural possibilities (e.g., see [29]). Two broad classes are the *solenoid*, a helical arrangement in which the linker DNA is bent (e.g., [9, 30]), and the *zigzag*, in which the linker DNA is mostly straight [18, 31]. Figure 15.2 illustrates some possible chromatin architectures with associated internucleosome patterns. Besides the classic solenoid and 2-start zigzag models, an interdigitated solenoid [9] is also shown. Many structural studies converge upon the irregular two-start *zigzag* which brings each nucleosome into closest contact with its second nearest neighbor (i±2). *Solenoid*-like structures, however, have dominant interactions between each successive pair of nucleosome (i ± 1) and/or involving (i ± 5) or (i ± 6) [32] depending on the repeated patterns (see Figure 15.2). Both monovalent and divalent ions as well as linker histones are known to affect these patterns significantly since, through electrostatic shielding, they can allow fiber segments to come into closer contact. In addition, it is likely that fiber structure is polymorphic [33, 34], so that several morphologies exist and interchange depending on detailed conditions in the cell. Chromatin modeling and simulation, thus, has the potential to provide important insights into these questions and add detailed structures for analysis.

15.3 The first-generation macroscopic chromatin models and results

Our first-generation "macroscopic" models treated the nucleosome and the wound DNA according to general mechanical and electrostatic properties, with the histone tails approximated as rigid bodies and linker histones neglected (Fig. 15.3) [24, 35, 36, 37, 38]. As Figure 15.3 shows, the nucleosome core is represented as a large regular disk and a short slender disk for part of the H3 histone tail resolved in the 1997 crystal structure [17], with discrete charges determined to approximate the Poisson-Boltzmann solution of the electric field using our program DiSCO (Discrete Surface Charge Optimization) (see Figure 15.4) [35]; the minimization is performed efficiently using our truncated Newton code TNPACK [39]. The linker DNA is denoted by charged beads using the well known wormlike/chain model for supercoiled DNA developed by Allison and coworkers (as applied in [20]) using the Stigter charged cylinder electrostatics formulation. For dynamics, variables r_i are defined on each core and each linker DNA bead, each associated with a local Euler coordinate frame (Fig. 15.5).

The advantage of this model is its relative simplicity and the fact that most associated energy parameters are taken directly from experiment. Figure 15.5 shows typical stretching, bending, twisting, electrostatic Debye-Hückel, and excluded volume terms used for the model. For Brownian dynamics, we use complete hydrodynamics to treat both the translation and rotation of nucleosome cores and linker DNA beads (Fig. 15.5, right), making possible nanosecond to microsecond simulations to capture folding/unfolding events of short polymers.

Specifically, translational diffusion constants (D_t) were verified against experimental values for dimers and trimers, with a gentle linker-DNA bending noted

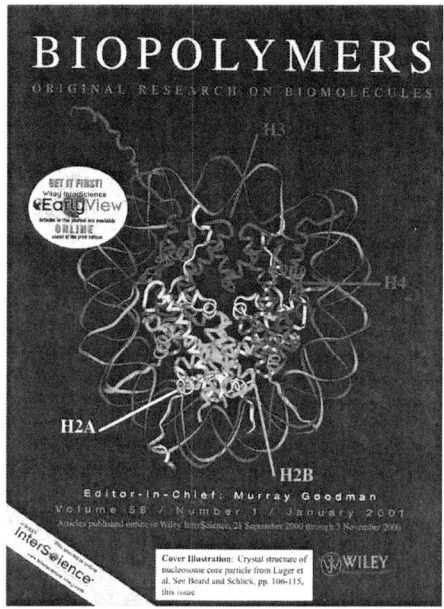
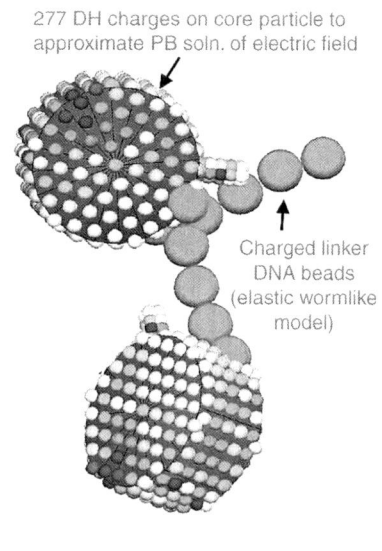

Figure 15.3: First-generation chromatin model. The nucleosome core (left) is represented by charged bodies denoting the nucleosomes with linker DNA represented as connecting beads (right). (See Plate 16).

for the latter, explaining the experimentally-noted sharper increase of D_t with the salt concentration [24]. The accordion-like opening at low salt of a polynucleosome can already be captured with this simple model (Fig. 15.6). Moreover, a 30 nm helical *zigzag* naturally emerges from the trimer geometry with a packing ratio of four nucleosomes per 11 nm (Fig. 15.6).

When the new crystal structure with full tails was resolved crystallographically in 2002 [40], we extended the DiSCO optimization approach to handle irregular surfaces (Fig. 15.7) [37]. With this new model, the salt-dependent folding of nucleosomal arrays was captured, along with explanations for the energetics involved (Fig. 15.7). Namely, at low salt, linker DNA repulsion triggers array unfolding while, at high salt, internucleosomal attraction from histone tail regions of the core triggers folding, with the H3 tails dominating internucleosomal attraction.

The importance of these histone tail interactions signaled to us that histone tail flexibility must be incorporated for further resolution and studies of chromatin structure and dynamics. Indeed, the histone tails are known to be crucial for compacting chromatin and, via biochemical modifications such as acetylation, methylation, and phosphorylation, can affect signaling pathways and transcriptional regulation. Detailed structures and physical insights are needed to interpret such observations.

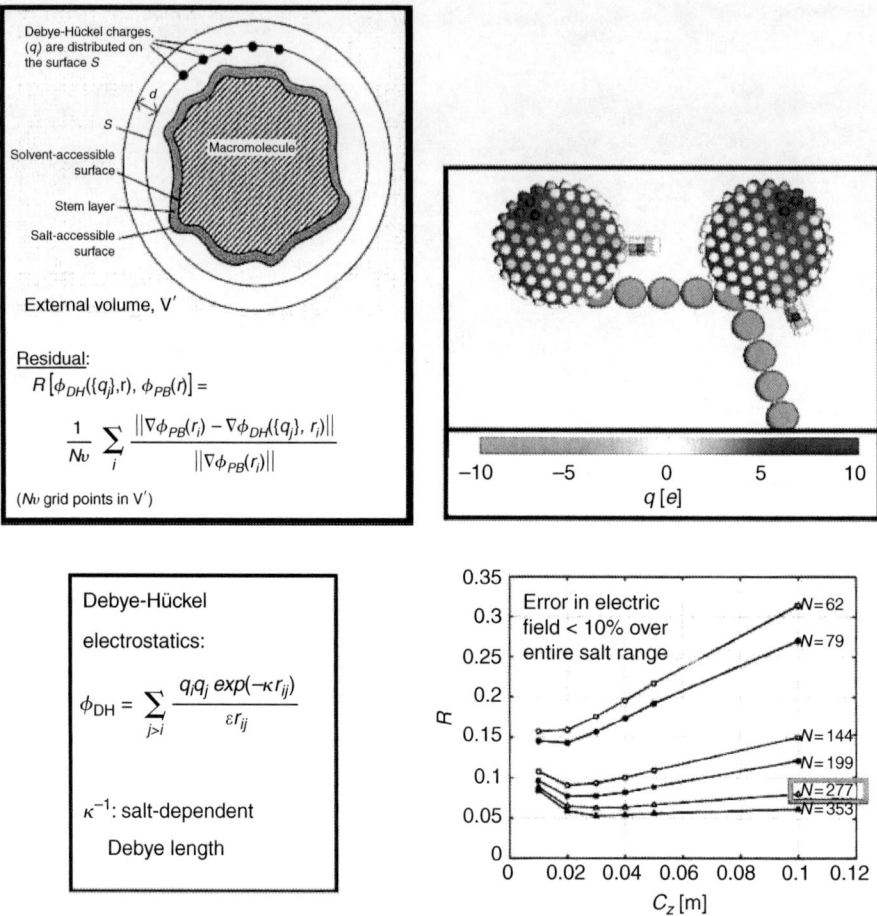

Figure 15.4: DiSCO optimization for the nucleosome core. To optimize the charges on the surface representing the histone octamer with wrapped DNA, we distribute charges homogeneously on the surface and choose charges to approximate the electric field of the atomistic core by a Debye-Hückel approximation in a region V', separated from the surface by a distance d, where the Debye-Hückel (linear) approximation is valid. Optimization is performed efficiently using our truncated Newton method TNPACK [39]. We found that 277 particles yielded an error of <10% are a large range of monovalent salt [35].

15.4 The second-generation mesoscopic chromatin model and results

To incorporate histone tail flexibility as well as linker histones, we applied coarse-graining to add components compatible with the nucleosome and linker DNA units (Fig. 15.8) [41]. Namely, starting from the amino-acid/subunit model of

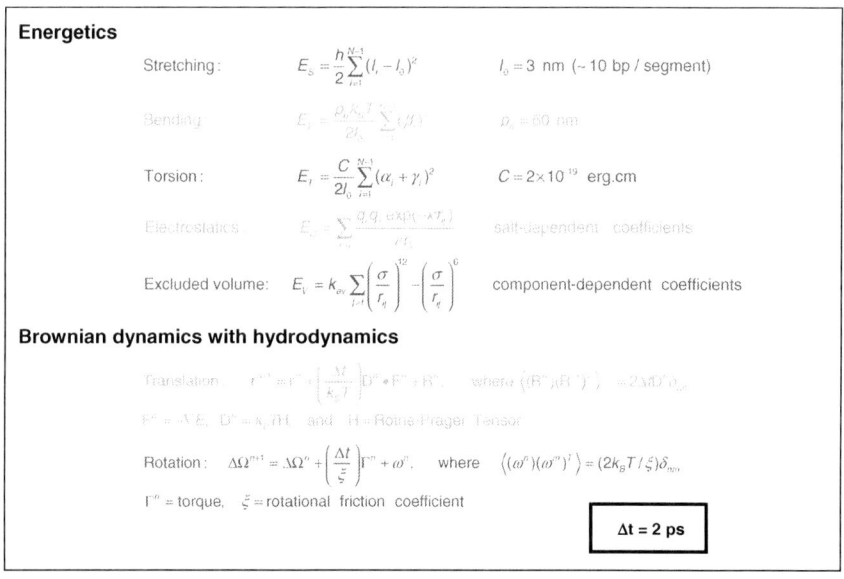

Figure 15.5: Model for chromatin dynamics and energetics. On the top, the basic unit along with position vector and coordinates for core and linker DNA beads are shown [24]. On the bottom, the governing energies and Brownian dynamics protocol are described.

Warshel and Levitt [42], for each histone tail (where each residue is a bead), we simulated Brownian dynamics of the tails and further coarse-grained the polymers to obtain parameterized protein beads (charges, excluded volume, harmonic stretching and bending) that reproduced configurational properties of the subunit model.

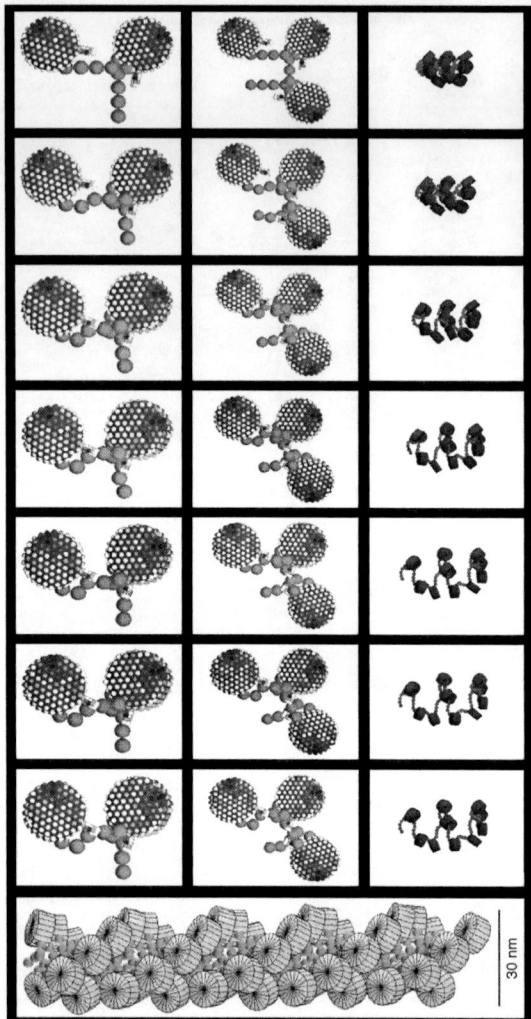

Figure 15.6: Nucleosome array trajectory snapshots from 10 ns Brownian dynamics simulations with the macroscopic model for dinucleosome, trinucleosome, and 12-nucleosome systems [35]. The linker DNA from the wormlike chain model is shown in red, and the nucleosomes are the cylindrically-shaped objects with electrostatic charges colored on a scale of red as negative to blue as positive. The dinucleosome and trinucleosome systems reflect compact structures under high univalent salt concentrations (50 mM). The 12-nucleosome array expands as the negatively-charged linker DNAs repel one another at low univalent salt concentrations (10 mM). At the bottom, the 30 nm fiber consisting of 48 nucleosomes was refined using Monte Carlo methods from a solenoid starting structure that was constructed using the dinucleosome folding motif.

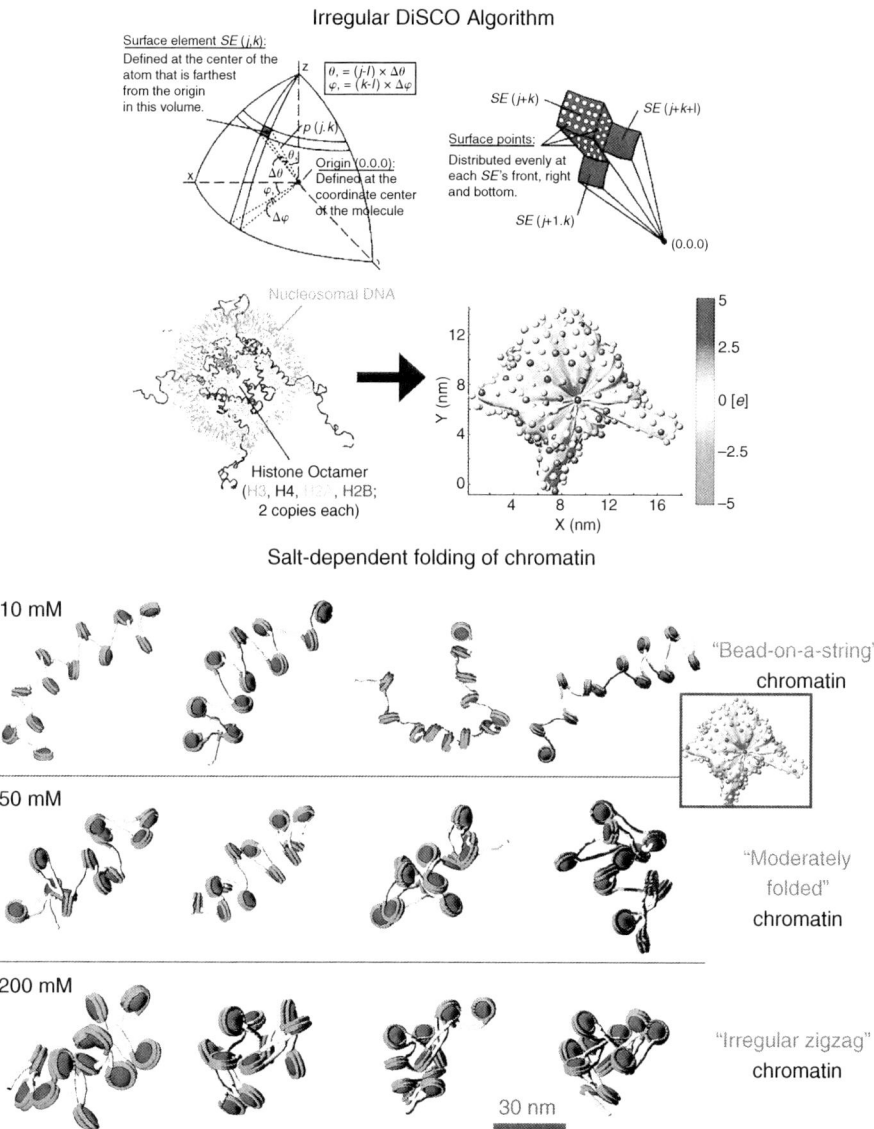

Figure 15.7: The irregular DiSCO algorithm and associated results. The algorithms are designed to create an irregular surface for the nucleosome so that tail geometries (though rigid) are captured (top) [37]. This model captured the salt-dependent folding and unfolding of chromatin (bottom) at different univalent salt environments [38].

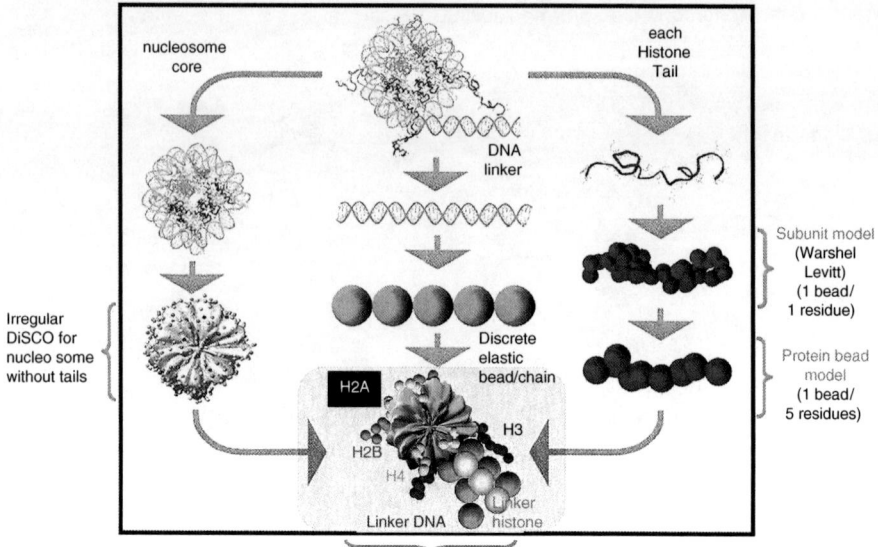

Figure 15.8: Mesoscopic oligonucleosome model. The irregular DiSCO model for the nucleosome without tails is combined with a coarse-grained representation of the histone tails and the linker histones [41].

To model the linker histone, we use the model of the rat H1d linker histone [43, 44] and represent the globular domain (76 residues) by one bead and the C-terminal domain (110 residues) by two beads, rigidly attached to the nucleosome core along the dyad as suggested experimentally (Fig. 15.8); we neglect the shorter (33-residue) N-terminal domain for simplicity.

The resulting oligonucleosome simulation model is extended accordingly, so that each independent variable (one per core, each linker DNA, linker histone, and histone tail bead) is associated with a Euler coordinate frame.

With this more complex, "mesoscopic" model, Brownian dynamics simulations were too computationally intensive. We therefore sampled oligonucleosome configuration space by Monte Carlo simulations with tailored local moves for translation and rotation of nucleosome core and linker DNA, global pivots of the polymer, and histone tail-regrowth moves; a new, end-transfer configurational bias method was developed to regrow tails at apposite ends so as to scale quadratically with polymer size rather than exponentially as in traditional configurational-bias MC [45].

The new chromatin model was validated against experiments as well as against the rigid-histone-tail model for configuration-dependent measurements such as sedimentation coefficient, $S_{20,W}$ (which measures how fast a polymer sediments in a fluid), translational diffusion constants (D_t), and tail-to-tail

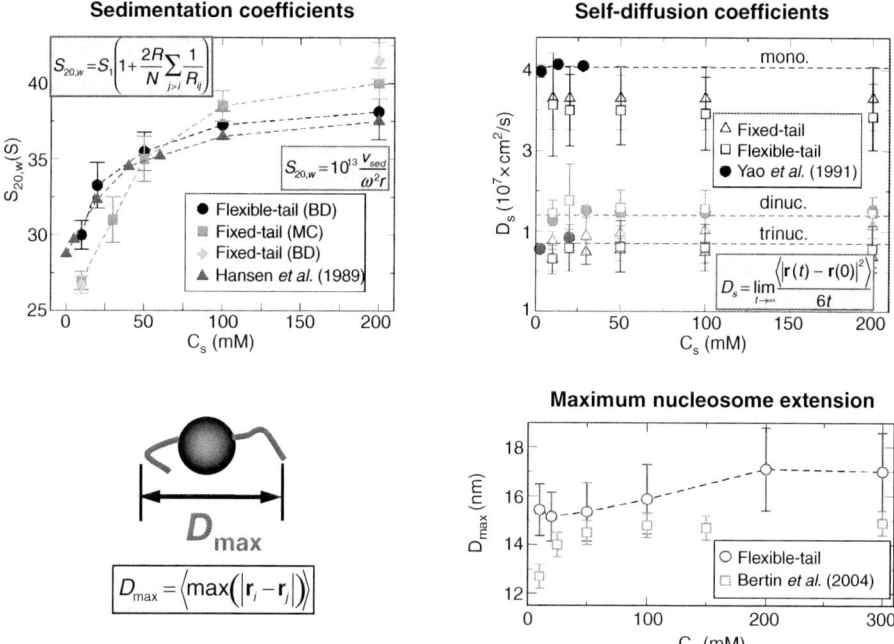

Figure 15.9: Experimental validation of a flexible-tail model for sedimentation coefficients, diffusion constants, and mono nucleosome extension [41]. Results show good agreement with experiment and better reproduction of the data compared to the rigid-tail model.

mononucleosome diameters (D_{max}) (see Figure 15.9) [41]. Not only did the refined model results compare well with experiment, results for the flexible-tail model improved upon results for the earlier, rigid-tail model.

Oligonucleosome dynamics (Fig. 15.10) reveal the dynamic nature of the histone tails, the strong repulsion among linker DNA at low salt, and the significant internucleosomal attraction mediated by the histone tails at high salt conditions. Note that *zigzag* topologies dominate at monovalent ion conditions, with fiber axis bending leading to fiber/fiber cross-interactions (Fig. 15.10). In fact, we found that each histone tail has a unique role in fiber organization because of its size and location [28, 52]: the H4 tails mediate the strongest internucleosomal interactions; the H3 tails bind strongly to parental DNA to screen the negatively-charged linker DNAs; and the H2A and H2B tails, located at the periphery of the nucleosome, mediate the most fiber/fiber interactions.

The most interesting aspect of our recent studies involves the dramatic effect of linker histone and divalent ions on fiber organization. The latter was modeled as a first approximation by setting the Debye parameter κ to the inverse DNA diameter $\kappa = 1/(2.5)$ nm, to allow the DNA beads to almost touch one another.

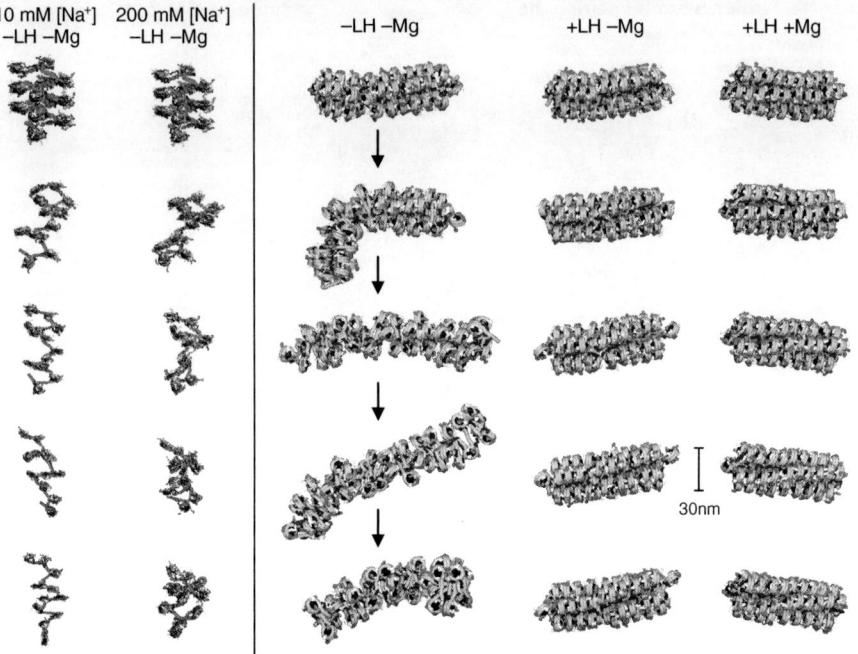

Figure 15.10: Nucleosome arrays with flexible histone tails simulated using the mesoscale chromatin model by Monte Carlo. The tails are colored as: H2A—yellow, H2B—red, H3—blue, and H4—green. The leftmost two columns illustrate behavior of 12-nucleosome arrays under low (10 mM) and high (200 mM) univalent salt concentrations. At low salt, the negatively-charged linker DNAs (red) repel one another and expand the array, while at high salt, the chromatin fiber condenses. The other columns show the behavior of 48-nucleosome arrays at high (200 mM) monovalent salt concentrations with linker DNA colored grey and nucleosome cores colored black at three compositions: without linker histones without magnesium (–LH–Mg); with linker histone H1 (cyan) without magnesium(+LH–Mg) ; and with linker histone H1with magnesium (+LH+Mg). Note the compaction introduced by linker histone as well as divalent ions. (See Plate 17).

Moreover, the linker DNA bending persistence length was reduced from 50 to 30 nm to mimick magnesium effects [46].

As Figure 15.11 shows [47], without linker histones and a monovalent concentration of $c_s = 0.15$ M, the fiber organizes as a classic open *zigzag* with a sedimentation coefficient $S_{20,W}$ of 39 ± 1.2 S (compared to 37 S experimentally) and a packing ratio of four nucleosomes per 11 nm. When the linker histone is added, sandwiching the entering and exiting linker DNA from each nucleosome, a rigid stem forms to allow closer contact; the fiber is markedly more compact

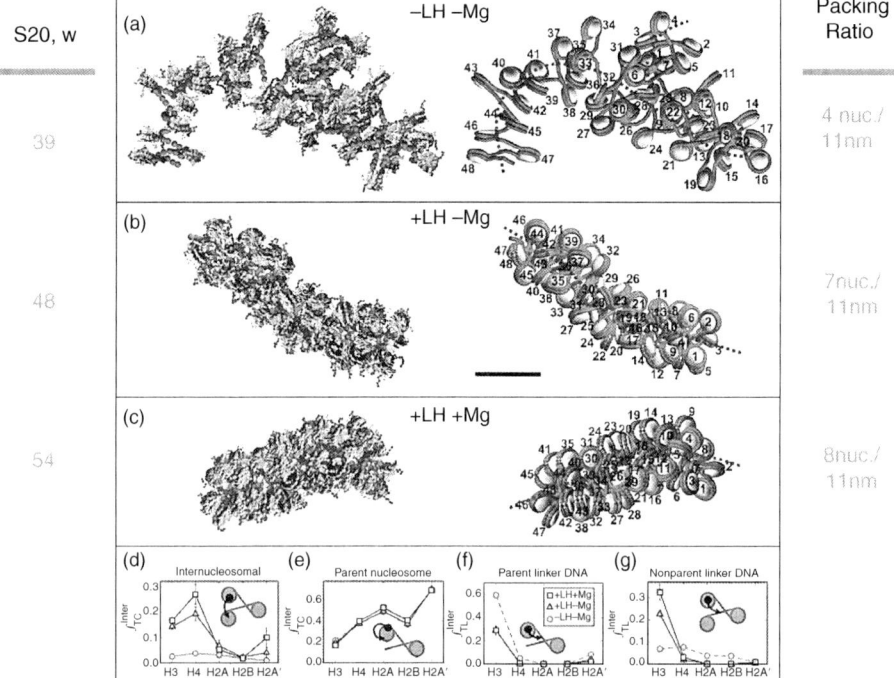

Figure 15.11: Representative fiber architectures without linker histone, without magnesium ions (top); with linker histone but without magnesium ions (middle); and with both linker histones and divalent ions (bottom). The histone tail analyses at the bottom reveal the important role of the tails in stabilizing structures, especially when linker histone is present.

($S_{20,W} = 49 \pm 1.7$ S, compared to 55.6 S experimentally, and packing ratio of seven nucleosomes per 11 nm). The linker DNAs remain relatively straight as in the classic *zigzag* (see below). When divalent counter ions are incorporated (to mimick 1 mM Mg^{2+}), further compaction follows: $S_{20,W} = 54 \pm 2.2$ S (compared to 60 S experimentally, and eight nucleosomes per 11 nm). Moreover, this further compaction is made possible by a tendency to bend a small proportion (e.g., 15%) of the linker DNAs rather than maintain all straight orientations. The dynamics associated with these fibers at these different conditions (Fig. 15.10) emphasize the greater flexibility and open structure without linker histones versus the more rigid and compact structures with linker histones.

This surprising result, of a compact fiber architecture at divalent ionic condition involving both straight and bent linker DNAs, was analyzed further using internucleosomal interaction patterns (Fig. 15.12). The entries of these interaction matrices, I'(i,j) measure the intensity of tail-mediated interactions between nucleosomes i and j (i.e., fraction of configurations over a 200 million

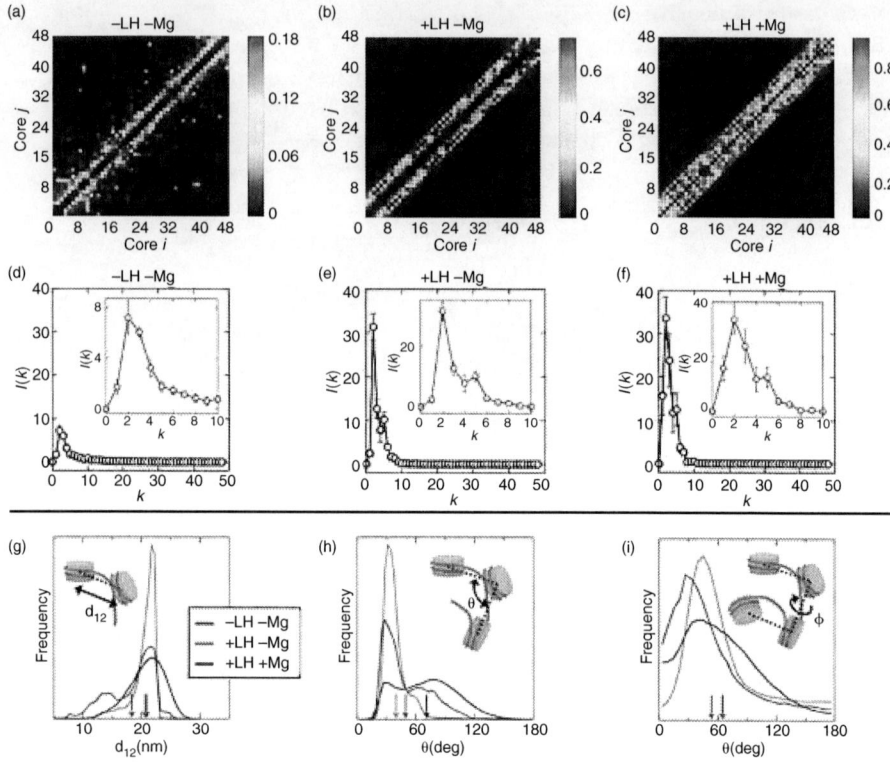

Figure 15.12: Internucleosomal nteraction patterns (averaged over 100,000 configurations from 200 million MC steps) for the three systems as in Figure 15.11 (LH = linker histone, Mg = magnesium ions). The pairwise pattern show the zigzag preference of the fiber without magnesium ions, and the introduction of solenoid-like (bent-DNA) features when divalent ions are introduced. The fiber compaction trends are also evident from analyses of the respective internucleosome distances, triplet, and quadroplet angles for successive nucleosomes [47].

MC ensemble where the two nucleosome cores are within 80% of the van der Waals radii). The one-dimensional plot I(k) decomposes the interactions (i, i±k) for k = 1, 2, 3. ...

Figure 15.12 shows that (i±2) and (i±3) interactions dominate without linker histone at monovalent salt (left column), reflecting *zigzag* configuration with some third-neighbor interactions. With the linker histone, (i±2) interactions dominate, producing the classic *zigzag* architecture (compare to Figure 15.2). In this structure, the nucleosome/nucleosome distances (d_{12}) and triplet angles (θ) formed by three consecutive nucleosome narrow considerably (Fig. 15.12, bottom). When, in addition, divalent ions are included (right column), (i ± 1) interactions indicative of *solenoid* configurations appear, with a bimodal θ distribution reflecting a mixture of mostly straight and some bent linker DNA.

This intriguing suggestion for a hybrid compact fiber in divalent ion conditions was recently verified by a new experimental technique termed EMANIC which uses formaldehyde cross-linking followed by unfolding and EM visualization to capture internucleosomal patterns [47]. Such an ensemble of interchanging configurations with straight and bent linker DNAs is energetically advantageous since linker DNA bending can minimize repulsion at the fiber axis. Moreover, it merges both classic models of *zigzag* and *solenoid* for optimal compaction!

15.5 Future perspective

Elucidating the structural details of the chromatin fiber and its dynamics at various physiological conditions and chemical compositions will undoubtedly remain an ongoing challenge for both modelers and experimental scientists. This is because of the magnitude of the folding problem, from the open "beads on a string"-like model at low salt of width 10 nm to very condensed states of mitotic chromosomes at "silent" stages of the cell cycle [2]. In addition, chromatin structure varies depending on the presence of specific linker histones (e.g., H1, H5), the length of the linker DNA (which roughly varies from 20 to 80 bp), the histone proteins themselves (which have well known variants), and the modified states of the tails. Four specific problems for modeling involving these components are described below, along with associated modeling development issues.

15.5.1 *Linker DNA length effects*

Though many experiments suggest typical two-start helix models for fiber organization based on *zigzag* arrangement of nucleosome as revealed by electron microscopy, *in vivo* studies of linker-length variations have not been systematic. Recently, Robinson *et al.* [9] have prepared *in vitro* assays of oligonucleosome with linker lengths varying from 10 to 70 bp. Surprisingly, they found two distinct fiber structures: a diameter of 33 nm and repeat pattern of 11 nucleosome per 11 nm for 10–40 bp linker lengths, and a diameter of 44 nm and 15 nucleosomes per 11 nm for the longer linker lengths (50–70 bp). Moreover, based on model building, they argue that a one-start helical model with nucleosome interdigitation and bent linker DNA explains these observations. However, this solenoid form differs from the favored zigzag configuration observed widely by X-ray crystallography [18].

These contradictory findings are puzzling, particularly since Robinson *et al.* [9] suggest that the linker histones alter the chromatin geometry to reconcile the findings. But our simulations and the hand-in-hand experiments by the Grigoryev group [47] show that linker histones favor the *zigzag* structure and that instead, divalent ions lead to some bending in linker lengths (Fig. 15.11). Thus, simulations with varied linker lengths are needed to reconcile these observations and provide atomic views and energetic estimates to interpret the linker-length effects. This application of linker-length variations is easily performed with our current mesoscale model of the chromatin fiber and is currently underway [53].

15.5.2 Histone variant effects

The naturally-occurring histone variant H2A.Z of H2A has been associated with assembly of specialized compacted forms of chromatin [4]. Remarkably, H2A.Z can alter the equilibrium dynamics of the fiber compared to H2A. It has been hypothesized that electrostatics differences between the two proteins are responsible for these profoundly different structural effects, and that different residues on H2A/H2A.Z interact differently with the H4 tails to impart these global effects. Other experiments reveal that a different variant, namely H2.Bbd has an opposite effect, promoting unfolding of condensed chromatin [48]. These problems form excellent questions for modeling. However, successful studies will likely require model enhancements since a 20–40% sequence homology difference of the variants compared to wildtype H2A may require more sensitive accounts than our current united-residue model. In fact, atomic-level modeling of the various histones H2A may be required to obtain potentials of mean force for each one and then incorporate these averaged force fields into the mesoscale oligonucleosome model. With this modification, varying the concentration of H2A and its variants (e.g., 60% H2A, 40% H2A.Z) and their distribution along the polymer chain is readily amenable for simulation.

15.5.3 Histone tail modifications

The histone-code hypothesis, in which different histone proteins are chemically modified to regulate transcription [49], is one of the fundamental tenets in modern biology. Because the N-termini of the histone tails are flexible and disordered in the crystal state, post translational modifications can trigger fiber changes and specific biological functions. However, how these chemical modifications contribute to fiber packaging and folding/unfolding events is not well understood. For example, acetylation (adding $-COCH_3$ units to arginine and lysine residues) can prevent condensation at high salt (e.g., when applied to lysine 16 of H4), while methylation (adding $-CH_3$ moieties to arginine and lysine residues) can cause folding when applied to lysine 9 of H3 but unfolding when applied to lysine 4 of H3.

In theory, modeling and simulation can systematically test these changes to interpret the structural effects. However, as above, a detailed residue model with accurate electrostatics may be needed on a fine level and integrated with a coarser model of oligonucleosome chains. A finer level of modeling would also require more careful parameterization of the nonelectrostatic terms.

15.5.4 Higher-order chromatin organization

Silent phases of the cell cycle are associated with heterochromatin—a highly condensed, highly ordered state of nucleosomal arrays, possibly compacted through clusters of repetitive DNA elements found at centromeres and telomeres [50]. Understanding heterochromatin formation and transformations is thought to be key to deciphering epigenetic control of the genome. Crucial to formation of heterochromatin are the binding of many chromatin-regulating (including linker) proteins as well as histone tail modifications, two elements of which may act in concert. From studies of mammalian nuclei, fibers of diameter ∼500–700 nm have been suggested. This high level of compaction must involve a decrease of the vacant spaces in the ∼30 nm fiber through interdigitation of nucleosomes, a generalization of single fiber inter-digitation, as shown in Figure 15.2. Fiber cross-linking must be enhanced by linker histones, chromatin-binding proteins like MeCP2 [51], and divalent ions. Modeling can in theory explore these potential interdigitated fiber states and study the dependencies on the above factors.

To a first approximation, the mesoscale model described here could be expanded to account for multiple oligonucleosome chains with attractive electrostatic interactions that mimick bridging effects of protein cofactors, linker histones, and divalent ions. Such studies could examine the feasibility of interdigitation of various solenoid, zigzag, or hybrid fiber models to exclude certain geometric possibilities and suggest favorable topologies. More detailed studies would require construction *de novo* of a new biomolecular network model appropriate for such complex investigations of multiple fiber/protein interactions.

15.6 Conclusions

As demonstrated in this overview, understanding chromatin structure, function, and dynamics is a challenging enterprise amenable to innovative multiscale models. As new developments in biology, modeling, and scientific computing are made available, more intricate models can be designed with more challenging questions addressed. In the not too distant future, I envision an all-atom simulation of an oligonucleosome (millions of atoms) and, within a few decades, a realistic simulation of the entire cell cycle of chromatin organization and associated epigenetic control! Biologists, chemists, and mathematics/physical scientists should continue to combine their expertise to achieve these exciting goals.

15.7 Acknowledgments

This work was supported by NSF award MCB-0316771 and NIH award R01 GM55164 to T. Schlick. Acknowledgment is also paid to the donors of the American Chemical Society Petroleum Research Fund for support of this research (Award PRF39225-AC4 to T. Schlick) and to Philip Morris USA and Philip Morris International. I thank Dr Sergei Grigoryev for valuable discussions, Yoon Ha Cha and Trang Nguyen for technical assistance with manuscript preparation, and Sean McGuffee for preparing three figures.

References

[1] Schlick T. and Dill K., Eds. (2006). "Multiscale modeling in biology", *Special Issue of SIAM J. Mult. Mod. and Sim.*, **5**(4), p. 1174–1366.

[2] Felsenfeld G. and Groudine M. (2003). "Controlling the double helix", *Nature*, **421**, p. 448–453.

[3] Segal E., Fondufe-Mittendorf Y., Chen L., Thåström A., Field Y., Moore I. K., Wang J. P., and Widom J. (2006). "A genomic code for nucleosome positioning", *Nature* **442**(7104), p. 772–778.
[4] Schlick T. (2002). *Molecular Modeling: An Interdisciplinary Guide*, Springer-Verlag, NY.
[5] Richmond T. J. and Davey C. A. (2003). "The structure of DNA in the nucleosome core", *Nature*, **423**(6396), p. 145–150.
[6] Bednar J., Horowitz R. A., Grigoryev S. A., Carruthers L. M., Hansen J. C., Koster A. J., and Woodcock C. L. (1998). "Nucleosomes, linker DNA, and linker histone form a unique structural motif that directs the higher-order folding and compaction of chromatin", *Proc. Natl. Acad. Sci. USA*, **95**(24), p. 14173–14178.
[7] Woodcock C. L. and Dimitrov S. (2001). "Higher-order structure of chromatin and chromosomes", *Curr. Opin. Gen. Dev.*, **11**(2), p. 130–135.
[8] Huynh V. A., Robinson P. J., and Rhodes D. (2005). "A method for the in vitro reconstitution of a defined '30nm' chromatin fiber containing stoichiometric amounts of the linker histone", *J. Mol. Biol.*, **345**(5), p. 957–968.
[9] Robinson P. J., Fairall L., Huynh V. A., and Rhodes D. (2006). "EM measurements define the dimensions of the '30-nm' chromatin fiber: Evidence for a compact, interdigitated structure", *Proc. Natl. Acad. Sci. USA*, **103**(17), p. 6506–6511.
[10] Grigoryev S. A. (2004). "Keeping fingers crossed: heterochromatin spreading through interdigitation of nucleosome arrays", *FEBS Lett.*, **564**(1–2), p. 4–8.
[11] Widom J. and Klug A. (1985). "Structure of the 300A chromatin filament – X-ray diffraction from oriented samples", *Cell*, **43**(1), p. 207–213.
[12] Bystricky K., Heun P., Gehlen L., Langowski J., and Gasser S. M. (2004). "Long-range compaction and flexibility of interphase chromatin in budding yeast analyzed by high-resolution imaging techniques", *Proc. Natl. Acad. Sci. USA*, **101**(47), p. 16495–16500.
[13] Caravaca J. M., Cano S., Gallego I., and Daban J. R. (2005). "Structural elements of bulk chromatin within metaphase chromosomes", *Chrom. Res.*, **13**(7), p. 725–743.
[14] Luger K. and Hansen J. C. (2005). "Nucleosome and chromatin fiber dynamics", *Curr. Opin. Struct. Biol.*, **15**(2), p. 188–196.
[15] Tremethick D. J. (2007). "Higher-order structures of chromatin: the elusive 30 nm fiber", *Cell*, **128**(4), p. 651–654.
[16] van Holde K. and Zlatanova J. (2007). "Chromatin fiber structure: Where is the problem now?", *Semin. Cell Dev. Biol.*, **18**(5), p. 651–658.
[17] Luger K., Mader A. W., Richmond R. K., Sargent D. F., and Richmond T. J. (1997). "Crystal structure of the nucleosome core particle at 2.8 Angstrom resolution", *Nature*, **389**(6648), p. 251–260.

[18] Schalch T., Duda S., Sargent D. F., and Richmond T. J. (2005). "X-ray structure of a tetranucleosome and its implications for the chromatin fiber", *Nature*, **436**(7047), p. 138–141.
[19] Hansen J. C. (2002). "Conformational dynamics of the chromatin fiber in solution: determinants, mechanisms and functions", *Ann. Rev. Biophys. Biomol. Str.*, **31**, p. 361–392.
[20] Jian H., Vologodskii A., and Schlick T. (1997). "A combined wormlike chain and bead model for dynamic simulations of long DNA", *J. Comp. Phys.*, **136**, p. 168–179.
[21] Schlick T. and Olson W. K. (1992). "Computer simulations of supercoiled DNA energetics and dynamics", *J. Mol. Biol.*, **223**, p. 1089–1119.
[22] Leuba S. H,. Yang G. L., Robert C., Samori B., van Holde K., Zlatanova J., and Bustamante C. (1994). "3-Dimensional Structure of Extended Chromatin Fibers as Revealed by Tapping-Mode Scanning Force Microscopy", *Proc. Natl. Acad. Sci. USA*, **91**(24), p. 11621–11625.
[23] Katritch V., Bustamante C., and Olson V. K. (2000). "Pulling chromatin fibers: Computer simulations of direct physical micromanipulations", *J. Mol. Biol.*, **295**(1), p. 29–40.
[24] Beard D.A. and Schlick T. (2001). "Computational modeling predicts the structure and dynamics of chromain fiber", *Structure*, **9**(2), p. 105–114.
[25] Wedemann G. and Langowski J. (2002). "Computer simulation of the 30-nanometer chromatin fiber", *Biophys. J.*, **82**(6), p. 2847–2859.
[26] Sun J., Zhang Q., and Schlick T. (2005). "Electrostatic mechanism of nucleosomal array folding revealed by computer simulation", *Proc. Natl. Acad. Sci. USA*, **102**(23), p. 8180–8185.
[27] Mozziconacci J., Lavelle C., Barbi M., Lesne A., and Victor J. M. (2006). "A physical model for the condensation and decondensation of eukaryotic chromosomes", *Febs Lett.*, **580**(2), p. 368–372.
[28] Arya G. and Schlick T. (2006). "Role of histone tails in chromatin folding revealed by a mesoscopic oligonucleosome model", *Proc. Natl. Acad. Sci. USA*, **103**, p. 16236–1624.
[29] Dorigo B., Schalch T., Kulangara A., Duda S., Schroeder R. R., and Richmond T. J. (2004). "Nucleosome arrays reveal the two-start organization of the chromatin fiber", *Science*, **306**(5701), p. 1571–1573.
[30] McGhee J. D., Nickol J. M., Felsenfeld G., and Rau D. C. (1983). "Higher order structure of chromatin: Orientation of nucleosomes within the 30 nm chromatin solenoid is independent of species and spacer length", *Cell*, **33**(3), p. 831–841.
[31] Woodcock C. L., Frado L. L., and Rattner J. B. (1984). "The higher-order structure of chromatin – evidence for a helical Ribbon Arrangement", *J. Cell. Biol.*, **99**(1), p. 42–52.
[32] Thoma F., Koller T., and Klug A. (1979). "Involvement of histone-H1 in the organization of the nucleosome and of the salt-dependent superstructures of chromatin", *J. Cell Biol.*, **83**(2), p. 403–427.

[33] Wong H., Victor J. -M., and Mozziconacci J. (2007). "An all-atom model of the chromatin fiber containing linker histones reveals a versatile structure tuned by the nucleosomal repeat length", *PLoS ONE*, **2**, e877:1–8.

[34] Gordon F., Luger K., and Hansen J. C. (2005). "The core histone N-terminal tail domains function independently and additively during salt-dependent oligomerization of nucleosomal arrays", *J. Biol. Chem.*, **280**, p. 33701–3706.

[35] Beard D. A. and Schlick T. (2001). "Modeling salt-mediated electrostatics of macromolecules: The algorithm DiSCO (Discrete Surface Charge Optimization) and its application to the nucleosome", *Biopolymers*, **58**, p. 106–115.

[36] Beard D. A. and Schlick T. (2003). "Unbiased rotational moves for rigid-body dynamics", *Biophys. J.*, **85**, p. 2973–2976.

[37] Zhang Q., Beard D. A., and Schlick T. (2003). "Constructing irregular surfaces to enclose macromolecular complexes for mesoscale modeling using the discrete surface charge optimization (disco) algorithm", *J. Comp. Chem.*, **24**, p. 2063–2074.

[38] Sun J., Zhang Q., and Schlick T. (2005). "Electrostatic mechanism of nucleosomal array folding revealed by computer simulation", *Proc. Natl. Acad. Sci. USA*, **102**(23), p. 8180–8185.

[39] Schlick T. and Fogelson A. (1992). "TNPACK – a truncated Newton minimization package for large-scale problems: I. algorithm and usage", *ACM Trans. Math. Softw.*, **18**, p. 46–70.

[40] Davey C. A., Sargent D. F., Luger K., Mader A. W., and Richmond T. J. (2002). "Solvent mediated interactions in the structure of the nucleosome core particle at 1.9 Å resolution", *J. Mol. Biol.*, **391**, p. 1097–113.

[41] Arya G., Zhang Q., and Schlick T. (2006). "Flexible histone tails in a new mesoscopic oligonucleosome model", *Biophys. J.*, **91**(1), p. 133–150.

[42] Warshel A. and Levitt M. (1975). "Computer simulation of protein folding", *Nature*, **253**(5494), p. 694–698.

[43] Bharath M. M., Chandra N. R., and Rao M. R. (2002). "Prediction of an HMG-box fold in the C-terminal domain of histone H1: insights into its role in DNA condensation", *Proteins*, **49**, p. 71–81.

[44] Bharath M. M. S., Chandra N. R., and Rao M. R. S. (2003). "Molecular modeling of the chromatosome particle", *Nuc. Acids Res.*, **31**, p. 4264–4274.

[45] Arya G. and Schlick T. (2007). "Efficient global biopolymer sampling with end-transfer configurational bias Monte Carlo", *J. Chem. Phys.*, **126**(4), p. 044107.

[46] Baumann C. G., Smith S. B., Bloomfield V. A., and Bustamante C. (1997). "Ionic effects on the elasticity of single DNA molecules", *Proc. Natl. Acad. Sci. USA*, **94**, p. 6185–6190.

[47] Grigoryev S. A., Arya G., Correl S., Woodcock C., and Schlick T. (2009). "Evidence for heteromorphic chromatin fibers from analysis of nucleosome interactions by modeling and experimentation", *Proc. Natl. Acad. Sci. USA*, In Press.

[48] Fan J., Rangasamy D., Luger K., and Tremethick D. (2004). "H2A.Z alters the nucleosome surface to promote HP1 alpha-mediated chromatin fiber folding", *Mol. Cell*, **16**(4), p. 655–661.

[49] Strahl B. D. and Allis C. D. (2000). "The language of covalent histone modifications", *Nature*, **403**, p. 41–45.

[50] Shiv G. and Jia S. (2007). "Heterochromatin revisited", *Nat. Rev. Gen.*, **8**, p. 35–46.

[51] McBryant S. M., Adam V., and Hansen J. C. (2006). "Chromatin architectural proteins", *Chrom.e Res.*, **14**, p. 39–51.

[52] Arya G. and Schlick T. (2009). "A tale of tails: How histone tails mediate chromatin compaction in different salt and and linker histone environments", *J. Phys. Chem. A.*, 113, p. 4045–4059.

[53] Schlick T. and Perisic O. (2009). "Mesoscale simulations of two nucleosome-repeat length oligonucleosomes", Submitted.

16

MULTISCALE NATURE INSPIRED CHEMICAL ENGINEERING

Marc-Olivier Coppens

16.1 Introduction

Adequate structuring at multiple length and time scales is crucial to processes in nature and in technology. In both cases there is a need to bridge microscopic building blocks with the macroscopic world in a way that does not destroy the desired functionality tied to the microscopic structure of the building block—a cell, a supramolecular complex, a nano-cluster, a gas bubble. In this chapter we draw particular attention to the interesting opportunities offered by designs that are inherently multiscale, inspired by useful multiscale architectures in nature. Examples are the fractal architecture of trees, lungs, and kidneys; the chaotic attractor of a gas-fluidized bed of solid particles; the broad power law distribution of water residence times in protein channels. Such patterns are not well described in terms of a discrete spectrum of characteristic scales. They involve a *continuum* of scales to interpolate between important discrete scales (a leaf and a tree-crown; an alveolus and the lung; a bubble and a fluidized bed), in order to preserve certain desired microscopic properties (gas exchange; chemical selectivity and activity) up to macroscopic scales of practical relevance.

The methodology discussed here aims to revive interest in describing phenomena in a *holistic* (but not necessarily phenomenological) way, as a powerful complement to methods that rely on an *atomistic*, bottom-up analysis and design, with successive, level-by-level coarse-graining. With "holistic", we refer to overall, collective properties, which embody inherent scaling symmetries, e.g., an underlying fractal scaling or an invariant characteristic distribution that captures a broad range of scales. Clearly, the word "holistic" carries its original meaning, without postmodern attributes or New Agey undertones.

Despite a surge of interest in the 1970s–90s—with renormalization group theory, percolation theory, chaos and fractals at its apex—holistic descriptions tend to be surpassed by atomistic descriptions (considered explicit, and therefore most rigorous), to replace phenomenological (and often empirical) engineering models. Likely this is due to a shift in focus to nanotechnology and to the availability of cheap computer power, but also because of deeply rooted historical and cultural reasons that associate ultimate control with atom-by-atom bottom-up construction. Atomistic and bottom-up approaches have proven to be immensely successful in synthesis (thanks to the possibilities offered by nano and microtechnology), in theory (coarse-graining of fundamental equations, molecular dynamics), and in computation (thanks to ever increasing computer power). Relatively simple systems are indeed describable in terms of discrete characteristic scales, possibly supplemented by linear, crystal symmetries (e.g., periodicity). Equations describing such systems can be readily derived and integrated to reveal macroscopically relevant information, precisely thanks to periodicity and/or the fast decay of interactions over short enough length or time scales, causing the underlying system to scale in a linear way.

With growing interest in complex systems, e.g., in the life sciences and in composite materials, this explicit atomistic approach is at least impractical, because multiple scales interact in a nonlinear way, and the equations are either incomplete or cannot be integrated. Furthermore, the fact that weeks of supercomputer time are needed to perform molecular dynamics on (bio)macromolecular systems of moderate complexity is unsatisfying and tells us that the underlying physics is not well understood. Nevertheless, as this book amply demonstrates, there has been significant progress in integrating over multiple scales in a smart way, such as multi-scale (adaptive) coarse-graining algorithms.

Further simplification is possible when a pattern or symmetry is recognized: discovering invariance under transformations has been key to solving many scientific problems. This is not always possible, yet it is common in nature, and it can be imposed by design in engineering. Properties of crystals are easier to calculate than those of amorphous materials, thanks to their periodicity and a range of invariances related to the crystal group. Similarly, in fractal systems the invariance of details under magnification, within a certain range of length scales, $[\delta_{min}, \delta_{max}]$, or time scales, $[t_{min}, t_{max}]$, allows for simplifications. Describing a fractal structure as a collection of subunits ignores its defining symmetry. Atomistic algorithms are ineffective within this range: it is a waste of resources to describe a tree in terms of its many individual branches, or a cloud in terms of all its billows.

This chapter focuses on recognizing multiscale patterns (Section 16.2) or imposing them (Section 16.3) to facilitate the problem of bridging scales, with particular emphasis on chemical engineering, the author's specialty. Scale-free, fractal architectures immediately jump to mind, but other examples involve chaotic attractors and invariant distributions over a broad range of scales. A chaotic attractor is a fingerprint for a chaotic system, containing a wealth of information. Imposing inherently multiscale patterns, with the desired bridging structure linking the micro with the macroworld, allows us to focus on rational design at the microscale, while bypassing complications at larger scales.

Overall, effective properties at the macroscale depend on the structure at the microscale (intrinsic properties), but also on the interpolating architecture between these sometimes vastly different scales. Right above the size of individual atoms and molecules, nonlinear interactions may generate properties in solid materials and certain complex fluids that are dependent on the atomic/molecular organization in a way that is different from what could be trivially inferred from the properties of the individual molecular or atomic units. Studying and designing the organization of those assemblies is the focus of nanoscience and nanotechnology. If the microstructure captures the property of interest, imposing it by design, there may be no need to explicitly resolve all details within the range of scales above that of the elementary building blocks that constitute the pattern. There are other situations where a wide range of scales cooperates to induce a phenomenon very different from what could be inferred by observing a single scale—turbulence, for example. Also these situations benefit from the multiscale methodology discussed in this chapter. Note that this approach may not be suitable for many engineering systems as they are typically conceived today, yet it is very often employed in nature, probably because it is a facile way to preserve microscopic properties up to macroscopic scales. Hence the proposal to use structures inspired by nature in engineering design.

In this chapter we do not focus on algorithms, for which we refer to specialized papers, but on examples illustrating the advantages of the multiscale methodology to a few selected applications in chemical engineering: reactors for the production of chemicals and power, fluid distributors, porous catalytic particles, and nature-inspired membranes for separation processes.

16.2 Resolving multiscale patterns in heterogeneous systems

16.2.1 *Bubbling gas-fluidized beds of solid particles*

Gas-fluidized beds of solid particles are used on a vast scale in the production of power (coal and biomass combustion and gasification), the industrial production of chemicals, and the drying and coating of particles. The solid particles react with the gas or catalyze gas-phase reactions. In such reactors, a gas flow is typically distributed through a distributor plate at the bottom of the bed at velocities higher than the so-called minimum fluidization velocity. When the particles are of the Geldart B-type ("sand-like"), gas bubbles appear, and the bed resembles a boiling liquid [1]. Bubbles rise, grow and coalesce, so that the bubble size distribution is broad. With increasing gas velocity, vortices appear. The hydrodynamics are very complex and depend on the geometry and size of the bed. As a result, fluidized beds are difficult to scale up. Furthermore, mass transfer between bubbles and the surrounding gas-solid suspension (also called emulsion phase) limits their overall performance, when gaseous reactants cannot sufficiently rapidly access the particles in the emulsion phase, with or on which they react.

With the advent of powerful computers, better algorithms, and better experimental tools to validate the simulations, models of fluidized beds have become more and more sophisticated. They have evolved from two-phase volume-averaged models to truly multiscale models that couple Lattice Boltzmann Methods at the single-particle level to discrete (Lagrangian) particle models to continuum (Eulerian) schemes to allow for the simulation of entire reactors [2]. Discrete particle dynamics can indeed only be used to study systems of several thousands to millions of particles at most, which is orders of magnitude less than in industrial (and even most lab-scale) fluidized beds. Despite the considerable progress made [3, 4, 5, 6, 7], computational fluid dynamics of fluidized beds is still very challenging, and care is needed in verifying the codes and validating results against experiments, a difficult task in itself [8].

Because of the extremely broad range of relevant length and time scales, Li and Kwauk [9] developed a multiscale method, which is based on energy minimization (EMMS). An alternative to these successive coarse-graining methods is to look at a fluidized bed as a complex system, and to search for overall characteristic signatures or patterns that directly account for the entire range of scales. If these patterns are truly representative for the multiscale hydrodynamics, then they can also be used to monitor them (safety, product quality), to preserve them (scale-up of a reactor to a larger scale), or to modify them (process control).

16.2.1.1 A fluidized bed as a chaotic system

Daw et al. [10], and Schouten et al. [11] showed that both bubbling and circulating gas-solid fluidized beds behave as low-dimensional chaotic systems. Their hydrodynamics can therefore be studied using tools from chaos theory, e.g., they have a Kolmogorov entropy K that is finite and nonzero ($0 < K < \infty$). Thanks to Takens's embedding theorem, the chaotic attractor of the fluidized bed can be reconstructed from a time-series of a single characteristic variable. In the case of a fluidized bed, local pressure fluctuation measurements are particularly convenient, since they are representative for the dynamics and easy to carry out even in industrial equipment. Assume that pressure measurements are taken every time $t_i = i\Delta t$ ($i = 1, \ldots, n$) at a high frequency $1/\Delta t$. Using time-delay coordinates, the pressure time series $\{P(t_i), i = 1, \ldots, n\}$ is converted into $n - m + 1$ delay vectors $\{[P(t_i), P(t_{i+1}), \ldots, P(t_{i+m-1})], i = 1, \ldots, n - m + 1\}$ in m-dimensional space. Takens [12] proved that this reconstructed attractor based on a single variable (pressure in this case), has the same dynamic characteristics as the "true" attractor obtained from all variables that govern the system.

This observation is extremely powerful, because the chaotic attractor is an excellent fingerprint of a fluidized bed as a multiscale system. Typical dimensionless hydrodynamic numbers, such as Re (Reynolds number) and Fr (Froude number), are related to characteristic length or time scales, and do not embody the inherent multiscale nature of the fluid system. The Kolmogorov entropy K (bits/s), tied to the chaotic attractor, is a direct representation of the information content in the bed. The difficult task of fluidized-bed scale-up can be simplified by preserving the normalized entropy, obtained by dividing K by the natural frequency of the entire bed. In addition, this normalized entropy can be shown to correlate with Fr and the bed dimensions in a particular way. A review by van den Bleek et al. [13] discusses how K or the attractor itself can be used as a pattern for the fluidized bed hydrodynamics, in such a way that hydrodynamic regime changes, scale-up and control are facilitated. The chaotic attractor can even be reconstructed from time series "measured" from a computer simulation of a fluidized bed, and compared to experiments. Furthermore, as van Ommen et al. [14] have shown, changes in the chaotic attractor are a powerful way to monitor hydrodynamic changes in real time, sending an early warning signal to operators that something is going on, such as undesired agglomeration or sticking of particles to each other. *What* exactly is changing may not be inferred by overall measures—for this, detailed simulations are needed—but the ability to capture the essence of the multiscale complexity in a sensitive and rapid way is remarkable, and pleads for using holistic measures, such as attractor comparison, as a complement to bottom-up simulations.

Clearly, the tools summarized here are not limited to fluidized beds or even other multiphase fluid systems (foaming, two-phase pipe flow, slurry reactors), but can be adopted to other situations with complex multiscale dynamics, as long as they can be probed by an observable representative parameter that can be measured at high frequency from experiments or simulations.

16.2.1.2 Polydispersity and the Student t-distribution as emergent pattern

On top of this, when pressure fluctuations are measured in a fluidized bed, their histogram is typically observed to be a Student t-distribution $\rho_\beta(x) \sim \left[1 + \alpha x^2/\beta\right]^{-\beta}$, where $x = \Delta P$ is the measured pressure fluctuation over a short, constant time delay Δt (e.g., ≤ 50 ms) in a point in the fluidized bed: $\Delta P = P(t + \Delta t) - P(t)$ [15]. The Student t-distribution deviates significantly from a Gaussian distribution for $x \gg 1$, since it decays with x as a power law $\sim x^{-2\beta}$, which is much more slowly than a Gaussian decay, $\exp[-x^2/(2\sigma^2)]$. For example, despite the complexity of gas-solid fluidized bed hydrodynamics, this distribution emerged over more than four orders of magnitude in a tall, large diameter (80 cm), experimental air-fluidized bed of sand particles. The origin of the Student distribution lies in the polydispersity of the bubble sizes, itself a result of bubble aggregation and breakup. The bubble size distribution $f(R) \sim R^{-\tau} \exp(-a/R)$. The pressure fluctuation range due to individual bubbles is Gaussian, i.e., $\rho_R(x) \sim \exp(-cx^2/R)$. The convolution of both generates the measured Student t-distribution.

This result is very general [16], because polydispersity can often be described by a power-law distribution with exponentional cutoff, as a result of similar Smoluchowski aggregation-fragmentation kinetics [17, 18]. If the phenomenon associated to a single scale R is fluctuating with Gaussian statistics, the overall result is a Student t-distribution. Classical diffusion and dispersion fall under this category (Brownian motion), as well as deterministic processes with a stochastic component that is well described by Gaussian noise due to a large number of independent interactions of finite variance on each object of size R (central limit theorem).

16.2.2 Diffusion in porous media

The modeling of diffusion in heterogeneous media is a problem rich in multiscale features. It is a field too vast to cover here in any detail, therefore we will only highlight how certain multiscale patterns arise, out of very different reasons. Again, bottom-up multiscale methodologies are usefully supplemented by an overall analysis.

Diffusion of fluids in biological and technical porous materials is important to molecular separation processes, via selective membranes or adsorbents. Diffusion is also crucial in many catalytic processes: molecules need to diffuse toward and away from the so-called "active sites" on the internal surface of porous catalysts. To achieve the large surface area needed for catalysis (typically hundreds of square meters per gram), the majority of the pores that traverse solid catalysts are typically only a few nanometers or less in diameter, i.e., they are "nanopores," so that the diffusing molecules almost constantly interact with the pore surface.

The nature of pores varies a lot. Sometimes they are straight and smooth, as in carbon nanotubes. Sometimes they are very rough and seemingly irregular, as in silica gels. Pores may be chemically homogeneous or heterogeneous, as in certain porous catalysts and in biological pores traversing a cell membrane. The latter consist of enzymes with a complex, partially charged surface. In such chemically heterogeneous materials, the potential felt by a diffusing molecule varies considerably from one point to another.

In nanopores, the environment is very different from that in a bulk gas or liquid. The mean free path of a gas is typically larger than the nanopore diameter, which means that diffusion is governed by molecule-surface, rather than by intermolecular collisions (as in bulk diffusion). As a result, Knudsen diffusion dominates: molecules collide with the surface, adsorb for a brief time and desorb again, leaving the surface in a random direction. Surface diffusion and slip may complicate the picture. Liquids in nanopores are also influenced by the walls, especially the molecules closest to the pore wall. For example, water forms a hydration layer along hydrophilic walls, and transport in this layer is different from that in the core zone. Molecules move from within a core zone in the pore into the surface layer, and back. The collective motion of molecules may be affected by the walls as well. The overall result of this complex process depends on the nature of the surface and the molecules.

Finally, there is the special, but practically important, case of diffusion in zeolites and other microporous materials [19], in which molecules constantly interact with the walls, so that diffusion is typically an activated process ($\sim \exp[-E/(R_g T)]$, with E the activation energy, R_g the gas constant, and T the temperature). Diffusion of tightly-fitting molecules in zeolites is best described in terms of a succession of hops, somewhat similar to defect diffusion in solids. Molecules cannot directly pass each other. At high molecular concentration, collective effects again start to dominate. The widely different time scales call for a multiscale approach, e.g., combining molecular dynamics with transition-state theory, dynamic Monte Carlo simulations, and (where applicable) continuum approaches. This subject has been reviewed by Keil *et al.* [20].

Pores may be parallel in some of the newest nanostructured materials, but more frequently form a network. Clearly, the network topology affects diffusion. If the connectivity of the network is high enough, this influence is only quantitative, and does not change Einstein's law, i.e., the mean-square displacement $\langle r^2 \rangle$ of the molecules is proportional to time t, and the self-diffusivity is

$D_s = \lim_{t \to \infty} \langle r^2 \rangle / (6t)$ in three dimensions. Only if the connectivity is critically low does the diffusion behavior change in a qualitative sense, so that Einstein's law breaks down.

In the case of a regular, periodic network, such as in zeolites, this only happens in the special case of single-file diffusion, i.e., there is no network at all: in non-intersecting pores that do not allow molecules to pass one another, the molecular motion is collective over the entire pore length. At time scales smaller than those required to cross the pore ends, Einstein's law breaks down and $\langle r^2 \rangle \sim \sqrt{t}$. In effect, there are multiscale features present at all time and length scales commensurate with the system size.

If the network is topologically disordered, certain nodes have few neighbors, while others may have several. Because of network bond or node blockage (e.g., due to deactivation processes during chemical reactions, the formation of solid deposits, poisoning, etc.), a pore network may become less and less accessible, until there is no path linking one side of the material to the other side: the percolation threshold is reached. Below the percolation threshold (i.e., if even more bonds or nodes are blocked), transport is localized. Transport on the percolation cluster is *anomalous* and sub-diffusive: $\langle r^2 \rangle \sim t^\beta$, where $\beta < 1$ is a critical exponent. This exponent is universal, in the sense that it only depends on the embedding dimension of the system (3 in real space), while the percolation threshold is dependent on the network details, e.g., its original topology [21, 22, 23].

The reason for diffusion to be anomalous on a percolation cluster is that the cluster is a self-similar fractal. Details of the cluster look similar to larger parts of it, so that it is invariant under magnification. More generally, diffusion on fractal networks, like the self-similar Sierpiński triangle or its three-dimensional extension (the Sierpiński gasket) is anomalous due to the loops-within-loops, which delay diffusion so much that it becomes localized. Within the fractal range, there is no characteristic scale, except for the smallest subunit (an "inner cutoff," $R_{\min}$) and the "outer cutoff" or correlation length (a maximum size $R_{\max}$, at most equal to the entire network size). Diffusion, conduction and reaction on fractal structures can therefore be studied with the same tools as critical phenomena, and renormalization group theory is particularly suited [24, 25, 26, 27].

The last examples showed the emergence of power laws and anomalous diffusion behavior as a result of geometrical invariance over a range of length scales, $[R_{\min}, R_{\max}]$, i.e., the pattern is tied to the geometrical structure of the material. Within the fractal scaling range, it is preferable to use tools from statistical physics that directly account for the continuous scale-free nature of such fractal materials or pore networks, which vary from rough solid surfaces to polymeric networks, while molecular dynamics are more suited at the smaller length scales. Extensive reviews on anomalous diffusion are provided by Havlin and Ben-Avraham [28] and Bouchaud and Georges [29].

Remarkably, however, also in some of the other mentioned diffusion problems power laws may emerge, but for quite different reasons—albeit that the unifying principle is that *a broad range of scales (in time, space or energy) is always at the origin of the power law*. The single-file diffusion case already illustrated this: here, it is the collective motion over many time scales that led to anomalous diffusion. Also, if the direction a molecule takes is dependent on its history in a power-law fashion, either in a persistent or an antipersistent way, Mandelbrot and Van Ness [30] showed that the overall diffusion process becomes anomalous, with $\langle r^2 \rangle \sim t^\beta$. This is called fractional Brownian motion. Long-range temporal rather than spatial correlations are at the origin of the anomaly. This effect may also be indirect, in a dynamic environment, such as in the single-file diffusion case or in confined concentrated media with strong intermolecular interactions (such as electrostatic and dipole-dipole interactions), so that molecules move in a coordinated fashion with neighboring molecules. The collective diffusivity, D_c, describing the mean-square displacement of the center-of-mass of groups of molecules differs from the self-diffusivity, D_s, as a result of the intermolecular correlations. This situation with dynamic—rather than static—heterogeneity is most difficult to simulate. Sometimes all interactions need to be simultaneously considered, sometimes a mean-field can be used from a certain distance on: likely, a hierarchy of simulation methods is necessary to obtain values for D_s and D_c.

In an environment in which molecules make long excursions l before changing direction, the distribution of these steps, $p(l)$, even when they are uncorrelated, determines the nature of the diffusion process. A typical case is the above-mentioned Knudsen diffusion process: the excursions are chords connecting points on the pore surface, through pore space. The trajectory of the molecules is a Lévy walk if $p(l) \sim l^{-\alpha}$, with $1 < \alpha < 3$ [29, 31, 32, 33], in which case the second moment of the chord length distribution, $\langle l^2 \rangle$, and therefore also $\langle r^2 \rangle$ diverges, and the walk becomes super-diffusive. If $\alpha > 3$, as in cylindrical pores ($\alpha = 4$), $\langle r^2 \rangle \sim t$, i.e., Einstein's law holds. For the marginal case of $\alpha = 3$, $\langle l^2 \rangle \sim \ln(t)$, and $\langle r^2 \rangle \sim t \ln(t)$ (Sinai's billiard without horizon; diffusion in infinite slit-shaped pores). For $\alpha = 2$, $\langle r^2 \rangle \sim t^2/\ln(t)$. Care has to be taken in molecular simulations, for convergence can be slow. A continuous-time random-walk (CTRW) description is well suited to model diffusion via jump processes in real time [34, 35], and anomalous super-diffusion ($\beta > 1$) in particular [33, 36]. Simulation algorithms that do not include the intrinsic power-law nature of the random walk waste a lot of resources; therefore it is advisable to combine short-term molecular dynamics simulations with, e.g., a CTRW description to obtain quantitative results within a reasonable amount of computational time.

Lévy walks and fractional Brownian motion are sometimes confused, because both lead to anomalous diffusion. However, anomalous diffusion is due to long-term correlations in the case of fractional Brownian motion, while no such correlations are invoked in Lévy walks. The origin of nonGaussianity is purely related to the geometry of the environment, i.e., it is due to extreme static heterogeneity, such as power-law polydispersity, and it occurs in the absence of any correlations between successive steps [37]. This is similar to the effect of polydispersity on the power-law tail observed in the Student t-distribution in Section 16.2.1.2 [16].

Confinement may not necessarily lead to such strong deviations from classical diffusion, but could still slow down diffusion in a way that is best captured by including the underlying multiscale nature of the medium in the simulation. Many amorphous porous materials, some of which are used as common catalyst supports, were shown to have a fractal surface over a range of scales $[\delta_{\min}, \delta_{\max}]$ encompassing molecular scales [38]. Coppens and Froment [39] showed analytically, and Malek and Coppens [40] confirmed by means of dynamic Monte Carlo simulations, that this leads to a molecular size-dependent decrease in the Knudsen diffusivity, via the expression: $D_s = D_0 / \left[1 + a\left(1 - \delta'^b\right)\right]$, or $D_s = D_0 \delta'^{2-D}$ to a first-order approximation. Here, a and b are constants that depend on the fractal dimension D of the surface ($2 \leq D \leq 3$) and its local accessibility; δ is the kinetic diameter of the molecule ($\delta_{\min} < \delta < \delta_{\max}$; $\delta' = \delta/\delta_{\max}$); and D_0 is the diffusivity in a porous medium of the same topology as the real medium, but with smooth walls.

Clearly, power laws frequently emerge in the analysis of diffusion in heterogeneous environments with multiscale features, but the origins are even more diverse [41], as dispersion in certain amorphous solids [42], and molecular simulations of water and ions in protein crystals [43] and protein channels in cell membranes [44] illustrate. In these more complex materials, the chemically heterogeneous environment, i.e., the broad energy landscape, is the main reason behind anomalous diffusion: the diffusing species is trapped for a time t that follows a very broad distribution $f(t)$. Note that the CTRW approach can be used to combine this trapping time distribution with the distribution of step lengths $p(l)$, in case of a jump-diffusion mechanism. A power-law distribution of trapping or waiting times, $f(t) \sim t^{-\gamma}$, induced by geometrical or chemical causes, may render diffusion anomalous for low values of γ (distribution with a "fat" tail). If the distribution does not have a fat tail, diffusion remains Gaussian and follows Einstein's law. For example, if $p(l) \sim l^{-4}$ (as in Knudsen diffusion in cylindrical pores), diffusion follows Einstein's law for $\gamma > 2$, otherwise the motion is subdiffusive [45]. Using molecular dynamics simulations, a power-law distribution of trapping times over at least three or four orders of magnitude was observed for water around proteins in solution [46] and in protein channels through cell membranes [44], as a result of the temporal fluctuations of the protein structure itself, which occasionally captures molecules for very long times, orders of magnitude longer (nanoseconds) than a normal vibration time (picoseconds).

In closing this paragraph, multiscale features are paramount and each one of them, individually or in combination, may influence diffusion to the extent that it becomes anomalous or is at least quantitatively affected. These multiscale effects are very diverse: they may be frozen in time, frozen in space or dynamic (due to a fluctuating environment, intermolecular correlations, and collective effects); they may result from memory effects; and they may result from the energy landscape, itself static or dynamic; they may even be caused by the vibration frequency of the material. Each of these effects could be studied by molecular simulations, but the inherent multiscale nature of the diffusion environment or the diffusion process itself demands tools that capture the overall behavior, which a single technique can rarely do. On the other hand, scaling laws are a powerful signature of multiscale behavior, but without good knowledge of the materials and the intermolecular interactions, scanned by experiments or by molecular simulations, false interpretations are risked. This has important consequences for the design of materials for catalysis and separations.

16.3 Imposing multiscale structure using a nature-inspired approach

16.3.1 *Hierarchical functional architecture in nature*

The previous section showed that an overall perspective helps to resolve heterogeneous, complex systems by discovering multiscale patterns. These patterns can be used in their analysis, aids in their understanding, and be utilized in applications. However, nowhere are scale-free and meticulously hierarchical structures as predominant as in biology. This leads us to postulate that natural structure-function relationships, as in the structure of enzymes for selective catalysis, go beyond the molecular and the nanoscale, and are truly multiscale. The interpolation between the micro and the macroscale is as important as structure at the microscale. By studying multiscale architectures in nature, and *imposing desirable multiscale patterns by design*, much more efficient materials and processes could be devised. Undesired complexity can be avoided by imposing scaling structures, or employing desired overall patterns to preserve properties up to macroscopic length or time scales. This section illustrates this point by means of a few examples.

Bone, for example, has a hierarchical structure of up to seven levels of organization, which has evolved to fulfill a variety of mechanical functions [47]. Rho et al. [48] note that mechanical properties of bone cannot easily be inferred from formulae using composite mixture rules, which points to important interactions in load transfer across the length scales. Fratzl and Weinkamer [49] point to the functional adaptation of the structure of biological tissues, such as wood, tendon, and bone, at all hierarchical levels, which is important to fracture healing. Insights from the hierarchical structure of biological functional materials like bone and tendon are a rich source of ideas for the design of synthetic materials [47, 50, 51, 52].

Inspired by construction methods used in gothic cathedrals, as well the shape of trees, Gaudí realized that trees are efficient distributors of weight, and he used branching tree-shaped columns to support the roof of the Sagrada Familia in Barcelona, claiming that "The architect of the future will build imitating Nature, for it is the most rational, long-lasting and economical of methods" [53]. Arguably, scale-free features are also seen in Hindu temples and Buddhist complexes (such as Borobudur in Java). It is not clear yet to this author whether this has a cultural/philosophical-religious or a functional basis, or both.

Mandelbrot [31] notes how Eiffel's insight of force balances on multiple lengthscales allowed the saving of an enormous amount of steel, yet enabled the building of a strong structure; indeed this is exemplified by the Sierpiński gasket, which, in the limit, only contains nodes where four branches meet, giving the structure stability and strength without any mass.

For chemical engineering purposes, fast transport, uniform distribution and collection of fluids, robustness, and easy scale-up are some of the most desirable properties in process and material design. In this respect, the multiscale features of trees, lungs, kidneys, the vascular network, and other biological networks are awe-inspiring. They form the basis for the *nature-inspired chemical engineering* applications discussed below.

Observing nature, *it is striking how a fractal structure is often interpolating between length scales $[R_{\min}, R_{\max}]$ over which the most relevant physics do not appreciably (i.e., qualitatively) change or should remain unchanged, while a transition in fundamental physics is accompanied by a radical change in structure.* In other words, characteristic length scales are tied to cross-overs in function, and fractal interpolation between these cross-over points is common, since it enables preservation of function and is accompanied by desirable features such as trivial scaling (also under organic growth), fast transport, mechanical strength, etc.

For example, transport through the 23 generations of the human airway tree [54, 55] changes from flow-dominated to diffusion-dominated around the level of the bronchioles, approximately 14–16 levels deep, counting from the trachea via the bronchi to the terminal bronchioles. The structure of the seven to nine spacefilling generations of acini, lined by the gas-exchanging alveoli, is very different from that of the impermeable upper generations. The mammalian airway tree is a fractal distributor/collector with a self-similar architecture that changes at the cross-over between flow and diffusion, to interpolate between the macroscopic scale of the entire organism and the gas-exchanging alveoli. The size of the alveoli themselves varies much less, even between organisms of widely different size, e.g., 31 (mouse, with a weight of 0.03 kg) and 133 micrometer (pig, with a weight of 50 kg) [56]. Sapoval *et al.* [57] demonstrated how the hierarchical structure of the lung, and the morphometric characteristics of the acinus are intimately tied to the efficiency of the lung as a gas exchanger. This is corroborated by Hou *et al.* [58] and Gheorghiu *et al.* [59], who showed that the lung operates in the "partial-screening" regime at rest and no diffusional screening at exercise (higher oxygen current), while its space-filling architecture is optimal with respect to active membrane surface area and minimum power dissipation, due to an equipartition of thermodynamic forces (pressure drop uniformly distributed over the branches of the bronchial tree; concentration drop uniformly over the membrane).

Similar arguments can be made about trees, which can be seen from a chemical engineering viewpoint as self-similar photosynthesis reactors. The tree crown serves as a support, and as a branching distributor of water. It grows thanks to the production of sugars in the leaves. The tree's self-similar architecture is perfectly suited to grow without change in functionality at the branch tips; the branches thicken, and the number of branching generations advances with age. The size of the leaves, on the other hand, does not change with the age of the tree. The veinal architecture of the leaves serves the final transportation stages to and from the photosynthesis complex, where the crucial reactions occur, and it forms the matrix for water, O_2/CO_2, nutrient, and biomass exchange. The tree and the lung are particularly inspiring to chemical reactor and catalyst design, subjects discussed in the next subsections.

Finally, several researchers have drawn a link between thermodynamic efficiency and the self-similar tree-like architecture within a finite range $[R_{\min}, R_{\max}]$ of vital transportation networks [60, 61, 62]. West et al. [61, 62] proposed thermodynamics as the explanation behind a ubiquitous allometric scaling law in nature, which states that the energy dissipation, $\dot{E}$, scales with body mass, M, to the 3/4 power, i.e., $\dot{E} \sim M^{3/4}$. Dissipation from a mass (proportional to a three-dimensional volume) via a classical two-dimensional surface would lead to $\dot{E} \sim M^{2/3}$, but flow through a space-filling, fractal vascular network and dissipation via its three-dimensional fractal boundary leads to the 3/4 law. This contention is under debate, however, as Banavar et al. [63] advanced that the properties of the transportation network could be relaxed from space-filling fractal to having some very general geometric properties. White and Seymour [64], on the other hand, even contest the generality of the long-standing 3/4 law, noting biological evidence for a bias resulting from different measurement conditions (e.g., body temperature and exercise versus rest). Nevertheless, the subject is worthy of further investigation, since any discovery of links between multiscale structure and thermodynamic efficiency is compelling not just to biologists, but for engineering applications as well, where minimizing entropy generation is a major objective [65]. Based on thermodynamic arguments, Bejan [62] used what he terms a "constructal" approach to explain why tree-like structures are so prevalent in nature and engineering. This bottom-up approach is very powerful, yet assumes no interactions between the building blocks from which the tree is recursively grown; hence, depending on the boundary conditions, the constructal approach could bypass the global optimum in favor of a tree-shaped, local optimum.

16.3.2 Nature-inspired fluid distributors and injectors

Consider a tree with n branching generations. Each branch of generation i ($1 < i < n$) has a diameter d_i and a length l_i. A branch of generation i splits, at its tip, into $m > 1$ branches, so that the first generation, just beyond the stem (generation 0), contains m branches, and the i-th generation m^i branches. There are m^n outlets, each one at the end of a twig, which is part of the final, n-th, generation. The path length, discounting the nodes, from the inlet (bottom of the stem) to each of the outlets is identical, and is given by: $L = \sum_{i=0}^{n} l_i$.

Such mathematically-constructed trees are rudimentary models for the many trees in nature, like botanical trees and their root networks, lungs, kidneys, and the vascular network, that serve to distribute or collect fluids. The equal (hydraulic) path length from the inlet to the multitude of outlets leads to uniform distribution over the volume or area covered by the branch tips of the final, n-th generation. Such a uniform distribution, while bridging an arbitrary number of scales, is highly desirable to chemical engineering applications. If the branching is performed in a self-similar, fractal way, then $l_i = l_{i-1}/\lambda$, and $d_i = d_{i-1}/\mu$, where λ and μ are constants, independent of the generation. The fractal dimension of the tree is $D = \log m / \log \lambda$, and the "diameter exponent" is $\Delta = \log m / \log \mu$ [31]. The tree is an excellent way to extrapolate or to interpolate between a broad range of scales. Either the stem is kept constant in size (l_0, d_0 constant) and recursion generates a tree with finer and finer twigs, with outlets that are closer and closer to each other, or the twigs are kept constant in size (l_n, d_n constant) and the tree grows, filling a larger and larger volume or area, while the outlets preserve their mutual distance, the stem becoming larger with increasing numbers of generation ($l_0 = l_n \lambda^n$, $d_0 = d_n \mu^n$).

Kearney [66] proposed to use such trees to distribute fluids in a uniform way as improved (two-dimensional) "shower caps", over fixed beds of particles, e.g., filling a chromatographic column or a silo. This stands in stark contrast to distributors consisting of a plate with equal-sized holes. Uniform distribution in this case is very difficult to achieve for large distributors, unless the pressure drop over the holes is excessively high. Even if structured columns are used [67], such as Sulzer packings (for distillation) or monolithic catalysts (for chemical reactors), the structure is underutilized by maldistribution if the structured packing or the monolith is not uniformly wetted. Using a *fractal distributor*, the distribution is guaranteed to be uniform. Even clogging of one of the branch tips (finest tubes) leads to a minimal influence to the overall flow, since the hydraulic path lengths from the inlet to each of the remaining outlets is still the same. In other words, such fractal distributors are an easy way to bridge multiple scales, they lead to uniform transport, and they are robust. Modular synthesis methods and (stereo)lithography can be used to construct fractal distributors.

In Section 16.2.1, the inherent multiscale nature of fluidized beds was discussed. Like turbulent single-phase systems, fluidized beds contain features (bubbles, vortices) over a broad range of length and time scales. The swirls may be advantageous to mix the fluidized bed contents, hereby promoting heat transfer to the reactor walls. However, scale-up and control are complex, and the presence of large bubbles and voids at greater heights above the bottom distributor plate decreases mass transfer and gas-solid contact. Similar remarks could be made about other multiphase fluid systems, such as slurry reactors.

Using a generalization of the nature-inspired fractal tree concept to inject gas throughout a three-dimensional reactor volume, a *fractal injector* provides a convenient way to impose an easier-to-scale structure on multiphase reactors [68, 69]. Since fluid is distributed throughout the volume, bubbles are smaller and local gas-solid contact is promoted, even for the same total throughput. The detailed mechanism of gas-solid mixing around a single fluid outlet can be studied using discrete particle simulations [70], while overall two-phase reactor models are sufficient to evaluate the benefits of injecting secondary gas injection via a fractal injector on the overall performance of fluidized bed reactors [71]. The staged fluid injection leads to higher chemical conversions, and, in certain cases, even to higher selectivities to desired reaction products.

This does not yet answer the question of what the optimal fractal injector geometry is, from the perspective of, e.g., maximum yield or uniform fluid distribution over the reactor contents. The translation of the fractal distributor idea from two to three dimensions, in order to feed a fluid in a uniform and self-similar way over a reactor volume, appears trivial at first [66]. If the reactor contents are *static*, as they are in a fixed reactor bed, and reaction can occur right at each of the outlets, similar principles hold as for the two-dimensional distributor ($D = 2$). Therefore, the optimal fractal dimension of the injector in this case is $D = 3$, since it should be space-filling, similar to the lung. However, also in nature, fractal transportation networks do not all have a dimension of three. For multiphase reactors, involving a fluid or fluidized reactor content, fed by another fluid that is injected into the reactor, the generalization to three dimensions is not obvious. In the *dynamic* environment of a fluidized bed, the reactor contents are mobile and one of the fluids will invariably rise with respect to the other one as a result of density differences, e.g., as gas bubbles in a bubbling fluidized bed [68, 72].

Since gases rise as bubbles and within the emulsion phase, an injector with $D = 3$ would typically lead to a surplus of reactants higher in the reactor. On the other hand, a classical two-dimension distributor plate (which is the other limit of the fractal dimension D, i.e., $D = 2$) might allow gas to access the entire reactor volume, but in a nonuniform way, as a result of depletion higher up in the bed. Therefore, the optimal fractal dimension is $2 < D < 3$ [72]. In the laboratory, a reactor with a fractal injector with $D = \log 6 / \log 2 \approx 2.6$ ($m = 6$, $\lambda = 2$) was tested and gave good results, even if this value was not optimized.

In summary, the fractal injector allows the replacement of an uncontrolled multiscale multiphase flow, with its complex hydrodynamics, with one where the behavior can be optimized at the micro-level (around a single outlet of the injector, surrounded by neighboring outlets), using lab-scale experiments and small-scale simulations, while scale-up to large-scale units has been simplified using a fractal tree, as frequently used in nature.

Tondeur and Luo [73] discuss fractal fluid distributors from the perspective of cost-minimization (pressure drop and viscous dissipation versus reactor volume). Bejan's [62] constructal approach can be used to optimize the trees in a thermodynamic sense, i.e., to minimize entropy generation [65]. Allometric scaling arguments could perhaps be an alternative way to optimize a fractal injector structure, inspired by the proposed link between efficiency and tree-architecture, as discussed in Section 16.3.1.

16.3.3 Optimal pore networks in catalysis

The high activity, selectivity, and stability of porous solid catalysts is primarily caused by the geometrical and chemical design of the "active sites," i.e., the nanoscale environment, which determine the *intrinsic* catalytic properties. Rational design of active sites [74] is a challenging multiscale quantum mechanical problem [75, 76]. For example, the atomic structure of nanoscale clusters of Pt and Re metal on porous Al_2O_3, and the chemical structure of the Al_2O_3 pore surface itself, determines the performance of a catalyst for the reforming of low-octane to high-octane gasoline. Faujasite is a zeolite, a crystalline aluminosilicate traversed by a regular array of 0.7 nm pores, and an excellent catalyst for catalytic cracking of gasoil, another refinery process; the highly selective cracking of gasoil into a desirable mixture of olefins and gasoline products is possible thanks to the molecular selectivity induced by the subnanometer pores, and the acidic Al-sites (balanced by counter-ions), i.e, the precise geometric and chemical environment of the zeolite at molecular scales. Similarly, the conversion of carbon dioxide into oxygen and sugars, using light, depends on the molecular structure of the photosynthetic complex. This example might not be one of a particularly active catalyst, especially in leaves of plants in temperate climates, but perhaps this is also unnecessary in this case. Enzymes are proteins that can be remarkably active and selective in catalyzing molecular conversions, a source of inspiration for the synthesis of biomimetic catalysts.

These examples have an element in common: the *effective*, overall catalytic properties depend not just on the nanoscale, but on the design of the catalyst and the reactor up to macroscopic scales [67, 77, 78]. Adequate distribution channels are required to transport molecules toward and away from the active sites. The previous section highlighted the ability of fractal networks to efficiently bridge meso to macroscopic length scales. Recalling the examples of trees and lungs, leaves and acini differ in structure from tree crowns and the upper respiratory tract. The size distribution of the transportation channels tends to be less broad at smaller length scales, especially when molecular diffusion, rather than flow, dominates the molecular transport properties.

How this observation in nature translates to catalyst design can be investigated by simulating diffusion and reaction in models of nanoporous catalysts. If no channels other than the nanopores are present, the situation is akin to a large metropolis with many narrow streets but with a lack of avenues and broad access roads: the city becomes congested, especially when traffic is dense during rush hour. Likewise, if the intrinsic activity of a catalyst is high, nanopores cannot transport the confined reactants rapidly enough. Large pore channels should be introduced inside the catalyst, connecting interior and exterior, but the question is how they should be distributed and what their optimal width (or width distribution) is? Analytical calculations [79] and multiscale computer simulations [80] demonstrate that the maximum overall yield of a chemical reaction (with first-order kinetics, easily generalized to arbitrary kinetics, see [78]) is obtained by introducing channels of a *constant* diameter, d, traversing the catalyst in such a way that the channels are separated by nanoporous catalytic walls or (sub)particles of size w. In doing so, the *local* effectiveness of the nanoporous walls is as high as possible, close to 100%, so that the intrinsic catalytic properties are preserved, while d is large enough to guarantee fast access, but not so large that too much of the available space is taken up by voids, rather than catalytic material. Details can be found in the previous papers, as well as in Wang *et al.* [81], who discuss the implementation using a combination of a gradient-based optimization package, NLPQL, a multigrid [82] solver, MGD9V, and a limited amount of in-house coding to tackle this multiscale simulation problem.

It is important to realize that a uniform, rather than a broad distribution of channels was found as a direct consequence of the optimization constraints. If, for example, the number of channels is kept constant, instead of allowing it to vary as an optimization variable, then broader distributions, $p(d)$ and $p(w)$, are found to be optimal, because finite-size effects such as corners and boundaries play a role [83]. If transport through the large channels is assumed to be extremely rapid, while transport limitations are constrained to the nanoporous material, fractal networks may be more efficient [84], similar to what happens at larger, flow-dominated length scales, such as those discussed in the previous section.

Another desirable property of fractal networks is their robustness to deactivation [85, 86, 87], i.e., the effective activity stays constant if the intrinsic activity decreases or diffusion is slowed down over a broad range, as a result of increasing blockage of the active sites or the pores. This can be interpreted in terms of increased screening of deeper generations in fractal trees, while the most accessible generations are more efficiently used [88, 89]. Sheintuch [90] and Coppens [91] have reviewed the topic of diffusion and reaction in fractal porous catalysts, including the possibility of influencing overall activity and product selectivity by changing the fractal surface roughness of porous catalysts.

These examples demonstrate opportunities to avoid some *undesired* multiscale features, which require extensive analysis, by instead *imposing* a particular multiscale architecture. Such calculations serve to guide the synthesis of catalysts that are optimal under those constraints most significant in a practical application.

16.3.4 *Nature-inspired membranes*

Progress in nanoporous material synthesis should allow us to design catalysts and also membranes with vastly improved properties. Membranes are increasingly attractive to carry out separation processes, because of their tunability and low energy requirements. The ideal membrane is strong and stable against fouling, and, of course, it should be ultra-selective, all while allowing the filtrate to permeate at high flux. Keeping the examples of the lung as well as the applications discussed in the previous section in mind, the geometrical and chemical structure at the microscale should induce the desired intrinsic properties of high throughput and selectivity, while an easily scalable structure, e.g., with fractal elements, would serve as a distributor or collector to interpolate between the macro and the microscale.

In this context, a source of inspiration even better than the lung is the vascular network and the membrane structure of erythrocytes (red blood cells), because the throughput/selectivity properties of the latter are so remarkable. Peter Agre received the Nobel Prize in Chemistry in 2003 for discovering that this was thanks to protein channels called aquaporins. These form hydrophilic, conducting pathways through the hydrophobic interior of the cell membrane, which is a thin lipid bilayer separating the inside from the outside of the cell. Aquaporin-1 (AQP1) in human red blood cells consists of proteins that form hourglass-shaped channels less than 1 nm in diameter at their narrowest point [92]. Remarkably, these channels allow the passage of up to 3×10^9 water molecules each second, at very low activation energies, while blocking protons [93]. Electrostatic interactions due to positive charges in the central part of AQP1 play an important role: when approaching this central zone, water molecules can reorient their dipole moments and pass, while protons and other cations are rejected. Water molecules maintain a mobility similar to that in bulk water. This ability to separate water and charged species at such high throughput is exactly what is needed in water desalination and purification. Likewise, other channels in the outer cell wall and in the nuclear membrane of eukaryotic cells are ultra-effective in the separations of ions or molecules, including proteins.

Atomistic molecular simulations are crucial in revealing the details of the separation mechanism, since the source may lie in very subtle features of the protein structure [94]. However, it takes many days of computer time to simulate transport of water through protein channels via molecular dynamics simulations, even over periods of only a few nanoseconds. Reasons are the long-range electrostatic and dipolar interactions, and the chemical heterogeneity of the structure. This problem only gets worse in the simulation of ion channels, containing a gate and a filter besides the channel itself—the opening and closing of the gate may occur over time scales of microseconds, inaccessible by molecular dynamics, with their femtosecond-resolution. Furthermore, transport may be mediated or "active", i.e., it involves energy exchange and chemical reactions to permit or to facilitate species transport even against gradients in chemical potential.

One way to approach this multiscale problem is to combine various coarse-graining techniques, e.g., by using Brownian dynamics and continuum simulation methods, or more sophisticated multiscale simulation methods discussed elsewhere in this book. As discussed in Section 16.2.2, despite the complexity of the transport phenomena in protein channels, multiscale patterns in the overall transport behavior may emerge. Molecular simulation of water diffusion in AQP1 and in the outer membrane porin OmpF revealed a power-law distribution of water trapping times over three orders of magnitude, from $\sim$10 ns to $\sim$10 ps [44]. The origin for this power law is the intrinsic protein fluctuations. However, power laws as universal patterns may emerge for the broad variety of reasons discussed in Section 16.2.2.

Key to the design of nature-inspired membranes is the identification of the principal actors in separation efficiency, as well as other important properties, like stability against fouling. It is unnecessary to mimic *all* molecular details needed in a biological context. Arguably, this is even counterproductive: the biological constraints may well be different from those in, e.g., water desalination or protein separation. However, the remarkable multiscale mechanisms used by nature to achieve properties superior to currently used zeolite and polymeric membranes in technical separations, give us pause. The *patterns* used to achieve molecular or ionic separations up to macroscopic scales are a useful source of inspiration to guide the overall multiscale architecture of a membrane unit and the chemical functionalization required to achieve high efficiency. This cannot be achieved using units that ignore the intrinsically multiscale workings of nature necessary to bridge scales, while maintaining nanoscale selectivity. Hence, while the concept of nature-inspired membranes is not yet in a stage of development comparable to that of the fractal injector or the hierarchically-structured porous catalysts discussed in the previous sections, it may be postulated that the feature-free, geometrically- or chemically-homogeneous environment of, e.g., pure carbon nanotubes or silicalite will be insufficient. The actors behind the generation of multiscale patterns, discussed in Section 16.2, in conjunction with molecular design, should be more effective.

16.4 Concluding remarks

This chapter discussed alternative "multiscale" analysis and design methodologies, for which bottom-up approaches and conventional coarse-graining techniques would be inefficient, because they ignore underlying symmetries, patterns, or invariant distributions. The holistic approach advocated here collectively studies (analysis) or imposes (design) a multitude of scales. Scale-free systems and architectures with a continuous spectrum of relevant length scales are often observed in nature, both in biology and in complex heterogeneous systems. A predominant example is that of fractal architectures, essential to the workings of nature and, if properly utilized, extremely useful to technology as well.

However, as this chapter illustrated, fractals are only one example of interesting natural patterns that we can learn from. Nature is full of patterns [95], many of which may not carry any function, but others do (superhydrophobicity, selective adhesion, strength, energy efficiency), and on top of that even emerge spontaneously by so-called "self-assembly". Such useful, often multiscale patterns are a source of inspiration for technology. Whitesides and Grzybowski [99] marvel on the ubiquity of self-assembling processes throughout nature and technology, and point out that life itself would be impossible without them. If multiscale patterns spontaneously emerge, reductionism and complexity are connected. Typically, as in life itself, pattern formation by self-assembly is not static but dynamic and out of equilibrium: it requires a constant flux of energy [97].

Although there is some interest in holistic approaches to harness complex (and, invariably, multi-scale) systems [98, 99], this chapter showed that, beyond the phenomenology of power laws, novel applications may emerge by combining holistic with atomistic approaches. Each approach offers different insights, which complement each other. The combination allows the design of truly innovative products and processes.

16.5 Acknowledgments

The Norwegian Academy of Science and Letters is gratefully acknowledged for support of a sabbatical stay at the Centre for Advanced Studies in Oslo. Prof. S. Kjelstrup and Prof. D. Bedeaux are warmly thanked for the hospitality in their home, which allowed me to prepare this chapter while recovering from surgery in the Netherlands.

References

[1] Kunii D. and Levenspiel O. (1991). *Fluidization Engineering*. Second ed., Butterworth-Heinemann, Boston.
[2] van der Hoef M. A., van Sint Annaland M., and Kuipers J. A. M. (2004). "Computational fluid dynamics for dense gas-solid fluidized beds: A multi-scale modeling strategy", *Chem. Eng. Sci.*, **59**(22–23), p. 5157–5165.
[3] Gidaspow D. (2004). *Multiphase Flow and Fluidization: Continuum and Kinetic Theory Descriptions*, Academic Press, New York.
[4] Fan L.-S. and Zhu C. (1998). *Principles of Gas-Solid Flow*, Cambridge University Press, Cambridge, U.K.
[5] Jackson R. (2000). *The Dynamics of Fluidized Particles*. Cambridge University Press, Cambridge, U.K.
[6] Sundaresan S. (2000). "Modeling the hydrodynamics of multiphase flow reactors: Current status and challenges". *AIChE J.*, **46**(6), p. 1102–1105.
[7] Curtis J. S. and van Wachem B. (2004). "Modeling particle-laden flows: A research outlook", *AIChE J.*, **50**(11), p. 2638–2645.
[8] Grace J. R. and Taghipour F. (2004). "Verification and validation of CFD models and dynamic similarity for fluidized beds". *Powder Tech.*, **139**(2), p. 99–110.
[9] Li J. and Kwauk M. (1994). *Particle-Fluid Two-Phase Flow: The Energy-Minimization Multi-Scale Method*, Metallurgical Industry Press, Beijing.
[10] Daw C. S., Lawkins W. F., Downing D. J., and Clapp N. E. (1990). "Chaotic characteristics of a complex gas-solids flow", *Phys. Rev. A*, **41**, p. 1179–1181.
[11] Schouten J. C., Takens F., and van den Bleek C. M. (1994). "Maximum-likelihood estimation of the entropy of an attractor", *Phys. Rev. E*, **49**, p. 126–129.
[12] Takens F. (1981). Detecting strange attractors in turbulence. *Lecture Notes in Mathematics, Vol. 898, Dynamical Systems and Turbulence*. Rand D. A. and Young L.-S. (eds). Springer Verlag, Berlin, Germany, p. 366–381.
[13] van den Bleek C. M., Coppens M.-O., and Schouten J. C. (2002). "Application of chaos analysis to multiphase reactors", *Chem. Eng. Sci.*, **57**(22–23), p. 4763–4778.
[14] van Ommen J. R., Coppens M.-O., Schouten J. C., and van den Bleek C. M. (2000). "Early warning of agglomeration in fluidized beds by attractor comparison", *AIChE J.*, **46**, p. 2183–2197.

[15] Gheorghiu S., van Ommen J. R., and Coppens M.-O. (2003). "Power-law distribution of pressure fluctuations in multiphase flow", *Phys. Rev. E*, **67**, p. 041305.
[16] Gheorghiu S. and Coppens M.-O. (2004a). "Heterogeneity explains features of "anomalous" thermodynamics and statistics". *PNAS*, **101**(45), p. 15852–15856.
[17] Chandrasekhar S. (1943). "Stochastic problems in physics and astronomy", *Rev. Mod. Phys.*, **15**, p. 1–89.
[18] Vicsek T. and Family F. (1984). "Dynamic scaling for aggregation of clusters", *Phys. Rev. Lett.*, **52**, p. 1669–1672.
[19] Kärger J. and Ruthven D. M. (1992). *Diffusion in Zeolites and Other Microporous Solids*. Wiley, New York.
[20] Keil F. J., Krishna R., and Coppens M.-O. (2000). "Modeling of diffusion in zeolites". *Rev Chem. Eng.*, **16**, p. 71–197.
[21] Gefen Y., Aharony A., and Alexander S. (1983). Anomalous diffusion on percolation clusters. *Phys. Rev. Lett.* **50**, 77–80.
[22] Stauffer D. and Aharony A. (1994). *Introduction to Percolation Theory.* Second edn, Taylor & Francis, Philadelphia.
[23] Sahimi M. (1994). *Applications of Percolation Theory*, Taylor & Francis, London.
[24] Gefen Y., Mandelbrot B. B., and Aharony A. (1980). "Critical phenomena on fractal lattices", *Phys. Rev. Lett.*, **45**, p. 855–858.
[25] Gefen Y., Aharony A., Mandelbrot B. B., and Kirkpatrick S. (1981). Solvable fractal family, and its possible relation to the backbone at percolation. *Phys. Rev. Lett.* **47**, 1771–1774.
[26] Given J.A. and Mandelbrot B. B. (1983). "Diffusion on fractal lattices and the fractal Einstein relation", *J. Phys. A: Math. Gen.*, **16**, L565–L569.
[27] Giona M., Schwalm W. A., Schwalm M. K., and Adrover A. (1996). "Exact solution of linear transport equations in fractal media – I. Renormalization analysis and general theory". *Chem. Eng. Sci.*, **51**, p. 4717–4729.
[28] Havlin S. and Ben-Avraham D. (2002). "Diffusion in disordered media". *Adv. Phys.*, **51**, p. 187–292.
[29] Bouchaud J.-P. and Georges A. (1990). "Anomalous diffusion in disordered media: Statistical mechanisms, models and physical applications", *Phys. Rep.*, **195**, p. 127–293.
[30] Mandelbrot B. B. and Van Ness J. W. (1968). "Fractional Brownian motions, fractional noises and applications", *SIAM Review*, **10**(4), p. 422–437.
[31] Mandelbrot B. B. (1983). *The Fractal Geometry of Nature*, Freeman W. H., San Francisco.
[32] Levitz P. (1997). "From Knudsen diffusion to Lévy walks", *Europhys. Lett.*, **39**(6), p. 593–598.
[33] Zumofen G. and Klafter J. (1993). "Scale-invariant motion in intermittent chaotic systems", *Phys. Rev. E*, **47**(2), p. 851–863.

[34] Montroll E. W. and Weiss G. H. (1965). "Random walks on lattices. II", *J. Math. Phys* **6**, p. 167–183.
[35] Klafter J. and Silbey R. (1980). "Derivation of the continuous-time random-walk equation", *Phys. Rev. Lett.*, **44**, p. 55–58.
[36] Shlesinger M. F., Zaslavsky G. M., and Klafter J. (1993). "Strange kinetics", *Nature*, **363**, p. 31–37.
[37] Mandelbrot B. B. (1998). *Multifractals and 1/f Noise*, Springer, New York.
[38] Pfeifer P. and Avnir D. (1983). "Chemistry in noninteger dimensions between two and three. I. Fractal theory of heterogeneous surfaces", *J. Chem. Phys.*, **79**, p. 3558–3565.
[39] Coppens M.-O. and Froment G. F. (1995). "Knudsen diffusion in porous catalysts with a fractal internal surface", *Fractals*, **3**, p. 807–820.
[40] Malek K. and Coppens M.-O. (2003). "Roughness dependence of self- and transport diffusivity in the Knudsen regime: dynamic Monte-Carlo simulations and analytical calculations", *J. Chem. Phys.*, **119**(5), p. 2801–2811.
[41] Coppens M.-O. and Dammers A. J. (2006). "Effects of heterogeneity on diffusion in nanopores – From inorganic materials to protein crystals and ion channels", *Fluid Phase Equil.*, **241**(1–2), p. 308–316.
[42] Scher H. and Montroll E. W. (1975). "Anomalous transit-time dispersion in amorphous solids", *Phys. Rev. B*, **12**, p. 2455–2477.
[43] Malek K., Odijk T., and Coppens M.-O. (2005). "Diffusion of water and sodium counterions in nanopores of a β-lactoglobuline crystal: A molecular dynamics study", *Nanotechnology*, **16**, S522–S530.
[44] van Hijkoop V. J., Dammers A. J., Malek K., and Coppens M.-O. (2007). "Water diffusion through a membrane protein channel: a first passage time approach", *J. Chem. Phys.*, **127**(8), p. 085101.
[45] Dammers A. J. and Coppens M.-O. (2005). Characteristics of Knudsen diffusion in channels with fractal wall roughness, *7th World Congress of Chemical Engineering*, Glasgow, U.K.
[46] Garcia A. E. and Hummer G. (2000). "Water penetration and escape in proteins". *Proteins*, **38**, p. 261–272.
[47] Weiner S. and Wagner H. D. (1998). "The material bone: Structure-mechanical function relations", *Ann. Rev. Mat. Sci.*, **28**, p. 271–298.
[48] Rho J. Y., Kuhn-Spearing L., and Zioupos P. (1998). "Mechanical properties and the hierarchical structure of bone", *Med. Engng and Phys.*, **20**(2), p. 92–102.
[49] Fratzl P. and Weinkamer R. (2007). "Nature's hierarchical materials", *Prog. Mat. Sci.*, **52**(8), p. 1263–1334.
[50] Baer E., Hiltner A., and Keith H. D. (1987). "Hierarchical structure in polymeric materials", *Science*, **235**(4792), p. 1015–1022.
[51] Lakes R. (1993). "Materials with structural hierarchy", *Nature*, **361**, p. 511–515.
[52] Mann S. (1993). "Molecular tectonics in biomineralization and biomimetic materials chemistry", *Nature*, **365**, p. 499–505.

[53] Wikiquote (2008). http://en.wikiquote.org/wiki/Antoni_Gaudí, 9 February 2008.
[54] Weibel E. R. (1963). *Morphometry of the Human Lung*, Springer, Berlin.
[55] Weibel E. R. (1984). *The Pathway for Oxygen*, Harvard University Press, Cambridge, MA.
[56] Lum H. and Mitzner W. (1987). "A species comparison of alveolar size and surface forces", *J. Appl. Physiol.*, **62**, p. 1865–1871.
[57] Sapoval B., Filoche M., and Weibel E. R. (2002). "Smaller is better – but not too small: A physical scale for the design of the mammalian pulmonary acinus", *PNAS*, **99**(16), p. 10411–10416.
[58] Hou C., Gheorghiu S., Coppens M.-O., Huxley V. H., and Pfeifer P. (2005). Gas diffusion through the fractal landscape of the lung: How deep does oxygen enter the alveolar system? In: *Fractals in Biology and Medicine*, Losa G. A., Merlini D., Nonnenmacher T. F., and Weibel E.R. (eds), Springer Verlag, Birkhäuser Basel, p. 17–30.
[59] Gheorghiu S., Kjelstrup S., Pfeifer P., and Coppens M.-O. (2005). Is the lung an optimal gas exchanger? In *Fractals in Biology and Medicine*, Losa G. A., Merlini D., Nonnenmacher T. F., and Weibel E. R. (eds), Springer Verlag, Birkhäuser Basel, p. 31–42.
[60] West G. B., Brown J. H., and Enquist B. J. (1997). "A general model for the origin of allometric scaling laws in biology", *Science*, **276**, p. 122–126.
[61] West G. B., Brown J. H., and Enquist B. J. (1999). "The fourth dimension of life: Fractal geometry and allometric scaling of organisms", *Science*, **284**, p. 1677–1679.
[62] Bejan A. (2000). *Shape and Structure, from Engineering to Nature*, Cambridge University Press, Cambridge, U.K.
[63] Banavar J. R., Damuth J., Maritan A., and Rinaldo A. (2002). "Supply-demand balance and metabolic scaling", *PNAS*, **99**(16), p. 10506–10509.
[64] White C. R. and Seymour R. S. (2005). "Allometric scaling of mammalian metabolism", *J. Exp. Biol.*, **208**, p. 1611–1619.
[65] Kjelstrup S., Bedeaux D., and Johannessen E. (2006). *Elements of Irreversible Thermodynamics for Engineers*. Second ed., Tapir Academic Press, Trondheim.
[66] Kearney M. M. (2000). "Engineered fractals enhance process applications". *Chem. Eng. Prog.*, **96**, p. 61–68.
[67] Cybulski A. and Moulijn J. A. (Eds) (2005). *Structured Catalysts and Reactors*, Second Ed. CRC Press, New York.
[68] Coppens M.-O. (2001). "Method for operating a chemical and/or physical process by means of a hierarchical fluid injection system", *U.S. Patent*, p. 6,333,019.
[69] Coppens M.-O. and van Ommen J. R. (2003). "Structuring chaotic fluidized beds", *Chem. Eng. J.*, **96**(1–3), p. 117–214.
[70] Christensen D., Vervloet D., Nijenhuis J., van Wachem B. G. M., van Ommen J. R., and Coppens M.-O. (2008a). "Insights in distributed

secondary gas injection in a bubbling fluidized bed via discrete particle simulations", *Powder Techn.*, **183**, p. 454–466.
[71] Christensen D., Nijenhuis J., van Ommen J. R., and Coppens M.-O. (2008b). "The influence of distributed secondary gas injection on the performance of a bubbling fluidized bed reactor", *Ind. Engng Chem. Res.*, **47**, p. 3601–3618.
[72] Coppens M.-O. (2005). "Scaling up and down in a nature inspired way", *Ind. Eng. Chem. Res.*, **44**, p. 5011–5019.
[73] Tondeur D. and Luo D. (2004). "Design and scaling laws of ramified fluid distributors by the constructal approach", *Chem. Eng. Sci.*, **59**(8–9), p. 1799–1813.
[74] Bell A. T. (2003). "The impact of nanoscience on heterogeneous catalysis", *Science*, **299**, p. 1688–1691.
[75] Van Santen R. A. and Neurock M. (2006). *Molecular Heterogeneous Catalysis: A Conceptual and Computational Approach*, Wiley, New York.
[76] Nilsson A., Pettersson L. G. M., and Norskov J. (Eds) (2007). *Chemical bonding at surfaces and interfaces*, Elsevier, Amsterdam.
[77] Froment G. F. and Bischoff K. B (1990). *Chemical Reactor Analysis and Design*, Second ed. Wiley, New York.
[78] Coppens, M.-O. and Wang, G. (2009). Optimal design of hierarchically structured porous catalysts. In press for: *Design of Heterogeneous catalysts*. Ozkan U., ed., Wiley, NY.
[79] Johannessen E., Wang G., and Coppens M.-O. (2007). "Optimal distributor networks in porous catalyst pellets. I. Molecular diffusion", *Ind. Eng. Chem. Res.*, **46**, p. 4245–4256.
[80] Wang G., Johannessen E., Kleijn C. R., de Leeuw S. W., and Coppens M.-O. (2007). "Optimizing transport in nanostructured catalysts: a computational study", *Chem. Eng. Sci.*, **62**, p. 5110–5116.
[81] Wang G., Kleijn C.R., and Coppens M.-O. (2008). "A tailored strategy for PDE-based design of hierarchically structured porous catalysts". *Int. J. Mult. Comp. Eng.* **6**, p. 179–180.
[82] Wesseling P. (1992). *An Introduction to Multigrid Methods*, Wiley, New York.
[83] Gheorghiu S. and Coppens M.-O. (2004). "Optimal bimodal pore networks for heterogeneous catalysis", *AIChE J.*, **50**, p. 812–820.
[84] Coppens M.-O. and Froment G. F. (1997). "The effectiveness of mass fractal catalysts", *Fractals*, **5**, p. 493–505.
[85] Villermaux J., Schweich D., and Authelin J.-R. (1987). "Transfert et réaction à une interface fractale représentée par le 'peigne du diable'". *C.R. Acad. Sci. Sér. 2*, **304**, p. 399–404.
[86] Mougin P., Pons M., and Villermaux J. (1996). "Reaction and diffusion at an artificial fractal interface: Evidence for a new diffusional regime", *Chem. Eng. Sci.*, **51**, p. 2293–2302.
[87] Sheintuch M. (2000). "On the intermediate asymptote of diffusion-limited reactions in a fractal porous catalyst". *Chem. Eng. Sci.*, **55**, p. 615–624.

[88] Pfeifer P. and Sapoval B. (1995). "Optimization of diffusive transport to irregular surfaces with low sticking probability". *Mater. Res. Soc. Symp. Proc.*, **366**, p. 271–277.
[89] Sapoval B., Andrade J. S., and Filoche M. (2001). "Catalytic effectiveness of irregular interfaces and rough pores: the 'land surveyor approximation'". *Chem. Eng. Sci.*, **56**, p. 5011–5023.
[90] Sheintuch M. (2001). "Reaction engineering principles of processes catalyzed by fractal solids", *Catal. Rev. Sci. Eng.*, **43**, p. 233–289.
[91] Coppens M.-O. (1999). "The effect of fractal surface roughness on diffusion and reaction in porous catalysts: from fundamentals to practical applications", *Catal. Today*, **53**, p. 225–243.
[92] Jung J. S., Preston G. M., Smith B. L, Guggino W. B., and Agre P. (1994). "Molecular structure of the water channel through aquaporin CHIP. The hourglass model", *J. Biol. Chem.*, **269**(20), p. 14648–14654.
[93] Zeidel M. L., Ambudkar S. V., Smith B. L., and Agre P. (1992). "Reconstitution of functional water channels in liposomes containing purified red cell CHIP28 protein", *Biochemistry*, **31**, p. 7436–7440.
[94] de Groot B. L. and Grubmüller H. (2001). "Water permeation across biological membranes: Mechanism and dynamics of aquaporin-1 and GIpF". *Science*, **294**, p. 2353–2357.
[95] Ball P. (1999). *The Self-Made Tapestry: Pattern Formation in Nature*, Oxford University Press, Oxford.
[96] Cross M. C. and Hohenberg P. C. (1993). "Pattern formation out of equilibrium", *Rev. Mod. Phys.*, **65**, p. 851–1112.
[97] Ottino J. M. (2003). "Complex systems", *AIChE J.*, **49**(2), p. 292–299.
[98] Barabási A.-L. (2003). *Linked*, Perseus, Cambridge, U.K.
[99] Whitesides G. M. and Grzybowski B. (2002). "Self Assembly at all scales", *Science* **295**, p. 2418–2421.

INDEX

a posteriori analysis 323, 334–51, 358, 359, 360, 364–7, 370, 378–82
   a priori hybrid error 372–6
a priori analysis 316–18, 320, 323
Aarnes, J. E. 3
ab initio calculations 481, 483
Abraham, F. F. 95, 111
accelerations 146, 195, 502
acetylation 519, 530
acini 546, 550
ACIS (modeling kernel) 403
Ackland-Thetford core repulsion 151
acoustic phonons 442, 469
acoustic wave
   frequency/propagation 143, 144
acoustic zone-center modes 490
action-reaction principle 255
active sites 541, 549–50, 551
adaptive mesh refinement 351–8, 385
adaptive modeling/procedures 286, 302, 400, 415–16
adaptive strategy 296–9
Adelman, S. A. 118
adiabatic system 428
adjoints 289, 293–4, 301, 302, 320, 323–51, 354–8, 362–7, 370–5, 378–9, 386, 387
   interpolation of 207, 208
admissibility 254, 257
   kinematic 252
   Lagrange multiplier 262
   macro forces 260–1
   static 252
advection 228, 231, 234, 237, 240, 242, 307
aero-gels 31, 33
affine shear 499
affine space 251
aggregation:
   adaptive smooth 197
   diffusion-limited 37
Agre, Peter 552
Airbus A380 plane 57
alcohol 31

Alexander, S. 42
algebra 4, 142, 226, 400
   linear 362–3, 370–1, 372
   see also AMG
algorithms 196, 230–1, 289, 311, 378, 384, 406
   adaptive 298, 353, 354, 537
   atomistic 537
   better 538
   biased Monte-Carlo 291
   coarsening 4
   computing surrogate models 292–3
   convergence of 262–3, 265
   coupling 286
   densification 292
   discretization 370
   efficient and numerically stable 147
   gap-tooth 226
   grid-based 404
   multiscale 193, 290–5
   Newton 296
   non-uniform coarsening 10, 11
   simulation 515, 543
   sublinear 409
   time-stepper-based bifurcation 220
   see also Goals Algorithm; Lanczos algorithm
aliasing 121
all-atom approaches 515
Allen-Cahn problem 348
Allison, S. A. 518
allometric scaling 547, 549
alpha iron 94
alveoli 546
AMG (Algebraic Multigrid) 194, 197, 200–1
amorphous nanocrystalline structures 491–2
amplitude 125, 156, 235, 499
   low 160
   negligible 498
   small 153, 160
   stochastic 208
   sufficiently large 159

angular momentum:
  balance of 47, 48
  conservation of 45, 115
  zero 124–5
anharmonic forces 143–4, 146, 159
anisotropies 4, 6, 11, 27, 43, 293
annealing 209–10
ansatz 137, 144
anterpolation 208
API (application programming interface) 399, 415
applied stress 486, 487, 500, 505
  effects on dislocation mobility 506
  vanishing 488
approximation:
  adaptive procedures to improve 400
  best 268, 273
  diagonal 149
  differential operator 350
  finite difference 216, 224, 228–9, 231, 238, 241
  finite element 108, 366, 368
  first-order 269, 544
  harmonic 146, 434–5, 441
  improved 203
  lumped-mass 145
  mass conservative 6
  mixed MsFEM 4
  numerical 349
  pointwise Lagrange multiplier 109
  polynomial 225
  possible at finite temperature 143
  QoI 285
  radial time-space 249, 265–77
  reinterpreted 41
  second-order 269
  semilocal 427, 441, 463
  third-order 269
  time-space 269
  time-stepper 219
  very good 469
  water-cut data 15
  see also Born; Boussinesq; Debye-Hückel; FAS; Galerkin; piecewise approximation; quasiharmonic approximation
aquaporins 552
arginine residues 530
argon 151
Arlequin methods 111, 167, 293, 294, 295, 296, 298, 299–300
Ashland Hetron 922 (epoxy) 82

asymmetric two-atom vacancy defect 126, 127
asymptotic expansion 59, 60, 222, 223
  multiple-scale 57, 196
AtC (atomistic-to-continuum)
  blending methods 165–6
  energy-based 167, 168, 181–5
  force-based 167, 168–170, 171, 172–82, 183, 184
athermal energy release rate 508–9
atomic indices 142
atomic interactions 125
  nearest neighbor 120
  stiffness that arises from 143
atomic lattice 428
atomic masses 140, 141
atomic scale 136
atomic shuffling details 498
atomistic approach 536, 537, 553
atomistic/continuum models 395, 398
  blended functionals 177–9
  concurrent adaptive multiscale simulation tool 410–17
  coupling of 93–133, 134–7, 165–89
atomistic scales 57, 428
atoms:
  bond energy between 431
  boundary 409
  bulk silicon 470
  carbon 125, 126, 426
  chains of 465
  coarse-grained 141, 204, 426, 425–6
  coarse-level locations 200
  computing forces between 408
  connected by covalent bonds 428
  constrained to move identically 102
  coupling between vibrations of 438
  dark 128
  defect 497
  defect mobility measures 500–10
  defining and storing information 406–8
  deformed configuration of 428
  described by empirical interatomic potential 426
  edges overlaid by 117
  enforcing constraints between continuum mesh and 114
  equally spaced 119
  far field 104
  fast relaxation scheme 408–10
  ghost force of 110
  handshaking domain 108, 110

high-energy 409
hydrogen 103, 129, 466, 467
independent 100
interior 452, 474
internal force of 124
last line of 105
layers of 463, 465, 467, 497, 501
light 128
link 102, 103, 104
local entropy of 456
local potential energy of 464
majority stored implicitly 406
mesh nodes behave as 135
movable 508
movements over simple shapes 402
multiple 148
neighboring 120, 151, 402, 408, 431, 438
potential energy of 140, 436, 438
removing 129
representative 137, 394, 426, 427
scaled masses of 119
simulation of systems 146
strength and deformation measures 486–500
surface 452, 453, 469, 471, 472, 473, 474
tetrahedrally bonded 428
thermal vibration of 426
virtual 100, 101
see also nuclei; position of atoms
attenuation 196, 482
attractive term coefficient 431
attractors 540
  chaotic 536, 537, 539
  strange 314
automated adaptive continuum simulations 414
automatic mesh generation 403
averaging principles/techniques 196, 286
axial tow 82

backward Euler method 312, 345
backward problem 301, 302
Bai, Dov 201
balance equations 41, 45
  see also force-balance equations
Banach spaces 326, 339
Banavar, J. R. 547
bandwidth 138, 161
  exceptional 485
Barcelona 545

base-catalyzed hydrolysis 31
basis functions 7–11, 18, 204, 257, 265, 324
  affine 258
  macro 261
  multiscale, boundedness of 17
  quadratic 258
Baskes, M. I. 94
Bauman, P. T. 287, 293
BCC (body-centered cubic) structure 496, 500–1
"beads-on-a-string"-like models 516, 518, 519, 521, 523, 524, 525, 529
behavioral models 396
Bejan, A. 547, 549
Belytschko, T. 94, 95, 116
Ben-Avraham, D. 542
Bensoussan, A. 223
bias tows 78, 82, 84
bifurcations 217, 220
bilinear form 359
bilinear identity 328, 329, 333, 339
biochemical modifications 519
biology 31, 545, 546
  fundamental processes 514, 515
bistable problem 348, 350, 351
blending constraint operators 175–7
blending methods, see AtC blending methods
Bloch's theorem 439
body forces 47, 48, 60, 413, 429, 452
  applied 451
  prescribed 249, 251, 254
Boltzmann constant 138, 435, 455
Boltzmann lattice problems/methods 219, 221, 538
  see also Poisson-Boltzmann
Boltzmann-type upscaling 208
bond angles 204, 431
  cosine of 434
bond breaking 113, 499
bond lengths 204, 434, 467
bonds 104, 119, 469
  atomistic 110
  blocked 542
  covalent 103, 290, 292, 296, 428, 466
  dangling 103, 129, 466
  dimer 466, 467
  electronic 134
  interatomic 135
  internal 154

bonds (*cont.*)
  Si-Si/Si-H 472, 473
  strong between neighboring atoms 402
bone 545
Bootstrap AMG 197
Born, M. 486, 488
  see also Cauchy-Born rule
Born approximation 158
Born-von Karman boundary conditions 439, 441
Borobudur 545
bottom-up approach 57, 425, 485, 536, 540, 547, 553
Bouchaud, J.-P. 542
boundary conditions:
  absorbing 145, 160
  adjoint 333, 334, 379
  arbitrary 217, 228
  artificial 173, 174, 205, 225
  asymptotic 156
  built-in 228
  cantilever 451
  complete control over 484
  complicated 34
  constant gradient 217, 223, 224, 242
  displacement 50, 52
  essential 412
  fine-level 206
  fixed-fixed 451
  homogenous 7–8, 25, 169, 222, 223, 333
  inhomogeneous 168, 169
  macroscopically inspired 228
  matching 96
  minimal 333
  mixed 48
  natural 52, 173, 174, 412
  no-flux 228, 232
  non-zero flux 9
  patch 242
  periodic 144, 151, 205, 342, 448, 490, 491, 494, 496, 505, 506, 507
  prescribed 252, 485, 507
  silent 95
  simple 44
  tooth 238–42
  unphysical interface 185
  well-defined 9
  zero flux 9
  see also Born-von Karman; Dirichlet; Neumann
boundary flux values 383

boundary representations 401, 402, 404
  nonmanifold geometric 403, 406
boundary value problems 33, 34, 35, 40, 41–53, 238, 334, 338
  elastic 61, 70, 77
  general second-order linear elliptic 336
  initial, in periodic media 196
  strong form of 58
Boussinesq approximation 307
Boussinesq equations 380
braid architecture 82, 83
branch tips 546–7, 548
Bravais lattices 433, 439, 440
  FCC (face-centered cubic) 432, 434
Brenner MM model 103
Brenner's potential 426
  see also MTB-G2
bridging scales 537
bridging-domain method 94, 95, 96, 97, 104–8, 111, 121, 129
  conservation properties of 115–17
  coupling used is stable 112
  effectiveness of 120
  ghost forces in 109, 110
  molecular/continuum dynamics with 119
Brillouin zone 143, 149, 150, 152, 153, 154
  first 439, 440, 442, 443
brittle material 499
bronchial tree 546
Broughton, J. Q. 94, 95, 139, 144, 146, 147, 150, 158
Brownian dynamics 515, 518, 521, 552
  simulations 522, 524
Brownian motion 145, 218, 540
  fractional 543
Brusselator problem 306, 369, 376
BSM (bridging scale method) 95, 97, 118–19, 139, 425
bubbles 536, 538–9, 540, 548, 549
buckling 84, 247
Buddhist complexes 545
buffers 227, 228–38
bulk modulus 488
bulk silicon structures 467–70
buoyancy term 307
Burgers equation 342
  one-dimensional 238

simulation of 238
viscous 221
Burgers vector 501, 505

CAD (computer-aided design) 395, 397, 401, 402, 403, 415
CADD (coupled atomistic/discrete-dislocation) 95
CAE (computer-aided engineering) systems 397
Cai, W. 95
calibration 300, 302
Cantor sets 35–6, 38, 45, 49, 51
carbide precipitates 134
carbon 125, 126
  Brenner MM model for 103
carbon black nanoparticles 32
carbon dioxide 550
carbon fiber 82
carbon nanotubes 541, 553
Carey, G. F. 383
Carpinteri, A. 43, 45, 46, 51
Cartesian geometry 10, 11, 15, 61, 293
catalysis 541, 544
  optimal pore networks in 549–51
  selective 545
catalyst design 547, 551
catalysts:
  biomimetic 550
  hierarchically-structured porous 553
  intrinsic activity of 550
  monolithic 548
  porous 538, 541, 549–50, 551
  solid 541
cations 552
Cauchy:
  discrepancy 447
  generalized inequality 325, 326
  sequences 329
  stress 249, 487, 494
Cauchy-Born rule 94, 112, 113, 137, 176, 177, 182, 428, 431–2, 433, 453, 454, 474, 475
cellular processes 514–15
cement-setting kinetics 510
center manifold theory 241
central difference method 114, 115, 119
ceramics 499, 510
CGMD (coarse-grained molecular dynamics) 94, 134–64, 425
Chaboche, J. L. 98

chain models:
  one-dimensional 496–7
  wormlike 522
chain rule 433, 460
channelized permeability fields 4, 7, 12
chaos theory 539
chaotic behavior 314, 315, 323, 348, 536, 537, 539–40
characteristic scales/variables 536, 537, 539, 542
chemical bonding 499
chemical engineering 536–59
chemical reactions 290, 510, 550, 552
  deactivation processes during 542
Chen, W. 197
Chiaia, B. 51
chord length distribution 543
chromatin folding 514–35
chromosomes 516, 529
chunkiness parameters 22, 25, 26
circular tube model 84–5
cleavage fracture 94
closed form solutions 95
CM (continuum models) 95, 98, 120, 125, 128, 129, 293, 415
  cutoff frequency of 113, 121, 122
  discretized 297, 298
  gradual transition from molecular model to 104
  nonlinear elastic 410
  QC method avoids ill-posedness of 113
  retaining the efficiency of 425
  see also MM/CM; QM/CM
CMC (compatible Monte-Carlo) 200, 208
coarse equations 58, 193, 242
  derivation of 57, 196, 197, 202–5, 206
coarse grids 3, 5, 8, 9, 15, 16, 21
  adaptive 10, 12
  Cartesian 10, 11
  highly anisotropic 4, 6, 11, 27
  large-scale changes calculated on 193
  non-uniform 4, 10, 11
  quasi-uniform 4, 17
  structured 13, 14
  unstructured 4, 6, 13, 26, 27
coarse-scale models 96, 97, 98, 102, 119, 286

coarse-to-fine transitions 195, 199–200
coarse variables 137, 199, 200, 209, 210
  choosing an adequate set of 195
  current-time and previous-time 207
  dependence of 202
  developing for high-Reynolds flows 201
  general equilibrium criterion for choosing 208
  neighboring 201, 202, 207
coarsening 198, 199–201, 204, 206
  equilibrium criterion 207
  general criterion 208
  localness of 202
  non-uniform 4, 5
  spatial 207
  stochastic 208–9
  unstructured 6, 14
coefficients 209, 229, 237, 345, 363, 371
  anisotropy 43
  attractive term 431
  continuous function 324
  diffusion 350
  drag 502, 503
  higher-order 458
  nonlinear diffusion 343
  potential energy 140, 142, 143–4, 146, 147
  reflection 155, 156, 158, 159
  repulsive term 431
  scattering 157
  sedimentation 515, 525, 526
  stiffness 488
  terminating 458
  thermal expansion 307
  transmission 155, 156
  unknown 50, 201
  see also recursion coefficients
coincident atoms/nodes 100, 101
collagen fibers 31
collective properties 536
compaction 516, 526–7, 528, 531
  optimal 529
compatibility:
  balance of 97, 98
  enforced 100, 108
  weak displacement 411
compatibility conditions 270, 271
  and equilibrium equations 251–2
compliance 48
component-based methods 399

compression 443, 444, 451, 488, 489
  effect of 448
compression operators 394
compression tests 449
compressive waves 160
computational domains 101, 307, 396, 397, 398, 405
  image data in construction of 402
  relationships between entities in 400, 403, 404
  topological and voxel representations that control 399
  topological entities and adjacencies for 415
computational fluid dynamics 482, 539
computational models 395, 396, 397, 398, 415, 516
  multiple 405
  transformation of mathematical to 400
computational procedure 76–8, 205, 395, 483
  accurate and efficient 308
  adaptive 396
  algorithms for surrogate models 292–3
  coarse-level computability of fine observables 210
  equation-free 216–46
  functionals and 323–34
  goals 309
  mechanics with time-space homogenization 247–82
  modeling tools 134–5
  multiscale strategy 258–65
  RG efficiency 195
  slowdown-free 198
  see also simulation; software
computer simulations 285
concurrent coupling 93–133
condensation 31, 530
conduction 542
conjugate heat transfer 307, 376–8
conjugate variables 251
connectivity 541, 542
conservation 119
  bridging-domain method 115–17
  properties of meshfree methods 116
  also energy conservation; momentum conservation
consistency 226–8, 232, 234–5

consistency tests 168, 177
constitutive behavior of phases 84, 86
constitutive equations 58, 60, 96, 216, 251
  appropriate 414
constitutive laws 394, 425, 446
  continuum 430
  material 426
constitutive model 109
constitutive relations 86, 255, 264, 272, 273, 274, 475
  computed using Helmholtz free energy density 426
  developing for continuum-level codes 484
  local 459
  material 428
  partitioned 76, 81
constraint equations 175, 177, 413
constraints 117, 118, 142, 173, 196
  algebraic 226
  biological 553
  blending 175–7, 184
  boundary 225–6
  compatibility 119
  consistent 121, 122, 124, 125
  diagonalized 121, 122, 123, 124, 125
  enforcing 106, 114, 174–7, 183–4
  mean field 138
  optimization 184, 352, 353, 551
  size 506
  trivially imposed 183
constructal approach 547, 549
construction methods 545
continua 33, 34, 35, 44, 46, 93, 414–15, 541
  generalized 185–6
  homogenized 45
  level description 429–30
  methods for coupling atomistic to 113
continued fraction 454
  GF expressed as 460
  infinite level 458
continuous function 38, 86
continuous linear functional 323
continuous variables 440
continuum discrete scales 394
continuum elasticity 136, 154, 165, 482
  linear 143
continuum equations 196, 425
  breakdown of 197

  domain definitions used with 402
  weak 173
continuum function spaces 412
continuum models 171–2
  defect 95
  discretized 295
  fictitious 103
  force-based 168, 170
  strong form of 170
  see also atomistic/continuum models
continuum scales 57, 394
control-based strategy 228
convection 319, 320, 343, 355
convergence 4, 5, 9, 11, 17, 18, 27, 43, 200, 201, 204, 231, 232, 268, 385, 386, 508
  a priori 337
  algorithm 262–3, 265
  approximation 275
  asymptotic range of 309
  iteration 378
  slow 543
  standard analysis techniques 369
  strong 223
convergence theorem 227
  for velocity 20
coordinate systems 47, 48, 110
  alternative 404
  Cartesian 61, 293
  fixed 46
  global 76
  local interface 76
Coppens, M.-O. 544, 551
copper 503
corner-point grids 5, 15
corrector calculation 115
correlations 209
  intermolecular 543, 544
  structure-property 484, 491
  temporal 37, 543
Cosserat material 262
Coulomb friction 264
coupled equations 305–6
coupling 95, 96, 97, 108, 119, 125, 126, 165–89, 248, 277, 503
  atomistic or quantum model with 103
  between vibrations of different atoms 438
  coincident 100, 101
  direct 98–9, 101, 112, 425
  finite 241

coupling (*cont.*)
  ideal 120–1
  length scales approach 134–7
  master-slave 94, 99–101, 117–18
  QM/MM 129
  strict 109, 110, 111
  weak 110, 111
crack front:
  advances 499
  extension 506–10
  initially straight 508
  sharp 506, 507
crack plane 94
crack propagation 159
  graphene sheet 129–30
crack tip 128, 197
  abrupt displacements 506
  compressive waves generated at 160
  cracked cylinder cut from 507
  elastic waves generated at 155
cracks 128, 247, 264, 265
  multiscale simulation 155
  semi-infinite 507
  strain localization around 492
Crank-Nicolson scheme 345
creep-resistant structures 484
critical values 341, 342, 490, 500, 508
crystal lattices 113, 134, 139, 146, 148, 435, 441, 474
  atoms in 428, 437, 438, 439, 440
  ductility or brittleness of 506
  formulation of CGMD relies on properties of 161
  mechanical stability of 486
  perfect 438, 439
  sharp crack evolves in 506
  underlying 430, 431
  unstrained 447, 448
crystalline aluminosilicate 550
crystalline materials 94, 402, 405–10
  mechanical properties of 500
CTRW (continuous-time random walk) 543, 544
current 306
Curtin, W. A. 109
cutoff frequencies 95, 113, 120, 121, 122
cylinder flow 307, 380–1, 382, 385
cylindrical waves 124

damage state variables 86
dampers 118
Daw, C. S. 539

Debye-Hückel model 518, 520, 525
decay of influence 320
decomposition 118, 264, 305–90
  additive 59
  bridging scale 165
  edge-to-edge to overlaid domain 117
  overlapping domain 95, 166, 185
  proper orthogonal 268
  spectral 267
  structure 252–6
defect configuration 496
defect singularity 494
defects 134, 495
  atomic configuration of 134
  atomistic measure of mobility 500–10
  continuum model 95
  equilibrium properties at finite temperature 426
  generic 467
  graphene sheets 119, 125–9
  local 493
  nucleation of 410
  point 489
  slit-like 130
  twin 497
decohesion:
  interface 134
  intergranular 492
deformation 46, 443, 444, 461, 485
  atomistic measures of strength and 486–500
  chemical reactivity and mechanics of 510
  continuous and atomistic fields 176
  elastic 493, 498
  finite 410
  finite-strain 494
  inelastic 59
  inhomogeneous 493
  nucleation of 499
  plastic 484, 493
  primary local 454
  reversible 498
  size effects of 45
  static 290
  sufficiently smooth 186
  uniaxial 442, 490
  see also eigendeformations; homogeneous deformation; shear deformations
deformation gradient 99, 293, 428, 429, 431–7, 442, 448, 486

constrained to be symmetric 488
  local 441
deformation space 487, 488
degrees of freedom:
  atomistic 94, 176, 413, 414
  coarse-grained 161
  continuum 176, 413–14
  fine-scale 95, 208, 294
  internal 137, 138, 141, 144
  larger-scale 209
  nodal 413
  primary 497
  simulated 136
  spatial 271
  systematic removal of 137
  time 271
  twinned crystal has too many 496
delta functions 138, 140, 469
  see also Dirac
dendritic structures 31
density 218, 250
  energy 434, 459, 460
  known average 205
  probability 199, 322
  reference values for 307
density functions:
  free energy 434, 436
  potential energy 447
  strain energy 430, 432, 434
dependence tables 202–3, 207, 208
dependent variables 404
derivation 139, 142, 143, 144, 488
  coarse equations 57, 196, 197, 202–5
  three-scale formulation 58
derivatives:
  bond length 434
  energy 126, 434, 447, 459
  fractional 37, 38, 46–7, 51
  GF 460
  Helmholtz free energy density 459
  LPDOS 460
  normal 382
  parametric 140
  partial 343
  potential energy density
    function 447
  spatial 59, 216, 224, 227, 228, 229, 230
  square-integrable 139
  strain energy 432, 434
  vibrational frequencies 437
  zero classical 51

  see also Gâteaux derivatives; Tersoff
    potential;
  time derivatives
desalination 552, 553
design problems 308
determinism and stochasticity 31, 196, 210
deterministic problems 196, 202
  relaxation in 199
Devil's staircase function 38, 45, 46, 51
DFT model 103
Dhia, H. B. 111, 112
diagonalization 115
diatoms 31, 32, 33
difference schemes 344
  backward Euler-second order
    centered 345
  space-finite 343
  third order sub-diagonal Padé 345
differentiability 40, 46, 47
differential equations:
  conservative 115
  linear 273, 330–1
  see also PDEs
differential operators 36, 37–9, 40, 41, 186, 330–1
  approximation of 350
  linear 335
differential self-consistent method 44
differentials 52
diffusion 196, 235, 241, 242, 306, 319, 349, 545
  anomalous 543, 544
  bulk 541
  classical, strong deviations from 544
  collective 543
  cross-over between flow and 546
  infinite slit-shaped pores 543
  localized 542
  negligible 146
  network topology affects 541
  porous media 37, 540
  single-file 542, 543
  surface 541
  water 551
  see also Dirichlet teeth; Knudsen;
    reaction-diffusion
diffusion constants 518–19, 524–5
diffusion problems 232, 234
  dissipative numerical methods for
    integrating 310
  linear 373

diffusion properties 515
dihedrals 201, 204
dimers 466, 467
  experimental values for 518
dinucleosome folding 522
Dirac delta function 62, 108, 412
Dirichlet boundary conditions 7–8, 170, 228, 232, 233, 238, 239, 307, 416
  homogeneous 169, 222, 231
  inhomogeneous 169
Dirichlet interface values 378
Dirichlet problems 327
  boundary value 334
Dirichlet teeth diffusion 239–40
DiSCO (Discrete Surface Charge Optimization)
  program 518, 524
discontinuities 248, 257, 264
  acceleration waves 493
  field 41
discrete compatibility equations 413
discrete-continuum method 293
discrete equations 106, 108
discrete particle techniques 483, 494, 538
discrete phase space 412
discrete scales 57, 394, 536
discrete-state variables 208
discrete systems 35
discrete (Lagrangian) particle
  models 538
discretization:
  coarse 264
  computational tensor fields
    constructed through 398
  displacement and stress fields 51
  double 400
  elliptic problems 336–7
  energy 100
  equilibrium equations 61
  evolution problems 312, 343, 345–6
  FEM 297
  fine-scale 238
  finite difference 239
  finite element 337, 343, 376, 415
  generalized 395
  holistic 240
  macroscopic equation 242
  microscale 239
  most commonly applied
    methods 400
  multiscale 359, 365

operator splitting 312, 370, 373
PDE 193, 197, 199, 205, 238
reaction-diffusion equation 367
refined 264
space-time 224, 343
spatial 10, 216, 234, 235, 257, 348, 357, 369, 376
time 229, 248, 257, 356
uniform refinement of 309
discretization enrichment problem 352
discretized continuum model 295, 297, 298
Diskin, Boris 207
dislocation 95, 484
  complicated 417
  continuously distributed 57
  core structures 135
  curved 508
  discrete 57
  dynamic/intrinsic properties of 500
  impediments to flow 134
  interacting 112
  mobility in silicon 503–6
  motion 410
  nucleated 410, 495, 508–9
  partial 503, 504, 508, 510
  screw 505
  single glide in a metal 500–3
  thermal release of 501
dislocation loops 417, 489, 499, 509
  nucleation of 508
  partial 508, 510
dislocation propagation 494
dislocation slip 495–8
dispersion 196, 544
  nonlinear 143
  phonon 443
dispersion curves 490, 491
displacement 94, 295, 408, 412, 487
  abrupt crack-tip 506
  average vertical 296
  coarse-scale 118
  continuity 41, 47
  continuum 106, 173, 183, 413
  discretization of 51, 413
  energy functional 183
  evolutions for 265
  fine-scale 118
  homogeneous 434
  incremental 494
  initial 120, 124
  inner 432, 433, 434, 435, 436, 437, 442, 447, 459, 468

interface 59, 254
LQHM model underestimates 451
mean-square 541, 543
nodal 100, 103, 104, 118, 137–8, 139, 146
particle 176, 498
peak 450, 451, 452
prescribed 171, 255
relative 432, 495, 497, 498
resolving as plane wave 494
shear 510
small 143
steady-state 306
thermal vibrational 428–9, 435
uniform 139
unknown site components 290
weak 411
zero 169
displacement control mode 494
displacement distribution 252, 509
shear 510
displacement fields 50, 61, 99, 139, 249, 251, 254, 255
asymptotic expansion of 59
asymptotic form of 156
continuous 176
corresponding space of 254
evolution over time 265
initial 300
optimizing reflection term added to 96
using higher-order gradients of 184
displacement vectors 105, 140, 296, 297, 300, 429
fixed 169
initial 301
nodal 103
prescribed 430
disregistry 502
dissipation 251, 252–3, 547
viscous 549
dissipative
method/forces/processes 115, 145, 159
distribution functions 404
piecewise 400
properly accounting for 405
divalent ions 518, 526, 527, 528–9, 531
dramatic effect on fiber organization 525
divergence expression 332
DNA (deoxyribonucleic acid) 519, 521, 524

duplex 516
eukaryotic 516
linker 515, 516, 518, 522, 525–7, 529–30
negatively-charged 515, 522, 525, 526
second code in 514
supercoiled 515, 518
wrapped 516, 520
see also linker histones
DNS (direct numerical simulation) 57
Dokainish, M. A. 94
Doll, J. D. 118
domains 12, 40, 61, 295
atomistic 104, 113, 117, 121, 167
bridge 167
coarse-scale 95
composite 49
continuum 98, 104, 117, 167, 294, 404
convex 337, 339, 343
coupling 109, 121, 125
decomposed into geometric grid or mesh 400
discretized 395
disjoint 403
edge-to-edge 117
explicit 396, 397, 398, 399, 402, 403, 404, 415
faceted crystal 406
finer-scale 403
finest-level 205
finite element 412
flow 200
fluid 380
fully discrete 296
domains geometric 402
geometric complexity of 34
globular 524
higher-scale 403
hybrid concurrent model 411
integration 440
mathematical 403
overlaid 117, 119
partitioned 296, 297, 298
periodic 209
polygonal 307, 337, 343
problem 41, 44
QM 126
solid 380
solving boundary value problems on 53
strain 98

domains (cont.)
  telescoping 403
  time-space 216, 248–9, 256–8, 262, 265, 266, 268, 272, 275, 334, 343, 395
  transport over 42
  see also bridging-domain method; computational domains; overlapping domains; spatial domains; subdomains
downscaling information 386
driving force 500, 506
dual spaces 324–5, 326
dual variational principle 50
duality 323, 329
dynamical evolution 498
dynamical matrix 140, 142, 143, 144, 439–40
  diagonal 147, 490
  eigenvalues of 442
  finite temperature 151
  inverse is trivial to compute 147
  nonzero terms of 149
  range of 151
  stiffness matrix differs from 156
dynamical problem 301
dynamical systems 207–8, 217, 242
dynamics:
  advection-dominated 231
  artificial 228
  chemical 306
  chromatin structure and 514
  coarse 208
  concurrent coupling methods for 95
  continuum 119
  controlled by spatial and/or temporal
    correlations 37
  discrete particle 538
  dislocation 484
  dislocation 93, 481
  lowsalt/high-salt unfolding/folding 515
  molecular/continuum 119
  oligonucleosome 525
  rigid-body 514
  scattering and 155–60
  small-amplitude 153
  tooth 240–2
  see also CGMD; computational fluid dynamics; Fitzhugh-Nagumo; Langevin; lattice dynamics; molecular

dynamics; patch dynamics
dynamics propagation 514
Dyskin, A. V. 44

E-admissibility conditions 254, 258, 261, 272
EAM (Embedded Atom Method) 410, 411
edge dislocation dipole 501
EDIP potential 505
effective coordination number 431
Eiffel, Gustave 546
eigendeformations 61, 63, 64
  independent partitioned 67, 68, 69, 70, 72, 76, 79, 80
eigenfunctions 42, 210, 268
  orthogonal 268
eigenmodes 241, 488
eigenseparation 63, 64, 87
  independent 72, 74
  partitioned 66
eigenseparation influence functions:
  governing equations for 62
  reduced-order 67, 77
eigensolutions 268
eigenstates 147
eigenstrain influence functions 61
  reduced-order 66, 77
eigenstrains 59, 63, 64, 72, 74, 84
  dependent 67, 68, 77
  partitioned 67, 68, 86
eigenvalues 148, 232, 233, 234, 268, 329, 437, 438, 440, 442, 455, 464, 489
  nonzero 147
  positive 268, 488
  zero 488
Einstein, Albert 58
Einstein's law 541, 542, 543, 544
elastic constants 143, 151, 430, 434, 438, 441, 451–2, 459, 467, 475, 486–9, 494
  adiabatic 437, 445, 447
  bulk 446–9
  effective 43
  evaluated at current state 489
  fourth-order tensor 487
  homogenized 44
  independent 459
  isothermal 436–7, 445, 447, 448
  isotropic 49
  predicting 426–7
  thermal softening of 144

third-order 410
vanishing of 490
variation of 473, 474
elastic theory 185
nonlocal 166
elastic waves 149, 151, 153, 161
generated at crack tip 155
long wavelength 145, 148
reflection from a CG region 157, 159
spurious reflection of 95
within CGMD 148
elasticity 50, 128
nonlocal theory 94
one of the most central issues in 486
see also continuum elasticity; linear elasticity
elastodynamics 196
elastostatic behavior 430
electric fields 518, 519
error in 520
electric vehicles 484
electron shells 143
electronic structure 101, 102, 103, 481, 482, 483, 491, 504
first-principles quantum mechanical calculations of 484
electrons 94, 210, 483
electrostatic current equations 306
electrostatics 515, 520, 521, 530, 531, 552
charges 207, 522
dipole-dipole interactions 543
shielding 518
elliptic equations 7–8, 17
elliptic problems 319, 327, 352, 353
a posteriori analysis for 337–8
discretization of 336–7
important stability property of 320
linear 372
triangular systems of 358–60
EMANIC (experimental technique) 529
embedding space 42, 46
Euclidean 36, 43
EMMS (energy minimization multiscale method) 539
empiricism 485
emulsion phase 538, 549
energy 94, 98, 102, 126, 408, 497
activation 541, 552
athermal release 508–9

atomistic 104, 112
bond 431, 472
characteristic 447
coarse-grained 138, 139, 140, 141, 142, 146
cohesive 140
constant flux of 553
continuum 104, 112
discretized 100
effective 426
elastic 139
entropic contribution 426
error in 125
exchange of 552
high frequency 121
initial 119
internal 138, 428, 471, 486
kink 505
linear scaling of 104, 112
migration 505
minimizing 209–10
minimum solution 409
quantum-mechanical zero point 435
relaxed 508
stacking 504, 508
strain 430, 432, 434, 489, 496, 498
thermodynamic 113
vibrational 455, 472, 473
zero point 447, 449
see also free energy; kinetic energy; MEP; potential energy
energy barriers 209, 506, 508
energy-based criterion 494
energy basins 209
energy conservation 115, 116–17, 144, 145
energy density 434, 459, 460
see also Helmholtz
energy dissipation 125, 547
high/low frequency 121, 123
energy equations:
linking to thermostats 113
steady-state 306
energy minimization principle 167
energy scaling function 105, 110, 114, 116
constant 118
Engquist, B. 227, 232
entropy 426, 456, 539
minimizing generation 547, 549
normalized 539
environmentally-durable structures 484

enzymes 541, 545
equal spring constants 296
equation-free framework 196, 197
  computation 216–46
equations of motion 95, 137, 142,
    155–7
  bandwidth of 138
  CGMD 146, 148
  explicit integration of 145
  linear 143
  Newton's 333–4
equidistribution 353, 366
equilibration 151, 200
equilibrium 48, 95, 149, 199, 254, 290,
    436, 487, 490, 501
  atom fluctuates around 428
  atomistic 209
  force 167
  general criterion 208
  mechanical 98–113
  molecular 210
  self-consistent multilevel 201
  small displacements from 143
  static 286
  thermal 137, 144, 155
equilibrium deformed
    configuration 171
equilibrium equations 60, 78, 102,
    104, 110, 137, 413
  compatibility conditions and 251–2
  discrete 106–7
  discretized 61
  ill-conditioning of 98
  standard 411
equipartition 146, 546
equiprobability 202
equivalence 142, 486
equivalent macroscale equations 426
error estimates 286, 287–9, 296–9,
    302, 305–90, 395
  finite element 394
error indicators 265, 415, 416
  good 263
errors 53, 103, 104, 123, 227, 231, 236,
    237, 520
  a posteriori 351–8, 378–82
  constitutive relation 272, 273, 274
  energy 125
  fractional flow 13
  greatly reduced 173
  help to reduce 174
  higher-order 155
  lifting 219

methods to control 395
minimized by nodal
    displacements 139
modeling 395
normalized 111
optimal 232
potential 405
relative 14, 150, 154, 268, 415
saturation 12, 13, 14, 15
smooth 193
water-cut 14
see also numerical error
erythrocytes 551–2
estimates/estimation 52, 285–304
  convergence 21
  production 11
  projection 22
  stability 22
  see also error estimates
estimators:
  gap-tooth 232, 235
  reference 231
  time-derivative 232
Euclid 35, 49, 326
Euclidean embedding space 36, 43
  fractal media 45, 47
eukaryotes 514, 516, 552
Euler coordinate frame 524
Euler equations 206–7
  see also backward Euler method;
    forward Euler
    scheme
Euler-Lagrange equations 142
evaporation 31
evolution law 252, 254
evolution(ary) problems:
  adaptive mesh refinement for 356–8
  analysis of stability for 318
  discretization of 312, 343, 345–6
explicit/implicit methods 117
exponentials 37, 540
external forces 99
extrapolation technique 458

Farrell, D. E. 95
FAS (full approximation storage)
    approach 197
fatigue resistance 510
faujasite 550
FBZ (first Brillouin zone) 439, 440,
    442, 443
FCC (face-centered cubic)
    structures 432, 434, 490, 499

FEM (finite element models) 43, 50, 52, 53, 65, 94, 99, 106, 109, 116, 150–1, 153, 294, 337, 434
  atomistic model coupled to 135
  bridging-domain method used to link MM to 129
  consistent-mass 141, 154, 158
  coupling an atomistic or quantum model 103
  cutoff frequency of 120
  discretization 297
  harmonic CGMD differs from 143
  hybrid scheme based on 136
  independent of and prior to nonlinear macro analysis 70, 77
  linear elastic 125, 143
  lumped-mass 149, 154, 157, 158
  mixed multiscale 3–30
  piecewise linear 64, 360, 367
  scattering properties of 156
  space-time 344
  well-known method in 98
  see also MsFEM
femtosecond-resolution 552
ferromagnetic material 348–9
Feyel, F. 98
Feynman, Richard 57
Feynman's path integrals 210
FFT (fast Fourier transform) techniques 147
field equations 45, 481–2
fields 42, 257, 404–5
  building 259
  CG mean 142
  coarse-scale 118
  continuum 118
  discontinuities at interfaces 41
  dual 250
  E-admissible 254
  interpretation of 138
  piecewise constant 241
  primal 250
  radiation 484
  temperature 307, 382
  variation of 394
  velocity 300, 307
  see also displacement fields; electric fields; force fields; strain fields; stress fields; tensor fields
fine grids 3, 6, 10, 11, 12, 14
  unstructured 5, 27
fine-level laws 196

fine-level patches 206
fine-scale/fine-level equations 57, 58, 197
fine-scale models 96, 97, 98, 119, 285, 286, 287, 295, 296, 403, 405
  atomistic 196
fine-to-coarse transformations 199–200
fine variables 201
  neighboring 199
finest scale of interest 58, 63, 69, 72
finite difference schemes 216, 217, 223, 224, 226–9, 231–3, 234, 236–8, 239, 240, 241, 242
finite element calculation 274
finite element codes 247
finite graphs 42
finite sum truncation 324
finite-volume method 10
Finnis-Sinclair many-body potential 151
first-order expansion 487–8
first-order relation 203
first-principle laws 195
first-principles calculations 486
Fish, Jacob 196, 197
Fitzhugh-Nagumo dynamics 221
five-spot problem 12
fixed-point theorem 43
flow equations 3, 4, 5, 6, 11, 14, 26–7
  mixed multiscale finite element method for 7–10
  two-phase 10
fluctuation-dissipation theorem 145
fluid distributors and injectors 547–9
fluid equations 378, 482
fluidized beds 536, 538–40
  inherent multiscale nature of 548
fluids 208, 382
  atomistic 200, 205
  complex 209, 537
  dense 219
  fast transport 546
  heat transfer problem between solids and 376–8
  incompressible two-dimensional 206
  reacting 308
  thermal conductivities of 307
  wave propagation and attenuation in 482
flux correction 382, 383
flux dependence 207
flux-divergence theorem 41, 47

Fokker-Planck equation 218
folding situations 205, 206, 514–35
force-balance equations 169, 173, 176
force balances 546
force-based models 167, 168–170, 171, 172–82, 183, 184
force constant matrix 435, 436, 437, 439, 462, 465
  eigenvalue of 464
  local 438, 464
  mass-weighted 455, 457, 458–9, 468
  tridiagonalized 458, 463, 464
  very large and computationally intractable 461
force distribution 252
force fields 255, 487, 488
  averaged 530
  derivation of 210
formal singularities 142
formaldehyde cross-linking 529
Fortafil 82
forward Euler scheme 224, 229, 230, 232, 233, 234, 236, 238, 312
forward problem 301, 302
fouling 551, 552
Fourier coefficients 324
Fourier representation 140, 142
Fourier series 324
Fourier transforms 119, 142, 144, 148, 149, 439
  fast 147
fractal distributors 546, 548, 549
fractal injectors 548, 549, 553
fractal networks 550, 551
  diffusion on 542
fractals 31–6, 38, 41–53, 536, 537, 544
  infinite 40
  porous 44
  self-similar 542, 548
  space-filling 547
fractional calculus 36–41, 42
fractional operators 36–41, 44, 45–53
fracture 37, 93, 97, 135
  brittle components 134
  cleavage 94
  defected graphene sheets by QM/MM and QM/CM 125–9
  study with MM/QM ONIOM schemes 102–3
Fratzl, P. 545
Frechet Differentiable 339, 341
free energy 138, 481

  see also Gibbs free energy; Helmholtz; QCFEM
frequencies 125, 154, 398
  atomistic 121
  consistent-mass 154
  cutoff 95, 113, 120, 121, 122
  lumped-mass 154
  phonon 441, 442, 445, 455, 456, 469, 471, 473
  true 154
  vibrational 435, 437, 438, 440, 470
frequency-wavelength relationships 148
frequency waves 120
  high/low 96, 123
Froment, G. F. 544
Froude number 539
fuel cells 484
functional properties 484
functional theory methods 506
functional values 326–7
functionals 219, 287, 289, 330, 334
  blended 172, 177–9
  coarse 209
  computing information and 323–34
  continuous 323, 329
  cost 300, 301
  energy 183, 185, 199
  global energy 182
  Hamiltonian-like 208
  interdependence between 203
  Lagrangian 184
  linear 288, 323–6, 328, 335, 363, 365, 366
  neighborhood 202, 203
  nonlinear 324, 365
  objective 302
  potential energy 182, 183
  residual 288, 298–9
  stored energy 293
  unique critical point of 50
functions 333, 343, 366, 399, 404, 502
  affine 257
  blending 172, 173, 174, 412, 414
  coarse-level 207
  continuous 324
  continuum blend 412
  cutoff 125, 400, 431
  error 193
  finite element 359
  interatomic potential 483
  irregular, randomly-obtained 269
  linear 488

linear interpolation 147
local enrichment 248
memory 145
nonlinear 310, 324
normalized and local 138
pairwise 151
periodic 59, 342
piecewise 337, 338, 344
polynomial 344
possibly discontinuous 257
potential 124
quadratic 338, 341
radial 269
real valued 39
scalar 7, 249, 268
scale-linking 405
smooth 7, 310, 331, 337, 359, 360
spectral 145
square root terminator 458, 459
stabilization 431
temperature 435
test 330–1, 413
time-space 9, 267, 268, 272–6
transformation 398
vorticity 206
weighting 137
see also basis functions; delta functions; density functions; distribution functions; eigen-functions; Green's functions; influence functions; lattice constant functions; partition functions; shape functions
fundamental equations 537

Galerkin methods 248, 257, 264, 343, 365
  approximation 344
Gangal, A. D. 37, 40, 45
gap-tooth scheme 216–17, 223–8
  and patch dynamics 220–1
gas constant 541
gas-exchanging alveoli 546
gas-fluidized beds 536, 538–40
gasoil cracking 550
Gâteaux derivatives 175, 289
Gaudí, Antoni 545
Gauss/Gaussian 77
  diffusion 544
  elimination 371
  quadrature 440
  statistics 540
Gauss integrals 140–1, 142

Gauss-Seidel relaxation/iteration 202, 363
gecko lizard 31
Geldart B-type particles 538
generic solutions 349
genome 515
  deciphering epigenetic control of 531
geology 3, 12
geometry 52, 378, 396, 400, 401, 403, 408, 415, 449–50, 451, 538, 542, 543
  complexity of 43, 44, 53
  configuration of a set of crystals 402
  corner-point grid 15
  deterministic 43
  device 430
  equilibrium 129, 469
  fractal 32, 33, 34, 35, 36, 42
  general properties 547
  loading and 263
  mathematical 48
  multiscale 36, 53
  optimized 126, 129
  self-similar 31, 33–4, 35–6, 49, 53
  statistically-based constructs 402
  unit cell 407
  see also Cartesian geometry
Georges, A. 542
Gheorghiu, S. 546
ghost forces 94, 108–11, 161, 165, 173
  eliminated 167
  help to reduce 174
  mitigated 185
  origin of 168
Gibbs free energy 487, 494
Gibbs integral 486, 487
global linear equations 259
Goals Algorithm 286, 294–5, 296, 298, 299, 302
governing equations 43, 66, 67, 411, 428, 430
  collocating at continuum nodes 434
  complex geometry in 34
  continuum elasticity 426
  continuum elastostatic 429
  reformulations of 44–53
  residual-free, at multiple scales 61–3
  upscaling of 58
GPGP (generalized phonon Grüneisen parameters) 436, 442–6, 475
gradient models 98

see also deformation gradient
Grafil 82
grain scales 57
grain structure 405–6, 408
  algorithmically-defined 407
graphene sheets:
  concurrent bridging-domain model of 104
  crack propagation in 129–30
  defected 119, 125–9
graphite 126
Green's functions 58, 61, 95, 119, 320, 327, 371, 372
  generalized 330, 335, 336, 337–8, 339, 343, 345, 373
  on-site phonon 454, 460
Green-Lagrange strain tensor 430, 459
grid-free solvers 194
Griffith's formula 129–30
Grigoryev, S. A. 529
Grüneisen parameters 426
  see also GPGP
Grunwald-Letnikov operators 36
Grzybowski, B. 553
Guidault, P. A. 95
Guinier-Preston zones 134

Hadjiconstantinou, N. G. 230
Hahn-Banach theorem 325
Hairer, E. 115
Hamiltonians 95, 113–14, 136, 137, 138, 145, 147, 194, 199, 201
  coarse 139–44, 203–5, 208, 209
  highly oscillating 208
handshake domains 95, 98, 104, 105, 111, 118
  atoms in 108, 110
harmonic coordinates 17
harmonic crystals 141–2, 143
harmonic potentials 109, 123–4, 292, 296
  analysis for blending linear elasticity and 167
Havlin, S. 542
helical models 529
Helmholtz free energy 439, 455–6, 460, 470, 471, 472, 474
  N-atom crystal lattice 440
  solid systems 438
  variation with lattice constant 462, 437, 468
Helmholtz free energy density 428, 434, 436, 437, 440, 454, 459

  carbon atoms 426
  static part 475
  vibrational 426, 427, 435, 441
heterochromatin
  formation/transformations 530–1
heterogeneities 41, 45, 59
  chemical 552
  dynamic 543
  finest scale 59
  macroscopic 293
  material 134
  static 543
heterogeneous systems 134, 135, 394
  resolving multiscale patterns in 538–45
hierarchical methods/models 96, 97, 98
  functional architecture in nature 545–7
hierarchy 411, 416
  topological 415
high-Reynolds flows 201
higher-order theory 101
Hilbert spaces 300, 330
  Riesz Representation for 326
Hill, R. 486, 493
Hill-Mandel conditions 57, 247
Hindu temples 545
histone octamers 523
  DNA wrapped around 515, 520
histone tails 516, 519, 520, 521, 527
  approximated as rigid bodies 518
  coarse-grained representation of 524
  dynamic nature of 525
  flexible, nucleosome arrays with 526
  modifications 530, 531
  stabilizing role of 515
  unique role in fiber organization 525
  see also linker histones
HMM (heterogeneous multiscale method) 196–7, 425
Hölder's inequality 325
holistic approaches 240, 540, 553
homogeneity 373, 494
  loss of 496
homogeneous deformation 293, 410, 439, 441, 486–9, 493, 494
  linear 108
homogenization 7, 34, 41, 196, 221
  computational 57
  hyperbolic 223
  mathematical 58–60
  parabolic 222–3, 227

robust theory of 96
  time-space 247–82
homogenized equations 227, 235, 237
Hooke's tensor 250
hopping properties 465
hot spots 97
Hou, C. 546
Huang, Z. 96
hybrids 294, 295
  a posteriori-a priori error
    analysis 372–6
  concurrent model domain 411
  discrete-continuum model 34
  FEM/MD schemes 155
  hierarchical/semiconcurrent
    methods 96, 97–8
hydration 541
hydraulic path lengths 548
hydrocarbon reservoirs 484
hydrodynamics:
  Brownian dynamics with 518, 521
  complex 538, 549
  fluidized bed 539, 540
hydrogen passivation 454, 466, 469,
  470, 471, 472, 473, 475
hydrophilic pathways 552
hyperelastic continuum 297
hyperspace 508

identity operator 263
imprint lithography 287, 289–90
incident waves 121, 124
inclusions 32
  see also matrix-inclusion interfaces
independent harmonic oscillators 438
independent variables 63, 404
  microscopic 218
inelastic material parameters 87
inequality 17, 19, 325
inertia 143
infinite series 324
influence functions 61, 62, 66, 67, 77
  numerical 58
inf-sup condition 175
  discrete 22, 378
inhomogeneities 134, 168, 169, 410,
  489, 493
initial conditions 226, 229, 251, 254,
  287, 331, 395
  complete control over 484
  prescribed 485
inseparable interaction scales 197
instability 207, 369, 506

elastic 490
  nano-indentation 493–5
  predictions of 494
  shear deformation 488, 495–8
  spinodal 488
  structural 486, 489, 490, 492,
    495
integers 208, 407
Integral Mean Value Theorem 341
integral operators 36, 39–41,
  44–5
integrals 338, 343, 344, 359–60, 366,
  440
  discretized 415
  see also Feynman's path integrals;
    Gauss
    integrals; Gibbs integral
integrated quantities 12
integrated responses 5, 27
integration 49, 248, 346
  diffusion 312
  projective 196, 221, 228
  reaction 312
  see also time integration
intellectual merits 483–4
interactions:
  atoms in layers 463, 465
  covalent bond 466
  cutoff functions to limit distance
    of 400
  dihedral 200, 201
  dipolar 543, 552
  domain 403, 415
  electrostatic 531, 543, 552
  extensions to 199
  finite variance 540
  force 165
  geometric 402
  indicative of solenoid
    configurations 529
  inseparable 197
  instantaneous 311, 312
  intermolecular 541, 543, 545
  internucleosomal 525, 527–8
  inter-scale 194, 199
  local 199, 207, 430
  long-range 207, 552
  many-body 430
  nearest neighbor 432, 439
  nonlinear 537
  non-local 198, 207
  scale 401
  simultaneously considered 543

interactions (cont.)
  Si-H/Si-Si 466, 468, 470
  spatial 402
  unique three-body 124
  see also Lennard-Jones interactions;
    Peach-Koehler interaction
interatomic potentials 143, 144, 146,
    161, 410, 469, 483, 488
  anharmonic characteristics/effects
    of 441, 451
  empirical 428, 466
  many-body 151, 508
  quasiharmonic approximation
    of 426, 428, 434
  static part of Helmholtz free energy
    obtained
    directly from 454, 475
  see also Lennard-Jones potentials;
    Tersoff potential
interface variables 253
interfaces:
  atomistic/continuum 95
  behavior of 255, 258
  characteristic length of 257, 263
  compatibility and momentum
    balance enforced
    across 97
  coupling 100
  crystal 406
  decomposition of structure into 253
  displacements at 59, 100
  field discontinuities at 41
  library 414
  MD/CG 139, 160
  meshed independently of
    subdomains 254
  neighboring 252, 256
  nonzero Kapitza resistance at 155
  quantum models often fractured
    at 103
  reflectivity and transmissity of 120,
    121
  suppressing spurious reflections
    at 118
  see also matrix-inclusion interfaces
internal variables 249, 250, 272, 273
internucleosomal
  attraction/patterns 525, 529
interphase 411, 413
  phase space of atoms in 412
interpolation 109, 110, 239, 240
  accuracy of 197, 203
  adjoint 207, 208
  error of 337
  fractal 546
  generalized 200, 206
  linear 147, 149
  micro and macroscale 545
  non-symmetric 224
  quadratic 242
  symmetric 217, 223, 224, 227
intervals 35, 37, 38, 39, 40
invariant distributions 537
inverse analysis 300, 302
inverse methods 84
inverse operators 371
inverses 142, 147
invertibility 341, 342
ions 515, 516
  divalent 518, 525, 526, 527, 528–9,
    531
  molecular simulations of water
    and 544
  monovalent 518, 525
  separations of 552
  simulation of channels 552
iron 94
isothermal bulk modulus 488
isothermal conditions 454
isothermal systems 426, 428, 434, 475
isotropic strain hardening 251
iteration:
  approximate solutions
    throughout 266
  convergence of 378, 386
  effect of 385–6
  fine/coarse 207
  linear stage 259, 260–1, 262, 271
  local stage 259–60
  nested procedures 102
  Newton-like 204
  preset number 295
  see also Gauss-Seidel

Jacobian 340, 341, 486
Java 545
Joos, B. 486
jumps 59, 504, 543, 544
  horizontal 493
  intermittent discontinuous 493
  vertical 494, 495

Kapitza resistance 155, 160
Karush-Kuhn-Tucker condition 117
Kearney, M. M. 548
Keil, F. J. 541

Kevrekidis, I. G. 223, 225, 238, 239, 240, 242
Khare, R. 93, 97, 103
Kigami, J. 42
kinematic strain hardening 251
kinematics 48
kinematics equations 45
kinetic energy 137, 140, 143, 146, 472, 473
kinetics 510, 540, 544
  arbitrary 550
  first-order 550
kink mechanism 503–6
Knap, J. 94
Knudsen diffusion 541, 543, 544
Kohlhoff, S. 94, 135
Kohn-Sham equations 210
Kolmogorov entropy 539
Kolwankar, K. M. 37, 40, 45
Kouznetsova, V. 98
Krenn, C. R. 486
Kronecker delta symbol 488
Kwauk, M. 539

Ladevèze, J. 266
Lagrange multipliers 42, 95, 104, 106–8, 115, 118, 142, 175, 176, 177, 184, 185, 253, 261, 288
  admissibility of 262
  discrete 110
  distribution of 274
  interpolated by finite element shape functions 109
  micro problems with 262
  normalized 111
  pointwise 109, 114
  stability of 111–12
  strict 110
  see also Euler-Lagrange equations
Lagrange polynomial 225
Lagrangian particle models 538
Lagrangian strain tensor 486
  see also Green-Lagrange strain tensor
LAMMPS (Large-scale Atomic/Molecular Massively Parallel Simulator) 409, 410, 414
Lanczos algorithm 458–9, 464
Langevin dynamics 145, 208
Laplace transform 95
  inverse 145
Laplacian operator 42
  boundary value problems for 334

LATIN method 248, 259, 266, 277
lattice constant functions 435, 436, 462, 468
lattice constants 94, 95, 98, 100, 103, 109, 119, 123, 441, 467, 469, 472, 473
  determined 436, 442, 443, 468
  equilibrium 151
  internodal spacing greater than 113
  zero temperature and zero pressure 143
lattice dynamics 154, 156, 434, 454–5
  acoustic and optical branches familiar from 147
  quantum-mechanical 426, 428
  soft phonon modes in 490
lattice models 290, 394
lattice spacing 110, 135
  atomistic 112, 506
least-squares 105, 118, 139
Lebesque Cantor staircase 51
Lefever, R. 306
length scales 226, 482–3, 537, 545, 548, 550, 553
  coupling 134–7
  flow-dominated 551
  force balances on 546
  geometrical invariance over a range of 542
  multiple 536, 546
Lennard-Jones interactions 204, 205
Lennard-Jones potentials 120, 292, 490, 491
Levitt, M. 521, 524
Lévy walks 543
Li, J. 539
Li, S. F. 96, 119
life sciences 537
lifting operators 219, 220, 221, 226, 230
limited global information 3–30
linear constants 96
linear elastic response 489
  expected 491
linear elasticity 48, 109, 125, 143
  analysis for blending harmonic potentials and 167
linear equations:
  elasticity 306
  evolution 261
  parabolic 24

linear momentum 120, 121, 124
  balance of 47, 48
  conservation of 45, 115, 116, 182
linearization 342, 347, 358, 364, 372, 373, 376, 378, 379, 387
  coefficients obtained by 345
linked scales 96
linker histones 515, 516, 518, 520, 524, 526–7
  chromatin structure and 529
  dramatic effect on fiber organization 525
  fiber cross-linking must be enhanced by 531
  interactions dominate (without) 528
lipid bilayer 552
Lipschitz continuity 369
lithography 548
Liu, W. K. 95, 118
loading 277, 506, 507, 509
  athermal 508
  critical 499, 506
  general condition 451
  geometry and 263
  hydrostatic 416, 417, 488
  periodic histories 248
  prescribed 489
local equations 199
Lomer-Cottrell junctions 417
long-range equations 199
Lorenz problem 313–17, 318, 321, 322, 323, 348
low order elements 113
LPDOS (local phonon density of states) approach 454–75
LQHM (local quasiharmonic) model 426–7, 437, 438–9, 441, 442, 445, 447–52, 461, 462, 467–9
LTOL (local error tolerance) 356
lungs 536, 546, 547, 549, 550, 551
Luo, D. 549
lysine residues 530

MAAD (macroscopic-atomistic-*ab initio* dynamics) 95, 97
macro problems 249, 262, 265
macro quantities 248, 262
  admissibility of 257
macroscale:
  effective properties at 537
  interpolation between micro and 545
macroscale response 410

macroscale structure 431
macroscopic continuous glide 500
macroscopic equations 216, 229, 231
  central finite difference schemes for 217
  discretization of 242
  higher-order or advection-dominated 228
  rigorous derivation of 196
  second-order 224
  unavailable 220, 242
  unknown 217, 220, 221–2
macroscopic flows 482
macroscopic models 196, 216, 220, 223, 239, 241
  chromatin 518–20
  dinucleosome, trinucleosome and 12-nucleosome systems 522
  elastic 515
  equivalent 218
macroscopic scales 57, 222, 536, 550, 553
  effective homogenized PDE on 223
  system behavior on 216
macroscopic variables 219, 220, 229
McWilliams, Jim 207
magnesium 526, 527, 528
Malek, K. 544
Mandelbrot, B. 35, 543, 546
marching cubes 406
marine micro-organisms 31, 32
mass matrix 142, 143, 155
  consistent 141, 149, 150
  diagonalized 114
  distributed 149
  Fourier transform of 148
  lumped 113
  symmetric 300
master-equation methods 514
master-slave coupling 94, 98–101, 117–18
material constants 293
material properties 57, 430, 439
  predicting accurately and efficiently 427, 475
material scales 57
material sciences 167
materials 31–56, 135, 481–513
  amorphous 537
  biological functional 545
  chemically heterogeneous 541
  composite 196, 247, 537

crystalline 94, 402, 405–10
  fractal 542
  hyperelastic 430
  inhomogeneous 134
  linear and nonlinear elastic 430
  microporous 541
  nanoporous synthesis 551
  nanostructured 541
  polymeric 402
  porous 544
  real 134
  solid 537
  synthetic 545
mathematical equations 404
mathematical models 308, 395, 396, 397, 398, 402, 415
  continuum 400
  transformation of computational to 400
matrices:
  cellular protein 515
  consistent constraint 115, 121
  constitutive 404
  diagonalized constraint 96, 115, 121
  epoxy 82
  identity 142
  interaction 527
  Jacobian 486
  modified GF 459
  non-collagenous 31
  orthogonal transformation 459
  orthonormal 463
  potential energy coefficient 140, 142, 143, 146
  projection 142, 148
  pseudo-stress 47
  random 363, 371
  singular values of 329
  transformation 76
  triangular 341
  water 547
  see also dynamical matrix; force constant
    matrix; mass matrix; stiffness matrix
matrix-inclusion interfaces 34, 44, 48
matrix inverse 139
  direct 457
MCTK (multiscale component tools kit) 415
MD (molecular dynamics)
  simulation 219, 441, 451, 454, 468, 489, 496, 498, 501, 502, 510

constant temperature 474
crack tip propagation 506
direct 506
equilibrium geometry of nanowire 469
modeling dislocation mobility using 484
nano-indentation 494, 495
short-term 543
simulation transport of water through protein channels 552
variation of virial stress at constant strain
  from 492
water through protein channels 544, 552
mean curvature 350
mean value theorems 339, 341
mean values 248, 257
mechanical equilibrium
  methods 98–113
mechanical functions 545
mechanical properties 454, 461, 467–8, 473–5, 518
  accurate results for 460
  bone 545
  crystalline materials 500
  local 459–60
  single-wall carbon nanotubes 426
mechanical variables 94
mechanics 31–56
  computational 247–82
  continuum damage 84
  nanoscale 510
  statistical 197, 198
  see also CM; MM; QM
membranes 538, 541, 544, 546, 551–3
MEMS (microelectronic mechanical switch) device 305
Menger sponge 44
MEP (minimum energy path) 506, 508, 509
mesh acceptance criterion 352
mesh generation 112, 406
mesh-independent specification 396
mesh modification procedures 415
  automated adaptive 403
mesoscale models 83, 84, 406, 481
mesoscopic chromatin models 530
  second-generation 520–9
metallurgy 134

metals 550
  crack front extension in 506
  failure processes 134
  plastic deformation in 484
  single dislocation glide in 500–3
metastable solutions 349, 350
methylation 519, 530
MGD9V (multigrid solver) 551
micro problems 261, 262, 271–2
  different loading cases for 277
  resolution of 249, 274–7
micro quantities 248
micro scales 83
microelectronic applications 395
micro/macro projectors 248, 265, 267
microns 483
microscale 95, 238, 264
  geometrical and chemical structure at 551
microscopic models 196, 216, 217, 218, 219, 223, 226, 227, 229, 242
  particle 228
  stochastic 220
  time integration of 230
microscopic scales 223, 227, 232
microscopic variables 218, 219
microscopy 485
  electron 515, 529
microsimulators 238, 239
microstructure 57, 483, 491–3
  evolving 484
  fine-scale 239
  hierarchical 31, 35, 41
  higher-order patterns in 134
  multiscale 31–56
  optimizing 134
  realistic 406
  underlying 484
microstructure evolution:
  mesoscale 481
  predictive models of 484
microtechnology 537
Miller, R. E. 109
Milstein, F. 486
minimization 494, 508, 518
  energy 539
minimization problems 50
  constrained 353
  unconstrained 184
mixed continuum-discrete models 410
mixed formulation 50
  numerical 26
  space-discrete 25

mixed variational principle 50, 51
MLS method 128
MM (molecular model) 93, 98, 99, 102–3, 106, 125, 293, 394
  consistent 109
  gradual transition to continuum model 104
MM/CM model 103, 111
  master-slave method can be applied to 117
model breakdown indicators 395
model prediction 308
modes 144
  internal 112
modular synthesis methods 548
moduli 43, 44
moisture effects 59
molecular chain 402
molecular components 31
molecular dynamics 95, 113–19, 136, 287, 414, 481, 482, 483
  combining with transition-state theory 543
  common assumptions for 409
  projection to longer timescales 514
  QC employed to develop coarse-grained alternative to 426
  supercomputer time needed to perform 537
  time-dependent problems in 300
  upscaling from quantum mechanics to 210
  see also CGMD; MD simulation
molecular scales 544, 550
molecular statics 286–7, 315, 414
  large-scale model 289–90
molecules 145, 200, 287, 290, 296
  adequate distribution channels required to transport 550
  complex 209
  cross-link 290
  diffusing 541
  greater complexity 515
  kinetic diameter of 544
  mean-square displacement of 541
  motion of 542
  neighboring 543
  separations of 552
  SiH 470
  tightly-fitting 541

trajectory of 543
water 552
moments 324
  higher-order 219
  irreducible chain 465
  low-order 219, 235
  recursion coefficients and 465
momentum 119, 124–5, 143
  atomic 137, 140
  balance of 97, 98
  initial 120
  see also angular momentum; linear momentum
momentum conservation 45, 115, 116, 307
  temporary violation of 120
momentum equations 307
monatomic lattices 147, 148
monomers 290, 293, 300
Monte-Carlo method 199, 209, 366, 425, 483, 515, 522, 525, 526, 527–8
  compatible 200, 208
  configurational-bias 524
  dynamic 541, 544
  kinetic 207, 228, 290–1, 481, 503, 505
  quasicontinuum 426
Mooney-Rivlin material 296
morphometric characteristics 546
Morris, J. W. 486
motion:
  continuous glide 500
  crack front 506
  dislocation 504
  jerky 501
  molecular 542
  sensitive to imperfections 501
MsFEM (mixed multiscale finite element methods) 3–30
MTB-G2 (modified Tersoff-Brenner) potential 125
Mullins, M. 94
multifractal geometry 33
multigrid bridging approaches 425
multigrid solvers 193, 195, 202, 551
multiphase fluid systems 548
multiphysics problems/models 248, 259, 305–90
multiple contacts 255
multiple masses 140
multiple scales 58, 59, 537
  asymptotic expansions 57, 196
  easy way to bridge 548

effective methods to relate domains across 403
  reduced order model 63–70
  residual-free governing equations at 61–3
multiplication 493
  simple matrix 465
multiplicative constants 37, 40
multipliers 400
  nodal 413
  see also Lagrange multipliers
multiscale modeling:
  ad hoc difference of systematic upscaling from 195–6
  general approach to 285–304
multiscale structure 545–53
multi-time-step methods 248

nanocrystalline structures 491–2
nano-indentation 493–5
nanomanufacturing 285, 286–7, 288
nanopores 550, 551
  liquids in 541
nanoscale 161, 510, 545, 550, 553
  crack front deformation at 506
  mechanical resonators 134, 135
nanostructures 134, 541
  LPDOS of 468–70
  mechanical behavior under external loads 449–54
  novel 484
  see also mechanical properties; NEMS; thermal properties
nanosystems 425, 426
nanotechnology 167, 536
  modeling software 395
  nanoscience and 538
  rapid advances in 425
nanovoids 416
national security 484
nature-inspired approach 536, 538, 545–53
Navier-Stokes:
  discretized equations 205
  simulations 206
nearest neighbors 109, 120, 161, 432, 439
  linear interpolation and 149
NEB (nudged elastic band) 497, 508
neighboring variables 199
NEMS (nanoelectromechanical systems) 136, 425–80

Network for Computational
    Nanotechnology (NSF) 395
network materials 33
Neumann boundary conditions 223,
    225, 307, 348
Neumann boundary value
    problems 334
Neumann series 371
neutral network-based law 96
Newton method 414
Newton-Picard method 217
Newton-Raphson method 292
Newtonian dynamics 208
    particle system governed by 210
Newtonian heat-conducting fluid 307
Newton's law:
    implicit discretization of 208
    second 167, 169
    third 168, 182, 185
NLPQL (gradient-based optimization
    package) 551
Nobel Prize in Chemistry (2003) 552
nodal indices 142
nodes 99, 117, 310, 383, 546
    blocked 542
    continuum 101, 104, 124, 428, 431,
        434
    distance between 113
    FEM 109
    finite element 413
    independent 100
    interface 119
    Lagrange multiplier 106, 107
    mesh 138
    neighboring 141
    neighbors 542
    overlapping 118
    time 345
noise drift 196
non-conservation form 236–8
non-derivability points 39
non-deterministic problems 194
non-dimensional heat equation 380
non-equilibrium effects 144–6
non-Gaussianity 543
non-intersecting pores 542
nonlinear equations 58
nonlinear macro analysis 70, 77
nonlinear problems:
    adjoint analysis for 339–40
    adjoint operator associated with
        error analysis for 358
    large-scale 57

nonlinear response 489, 491
nonlinearity 372, 499
    onset of 500
nonlocal effects 4, 12, 37
    strong 6
nonlocal models 113
non-Markovian dynamics 37
nonperiodic case 59
normalization 110–11, 119
norms 324, 352
    dual 325
    Euclidean 326
    space 349
nucleation 410, 415, 489, 499, 505
    dislocation 493, 508
    double-kink 503
    elliptical microcrack 491–2
    homogeneous 493
    kink 505
    partial 504
    spatially-dependent criterion 494
nuclei 100, 108
    displacements of 102
    mammalian 531
    need not be coincident with nodes
        of continuum mesh 104
    positions updated to bring towards
        equilibrium 102
nucleosomes 514, 516–20, 522–8, 530
    detailed structures of building
        block 515
    interdigitation of 529, 531
numerical error 338, 370
    pointwise 316
    quantification and control of 309
numerical methods 57
numerical micromodels/problems 96,
    98
numerical upscaling 196–8

observables 208
    coarse-level 203
    correlation 209
    fine, coarse-level computability
        of 210
    interdependence of 204
    product 204
    still-higher-moment 204
observation 53, 339, 427, 492
    experimental 195, 300, 485
octree structures 402, 404, 406
oil and water phases 5
oilfield exploration 484

olefins 550
oligonucleosomes 524, 530
　in-vitro assays of 529
　multiple chains 531
one-dimensional studies 119–23
ONIOM method 101–4, 127
　modified 126, 129
open-source programs 395
operator splitting 310, 368, 369, 370, 371, 373, 374, 376
　abstract 311, 372
　effects of 375
　numerical 372
optical modes 445
optical phonons 442, 460
optimal control theory 300, 301
optimality condition 171
optimization:
　control and 220
　parameter 308
　performed efficiently 520
　see also DiSCO; NLPQL
optimization problems 292
　constrained 184, 352, 353, 551
　global 210
Orbach, R. 42
Ornstein-Uhlenbeck process 218
orthogonal projection 22, 105, 255, 360, 365
Ortiz, M. 94
oscillations 313
　nonphysical 369
overlapping domains 95, 98, 104, 105, 106, 124, 403
　carbon atoms in 126
　constant scaling in 111, 118
　decomposition schemes 95, 166, 185
　joining coarse-scale to fine-scale on 119
Owhadi, H. 7
oxidation resistance 510
oxygen 546, 550

pad atoms 94, 95
Padé approximant 150
Padé third-order sub-diagonal difference scheme 345
Panagiatopoulos, P. D. 43
Panagouli, O. K. 43
parabolic equations 4, 24, 222
parabolic problems 333, 357
　homogenization 222–3, 227
　linear 345

parallelism 259
parameter identification process 83
Parareal approach 248
Parasolid (modeling kernel) 403
Park, H. S. 95
Parrinello-Rahman procedure 501
particles 205, 295, 496, 520
　behavior of 218
　colloidal 31
　drying and coating of 538
　evolution of the probability density of 218
　interacting 219
　porous catalytic 538
　solid 536, 538–9
　sticking to each other 540
　trajectories at temperature of interest 489
　trees to distribute fluids over fixed beds of 548
partition functions 138, 141, 142
　constrained 140
partition-of-unity property 116, 139, 166
partitions 83, 167, 253, 295, 296, 297
　coarse 257
　interface 64, 65
　phase 84
　surface 65
　volume 65
patch dynamics 216–46
patch tests 167, 168, 172, 177, 181, 185
　satisfying 110, 111, 180
PDEs (partial differential equations) 37, 194, 206, 216, 220, 221, 227, 313, 375, 402
　continuum 394, 397, 399, 405
　diffusion 240
　discretized 193, 197, 199, 205, 238, 239, 394, 395
　estimating error of finite element solution 334
　homogenized 222, 223
　hyperbolic 223
　parabolic 222
　solving over continuum domains 404
PDOS (phonon density of states) 426, 441–6
　see also LPDOS
Peach-Koehler
　stresses/interaction 504, 505

Peierls barrier 503, 504
percolation theory 536, 542
perfect crystals 486
performance tests 148–60
periodic media:
  initial-boundary-value problems in 196
  linear 247
periodic solutions 217, 222
periodic theory 59
periodicity 247, 439, 466, 537
  fine-scale 196
permeability fields 5, 10, 15
  channelized 4, 7, 12
  fine-scale 12
  random 7
perturbations 315, 318, 319, 321, 330, 350, 487
  localized 320
  modulating wave 498, 500
  polynomial rate accumulation of 376
  sensitivity to 323
  small 249
petroleum applications 5
Petrov-Galekin forms 415
phase damage evolution 86
phase transformation/transition 59, 198
phonons 113, 148–50, 151, 152, 456–9
  longitudinal and transverse 147
  nearly perfect transmission of 121
  soft modes 490
  suppressing all reflections of 120–1
  see also GPGP; LPDOS; PDOS
phosphorylation 519
photosynthesis 546, 547, 550
physics 308, 358, 537, 546
  atomistic 425
  computational 481
  diffusion 306
  fundamental problems 268
  mathematical 287
  nanoscale 510
  reaction 306
  statistical 197, 500, 542
physiological conditions 515, 516, 529
piecewise approximation:
  continuous finite element 368
  linear discrete 312
  polygonal 337

Piola-Kirchhoff stress 101, 429, 430, 433, 434, 436, 437, 438, 440, 459, 487
PITA approach 248
Planck's constant 435, 455
  see also Fokker-Planck equation
plane waves 147, 149, 153
  resolving displacement as 494
plankton 31
plasticity:
  crystal 57, 500
  phemenological 57
  visco- 267
PM3 quantum model 103, 125, 129
PML (perfectly matched layer) method 119, 121
point group operations 151
poisoning 542
Poisson's ratio 43, 320
Poisson-Boltzmann solution 518
polarization 196
polycrystalline structure 406
polydispersity 540, 543–4
polymer 200, 204, 205, 206, 207, 300, 515
  coarse-grained 521
  creating through chemical process 290
  different realizations of structure 300
  equilibrium problem 287
  filled 31
  lattice-based models of 286
  relaxation processes of 34
  sediments in fluid 524
  short, folding/unfolding events of 518
  stamping etch barriers at room temperature 289
polymer chains 290, 402, 530
  covalent bonds on 292
polymer densification 290–5, 296
polymerization 287
  kinetic Monte-Carlo method 290–5
polymethylene 204
polynomials 203, 257, 344, 377
  approximating 225
  quadratic 226
  symmetric interpolating 223, 227
  Taylor-like 228
polynucleosomes 522
pore networks 542, 549–51
porosity:

assumed constant 5, 15
  high 31
porous media 44
  diffusion in 37, 540
  transport on 37
position of atoms:
  arrangement and 134
  equilibrium 429
  instantaneous 429, 435, 438, 455
  relative 435
  unit cell 407
positive definite operators 261
potential energy 99, 104, 105, 142,
    143, 146, 147, 182, 430–1, 454, 503
  atomistic 140, 171, 183, 185, 436,
    438
  continuum 183
  derivatives of density function 447
  harmonic and anharmonic
    parts 140, 141
  lattice 447
  local 438, 464
  mechanical definition of 185
  minimized 170
  static 435, 455, 456, 472
  sum of kinetic and 137
potentials 210
  angle-bending 116
  applied to discrete entities 400
  atomistic 112, 113
  electric 306
  empirical bond-order 491
  governing 306
  mean force 530
  molecular 292, 300, 394
  pair 113
  quadratic 143
  thermodynamic 487
  see also EDIP potential; harmonic
    potentials; Finnis-Sinclair;
    interatomic potentials;
    Lennard-Jones; MTB-G2; REB;
    Tersoff
Poutet, J. 44
power law 37, 38, 39, 41, 44, 543,
    553
  scaling of elastic constants given
    by 43
power law distribution 536, 540, 544,
    552
power production 538
practical resolution technique 273–4
predictions and experiments 505

predictive models 484
pressure measurements 539
Prigogine, I. 306
primal variational principle 50
probability 199, 208, 322
problem-setting 285–7
process control 539
projection 139, 358, 366, 377
  global operators 344
  least-square 105, 118
  orthogonal 22, 105, 255, 360, 365
  recursive 217
Prony series 37
proteins 514, 516, 521, 531, 536
  channels in cell membranes 544
  intrinsic fluctuations 552
protons 552
pseudo-potentials 251
pseudo-strain 48
pseudo-stress 47, 48, 53
PVI (pore volume injected) 11–12, 13,
    16

QC (quasicontinuum) method 34, 94,
    109, 112–13, 135, 165, 394, 425,
    426
  AtC blending contrast with 182
  combined with continuum defect
    models 95
  early versions of 197
  finite temperature 427–54, 475
  fundamental principle of 137
QCFEM (QC free energy
    minimization) method 426
QCMC (QC Monte-Carlo)
    method 426
QHM (real space quasiharmonic
    model) 143, 144, 426, 437
  see also LQHM; QHMK
QHMG-n method 460–3, 464, 467–9,
    471–2, 474, 475–6
QHMK (k-space quasiharmonic)
    model 426, 427, 437, 439–41,
    442–4, 447–52, 454, 460–2, 467,
    468, 469, 471
QM (quantum mechanics) 93, 95, 97,
    98, 125, 435, 447
  lattice dynamics 426, 428
  multiscale problem 550
  Schrödinger equation of 483
  upscaling to molecular
    dynamics 210

QM/CM (quantum
   mechanics/continuum mechanics)
   coupling 98, 101, 119
   fracture of defected graphene sheets
     by 125–9
QM/CM/MM calculation 130
QM/MM (quantum
   mechanics/molecular mechanics)
   coupling 98, 101, 102–3, 104, 119,
     125, 126, 128–9
QM/MM/CM model of graphene
   sheet 129
QoIs (quantities of interest) 248, 272,
   285–8, 293–6, 298–300, 318,
   321–3, 335–7, 343, 345, 347, 350,
   351, 354–6, 360, 362–7, 369–71,
   373, 378, 381, 382, 385
QtMMOD (modified ONIOM)
   scheme 126, 127
quadrature 57, 360
   errors arising from 338
   formula 235
   Gaussian 440
   lumped-mass 345
quantifiability 485
quartz 290
quasiharmonic approximation 435,
   436, 441, 454, 468
   interatomic potentials 426, 428, 434
   local 426, 438, 464
quasiharmonic models, see QHM
quasi-static problems 42, 48, 249
quasi-uniform meshes 22

radial hyperreduction 267
radial local functions 206
radiation 160, 484
random arrays 134
random colors 6, 12, 15, 16
random forces 145
random number generations 197
random variables 285, 287
   small set of moments of 324
random walks 543, 544
Rateau, G. 111
Rayleigh quotient 268
reactants 290, 291
   confined 550
   gaseous 538
reaction 349, 542
reaction-diffusion equations 306, 308,
   310, 312–13, 343
   discretizing 367

multiscale operator splitting for 368
   operator splitting for 311
   problems 367–70, 373
reaction pathway sampling 506, 508
reactors 539, 550
   chemical 547, 548
   fluidized bed 549
   multiphase 548, 549
   scale-up to larger scale 539
   self-similar photosynthesis 546
   simulation of 538
   slurry 540, 548
   two-phase 548–9
real space 140, 437
   averaging in 147
REBO (reactive empirical bond order)
   potential 125
reciprocal space 142, 153, 439
   shape functions in 147–8
reconstruction operators 394
recursion 460, 468, 469, 548
   phonon GF and 456–9
recursion coefficients 459, 460, 461,
   463, 464
   asymptotic 462
   moments and 465
recursion constants 465–6
recursive projection method 217
red blood cells 551–2
reduction theory 57–90
reference frame 46, 49
reference points 273
reference problem 249–52
   reformulations of 252–6
reference solution 265
reference times 270, 274
refinement procedures 394
reflections 96, 119, 157, 159
   phonon, substantial 121
   wave 145
   see also spurious reflections
relativistic effects 501
relaxation 34, 193, 199, 202, 378, 508
   compatible 200, 201
   fast 405, 406, 408–10
   full 409
renormalization 44, 135, 194–5, 199
   see also RG; RMG
repeated indices 140
replicas 220, 506, 508
repulsion 143, 151, 525, 529
reservoirs 10
   synthetic 15

residual force calculation 293
residuals 110, 288, 289, 294–5, 348, 349, 362, 366
  approximating 298
  conjugate gradient minimization of 414
  finegrid 193
  single physics 363
  space 349, 350
  time 349, 350
  time-dependent 302
  weak 364
resistance:
  creep 484
  drag 501
  fatigue 510
  glide 500, 501
  lattice 506, 509
  obstacle 501
  oxidation 510
  see also Kapitza resistance
restriction operator 219, 220, 221
restrictive assumptions 4
restrictive hypothesis 44
Reynolds number 382, 539
  see also high-Reynolds flows
RG (renormalization group) method 43, 195, 196, 202, 536, 542
rheologic behavior 37
Rho, J. Y. 545
Rice, J. R. 493
Riemann sums 39, 40
Riemann-Liouville operators 36, 45
Riesz Representation 45, 326, 328
rigid bodies 43, 256, 432, 514
  histone tails approximated as 518
RMG (Renormalization Multigrid) method 197
Roberts, A. J. 225, 238, 239, 240
Robinson, P. J. 517, 529
room temperature 289, 293, 474, 502
Rudd, R. E. 94, 139, 144, 145, 146, 147, 150, 158
Runge-Kutta time integration 235–6
Rustad, Alf B. 5, 15
RVE (Representative Volume Element) 291, 292–3, 494

saddle-point state 50, 184, 493, 496, 497, 506, 507
  contours across slip plane at 509
  difference between initial state and 508
  shear displacement distribution along crack front at 510
Sagrada Familia 545
Saint Venant principle 248
salt concentrations 516, 519
  monovalent 526
  univalent 522, 526
Samaey, G. 226, 232, 234, 235
Sandia National Laboratories 414
Sapoval, B. 546
saturation 5, 12, 14, 129
  coarse-scale 10
  upscaling 14, 15
saturation equations 5, 13
  upscaling of 11
SBED^{TM} 5, 15
scalable parallel code 298
scalars 7, 170, 249, 268, 251, 271, 412, 465
  constant 264
scale decoupling 34, 41
scale-free features 545
scale-linking methods 394, 395, 398, 399, 400, 401, 405
scattering 33
  dynamics and 155–60
Schouten, J. C. 539
Schrödinger operator/equation 210, 483
Schwarz inequality 19
scientific computing 514
search direction 260, 261, 264, 271, 274, 275
secular equations 148
self-assembling processes 553
self-diffusivity 541, 543
self-similarity 43, 44, 48, 51
  deterministic 35, 49
  geometry of 31, 33–4, 35–6, 49, 53
  hierarchical 42
  multiple spatial ranges of 31
  stochastic 31, 34, 35
semiconcurrent methods 97–8
semiconductors 499
  crack front extension in 506
  high-precision features of 289–90
  nanomanufacturing 285, 286–7
semiempirical methods 125
semilocal models 460–6, 467, 468, 475–6

sensitivity analysis 308
separation 538, 545, 551, 552
  molecular or ionic 553
  protein 553
  scale 6–7, 57
  technical 553
  vast-scale 196
Seymour, R. S. 547
shape functions 51, 99, 106, 137, 138, 147–8
  continuum 143
  finite element 109, 413
  Lagrange multipliers interpolated by 110
  linear 64, 65, 149
  low-order 415
  partition-of-unity property of 116, 139
  piecewise constant 403
  tensor field discretized into a set of 405
shear deformations 442, 448, 449, 492, 501, 502, 509, 510
  instability in 495–8
shear localization 493–4, 495, 499
shear moduli 488, 510
shear tests 442
shear viscosity 482
Sheintuch, M. 551
Shilkrot, L. E. 95
shock waves 499
  dispersed 144–5, 160
  smoothed out 155
  step-like 159
Shockley partial dislocation 508, 510
short-wavelength phenomena 247
SIAM journals 514
Sierpinski gasket/carpet/triangle 42, 43, 542, 546
silica gels 541
silicalite 553
silicon:
  diamond structure 469
  dislocation mobility 503–6
  nanowire 454, 467, 469–75
Simmetrix Simulation Applications Suite 401
simplex triangulation 337
simulated load-displacement curve 494
simulation:
  accelerating 195

atomistic 196, 448, 483, 484, 486, 488, 489, 552
automated adaptive-continuum 414
better experimental tools to validate 538
bottom-up 540
chromatin 518
coarse-level 195, 203, 206
computer 285
concurrent at finite temperature 134–64
constant temperature 138
continuum 414, 552
coupled atomistic-continuum 219
deformation 491
direct flow 3
discrete particle techniques 483
dislocation dynamics 484
fine-scale/fine-level 195, 196, 197, 201, 203, 205, 206, 208, 210
hierarchy of methods 543
local 216
long-term 234
macroscopic 197, 216, 217
materials modeling and 481–6
microscale 95
microscopic 216, 217, 220, 223, 225, 228, 229, 233–6, 238, 239
molecular 543, 544, 545, 552
multiple replica 220
numerical 236
particle-based 221
predictive 300
problems now amenable to 481
reactor 538
resonator 136
single-scale 396, 397, 400–1
small systems 483
structure-property correlation through 484
two-phase flow and transport 10
unprecedented computing power through 485
see also Boltzmann; DNS; MD; Monte-Carlo; Navier-Stokes
Sinai's billiard without horizon 543
single-phase pressure equations 7
single physics paradigm 362–3, 386
sinusoidal wave 498
slip plane 509
slit defects 129
slurry reactors 540, 548

Smoluchowski
    aggregation-fragmentation
        kinetics 540
smoothness 194, 223
Sobolev embedding theorem 25
Sobolev spaces 7, 139, 169
soft modes 489–91
software 98, 101, 393–421
    commercial geomodeling 15
solenoid model 522, 528, 529, 531
    interdigitated 517, 518
solid deposits 542
solids 151, 382, 536
    ab initio calculation of
        properties 481
    amorphous 544
    conjugate heat transfer problem
        between
        fluids and 376–8
    crystalline 483
    deformation response of 499
    diffusion in 541
    elastostatic behavior of 430
    finite temperature without
        thermostat 146
    harmonic 141
    measure of theoretical strength
        of 489
    monatomic 149
    structural transitions in 486
    thermal conductivities of 307
    zero external stress 486
Song, J.-H. 97
space variables 249, 267
spatial domains 310, 401
    explicit and computational 403
spatial indices 142, 143
spatial scales 58, 514
SPE 10 (Tenth SPE Comparative
    Project) 4, 10, 12, 15
specific heat 307
spectra 241
    CGMD 142, 143, 151–5
    elastic wave 151
    exact MD 150, 151
    FEM 150, 153–4
    high frequency part of 125
    phonon 148–50, 151
    wave 153
spherical/ellipsoidal pores 44
splitting of degeneracies 442
spurious reflections 95, 98, 113
    suppressing 115, 118, 119, 120–1

SRT (square root terminator)
    function 458, 459
stability 306, 310, 311, 313–34, 342,
    358, 387, 552
    complex 308
    consistency and 226–8, 232
    elastic 494
    global information 370
    importance of obtaining accurate
        information
        about 386
    Lagrange multiplier
        methods 111–12
    lattice 487, 488
    mechanical 486
    multiscale operator decomposition
        affects 369
    structural 486
    structural 546
    vibrational 489
stability conditions 233, 486, 488, 494
    load-dependent 489
stability factor 336
stacking fault 489, 504, 505
    tetrahedra 417
state laws 250–1
state variables 84
static systems 137
stationarity 268
stationary points 100
stationary problems 318
    adaptive mesh refinement for 352
STATOIL 5, 15
steady states 206–7, 216, 217, 306,
    314, 349, 373
steel 134, 510, 546
    stainless 380
stencil width 242
step lengths 544
stiffness matrix 142, 143, 145, 146,
    150, 154, 344, 483
    bandwidth reduction for 161
    explicit formula for 147
    FEM 151
    Fourier transform of 148
    region where it differs from the MD
        dynamical matrix 156
    subtle cancellation between mass
        matrix and 155
    temperature-dependent 144
Stigter charged cylinder
    electrostatics 518
Stillinger, F. H. 508

stochastic models 285
stochastic problems 196, 199, 202, 267
stochastic sampling 505
stochasticity:
  added 208
  determinism and 31, 196, 210
strain 48, 61, 70, 98, 139, 417, 427, 434, 441–6 449, 484, 487
  applied 126
  asymptotic expansion of 59
  atomistic 128
  constant 490, 492
  critical 489, 491, 492
  damage equivalent 86
  effects on PDOS 426
  energy density 99
  finite 486
  fixed 489, 491, 494
  fracture 103
  hardening 251
  hydrostatic 490
  inelastic 273
  large shear 495
  leading order 60
  macro 72, 262
  small 489, 491
  tensile 490
  uniaxial 490
  vanishing 66
  zero 498
  see also pseudo-strain
strain fields 50, 254
  expected analytic piecewise constant 52
strain localization 492, 498–500
strain tensors 46, 47, 172, 434, 468
  see also Green-Lagrange strain tensor; Lagrangian strain tensor
strength:
  atomistic measures of 486–500
  buckling 84
  limits to 486–9
  material 486
  mechanical 546
  structural 546
  tensile 489
stress 48, 61, 449
  complementary 47
  compressive 493
  conjugate 487
  constant 488
  continuum 398, 411
  critical 489, 492

discrepancies tend to zero 265
discretization of 51
dramatic reduction 489
external 486, 487, 493
fracture 130
macro 69, 262
residual-free 58, 62, 65, 70, 80
shear 496, 501
single-frequency waves 196
tensile 126
true 487
virial 489, 491, 492
  see also applied stress; Cauchy stress; Griffith's formula; Peach-Koehler stresses; Piola-Kirchhoff; pseudo-stress
stress calculations 126
stress dependence 502
  highly nonlinear 501
stress fields 48, 50–1, 52, 53, 249, 251, 284
  asymptotic expansion of 60
  complementary 47
stress intensity factor 128
stress-strain curves 125, 126
stress-strain law 96
stress-strain relations 430
stress-strain responses 486, 489, 490, 491
  qualitative behavior of 492
stress tensors 172, 429, 430, 434, 442, 459
stretch 442, 449, 474
stretch and angle bending 123–4
Strichartz, R. S. 42
Stroh solution 508
structural scales 57
structure-function relationships 545
Student t-distribution 540, 544
subdomains 41, 98, 101, 119, 128, 197, 201, 252, 255, 294, 416
  arbitrary 47
  atomistic 94, 110, 113, 165
  bond-breaking 95, 97
  coarse 206
  continuum 99, 110, 113, 165, 411
  coupled 116, 121
  meshed independently of interfaces 254
  molecular 99, 104
  nonoverlapping 64
  overlapping 105, 111, 120
  quantity of interest 296

refined 206
  residual contributions 295
  small representative 195
  time-space 270
  tiny 196
sub-grain scales 57
subnanometer pores 550
substitutions 141, 148, 333, 436
substructures 248, 249, 252–6, 260–6, 271, 273, 274
subunit model 521, 524
sugars 546, 550
Sulzer packings 548
supercomputers 193
super-diffusion 543
superposition methods 37, 95
surface models 466–7
surrogates 289, 292–3, 294, 296, 297, 298, 299, 302
Svensson, M. 101
Swendsen, R. H. 203
symmetric properties 449
symmetry 153, 155, 486, 537
  broken 495
  cubic 43, 151, 459, 488
  four-fold 124
  perfect crystal space group 142
  scaling 536
  translational 34
  underlying 553
synthesis 551

Tadmor, E. B. 94, 112, 135
Takens's embedding theorem 539
tantalum 151–5
TAO/PETSc codes 292
Tarasov, V. E. 43, 44, 45
Taylor-Hood finite element pair 378
Taylor series/expansion 144, 228, 229, 435, 454
temperature 208, 209, 380, 381, 385, 489, 490, 541
  finite 134–64, 196, 425–80, 500
  governing 306
  high 209
  low(er) 200, 209, 503
  nonzero 137
  reference values for 307
  ultra-high 510
  velocity decreases with 501, 502
  see also room temperature; zero temperature
temperature dependence 502

positive 503
temporal scales 514
tendons 31, 32
tensile pressure/strength 450, 489
tension:
  hydrostatic 488, 491, 492
  maximum 492
  thermodynamic 487
  uniaxial 489, 499, 507
tensor fields 397–8
  computational 399, 400, 404
  discretized 398, 405
  explicit 399, 404
tensors 140, 396, 488
  coefficient 70, 72, 74, 76–7, 494
  constant, symmetric and positive definite 250
  continuum elastic constant 143
  deformation 293
  elastic constant 494
  fourth-order/fourth-rank 487, 488
  identity 429
  scalar 170
  second-order 251
  stiffness 494
  see also Green-Lagrange; Hooke's tensor; Piola-Kirchhoff; strain tensors; stress tensors
TEOS (tetraethoxysilane) 31
Tersoff-Brenner potential 125
Tersoff potential 430, 431, 432, 434–5, 439, 465, 467, 468
  interatomic 466, 475
Tersoff silicon 432, 449, 462, 467
  dynamical matrix for 439
  elastic constants of 446, 447, 475
  generalized Grüneisen parameter for 444, 445, 446
  lattice constant for 468, 469
  unstrained 442–3, 446
tetrahedral structure 417, 428, 431, 440
  shear instability 488
tetranucleosomes 515
thermal effects 290
  activation region 502
  actuators 305–6
  changes 59
  conductivity 382, 482
  flip-flop motion 466
thermal expansion 144, 441
  coefficient of 307
  shifts due to 445

thermal fluctuations 137
  eliminating 144
thermal properties 461, 467–8
  local 470–3, 475
  single-wall carbon nanotubes 426
thermodynamic properties 454
  accurate results for 460
  local, LPDOS and 455–6
  silicon nanowire 467
thermodynamics 143, 434, 486, 546, 547
  feasibility of calculating limits 197
  self-consistent 146
thermostats 146, 160
  linking energy equation to 113
three-scale formulation/analysis 58, 78–86
three-body potential model 508
tight-binding descriptions 428
tiling procedure 498
time-dependent memory kernel 95
time-dependent problems 24–6, 216
time derivatives 115, 116–17, 121, 220, 232, 233, 249, 300, 333
  finite difference 231
  macroscopic 216, 217, 228, 229, 230, 235–6
time differentiation 47
time history kernel function 95, 96
time integration 114, 119, 146, 226, 228, 229, 230
  accelerated 220
  parallel, time-decomposed 248
  projective 239
  see also Runge-Kutta
time scales 482–3, 536, 539, 542, 548
  separation of 219, 220
time-steppers 215, 229
  coarse 217, 218, 219–20, 221, 239
  finite-difference 227, 232
  forward Euler 232, 233, 234, 236
  gap-tooth 226, 231, 232
  microscopic 221
  upwind 234, 236
time steps 196–7, 207, 236, 343, 358, 501
  coarse 218
  diffusion 376
  error versus 369
  gap-tooth 230, 234
  large 206, 208
  maximum length of 257
  microscale 238

microscopic 217, 218, 220
  reaction 375, 376
time variables 267
time-series 539
tire rubber 31, 32
TMOS (silicon tetramethoxide) 31
TNPACK (truncated Newton code) 518, 520
To, A. 119, 121, 122
tolerance 350, 352, 353, 356, 358, 378, 382, 384
  error 286
  see also LTOL
tolerance values 416
Tondeur, D. 549
tooth boundary conditions 238–42
top-down approach 425, 426, 428
topological entities 399, 402, 403, 404, 415
torsion 200, 204
traction conditions 41, 47, 52, 404
  continuity 48
transition-state theory 543
translational diffusion constants 518, 525
transmission factor 121
transport 10, 37, 42, 552
  adequate distribution channels 550
  chemical reactivity and mechanics of 510
  fast 546
  localized 542
  uniform 548
transport equations 3–4, 11, 14
  upscaling of 5, 6, 26, 27
transportation networks 547
trapping time distribution 544
trees 536, 537, 545, 548, 549, 550, 551
  mammalian airway 546
  mathematically-constructed 547
Trellis (object-oriented framework) 414–15
trial and error 32
trial problem 48, 49, 50, 51, 52
trial solution 53
trial space 294, 300
trimers 518
trinucleosomes 522
trust region method 292
tube crush tests 58, 82
tungsten 94
turbulent flows 37
twinning 495–8

two-dimensional
    studies/problems 119, 123–5, 205
two-phase flow 7
two-scale description of
    unknowns 256–7
two-scale model reduction
    approach 58

ultraviolet light 290
uncertainty 309
    analyzing the effects of 323
uncertainty quantification 285, 299,
    303, 309, 313, 330
uncoupling 250
    micro-macro 262
unit cell problems:
    finest scale 57
    multiple-scale 58, 62, 63
    nonlinear 57, 58
    reduced-order 70–6, 78, 81
unit cells 148, 432, 448, 469, 507
    geometry of 407
unity 158
    partition of 116, 139, 166
unstructured grids 4, 6, 13, 27
    analysis of global mixed MsFEM
        on 16–26
upscaling 3, 4, 57, 58, 386
    saturation 11, 13, 14, 15
    systematic 193–215
    transport equation 5, 6, 26, 27
upwind schemes 234, 235–6, 237, 238,
    239
    theoretical support for 242
Usher, M. 42

vacancies 126, 127, 409
validation 58
    braid architecture considered for 82
    experimental 525
    three-scale reduced-order
        formulation 82
    verification and 398
Van den Bleek, C. M. 539
van der Waals radii 528
Van Ness, J. W. 543
variable order p-version analysis
    procedures 403
variational analysis 344
variational equations 169
    weak 172
vascular network 546, 547, 551–2
vectors 99, 100, 255, 310
    adjoint 295, 296
    basis 407
    body force 429
    boundary traction 47
    constraint 106
    delay 539
    equilibrium position 429
    lattice 407
    normalized 459
    normed 324
    one-dimensional snapshot of 324
    polarization 494
    random 363, 371
    relaxed 197
    surface traction 430
    time-dependent 300
    unit outward normal 429
    velocity 300, 301
    wave 149, 439, 443, 444, 494
    see also Burgers vector;
        displacement velocity
velocity 4–11, 13, 16–18, 22, 24, 119,
    120, 124, 146, 151, 208, 350
    convergence theorem for 20
    dislocation 500–1, 502, 505
    flux averaged inlet 382
    gas 538
    inflow 380
    nodal 138
    prescribed 257
    wave 113, 159
verification 58, 489
    error estimator and adaptive
        strategy 296–9
    validation and 398
Verlet method 114, 146
    symplectic 115, 119
vibration 153, 461
    different atoms 439
    localized modes 42
vibrational frequencies 435, 437, 438,
    440
    stretching 470
virtual experiments 293
viscoelastic/viscoplastic behavior 37,
    251
viscosities 5, 12
    molecular 307
    shear 482
visual insights 485–6
voids 43, 44, 135, 548, 551
    dislocation structures around 416,
        417

Voigt notation 459
volumetric PGPs (phonon Grüneisen parameters) 442, 444, 445, 446
Von Mises stress field 265, 275
vortices 207, 538, 548
  idealized/strong 206
voxels 399, 402, 404
  multimaterial 406

Wagner, G. J. 95, 118
Wang, J. 486
Warshel, A. 521, 524
water 5, 544
  branching distributor of 546
  desalination 552, 553
  hydration layer along hydrophilic walls 541
  matrix for 547
  purification 552
  residence times 536
water-cut curves 5, 11, 12, 14, 15, 27
  reference and multiscale solutions 13, 16
wave distortions 500
wave equations 4, 24, 120
  multiple coarse-level representations for 194
wave forms 500
wave propagation 42, 144
  and attenuation 482
wave reflection 144, 145
  potential problems associated with 155
  unphysical 155
wavelengths 33, 143, 148, 149, 150, 158, 160
  longest 155
  shortest 144

wavenumbers 143, 149, 153, 154, 157, 158, 241
wavespeed 113, 120, 143
weak continuum equations 173
weather patterns 321
Weber, T. A. 508
Weinkamer, R. 545
well-posedness 313
West, G. B. 547
Wheeler, M. F. 383
White, C. R. 547
Whitesides, G. M. 553
Wilson, K. G. 203
windows 194, 195, 198, 205–6, 207, 208
  relatively small 201, 210
wormlike models 519, 522

Xiao, S. P. 95
X-ray crystallography 515

Young, Ju Lee 207
Young's modulus 44, 120, 129, 263, 264, 474

zeolites 541, 542, 550, 553
zero-displacement reference state 144
zero-energy mode 139
zero temperature 96, 143, 286, 431–4, 490, 500
  QC energy 426
Zhang, L. 7
Zhou, Z. 486
zigzag models 525, 526, 527, 528, 529, 531
  helical 519
  irregular two-start 518
zinc-blends structure 490, 491